PLATE 2 *Cepaea nemoralis* shell color and banding polymorphisms in snails from a single locality in France (original photograph by John Maltby).

Elements of Evolutionary Genetics

Elements of Evolutionary Genetics

Brian Charlesworth

and

Deborah Charlesworth

GEORGE GREEN LIBRARY OF
SCIENCE AND ENGINEERING

Roberts and Company Publishers
Greenwood Village, Colorado

Roberts and Company Publishers

4950 South Yosemite Street, F2 #197
Greenwood Village, Colorado 80111 USA
Internet: www.roberts-publishers.com
Telephone: (303) 221-3325 _10 06382923_
Facsimile: (303) 221-3326

Publisher: Ben Roberts
Artist: Emiko-Rose Paul at Echo Medical Media
Copyeditor: Lee Young
Production Manager: Betty Gee at Side By Side Studios
Cover and Interior Designer: Mark Ong at Side By Side Studios

© 2010 by Roberts and Company Publishers

Reproduction or translation of any part of this work beyond that permitted by Section 107 or 108 of the 1976 United States Copyright Act without permission of the copyright owner is unlawful. Requests for permission or further information should be addressed to the Permissions Department at Roberts and Company Publishers.

Library of Congress Cataloging-in-Publication Data

Charlesworth, Brian.
 Elements of evolutionary genetics / Brian Charlesworth and Deborah Charlesworth.
    p. cm.
   ISBN 978-0-9815194-2-5
   1. Evolutionary genetics.   I. Charlesworth, Deborah.   II. Title.
   QH390.C43 2011
   572.8'38—dc22

                                                    2009042482

To James F. Crow

Contents

5 The evolutionary effects of finite population size: basic theory

9   The evolution of breeding systems, sex ratios, and life histories

Preface

The project of writing a textbook on evolutionary genetics was suggested to us by Ben Roberts over three years ago. We are very grateful to him for his support and encouragement over the seemingly endless process of writing and revision, and to Emiko Paul for turning our crude sketches and blurred images into handsome illustrations. We also thank Jane Charlesworth for assistance with the index, and Betty Gee and Mark Ong for overseeing the production of the book.

We especially thank the people who generously gave their time to read and comment on some or all of the draft chapters: Hiroshi Akashi, Jane Charlesworth, Jim Crow, George Gilchrist, Philip Johnson, Laurence Loewe, Christina Muirhead, Sally Otto and her lab group, Manus Patten, Jitka Polechova, Suo Qiu, Sohini Ramachandran, Monty Slatkin, and Kai Zeng. They corrected numerous errors and made many suggestions for improving the contents of the book. We have not always taken their advice, and the responsibility for the remaining errors and lapses from clarity is entirely ours.

Edinburgh, November 2009

Introduction

"Evolution" means cumulative change over time in the characteristics of a population of living organisms. Evolutionary changes have been in progress at all times, including the present. Some can be very fast. A pest or bacterial population susceptible to insecticide or antibiotic can be transformed into a resistant population in a few generations, under the strong selection pressures that operate when a pesticide or drug is introduced into use. Many evolutionary changes, however, are very slow, and take place over millions of years; for example, the emergence of mammals from reptiles. Events such as the evolution of a new species lie in the middle of this time spectrum. The types of traits involved in evolutionary change are enormously varied, ranging from easily observable characteristics of organisms, such as visible color patterns or behavior, to the components of the genome itself. Understanding evolutionary change in the sequences of the bases that make up the genetic information carried by DNA is a major focus of modern biology.

All evolutionary changes require initially rare genetic variants to spread among the members of the population, rising to a high frequency, so that the population becomes genetically different from its ancestral condition. The goal of evolutionary geneticists is to understand such changes, and to explain what has happened in evolution. We now largely understand the origin of the variants that are the basis of evolutionary changes, and the mechanisms that can cause variants to spread through populations. As in other branches of science, we assume that the properties of organisms living in the past were fundamentally the same as for present-day living organisms, so that even remote evolutionary events were subject to the rules revealed by experimental studies today. This assumption is supported by the basic similarities of the genetic material and the genetic code across all types of organisms alive today, even ones whose last common ancestor lived more than two billion years ago.

Evolutionary genetics differs from most other branches of biology in its strong theoretical structure. The rules of transmission of genetic information from parents to offspring, uncovered by experimental genetics, impose important constraints on what can happen in evolution. From the very beginnings of genetics, early in the 20th century, mathematical models of genetic processes in populations have been used to study evolution. A major goal of this book is to show how these models help us understand evolution, and how they can add insights beyond those that we can gain intuitively, but sometimes giving counterintuitive results. In addition, they can provide results that are far from obvious, and provide a quantitative understanding that leads to new ways of asking and answering important biological questions.

In this book, therefore, we present the models in the context of biological questions, and analyze them only to a level of detail needed to illuminate these questions. Our aim is keep our book as short as possible, and to provide a foundation of knowledge of the main concepts of evolutionary genetics for use by people with a variety of biological interests. We assume familiarity with the basic principles of genetics and molecular biology; readers with backgrounds in fields other than biology will first need to learn these principles from an introductory text.

In each chapter, the simpler and more basic concepts are described first, and advanced topics are usually presented either at the ends of some chapters (where, as we indicate, they can be skipped, because most are not essential for understanding later chapters), or in **Appendices** (**Chapters 5**, **6**, and **8**). Some special topics illustrating important principles are discussed as separate **Case Studies**. Each major section of the book (indicated by arabic numerals) could have been a separate chapter, but we have grouped related topics into chapters, with divisions into major sections providing natural breaks that we hope will help readers to pause and think, and make sure that they have understood the ideas they have been reading about. Slow reading and working through many sections will be necessary, because they contain concepts and approaches that will be new to most readers.

Our style of derivation is not rigorous. As J. B. S. Haldane (1964) wrote in his famous essay *A defense of beanbag genetics*:

> . . . Wright, Fisher, and I all made simplifying assumptions which allowed us to pose problems soluble by the elementary mathematics at our disposal. . . . Our mathematics may impress zoologists, but do not greatly impress mathematicians.

To avoid disrupting the flow of the verbal arguments, the mathematical derivations are separated into boxes, while the text discusses and illustrates the main conclusions that can be drawn from them. To help readers understand the approaches and concepts, problems are also provided, with worked answers at the end of the book. The problems range from simple numerical examples (intended to help readers see how a concept is used), to short derivations (to help readers understand how the derivations work). Working through the boxes and the problems will give a good understanding of the basics of the theory, but the text can be read without reading the boxes (e.g., when first reading a section of the book). Most of the boxes and problems are accessible to any reader who can do simple algebra, but some (indicated by stars) require basic calculus or matrix algebra. A summary of some useful basic results in mathematics and statistics is given in the final **Appendix**.

This book is appropriate for graduate students in evolutionary biology and human genetics (or students in advanced level undergraduate courses), and for researchers in related fields who wish to become acquainted with evolutionary genetics. Because of the emphasis on specific biological questions, individual sections of the book can also be used in general evolution or genetics courses.

We illustrate the questions illuminated by the theory with examples drawn from a wide variety of organisms, ranging from viruses to humans. In the earlier, more basic, chapters we emphasize the historical development of the field. Classical population genetics is classic science, with beautiful and often simple results, providing the basis for an understanding of evolution and its genetic basis that should be part of every biologist's consciousness. **Chapter 1** describes the evidence accumulated from over a century of work on the nature and amount of genetic variability in populations. Recent advances in DNA sequencing are revolutionizing the collection of data on differences between species and among individuals within species. This is providing evolutionary genetics with an unprecedented wealth of material, to which the models developed by the pioneers of population genetics are often surprisingly relevant. Furthermore, the great power of genome-wide surveys of populations to detect associations between genetic variants and diseases or complex traits, and to test for the effects of selection, means that evolutionary genetics is more relevant than ever before to practical problems, such as animal and plant improvement and human health.

The major processes that cause variability within species and populations to be incorporated into evolutionary change are *natural selection* (differences in reproductive success among individuals with different characteristics) and *genetic drift* (the random sampling of variants during the transmission of genes from one generation to the next). These processes, and their interactions with other evolutionary forces such as *mutation*, *migration*, and *recombination*, are described in **Chapters 2–8**. The final two chapters (**9** and **10**) use these concepts to study some evolutionary questions for which genetic approaches are central.

The book does not comprehensively cover the whole of evolutionary genetics, which has now developed into a major research area in biology, some of it highly mathematical. Rather, as explained above, our aim is to describe basic population genetic processes and to illustrate their relevance to interesting evolutionary situations. For this reason, we only touch on the rapidly growing field of statistical inference in population genetics. Many other interesting results in population genetics and evolution are also omitted, as are details of applications to medical genetics and plant and animal improvement. We hope that our choice of topics represents the major concepts in the field, but such judgements are subjective, and reflect our own work and interests. Where possible, we provide references to further work, either work on topics related to those we discuss, or more mathematical developments of the same topics.

CHAPTER ONE

Variability and Its Measurement

CHAPTER SUMMARY

Genetic variability is essential for evolutionary change. Many observable traits (phenotypes) of individuals are influenced by their genetic makeups (genotypes), as well as by the environment. This chapter reviews variability within populations. We first describe phenotypic variability, ranging from discrete polymorphisms to continuous variation in morphology, behavior, and physiology (quantitative variability). We also describe how inbreeding reveals the presence of rare variants with deleterious effects on fitness, i.e., survival or fertility.

Genetic variability affecting the phenotypes of individuals is caused by differences between individuals in the DNA sequences that encode proteins and functional RNA molecules, and in noncoding DNA sequences that control gene expression, as well as some heritable variation that does not involve sequence differences (epigenetic variation). Some sequence differences may, however, have little or no effect on any phenotype, so that natural selection acting on them may be weak, or even nonexistent (selective neutrality). In this chapter, we focus on DNA sequence differences revealed by comparing individuals of the same species (the same types of differences are also found between members of different species). Among normal individuals in natural populations of most organisms, there are typically millions of DNA sequence variants in the genome as a whole. We introduce the terminology used for describing variability and the methods for quantifying sequence variability within a population.

We then start the process of understanding the causes of this variability, using models of very large populations (in which the frequencies of selectively neutral variants in each new generation are close to those in the parental generation). We derive the Hardy–Weinberg equilibrium genotype frequencies under Mendelian inheritance in a diploid, randomly mating population, for which the genotypes of mates are independent. The results show that neutral variability is maintained indefinitely in this case. The same is true for uniparental transmission of variants in mitochondrial or chloroplast genomes, and with asexual reproduction.

We also determine the genotype frequencies expected in populations that do not mate at random. Inbreeding is a common and important type of nonrandom mating. Inbreeding populations have higher frequencies of homozygotes than with random mating, due to the descent of an individual's maternal and paternal alleles from a recent common ancestor. We use the concept of identity by descent to quantify this effect; this concept is centrally important in population genetics and will appear again in later chapters. In the present chapter, we use it to show how inbreeding can be measured, and prove that variability is also maintained in large inbreeding populations, despite a reduced frequency of heterozygotes.

New, stably transmitted variants that are found in natural populations originate by mutation, and **Chapter 1** ends with an outline of current empirical evidence on the nature and rate of occurrence of new mutations.

> Any variation that is not inherited is unimportant for us. But the number and diversity of deviations of structure . . . is endless.
>
> Charles Darwin, *The Origin of Species*, Chapter 1

INTRODUCTION

Evolutionary change requires heritable variability among individuals within a population. For example, the process of natural selection on a trait involves differences in survival or fertility (*fitness differences*), related to the trait values of individuals. Similarly, artificial selection involves the human choice of certain individuals for breeding, and the rejection of others. If differences among individuals in a trait are not transmitted to their offspring, the offspring of the individuals with the highest fitness will be similar to the offspring of parents taken randomly from the population. Selection will then cause no change in the composition of the population. In contrast, if genetic differences contribute to variation in a trait, the parents with higher fitness values will have offspring that resemble them to some extent. There can then be a corresponding change in the population, in the direction of trait values associated with higher fitness, as we describe more fully in **Chapters 2** and **3**. Similar principles apply to changes caused by other evolutionary processes such as *genetic drift*, which involves random changes in the genetic composition of a population caused by the sampling effects of finite population size (described in **Chapters 5** and **6**).

All of the diversity of life on earth must therefore result from the transformation of within-population genetic variability into differences between populations, species, and higher taxonomic groups. Understanding the nature and causes of genetic variability is therefore fundamental to our understanding of evolution. It is also important for the informed improvement of domestic animals and plants by selective breeding, for determining the genetic basis of human diseases, and for the control of pathogenic organisms and pests. Although Charles Darwin fully understood the importance of genetic variability for evolutionary change, and emphasized it strongly in *The Origin of Species* (Darwin 1859), his understanding of the mechanisms of evolution was seriously hampered by not knowing how inheritance works, as we discuss in more detail at the beginning of **Chapter 2**. Modern evolutionary thought is based on genetic knowledge, which has confirmed the basic principles of evolution by natural selection outlined by Darwin and Alfred Russel Wallace (1858). It has also greatly extended our ideas about the ways in which evolution works, introducing processes unknown to Darwin and Wallace.

Because of the importance for evolutionary change of genetic variability within a population, a major task of evolutionary genetics is to investigate its nature and to measure its extent. In this chapter, we shall see how our understanding of variability has been progressively refined by advances in genetics and molecular biology, most recently by studies of variation in DNA sequences. We will also begin to interpret data on variability in terms of models of the evolutionary processes acting within populations; these models lay the foundations for later chapters.

1.1. CLASSICAL AND QUANTITATIVE GENETIC STUDIES OF VARIABILITY

We first describe variability that can be scored or measured without the use of elaborate molecular methods. This type of variability provided the materials for genetic and evolutionary studies from the time of Darwin and Mendel until the 1950s, and many important questions about variability that are being answered by molecular approaches were first raised by studies of such traits. It should not be forgotten that many aspects of organisms that are of most interest to us, such as behavior, morphology, and many human diseases, represent traits whose variability is studied in the ways we describe here, although there is increasing emphasis on relating this variability to its underlying causes at the level of DNA sequences.

1.1.i. Discrete variability

A casual glance at the occupants of a classroom or bus reveals many differences in appearance, some of which can be classified into clear-cut, discrete alternatives, such as hair or eye color, while others are more continuous (e.g., height). The presence of alternative discrete variants in a population is called *polymorphism*. Polymorphisms for traits that are readily scored by simple visual methods represent, however, only a minor component of the variability within a species; after briefly discussing such polymorphisms, we will discuss the much more abundant *quantitative variability* and then introduce molecular variability.

Polymorphisms often involve differences in conspicuous characteristics, such as the polymorphisms for mimetic forms of butterflies in the genus *Papilio* (**Plate 1**, front end paper) and for shell color and band number in the land snail *Cepaea nemoralis* (**Plate 2**, front end paper), reviewed by Sheppard (1975, Chapters 5 and 11). In both *Papilio* and *Cepaea*, the color differences are inherited as variants at a single genetic locus with several alleles; in *Cepaea*, another gene controls the number of bands on the shell (Sheppard 1975, pp. 87–90). Genetic polymorphisms for size and shape characters are less common than color differences, but they sometimes occur—for example, the beak size dimorphism of the African finch *Pyrenestes* (Smith 1993). Similar kinds of variation can be caused by the environment and are not genetically determined—for example, the seasonal differences in wing patterns displayed by African butterflies of the genus *Bicyclus* (Roskam and Brakefield 1996). Some polymorphisms involve the joint effects of genes interacting with effects of the environment (Roff 1996).

Discrete polymorphisms have also been discovered by microscopic and biochemical studies of organisms; these are sometimes called *cryptic polymorphisms*. The human blood group polymorphisms are an example. Blood groups were discovered between the 1910s and 1950s by immunologists studying the antigenic properties of blood cells. Several dozen different polymorphic blood group genes are now known (each with two or more alleles). The first of these were discovered early in the history of blood transfusions, because transfusions between individuals differing in some blood group types are incompatible and cause clotting. These polymorphisms have been used extensively to study variation within and relationships between different human populations, and some are shared with some of our closest relatives, such as gorillas and chimpanzees (Race and Sanger 1975).

Chromosomal rearrangements, such as inversions, are another important category of cryptic polymorphism (**Figure 1.1**). These have been most studied in species of the fruitfly *Drosophila*, where they are easily detected by microscopic examination of the giant larval salivary gland chromosomes (Krimbas and Powell 1992a). Chromosomal polymorphisms also occur in other organisms, including humans, where numerous inversions have recently been uncovered by studies of DNA sequence variation.

1.1.ii. Quantitative variability

1.1.ii.a. *Describing quantitative variability*
Most easily observable variability within populations is not discrete. Unlike the discrete polymorphisms just described, *quantitative traits*, such as height differ-

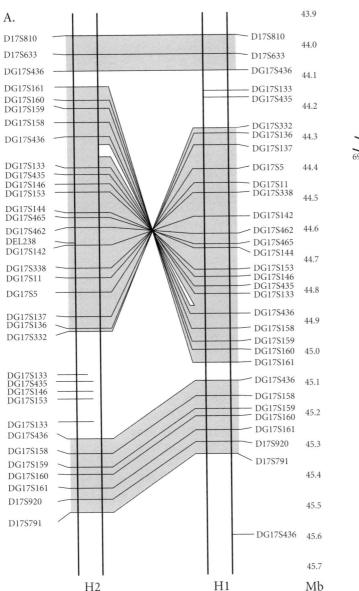
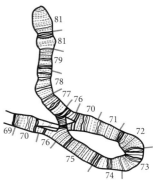

FIGURE 1.1 Inversions in humans and *Drosophila*. **A.** A 900-kb inversion of part of human chromosome 17 (the 17q21.31 locus). This inversion is polymorphic in the human population, with both arrangements being common in Europeans. The right-hand chromosome, H1, is the same as in the assembly of the human genome. The left-hand chromosome, H2, is inverted in this region. [Adapted from Figure 1 of Stefansson et al. (2005).] **B.** The polytene chromosomes of a *Drosophila pseudoobscura* individual heterozygous for the Arrowhead and Standard inversions in chromosome 3. The inverted and noninverted arrangements have paired in the region of the inversion, forming a loop structure. [Adapted from Figure 1 of Sturtevant and Dobzhansky (1938).]

ences among humans, have a roughly continuous distribution of values. Many characters used by systematists and paleontologists to study differences between living species, or changes over time, are of this kind, as are many traits of economic and medical importance. *Metrical* traits are those that can be measured on a continuous scale, and generally have a bell-shaped probability distribution approximated by the *normal* or *Gaussian* distribution (**Figure 1.2** and **Appendix A2.v.e**). With the availability of data on levels of gene expression across the genome, it is increasingly common to treat messenger RNA abundance levels as quantitative traits (Gibson and Weir 2005). *Meristic* characters are ones with several discrete trait categories, such as vertebra number in fish and snakes, or the number of rows of kernels on a maize cob (**Figure 1.3**). These often have a unimodal distribution, with the intermediate classes being the most frequent. A related category of trait, called *threshold traits*, includes traits with two discrete alternatives, where the expression of one of the two alternatives is determined by whether or not a threshold in an underlying continuously distributed trait is exceeded; some human diseases, such as diabetes, behave in this way (Falconer and Mackay 1996, Chapter 18).

Variability for either meristic or metrical traits is quantified by standard statistical measures such as the dispersion of values around the mean (the *variance*) or its square root (the *standard deviation*—see **Appendix A2.iii**). A useful measure is the *coefficient of variation* (CV), which is the standard deviation of the trait divided by its mean; the CV removes the scale of measurement, and is approximately equivalent to using the natural logarithm (**Appendix A1.ii.c**) of the trait value instead of the original scale. Coefficients of variation for many traits are often 10% or more (Wright 1968, Chapter 6; Yablokov 1974). With a normal distribution, a CV of 10% means that 16% of the population

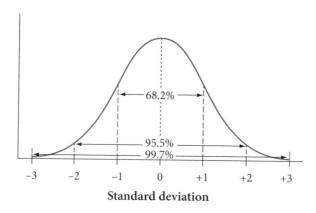

FIGURE 1.2 The Gaussian or normal distribution, showing the proportions of the values lying within one or more standard deviations from the mean.

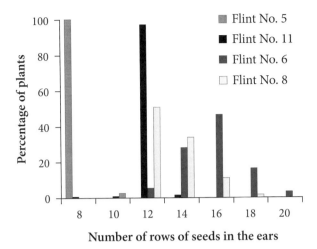

FIGURE 1.3 An example of a meristic trait. The figure shows the distributions of numbers of rows of seeds in ears of maize from different strains. [From Table 6 of East (1910).]

have values greater than 110% of the mean, and 2.3% have values of 120% of the mean or more.

1.1.ii.b. *Genotypes and phenotypes*

Selection on a trait in a genetically variable population is usually highly effective, and can rapidly change its mean value. Laboratory or domesticated populations of animals, plants, and microbes have successfully been subjected to artificial selection for changed trait values in many hundreds of selection experiments (Falconer and Mackay 1996, Chapter 11). This demonstrates the presence of genetic variability, as Darwin realized from his studies of domesticated animals and plants (Darwin 1859, Chapter 1; Darwin 1868).

An important advance early in the history of twentieth-century genetics was the proof that a response to selection requires genetic variability that exists independently of any effects of the environment on the trait in question. Previously, it had seemed possible and even likely that parents often transmit to their offspring variation that had been acquired during their lifetime; this was believed by Darwin (1868). One way of testing for this "inheritance of acquired characters" is to conduct selection experiments on inbred lines, made by many generations of matings between close relatives. All individuals from the same inbred line will then be genetically nearly identical and will be homozygous for most loci, i.e., there will be no genetic variability (see **Section 1.3.ii**). Alternatively, in some organisms, genetic identity can be achieved by clonal propagation, e.g., by cuttings of plants or by cell divisions of unicellular organisms. In contrast to the effectiveness of selection on non-inbred populations, selection on these genetically uniform lines is almost always ineffective in changing their means, as was first shown by Johannsen (1909). For selection to be effective, enough time must have elapsed for new genetic variants to arise by mutation (Lynch 1988; Falconer and Mackay 1996).

However, highly inbred or clonally propagated lines do not lack visible variability or variability in quantitative traits. This variability must be due to nongenetic factors influencing the trait, including effects of environmental conditions, and also internal accidents of development (often, confusingly, these are collectively called *environmental effects*: Falconer and Mackay 1996, p. 138). In contrast to what is found in natural populations, within a line there is generally no correlation between the values of parents and offspring (when precautions are taken to exclude environmental causes of such correlations), even when there is extensive variability in the trait. No immediate response to selection is therefore possible (see **Section 3.3.iii**). A classic example of this is the study by Sewall Wright of the piebald pattern of guinea pigs, for which the amount of white in the coat is variable between individuals (Wright 1920). A stock that had been maintained for 20 generations by brother–sister mating remained variable, but showed no evidence for parent–offspring correlations, although significant correlations were detected in a control, random-bred stock.

These experiments show that, if the possibility of genetic variation is excluded, variation in a trait does not allow a response to selection. This result yields the important conclusion that parents do not transmit nongenetic variation to their offspring. This is the reason why biologists emphasize the distinction between an individual's measurable *phenotype* (the value of a trait of interest) and its *genotype*. It implies that phenotypes are controlled by the joint effects of genetic factors (which can be transmitted to the offspring) and nongenetic factors (which are usually not transmissible), i.e., nature together with nurture (first emphasized by Johannsen 1909). If we measure many individuals with the same genotype, nongenetic effects that increase or decrease the values of individual phenotypes should average out, and their mean provides an estimate of the individuals' *genotypic value*.

1.1.ii.c. *The genetic basis of quantitative variability*

What is the basis of genetic variability in quantitative traits? In Darwin's time, it was almost universally believed that inheritance worked by the blending of material from the two parents when offspring are formed. Mendel (1866) disproved this for traits controlled by single gene differences, and realized that it was also wrong for a quantitative trait (flower color) that he studied. His ideas were, however, ignored. The observation that the F1 hybrid progeny of a cross between two inbred lines are often intermediate in mean value between the means of the two parental lines seemed to support blending inheritance. Mendel's quantitative approach, including the study of later generations as well as the F1, established that inheritance is particulate (nonblending), with the full reappearance of recessive parental traits in the F2 generation produced by intercrossing F1 individuals.

After Mendel's work was rediscovered in 1900, most people studying quantitative traits, notably Karl Pearson and his associates, did not accept that it could explain the observed quantitative trait variability, although they stressed the evolutionary importance of natural selection on quantitative traits (Provine 1971). In contrast, geneticists working on simple traits obeying Mendel's laws tended to downplay the importance of quantitative traits and natural selection, emphasizing instead mutations creating discrete new character states (Provine 1971).

There is, however, no contradiction between Mendelian inheritance and quantitative inheritance. Studies of quantitative traits in crosses between inbred lines of plants, such as wheat, oats, and maize, showed greater variability in the F2 than in the F1 or either parental line (Nilsson-Ehle 1909). This is impossible with blending inheritance, but is expected when one or more Mendelian genes control the difference between two lines (**Figure 1.4**). Indeed, the increase in variability in the F2 generation provides a way to measure the number of different genes distinguishing the parental lines (Wright 1968, Chapter 15).

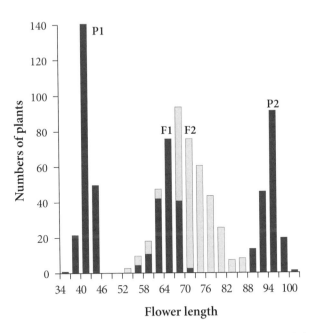

FIGURE 1.4 Quantitative variability in parental, F1, and F2 generations made by crossing two inbred strains (homozygous parental genotypes denoted by P1 and P2) of the plant *Nicotiana tabacum*. The figure shows the distributions of flower size. The parental strain and F1 flower sizes are in black, whereas the F2 are in gray. [From Table 1 of East (1916).]

This experimental evidence established firmly that the genetic factors under-lying quantitative trait variability are just like those underlying Mendelian traits (Wright 1968, Chapter 15); the only difference is that the segregation of individ-ual genes cannot be detected directly, because their contributions are obscured by nongenetic variation and by the effects of other genes. Moreover, linkage can be detected between Mendelian marker genes and genes affecting quantitative traits. By the 1920s, genes affecting traits such as bristle numbers in *Drosophila melanogaster* had been roughly localized to defined positions on the fruitfly's chromosomes (Muller and Altenburg 1920; Payne 1920). The mapping of these genes was limited only because few genetic markers were then available. Modern studies, using molecular markers scattered across the genome, allow fine-scale mapping of *quantitative trait loci* (QTL). Rapid progress is being made in the study of QTL in model organisms such as *Drosophila* and mice, in domestic animals and plants, and in humans (Mackay 2004; Mott 2006; Flint and Mackay 2009).

1.1.ii.d. *Modeling quantitative trait inheritance*

Even today, however, much work on quantitative traits uses statistical descrip-tions of their population properties, since the details of the underlying genes and their properties are mostly unknown. These descriptions rely on mathe-matical models that were developed independently in the 1910s by Fisher, Weinberg and Wright (Provine 1971), who showed how to interpret the statisti-cal appearance of blending in terms of Mendelian inheritance, taking into account the joint effects of multiple genes and nongenetic sources of variability.

We shall describe this theory in more detail in **Section 3.3**. For now, we need only note that estimates of the correlations (see **Appendix A2.iv**) between close relatives, such as parents and offspring, or siblings, can be used to infer the pro-portion of the variance in a trait in a sexually reproducing population that is caused by *additive genetic* effects. The additive genetic value of an individual can be thought of as measuring the individual's genetic contribution to the mean value of the offspring that it would produce if mated to a large number of other individuals chosen randomly from the population (Falconer and Mackay 1996, pp. 135–136). In other words, the additive genetic value is the component of the genotype that is transmissible to the offspring (in organisms in which mothers provision their young, it is important to exclude nongenetic maternal influences).

Estimates of individuals' additive genetic values are useful in animal breeding, where artificial insemination can be used to produce many progeny from a single male (Brotherstone and Goddard 2005). In most other situations, we can only measure the proportion of the total phenotypic variance in a trait contributed by the variance in additive genetic values (the *heritability*). This is an important determinant of the rate of response to selection on the trait (see **Section 3.3**). Heritability values are usually in the range 0.2–0.8 (see Table 10.1 of Falconer and Mackay 1996), and very few traits have zero heritability (some examples are described by Blows and Hoffman 2005 and Toro et al. 2006).

1.1.iii. Concealed variability and mutational variability

1.1.iii.a. *Introduction*

Close *inbreeding* (e.g., brother–sister mating in animals or self-fertilization in plants) reveals *concealed variability*, which is not normally detectable in an out-breeding population. The harmful effects of inbreeding on traits related to fitness were first systematically investigated by Darwin (1876). Genetically, these effects are interpreted as reflecting the presence of recessive deleterious alleles, which, when they are made homozygous, reduce the values of traits such as size and survival (see **Section 4.4.i**). Pervasive deleterious effects of inbreeding are documented by numerous experiments on domestic and laboratory populations (Wright 1977, Chapter 1; Charlesworth and Charlesworth 1987), and more recently by field studies (Keller and Waller 2002). This work shows the following:

1. The mean values of traits like survival, fertility, and growth rate are reduced in inbred lines, compared with either the outbred initial population, or with F1 crosses between inbred lines (*inbreeding depression*). We will describe quantitative measures of inbreeding depression in **Section 4.4.ii**.
2. Phenotypic variability among different lines derived from the same source population is increased as inbreeding progresses, and abnormalities caused by homozygosity for rare single-gene major mutations are sometimes seen in some of the lines.

1.1.iii.b. *The genetics of inbreeding depression*

Severe inbreeding depression for early life survival is observed in most outbreeding species of plants and vertebrates so far studied, but the detailed analysis of its genetic basis is difficult. In *D. melanogaster*, however, powerful genetic techniques were developed by Muller (1928), which allow entire chromosomes to be made homozygous and "cloned" (**Figure 1.5**). This species has three major chromosomes (X, 2, and 3), and **Figure 1.5** gives the breeding scheme for isolating a single second chromosome, using the *Cy* major gene mutation as a genetic marker, as explained in the figure legend.

If a set of wild-type second chromosomes is isolated from different wild-caught males, and each chromosome is then made homozygous as shown in **Figure 1.5**, an average of about 30% of chromosomes prove to be *lethal* (Crow 1993), i.e., no wild-type adult flies are found in the final generation of the crosses carrying them. A similar number is carried on the third chromosome. Very few lethals are present on the X chromosome, because X-linked lethal

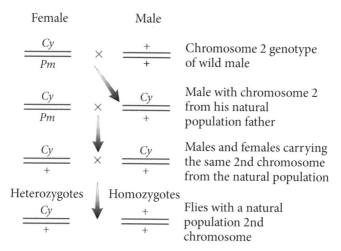

FIGURE 1.5 How balancer chromosomes can be used to breed multiple *Drosophila melanogaster* individuals that are all homozygous for the same chromosome present in a wild fly captured in nature. The diagram shows a wild-caught male at the top right (whose two chromosomes are denoted by +). The balancer chromosome carries several inversions (**Section 1.1.i**) that suppress crossing over with the normal second chromosome in heterozygous females (*Drosophila* males have no crossing over). The balancer also carries a dominant marker gene (*Cy* in this example, which gives the flies curled wings), detectable in carrier flies. The marker is homozygous lethal, and the balancer is kept in a stock (the left-hand genotype in the top two generations) whose other second chromosome carries another dominant marker, *Pm* (giving dark red eyes, instead of the usual bright red), which is also lethal when homozygous. Only *Cy/Pm* individuals survive, so the heterozygous balancer stock breeds true.

variants kill their male carriers (males are XY, and the Y chromosome lacks nearly all the genes present on the X), and are immediately eliminated from the populations. A fly carries an average of about 0.6 recessive lethal chromosomes per haploid genome, or 1.2 per diploid genome.

Genetic mapping of lethal chromosomes shows that these nearly always carry variants in single genes (rather than having sets of genes that collectively reduce viability to zero). In addition, crosses between two lines, each carrying different lethal chromosomes, show that about 98% of the heterozygotes produced are viable, indicating that different lethal chromosomes usually carry recessive lethal variants in different (*nonallelic*) genes. It follows from this result that lethal variants are individually rare in the population, i.e., the high frequency of lethal chromosomes is due to individually rare variants distributed over many different genes (Crow 1993). In **Section 4.2.ii**, we describe how these can be maintained in the population by a balance between input by *mutation* (the origin of new variants by accidents in the replication and transmission of the genetic material) and elimination by selection.

Furthermore, among nonlethal chromosomes, the ratio of wild-type homozygotes (+/+) to marker heterozygotes (*Cy*/+ in the example in **Figure 1.5**) is usually less than the expected 1:2 Mendelian ratio, implying that the homozygotes for nonlethal chromosomes have lower viability from the egg to the adult stage than *Cy*/+ (on average, about 20% lower: Crow 1993). This indicates the presence of variants with minor effects on egg-to-adult survival. More detailed investigations, where a nonlethal wild-type chromosome is allowed to compete with a balancer chromosome in a laboratory population of flies, allow estimates of the net effect on fitness throughout the flies' lives of homozygosity for a whole chromosome (Sved 1971; Latter and Sved 1994). The theory underlying these experiments is described in **Section 2.1.ii.c**; for now, it is sufficient to note that the final frequency reached by a wild-type chromosome in a competition experiment will be highest if its effect on fitness is slight. The mean fitness reduction caused by homozygosity for a whole *D. melanogaster* second chromosome is about 84%. A *Drosophila* individual that is homozygous at all loci effectively has zero fitness.

In species other than *D. melanogaster*, it is still usually unknown how much inbreeding depression in viability is due to lethal mutations rather than to many genes with minor effects. Controlled crosses have recently been done in zebrafish and killifish between individuals taken from natural populations to produce large F2 families from many different sib-matings (McCune et al. 2002). These experiments detected the segregation of recessive alleles with lethal effects on early development. They indicated, however, that fish caught in nature carried numbers of heterozygous lethal mutations similar to those in *Drosophila*. This is surprising, since fish genomes have around 25,000 genes, many more than those of flies (which have around 14,000 genes).

1.1.iv. Limitations of the classical and quantitative genetic approaches

While validating Darwin's view that there is plenty of genetic variability available for use in evolution, the evidence obtained by the methods we have just described raises two important questions, which were clearly posed by the end of the 1950s when much of the information described above was available:

1. How much variation is there in an average gene within a natural population? The classical and quantitative genetic methods provide no means of sampling variation randomly from the genome.
2. Are the frequencies of variants within populations controlled by natural selection, or are variants selectively neutral? What is the role of other potentially important evolutionary factors, such as mutation and genetic drift?

The "classical" school of population geneticists predicted that most genes would have a high frequency wild-type allelic form, with rare deleterious variants due to mutations, like the recessive lethals we have just discussed (Muller 1950). In contrast, the then newly developed "balance hypothesis" proposed that genes typically have several different forms, like the polymorphisms discussed at the beginning of this chapter, and suggested that these variants are maintained in populations by selection (Dobzhansky 1955). Various forms of such *balancing selection* will be discussed in **Sections 2.2** and **2.3**.

With the techniques available at the time, no clear-cut empirical test to distinguish between these two views of the causes of natural variability was possible. Classical genetic techniques can ascertain the existence of a gene only if at least two alternative alleles exist, so that genes lacking phenotypic variation go undetected. Studies of quantitative and concealed variability, with the exception of lethals in *Drosophila*, do not allow the identification of individual genes. It was therefore impossible to estimate how much variability existed in a typical gene. This requires methods based on molecular genetics, as we will now describe.

1.2. MOLECULAR VARIABILITY

Discoveries about the nature of genes and their relationships with proteins led to the development of methods for cloning and sequencing portions of the genetic material, making it possible to study variation and evolution in the sequences of proteins, and in the DNA sequences of the genes that encode them, as well as in noncoding DNA. The new kinds of data stemming from these methods have conclusively answered question (1) of the previous section, while question (2) is still an active area of research.

Although the study of DNA sequence variants is now very important, and will certainly continue to help to answer the questions just posed, it is helpful to start with protein variants—the first molecular variants that were discovered—in order to appreciate the historical development of our ever more detailed knowledge of variability in populations.

1.2.i. Gel electrophoresis of proteins

1.2.i.a. *Introduction*

Variation within natural populations in single genes (question 1 above) was first estimated by using the knowledge that most genes correspond to stretches of DNA that code for polypeptides. By studying the polypeptide sequence encoded by a gene, we can "see" some of the variation in the gene itself. This was first achieved in the mid-1960s by Richard Lewontin and John Hubby working in Chicago with *Drosophila pseudoobscura* (Hubby and Lewontin 1966; Lewon-

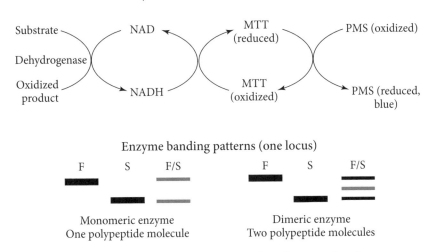

FIGURE 1.6 Staining for a dehydrogenase enzyme, such as glucose-6-phosphate dehydrogenase, after gel electrophoresis. The figure shows the reactions used to stain for the enzyme, and an example of fast (F) and slow (S) variants seen on a gel following electrophoretic separation of the proteins according to their charge; the patterns in heterozygotes are also shown. The abbreviations MTT and PMS refer to chemicals that form part of the staining reactions, and NAD is a cofactor necessary for dehydrogenase reactions.

tin and Hubby 1966), and by Harry Harris in London, studying variability in humans (Harris 1966). Both groups used the technique of *gel electrophoresis* to screen populations for variants affecting protein migration rates in an electric current (**Figure 1.6**). They studied many different soluble proteins controlled by independent genes, mostly enzymes with well-understood metabolic roles. The proteins were chosen purely because they could be studied easily, with no bias as to their level of variability. Enzyme loci detected by this method are known as *allozyme* loci.

This method is outlined in **Figure 1.6**. Polypeptides encoded by different alleles at a locus may differ by one or a few amino acids in their sequences. Some amino acid differences involve residues whose electric charges differ, causing different mobilities in gels exposed to electric currents, resulting in detectable differences in band positions. In the example, two types, fast (*F*) and slow (*S*), are distinguishable. Such variants are generally inherited as alleles at a single locus. In the simplest situation of a monomeric enzyme (lower left-hand diagram in **Figure 1.6**), fast individuals are homozygotes for two *F* alleles, slow individuals are homozygous for two slow alleles (*S*), and individuals with two bands are heterozygous, *F/S*. When the two alleles in heterozygotes are detectable, the variants are said to be *codominant*, as is usually the case for allozyme variants. Sometimes multiple different allelic forms of the polypeptide are controlled by the same gene, or more than one polypeptide contributes to the same protein (lower right-hand diagram in **Figure 1.6**), but the basic principle remains the same. Gel electrophoresis provides a rapid, cheap way to find variants in natural populations of virtually any species.

1.2.i.b. *The discovery of high molecular diversity*

The introduction of this technique into population genetics resulted in hundreds of "find 'em and grind 'em" surveys of variability (reviewed by Lewontin 1974, 1985). These electrophoretic surveys estimated that a large fraction (as high as 30% in *D. pseudoobscura*) of loci are polymorphic; most natural populations of many other species subsequently studied are highly variable with respect to alleles detected by electrophoresis, apparently overthrowing the "classical" view of genetic variability.

To compare different populations and loci, and to make progress in understanding the factors that control levels of variability, variability must be quantified. An electrophoretic or any other kind of molecular survey of genetic variation yields data on the numbers of each genotype in a sample of individuals from a population. These are summarized in terms of the *genotype frequencies* for each locus, as shown in **Box 1.1**. For codominant alleles, the *allele frequencies* at each locus can be inferred directly from the genotype counts (**Box 1.1**).

Box 1.1 GENOTYPE AND ALLELE FREQUENCIES AT A POLYMORPHIC LOCUS

Assume that there are three variants, S (slow), I (intermediate), and F (fast), at an autosomal allozyme locus in a diploid species. If these represent three different alleles, six genotypes are possible. Because such electrophoretic alleles are *codominant*, heterozygotes and homozygotes for each allele combination can be scored directly from the gels and counted. A specimen set of results is as follows:

Genotype	SS	II	FF	SI	SF	IF	Total
Number	12	21	60	34	46	77	250

These genotypic data can be reduced as follows to a set of the three *allele frequencies*. A given allele (e.g., S) is present in two copies in the homozygotes (*SS*, etc.) and one copy in heterozygotes (*SI*, etc.). With diploidy and autosomal inheritance, every individual carries a paternal and a maternal copy at each locus. With k individuals in a sample (250 in the present case), there are $2k$ gene copies. The allele frequencies are thus

allele S: $(2 \times 12 + 34 + 46)/500 = 0.208$
allele I: $(2 \times 21 + 34 + 77)/500 = 0.306$
allele F: $(2 \times 60 + 46 + 77)/500 = 0.486$

Equivalently, the calculation for a given allele can instead be done by adding the frequency of the homozygote to half the frequencies of the relevant heterozygotes, and dividing by the sample size; which method to use is merely a matter of convenience.

It is easily verified that the allele frequencies sum up to one. Thus only two of them are needed to describe the data. In general, with i alleles, there are i homozygous and $i(i - 1)/2$ heterozygous genotypes to be enumerated, but the same principles as above can be used to derive the allele frequencies. Similar principles apply to data from microsatellite or minisatellite loci (**Section 1.2.iii.c**), where alleles are also codominant.

In haploid organisms, such as bacteria, allele frequencies are directly obtained from counts of the numbers of each genotype. With sex-linked inheritance, males and females must be distinguished, since the *heterogametic* sex is effectively haploid rather than diploid, so that its set of genotypes corresponds to the set of alleles. Allele frequencies can then be averaged across males and females.

For non-codominant alleles, like blood group genes and several types of DNA-based markers such as RAPDs and AFLPs (Ouborg et al. 1999; Schlötterer 2004), genotypes cannot be inferred directly from counts of the numbers of the different phenotypes. Computations of allele frequencies then require some assumptions about the relation between allele frequencies and genotype frequencies, such as random mating and Hardy–Weinberg genotype frequencies (see **Section 1.3.i.a**), and special statistical methods have been developed for these cases (Weir 1996, Chapter 2).

The allele frequencies provide a useful reduced description of the variation at each locus, but are still cumbersome if many loci are studied. "Summary statistics" have thus been developed to measure variability in samples, which are explained in **Box 1.2**. The three main measures are

1. The *proportion of polymorphic loci* (P).
2. The number of different alleles at the locus (A).
3. The *gene diversity* (H). H is also called the *heterozygosity*.

Note that, to measure diversity, all loci surveyed must be included, not only polymorphic ones; if, at a given locus, all individuals in a population or species

Box 1.2 DESCRIBING LEVELS OF VARIABILITY IN A SAMPLE OF ALLELES AT A LOCUS

Given a set of loci with known allele frequencies, we can summarize the level of variability in a sample. Three main measures are in common use.

B1.2.i. Proportion of polymorphic loci (P)

A locus is generally classified as polymorphic if there is at least one minority allele with a frequency greater than some cut-off value, e.g., 0.01 or 0.05. The rationale for this is that rare alleles do not contribute much variability to the population, compared with alleles at intermediate frequencies. A survey that finds variants at 7/30 loci, two of which are not considered polymorphic by this criterion, yields $P = 5/30 = 17\%$ (**Problem 1.1**). While this measure is quick and convenient to calculate, the arbitrary cut-off criterion, which depends entirely on the taste of the investigator, is unsatisfactory and makes it difficult to compare different studies.

For this reason, the following methods are preferable.

B1.2.ii. Allele number (A)

For a given locus, the number of different alleles in a sample is counted, and the mean can be taken over the set of loci that are studied. For the example of allozyme variants in **Figure 1.6**, there are two alleles. Variation at microsatellite loci (**Section 1.2.iii.c**) is often quantified in terms of allele numbers, but variances in repeat numbers have better statistical properties (Goldstein and Schlötterer 1999). This method suffers from the problem that A is highly dependent on the sample size.

B1.2.iii. Gene diversity (H)

For a given locus, a useful diversity measure is the frequency with which a pair of randomly chosen alleles from the sample differ in state (Nei 1973). Similar measures are used in other branches of science, e.g., for species diversity in ecology. For allelic variants, such as allozymes or microsatellites, this measure is called H. In the example in **Box 1.1**, by taking all three pairs of different alleles and finding the sum of twice the products of their frequencies, we obtain

$$H = 2 \times [(0.208 \times 0.306) + (0.208 \times 0.486) + (0.306 \times 0.486)] = 0.627$$

The factor of 2 arises because the order in which the two alleles are chosen is irrelevant.

A quicker way of obtaining the same result is to subtract from 1 the frequency with which a pair of alleles is the same, i.e., to calculate

$$H = 1 - (0.208)^2 - (0.306)^2 - (0.486)^2 = 0.627$$

To summarize the results for a set of loci, we simply take the mean of the H values for each locus (**Problem 1.1**).

H measures the level of variability in a convenient way, since alleles present at low frequencies contribute little to its value; this measure is also not strongly dependent on the sample size. For a single locus with two alleles, H reaches a maximum of 0.5 when each allele has a frequency of 0.5, and is close to zero when either allele is at a low or high frequency.

If the sample came from a randomly mating population, the frequencies of heterozygotes at each locus should be the same as the H values predicted by the Hardy–Weinberg formula (**Section 1.3.i.a**), within the limits of statistical error, since individuals are produced from eggs and sperm that have been combined at random. H is therefore often referred to as the expected *heterozygosity*, H_e, or sometimes just the *heterozygosity*. This usage is not recommended. It is liable to cause confusion, because in non-randomly mating populations, H calculated as above is not the expected heterozygote frequency (see **Section 1.3.ii.b**).

By viewing H simply as a measure of allelic diversity derived directly from the allele frequencies, it can be applied to any type of mating system, and indeed can be used for haploid or X-linked loci. It is advisable to use "heterozygosity" to mean the frequency of heterozygotes (often denoted by H_o, for the observed frequency of heterozygotes), and "gene diversity" or H for the diversity measure.

have the same allele, the locus is said to be *monomorphic* (and the allele is said to be *fixed* in the population). In humans, the mean gene diversity at a locus, estimated from electrophoretic surveys, is about 7%, and the proportion of polymorphic loci is about 30%. Much higher variability is found in insects and bacteria, and species or populations lacking variability are rare (Lewontin 1974, p. 117).

However, these results have several limitations. First, only soluble proteins can easily be studied by standard gel electrophoresis. More importantly, the method detects only amino acid changes that affect the mobility of proteins in gels (mostly only those causing charge changes); these variants probably represent only about one-third of the total possible mutational changes in the amino acid sequence of a protein (Lewontin 1985). Finally, changes in the DNA that do not affect the protein sequences are not detected. Technical improvements have increased the sensitivity of electrophoresis (Lewontin 1985), and allozymes still provide very useful genetic markers, but it was clear by the end of the 1970s that eliminating these limitations in quantifying variability in natural populations would require studies of DNA sequences.

1.2.ii. Detecting variation by restriction mapping

Even before the development of the *polymerase chain reaction* (PCR) made rapid DNA amplification and sequencing possible, the laboratory of Charles Langley pioneered surveys of variation in DNA sequences in natural *D. melanogaster* populations by using *restriction enzymes* (explained in the next paragraph). The variants detected are called restriction fragment length polymorphisms (RFLPs), and they are widely used in genetic mapping. *D. melanogaster* was mostly used in the early surveys of DNA variation, because of the ability to create flies homozygous for any of the three major chromosomes as explained in **Section 1.1.iii.b**. In such homozygotes, only one allele is present at each locus, allowing haploid genotypes to be determined using restriction mapping (e.g., Aquadro et al. 1986).

For studying variability, this approach has largely been superseded by DNA sequencing, so we will not describe it in detail here. In brief, DNA from several independent homozygous lines isolated from a population is subjected to digestion by restriction enzymes, which cut the DNA at short sequences specific to the enzyme in question (usually four or six basepairs (bp) long). If different alleles differ in the presence or absence of the sequence recognized by a given restriction enzyme, the sizes of the bands on gels from different homozygous genotypes will differ, and the numbers and types of differences between the sequences can be inferred. **Figure 1.7** shows some results of a study using this approach.

Studies of several *D. melanogaster* genes found that most of the variants are single nucleotide site changes that simply add or remove a restriction enzyme recognition site, either inside regions coding for polypeptides or in a *noncoding* region (an intron or intergenic region). Short insertions or deletions (*indels*), usually a few bases in length, are also found, usually in the noncoding regions (**Figure 1.7**). Finally, there are occasional large insertions of transposable elements; these are almost invariably found outside coding regions, indicating that natural selection removes such insertions from the coding sequences (see **Section 10.3.iv.d**).

Similar results were obtained for globin genes in humans, although this was technically more difficult than the *Drosophila* work, because diploid individuals

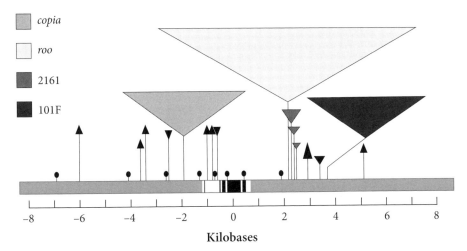

FIGURE 1.7 Insertional and single-nucleotide site variation around the alcohol dehydrogenase (*Adh*) gene in a sample of 49 chromosomes from a *Drosophila melanogaster* population. The noncoding sequences surrounding the gene are in grey. The gene's three protein coding exons are shown in black, and the introns in white (these include all parts present in the messenger RNA but not translated into the protein encoded by the mRNA). The position of the difference in the DNA sequence that causes the amino acid difference that leads to the fast and slow electromorphs (allozyme variants) is in the last exon. To symbolize the different kinds of variants found in this sample of sequences, circles indicate single-nucleotide polymorphisms (explained in the next section), small triangles indicate small indels, and large triangles are transposable element insertions (TEs, see **Section 10.3.iv**, with different types indicated by different shading). [Adapted from Figure 1 of Aquadro et al. (1986).]

rather than homozygous chromosomes were studied (Kazazian et al. 1983). However, reliable diversity estimates could be obtained, because there was a low level of variability and no transposable element insertion variants were present. These differences between flies and humans were confirmed when their DNA was sequenced, as we discuss next.

1.2.iii. DNA sequencing

1.2.iii.a. *Introduction*

Restriction mapping is generally limited to surveying genomic regions in or close to coding sequences that have been cloned. Another limitation of this method is that many variants remain undetected, since the method mainly detects variants within short restriction enzyme recognition sites, which are sparsely scattered over the genome. Sequencing, however, can provide complete information about any part of a genome.

Before we describe the results of DNA sequence-based methods for studying natural variability, we note that their introduction has resulted in a change in

TABLE 1.1 Terminology used to describe different kinds of molecular variants and the diversity measures used for quantifying them

Scale at which data are collected	Terms used	Examples	Diversity measures
Gene, locus	Alleles[1]	Allozyme variants	H, A, P[2] (see **Box 1.2** in **Section 1.2.i.b**)
		Microsatellites	H, variance in repeat number
Sequence from a defined genomic region (sometimes called a "locus")	Haplotypes (also often called "alleles")	Variable DNA sequence, restriction enzyme band	Number of haplotypes (also sometimes denoted by H), haplotype diversity
Single nucleotide sites	Variants, sequence variants	Single nucleotide polymorphisms (SNPs)	Per nucleotide site diversity measures, π or θ_w (see **Box 1.3** in **Section 1.2.iv.b**)

[1]Note that the word "allele" is sometimes used to mean an identified "type" at a locus (e.g., when we refer to the A, B, and O blood group alleles, or in the phrase "allele frequency," which is often also called "gene frequency"), but sometimes it is used to mean one of the sequences in a sample (e.g., when five diploid individuals are studied, we say that the sample consists of 10 alleles).

[2]An allozyme locus is considered polymorphic only if the rarest allele is above a predetermined frequency (often 1% or 5%).

terminology from that used for describing the electrophoretic variation of proteins. To help avoid confusion, **Table 1.1** summarizes the terms used to describe different types of molecular variability. In particular, note that, in the context of DNA sequences, the word "polymorphism" is used for variants that are present at any frequency, whereas for electrophoretic variation only alleles present at intermediate frequencies are classed as polymorphisms. As we shall see in **Section 6.3.ii**, most nucleotide variants are present at low frequencies at any given site.

1.2.iii.b. *DNA sequence polymorphisms; SNPs*
With the invention of PCR amplification for isolating specific regions of DNA, and with modern automated sequencing methods, DNA sequencing of multiple copies of the same region of the genome, sampled from a population or set of populations, has become the most commonly used method for surveying DNA sequence variation. **Figure 1.8** shows an example of a set of allelic DNA sequences sampled from the human population. Such a study is called a *resequencing* study.

Exon 6 — Exon 7 — Exon 8

Homo sapiens Haplotype group	Number in sample	29	215	278	340	463	498	514	547	712	823	906	933	940	945	948	1073	1087	1093	1129	1276	1309	1329	1541	1591	1596	1636	1656	1799	1970	2015	2031	2059	2166	2178
B	24	A	A	A	G	A	C	T	C	G	G	A	T	C	C	G	A	G	A	G	G	C	G	T	A	C	C	C	C	C	G	T	A	T	C
B	4	A	A	A	G	A	C	T	C	G	G	A	T	C	C	G	A	G	A	G	G	C	G	T	A	C	C	C	C	C	G	T	A	T	C
B	2	A	A	A	G	A	C	T	C	G	G	A	T	C	C	G	A	G	A	G	G	C	G	T	A	C	C	C	C	C	A	T	A	T	T
B	2	A	A	A	G	A	C	T	C	G	G	A	T	C	C	G	A	G	A	G	G	C	G	T	A	C	C	C	C	C	A	T	A	T	T
B	2	A	A	A	G	A	C	T	C	G	G	A	T	C	C	G	A	G	A	G	G	C	G	T	A	C	C	C	C	C	G	T	A	T	C
B	4	A	A	A	G	A	C	T	C	G	G	A	T	C	C	G	A	G	A	G	G	C	G	T	A	C	C	C	C	C	G	T	A	T	C
B	2	A	A	A	G	A	C	T	C	G	G	A	T	C	C	G	A	G	A	G	G	C	G	T	A	C	C	T	C	C	G	T	A	T	C
B	1	A	A	A	G	A	C	T	C	G	G	A	T	C	C	G	A	G	A	G	G	C	G	T	A	C	C	C	C	C	G	T	A	T	C
B	1	A	A	A	G	A	C	T	C	G	G	A	T	C	C	G	A	G	A	G	G	C	G	T	A	C	C	C	G	C	G	T	A	T	C
A–	7	G	A	A	A	A	C	T	C	G	G	A	T	C	C	G	G	G	A	G	G	C	G	T	A	C	C	C	G	C	G	T	A	T	C
A+	2	G	A	A	G	A	C	T	C	G	G	A	T	C	C	G	G	G	A	G	G	C	G	T	A	C	C	C	G	C	G	T	A	T	C
Chimpanzee	1	A	G	G	G	G	T	C	T	C	T	C	G	G	T	C	A	T	G	C	A	T	A	G	C	T	T	C	C	T	G	G	G	G	C

Exon 11 — Exon 12 — Exon 13 — Exon 14

Homo sapiens Haplotype group	Number in sample	2190	2390	2397	2450	2495	2660	2683	2886	2906	2942	2977	3006	3055	3212	3244	3268	3549	3617	3846	3916	4010	4062	4067	4075	4141	4192	4380	4653	4712	4715	4907	4938	4974	5022
B	24	T	C	A	A	C	C	C	G	C	C	G	C	G	T	C	T	G	G	A	C	T	C	C	C	C	A	C	C	A	G	C	G	C	A
B	4	T	C	A	A	C	C	C	G	T	C	G	C	G	T	C	T	G	G	A	C	T	C	C	C	C	A	C	C	A	G	C	G	T	A
B	2	T	C	A	A	C	C	C	G	C	C	G	T	G	T	C	T	G	A	A	C	C	C	C	C	C	A	C	C	G	G	C	G	C	A
B	2	T	C	A	A	C	C	C	G	C	C	G	C	G	T	C	T	G	G	A	C	C	C	C	C	C	A	C	C	G	G	C	G	C	A
B	2	T	C	A	A	C	C	C	G	C	C	G	C	G	T	C	T	G	G	A	C	C	C	C	C	C	A	C	C	G	G	C	G	C	A
B	4	T	C	A	A	C	C	C	G	C	C	G	C	G	T	C	T	G	G	A	T	C	C	C	C	C	A	C	C	G	G	C	G	C	A
B	2	T	C	A	A	C	C	C	G	C	C	G	C	A	T	C	T	G	G	A	C	C	C	C	C	T	A	C	C	G	G	C	G	C	A
B	1	T	C	A	A	C	C	C	G	C	C	G	C	G	T	C	T	G	G	A	C	C	C	C	C	T	A	C	C	G	G	C	G	C	A
B	1	T	C	A	A	C	C	C	G	C	C	G	C	G	T	C	T	G	G	A	C	C	C	C	C	T	A	C	C	G	G	C	G	C	A
A–	7	T	C	A	A	T	C	C	G	C	T	G	C	G	T	C	T	G	G	A	C	C	C	C	C	C	A	C	C	G	G	C	G	C	A
A+	2	T	C	A	A	C	C	C	G	C	T	G	C	G	T	C	T	G	G	A	C	C	C	C	C	C	A	C	C	G	G	C	G	C	A
Chimpanzee	1	C	G	T	G	C	T	T	A	C	C	A	C	G	C	T	C	A	G	G	C	A	C	T	T	C	G	T	G	G	A	A	C	C	G

FIGURE 1.8 Sequences of part of the gene encoding the glucose-6-phosphate dehydrogenase enzyme in a worldwide sample of humans and a single chimpanzee. A region of just over 5 kilobases (kb) was sequenced in a sample of 51 humans, including three types of allele (B, A–, and A+); these differ at two amino acids. B alleles encode a protein with high enzyme activity. A– alleles are quite common and encode an allele with only 12% of normal activity (a deficiency allele), whose presence confers malaria resistance, with a 50% reduction in the risk of severe malaria in heterozygous females and in males. A+ is a mild deficiency allele (85% of normal activity and no malaria resistance), which is rarer than A– in human populations. Exons are marked with solid boxes. The diagram shows only sites with variants, either within humans (18 sites), or between the human and chimpanzee sequences. Their positions from the start of the region sequenced are given at the top of each column. SNPs are highlighted and the two amino acid variants are boxed. [Data from Saunders et al. (2002).]

The first thorough comparison of DNA sequences of alleles at the same locus from samples of multiple individuals within a species was done in *D. melanogaster* by Martin Kreitman (1983), using sequences of each of 11 different alleles of the *Adh* gene isolated from *D. melanogaster* collections from around the world, five with the *F* electrophoretic variant of the protein and six with the *S* variant (**Table 1.2**). This study demonstrated high variability per nucleotide site, consistent with the surveys using restriction mapping. Most variants are changes from one nucleotide base to another (called *single nucleotide polymorphisms* or SNPs), although a total of six indels were also found, all in noncoding sequences.

Studies of further loci and species show that, as well as containing many SNPs, noncoding sequences also include indel variants (these are about 20% of all variants in *Drosophila* and 10% in mammals); indels are much less common in protein coding sequences, where they may alter the reading frame, causing loss of the protein's function. As already mentioned, some variation is also contributed by transposable element insertions, at least in some species like *Drosophila*. It is hard to quantify the contributions from these types of variants, and they are often just listed and omitted from quantitative measures of DNA sequence diversity.

A very interesting finding in the *D. melanogaster Adh* gene study (Kreitman 1983), also confirmed by later resequencing work, was that most of the variability involved *silent* changes that do not affect the protein sequence: the sequence differences were either in noncoding regions that do not code for amino acids, or were changes that do not affect the amino acid sequence (*synonymous variants*). In the *D. melanogaster Adh* gene survey, only one amino acid polymorphism was detected—the one already known to cause the difference between the fast (*F*) and slow (*S*) electrophoretic alleles. About 39 amino acid variants would have been found if the same level of variability applied to silent and *non-*

TABLE 1.2 DNA sequence variation at the *Adh* locus of *Drosophila melanogaster*[1]

	Sequence region		
	Intron	Coding	Nontranslated, transcribed
Number of sites sequenced	789	765	335
Number of silent sites	789	192	335
Number of polymorphic silent sites	18	13	3
Fraction of silent sites that are polymorphic	2.2%	6.7%	0.9%

[1]From Kreitman (1983).

synonymous variants that affect the amino acid sequence of a protein (Kreitman 1983).

This observation implies that protein sequences are constrained by natural selection; most nonsynonymous mutations must cause harmful changes to their functioning and be eliminated from the population, so that such variants are rarely found in samples. In other words, only a fraction of nonsynonymous mutations cause "acceptable" changes in proteins. Most variants that *are* detected in coding sequences are, therefore, synonymous ones. Similarly, variability is usually higher in noncoding sequences (most of the genome in higher eukaryotes) than in coding sequences, except for noncoding sequences with important functions, such as regulating gene activity. Similar patterns are also seen in between-species comparisons of DNA sequences (see **Section 6.1.ii.e**). These patterns tell us that *silent* mutations (i.e., mutations in noncoding sequences or synonymous variants in coding sequences) mostly have much smaller effects on fitness than nonsynonymous mutations. In **Chapters 6** and **8** we describe in more detail how this helps us to detect natural selection acting on DNA and protein sequences.

These findings on DNA sequence variability have been confirmed in many different species (see **Section 1.2.iv** below). In species where sequence variability is being intensively studied, such as *Drosophila* and humans, it is now commonplace to survey many different stretches of DNA around the genome for their patterns of variability. For human populations, and for organisms of economic or medical significance, vast resources have been made available to characterize DNA sequence variability. Several million SNPs are now available for use in studies of humans, and special techniques are available that avoid the labor of resequencing the same pieces of DNA multiple times, but instead provide efficient means of detecting known SNPs. The principal methods of analysis remain the same, however, regardless of the sizes of the studies and the technologies used.

There is, however, an important difference between resequencing studies and SNP detection: the variants detected in preliminary surveys to identify SNPs for use in later studies of larger samples tend to be common variants, since a small sample is unlikely to contain rare variants. SNP detection can thus be useful for comparing diversity levels (we shall see in the next section that rare variants contribute little to diversity measures). But for full population genetic studies, this frequency bias causes serious problems. As we show in **Section 6.4**, information about variant frequencies can be very useful for inferring whether selection has been acting on a gene, and what kind of selection has acted or is acting. Resequencing of unbiased samples is the only reliable source of such data, and the development of new high-throughput sequencing methods, such as 454 sequencing (Green et al. 2006), will soon greatly increase our knowledge of DNA sequence variability and variant frequencies, with the prospect of resequencing large numbers of whole genomes (http://www.1000 genomes.org/page.php; http://1001genomes.org/).

1.2.iii.c. *Other types of DNA sequence variants: repeat number polymorphisms and copy number polymorphisms*

In addition to variability arising from "point mutations," many organisms also have loci with highly variable numbers of copies of members of *tandem arrays* of repeated sequences (**Figure 1.9**). There are two such types of variable repeats: *microsatellites* and *minisatellites*. There is no sharp distinction between them; microsatellites involve quite large numbers (sometimes as many as a hundred or

A.

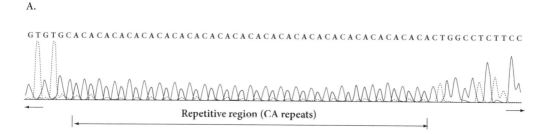

B.

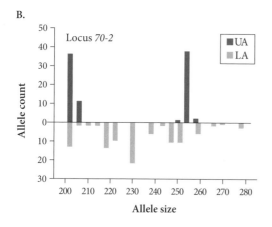

FIGURE 1.9 Microsatellite variability. **A.** Sequence read of a single allele of a microsatellite repeat region in the sea louse *Lepeophtheirus salmonis*. The figure shows the output of a DNA sequencing run. This allele has a succession of alternating CA bases, repeated 12 times (center part of the figure), while at the two ends some unique, nonrepetitive flanking sequence is seen. [Adapted from www.st-andrews.ac.uk/ ~merg/sea%20lice.htm.] **B.** Distribution of repeat numbers of a tetranucleotide microsatellite repeat in alleles sampled from natural populations of a fish (the guppy, *Poecilia reticulata*). The black bars show results from the upper Aripo River (Trinidad), and the grey bars show data from the lower river. [Adapted from Figure 1 of Oosterhout et al. (2006).]

more) of short repeats, usually 2–5 nucleotides long, whereas minisatellites have unit lengths of 15 nucleotides or more (Charlesworth et al. 1994). There are far fewer variable loci of this type than SNPs (about 30,000 variable microsatellite loci are estimated to exist in the human genome), but they often have large numbers of alleles because they mutate at high rates, due to processes such as slippage during replication that can add or lose repeat units. This makes them useful for genetic mapping (Schlötterer 2004), and as genetic markers when it is important to distinguish different individuals, e.g., when determining the source of animal and plant materials (Fang et al. 1997) and in forensic genetics (Goldstein and Schlötterer 1999).

A recent, unexpected discovery is the existence of a substantial variation for the presence and absence of sizeable tracts of DNA sequences in organisms with very large genomes, such as humans, *Drosophila melanogaster* and maize (Iafrate et al. 2004; Morgante et al. 2005; Sebat 2005; Emerson et al. 2008). At least 11 variants of more than 40 kb in size have been found in humans (Kidd et al. 2008).

1.2.iv. Measuring DNA sequence variability

1.2.iv.a. *Introduction*

In the preceding paragraphs, we gave several reasons why DNA sequence data give a more complete picture of the variants present in a sample than other types of diversity data. Another important advantage is the clear relationship between the measures used to quantify diversity for SNPs and the parameters of population genetic models (see **Sections 5.1.ii** and **5.2.iii.e**). This relationship means that we can understand quantitative diversity patterns across populations and species in terms of biological differences between them. Here we show how to quantify DNA sequence diversity, a necessary first step in understanding diversity patterns.

1.2.iv.b. *Statistics for describing variability*

As with electrophoretic variability, we need summaries to describe the large amounts of data generated by surveys of DNA sequence variation. To compare different types of sequences and different populations, we usually measure variability per nucleotide site. **Box 1.3** describes two major "summary statistics," π and θ_w, which are commonly used to measure variability for the SNPs found in a sample of allelic sequences. An alternative approach, which we will not explain in detail here, uses maximum likelihood methods (see **Appendix A2.vi.b**) (Kuhner et al. 1995; Wright and Charlesworth 2004; Felsenstein 2006).

The results for a set of genes or noncoding regions, each of whose variability has been characterized, can be summarized by taking the mean of these statistics over all the loci surveyed, as already explained for electrophoretic loci. A set of sequences taken randomly from across the genome provides an estimate of the genome-wide diversity. Estimates of π and θ_w have now been obtained from a number of different organisms (although, apart from humans, *D. melanogaster*, *Arabidopsis thaliana*, and some bacteria and viruses, only a few genes or genomic regions per species have mostly been studied). When comparing estimates of genome-wide diversity in different populations or species, it is best to use data on synonymous or silent site variability, because such variants are less affected by selection than nonsynonymous variants, as explained above. Many surveys therefore focus on noncoding sequences (although it should not be assumed that these are always free from selection, as they contain many functional sequences, including those controlling intron removal from messenger RNA and sequences controlling gene expression levels).

Box 1.3 DESCRIBING AND QUANTIFYING VARIABILITY IN A SET OF HOMOLOGOUS DNA SEQUENCES

In studying DNA sequence variation, we often refer to a stretch of DNA from a region of the genome as a *locus* (even if only part of a gene was sequenced, or even if the sequenced region is noncoding), and the independently sampled sequences as *alleles*. This is very different from the terminology used for allozyme variability, where locus was used to refer to a gene coding for a polypeptide identified on an electrophoretic gel, and alleles mean variants at such a locus that segregate from each other in a Mendelian fashion (see **Table 1.1** of **Section 1.2.iii.a**). Two summary statistics are widely used for sequence data.

B1.3.i. Nucleotide diversity (π) in a sample of k homologous sequences (alleles)

The first measure that we describe is analogous to the gene diversity, H, described in **Box 1.2** in **Section 1.ii.b** for allelic variants that were scored in a sample (e.g., allozymes), but is applied to single nucleotide sites rather than whole genes. It measures the frequency with which a randomly chosen pair of alleles from the set differ at a given nucleotide site. Values for individual sites are averaged over all sites in the sequence to provide a summary statistic for the whole sequence.

Nucleotide diversity can be calculated from data on a sample of alleles by determining the sum of the numbers of differences between all possible pairs of alleles and then dividing by the number of sequence pairs that were compared [with k independent alleles, this equals $k(k-1)/2$], and by the number of bases studied. In the example from **Figure 1.8** (**Section 1.2.iii.b**), the last three alleles, labeled B, A– and A+, are 5022 bases long. Of these, the B allele differs at seven nucleotide sites from the A– allele, and at four (different) sites from the A+ allele, and the A– and A+ alleles differ at three sites. The sum of the pairwise differences between all three combinations of sequences is $7 + 3 + 4 = 14$, so that the mean difference per site per allele pair is just under 0.1%:

$$\pi = 14/(3 \times 5022) = 0.00093$$

B1.3.ii. Watterson's θ (θ_w) for a sample of k DNA sequences

Another way to measure diversity would simply be to count the number of polymorphic sites (sites in the sequence that have variants in the sample). This is sometimes denoted by S_k, and is used by human geneticists (in statements such as "two variants per kilobase"). This is clearly unsatisfactory, because these numbers will be larger in larger samples. However, such counts can be corrected to take into account the number of sequences in the sample, k. This uses the quantity a_k, which is given by **Equation B1.3.1**:

$$a_k = 1 + 1/2 + 1/3 + \cdots + 1/(k-1) \qquad (\text{B1.3.1})$$

If k is large, a_k is approximately equal to $\ln(k)$, the natural logarithm of k.

When S_k is divided by the "correction factor" a_k and by the number of bases in the sequence (to get a per nucleotide diversity estimate), we obtain the statistic called *Watterson's* θ, or θ_w. If the population is at equilibrium, and there is no selection on the variants at this locus, the expected value of $\theta_w = S_k/a_k$ is the same as that of π (the basis for this is derived in **Section 5.1.iii.c**, but here we simply state the result). When divided by the sequence length, this quantity measures diversity per nucleotide site.

For the data in the example just explained, we have $S_k = 7$ and $a_k = 1.5$. In this case, θ_w is exactly the same as π:

$$\theta_w = 7/(5022 \times 1.5) = 0.00093$$

Figure 1.10 shows some representative diversity results. The highest genetic diversity values are found in microbial populations, with their vast numbers of individuals; insects and outcrossing species of plants are in the middle of the range, and humans and inbreeding species of plants and animals are at the low end. The reasons for these patterns will be discussed later, mainly in **Section 5.2.iii**.

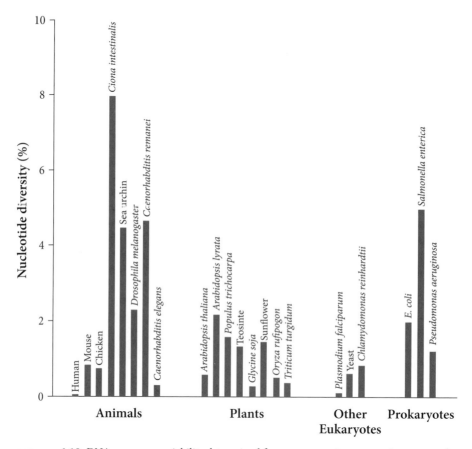

FIGURE 1.10 DNA sequence variability determined from resequencing surveys in a range of different species. The *y* axis shows nucleotide diversity estimates (for synonymous sites where available, or, in a few cases, noncoding sites) from the species named on the *x* axis. [Data are for species with at least six loci sequenced and come from the following sources: Asthana et al. (2007), Berlin et al. (2008), Caicedo et al. (2007), Charlesworth and Eyre-Walker (2006), Cutter (2005), Cutter et al. (2006), Haudry et al. (2007), Hyten et al. (2007), Ingvarsson (2005), International Chicken Polymorphism Map Consortium (2004), Jolley et al. (2005), Kolkman et al. (2007), Moeller et al. (2001), Mu et al. (2007), Nordborg et al. (2005), Ross-Ibarra et al. (2008), Schacherer et al (2009), Schmid et al. (2005), Sea Urchin Genome Sequencing Consortium (2006), Shapiro et al. (2007), Small et al. (2007), Smith and Lee (2008). The unpublished mouse data were kindly provided by D. L. Halligan and P. D. Keightley.]

Although levels of diversity per nucleotide site are small, even the small, gene-dense genome of the bacterium *Escherichia coli* has 4.2 million nucleotide sites, including around 900,000 silent or synonymous sites (Blattner et al. 1997). The mean value of silent θ_w in *E. coli* is about 0.02 (**Figure 1.10**), so that the formulae in the second part of **Box. 1.3** imply that the total number of nucleotide sites segregating in a sample of 100 *E. coli* genomes is approximately $9 \times 10^5 \times 0.02 \times \ln(100)$, i.e., about 82,893, or 2% of the total number of sites. Similar calculations show that several million SNPs are present genome wide in the populations of most multicellular organisms, which have much larger genomes.

1.2.v. Haplotypes

1.2.v.a. *Definition and use in measuring variability*
The number of distinct alleles possible at a given locus (a defined region of the genome) depends on the number of variants at each nucleotide site in the stretch of genome that has been sequenced multiple times, and on the number of combinations of these different variants. In a sample from a population, the term *haplotype* is commonly used to refer to a sequence that carries a particular set of variants (**Figure 1.11**), and a sample can be described by listing the different haplotypes that it contains, and their frequencies. Each site in a sequence

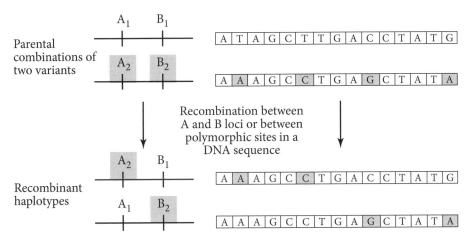

FIGURE 1.11 Diagram to show the effect of reciprocal recombination (crossing over). The left-hand part shows recombination between two genes, A and B, each with two different alleles (indicated by subscripts and shading). The right-hand part shows a crossover between two different alleles of a gene with several sequence variants, each with two different nucleotides (indicated by shading). The crossover event illustrated brings the two left-hand sequence variants into a new haplotype that does not have the two right-hand variants, and the reciprocal haplotype is also produced.

usually has just two alternatives, since variable sites are infrequent, and the presence of more than two variants at the same site is thus fairly unlikely. (Since there are four different bases, it is in reality possible to have three, or even all four, possible variants at a given nucleotide site, and this is occasionally seen in species with very high levels of variation.)

Nevertheless, two haplotypes from a population will usually differ at multiple sites, unless only a short length of sequence is studied. Sequence analyses of allozyme variants have produced many examples in which the alleles identified by electrophoresis differ by multiple amino acid and silent site differences (Eanes 1999; Filatov and Charlesworth 1999; Wheat et al. 2006), including the human example shown in **Figure 1.8**.

As in the case of data that are described simply in terms of allele frequencies, such as allozymes (see **Box 1.2** in **Section 1.2.i.b**), a simple way to measure DNA sequence variability is just to count the number of different haplotypes in a sample. Variability in human genes is often described in this way; for instance, the high allele numbers at the *major histocompatibility* (MHC) loci of mammals, with as many as 500 haplotypes at the HLA-B locus, are often used to illustrate the high diversity of these genes (Garrigan and Hedrick 2003). This is, however, a poor measure of diversity. It is highly sample-size-dependent, like the counts of numbers of polymorphic sites in a set of sequences (**Box 1.3**), and it also increases with the length of genome sequenced. One can also measure haplotype diversity, just as with gene diversity (see **Box 1.2**); this takes the sample size into account, but it again depends on the length of the region sequenced. It has the further disadvantage of ignoring all the information from the numbers and positions of sequence differences between the haplotypes (nonidentical haplotypes are treated in the same manner, whether there is just one nucleotide difference, or many).

1.2.v.b. *The effect of recombination on haplotype variability*
Another fundamental problem with measuring DNA sequence diversity by counting haplotypes (at least in nuclear genes) is that genetic recombination among different haplotypes affects haplotype numbers. Recombination does not affect diversity per nucleotide site, but merely distributes variants differently among haplotypes, in a process in which existing variants "migrate" between them (**Figure 1.11**). The same number of polymorphic nucleotide sites can thus be present in different numbers of haplotypes. With two alternative variants at each variable nucleotide site in a sequence, two polymorphic sites can generate four haplotypes (**Figure 1.11**); if there are S variable sites, the maximum number of possible combinations is 2^S. Usually, only a subset of the possible haplotypes is observed in a sample of the size commonly used in surveys of DNA sequence variation. If $S = 10$, for example, $2^{10} = 1024$ different sequences

are possible, and only a fraction of these can be found in a sample of 50 diploid individuals (with 100 alleles, and thus at most 100 different haplotypes).

An important use of haplotypes is to compare their frequencies with those predicted from population genetic models. If variants at different polymorphic nucleotide sites in the sample are distributed into haplotypes independently of each other, the expected frequency of a haplotype is simply the product of the frequencies of the variants at each site in the sequence, from the rule for combining independent probabilities (**Appendix A2.ii**). Departure of the frequencies of combinations of polymorphic variants from this random expectation is called *linkage disequilibrium* (often abbreviated to LD). The problem of describing LD and explaining patterns of LD is discussed in detail in **Chapter 8**. We simply note here that genetic recombination among different sites eventually eliminates LD in a very large randomly mating population if no selection is acting to build it up. The speed of this process is proportional to the recombination frequencies among the sites. It is very slow for sites within the same gene, but fast for sites more than a few kilobases apart (**Section 8.2.ii**); for non-recombining genomes, like the mitochondria of many organisms, or Y chromosomes, LD cannot break down in this way. However, in species whose mitochondria can be passed through the male parent, individuals can occasionally become "heteroplasmic" (i.e., heterozygous) and recombination in such individuals can be detected in their progeny, and from the presence of recombinant haplotypes in the population (McCauley and Ellis 2008).

1.3. THE CAUSES AND MAINTENANCE OF VARIABILITY

The evidence for abundant genetic variability in natural populations, from the level of the measurable phenotype down to the nucleotide site, prompts the question of where it comes from and what maintains it.

1.3.i. Conservation of variability by the mechanism of inheritance

One of the most important results of early population genetics was the answer to the question: *what happens in the absence of any evolutionary forces?* In other words, what happens to the frequencies of variants in a population if selection is not acting on them (i.e., they are selectively neutral), there is no mutation or migration to bring in new variants, and the population is so large that random fluctuations in genotype frequencies can be neglected? (The effects of population size are treated more formally in **Chapter 5**.) Contrary to what was believed by Darwin and his contemporaries, *nothing changes* under these conditions.

This is because inheritance is particulate rather than blending in nature, i.e., the contributions from either parent of an individual at a single site in the genome are transmitted intact through the individual's gametes. With Mendelian inheritance in a sexual population, variants present in a very large parental population under the conditions just specified will also be present in their progeny at the same frequencies as in the parents; variation is therefore conserved. The same holds for asexual reproduction, since each different genotype reproduces itself exactly. In contrast, as discussed in more detail at the beginning of **Chapter 2**, if inheritance were blending, as believed by Darwin and most of his contemporaries (see **Section 1.1.ii.c** above), populations would lose their variability (Fisher 1930b, Chapter 1).

1.3.i.a. *Hardy–Weinberg equilibrium*

The mathematical formulation of this conservation of variability is embodied in the *Hardy–Weinberg equilibrium* (Hardy 1908; Weinberg 1908). Consider a very large, diploid, sexually reproducing population with *discrete* (nonoverlapping) generations (e.g., an annual species of plant or insect). Parents in each generation reproduce simultaneously and then die, leaving offspring behind to start the next generation. Imagine two alternative autosomal variants (e.g., two different nucleotides at a particular site in the DNA), A_1 and A_2. Their frequencies among the breeding parents can be written as p and q (if the variant frequencies differ initially between males and females, they will become equal after one generation, since males and females contribute equally to each offspring).

We assume that there is *random mating*—individuals choose mates without regard to their genotypes. The new genotype frequencies in the zygotes are then produced by randomly combining the alleles in proportion to their frequencies in the parents (**Problem 1.2**), i.e., we have:

Genotype	A_1A_1	A_1A_2	A_2A_2	
Frequency	p^2	$2pq$	q^2	(1.1)

Clearly, the same genotype frequencies will be produced in each succeeding generation. Given the assumption that no evolutionary forces are acting on the population, the genotype and allele frequencies remain constant.

Note that this instantaneous achievement of equilibrium with respect to the genotype frequencies is valid only for a discrete generation population. In populations with *overlapping generations*, where individuals born at different times can reproduce at the same time, the approach to constant allele frequencies and Hardy–Weinberg equilibrium is only asymptotic, although rapid (Norton 1928; Charlesworth 1994b, Chapter 2). Even in a discrete generation population, an X-linked gene will approach Hardy–Weinberg equilibrium via an oscillatory path if the allele frequencies in males and females are initially different (**Problem 1.3**).

The genotype frequencies at single loci or nucleotide sites in many natural populations and human populations show close agreement with Hardy–Weinberg expectations, which can be evaluated by simple statistical tests such as the chi-squared goodness-of-fit test (**Appendix A2.v.f**); an example is given in **Problem 1.4**.

1.3.i.b. *Gene conversion and mutation*

The conservation of neutral variability due to discrete inheritance is not quite absolute, however, since Mendel's genes are not completely unchanging because of the occasional occurrence of *mutations*. We discuss mutation in **Section 1.3.iii** below. Here we deal with another process that alters the states of genes, called *gene conversion*. To a very good approximation, heterozygous parents transmit their two alternative variants to their gametes in a Mendelian 1:1 ratio, but this is not always true for DNA sequence variants. During meiosis, the two alleles at a locus can exchange short tracts of DNA (Hilliker and Chovnick 1981). Sometimes this exchange is *biased* and one of the two variants present in a heterozygote has a slightly greater probability of transmission to the gametes than the other (Marais 2003). The resulting slight departures from the expected 1:1 ratios are usually undetectable, except in very large breeding experiments, but they will eventually lead to replacement of the variant that does not benefit from this bias by the alternative variant that benefits (even though no new variants arise). Thus, a variant can spread throughout a population because of biased gene conversion, just as if it had a selective advantage (when a variant has increased and is present in all individuals in a population or species, it is said to be *fixed*). Biased gene conversion therefore leads to fixation and loss of variability.

Subject to these qualifications, Mendelian inheritance means that populations retain variability unless some evolutionary force acts on them. Furthermore, variability is conserved even in populations that do not mate at random, as we will now discuss.

1.3.ii. Nonrandom mating and identity by descent

Most real populations are not random mating. Inbreeding, i.e., matings between close relatives, is common in hermaphrodite organisms (see **Section 9.1.iii.d**), and limited dispersal means that, even in species with separate sexes, relatives may mate more often than if mating were random in the species as a whole. A different form of nonrandomness involves preferential mating between individuals with similar genotypes (*positive assortative mating*) or between individuals of unlike genotypes (*negative assortative mating*; an extreme example is chromosomal sex determination, e.g., when all matings involve an XY male and an XX female, as in humans and *Drosophila*).

1.3.ii.a. *Identity by descent and coalescence of alleles in common ancestors*
We concentrate here on the effects of inbreeding. Methods for quantifying the
degree of inbreeding of individuals and populations were invented early in the
history of population genetics by Sewall Wright, using "correlations between
uniting gametes," together with methods for calculating these correlations from
information on pedigrees, or from the population's mating system (Wright
1921). This approach has largely been replaced by the concept of *identity by
descent*, devised independently by Charles Cotterman (1940) and Gustave
Malécot (1948, 1969). We will use this method, which is simpler and has wide
applications in population and statistical genetics. This important concept
appears again in **Section 3.1.v.d**, when dealing with kin selection, and in
Section 5.1.i.b, when we describe the evolutionary effects of finite population
size and consider the ancestry of sequences at a locus. Identity by descent is also
the basis for coalescence theory, which is now widely used for modeling molecular variation, as discussed in **Chapters 5** and **6**.

Two alleles are said to be identical by descent or *i.b.d.* if they trace their
ancestry back (or *coalesce*) to the same ancestral allele (**Figure 1.12**); this
includes the case when one allele is descended from the other. We are not concerned here with the states of these alleles with respect to their DNA sequences
(if mutations occurred during the time interval between the present time and
the time when the two alleles coalesce into their ancestral allele, they might differ in sequence, but, according to the definition just given, the alleles are still
i.b.d.). If we take two individuals C and D from a pedigree (**Figure 1.12A**), we
can calculate the probability, f_{CD}, that an allele at an autosomal locus, chosen
randomly from individual C, is i.b.d. with an allele chosen randomly from D.
This is called the individual's *coefficient of kinship* (sometimes known as the

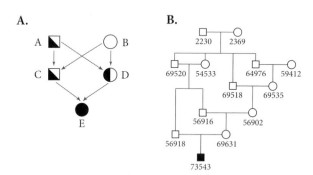

FIGURE 1.12 **A.** Pedigree of a full-sib family. **B.** The pedigree of a male song sparrow
(*Melospiza melodia*) hatched on Mandarte Island, near Vancouver, British Columbia, Canada, in
1992. Males are represented as squares, whereas females are represented as circles. This bird's
maternal grandfather is also his paternal uncle. Pedigrees such as this allow the calculation of
inbreeding coefficients. [Adapted from Keller and Waller (2002).]

coefficient of consanguinity or, alternatively, the *coancestry*). The *inbreeding coefficient* of a progeny individual E produced by a mating between C and D is defined as the probability f_E that the two alleles at a locus in E are i.b.d. Since E's alleles are received randomly, one from each parent, f_E is the same as f_{CD}.

We can work back along a given pedigree from a starting pair of individuals, C and D, since their coefficient of kinship can be related to those of their four parents: these are denoted by A and B for C, A' and B' for D (in the example in **Figure 1.12A**, A and A' are the same individual, and B is the same as B'). For an autosomal locus, we use the fact that one allele of an individual comes from its mother and the other from the father, with equal probability under Mendelian rules. Taking into account all four equally probable combinations of origins of alleles, we get:

$$f_{CD} = \frac{1}{4}\left(f_{AA'} + f_{AB'} + f_{BA'} + f_{BB'}\right) \tag{1.2}$$

Applying this rule consistently to a specified pedigree, we can obtain a general formula for the f value of a pair of individuals in terms of the inbreeding coefficients of all their common ancestors in the pedigree (**Box 1.4**). **Figure 1.12B** shows an example of a pedigree for a natural population.

Box 1.4 COEFFICIENTS OF KINSHIP IN A PEDIGREE

In an arbitrary pedigree, we can designate the two parents of an individual E as C and D (e.g., **Figure 1.12A**). C and D may trace their descent back to a number m of common ancestors. Denote these common ancestors by A_i ($i = 1, 2, ..., m$; it is, of course, possible, for one or more of these ancestors to be an ancestor of one or more of the others). By examining the pedigree, we can identify each path that connects C and D to A_i.

Looking down one of these paths, there is a probability of one-half (for an autosomal locus) that the same allele of A_i is transmitted to the two progeny of A_i that constitute the next step in the path. There is also a probability of one-half that a maternal allele is transmitted to one of these progeny and a paternal allele to the other. In this case, the probability that they are i.b.d. is the inbreeding coefficient of A_i, denoted by f_{A_i}. The net probability that the two genes are i.b.d. is thus $(1 + f_{A_i})/2$. This is equivalent to the inbreeding coefficient of an individual produced by self-fertilization of A_i.

Tracing the transmission of alleles from these two progeny individuals, let there be $n_i - 1$ further links in the path connecting A_i to C and D, where n_i is the total number of individuals in the path, including A_i, C, and D. At each of these links, there is a probability of one-half of handing on an allele inherited from A_i. The net probability that an allele sampled from C is identical with one from D, as a result of transmission down this path, is thus:

$$\left(\tfrac{1}{2}\right)^{n_i}\left(1 + f_{A_i}\right)$$

The coefficient of kinship of C and D can be found from this formula by adding up the contributions from all possible mutually exclusive paths leading from the common ancestors, as shown in **Equation B1.4.1**:

$$f_{CD} = \sum_i \left(\tfrac{1}{2}\right)^{n_i}\left(1 + f_{A_i}\right) \tag{B1.4.1}$$

A similar result can be obtained for an X-linked locus (with male heterogamety) by noting that males have only one X chromosome, and therefore transmit a given allele with probability one, so that a contribution from a male to his daughter is not discounted by one-half. Paths with two successive males are ignored, since males do not transmit X chromosomes to their sons. Transmission through females is the same as for the autosomal case, and so the numbers of females in the paths are counted in the same way as for the autosomal case. Males and females are interchanged for the case of female heterogamety.

For further details on inbreeding in pedigrees, see Crow and Kimura (1970, Chapter 3), Nagylaki (1992, Chapter 9), and Falconer and Mackay (1996, Chapter 5).

The earliest individuals for which we have pedigree information are descended from individuals that were members of the population, but are not themselves represented in the pedigree. To deal with such unknown ancestry, we can define an initial generation of the population for which all alleles are considered *not* to be i.b.d., i.e., for which $f = 0$. Alternatively, we can think of $1 - f$ for an individual as the probability that two alleles at a locus are non-i.b.d., *relative* to the probability for a random pair of alleles drawn from the initial generation; this value tends to decrease over time.

1.3.ii.b. *Genotype frequencies with inbreeding and assortative mating*
The measures just defined can be extended to a population. If we sample two
alleles from a randomly chosen individual, the probability that these are i.b.d.
defines an inbreeding coefficient, f, for the population. As we shall soon see, this
probability can sometimes be calculated from the population's mating system,
but for now we simply treat it as a parameter that describes the population's
level of inbreeding. The genotypic makeup of an inbred population is determined jointly by the variant (or allele) frequencies and f, as shown in **Box 1.5**.
Instead of the Hardy–Weinberg frequencies, the genotype frequencies for a site
with two variants are now the following:

$$\begin{array}{cccc}
\textbf{Genotype} & A_1A_1 & A_1A_2 & A_2A_2 \\
\textbf{Frequency} & p^2 + fpq & 2(1-f)pq & q^2 + fpq
\end{array} \qquad (1.3)$$

As one would expect, when there is random mating ($f = 0$), we recover the
Hardy–Weinberg frequencies.

Box 1.5 GENOTYPE FREQUENCIES IN INBREEDING
POPULATIONS

Consider an autosomal site or locus with two variants, A_1 and A_2, with
frequencies p and q. By definition, if we sample one of the two alleles of an
individual taken from the population, the probability that it is A_1 is p; the
chance that the allele is i.b.d. with the other allele of the individual is f.
Neglecting mutation over the (relatively short) time that it takes for the
two alleles to coalesce into their ancestral allele, there is a probability f that
the second allele also has A_1 at the site in question. Thus, there is a net
contribution fp to the probability that the individual has genotype A_1A_1.

There is a probability $1 - f$ that the other allele is nonidentical, in which
case the probability that the individual is A_1A_1 is p. This means that the
frequency of A_1A_1 is equal to $fp + (1 - f)p^2 = p^2 + fpq$. Similarly, the frequency of A_2A_2 is $q^2 + fpq$; by subtraction from 1, the frequency of A_1A_2 is
$2(1 - f)pq$.

With two polymorphic variants at a site in a genome, the three genotype frequencies sum to one, so there are only two independent frequencies (two
degrees of freedom). It follows that, if we specify the variant frequency q, only
one other variable is needed to describe the population. The representation
above is, therefore, much more general than for just the case of inbreeding.

Indeed, f can be viewed as a measure of departure from random mating (Wright 1951), which can take values from –1 to +1. This is called the *fixation index*, often denoted by F in order to distinguish it from the more restricted concept of identity probability (see **Section 7.1.i.a**). Positive F represents an excess of homozygotes over the random-mating expectation, and negative F represents heterozygote excess.

It is obvious that inbreeding generates an excess of homozygotes in a population, with a maximum value of $F = 1$. It is important to understand that homozygosity is not the same as an absence of genetic diversity in the population. **Equation 1.3** shows that it is quite possible for highly inbred populations to have nonzero diversity (**Problem 1.5**).

Positive assortative mating (i.e., preferential mating between individuals with similar genotypes) also generates homozygote excess, but this is caused by identity in state of alleles, and not directly by identity by descent. Negative assortative mating has the opposite effect. We do not deal with the detailed theory of assortative mating in this book (see Crow and Kimura 1970, Chapter 4; Bulmer 1980, Chapter 8; Nagylaki 1992, Chapters 5 and 10). However, it is worth pointing out an important difference between the effects of inbreeding and assortative mating; inbreeding leaves allele frequencies unchanged, so that a change in the mating system to a higher frequency of matings between relatives simply increases the value of F, i.e., it changes the genotype frequencies but not the allele frequencies. Assortative mating, however, can affect the fertilities of different genotypes, and may thus act as a form of selection that changes allele frequencies.

1.3.ii.c. *Regular systems of mating*

In some simple cases, the inbreeding coefficient can be calculated from a set of rules for the frequencies of the different types of matings in a population. The simplest case is when there is a mixture of random mating and self-fertilization with frequency S (the probability that a zygote is the product of self-fertilization, or the *selfing rate*), as in many species of hermaphroditic organisms, such as many flowering plants. This is called the "mixed mating model" (Brown and Allard 1970; Ritland 1990b). We assume that the progeny of a cross between unrelated individuals has an f value of 0, i.e., we ignore inbreeding other than selfing. **Box 1.6** shows that the population approaches an *equilibrium* value, f^*, given by **Equation 1.4**:

$$f^* = \frac{S}{(2 - S)} \tag{1.4}$$

If a population is 100% selfing, the inbreeding coefficient quickly approaches unity, and all individuals will become homozygous. Many natural populations of highly selfing animal and plant species are close to this state

(Vogler and Kalisz 2001), including two important model organisms whose genomes have been sequenced, the weedy plant *Arabidopsis thaliana* and the nematode worm *Caenorhabditis elegans* (Abbott and Gomes 1988; Barrière and Félix 2005).

Box 1.6 INBREEDING COEFFICIENTS IN A SELF-FERTILIZING POPULATION

Assume that a proportion S of zygotes are produced by self-fertilization and a proportion $1 - S$ by outcrossing, i.e., by matings between different individuals drawn randomly from the population. S could arise in one of two ways. Either all individuals might self-fertilize a proportion S of their female gametes, or the environment might cause a proportion S of the individuals in the population to reproduce by self-fertilization, while the others outcross; these represent extreme assumptions, and biological reality is probably in between.

Let the inbreeding coefficient in a given generation be f. The inbreeding coefficient of the progeny of self-fertilization is $(1 + f)/2$, by the result derived in **Box 1.4** of **Section 1.3.ii.b** above. The inbreeding coefficient of a zygote produced by outcrossing is set to zero. The inbreeding coefficient in the next generation, f', is therefore given by **Equation B1.6.1**:

$$f' = \frac{S}{2}(1 + f)$$

(B1.6.1)

The change in inbreeding coefficient, Δf, is:

$$\Delta f = \frac{S}{2} - \left(1 - \frac{S}{2}\right)f$$

(B1.6.2)

For small f, $\Delta f > 0$, unless $S = 0$. Also, unless $S = 1$, at large f values we must have $\Delta f < 0$. The inbreeding coefficient therefore approaches an equilibrium at which $f' = f$. Inserting this into **Equation B1.6.1** and rearranging, we find that the equilibrium inbreeding coefficient f^* for a population with a given value of S is given by **Equation 1.4** of the main text.

In reality, S may change over time, due to environmental changes (e.g., changing pollinator abundance in a plant, or changing density of individuals available for outcrossing), or there may be genetic differences between individuals, so that not all have the same selfing rate. Evolutionary changes in selfing rates will be discussed in **Section 9.1**.

1.3.iii. Mutation: the source of new variants

1.3.iii.a. *The nature of mutations*

We have seen how genotype frequencies at variable sites in the genome are generated, and that these variants are conserved in the absence of evolutionary forces such as selection and genetic drift. This does not, however, answer the question of where the variants at these sites come from in the first place. The answer is *mutation*. Mutations are alterations in the genetic material, and are the ultimate source of the genetic variability that allows evolutionary responses to natural selection and is used in animal and plant improvement.

Before discussing mutations, we note that the expression of a particular gene in a given cell can be modified by signals from other cells or environmental factors. Alterations in gene activity can be transmitted to daughter cells, as regularly happens in the differentiation of cell types during development. These alterations, however, are rarely passed between generations. Some examples of such *epigenetic* inheritance have been discovered in plants (Henderson and Jacobsen 2007) and mammals (Reik 2007), mostly involving changes in the methylation of nucleotide bases. Such epigenetic changes are, however, usually less stable than mutational changes in nucleic acid sequences, and hence are unlikely to be important in evolution. If they are stably inherited, they will obey Mendelian rules, and hence can be treated in a similar way to DNA sequence variants; for a general review of this question, see Richards (2006).

Mutational changes to DNA sequences are random events caused by damage in resting cells or errors in replication. These changes are usually transmitted stably across the generations. In animals with a separate "germ line" (cells in the egg or sperm cell lineages), only mutations in germ-line cells are transmitted to the progeny. Mutational changes range from gross changes in genome organization, such as *polyploidy* (multiple complete sets of chromosomes), chromosome rearrangements (*deletions, duplications, inversions,* and *translocations*), down to smaller-scale changes, such as insertions of transposable elements, and the so-called *point mutations*, which are the most frequent. These involve any of the categories of variants defined above, including substitutions of one nucleotide base for another (*nucleotide substitutions*), or additions or deletions of small numbers of bases (*indels*). A final category of mutation is the transfer of a stretch of DNA sequence from one species to another, or from the mitochondria or plastids to the nucleus (Lin et al. 1999; Race et al. 1999; Abdallah et al. 2000). Cross-species transfers have played an important role in bacterial variation and evolution (Hao and Golding 2006), but seem to be very infrequent in the nuclear genomes of eukaryotes, except for some cases involving transposable elements (Kidwell 1992; Robertson and Lampe 1995; Bartolomé et al. 2009).

1.3.iii.b. *Mutation rates*

Mutation rates are usually extremely low; the frequencies of single nucleotide changes have been estimated to be about 10^{-10} per nucleotide site per cell division in the bacterium *E. coli* (Drake et al. 1998), and 3.3×10^{-9} per cell division in the yeast *Saccharomyces cerevisiae* (Lynch et al. 2008). In multicellular organisms, nuclear genes have been estimated to have mutation rates of 2.1×10^{-9} per nucleotide site per generation in the nuclear genome of *C. elegans* (Denver et al. 2009), 5×10^{-9} per nucleotide site per generation in the nuclear genome of *D. melanogaster* (Haag-Liautard et al. 2007; Keightley et al. 2009), and 2×10^{-8} per nucleotide site per generation in humans (Nachman and Crowell 2000; Kondrashov 2003). The results for humans were obtained in two ways; the first was to use comparisons of noncoding DNA sequences between humans and chimpanzees, using the principles described in **Section 6.1.ii** (Nachman and Crowell 2000). The second used data on the rates of occurrence in the human populations of disease-causing new mutations with known changes in DNA sequences; mutations in 20 genes were studied (Kondrashov 2003). The two methods gave very similar results. The data for other organisms come from experiments in which the rates of occurrence of mutations changing sequences were measured. **Figure 1.13** shows how the *D. melanogaster* nucleotide mutation rate was estimated. The yeast and *C. elegans* estimates were based on data from sequences of the genomes of individuals allowed to accumulate mutations in similar experiments.

Mutation rates in most DNA-based genomes are low because there are elaborate molecular mechanisms for correcting errors in DNA replication and repairing DNA damage, and mutation rates are greatly increased by mutations in the genes controlling them (Drake et al. 1998). Genomes that lack such mechanisms, such as RNA viruses (Drake et al. 1998) and animal mitochondrial DNA (Haag-Liautard et al. 2008), have much higher mutation rates. In multicellular organisms with a germ line, mutations can arise in germ-line cell lineages before meiosis. A cluster of identical mutations may therefore appear in a brood of offspring, rather than just a single mutation (Woodruff and Thompson 1992, 2005). The above mutation rates for multicellular organisms effectively count each mutation in a cluster when estimating the mutation rate, as is appropriate for understanding the contribution of mutations to the rate of DNA sequence evolution (**Section 6.2.ii.b**).

1.3.iii.c. *Mutation and evolutionary change*

Although mutations are important as the source of variation for evolution, their low rate of occurrence means that mutation is an ineffective force for evolutionary change at an individual site in the genome. People unaccustomed to thinking about evolution in terms of populations often forget that a new mutation is

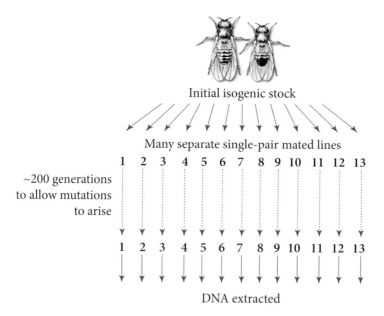

Initial isogenic stock

Many separate single-pair mated lines

1 2 3 4 5 6 7 8 9 10 11 12 13

~200 generations
to allow mutations
to arise

1 2 3 4 5 6 7 8 9 10 11 12 13

DNA extracted

FIGURE 1.13 Method for using mutation accumulation lines to estimate the mutation rate in *Drosophila melanogaster* (Haag-Liautard et al. 2007). After many generations of mutation accumulation, mutations in a given sequence can be detected by sequencing parts of the genome in flies from all of the lines, or by methods that detect that the sequence of a genomic region has changed from that in the initial inbred strain.

present initially in only a very small number of individuals. Most species with any prospect of continued survival contain an average of many thousands of individuals in each generation, so that the injection of at most a handful of gene copies into the population by a mutational event is an insignificant evolutionary change. To contribute to evolutionary change, a mutation must increase to at least a moderate frequency in the population.

Given enough time, however, a given mutational change may occur repeatedly at a site in a genome in a large population, so that the frequency of the mutation is driven up by *mutation pressure*. Consider the simple case of two alternatives, e.g., nucleotide pairs GC and AT at a particular site in a DNA sequence. Let GC mutate to AT at a rate u per generation, and AT mutate back to GC at rate v. If the population is very large, so that random changes in frequencies can be ignored, the frequency of GC in the next generation, p', is given by **Equation 1.5**:

$$p' = v(1 - p) + (1 - u)p \qquad (1.5)$$

The change in frequency, Δp, is given by **Equation 1.6**.

$$\Delta p = p' - p = v(1 - p) - up \qquad (1.6)$$

This shows that the rate of change in variant frequencies due to mutation pressure is of a similar magnitude to the mutation rate: if GC is initially very rare (i.e., p is close to 0), it will increase in frequency at a rate of approximately v. **Equation 1.6** also shows that there is an equilibrium in allele frequency at $p^* = v/(u + v)$ when $\Delta p = 0$ (**Problem 1.6**). This equilibrium is *stable*, i.e., it is approached when p is either above or below p^*. This model is easily generalized to allow for mutations among all four possible nucleotides at a site, but then the methods of matrix algebra must be used to determine the composition of the equilibrium population (**Problem 1.7***).

In principle, therefore, two or more alternative nucleotides can be maintained at intermediate frequencies in the population, just by mutation pressure, if other evolutionary factors are absent. The approach to equilibrium is, however, very slow; with a mutation rate like the human one (about 2×10^{-8} per generation), it would require a time on the order of 50 million generations (1.25 billion years, assuming 25 years per human generation) to come close to the equilibrium point from an initial state with no variability. For this reason, it is thought that forces other than mutation are needed to explain most patterns of variation that we see in populations, as well as evolutionary change over time. We shall start to examine these forces in **Chapter 2**.

PROBLEMS

1.1 A survey of variation at eight protein and enzyme loci of *Drosophila landei* showed that six loci had a predominant allele with frequency of at least 0.99. The frequencies at the remaining two loci were as follows:

Locus	Allele	Frequency
6-Pgd	S	0.79
	F	0.21
Pt-8	S	0.020
	I	0.876
	F	0.104

Using the formulae in **Box 1.2** of **Section 1.2.i.b**, calculate

i. The proportion of polymorphic loci.
ii. The mean gene diversity (heterozygosity).

1.2 Show how the Hardy–Weinberg frequencies (see **Section 1.3.i.a**) of the three genotypes at an autosomal locus with two alleles arise, assuming a randomly mating population that is so large that random fluctuations in allele frequencies can be ignored.

1.3　Determine how the frequencies of two alleles, A_1 and A_2, at an X-linked locus change in a large, randomly mating population. Assume that the initial frequency of A_1 is p_f in eggs, and p_m in sperm, and derive an expression for the change in the difference between p_f and p_m. Note that, with X-linkage, each female zygote receives one allele from her mother and one from her father, but each male zygote receives an allele only from his mother.

1.4　A survey of a sample from a European–American population in Utah for SNP variants at a nucleotide site in the promoter of the *EPO* gene (suspected of being involved in complications of diabetes) gave the following results for nondiabetic individuals (the two alternative variants at the relevant site on the coding strand are G and T):

SNP Genotype	GG	GT	TT	Total
Number of individuals	46	127	66	239

Conduct a statistical test using the chi-squared distribution (**Appendix A2.v.f**) to determine whether there is any evidence for a departure of these genotype frequencies from the Hardy–Weinberg expectations.

1.5　A sample of individuals from a population of the highly self-fertilizing plant species, *Arabidopsis thaliana*, was found to exhibit the following genotype frequencies at an electrophoretic locus: 45 FF, 52 SS, 2 FS. Calculate estimates of

   i.　The gene diversity and the frequency of heterozygotes at this locus (see **Section 1.2.i.b**).

   ii.　The inbreeding coefficient, f (see **Section 1.3.ii.b**).

   iii.　The frequency of self-fertilization (assume equilibrium under a mixture of random mating and selfing; see **Section 1.3.ii.c**).

1.6　Show that the equilibrium frequency of GC in an infinitely large population under the reversible mutation model of **Equation 1.5** (**Section 1.3.iii.c**) is $v/(u + v)$, and that this is a stable equilibrium.

1.7*　Using matrix algebra (**Appendix A1.iv**), find an expression for the equilibrium frequencies of the four possible bases on a single strand of DNA (G, C, A, and T) under mutation pressure, allowing different rates for all of the 12 types of mutations.

CHAPTER TWO

Basic Selection Theory and the Maintenance of Variation

CHAPTER SUMMARY

We now examine some of the most important theoretical and experimental results concerning the action of selection on a single locus or nucleotide site. We explain how selection is measured, define selection coefficients and dominance coefficients, and derive the equations for changes in allele frequencies at a locus for the simplest possible case (a single population with discrete generations). If genetic variants affect survival or reproduction in the same way in all environments and at all times (constant fitnesses), the relative fitnesses of the possible genotypes determine the outcome of selection in a simple fashion.

We first consider the fate of a new, unconditionally advantageous genetic variant. There is then "directional selection" for the variant versus a wild-type allele, the simplest translation into modern genetic terms of Darwin's and Wallace's concept of how natural selection causes evolutionary change. In a diploid, the favored variant always spreads through the population if one homozygote is superior to the other two genotypes.

In contrast, when the heterozygous genotype has the highest fitness, selection can maintain variability within randomly mating populations, causing a balanced polymorphism. The discovery of balanced polymorphism was an important early contribution of population genetics to the understanding of evolution; no such possibility was envisaged in classical evolutionary thought. We describe some biologically and medically important examples of this form of selection. In contrast, with heterozygote disadvantage and constant fitnesses, variability is eliminated from the population. With constant fitnesses, selection acting at a single locus always maximizes the mean fitness of a randomly mating population, even though homozygotes with lower fitness are present when there is heterozygote advantage.   Other forms of balancing selection can maintain variability when fitnesses are not constant, even if there is no heterozygote advantage. For example, the fitnesses of different genotypes may depend on the frequencies of the other genotypes in the population. We describe some well-studied examples of such frequency-dependent selection (including mimicry in butterflies, self-incompatibility and host-parasite systems). We then outline the theory for when fitnesses vary in time or space, showing when such fluctuating selection maintains variability.

Although it is very useful to model selection using fitnesses assigned to genotypes without specifying the underlying causes of fitness differences, it is important to consider what aspects of an organism's biology may determine its fitness. We therefore examine the biochemical basis for the relationship between the activities of enzymes in a biochemical pathway and organisms' fitnesses, and we explain how the predicted relationships can help us to understand protein polymorphisms.

> This preservation of favourable individual differences and variations, and the destruction of those which are injurious, I have called Natural Selection . . . .
>
> Charles Darwin, *The Origin of Species*, Chapter 4

INTRODUCTION

Natural selection is the only evolutionary force that can cause adaptation: the appearance of good engineering design that is so pervasive among living organisms. It is therefore widely accepted as being the most important and interesting of the causal processes of evolution. Evolution by natural selection was first proposed as a detailed scientific theory by Charles Darwin and Alfred Russel Wallace (Darwin and Wallace 1858), although the idea of selection had been suggested previously by several people, including the Roman writer Lucretius in his philosophical poem *De Rerum Naturae*. In *The Origin of Species*, Darwin (1859) critically evaluated the hypothesis of natural selection, and presented a large amount of supporting evidence from many different organisms. There were, however, two gaps, and many biologists doubted the importance of selection as a cause of evolution until well into the twentieth-century (Provine 1971). The first gap was the lack of direct evidence for the action of selection in nature; this evidence has since been supplied by the work of numerous field naturalists (reviewed by Sheppard 1975; Endler 1986; Kingsolver et al. 2001; Bell 2008). The second was the lack of understanding in the nineteenth-century of the mechanism of inheritance, which we discussed in **Section 1.1.ii.c**.

The nineteenth-century view of inheritance as a process of blending of material from the two parents implies that variability is constantly being lost from populations (Fisher 1930b, Chapter 1). It will thus not be available for use by selection, as was pointed out in a review of *The Origin of Species* by the Edinburgh engineering professor Fleeming Jenkin (Jenkin 1867). Since variability is in fact commonly observed, Darwin inferred that the depletion of variation must be counteracted in some way. He proposed in his theory of pangenesis that new heritable variation was generated by direct effects of the environment on the parents, which are transmitted to their offspring (Darwin 1868)—so-called *Lamarckian inheritance*. Even after the rediscovery of Mendelian genetics in 1900, many biologists continued to believe that quantitative traits showed blending inheritance (Provine 1971), and several prominent evolutionists continued to accept the "inheritance of acquired characters" until the 1920s (Wright 1978, p. 324).

As we discussed in **Section 1.1.ii.c**, belief in blending inheritance ended after the discovery that quantitative traits are controlled by multiple Mendelian genes. By the early 1920s, the investigation of various unusual cases of inheritance that appeared superficially to defy Mendelian rules led to the conclusion that Mendelian inheritance was involved in virtually all types of variation (Muller and Altenburg 1920). Certain maternally inherited traits, mostly in plants, provide an exception (Wright 1968, Chapter 4), but even these exhibit discrete rather than blending inheritance. As we saw in **Section 1.3.i**, discrete Mendelian inheritance means that variability in populations is conserved, and is thus available for use by natural selection and other evolutionary forces.

These advances in genetics set the stage for a revival of interest in natural selection as a major evolutionary force, and the realization that evolutionary processes can be understood in detail by examining the behavior of genes in populations. In particular, theoretical studies of selection by Fisher, Haldane, Norton, and Wright between 1910 and 1930 showed that natural selection acting on Mendelian variants can change the genetic composition of a population over short evolutionary time scales (Provine 1971), as we will discuss in detail in **Section 3.1**.

2.1. BASIC SELECTION THEORY: CONSTANT FITNESSES

We start by examining the substitution of one variant at a site or locus by a new, selectively advantageous alternative variant at a single site in the genome, first in the simplest situation of a haploid population, and then in diploid populations.

2.1.i. Haploid populations

To see how selection can be modeled, we first consider one complete generation of a population, starting with the offspring of one generation, and ending with their offspring. The easiest type of system to model is a haploid population with just two alternative genotypes, A_1 and A_2. In this chapter, we will refer to these variants as alleles, as there is no need to specify the exact nature of the genetic differences; the theory applies equally well to single-nucleotide variants at a particular site on the DNA sequence, to classical alleles at a single locus, which may differ at several different nucleotide sites, or to different arrangements of a chromosome. Many microbial species, such as bacteria, are haploid, so this model applies directly to them. It is also valid for several other important situations, including strictly maternally transmitted organelle genomes like plastids and mitochondria, and bacterial endosymbionts, such as *Wolbachia* in multicellular organisms (O'Neill et al. 1997). In species with sex chromosomes, it applies to Y-chromosomal genes without X-linked counterparts. The model also applies to asexually reproducing or completely self-fertilizing populations with any level of ploidy; haploid genotypes are simply replaced by diploid or polyploid genotypes that reproduce themselves exactly. It is also possible to generalize the model to multiple alternative alleles instead of just two (e.g., Nagylaki 1992, Chapter 2).

As described in **Section 1.3.i**, we assume that generations are discrete, and that the population size is so large that random fluctuations in genotype frequencies can be disregarded (for convenience, this is called an "infinite" population). Let the frequencies of A_1 and A_2 at the start of a generation be p and q, respectively. Since $p + q = 1$, only one of the two allele frequencies is needed to describe the state of the population. We will usually use the frequency of A_2, which we will often imagine to be a new variant produced by mutation.

We can model selection by assigning different probabilities of survival to adulthood to the two different genotypes (**Box 2.1**). The frequencies of the genotypes among the mature adults will then become different from those of the offspring at the start of a generation. The survival probability of genotype A_i is denoted by w_i ($i = 1$ or 2). In the absence of fertility differences, this represents the *fitness* of the genotype. If selection acts on fertility differences, the w_i values also take each genotype's number of offspring into account.

It is important to note that we are not necessarily assuming that the population is segregating only for variants at the site being considered. If there is variation at many other sites in the genome (as is usually the case—see **Section 1.2**), the fitnesses of A_1 and A_2 represent the average values for each of these two genotypes over all possible genetic backgrounds (i.e., the genotypes at other sites). We assume here that the fitness effects of A_1 and A_2 are independent of the effects of variants at other sites, so we can treat w_1 and w_2 as constants; this is of course an oversimplification, and the effects of relaxing this assumption will be considered in **Section 8.4**.

The algebra in **Box 2.1** yields the following equation for the change in the frequency of allele A_2 over one generation:

Box 2.1 SELECTION ON A HAPLOID POPULATION

At the start of a generation, we have frequencies p and q of alleles A_1 and A_2, which we assume to affect survival. The frequencies of A_1 and A_2 among the survivors are obtained by multiplying their frequencies before selection by their probabilities of survival, giving pw_1 and qw_2, respectively, and dividing these by the proportion of the population that has survived. This ensures that the frequencies after selection add up to 1. The proportion of survivors is the *mean fitness* of the population, and is given by **Equation B2.1.1**:

$$\bar{w} = pw_1 + qw_2 \qquad (B2.1.1)$$

The new frequencies of A_1 and A_2, which correspond to the frequencies before selection in the next generation, are therefore given by:

$$p' = \frac{pw_1}{\bar{w}} \qquad (B2.1.2a)$$

and

$$q' = \frac{qw_2}{\bar{w}} \qquad (B2.1.2b)$$

The change in allele frequency, Δq, is derived as follows:

$$\Delta q = q' - q = \frac{qw_2}{\bar{w}} - q$$

$$= \frac{qw_2 - q\bar{w}}{\bar{w}} = \frac{qw_2 - pqw_1 - q^2 w_2}{\bar{w}}$$

$$= \frac{q(1-q)w_2 - pqw_1}{\bar{w}}$$

Noting that $p = 1 - q$, this simplifies to **Equation 2.1** of the main text.

$$\Delta q = \frac{pq(w_2 - w_1)}{\bar{w}} \qquad (2.1)$$

The frequency of A_1, of course, undergoes the opposite change, since $p + q = 1$ ($\Delta q = -\Delta p$). If the fitnesses remain constant over time (and if q is initially between 0 and 1, as must be the case if the population contains genetic variability), it is obvious that A_2 increases in frequency when $w_2 > w_1$, and decreases if the opposite is true. Natural selection thus causes the allele associated with higher fitness to spread through the population towards *fixation*—that is, to a frequency of one. In an infinite population, this state can never be attained, but only closely approached; fixation can, however, happen in the real world of finite populations.

The main use of this simple model is to determine how fast the composition of a population changes under selection. This is a crucial question in relation to the effectiveness of selection as a force in evolution, and we will use the model for this purpose in **Section 3.1.ii**, where we apply it to some real-life examples.

2.1.ii. Diploid populations

2.1.ii.a. *Introduction*

Many organisms of interest to biologists are diploids, and diploidy makes the model more complicated. The simplest system for studying selection in a sexually reproducing, diploid population is an autosomal locus with two alleles, A_1 and A_2, just as for haploids in **Section 2.1.i** above. As in that case, we need only use one of the two allele frequencies to describe the change in the genetic composition of a population.

To begin the analysis, we need the frequencies of the three possible genotypes at the start of a generation. We apply the assumptions used in **Section 1.3.i.a** to derive the Hardy–Weinberg genotype frequencies: random mating, infinite population size, and discrete generations. At the start of a generation, the frequencies of the three zygote genotypes, A_1A_1, A_1A_2, and A_2A_2, are then p^2, $2pq$, and q^2, where p and q are the frequencies of A_1 and A_2 among the gametes of the previous generation (for the moment, we ignore differences in fitnesses between the sexes, so that there is no need to distinguish between frequencies in eggs and sperm).

To model selection, we will assume for simplicity that it involves only survival between the zygote and adult stages. Identical results are obtained for the simplest type of model of selection on fertility, where the fitness of a genotype is proportional to its contribution of gametes to the offspring of the next generation, with no sex differences in the relative values of the fertilities of different genotypes. We also assume that the fitness of an individual is determined solely by its own genotype, i.e., we are ignoring complications such as effects of the maternal genotype on the survival or fertility of her progeny. **Box 2.2** shows the model in detail.

Box 2.2 SELECTION ON A DIPLOID POPULATION: BASIC FORMULATION

In this case, the mean fitness of the population is given by:

$$\bar{w} = p^2 w_{11} + 2pq w_{12} + q^2 w_{22} \qquad \text{(B2.2.1)}$$

The frequencies of $A_1 A_1$, $A_1 A_2$, and $A_2 A_2$ among the survivors are therefore:

$$p^2 w_{11}/\bar{w}, \ 2pq w_{12}/\bar{w}, \ \text{and} \ q^2 w_{22}/\bar{w}$$

The new frequencies of A_1 and A_2 can be found by noting that the frequency of an allele among its homozygous carriers is 1, and the frequency among heterozygotes is 1/2. Hence, by summing the contributions from the relevant genotypes for each allele, we get:

$$p' = \frac{p\left(p w_{11} + q w_{12}\right)}{\bar{w}} \qquad \text{(B2.2.2a)}$$

and

$$q' = \frac{q\left(p w_{12} + q w_{22}\right)}{\bar{w}} \qquad \text{(B2.2.2b)}$$

It is convenient to use the notation $w_{1.} = p w_{11} + q w_{12}$ and $w_{2.} = p w_{12} + q w_{22}$. These are called the *marginal fitnesses* of the alleles A_1 and A_2, respectively. As the formulae show, the marginal fitness of allele A_i is a measure of its average fitness, taking into account the frequencies of the other alleles present in the genotypes in which A_i is present. By using the marginal fitnesses, we can write the new allele frequencies (after selection) in the same form as in the haploid case (**Equations B2.1.2a** and **B2.1.2b**), replacing the constants w_1 and w_2 by $w_{1.}$ and $w_{2.}$ to give **Equations 2.2a** and **2.2b** of the main text. The argument used to obtain **Equation 2.1** for the haploid case then gives the allele frequency change equation for diploids as:

$$\Delta q = \frac{pq\left(w_{2.} - w_{1.}\right)}{\bar{w}} \qquad \text{(B2.2.3)}$$

As shown in **Box 2.2**, the survival probability for a genotype A_iA_j is denoted by w_{ij}; combining this with the frequency of A_iA_j zygotes allows us to calculate the frequency of A_iA_j among the surviving adults. From **Equations B2.2.2a** and **B2.2.2b** in **Box 2.2**, and using the *marginal fitnesses* of the two alleles, w_1 and w_2, defined in **Box 2.2**, we obtain the frequencies of A_1 and A_2 among the offspring in the convenient forms:

$$p' = \frac{pw_{1\cdot}}{\bar{w}} \tag{2.2a}$$

$$q' = \frac{qw_{2\cdot}}{\bar{w}} \tag{2.2b}$$

These relations lead to an equation for allele frequency change, Δq, which is similar to **Equation 2.1**, except that the marginal fitnesses replace the haploid fitnesses (see **Box 2.2**).

Although we are still assuming that the genotypic fitnesses w_{ij} are constant over time, the marginal fitnesses depend on the allele frequencies. This means that it is not inevitable that allele frequencies will approach zero or one, as in the haploid case—it is also possible for the two marginal fitnesses to become equal. In this case, allele frequency change will cease, even if p and q are between 0 and 1.

Equations 2.1 and **2.2** also show that multiplying the fitness of all genotypes by a factor C has no effect, since it cancels from top and bottom (because the $\bar{w}$ value in the denominator is calculated from the three w_{ij} values). Thus, only the *relative* values of the fitnesses of the three genotypes enter into the selection equations, in both the haploid and diploid cases. We can thus select one genotype as a standard, and conveniently express all the fitnesses relative to this one by dividing them by the value for the standard genotype. The fittest genotype is often chosen as the standard, so that its relative fitness is 1, and all the other fitnesses are then less than 1, but there are many situations in which this convention is not obeyed (e.g., **Section 3.1**).

2.1.ii.b. *Directional selection*

Section i of **Box 2.3** models directional selection, which is the situation when one allele (A_1 in the example) confers the highest fitness when homozygous; using relative fitnesses, as just described, this homozygote's fitness is given the value 1. The alternative homozygote is assigned fitness $1 - s$, and s is called the *selection coefficient* of the A_2 allele; it measures the fitness reduction experienced by A_2A_2 homozygotes. The heterozygote A_1A_2 is

Box 2.3 SELECTION ON A DIPLOID POPULATION: THE USE OF SELECTION COEFFICIENTS

B2.3.i. Directional selection

As noted in the main text, only the relative fitness values of the genotypes enter into the selection equations. If there is *directional selection*, with A_2 being a deleterious variant, so that A_1A_1 is the fittest genotype of the three, we can divide w_{12} and w_{22} by w_{11}. Using these relative fitnesses, we then represent the fitnesses of A_1A_2 and A_2A_2, relative to a value of 1 for A_1A_1, as $1 - hs$ and $1 - s$, where $s > 0$ and $0 \le h \le 1$.

If these are substituted into the expressions for the marginal fitnesses (**Box 2.2**), we have $w_{1.} = 1 - qhs$ and $w_{2.} = 1 - phs - qs$. From **Equation B2.2.3**, the sign of the allele frequency change, Δq, is the same as that of $w_{2.} - w_{1.} = qhs - phs - qs = -qs(1 - h) - phs$. Provided that pq is nonzero, so that there is variability at the locus, Δq will always be negative, given the conditions on s and h. A_2 will thus always decrease in frequency. Its rate of decrease is determined jointly by the amount of variability, as measured by pq, and the strength of selection, as measured by s and h.

B2.3.ii. Heterozygote advantage

Here, A_1A_2 is the fittest genotype, and the relative fitnesses of A_1A_1 and A_2A_2 are conveniently written as $1 - s$ and $1 - t$, respectively. We now have $w_{1.} = 1 - ps$ and $w_{2.} = 1 - qt$, so that:

$$w_{2.} - w_{1.} = ps - qt \tag{B2.3.1}$$

This shows immediately that $\Delta q = 0$ when $ps = qt$. This corresponds to an equilibrium with both alleles present in the population. Using the fact that $p + q = 1$, and rearranging, we obtain **Equation 2.3** of the main text for the equilibrium allele frequencies, p^* and q^* (**Problem 2.1**).

We can also see that both terms on the right-hand side of **Equation B2.3.1** decrease as q increases, so that $w_{2.} - w_{1.} > 0$ when $q < q^*$, and $w_{2.} - w_{1.} < 0$ when $q > q^*$, with the equality being satisfied only at $q = q^*$ (**Problem 2.1**). This leads to the conclusion that the equilibrium is approached from either side, as outlined in the main text.

assigned fitness $1 - hs$, where h is the *dominance coefficient* of the A_2 allele; h is the fitness reduction of the heterozygote relative to s. If $h = 0$, the A_2 allele is recessive in its effect on fitness; if $h = 1$, A_2 is dominant (and A_1 is recessive). Intermediate h values indicate that the heterozygote's fitness lies between those of the two homozygotes; $h = 1/2$ corresponds to exactly intermediate dominance (often called *semidominance*, or additivity). This terminology is used throughout the book.

Section i of **Box 2.3** shows that, provided that h is in the range 0–1, the marginal fitness of the advantageous allele A_1 is always higher than that of A_2. A_1 will then always increase towards fixation from any nonzero initial frequency (just as for the haploid case). Thus, as Darwin and Wallace realized, natural selection can cause a selectively favored type to spread through a population and replace its alternative (the speed of the change is calculated in **Section 3.1.iii.b**).

2.1.ii.c. Heterozygote advantage and the maintenance of variation
Selection can, however, have other outcomes, unsuspected by Darwin and Wallace. Specifically, it may actively maintain variation in the population, a situation referred to as *balancing selection* or *balanced polymorphism*. **Section ii** of **Box 2.3** describes the simplest such situation, involving *heterozygote advantage* (sometimes called *overdominance*). In this case, we assign a fitness of 1 to the heterozygote, and represent the lower fitnesses of the two homozygotes A_1A_1 and A_2A_2 as $1 - s$ and $1 - t$, respectively.

Section ii of **Box 2.3** also shows that the two marginal fitnesses can become equal, and that this happens when $qt = ps$. There is then a *polymorphic equilibrium*, with no change in allele frequency but with both alleles present. It can then be shown (**Problem 2.1**) that the equilibrium allele frequencies, which we denote by p^* and q^*, are given by:

$$p^* = \frac{t}{(s + t)}, q^* = \frac{s}{(s + t)} \tag{2.3}$$

The equilibrium state is of little interest, however, unless the population moves towards it from other allele frequencies. Examining the behavior of the two marginal fitnesses shows that $w_2 > w_1$ when $q < q^*$; the reverse is true when $q > q^*$ (**Problem 2.1**). The allele frequency change per generation, Δq, has the same sign as $(w_2 - w_1)$. Thus, when $q < q^*$, Δq is positive (> 0), and is negative when $q > q^*$. The relationship between Δq and q is sketched in **Figure 2.1A**, which suggests that q^* is approached from either higher or lower allele frequencies, i.e., q^* is a *stable* equilibrium.

A.

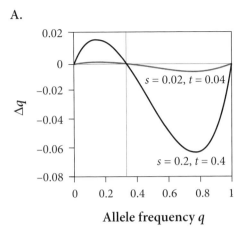

B.

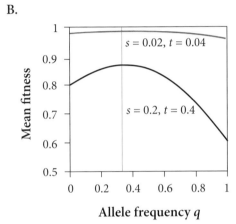

FIGURE 2.1 Allele frequency changes, Δq, from **Equations B2.2.3** and **B2.3.1**, in a diploid population when there is heterozygote advantage. **A.** Polymorphic equilibrium is approached from either higher or lower frequencies. **B.** This shows that stable equilibria under selection correspond to local peaks of mean fitness.

This is not, however, a complete examination of the stability of the equilibrium. We have not ruled out the possibility that, near q^*, Δq could be so large that the population jumps across the equilibrium, and then jumps back again, so that the allele frequencies oscillate. To examine this possibility, it is necessary to analyze the *local stability* of the equilibrium by looking at Δq after small perturbations of allele frequencies away from equilibrium. This is described in **Box 2.4***. It turns out that, with heterozygote advantage, there is always a smooth approach to q^*.

Box 2.4* LOCAL STABILITY OF AN EQUILIBRIUM WITH HETEROZYGOTE ADVANTAGE

For a system changing in discrete time, which can be described by a single variable such as the allele frequency q, we can write the new value of the variable q' in terms of its old value as:

$$q' = f(q) \qquad \text{(B2.4.1)}$$

To evaluate the stability of an equilibrium, we examine the behavior of the system after a small perturbation away from the equilibrium frequencies, $q - q^*$, by using the first two terms of the Taylor's series expansion (**Appendix A1.ii.a**) of **Equation B2.4.1**; i.e., we neglect squared and higher-order terms in $(q - q^*)$:

$$q' \approx f(q^*) + (q - q^*)\left(\frac{df}{dq}\right)_{q^*} \qquad \text{(B2.4.2a)}$$

At the equilibrium (i.e., q^*), $q^* = f(q^*)$, so that **Equation B2.4.2a** simplifies to:

$$q' - q^* \approx (q - q^*)\lambda(q^*) \qquad \text{(B2.4.2b)}$$

where $\lambda(q) = df(q)/dq$. Geometrically, this is equivalent to approximating the change in $f(q)$ in the neighborhood of q^* by its slope at q^*, multiplied by the deviation of q from q^*.

Equation B2.4.2b is completely general, and tells us that the approach to equilibrium is determined by the value of $\lambda(q^*)$. For a fuller account, see Otto and Day (2007, Chapter 5).

We can apply it to the case of heterozygote advantage by using the expressions for $f(q)$ and q^* in terms of the selection coefficients s and t, which we have already developed. The derivation is simplified by using the fact that $f(q) = q + \Delta q$, so that $\lambda(q) = 1 + (d\Delta q/dq)$. Furthermore, Δq is given by **Equation B2.2.3**, in which the term $w_{2.} - w_{1.} = 0$ when $q = q^*$. From the rule for differentiating a product (**Appendix A1.i**), this means that the only nonzero term in $d\Delta q/dq$ at $q = q^*$ is:

$$\frac{p^* q^*}{\overline{w}^*}\left(\frac{d(w_{2.} - w_{1.})}{dq}\right)_{q^*}$$

Substituting the equilibrium allele frequencies into this expression, and carrying out some algebra (**Problem 2.4***), we obtain **Equation B2.4.3**:

$$\lambda(q^*) = \frac{1 - \dfrac{2st}{(s+t)}}{1 - \dfrac{st}{(s+t)}} \qquad (\text{B2.4.3})$$

The numerator is always smaller than the denominator, so that $\lambda(q^*) < 1$. In addition, the minimum value of the numerator is reached when $s = t = 1$, in which case $\lambda(q^*) = 0$. This corresponds to the case when each allele is homozygous lethal, as in the case of the *Drosophila* balancer stocks we discussed in **Section 1.1.iii.b** (**Figure 1.7**). In this situation, the population can never depart from equal frequencies of the two alleles.

In all other cases with heterozygote advantage, $\lambda(q^*)$ lies between 0 and 1, so that the new value of $q - q^*$ given by **Equation B2.4.2b** is always less in magnitude than the old value, and has the same sign. This implies that the departures from equilibrium diminish steadily in size, so that the equilibrium is said to be *locally stable*.

The *Drosophila* experiments described in **Section 1.1.iii.b**, where a balancer chromosome is allowed to compete against a wild-type chromosome in an artificial population, provide an example of the usefulness of these results. Since the balancer is homozygous lethal ($t = 1$), the equilibrium frequency of the wild-type chromosome is determined purely by this chromosome's homozygous effect on fitness (see **Problem 2.2**). This makes it possible to estimate the selection coefficient, s, against flies homozygous for such chromosomes.

2.1.ii.d. *Biological implications of heterozygote advantage*

The model just described shows that natural selection can lead, not just to replacement of one allele at a locus (or variant in a sequence) by a better one, but also to the maintenance of variation. This was first discussed by Fisher (1922). As we will see in **Section 2.2** below, heterozygote advantage is just one form of balancing selection.

A useful way of understanding heterozygote advantage intuitively is to think about mutation or migration introducing one allele, say A_2, at a low frequency q into a population initially fixed for A_1. Because the initial q value is low, the frequency of A_2A_2 homozygotes (q^2) produced by random mating is small compared with the frequency of heterozygotes (with $p \approx 1$ and $2pq \approx 2q$). Thus, if $q = 10^{-3}$, the frequency of A_2A_2 is 10^{-6} and the frequency of A_1A_2 is approximately 2×10^{-3}, i.e., it is 2000 times larger than that of A_2A_2. Because A_2A_2 homozygotes are so rare, the condition for A_2 to increase in frequency is simply that A_1A_2 heterozygotes have a higher fitness than the initial A_1A_1 genotype, even if A_2A_2 individuals have severely reduced fitness. As A_2 rises in frequency, however, the frequency of A_2A_2 homozygotes will increase; if these have low fitness, this reduces the marginal fitness of the A_2 allele, and will eventually check its spread. A symmetrical argument applies to the introduction of A_1 into a predominantly A_2 population.

It is therefore tempting to conclude that many natural polymorphisms, of the kind discussed in **Sections 1.1** and **1.2**, are maintained by heterozygote advantage, especially as it is easy to imagine that a mutational change to a protein or RNA molecule may sometimes impair its function when homozygous, even if there is a fitness advantage to the heterozygote. In practice, however, unless selection is strong, it is difficult to determine the nature of selection acting on a polymorphism without very extensive data on the relative survival and fertility rates of the different genotypes.

Suppose, for example, that we are concerned with selection acting on survival between birth and adulthood. Assume that we score the adult offspring for heterozygotes versus homozygotes for a pair of equally frequent alleles that are polymorphic in the population ($p = 0.5$); the expected frequency of heterozygotes among the progeny under random mating is then 0.5. If n individuals are counted, the binomial probability formula (**Appendix A2.v.a**) gives the standard deviation of the estimate of heterozygote frequency as $\sqrt{0.25/n}$. With $n = 1000$, the standard deviation is thus 0.016. On the null hypothesis that the fitnesses of heterozygotes and homozygotes are equal, a difference (in either direction) of more than 1.96 standard deviations has a probability of less than 5% of occurring by random sampling. In this example, only deviations of more than 0.032 from 0.5 would be regarded as statistically significant, even with this fairly large sample. With 10,000 individuals, a deviation of 0.01 would be significant. As we will discuss in **Section 6.4.iv**, many protein sequence polymorphisms probably have selection coefficients that are one or two orders of magnitude smaller than 0.01, so that direct estimates from survival or fertility data are not feasible in most cases.

There are thus only a few cases where heterozygote advantage has been firmly established. The classic case is *sickle cell hemoglobin* (Allison 1964, 2004), where heterozygotes for the A allele conferring normal hemoglobin and the S mutation in the β-globin gene (a change of glutamine to valine at position 6 in the polypeptide chain) are protected against malaria, but SS homozygotes suffer a near-lethal disorder, *sickle cell anemia*, caused by the clogging of blood vessels by misshapen red blood cells. The cause of the heterozygote advantage was established by comparing the malarial infection rates of AA and AS genotypes in African populations, where it is known that malaria is an important source of mortality (Allison 1964, 2004). In addition, the frequency of AS individuals among adults is consistently higher than among infants in these populations, implying a survival advantage over AA individuals of the order of 10–20% (Allison 1964). These observations resolved the paradox that a highly debilitating condition caused by a recessive allele occurs at a high frequency in these populations (the frequency of homozygotes can be as high as 2% among infants in some populations). Numerous other human molecular variants are associated with malaria resistance. In several cases these have severely harmful effects when homozygous, but heterozygote advantage has not been detected in all cases (Kwiatkowski 2005).

2.1.ii.e. *Mean fitness*

Does the maintenance of deleterious homozygous genotypes reduce the fitness of the population? It turns out that the mean fitness of a randomly mating population with heterozygote advantage is at a maximum at the equilibrium under heterozygote advantage. This is shown in **Figure 2.1B**, and a proof is given in **Box 2.5***. Thus, while the presence of homozygotes with low fitness in cases of polymorphisms like sickle cell hemoglobin is unfortunate for the individuals concerned, the action of selection has increased the mean fitness of the population as a whole by maintaining malaria-resistant individuals in the population.

Box 2.5* MAXIMIZATION OF MEAN FITNESS BY SELECTION

We will first show that, with constant fitnesses and random mating, the change in allele frequency per generation is always proportional to the derivative of mean fitness with respect to allele frequency, so that a change in allele frequency is in the same direction as the slope of the graph of mean fitness against allele frequency (Wright 1937). In the present case, this can be used to show that the stable equilibrium under heterozygote advantage corresponds to the maximum in mean fitness.

We note first, from **Equation B2.2.1** of **Section 2.1.ii.a** and the definition of marginal fitness, that the mean fitness is equal to the mean of the two marginal fitnesses, i.e.:

$$\bar{w} = pw_{1\cdot} + qw_{2\cdot}. \tag{B2.5.1}$$

With constant genotypic fitnesses, this yields **Equation B2.5.2a**:

$$\frac{d\bar{w}}{dq} = \left(w_{2\cdot} - w_{1\cdot}\right) + p\frac{dw_{1\cdot}}{dq} + q\frac{dw_{2\cdot}}{dq} \qquad \text{(B2.5.2a)}$$

By evaluating the derivatives of the marginal fitnesses in this expression, we find that:

$$\frac{d\bar{w}}{dq} = 2\left(w_{2\cdot} - w_{1\cdot}\right) \qquad \text{(B2.5.2b)}$$

This can be substituted into **Equation B2.2.3** to obtain:

$$\Delta q = \frac{pq}{2\bar{w}}\frac{d\bar{w}}{dq} = \frac{pq}{2}\frac{d\ln(\bar{w})}{dq} \qquad \text{(B2.5.3)}$$

This implies that an equilibrium with q^* between 0 and 1 corresponds to a state in which the derivative of mean fitness is zero, a *stationary point* of mean fitness. This will be a *maximum* if and only if the second derivative of mean fitness is negative (**Appendix A1.i**). Using **Equation B2.5.2b**, it can be shown that:

$$\frac{d^2\bar{w}}{dq^2} = 2\left(w_{11} + w_{22} - 2w_{12}\right) \qquad \text{(B2.5.4)}$$

This is negative if there is heterozygote advantage, and positive if there is heterozygote disadvantage, so that an equilibrium with heterozygote advantage corresponds to a maximum in mean fitness, and an equilibrium with heterozygote disadvantage to a minimum.

This illustrates the fact that natural selection is a blind force, which acts solely on the relative fitnesses of genotypes thrown up by mutation, without regard to the sufferings of the individuals concerned. An understanding of balancing selection thus suggests caution about removing deleterious genetic variants from human and domestic animal or plant populations. If variants are maintained by heterozygote advantage, the fitness of the population as a whole might be reduced, since allele frequencies will be pushed away from the point of maximum fitness. Overall, therefore, more health problems would be caused by such intervention.

A more general result (which we will not derive) is that a single locus system in a randomly mating population, with constant fitnesses and any number of alleles, always changes in such a way that its mean fitness increases from one generation to the next (Kingman 1961; Ewens 2004, Chapter 2). This implies that stable equilibria under selection correspond to local peaks of mean fitness, as in the simple model we have just examined (**Figure 2.1B**). This result is the formal justification for Wright's concept of an *adaptive landscape*, according to which populations move towards local peaks of mean fitness (Wright 1932). As we shall see in **Sections 2.2.i** and **8.4.ii**, this is often not true without the assumptions of constant fitnesses and a single locus. However, it still provides a useful way of visualizing the action of selection in complicated situations.

One might think that any number of alleles could potentially be maintained at a locus by selection with constant fitnesses, given suitable fitness relations among the different genotypes. The detailed mathematical conditions for stability of multiple alleles at a single locus in a randomly mating population are described by Crow and Kimura (1970, pp. 227–272). It can be shown, however, that polymorphisms with large numbers of alleles are unlikely to be maintained, except under conditions that seem implausible for most biological situations, such as when all heterozygous genotypes have the same fitnesses, and all homozygotes have lower fitnesses. A locus where this is known to happen is the sex-determining locus of honeybees (called the *csd* locus). Heterozygotes for different alleles at the locus are female, while males are produced by females laying unfertilized eggs, and are haploid. Homozygotes for *csd* alleles develop as males (since the signal for female development is heterozygosity), but are killed by the worker bees (Yokoyama and Nei 1979; Bull 1983, pp. 150–152). However, this is a special and almost unique situation. Computer experiments in which sets of fitnesses are generated randomly for multi-allelic systems show that it is very unlikely that large numbers of alleles will be present in a stable equilibrium situation (Gillespie 1977; Lewontin et al. 1978; Boer et al. 2004). Single-locus systems where many alleles appear to be maintained by selection are thus likely to involve some form of selection in which fitnesses are not constant (i.e., some form of frequency-dependent selection, as explained in **Section 2.2.i**).

2.1.ii.f. *Heterozygote disadvantage*

When the heterozygote is inferior to both homozygotes (*heterozygote disadvantage* or *underdominance*), the opposite of balancing selection occurs: selection drives the population to either of two alternative states of fixation. Such selection can be analyzed in the same way as before, by making both s and t negative instead of positive. There is an equilibrium allele frequency of the same form as with heterozygote advantage. But q^* has the minimum mean fitness, and the equilibrium is now *unstable*, i.e., the population moves away from q^* towards 0 or 1, depending on which allele is initially commonest in the population (**Problem 2.3**).

Heterozygote disadvantage might seem biologically implausible. It is likely to be uncommon in nature, at least for randomly mating populations, since, as explained above, a mutation that suffers heterozygote fitness loss will be eliminated, i.e., polymorphisms cannot be maintained. The theory is, however, relevant to an understanding of the establishment of chromosome rearrangements, such as inversions and translocations, which are often disadvantageous in heterozygotes because disturbances to regular meiotic segregation in heterozygotes can lead to offspring with harmful deletions of portions of their genome (e.g., White 1973, Chapter 7). The above results suggest that such chromosomal mutations should be removed by selection, yet rearrangements occur frequently in evolution. For example, humans differ from chimpanzees by 11 major chromosome rearrangements (Yunis et al. 1980). One solution to this puzzle invokes the effects of random sampling of allele frequencies in finite populations, causing a rearrangement to spread to a higher frequency than the equilibrium, in opposition to selection, after which it can be fixed by selection (this process is discussed further in **Section 8.4.v.b**).

Another example of heterozygote disadvantage is provided by the Rhesus blood group system of humans, which is controlled by a pair of closely linked genes, *D* and *C–E* (reviewed by Avent and Reid 2000). Hemolytic disease of newborn babies is frequently associated with the development of a fetus that expresses the Rhesus D antigen when the mother is D-negative. The antigen can cross the placenta, causing the mother to produce anti-D antibodies that react with the fetal antigen. This results in a form of heterozygote disadvantage, since the affected individuals are always heterozygous for the allele conferring the D-negative state (Haldane 1942). Most human populations have relatively low frequencies of the D-negative allele, with the exception of the Basque population on the borders of Spain and France, where its frequency is about 50% (Bauduer et al. 2005). Even after 70 years of research on the Rhesus system, it is not known whether there are compensating selection pressures at the locus controlling the presence or absence of the D antigen, which help to maintain it in the population, and whether the high frequency of D-negative alleles in Basques is caused by chance or selection.

2.2. VARIABLE FITNESSES

The conclusions above change significantly if we remove the biologically unrealistic assumption that fitnesses remain constant. In particular, variability can be maintained by selection without any heterozygote advantage, as we show in the next section.

2.2.i. Frequency-dependent selection

2.2.i.a. *Theory*

Relative fitnesses may often depend on the frequencies of the different geno-types in the population. For example, assume that there is a dominant allele A_1, such that A_1A_1 and A_1A_2 (collectively $A_1/-$) have the same fitness, which is different from that of A_2A_2. If the fitness of $A_1/-$ is higher than that of A_2A_2 when A_1 is rare, but becomes lower when A_1 is frequent, a stable polymorphism can exist (see **Figure 2.2**). This is an example of *negative frequency dependence* (rare allele advantage). Frequency dependence can also operate in such a way that the fitness of a genotype increases with its frequency in the population (*positive frequency dependence*), but this cannot maintain variation.

When a polymorphism exists in the example above, all three genotypes must have equal fitnesses. There is thus no heterozygote advantage. If dominance is incomplete, however, biologically realistic examples of frequency-dependent selection are possible where the heterozygote's fitness is either *higher* or *lower* than that of either homozygote at or near the equilibrium point, although the heterozygote must have an advantage over the prevailing homozygote for either allele to invade a randomly mating population from a low initial frequency (Charlesworth and Charlesworth 1975a; Wilson and Turelli 1986). Haploid or asexual diploid populations can also have variability maintained by frequency-dependent selection, in contrast to the situation with constant fitnesses. Furthermore, the dependence of fitnesses on genotype frequencies means that the argument concerning the maximization of mean fitness used in **Box 2.5*** breaks down; it is no longer necessary for a stable equilibrium to coincide with a maximum in mean fitness (**Box 2.6***).

Another important property of frequency dependence is that it is no longer true that a polymorphic equilibrium is locally stable, even when selection tends to drive the population towards it. If frequency dependence is sufficiently strong, the population can overshoot the equilibrium from a nearby point and *limit cycles* or even *chaos* can result, with perpetual oscillations in allele frequencies (May and Anderson 1983). The mathematical basis for this is explored in **Box 2.6***.

2.2.i.b. *Examples of frequency dependence*

Biological situations in which negative frequency-dependent selection occurs are probably common (Clarke 1979). One example is *apostatic selection*, in which there is selection by predators that learn while they hunt; this favors rare forms of the prey (Clarke 1969). Vertebrate predators, such as birds, tend to form search images of their prey. For example, if a snail with a particular shell color is encountered, and is found to be good to eat, the bird will look out for more of the same. Rare color variants that differ from the majority of the population will tend to be ignored, and experience lower predation rates. This will

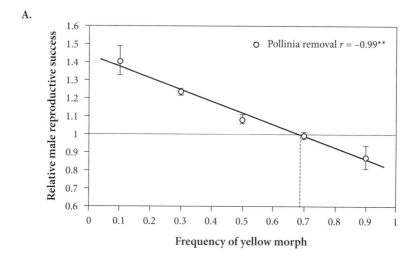

A.

Relative male reproductive success (y-axis: 0.6 to 1.6)

Frequency of yellow morph (x-axis: 0 to 1)

○ Pollinia removal $r = -0.99^{**}$

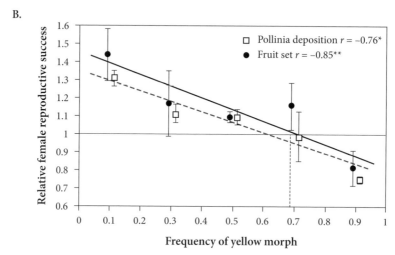

B.

Relative female reproductive success (y-axis: 0.6 to 1.6)

Frequency of yellow morph (x-axis: 0 to 1)

□ Pollinia deposition $r = -0.76^{*}$
● Fruit set $r = -0.85^{**}$

FIGURE 2.2 An example of negative frequency-dependent selection (i.e., the advantage of a rare phenotype) in an orchid species. In a plant whose flowers do not provide any nectar reward, pollinating insects visiting flowers may switch from plants resembling those they initially visited and try phenotypically different plants. In a population with different flower phenotypes, the rarer of two phenotypes will thus have an advantage, because it will generally not be the first type visited and the insects will not learn that its flowers are unrewarding. Pollinating insects might then give an advantage to the rarer of two flower color types. The figure shows the results of an experiment designed to test this possibility in a species whose flower color is polymorphic. Orchids are convenient for testing such hypotheses, because pollen movements can be measured, since the pollen is not loose but is transferred in clumps (pollinia). The frequencies of two different flower color types (yellow and purple) were manipulated experimentally, and the expected advantage is seen in terms of the relative male and female reproductive success of the yellow morph (measured as numbers of pollinia removed per flower, and the number deposited per flower, respectively). [Adapted from Figure 3 of Gigord et al. (2001).]

cause them to rise in frequency, but the birds will then start to encounter them more frequently. The selective advantage will then decrease, possibly resulting in a stable equilibrium with both forms present. There is evidence that the different shell colors and patterns of the snail *Cepaea nemoralis*, shown in **Plate 2**, front endpaper, are subject to selective pressures imposed by predation (Sheppard 1975, Chapter 11), and it is likely that apostatic selection is involved in the maintenance of these polymorphisms (Clarke 1969), as well as some other examples of color polymorphisms (Olendorf et al. 2006). Similar selection can operate in plant–pollinator interaction systems (**Figure 2.2**).

Box 2.6* LOCAL STABILITY WITH FREQUENCY-DEPENDENT SELECTION

Equation B2.4.2b of **Section 2.1.ii.c** can be used in this case, but now the derivatives of the genotypic fitnesses with respect to p must be included in the expression for $\lambda(q^*)$, so that the final expression is:

$$\lambda(q^*) = 1 + \left[\frac{p^* q^*}{\bar{w}^*} \right] \times$$

$$\left[w_{11} + w_{22} - 2w_{12} + p^* \left(\frac{d(w_{12} - w_{11})}{dq} \right)_{q^*} + q^* \left(\frac{d(w_{22} - w_{12})}{dq} \right)_{q^*} \right] \quad \text{(B2.6.1)}$$

(Note that the fitnesses are functions of q, so that the values of the two derivatives with respect to q are for q^*, the equilibrium value of q.)

The possibilities are now much richer than in the case of constant fitnesses. In the first place, heterozygote advantage is no longer required, so that the sum of the first three terms in braces is not necessarily negative. For example, with complete dominance of A_1 or A_2, the three fitnesses are all equal at equilibrium, so that this sum vanishes. Second, we now have to reckon with additional contributions from the terms involving the derivatives of the fitnesses.

With a dominant A_1 allele, we have the simplest situation, in which we need consider only the derivative of w_{22}, if fitnesses are measured relative to those of carriers of A_1, so that $w_{11} = w_{12} = 1$. If we write $w_{22} = 1 - s(q)$ and note that $s(q^*) = 0$, we obtain:

$$\lambda(q) = 1 - p^* q^{*2} \left(\frac{ds}{dq} \right)_{q^*} \quad \text{(B2.6.2)}$$

where ds/dq is always positive if there is negative frequency dependence, implying that $\lambda(q^*) < 1$. The possibility is, however, left open that $\lambda(q^*) < 0$, if ds/dq is sufficiently large, in contrast to what is found with constant fitnesses.

If $-1 < \lambda(q^*) < 0$, the general analysis presented in **Box 2.4*** implies that the value of a small deviation $q - q^*$ from equilibrium will be of opposite sign, but smaller in magnitude, than its value in the previous generation. This means that the population will converge to the equilibrium, but will oscillate from one side of the equilibrium to the other on its approach.

If $\lambda(q^*) < -1$ (i.e., if ds/dq is sufficiently large), the deviations will change sign each generation, and increase in magnitude (**Problem 2.5***). This means that the equilibrium is *unstable*. But, in contrast to the case of heterozygote disadvantage, the boundaries of allele frequencies $q = 0$ and $q = 1$ are also unstable, since selection pushes allele frequencies away from them in this model. This suggests that the population will end up oscillating permanently around the equilibrium point q^*.

Deeper analyses using the general properties of dynamical systems confirm this conclusion (May and Anderson 1983). If ds/dq is just large enough that $\lambda(q^*) < -1$, then the population will exhibit a *two-point limit cycle*, in which it swings backwards and forwards between two allele frequencies on either side of q^*. Progressively larger values lead to limit cycles involving 4, 8, 16, ... successive allele frequencies. Eventually, a region of *chaos* will be reached, in which the population fluctuates endlessly, with no apparent repetition of a cycle of change. Otto and Day (2007, Chapter 5) provide a detailed treatment of limit cycles and chaos.

Frequency-dependent selection resulting from predation is also important in *Batesian mimicry*, in which a distasteful or dangerous "model" species is mimicked by an edible species subject to predation (Fisher 1930b, Chapter 7). Here, the mimicking species is edible, and so will suffer mortality from predation. Inexperienced predators that capture and eat members of the model species will have an unpleasant experience that teaches them to avoid other, similar individuals, leading to a relatively low rate of predation on the model species. A rare variant of the mimic species that resembles the model in color or pattern will therefore share in its protection from predation, and tend to increase in frequency. Just as with apostatic selection, the increase in frequency of the mimetic form leads to an increased chance that naïve predators eat it, and so come to associate this form with a pleasant rather than unpleasant experience, so that its selective advantage dwindles. As a result, Batesian mimetic forms often show polymorphisms, as illustrated in **Plate 1**, front endpaper (Fisher 1930b, Chapter 7; Sheppard 1975, Chapter 11). Balanced polymorphisms are, however, not an inevitable outcome of either apostatic selection or Batesian mimicry. The maintenance of polymorphism requires an appropriate relationship between fitness and frequency, and a new allele can spread to fixation (and thus not come to a polymorphic equilibrium) if its selective advantage still exists at high frequencies, even when the fitnesses of its carriers decline with frequency (Charlesworth and Charlesworth 1975a).

A less obvious, but very important, system in which frequency-dependent selection causes the maintenance of polymorphisms is *self-incompatibility* in hermaphroditic species of plants (Wright 1939; Vekemans and Slatkin 1994; Schierup et al. 1998). Genetic studies of plant species with an inability to self-fertilize have shown that these are often controlled by one or a small number of "S loci," which are among the most highly polymorphic genes known in terms of numbers of alleles (Richman et al. 1996; Sato et al. 2002). The simplest such system involves a single *gametophytic self-incompatibility* locus, found in plants such as the relatives of tobacco and in apples. Pollen grains carrying an allele S_i at the S locus cannot fertilize ovules of plants that also carry S_i, and these grains' tubes usually fail to grow down the stigma to the ovary (**Figure 2.3**). This means that all plants in the population are heterozygous for two different S alleles, and that there must be at least three of these for the plants to be able to cross-fertilize each other and allow the population to survive. Usually there are at least a dozen S alleles in a population, and often many more. A mutation that creates a new S-allele will have a selective advantage, since pollen carrying a new variant can fertilize any plant in the population, whereas pollen carrying an established allele cannot fertilize the portion of the population that carries that

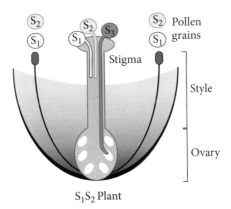

FIGURE 2.3 Diagram showing gametophytic self-incompatibility, which is found in several flowering plant families. The pollen expresses its own haploid genotype and pollen grains carrying an allele S_i at the S locus cannot fertilize ovules of plants that also carry S_i.

allele. This selective advantage will decline as the new allele increases in frequency; if the alleles have no other effects on fitness, equilibrium will be achieved when all possible heterozygotes have equal frequencies (Wright 1939).

Frequency-dependent selection also arises from the relations between hosts and parasites. The basic idea was sketched in a seminal paper by Haldane (1949b). The transmission of parasites among hosts, and hence the survival of a parasite population, requires encounters between infected and susceptible, uninfected host individuals. The rules governing the ecological relationships between hosts and parasites are the subject of the well-established science of mathematical epidemiology (May and Anderson 1983; Anderson and May 1991). For our purpose, we only need to consider only a system in which the population sizes of the host and its parasite have come into equilibrium, so that the numbers of each are stable. Now imagine a new mutation in the host, which confers greater resistance to the parasite compared to the rest of the host population. This will have a selective advantage, and the mutation will start to spread. As it becomes more common, the parasites find fewer new hosts to infect, and so their abundance decreases. The chance that nonresistant hosts acquire new infections also declines, and so the advantage of resistance falls. If resistance comes with a *cost*, so that the fitness of resistant individuals (in the absence of the parasites) is lower than that of nonresistant individuals, there is a potentiality for a polymorphic equilibrium, with resistant and nonresistant hosts coexisting.

The parasite may also experience frequency-dependent selection. This can happen either if there is genetic variation in resistance to the parasite, arising from the type of mechanism we have just discussed, or if host individuals can acquire immunity to infection to specific types of parasite (this can occur in vertebrates, since they have the ability to develop lasting, specific immune responses to antigens produced by parasites). In the first case, a new form of parasite that can infect the previously resistant host genotype has a selective advantage, and will therefore increase in abundance. As a result, the frequency of this host genotype will decline, reducing the selective advantage to the new type of parasite. If it suffers a fitness cost, perhaps because it is not so good at infecting the old host genotype as the alternative parasite genotype, then its frequency may be stabilized in the population. This type of interaction between host and parasite genotypes can give rise to violent cycles of allele frequency changes in both host and parasite (May and Anderson 1983). Such situations are likely to be common in the dynamics of host-pathogen systems (May and Anderson 1983; Seger 1988; Frank 1996; Brown 2003), as shown in **Figure 2.4**. Relations between plants and their pathogens often involve "gene-for-gene" interactions of the kind often assumed in these models, and there is suggestive evidence in some cases that these have resulted in the long-term maintenance of variability in DNA sequences, at least in the plant resistance genes (Rose et al.

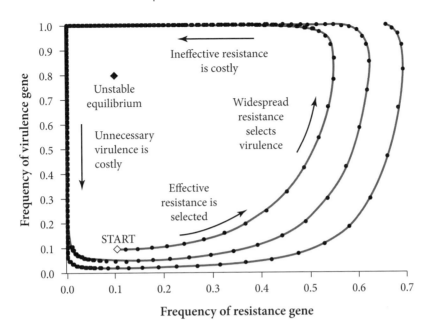

FIGURE 2.4 Dynamics of allele frequencies in a simple model of gene-for-gene interactions between a plant and a parasite. In each generation, each plant is attacked by one parasite and all parasites have an equal chance of attacking each plant. A resistance allele for a common parasite genotype has a selective advantage because it protects the host against the majority of parasites that lack the ability to attack resistant hosts (the lower part of the cycle). As it increases in frequency, it selects against this avirulent parasite genotype (the right part of the cycle). If the resistance gene has a fitness cost, the resistance alleles will start to decline when the virulent parasite (which is able to overcome host resistance) is common (the top of the cycle). If virulence is costly to the parasite, the loss of host resistance causes the virulent parasite genotype to decrease in frequency (the left part of the cycle). In the example plotted, the fitness cost (the relative reduction in fitness) to the plant of being diseased (s) is 0.25 and the cost of having the resistance gene (t) is 0.05. The cost to an avirulent pathogen, which is unable to infect a resistant host, (c) is 0.9 and the cost of being virulent (d) is 0.09. The trajectories of gene frequencies spiral around and away from an unstable equilibrium point where the frequency of resistance is $d/c = 0.1$ and the frequency of virulence is $(s - t)/s = 0.8$. The model was run for 500 generations. [Adapted from Figure 5 of Brown (2003).]

2004). The nature of the evidence for long-term maintenance of sequence variability will be explained in **Section 8.3.ii**.

With acquired immunity, a common type of parasite will probably have infected a large fraction of the hosts, which become immune to further infection. A new antigenic type of the parasite will be capable of infecting these individuals, as well as the rest of the population, and therefore gains a frequency-dependent advantage. In principle, this could maintain a large number of different antigenic types in the parasite population. There are, in fact, many examples of highly polymorphic genes in pathogens, such as malarial parasites, which control proteins involved in antigenic reactions with the hosts (Polley and Conway 2001). It seems highly likely that these polymorphisms are maintained by this type of frequency dependence, especially as these are predominantly haploid species where heterozygote advantage is impossible.

It is also possible, although far from certain, that many amino acid polymorphisms in the *major histocompatibility complex* (MHC) loci of vertebrates are maintained by frequency-dependent selection involving interactions with pathogens. These genes control molecules that are involved in presenting antigenic polypeptides, derived from the degradation of proteins of pathogens, to cells of the immune system. Different variants at a given MHC locus are known to have different specificities with respect to antigen presentation, and it seems plausible that relative abundances of pathogen types and MHC type could cause frequency-dependent selection, especially because the MHC polymorphisms mostly involve the amino acids involved in antigenic recognition (Garrigan and Hedrick 2003).

Finally, classical Lotka–Volterra ecological models of the coexistence of two species that compete for overlapping, but somewhat different, resources imply negative frequency dependence, since a system in which a species has reached equilibrium population size can be invaded by another, rarer species that uses a different resource spectrum (Begon et al. 2006, Chapter 8; Crow and Kimura 1970, Chapter 1). The second species will eventually lose its competitive edge when it becomes sufficiently abundant that it uses up its resources to a similar extent to the first species. This is formally equivalent to selection in a haploid or asexual population of a single species, and studies of bacterial systems have shown how several different genetic variants can coexist by this type of mechanism (Goymer et al. 2006; Bell 2008, pp. 349–358). Diploid, sexual populations can behave in the same way (Wright and Dobzhansky 1946). A recent possible example of this is the behavioral polymorphism of *D. melanogaster* larvae, with two types distinguished by their levels of activity: "rovers" and "sitters," controlled by a mutation in a single gene. The relative fitnesses of the two morphs exhibit negative frequency dependence at low larval densities in laboratory cultures (Fitzpatrick et al. 2007).

2.2.ii. Fitnesses that vary in time

One might also expect that temporal variation in the relative fitnesses of different genotypes could preserve variability, since an allele that does badly in one generation may gain an advantage in subsequent generations, and hence avoid elimination. We can study this possibility by modifying our model to allow the relative fitnesses of the various genotypes to vary between generations (see **Box 2.7**). For haploid populations, an argument based on **Equations B2.1.2a** and

Box 2.7 SELECTION WITH TEMPORAL VARIATION IN FITNESSES

B2.7.i. Haploids

In the case of haploids, we can use **Equations B2.1.2a** and **B2.1.2b** of **Box 2.1** in **Section 2.1.i** to write:

$$p_{t+1} = \frac{p_t w_{1,t}}{\overline{w}_t} \qquad (B2.7.1a)$$

and

$$q_{t+1} = \frac{q_t w_{2,t}}{\overline{w}_t} \qquad (B2.7.1b)$$

where the subscript t denotes values in generation t.

For the next step, it is more convenient to use the ratio $u = q/p$ instead of the variant frequencies themselves (because it eliminates the denominator, $\overline{w}$). We then obtain, following Haldane (1924):

$$\frac{u_t}{u_{t-1}} = \frac{w_{2,t-1}}{w_{1,t-1}}$$

By iterating this relation over successive generations, we find that:

$$\frac{u_t}{u_0} = \frac{w_{2,t-1} w_{2,t-2} \cdots w_{2,0}}{w_{1,t-1} w_{1,t-2} \cdots w_{1,0}}$$

This can be written as:

$$\frac{u_t}{u_0} = \frac{\prod_{v=0}^{t-1} w_{2,v}}{\prod_{v=0}^{t-1} w_{1,v}} \qquad (B2.7.2)$$

where the Π symbol denotes a product, and v denotes a generation between 0 and t.

If the fitness product involving $w_{2,v}$ exceeds that for $w_{1,v}$, A_2 will increase in frequency towards fixation; A_1 will become fixed if the opposite is true. It is possible for intermediate allele frequencies to be maintained if the ratio of the fitness products is cyclical (Nagylaki 1975), but the values of these frequencies depend on their initial values, and so there is no tendency for rare alleles to increase in frequency. In a randomly varying environment, the mean value of the fitness product will either be greater or less than 1, so that one or other allele will eventually be fixed.

B2.7.ii. Diploids

In the case of diploids, we consider first the situation when q is close to 0. In this case, A_2 is carried mainly in heterozygotes if the population is randomly mating, and so its marginal fitness is close to the fitness of the heterozygotes, deviating by an amount of order q. Similarly, the marginal fitness of A_1 is equal to the fitness of A_1A_1 homozygotes, deviating by a term of order q. We can then use **Equations B2.2.2a** and **B2.2.2b** in **Section 2.1.ii.a** to establish that:

$$\frac{u_t}{u_{t-1}} \approx \frac{w_{12,t-1}}{w_{11,t-1}}$$

(B2.7.3a)

The approximation is accurate to the order of terms in q^2, and thus becomes increasingly accurate as q approaches 0.

Iterating this over successive generations, we get:

$$\frac{u_t}{u_0} \approx \frac{\prod_{v=0}^{t-1} w_{12,v}}{\prod_{v=0}^{t-1} w_{11,v}}$$

(B2.7.3b)

There will be a net increase in the frequency of A_2 over the set of generations if

$$\frac{\prod_{v=0}^{t-1} w_{12,v}}{\prod_{v=0}^{t-1} w_{11,v}} > 1$$

(B2.7.4a)

This implies that the product of the fitnesses of the A_1A_2 heterozygotes must be larger than the product of the fitnesses of the A_1A_1 homozygotes.

If we take the tth root of the denominator and numerator of the left-hand side of this expression, we obtain the ratio of the *geometric mean fit-*

nesses of the two genotypes. (The geometric mean is an alternative measure to the standard arithmetic mean defined in **Appendix A2.iii.**) The ratio of the geometric fitnesses of A_1A_2 and A_1A_1 can exceed one only when the ratio of the corresponding products is greater than one, so that this provides a convenient way of representing the condition for invasion.

Exactly the same analysis can be carried out for the case when A_1 is the rare allele. Changing subscripts appropriately, the condition for its increase is:

$$\frac{\prod_{v=0}^{t-1} w_{12,v}}{\prod_{v=0}^{t-1} w_{22,v}} > 1 \qquad (B2.7.4b)$$

B2.1.2b (see **Section 2.1.i**) shows that one genotype will eventually prevail, except under rather restricted conditions, and that it will be the one with the higher product across generations of the fitness values (Dempster 1955; Nagylaki 1975). Thus, perhaps counterintuitively, temporal variation in fitness is unlikely to help maintain variability in haploids (Gillespie 1991, pp. 146–149); the same is true for diploid asexual populations.

The case of diploidy can be studied by examining the conditions under which each allele invades the population when rare, avoiding the difficult task of examining the full trajectory of the population when fitnesses vary in time (Haldane and Jayakar 1963). This situation is referred to as a *protected polymorphism* (Prout 1968). **Box 2.7** shows that each allele can invade a population in which the other allele is initially fixed—that is, there can be a protected polymorphism—if the product across generations of the heterozygote fitnesses exceeds the corresponding product for either homozygote. The following table of fitnesses gives an example:

	A_1A_1	A_1A_2	A_2A_2
Environment 1	0.4	1	2.0
Environment 2	2.0	1	0.4

The fitness products for the two homozygotes are both 0.8, whereas the heterozygote's value is 1. The means of both homozygotes' fitnesses are the same, 1.2, so that on average there is heterozygote disadvantage.

The situation when one allele is completely dominant over the other must be considered separately; the condition for increase of the recessive allele when rare is then that the arithmetic mean fitness of homozygous carriers relative to

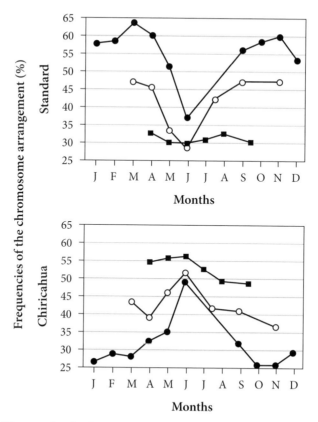

FIGURE 2.5 The annual cycle of changes in the frequencies of two chromosome arrangements in *Drosophila pseudoobscura*, Standard and Chiricahua, at three different locations on Mt. San Jacinto, California. The letters on the *x* axis refer to months of the year. [Adapted from Figure 2 of Wright and Dobzhansky (1946).]

that of the fitness of carriers of the dominant allele exceeds 1, while the condition for increase of the dominant allele is the same as above (Haldane and Jayakar 1963).

There are many examples of natural polymorphisms where the relative fitnesses of genotypes vary in time. A classic example is that of the inversion polymorphisms in *Drosophila pseudoobscura*, studied by Dobzhansky (1943) and Wright and Dobzhansky (1946), whose seasonal frequency changes are shown in **Figure 2.5**. Other examples are reviewed by Bell (2008, pp. 242–245). So far, however, there is little evidence that polymorphisms are maintained in this way, in the absence of heterozygote advantage or frequency-dependent selection. A possible example is the flower color polymorphism in the plant *Linanthes parryae*, which grows in the deserts of the southwestern United States of America (Turelli et al. 2001).

2.2.iii. Spatial variation in fitnesses, hard versus soft selection, and speciation

A protected polymorphism can also be maintained in spatially variable environments, under the conditions assumed in *Levene's model* (Levene 1953). This model (**Box 2.8**) assumes that a population encounters two or more environmental patches or *niches*, within which all selection takes place. The relative fitnesses of the genotypes vary across niches. The survivors emerge from their niches, to mate at random without regard to their niche of origin, and each niche contributes a fixed proportion of adults to the mating population—that is, there is complete density-dependent regulation of population size within niches. This model is formally very similar to the maintenance of two forms by the frequency-dependent exploitation of different niches that we discussed above, and works perfectly well in haploids. The relevant equations are derived in **Box 2.8**.

Box 2.8 SELECTION WITH SPATIAL VARIATION IN FITNESSES (LEVENE'S MODEL)

We assume that there is a set of n niches, such that the kth niche contributes a fixed fraction c_k to the gametes of the next generation (i.e., soft selection—see the main text), with completely random mating among individuals from different niches. The fitness of genotype A_iA_j in niche k is denoted by $w_{ij}^{(k)}$. Given the assumption of random mating across niches, each niche will have the same allele frequency q at the start of a generation. After selection within a given niche k, the frequency among individuals in the niche will be changed to a new value, given by the equivalent of **Equations B2.2.2a** and **B2.2.2b** of **Box 2.2** in **Section 2.1.ii.a** but using the fitness values $w_{ij}^{(k)}$.

We can study the case when A_2 is rare (q is close to 0) by using similar reasoning to that for the case of temporal variation in fitness; the mean fitness of the population for niche k is close to $w_{11}^{(k)}$ and the marginal fitness of A_2 is close to $w_{12}^{(k)}$. Summing the contributions from each niche, we obtain the approximate new value of q as:

$$q' \approx q \sum_{k=1}^{n} c_k \frac{w_{12}^{(k)}}{w_{11}^{(k)}} \tag{B2.8.1a}$$

A similar argument can be used for the case when A_1 is rare:

$$p' \approx p \sum_{k=1}^{n} c_k \frac{w_{12}^{(k)}}{w_{22}^{(k)}}$$ (B2.8.1b)

It follows from these relations that both alleles can invade when rare when the following two relations are satisfied:

$$\sum_{k=1}^{n} c_k \frac{w_{12}^{(k)}}{w_{11}^{(k)}} > 1$$ (B2.8.2a)

and

$$\sum_{k=1}^{n} c_k \frac{w_{12}^{(k)}}{w_{22}^{(k)}} > 1$$ (B2.8.2b)

This situation is likely to occur in insect species that eat plants and spend their entire larval life on one food plant. If different species of plants are exploited by the same species of insect, genotypes may differ in their performance on the respective plant hosts. Some *host races* of insect species show adaptations to different food plants, as assumed in this model (Via 2001). Further theoretical work shows that this type of selection can lead to the evolution of reproductive isolation between genotypes adapted to different niches, i.e., to *sympatric speciation* (Maynard Smith 1966), because this reduces the frequency of heterozygotes, which have low fitness. There is evidence that this may have happened in species such as the pea aphid, where there are two closely related forms adapted to different host plants (Via 2001; Via and West 2008).

This situation is often referred to as *soft selection*, because there is no relationship between the mean fitness of the individuals born into a given niche and the niche's contribution to the mating population, i.e., the extent to which individuals are eliminated by selection in a niche has no effect on its output. The extreme alternative is *hard selection*, in which each niche contributes to the mating population in proportion to its mean fitness (Dempster 1955). A similar analysis to that in **Box 2.8** shows that in this case variation can be maintained if there is an average advantage to heterozygotes (**Problem 2.6**). Intermediate situations can be imagined, where there is partial density regulation within niches, and these give results in between the two extremes (Arnold and Anderson 1983).

A fitness difference between the two sexes is a form of soft selection. For an autosomal locus, each sex contributes equally to the offspring, so that males and females are equivalent to two niches, each with a frequency of one-half. If alleles A_1 and A_2 have opposite effects on fitness in males and females (*sexual antagonism*), polymorphism can be maintained without heterozygote advantage (Prout 2000). Examples of polymorphisms with different variant frequencies in males and females are known, and are candidates for this form of selection (e.g., Marshall et al. 2004).

Unless there is high variability in relative fitnesses among environments, however, polymorphism is maintained by either temporal or spatial variation if the average fitness over the different environments is highest for the heterozygotes (Gillespie 1991, Chapter 4). This is shown in **Box 2.9***, which proves that the condition for the spread of either allele when rare depends on both the mean and variance of the selection coefficient for the heterozygote relative to that of the prevailing homozygote. If the variance for the heterozygote is sufficiently large (implying strongly fluctuating selection coefficients), an allele can spread even if it causes a mean reduction in fitness to the heterozygote. Mean heterozygote advantage is, however, always sufficient for a protected polymorphism.

Box 2.9* RELATION OF MEANS AND VARIANCES OF SELECTION COEFFICIENTS TO THE SPREAD OF ALLELES IN VARIABLE ENVIRONMENTS

B2.9.i. Temporal variation

We can conveniently write the ratio of heterozygote fitness to the fitness of A_1A_1 homozygotes in generation v as $w_{12,v}/w_{11,v} = 1/(1-s_v)$, where s_v is the selection coefficient against A_1A_1 homozygotes in generation v. Taking natural logarithms of the quantities on both sides of **Equation B2.7.4a** in **Section 2.2.ii**, the condition for the spread of allele A_2 is equivalent to:

$$\sum_{v=0}^{t-1} \ln\left(\frac{1}{(1-s_v)}\right) > 0$$

If we assume that the fluctuations in fitnesses over a sufficiently long period of time follow a distribution with mean $\bar{s}$ and variance V_s, we can replace the sum by t times the expectation (mean) of the bracketed term, so the condition for spread is equivalent to:

$$E\left\{\ln\left(\frac{1}{(1 - s_v)}\right)\right\} > 0 \text{ or } E\{\ln(1 - s_v)\} < 0 \qquad (B2.9.1)$$

where E denotes the expectation of the quantity in braces (**Appendix A2.iii**).

Using the first three terms of the Taylor's series expansion of the logarithmic expression (**Appendix A1.ii.c**), we can approximate this by:

$$E\left\{s_v + \frac{s_v^2}{2}\right\} > 0 \text{ or } \bar{s} + \frac{(\bar{s}^2 + V_s)}{2} > 0 \qquad (B2.9.2a)$$

If selection is relatively weak, the square of $\bar{s}$ can be neglected, and the condition simplifies to:

$$\bar{s} + \frac{1}{2}V_s > 0 \qquad (B2.9.2b)$$

A similar expression can be derived for the spread of A_1, substituting the selection coefficient against A_2A_2 homozygotes for that against A_1A_1 homozygotes.

B2.9.ii. Spatial variation (Levene's model)

Write $w_{12}^{(k)}/w_{11}^{(k)} = 1/(1 - s^{(k)})$, and let the mean and variance of $s^{(k)}$ across the environments encountered by the population again be $\bar{s}$ and V_s, respectively. By means of a similar argument to that for temporal fluctuations, but this time evaluating the Taylor's series expansion of $1/(1 - s^{(k)})$ (**Appendix A1.ii.c**), the condition for the spread of A_2, given by **Equation B2.8.2a**, can be approximated by:

$$E\{s + s^2\} \approx \bar{s} + \bar{s}^2 + V_s > 0 \qquad (B2.9.3a)$$

As with temporal variation, if selection is weak, this simplifies to:

$$E\{s + s^2\} \approx \bar{s} + V_s > 0 \qquad (B2.9.3b)$$

2.2.iv. The biochemical basis of fitness differences

We have seen that both temporal and spatial environmental variability can help to maintain polymorphism, even if selection is weak, but we have not explained when this is likely to occur in reality. One situation in which this happens is when fitness is related to a causal variable, x, like the activity of an enzyme, through a *concave* function of x (i.e., one with diminishing returns relation to fitness; see **Figure 2.6**) (Gillespie and Langley 1974; Gillespie 1991, Chapter 4). This occurs when fitness is determined by the output of a metabolic pathway, of which the enzyme is one component (Wright 1934; Kacser and Burns 1981; Dykhuizen and Dean 1990). In addition, environmental differences may affect how well different genotypes can metabolize and function, causing differences in their fitnesses.

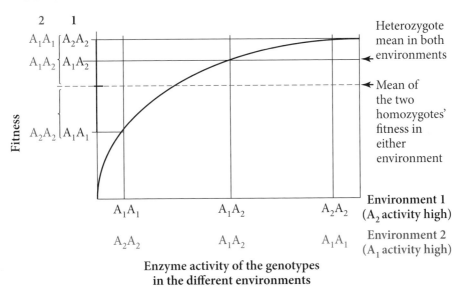

FIGURE 2.6 Heterozygote advantage resulting from the concavity of fitness (the y axis) as a function of enzyme activity (the x axis). The fitness of the heterozygote in a given environment (say, environment 1) will be close to that of the fitter of the two homozygotes. Dominance with respect to fitness in a single environment thus arises naturally in this model. When there is another environment (say, environment 2) in which the other homozygote has the higher fitness, the mean heterozygote fitness is higher than that of either homozygote averaged across the two environments (indicated by the horizontal dashed line, which is exactly half way between the two homozygote values). For clarity, the example shows a case in which the homozygote fitnesses in the two environments are symmetrical, so that heterozygotes have the same fitness in both, but the same result arises in more realistic situations.

As a simple but extreme example, consider the case of two equally frequent environments, and a gene with two alleles whose products affect enzyme activity, x, differently. Suppose that the differences between the activities for the two homozygotes A_1A_1 and A_2A_2 are in opposite directions in the two environments (A_1 is associated with high x values in one environment and A_2 in the other; see **Figure 2.6**). In each environment, we assume semidominance, so that x for the heterozygote lies exactly between the two homozygous values, as is usually the case for alleles that affect enzyme activity (Gillespie and Langley 1974).

Some examples of the concave relationship between enzyme activity and a measure of fitness for bacterial enzymes are shown in **Figure 2.7**. As a consequence of this property of the fitness–activity relation, the fitness of the heterozygote in a given environment will be closest to that of the fitter of the two homozygotes (**Figure 2.6**). Some degree of dominance with respect to fitness of the allele with the higher fitness in a single environment thus arises naturally in this situation (Wright 1934; Kacser and Burns 1981). As discussed in **Section 4.4.v**, this sheds light on the observation that mutations are usually recessive with respect to their effects on the phenotype.

In the model of **Figure 2.6**, the mean of the fitness of A_1A_2 over the two environments is higher than that of either A_1A_1 or A_2A_2 (**Figure 2.6**), so that net heterozygote advantage for fitness arises from the combination of geno-

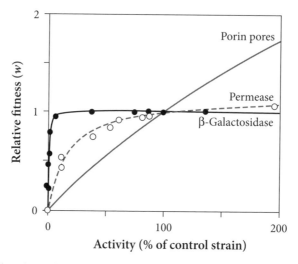

FIGURE 2.7 The relationship between activity and a measure of fitness for two bacterial enzymes and for the cellular structure, porin pores. For β-galactosidase (black dots), the relationship is strongly concave, whereas for permease (open circles), it is concave, but not so strongly. For porin pores, there is little such concavity. [Adapted from Figure 5 of Dykhuizen and Dean (1990).]

type–environment interactions on enzyme activity, x, and the nonlinear relation between x and fitness. This forms the basis of the *stochastic additive scale-concave fitness function* (SAS-CFF) model of John Gillespie, which he developed into a detailed mathematical theory of polymorphism in arbitrary numbers of different environments, or with randomly distributed environmental changes (Gillespie and Langley 1974; Gillespie 1991, Chapter 4). The basic idea is that the value of x associated with a given allele when homozygous may vary in time or space. As a result of the concave relation between fitness and x, the resulting means and variances of the fitness of the diploid genotypes can easily satisfy the conditions for maintaining polymorphism in variable environments, which we have already discussed in **Sections 2.2.ii** and **2.2.iii** (**Problem 2.7***). An attractive feature of this model is that the conditions for maintaining polymorphism arise directly from the few, quite realistic, assumptions just outlined. In particular, the model assumes that the allele frequencies of protein variants, including those studied as allozymes (**Section 1.2.i**), are sensitive to environmental conditions, as is often found to be the case (Gillespie 1991, Chapter 1). We describe an example in **Section 4.1.ii.a**.

A conceptually similar model to that shown in **Figure 2.6** is often called *antagonistic pleiotropy* (Rose 1982; Charlesworth and Hughes 2000; Prout 2000). Instead of two different environments, there are two or more different *components of fitness*, such as fertility and survival. If the homozygous effects of an allelic difference are in opposite directions for the two characters, and the favorable allele with respect to each character is dominant over the harmful allele, heterozygote advantage for fitness results. We discuss this idea again in **Section 9.5.v.d** in relation to the evolution of aging.

PROBLEMS

2.1.   **i.** Using **Equation B2.3.1** from **Box 2.3** in **Section 2.1.ii.b**, show that the equilibrium frequencies of A_1 and A_2 with heterozygote advantage are $p^* = t/(s + t)$ and $q^* = s/(s + t)$, respectively.

   ii. Show that $w_{2.} - w_{1.} > 0$ when $q < q^*$, and that $w_{2.} - w_{1.} < 0$ when $q > q^*$.

2.2   A laboratory population of the fruitfly *D. melanogaster* was maintained in a cage, with a second-chromosome balancer marked with the dominant mutation Cy, which is lethal when homozygous. The balancer was competed for many generations against a single wild-type second chromosome (+), derived from a natural population. The final frequency of Cy individuals among adult flies was 0.6. In a separate experiment, the egg-to-adult viability of +/+ flies was found to be 0.8, relative to that of Cy/+.

   i. Determine the fitness of +/+ flies relative to Cy/+, using the theory for equilibrium under heterozygote advantage.

      ii. Discuss possible reasons for why the fitness and viability values differ.

2.3   Show that the equilibrium with heterozygote disadvantage in a randomly mating population is unstable.

2.4* Use **Equations B2.4.1** and **B2.4.2b** of **Box 2.4** in **Section 2.1.ii.c** to derive **Equation B2.4.3** for the local stability of an equilibrium with heterozygote advantage.

2.5.* **i.** Obtain an expression for the stability criterion, $\lambda(q^*)$, with frequency-dependent selection (**Box 2.6*** in **Section 2.2.i.b**), in the case of a dominant allele when the selection coefficient against A_2A_2 homozyotes given by $s(q) = bq - a$, where a is a measure of the fitness cost to A_1 when A_2 is rare, and b measures the dependence of s on q.

     ii. Writing $k = b/a$, find a necessary condition for the existence of limit cycles or chaos in terms of k and b.

2.6   Construct a model of hard selection in a heterogeneous environment, similar to Levene's model of soft selection (**Box 2.8** in **Section 2.2.iii**). Show that polymorphism at a biallelic locus is maintained only if the average fitness of each genotype across niches is highest for the heterozygote.

2.7.* **i.** Use the model in **Figure 2.6** of **Section 2.2.iv**, which describes a locus with two alleles A_1 and A_2 that affect a variable x with a concave relation with fitness, to determine the conditions for net heterozygote advantage in a varying environment. Assume semidominance of the effects of the alleles on x, and use the first three terms of a Taylor's series to obtain expressions for the fitnesses of the three possible genotypes in terms of the means and variances of x for A_1A_1 and A_2A_2.

     ii. Relate this result to the conditions for the maintenance of variation in variable environments in **Box 2.9*** in **Section 2.2.ii**, assuming that the means and variances of x are the same for A_1A_1 and A_2A_2, and that fitness is a concave function of x.

CHAPTER THREE

Directional Selection and Adaptation

CHAPTER SUMMARY

In order to understand evolution by natural selection, it is important to determine the rates of evolutionary change that are possible under selection. We first calculate the numbers of generations required for a selectively favorable mutation to reach different frequencies within a large population. For haploid populations, or for nonrecessive mutations in randomly mating diploid populations, these times depend only weakly on the variant's initial and final frequencies. In these cases, even very rare mutations can spread to fixation in a time that is only a small multiple of the inverse of the selection coefficient. However, rare, favorable recessive mutations spread very slowly in diploid, randomly mating populations, explaining the fact that most recent evolutionary changes in such populations involve nonrecessive mutations. When the frequencies of favorable mutations are monitored during laboratory experiments, or in natural populations, the theoretical results can be used to estimate the strength of selection involved.

The theory is extended to fitness differences between the two sexes, when autosomal and sex-linked variants behave differently, and to inbreeding populations. We also discuss selection on mutations that cause individuals to benefit their relatives, even at a cost to their own fitness, and show how such kin selection can help to explain the "altruistic" behavior of social organisms.

We then introduce the topic of randomness (stochasticity) in the transmission of variants from one generation to the next, due to the fact that offspring numbers necessarily vary among individuals. Even in a very large population, random changes in the number of copies of a rare, favorable mutation can result in its loss from the population. We calculate the probability that a mutation becomes established at a frequency high enough that it is unlikely to be lost, i.e., its survival probability. The results show that most favorable mutations will be lost, even from large populations. Isolated populations that are exposed to the same type of selection may therefore diverge genetically by chance, and we discuss evidence for this in the case of the malaria resistance variants of humans.

We next outline some theory for quantitative trait variation. This is important for predicting the outcomes of both natural and artificial selection, where the details of the genetics of the traits involved are usually unknown, so that only statistical descriptions of the state of the population can be used. The theory shows how continued selection can produce combinations of variants that create phenotypes far outside the range of variation in the original, unselected population, even in evolutionarily very short times. Such rapid changes are indeed observed in long-term selection experiments. We briefly discuss correlations between different characters, and outline how the theoretical results can be used to estimate the strength of natural selection in the wild.

Finally, we show how the theory developed in this chapter gives insights into the expected sizes of mutational changes that become fixed in populations during the course of the adaptive evolution of complex phenotypes, including both morphological traits and protein sequences. The theory, and empirical evidence, support the view that the adaptive evolution of a multi-component structure usually involves a succession of genetic changes affecting separate components, rather than single "macromutations" simultaneously changing whole sets of characters, although individual mutations may contribute a significant fraction of the total change involved.

> Natural Selection is the only means known to biology by which complex adaptations of structure to function can be brought about.
>
> R. A. Fisher, *The Genetical Theory of Natural Selection*, Chapter 7

> A satisfactory theory of natural selection must be quantitative. In order to establish the view that natural selection is capable of accounting for the known facts of evolution, we must show not only that it can cause a species to change, but that it can cause it to change at a rate which will account for present and past transmutations.
>
> J. B. S. Haldane (1924)

INTRODUCTION

As made clear by Haldane's statement, we need to know if natural selection is capable of producing change over the time scales observed in the fossil record, or inferred from observations on living species. If, for example, it would take a billion generations for a population to become fixed for a new, selectively favorable mutation, we would severely doubt the effectiveness of selection.

Much early work in population genetics was therefore devoted to calculating the time needed for selection to change the frequency of an allele at a single locus. This started with studies by the English mathematician H. T. J. Norton, published as an appendix to a book on mimicry in butterflies by the early geneticist R. C. Punnett (1915), followed by the classic paper of Haldane (1924). This work, and its later extensions, provide a framework for understanding, among other things, the very simple type of adaptive evolution that is often seen in the responses of organisms to selection pressures resulting from human activities, such as antibiotic resistance in bacteria and the resistance to insecticides of many insect pests.

Changes in variant frequencies at individual sites in the genome, either single-nucleotide mutations or more complex mutational changes, must underlie nearly all adaptive evolution, even if many such sites are involved in the control of a trait under selection. In most cases of interest, however, we only know about changes in quantitative traits, such as the cranial capacity of humans, without any details of their genetics. For the purposes of understanding how fast natural selection can change quantitative traits, and for designing artificial selection programs for animal and plant improvement, models of how selection causes changes in quantitative traits are essential.

As we will see, the two types of selection models that we shall discuss (involving changes in the frequencies of allelic variants at single locus or site, and models of changes in the means of quantitative traits) have proved very successful in helping us to understand the evolutionary changes that constitute what is often called *microevolution*, i.e., changes in the genetic makeup of a single population or species. The theoretical results confirm what Darwin called "the power of selection". But there has been a debate about the causes of the evolution of species differences and "major" evolutionary innovations. Was Darwin right to view these as the cumulative outcome of individually small steps, each of which is selectively advantageous, or do they require unusual genetic processes?

In this chapter, we shall mainly be concerned with *directional selection*, the situation in which variant frequencies or trait means are subject to pressures of selection that cause them to change in a consistent direction for many generations. As we will show, selection is very effective at bringing about such changes in sufficiently large populations. Such populations will thus be close to equilibrium for most traits, unless there has been a recent change in the environment, so that one or more traits are no longer near their "optimum" values. An environmental change will probably be followed by a rapid readjustment of the composition of the population towards the new optimum. After such an event, further evolutionary changes will then cease, unless the environment continues to change (ignoring, for now, the continual removal of deleterious mutations by purifying selection). As was stressed by the founders of the Modern Synthesis of evolutionary biology, including the paleontologist George Gaylord Simpson (1953, Chapter 7), this means that episodes of rapid evolutionary change are likely to be correlated with shifts in the environment, including shifts imposed by interactions with other species. Environmental changes are involved in the evolution of drug and insecticide resistance, in the proliferation of new species on oceanic islands, and in the emergence of higher taxonomic groups, such as the radiation of the mammals after the extinction of the dinosaurs. Models of directional selection are therefore most relevant to situations in which an environmental change gives a selective advantage to new genotypes or phenotypes over the ones prevailing in the population (a situation often referred to as *positive selection*). Bell (2008) provides numerous examples of studies of selection of this kind, for both experimental and natural populations.

3.1. DIRECTIONAL SELECTION ON A PAIR OF ALLELIC VARIANTS

3.1.i. Introduction

In studying directional selection of this kind, we proceed as follows. The model of selection specifies the relative fitnesses of individuals carrying alternative variants at a particular nucleotide site or locus, as in **Section 2.1**. The population size is assumed to be so large that random changes in variant frequencies can be ignored, i.e., it is infinite in the sense of **Section 2.1.i**. We then try to answer the question: given the initial frequency of a variant, how long does it take for its frequency to change to a specified new value? An example would be a mutation that causes resistance to an insecticide; molecular genetic studies have revealed many cases in which a single mutation, causing a change in the amino acid sequence of a protein involved in the biochemical pathway affected by an insecticide, has risen to a very high frequency, when initially it was virtually absent from the population (French-Constant et al. 2004).

3.1.ii. Haploid populations

First consider haploid populations (or the equivalent situations mentioned in **Section 2.1.i**). An exact solution for the time needed for a change in allele frequency is derived in **Box 3.1**, using elementary algebra, with the final result given by **Equation B3.1.3**. **Equations B3.1.4** and **B3.1.5** give good approximations to this when the selection coefficient s is sufficiently small that s^2 can be neglected compared with s (*weak selection*). This will generally be a reasonable approximation if $s < 0.05$, and a good one for $s < 0.01$. **Equation B3.1.4** shows

Box 3.1 DIRECTIONAL SELECTION ON HAPLOIDS

We use the same type of discrete generation model as in **Box 2.1** in **Section 2.1.i**, but we now consider A_2 to be the variant favored by selection and assign it a selective advantage s over its alternative, A_1. The relative fitnesses of A_1 and A_2 are thus 1 and $1 + s$.

If the frequency of A_1 in a given generation is p and that of A_2 is $q = 1 - p$, the argument used in **Box 2.1** shows that the values in the next generation are:

$$p' = p/\overline{w}, \qquad q' = q(1 + s)/\overline{w}$$

(Note that, with the present notation, $\overline{w} = 1 + sq$.)

As in **Box 2.7** in **Section 2.2.ii**, we can get rid of the factor of $\overline{w}$ by taking the ratios of the two equations, writing u for the ratio q/p. This gives:

$$u' = u(1 + s) \tag{B3.1.1}$$

When selection is continued over many generations, we can write the value of u in a given generation, t, in terms of its value in the initial generation, 0, as:

$$u_t = u_0(1 + s)^t \tag{B3.1.2}$$

Taking natural logarithms gives:

$$\ln(u_t/u_0) = t \ln(1 + s) \tag{B3.1.3}$$

If s is so small that terms in s^2 can be neglected in comparison with s, the right-hand side of this relation is approximately ts (**Appendix A1.ii.c**), so that the time to change the variant frequencies by a specified amount can be written as:

$$t \approx \frac{1}{s}\ln\left(u_t/u_0\right) \tag{B3.1.4}$$

This can be rearranged to give a formula for the frequency of A_2 at time t, which is similar to that for the logistic growth equation of ecology (Begon et al. 2006, Chapter 5; Crow and Kimura Chapter 1):

$$q_t \approx \frac{1}{1 + \dfrac{p_0}{q_0}\exp(-ts)} \tag{B3.1.5}$$

that the time, t, for a frequency change is then inversely proportional to s; in other words, the product ts is invariant with respect to s, provided that the weak selection assumption is met.

Equations B3.1.3 and B3.1.4 yield another very important conclusion—the time it takes for the frequency to change under selection in this model depends only *logarithmically* on the variant's initial and final frequencies. Since the logarithm of a number increases much more slowly than the number itself, this means that the time is not very sensitive to these frequencies. For example, if the initial frequency of a favorable mutation is one in a million, **Equation B3.1.5** shows that it takes approximately $(1/s) \ln(10^3/10^{-6}) = 20.7/s$ generations to reach a frequency of 0.999. (This approximation has the useful property that, if the selection coefficient is multiplied by a factor C, the time taken to reach a given final allele frequency from a given initial one is divided by C.) If the initial frequency is one in 10^9, the time is increased to only $27.6/s$ generations. On the geological time scale, this is not a long time, unless s is extremely small; for example, it is 27,600 generations when $s = 0.001$, and 2760 generations when $s = 0.01$. This implies that selection is a powerful force in causing the spread of favorable variants, even from very low initial frequencies.

3.1.iii. Autosomal inheritance with random mating

3.1.iii.a. *Introduction*

The theory for diploid populations is more difficult, even with random mating and autosomal inheritance. An exact solution can be found for the case of a recessive lethal, and for one other situation (**Problem 3.1**); in other cases, numerical iteration of the allele frequency change equation (see **Boxes 2.2** and **2.3** in **Section 2.1.ii**) is necessary. To study selection on an advantageous mutation in a diploid population, it is most convenient to use a notation similar to that in **Box 2.3**, but changing the sign of the selection coefficient, so that s is now the fitness advantage to the A_2A_2 homozygote for the new allele, and hs is the advantage to A_1A_2.

Some examples of exact calculations of the trajectories of the frequency of a selectively favored variant (with $s = 0.1$), starting at a low initial frequency, are shown for different dominance coefficients in **Figure 3.1A**. Examples of the times taken to change between different initial and final frequencies are shown in **Figure 3.1B** for the case of semidominance.

3.1.iii.b. *Results for weak selection*

In all other cases with diploid, autosomal inheritance, we have to resort to approximations, which are valid only for weak selection. We first note that the denominator $\bar{w}$ in the equation for frequency change (**Equation B2.2.3 of Box 2.2** in **Section 2.1.ii.a**) can be replaced by 1, to a good approximation, when

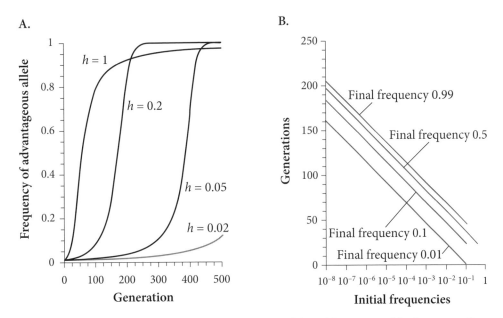

FIGURE 3.1 The increase in frequency of an advantageous allele. **A.** Trajectories of the frequency of a selectively favored variant (with $s = 0.1$), starting at a low initial frequency, are shown for different dominance coefficients (h, see **Box 2.3** of **Section 2.1.ii.b**). **B.** Examples of the times taken to change between different initial and final frequencies, for the case of semidominance ($h = 1/2$).

selection is weak. This is because $\bar{w}$ is equal to 1 plus terms in the selection coefficient s, so that $1/\bar{w}$ also equals 1 plus terms involving s and powers of s (see **Appendix A1.ii.c**). The difference between the marginal fitnesses of the two variants, $w_{2.} - w_{1.}$, involves only terms in s. We can thus ignore the terms involving s^2 and higher powers of s in $(w_{2.} - w_{1.})/\bar{w}$, leaving $(w_{2.} - w_{1.})$ as the main term.

The frequency change, **Equation B2.2.3**, is then approximated by **Equation 3.1**:

$$\Delta q \approx pq(w_{2.} - w_{1.}) \tag{3.1}$$

or **Equation 3.2**:

$$\Delta q \approx spq\,[ph + q(1 - h)] \tag{3.2}$$

The inaccuracy in this expression involves terms in s^2 and higher powers of s. This method of approximating changes in allele frequencies when the evolutionary forces acting on them are weak will be used many times in the rest of the book.

Further results require the use of differential calculus (see **Appendix A1.i**), and are derived in **Box 3.2***. For semidominance ($h = 0.5$), **Equation B3.2.3**

Box 3.2* APPROXIMATING THE EQUATIONS FOR DIRECTIONAL SELECTION IN DIPLOIDS

We replace the expressions for changes in frequencies between successive generations by differential equations (Haldane 1924). Time t is now a continuous variable, measured in units of generations. If selection is weak, this should be a good approximation to the discrete generation model, since the jumps in frequencies between successive generations are small, and the graph of frequency over time is close to a continuous curve (see **Figure 3.1A**). Using this idea, we can replace **Equation 3.2** with **Equation B3.2.1**:

$$\frac{dq}{dt} \approx spq\big[ph + q(1 - h)\big] \tag{B3.2.1}$$

On integration, this gives:

$$\int_{t_0}^{t_1} dt = \big(t_1 - t_0\big) \approx \frac{1}{s}\int_{q_0}^{q_1} \frac{dq}{pq\big[ph + q(1 - h)\big]} \tag{B3.2.2}$$

where the subscripts 0 and 1 represent the initial and final values for t and q.

Provided that $0 < h < 1$, evaluation of the right-hand side using standard integral formulae gives:

$$
\begin{aligned}
t_1 - t_0 &\approx \frac{1}{(1 - h)s}\ln\!\left(\frac{q_1\big[1 - h + q_1(2h - 1)\big]}{q_0\big[1 - h + q_0(2h - 1)\big]}\right) \\[2mm]
&+ \frac{1}{hs}\ln\!\left(\frac{p_0\big[h + p_0(1 - 2h)\big]}{p_1\big[h + p_1(1 - 2h)\big]}\right)
\end{aligned}
\tag{B3.2.3}
$$

The cases of complete dominance or recessivity of the favored variant ($h = 1$ and $h = 0$) can be solved directly from **Equation B3.2.2** (**Problem 3.2***):

$$t_1 - t_0 = \frac{1}{s}\left[\ln\!\left(\frac{p_0 q_1}{p_1 q_0}\right) + \frac{1}{p_1} - \frac{1}{p_0}\right] \qquad (h = 1) \tag{B3.2.4a}$$

$$t_1 - t_0 = \frac{1}{s}\left[\ln\!\left(\frac{p_0 q_1}{p_1 q_0}\right) + \frac{1}{q_1} - \frac{1}{q_0}\right] \qquad (h = 0) \tag{B3.2.4b}$$

gives the time required to change variant frequencies from q_0 to q_1, yielding the following expression:

$$t_1 - t_0 \approx \frac{2}{s}\ln\left(\frac{p_0 q_1}{p_1 q_0}\right) \tag{3.3}$$

This is the same as the result for haploids (**Equation B3.1.4** in **Section 3.1.ii**), but with selection coefficient $s/2$.

3.1.iii.c. *Behavior at extreme frequencies*
Comparisons of different selection models bring out some interesting points (**Figure 3.1A**). Most importantly, the response to selection for a rare *favorable recessive* mutation ($h = 0$) is extremely slow. This reflects the point made in connection with heterozygote advantage (**Section 2.1.ii.d**): with random mating, the frequency of the homozygotes for the rarer variant A_2 is q^2, which is very small when q is small, making it hard for selection to increase its frequency. More formally, **Equation 3.2** in this case becomes:

$$\Delta q \approx spq^2 \tag{3.4a}$$

so that the ratio of Δq to q (the proportional change in frequency) tends towards sq as q approaches zero, i.e., it also approaches zero.

In contrast, if $h \gg q$, **Equation 3.2** gives:

$$\Delta q \approx hsq \tag{3.4b}$$

so that $\Delta q/q$ tends towards the constant value hs as q approaches zero.

A similar argument also implies that the change in frequency of a fully dominant favorable mutation ($h = 1$) becomes very slow as the allele becomes abundant. This means that it is much more difficult for selection to eliminate recessive, autosomal deleterious mutations from a randomly mating population than to remove deleterious mutations that affect the phenotype of the heterozygotes. We return to this point in **Section 4.2.ii.a**.

These equations for extreme variant frequencies apply even for strong selection (**Problem 3.3**).

3.1.iii.d. *Continuous time*
The differential equation approach described in **Box 3.2*** in **Section 3.1.iii.b** can also be applied to a model of selection acting on a population reproducing in continuous time, where there are no distinct generations or age classes, as with the models used in ecology to represent the dynamics of numbers of individuals in a population (Crow and Kimura 1970, Chapter 1; Begon et al. 2006, Chapter 5). This is most appropriate for modeling single-celled organisms that reproduce by simple cell division, with given rates of cell division ("birth") and cell death per unit time. The fitness of a genotype can then be regarded as the

per capita rate of growth in size N of a population that consists solely of individuals with the birth and death rate of that genotype, and growing without any density-dependent restraints. This is called the *Malthusian parameter* of the genotype (Fisher 1930b, Chapter 2), and is equivalent to $d\ln(N)/dt$, the rate of change of the natural logarithm of N.

Selection coefficients are then measured as differences in the Malthusian parameters from the value for a standard genotype, and can be used to predict the rates of change of genotype frequencies (**Problem 3.4***). With asexually reproducing, single-celled organisms, it is possible to measure the rates of growth of different genotypes in laboratory cultures (expressed as changes in $\ln(N)$ values), yielding direct estimates of their fitnesses. An important application has been to estimate the fitness effects of deleting entire genes in budding yeast (Giaever et al. 2002; Deutschbauer et al. 2005). The results show that the complete loss of a gene often impairs fitness by only 1% or less under standard laboratory growth conditions (**Figure 3.2**). This is partly because some yeast genes are duplicates, so that deleting just one copy reduces growth only slightly, whereas deleting both copies has effects similar to deleting genes that have no duplicates (Deluna et al. 2008). Nevertheless, most deletions of single-copy genes reduce growth by only around 1%.

With sexual reproduction, however, there are difficulties in assigning Malthusian parameters to genotypes, since a genotype does not propagate itself exactly, but produces a mixture of offspring genotypes. As we shall explain in **Section 9.5.iii.d**, we can get good approximations for this case when selection is weak. For most practical purposes, it is adequate to use discrete generation models, because the main results from these also apply to more complex types of demography, at least when selection is weak. This is what we will do throughout most of this book.

3.1.iv. Biological applications

3.1.iv.a. *Estimating the strength of selection*

The equations for the rate of allele frequency change under selection can be used to estimate the strength of selection from observed frequency changes. Haldane (1924) used his equation for a dominant, favorable variant (**Equation B3.2.4a** of **Box 3.2*** in **Section 3.1.iii.b**) to estimate the selection coefficient for industrial melanism in the peppered moth, *Biston betularia*. The frequency of the black (melanic) form of this moth in industrial areas of Britain increased from near zero in the early nineteenth century to near fixation by the beginning of the twentieth century. Knowing that melanism is controlled by a single dominant allele, and that the moth has one generation per year, Haldane estimated s as about 33% (the assumption of weak selection is untrue in this case, so that this is only a rough approximation). Mark–release–recapture experiments,

A.

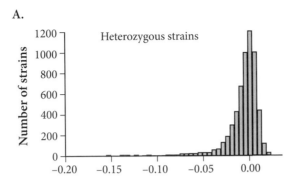

B.

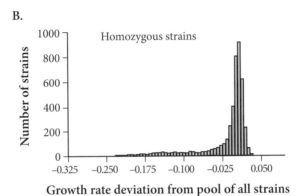

Growth rate deviation from pool of all strains

FIGURE 3.2 The distribution of the fitness values of deletions in yeast. The fitnesses are estimated from the rates of growth of log population size of mutant strains (determined by growing each strain for 20 hours). Each mutant strain has a deletion in 1 of 4624 genes tested. The *x* axis values are the fitness estimates for each strain, expressed as the deviation from the growth rate estimated for a pool containing all the strains. **A.** Results for heterozygous deletion strains. **B.** Results for homozygous deletions. [Adapted from Figure 1 of Deutschbauer et al. (2005).]

involving the release of melanic and wild-type forms of the moth in polluted woodlands, confirmed that there is a selective disadvantage to the light-colored form of this order of magnitude, reflecting lower predation on the melanic forms in polluted areas where trees lack lichens (Kettlewell 1956; Cook 2003).

3.1.iv.b. *Experimental evolution*

A refined version of this approach has been used for studying asexually reproducing microbial populations, where the relative frequencies of two variants at a locus under selection can be measured by using an easily scored marker (e.g., an antibiotic resistance gene), introduced into one of the two strains distinguished by the variants of interest. If the marker is selectively neutral under the experimental conditions, the marker frequency accurately reveals the trajectory for the selected variant (see **Section 8.3.v.a**). Large samples can be scored with this method, so that s can be very accurately estimated (down to s values as low as 0.001) using either **Equation B3.1.4** of **Section 3.1.ii** or else the continuous-time equivalent version derived in **Problem 3.4*** (Dykhuizen 1990). A very interesting result from such experiments is that two variant forms of a meta-bolic enzyme may behave as selectively neutral when resources are plentiful. If the population is placed in a more challenging environment, however, signifi-cant selective differences between the same variants are often detected (Hartl and Dykhuizen 1981; Dykhuizen 1990). This illustrates the point mentioned in the introduction to this chapter, that episodes of directional selection are often triggered by environmental changes.

3.1.iv.c. *Haldane's sieve*

We have already stressed that recessive favorable variants increase very slowly when present at a low initial frequency in a randomly mating population. Hal-dane (1924) pointed out that this implies that dominant or partially dominant mutations are more likely than recessive ones to rise quickly to high frequen-cies. Turner (1977) called the process of differential success of different kinds of mutations a *selective sieve*; in the case under consideration here, *Haldane's sieve* favors dominant or partially dominant mutations. Haldane proposed that this explains why, in most cases of recent evolution where single gene changes are involved, such as *Biston betularia*, dominant or partially dominant mutations have spread (Lees 1981). This pattern has been confirmed by later research on the evolution of pesticide resistance in insects: when there is simple genetic control, the selected forms overwhelmingly show significant phenotypic effects on their heterozygous carriers (Wood 1981; Bourguet and Raymond 1998). This contrasts with the well-known finding that the majority of laboratory mutations with a major effect on a visible phenotype are recessive in their qual-itative effect on the phenotype (Fisher 1930b, Chapter 3), although careful measurements often reveal slight heterozygous effects.

3.1.iv.d. *Developmental constraints on evolution*

This example provides a test of the idea that the directions of evolutionary changes are strongly constrained by the rules of developmental and functional biology, a long-debated idea in evolutionary biology (Maynard Smith et al. 1985). The recessivity of most mutations probably arises from the behavior of metabolic and developmental pathways (**Sections 2.2.iv and 4.4.v**). The obser-vation that alleles that have been driven to high frequency by selection are mostly dominant contrasts strongly with the mainly recessive mutational input of new alleles. This shows that selection can seize on the occasional nonreces-sive mutation, despite the rules of functional biology that make recessive ones far commoner. In this case, natural selection, not developmental constraints, predominates. There is thus no reason to invoke a general role for developmental constraints, rather than natural selection, in controlling the outcome of evolution (Charlesworth et al. 1982).

Nonetheless, there are clearly restrictions on the phenotypes that can be produced by mutation, governed by the properties of the developmental system characteristic of the species in question. As H. J. Muller once wrote,

> . . . the organism cannot be considered as infinitely plastic and certainly not as being equally plastic in all directions, since the directions which the effects of mutations can take are, of course, conditioned by the entire developmental and physiological system resulting from the action of all the other genes already present. Muller (1949, p. 427)

Since we still lack a truly predictive theory of development, it is usually impossible to know which types of changes are allowed or excluded by the rules of development and biochemistry. Progress is being made on this problem, which promises to increase our understanding of the possible range of phenotypes that can be generated by mutation. For example, Prusinkiewicz et al. (2007) have suggested that a model of inflorescence development supported by molecular genetic analyses of *Arabidopsis thaliana* can account for the relatively restricted number of inflorescence types observed in taxa of flowering plants. However, the example of dominance shows that, even when such rules appear to be of widespread applicability, the outcome of evolution is not completely restricted by them.

3.1.v. More complex selection models: relaxing some of the assumptions

3.1.v.a. *Autosomal inheritance with sex differences in selection*

Recall from **Section 2.2.iii** that selection acting in opposite directions in the two sexes can maintain polymorphisms without heterozygote advantage if selection is sufficiently strong. We will now show that, with weak selection, a simple average over males and females gives a good approximation to the fitness of a genotype. In particular, heterozygote advantage for the sex average of fitness is required for stable polymorphism.

First, selective differences between the sexes cause different variant frequencies in males and females. In a given generation, let the frequency of A_1 among reproducing females be $p + \delta$, and the corresponding male frequency be $p - \delta$. The frequency of A_1A_1 among the progeny will thus be $(p + \delta)(p - \delta) = p^2 - \delta^2$. Since δ must be of similar magnitude to the strength of selection causing the difference between the sexes, this means that δ^2 can be neglected in a weak selection approximation. A similar argument applies to each of the other genotype frequencies, and shows that the Hardy–Weinberg frequencies hold to the required level of approximation.

Given this result, it is possible to show that the standard selection equation applies if we treat the fitness of a genotype, w_{ij}, as the mean over the female and male fitnesses, w_{ij}^f and w_{ij}^m (Nagylaki 1979). There is, however, one difficulty. If, as is likely, differences in male fitnesses often involve differences in success in competition for mates, can we safely assume constant male fitnesses? It turns out that this assumption is legitimate when the fertility of matings depends solely on the female partner's genotype, and differences in male reproductive success are caused only by their ability to obtain mates, provided that there are fixed relative probabilities of mating success for different male genotypes (Charlesworth 1994b, pp. 143–144). If the fertility of a mating depends jointly on the genotypes of both partners, however, more complex models are needed (see Prout 1971a,b; Nagylaki 1987).

3.1.v.b. *Selection on an X-linked gene: Haldane's sieve and sexual antagonism* A similar argument can be used to model weak selection on an X-linked variant, or selection in species with *haplodiploidy* (male haploidy and female diploidy); Z-linked variants in a species with female heterogamety can be treated similarly by switching males and females. With weak selection, we can again use mean frequencies over males and females to represent the state of the population. For an X-linked gene, two-thirds of its copies are present in females and one-third in males (we only count genes that are transmitted to the offspring; because each offspring has one parent of each sex, but males inherit only their mother's X chromosome, the result is valid regardless of the primary sex ratio and the population's mating system—e.g., polygyny versus monogamy). These are, therefore, the appropriate weights to use in calculating the mean allele frequency. This gives **Equation 3.5** for a biallelic system:

$$\Delta q \approx \frac{pq}{3}\left[2\left(w_{2\cdot}^f - w_{1\cdot}^f\right) + \left(w_{2\cdot}^m - w_{1\cdot}^m\right)\right] \tag{3.5}$$

where $w_{i\cdot}^f$ and $w_{i\cdot}^m$ are the marginal fitnesses (**Section 2.1.ii.a**) of variant A_i in females and males, respectively.

With directional selection and a low frequency of A_2, we can simplify as for **Equation 3.4b** in **Section 3.1.iii.c**:

$$\Delta q \approx \frac{q\left(2hs_f + s_m\right)}{3} \tag{3.6}$$

where s_f and s_m are the selection coefficients for A_2A_2 and A_2 in females and males, respectively, and h is the dominance coefficient.

The equivalent equation for autosomal inheritance is:

$$\Delta q \approx \frac{qh\left(s_f + s_m\right)}{2} \tag{3.7}$$

Comparing these two equations shows that, if $h < 1/2$, so that A_2 is at least partially recessive, its rate of spread will depend equally on its effect on male and female fitnesses if it is autosomal (as one would expect intuitively), whereas, if it is X-linked, its rate of spread will depend more strongly on its effect on male than female fitness. A completely recessive, favorable mutation with fitness effects on males can thus spread if it is X-linked, whereas its rate of spread would only be of order q^2 in the autosomal case—see the discussion of Haldane's sieve in **Section 3.1.iv.c** above.

This suggests that there should be faster adaptive evolution of genes on the X chromosome compared with the autosomes, if favorable mutations tend to be recessive. This suggestion has been confirmed by detailed theoretical calculations of evolutionary rates, using the methods described in **Sections 3.2** and **6.2.iii** to include the effects of finite population size (Charlesworth et al. 1987).

If variants have *sexually antagonistic* fitness effects (Rice 1984, 1987a), so that s_f and s_m have opposite signs, it can be seen from the equations that an X-linked mutation A_2 that is advantageous in males can spread even if it has a similar-sized or somewhat larger disadvantage in females, provided that h is sufficiently small, but that an autosomal mutation with such a pattern of effects might fail to spread (**Problem 3.5**). The opposite pattern of sex difference in fitness effects promotes the faster fixation of female-advantageous X-linked mutations relative to autosomal ones, if h is large. Thus, recessive or partially recessive variants benefiting males but harming females should accumulate preferentially on the X chromosome; conversely, recessive or partially recessive variants that benefit females will accumulate relatively more on the autosomes than the X, if males are harmed. The opposite would be seen for partially dominant variants, or with female heterogamety (Z and W chromosomes instead of X and Y).

Once mutations with these sex-specific effects have become fixed in the populations, there will be selection on variants that regulate gene expression at the loci concerned, such that genes with beneficial effects on females but harmful effects on males might evolve reduced levels of expression in males relative to females, with the opposite holding for genes that are beneficial in males and harmful in females (Rice 1987a; Vicoso and Charlesworth 2006). This implies that selection on dominant or partially dominant favorable mutations should result in a deficiency of genes expressed in males on the X relative to the autosomes, and an excess of genes expressed in females. The opposite would be true if favorable mutations were partially recessive.

Is there indeed evidence for faster adaptive evolution of genes located on the X or Z chromosome compared with the autosomes, and are there differences in the gene contents of the X chromosome versus the autosomes in relation to their sex specificity of expression? The evidence for faster X- or Z-evolution of protein sequences is mixed, with more compelling evidence for its existence in mammals than *Drosophila*, and the strongest evidence coming from genes with male-biased expression (reviewed by Vicoso and Charlesworth 2006; see also

Baines et al. 2008). For *Drosophila* and *C. elegans*, the X chromosome is depauperate in genes that are expressed more highly in males than females, as predicted when favorable mutations are dominant or partially dominant (Vicoso and Charlesworth 2006; Sturgill et al. 2007). In birds, the Z chromosome is enriched in genes with male-biased expression, consistent with the prediction for this case for a species with female heterogamety (Kaiser and Ellegren 2007; Mank et al. 2007). The pattern in mammals is, however, the opposite, with relatively more male-biased genes on the X. The reasons for these differences are unclear (Vicoso and Charlesworth 2006).

3.1.v.c. *Selection in inbreeding populations*
The exposure of X-linked recessive variants to selection when hemizygous has a parallel in inbreeding populations, where homozygotes for recessives appear at higher frequencies than with random mating. We can model this by means of the inbreeding coefficient, f, introduced in **Section 1.3.ii**. For given values of f and the variant frequency q, the genotype frequencies before selection are given by **Equation 1.3** for an autosomal variant. This enables us to partition an inbred population into two components: an "outbred" one, with genotype frequencies in Hardy–Weinberg proportions (**Section 1.3.i.a, Equation 1.1**), and an "inbred" component, with only A_1A_1 and A_2A_2 homozygotes at frequencies p and q, respectively. These components are given weights of f and $1 - f$, respectively. It is then straightforward to show that:

$$\Delta q = \frac{pq\left[\left(1 - f\right)\left(w_{2\cdot} - w_{1\cdot}\right) + f\left(qw_{22} - pw_{11}\right)\right]}{\bar{w}} \tag{3.8}$$

where $w_{1\cdot}$ and $w_{2\cdot}$ are the marginal fitnesses used previously, and

$$\bar{w} = \left(1 - f\right)\left(pw_{1\cdot} + qw_{2\cdot}\right) + f\left(pw_{11} + qw_{22}\right) \tag{3.9}$$

With directional selection and small q, **Equation 3.8** becomes:

$$\Delta q \approx qs[h(1 - f) + f] \tag{3.10}$$

If f is large compared with q, then even a completely recessive mutation will spread at a rate similar to a dominant one, so that no Haldane's sieve (**Section 3.1.iv.c**) operates. This suggest that adaptive evolution in highly inbreeding populations should often involve the fixation of mutations with recessive or nearly recessive effects on the phenotypic target of selection, in contrast with what is seen for randomly mating populations. The scarce evidence available on this point seems to be consistent with this prediction (Charlesworth 1992; Schemske and Bradshaw 1999; Zeyl et al. 2003).

Inbreeding also increases the effectiveness of directional selection, unless the favorable mutation is completely dominant. Indeed, with $f = 1$, the selection equation is equivalent to that for haploids, if we substitute the fitnesses of the two homozygotes for the fitness of the corresponding haploid genotypes. In contrast, if there is heterozygote advantage, a high level of inbreeding reduces the ability of selection to maintain variability, since heterozygotes are rare and the outcome of selection is largely controlled by the ratio of the two homozygotes' fitnesses. Unless this ratio is close to 1, selection favoring heterozygotes cannot maintain a polymorphism when f is higher than a threshold value (Kimura and Ohta 1971, Appendix 2). Polymorphisms in highly inbreeding populations are therefore unlikely to be caused by heterozygote advantage.

One caveat is that we have taken f to be unchanging. With selection acting, f will, however, generally change as the composition of the population changes during the course of a generation. It is thus not strictly correct to use the results for neutrality; for example, to use the equilibrium f under partial self-fertilization given in **Section 1.3.ii.c**. Rigorous models for special cases such as partial sib-mating or selfing can be constructed (Karlin 1968; Nagylaki 1992, pp. 103–105), but we will not describe them here. With weak selection, however, the deviation of f from the value expected under neutrality is of order s; this deviation can then be neglected in equations such as **Equation 3.10**, leaving an error of order s^2. To this level of approximation it is, therefore, legitimate to use the neutral formulae for f.

3.1.v.d. *Kin selection*

The concept of *kin selection* (Maynard Smith 1964) is a very important extension to the theory of natural selection. It helps to explain many phenomena that appear to be inconsistent with the outcome of selection at the level of differential survival or fertility of individuals. The most spectacular of these is *eusociality*: the existence of sterile castes of individuals in insects like the Hymenoptera (bees, wasps, and ants) and termites, and in a few mammals such as the naked mole rat (Frank 1998). The sterile individuals maintain and provision the colony, providing substantial benefits to the other colony members, at the expense of their own ability to reproduce. This is an extreme example of what Haldane (1932, pp. 207–208) called *altruism*: behavior in which individuals give up their own fitness to benefit other members of their social group. As Darwin himself pointed out in *The Origin of Species* (Chapter 7), these examples appear at first sight to contradict the theory of natural selection.

However, the theory of kin selection shows that a variant that causes its carriers to lose some of their own fitness can have a net advantage, provided that it sufficiently increases the fitness of related individuals (Fisher 1930b, p. 159; Haldane 1932, pp. 207–208; Hamilton 1963). This is because related individuals have a higher than average chance of carrying a copy of the same variant. We will only describe a simple, general framework for a general understanding of kin selection. More detailed developments of the theory, and discussions of its relation to data on social behavior, are given in books such as those by Crozier and Pamilo (1996) and Frank (1998).

We assume a randomly mating population with an autosomal gene that affects the behavior of its carriers. During the course of a generation, individuals engage in pairwise interactions, in which *donors* sacrifice some fitness and *recipients* gain some fitness. Let the fitness cost of an interaction to a donor of genotype A_iA_j be c_{ij}, and the benefit to the recipient be b_{ij}, the fitness of each genotype in the absence of interactions being 1. Let the expected number of donor acts per generation for an A_iA_j individual be a_{ij}. If the interactions are weak, so that the products $a_{ij}b_{ij}$ and $a_{ij}c_{ij}$ for each genotype are small (i.e., less than some quantity ε, such that $\varepsilon^2 << \varepsilon$), selection must also be weak. We can thus use Hardy–Weinberg frequencies to approximate the genotype frequencies among the donors and recipients.

We also need to define a measure of *relatedness*, r, between the donors and recipients. The appropriate definition of r in this context is the probability that a recipient carries an allele identical by descent with an allele sampled randomly from the donor (**Figure 3.3**). This is not the same as the coefficient of kinship discussed in **Section 1.3.ii.a**; with autosomal inheritance, r is twice the coefficient of kinship, since either one of the recipient's two alleles could be identical to the allele from the donor, so that both possibilities must be added to obtain r.

An intuitive justification for the use of r is as follows. Suppose that a mutation is introduced into a nonaltruistic, randomly mating population, and causes its heterozygous carriers to suffer a reduction c in fitness as a result of behavior that increases the fitness of its relatives by an amount b. The mean number of copies of the mutation among the recipients is r, so that each altruistic act increases the copy number by rb, while simultaneously decreasing it by c. The mutation will therefore increase in frequency if $rb > c$. This is known as

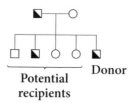

FIGURE 3.3 The probability that a recipient of an altruistic act carries an allele identical by descent with an allele sampled randomly from the donor, given a known relationship between the two. Autosomal inheritance is assumed. The donor carries the allele of interest (inherited from the left-hand parent), which is indicated by a black filled symbol. The potential recipients are its full-siblings. The offspring individual on the left has not received the black allele, while that to its right has done so. Overall, the probability of a recipient carrying the allele is 1/2.

Hamilton's rule (Hamilton 1963), first stated in a less precise, but more dramatic, form by Haldane (1955):

> Let us suppose that you carry a rare gene which affects your behaviour so that you jump into a flooded river and save a child, but you have one chance in ten of being drowned . . . . If the child is your own child or a brother or sister, there is an even chance that the child will also have this gene, so five such genes will be saved in children for one lost in an adult. If you save a grandchild or nephew, the advantage is only two and a half to one. If you only save a first cousin, the effect is very slight . . . . It is clear that genes making for conduct of this kind would only have a chance of spreading in rather small populations where most of the children were fairly near relatives of the man who risked his life . . . . Of course, the conditions are even better in a community such as a beehive or ants' nest, whose members are all literally brothers and sisters.

Given this model, the algebra in **Box 3.3** shows that the fitnesses in the standard selection equation can be replaced by *inclusive fitnesses* (Hamilton 1964a,b), defined as

$$\tilde{w}_{ij} = 1 + a_{ij}\left(rb_{ij} - c_{ij}\right) \tag{3.11}$$

Box 3.3 ALLELE FREQUENCY CHANGE UNDER KIN SELECTION

The following derivation is based on Hamilton (1964a) and Charlesworth (1980b). Consider the case of two variants at an autosomal site, following the rules of behavior described in the text. Let there be n individuals in the population at the start of the behavioral interactions among individuals in a given generation. Let the frequency of variant A_1 be p and A_2 be q, as usual, so that there are $n_1 = 2np$ and $n_2 = 2nq$ copies of A_1 and A_2, respectively. In the absence of any interactions, let the overall fitness of these individuals be w, such that $2nw$ genes would be transmitted to the next generation. The true mean fitness, $\bar{w}$, is equal to w plus terms of order s.

In the absence of selection, the frequency of A_2 among individuals who are related to a given individual X by coefficient of relatedness r is given by

$$q_r = r\,q_X + (1 - r)q \tag{B3.3.1}$$

where q_X is the frequency of A_2 in X (either 0, 1/2, or 1). In the presence of selection, this expression must be modified by a term of order s, since the transmission probabilities of alleles down a pedigree will be changed by selection from their neutral values.

This expression can be used as follows. Consider, for example, a heterozygote A_1A_2, for which $q_X = 1/2$. It behaves as a donor an average of a_{12} times; each interaction involves a change of $-wc_{12}$ in the number of copies of A_2 that it transmits to the next generation, and of $2wb_{12}q_r = wb_{12}[r + 2(1 - r)q]$, plus an error of order s, in the number of copies of A_2 transmitted by the recipients. Using our assumption of small values of the a_{ij}, we can ignore the contributions from situations in which individuals act both as donors and recipients, again with accuracy of order s.

We can therefore sum up the contributions from all three genotypes, which are given by terms similar to this one. Neglecting terms in s^2 (which include products of the type $a_{ij}b_{ij}r$ and $a_{ij}b_{ij}r$), the new values of the number of copies of A_1 and A_2 are thus:

$$n_1' \approx 2nw \times$$

$$\left[p + p\left[pa_{11}\left(rb_{11} - c_{11}\right) + qa_{12}\left(rb_{12} - c_{12}\right)\right] \atop +p(1 - r)\left[p^2a_{11}b_{11} + 2pqa_{12}b_{12} + q^2a_{22}b_{22}\right] \right] \quad \text{(B3.3.2a)}$$

$$n_2' \approx 2nw \times$$

$$\left[q + q\left[pa_{12}\left(rb_{12} - c_{12}\right) + qa_{22}\left(rb_{22} - c_{22}\right)\right] \atop +q(1 - r)\left[p^2a_{11}b_{11} + 2pqa_{12}b_{12} + q^2a_{22}b_{22}\right] \right] \quad \text{(B3.3.2b)}$$

The new total number of alleles, n', in the population is the sum of these two equations. We can obtain the expression for the change in frequency of A_2, to the desired level of approximation, by subtracting qn' from **Equation B3.3.2b**, and dividing by $2nw$. The second term in braces on the right-hand side of **Equation B3.3.2b** can be shown to be cancelled by this operation (**Problem 3.6**), and so we arrive at an equation identical in form to **Equation B2.2.3** of **Box 2.2** in **Section 2.1.ii.a** for allele frequency change with diploidy, replacing fitnesses by the inclusive fitnesses of **Equation 3.11** of the main text.

It is possible to include the possibility that different types of relatives interact with each other by including specific terms for interaction probabilities for each degree of relationship, and summing over all types of relationship (Hamilton 1964a,b). This shows that variants are likely to spread if they increase the probability of interacting with close relatives, at the expense of interactions with more distant ones.

Eusociality has evolved independently more than 10 times within the Hymenoptera (Crozier and Pamilo 1996). The theory of kin selection we have just outlined has therefore been extended to haplodiploid sex determination (diploid females and haploid males), the system in all Hymenoptera. In this case, r depends on the sexes of the interactants, and is not necessarily equal to twice the coefficient of kinship (**Table 3.1**). However, a similar argument to that used for the autosomal case enables us to define separate inclusive fitnesses for females and males (Hamilton 1964a,b; Charlesworth 1980b). These can be used instead of the usual expressions for male and female fitnesses in **Equation 3.5** of **Section 3.1.v.b**, to give a weak selection approximation for haplodiploidy, or for the case of X linkage.

The fact that, with haplodiploidy, sisters have an r of 3/4 instead of 1/2, whereas brothers have an r of only 1/2, led Hamilton (1964b) to suggest that this explains why sociality has evolved repeatedly in the Hymenoptera, and involves altruistic behavior of females only. It is now thought that these properties of relatedness coefficients with haplodiploidy are probably not the correct explanation for the initial evolution of sociality, but are nonetheless very important in explaining many features of the behavior of advanced Hymenopteran societies (Crozier and Pamilo 1996; Frank 1998; Wenseleers and Ratnieks 2006).

TABLE 3.1 Coefficients of relatedness for some important types of relationships

Relationship	Autosomal inheritance	Haplodiploid inheritance
Mother–daughter	1/2	1/2
Mother–son	1/2	1/2
Father–daughter	1/2	1
Father–son	1/2	0
Sister–sister	1/2	3/4
Sister–brother	1/2	1/4
Brother–brother	1/2	1/2

Hamilton's rule for the spread of a rare altruistic variant can easily be derived from the equation for inclusive fitness (**Problem 3.7**). It should, however, be clear from the derivation that this rule is an approximation, valid only for weak behavioral interactions. For making testable predictions about social behavior, there are other ways in which kin selection can be modeled, some of which are more useful than this approach (Crozier and Pamilo 1996; Frank 1998). When the assumption of weak interactions among participants is invalid, more exact approaches must be used, and **Equation 3.11** can then be violated (e.g., Cavalli-Sforza and Feldman 1978; Charlesworth 1980b; Grafen 2006; Gardner et al. 2007). However, the inclusive fitness approach shows that, with weak effects of individual variants on behavior, the equations for kin selection reduce approximately to the standard selection equations. It is, of course, always possible to determine the conditions for spread of a variant affecting behavior by the standard fitness approach described earlier in this chapter, taking into account the effects of the acts of all members of a social group on the fitnesses of individuals carrying the variant in question. Providing the assumptions just mentioned are valid, this will give the same results as the inclusive fitness approach (Hamilton 1964a; Cavalli-Sforza and Feldman 1978; Maynard Smith 1983). In practice, the inclusive fitness approach is more widely used, because it is easier to apply to specific problems.

3.2. SURVIVAL AND FIXATION PROBABILITY OF A FAVORABLE MUTATION

3.2.i. The problem

We have assumed up to now that a new advantageous mutation will spread smoothly through the population in a completely deterministic way. In reality, even if a population is very large, a new mutation will initially be present in just a few individuals. Individuals vary in the numbers of genes that they transmit to the next generation; even with no fitness differences, some individuals fail to leave offspring purely by chance, while others leave many. This variability can be described by a probability distribution of offspring numbers, and means that there is a chance that, in any one generation, all the mutants will fail to leave any offspring, in which case the mutation becomes extinct (**Figure 3.4**).

This process of random fluctuations in the number of copies of a mutation is an example of a *branching process* or *Galton–Watson process* (named after the two nineteenth-century scientists who used it to study the extinction of surnames). Branching processes were first used in genetics by Fisher (1922), who assumed that the wild-type population is so large that random changes in its size can be ignored when considering the fate of a mutation (in **Section 6.2.iii** we discuss what happens when this is not true).

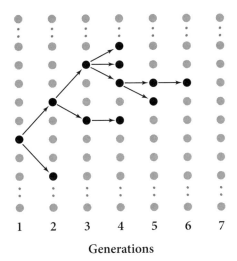

Generations

FIGURE 3.4 The process of chance extinction of a new mutation (dark circles) in a population of nonmutant alleles (lighter circles); the nonmutants will make up the great majority, but the figure shows only a few of them, as part of a larger population (indicated by dots). When offspring numbers per individual vary, and are not very large, a new advantageous mutation may fail to leave descendants purely by chance. The population is assumed to be stationary in size, i.e., the mean number of offspring for the nonmutants is 1 (the lines of descent for these alleles are not shown). In this case, the mutant fails to establish itself in the population, and is extinct by generation 7.

3.2.ii. The solution

Fisher investigated the *survival probability* of a mutation, i.e., the chance that it becomes established in the population without significant risk of future chance loss. The details are given in **Box 3.4** for the case of a haploid population. If the population is stationary in size, and there is a Poisson distribution of offspring number, the survival probability is approximately equal to twice the selective advantage, s, of the favorable mutation (Haldane 1927b).

In a diploid, randomly mating population, the same approach can be used, with hs now representing the selective advantage to the heterozygous carriers of the mutation. Most matings are between mutant heterozygotes and wild-type homozygotes; one-half of the offspring of such matings carry the mutation. If the wild-type population is stationary in size, the mean number of offspring must be two per parent; the expected number of mutant offspring per heterozygous carrier is then $2(1 + hs)/2 = 1 + hs$. The change in numbers of copies of a rare mutation can thus be treated in exactly the same way as for haploids. With a Poisson distribution of offspring number, the probability of survival for a weakly selected mutation is approximately $2hs$.

If a favorable mutation survives early random loss, its carriers will eventually become sufficiently numerous that there is effectively no chance that the

Box 3.4 SURVIVAL PROBABILITY OF A NEW MUTATION

We first describe the general model, which makes it clear why new favorable mutations will not always spread in a population, and then derive Haldane's (1927b) very useful approximate result for the chance of survival of a new favorable mutation.

B3.4.i. General formulation

We assume that there is a distribution of the number of adult individuals contributed by an adult to the next generation, writing P_i for the probability that an individual contributes i offspring ($i = 0, 1, 2, \ldots$), and E for the probability of ultimate extinction of the mutation, i.e., the probability that no descendants are ultimately left by the initial copy. This can happen if it leaves no offspring in the first generation, or if it leaves one offspring in the first generation, but the descendants of this individual ultimately go extinct (probability E), and so on.

In general, if i offspring are contributed to the first generation, the chance that they all ultimately fail to leave descendants is E^i. The net chance of extinction thus satisfies the following relation:

$$E = P_0 + P_1 E + P_2 E^2 + \cdots + P_i E^i + \ldots \qquad (B3.4.1)$$

Equation B3.4.1 can be solved numerically for a given assumed distribution of offspring number—for example, by using Newton's method of iteration (**Appendix A1.ii.b**). One general result that can be derived is that $E < 1$ only if the mean of the offspring distribution exceeds 1, i.e., a mutation is doomed to stochastic extinction if its *absolute* fitness is less than or equal to 1. The mutation will have a nonzero chance of survival if this is greater than 1, regardless of the performance of the rest of the population (**Problem 3.8***).

B3.4.ii. Haldane's approximation

In the case of weak selection and a constant population size, and assuming that a haploid asexual population is initially fixed for a variant A_1, we can obtain a useful approximation for the chance of survival of a new favorable mutation A_2. If the population size of the A_1 population is constant, it has a mean number of offspring of one per individual. If the mutation is neutral, it follows that its mean offspring number is also one, so that it is doomed to extinction, using the result stated above. But if A_2 is favored by selection, we can write the mean offspring number of A_2 individuals as $1 + s$, where s is assumed to be small.

Assume further that the offspring distribution is Poisson (**Appendix A2.v.d**). This will be true if there is purely random variation in reproductive success for a given genotype. In this case, the probability that a given individual is the parent of a given member of the offspring generation is very small, so that the Poisson distribution is a good approximation to the binomial distribution that describes the number of offspring contributed by a parent. This means that the offspring distribution for the mutants is given by:

$$P_i = (1 + s)^i \exp[-(1+s)]/i! \tag{B3.4.2}$$

Substituting this into **Equation B3.4.1**, we get:

$$E = [1 + (1 + s)E + (1 + s)^2 E^2/2 + (1 + s)^3 E^3/3! \dots] \exp[-(1 + s)]$$
$$= \exp[(1 + s)\,E]\,\exp[-(1 + s)] = \exp[(1 + s)(E - 1)] \tag{B3.4.3a}$$

Let the probability of survival of the mutation be $Q = 1 - E$. **Equation B3.4.3a** becomes:

$$1 - Q = \exp[-Q(1 + s)] \tag{B3.4.3b}$$

By using the formula for $\exp(-x)$ as a series in x (**Appendix A1.ii.c**), the right-hand side of **Equation B3.4.3b** can be written as $1 - Q(1 + s) + Q^2(1 + s)^2/2$, plus higher-order terms in Q and s.

Since Q and s must be similar in size, this can be approximated further by $1 - Q(1 + s) + Q^2/2$. Equating this to the left-hand side of **Equation B3.4.3b**, we get:

$$Q \approx 2s \tag{B3.4.4}$$

mutation will be lost, and the infinite population equations then apply. For advantageous mutations subject to simple directional selection, the probability of survival is equivalent to the probability that the mutation becomes fixed in the population, so that the term *fixation probability* is often used. For a mutation that does not become fixed, but remains polymorphic due to balancing selection (**Section 2.1.ii.d**), the probability of survival is the chance that it becomes permanently established in the population.

These results reinforce the conclusion that there is a selective sieve for dominant or partially dominant, favorable autosomal mutations in very large, randomly mating populations, since hs is 0 for fully recessive mutations and they accordingly behave as selectively neutral, with a survival probability of 0. Of course, this result is not quite accurate, because populations are not infinite in size. A recessive, favorable mutation arising in a randomly mating population can be shown to have a survival probability greater than that of a neutral mutation (using the methods described in **Section 6.2.iii.a** and **Appendix 6A.1** of **Chapter 6**), although it is much smaller than for dominant mutations. With a Poisson distribution of offspring number, as assumed in deriving **Equation B3.4.4**, the fixation probability is approximately $1.13\sqrt{s/(2N)}$, where s is the selective advantage to homozygotes for the favorable allele, and N is the number of breeding adults in the population (Kimura 1962, 1964).

3.2.iii. Biological implications

We have seen that a favorable mutation has only a small chance of becoming established in a population, even if it has a sizeable selective advantage in heterozygotes, and the population is extremely large. This reflects the fact that new mutations are initially present in low copy numbers, which means that the numbers of carrier individuals fluctuate from generation to generation, leading to a high chance of random loss. It follows that many independent mutations must occur before there is a good chance that at least one of them becomes established in the population (**Problem 3.9**).

This means that the process of directional selection has a strong random element, in terms of the precise identity of the mutations that become established in a population (unless the population is so large that all possible beneficial mutations arise repeatedly). In particular, isolated populations will tend to diverge genetically, even if they are exposed to the same selection pressure. If we examine the evolution of traits for which several alternative genetic mechanisms can produce similar phenotypes (Turner 1977 called this the "largesse of the genome"), we would therefore expect variants established by the same type of selection (e.g., selection for resistance to a pathogen) to differ among different populations, since it is a matter of chance which particular mutation is the first one that survives extinction. Many examples are known of such *convergent evolution* at the genetic level. For instance, different human populations have evolved malaria resistance by several different genetic mechanisms (**Table 3.2**). Similarly, four different regulatory sequence mutations that confer lactose tolerance appear to have been selected in different human populations that consume milk, (Tishkoff et al. 2007), and several independent mutations that cause loss of self-incompatibility (see **Section 2.2.i.b**) in the self-fertilizing plant *Arabidopsis thaliana* have spread in different populations of this species (Sherman-Broyles et al. 2007). This process of essentially random divergence among isolated populations probably plays an important role in speciation (see **Section 8.4.vi**).

TABLE 3.2 The range of proteins for which polymorphic variants are known to affect resistance to malaria[1]

Protein	Function	Effects on malaria
Common blood cell and plasma variants		
Duffy antigen	Chemokine receptor	FY*O allele completely protects against *P. vivax* infection
Glucose-6-phosphatase dehydogenase	Enzyme	G6PD deficiency protects against severe malaria
Glycophorin A, B, and C	Sialoglycoprotein	GYPA-deficient erythrocytes are resistant to invasion by *P. falciparum*
HBA α-globin	Component of adult hemoglobin	Deletion of one copy of the two α-globin genes (α-thalassemia) protects against severe malaria environments
β-Globin component of hemoglobin	Component of adult hemoglobin	HbS and HbC alleles protect against severe malaria; HbE allele reduces parasite invasion
Haptoglobin	Hemoglobin-binding protein present in blood plasma	Haptoglobin 1-1 genotype is associated with susceptibility to severe malaria in Sudan and Ghana
CD233, erythrocyte band 3 protein	Chloride/bicarbonate exchanger	Deletion causes ovalocytosis but protects against cerebral malaria
Cytoadherence: adhesion molecules expressed on endothelium, platelets, macrophages, and erythrocytes		
CD36 antigen	Thrombospondin receptor PE-binding receptor on endothelium and dendritic cells	Homozygotes for a nonsense polymorphism, the CD36_1264G allele, were susceptible to cerebral malaria in Gambin and Kenya, but heterozygosity for this allele in Kenya was associated with protection against severe malaria; in Thailand, a dinucleotide repeat sequence in intron 3, causing alternative splicing, may protect against cerebral malaria
CR1	Complement receptor 1 PE-binding receptor on erythrocytes	Variable associations with severe malaria in the Gambia, Thailand, and Papua New Guinea
CD54	Intercellular adhesion molecule-1 PE-binding receptor on endothelium	*ICAM1* polymorphisms show variable associations with severe malaria in Kenya, Gabon, and the Gambia

(continued)

TABLE 3.2 (continued)

Protein	Function	Effects on malaria
PECAM1	CD31, platelet–endothelial cell-adhesion molecule PE-binding receptor on endothelium	Variable associations with severe malaria in Thailand, Kenya, and Papua New Guinea

Immune genes

Protein	Function	Effects on malaria
CD32	Low affinity receptor for Fc fragment of immunoglobulin G	*FCGR2A* associated with severe malaria in the Gambia
HLA-B	Component of MHC class I antigen presentation involved in cytotoxic T cell production	HLA-B53 associated with severe malaria in the Gambia
HLA-DR	Component of MHC class II antigen presentation involved in antibody production	HLA-DRB1 associated with severe malaria in the Gambia
Interferon-a receptor component	Cytokine receptor	*IFNAR1* associated with severe malaria in the Gambia
IFNG Interferon g	Cytokine with antiparasitic and pro-inflammatory properties	Weak associations with severe malaria in the Gambia
Interferon-g receptor component	Cytokine receptor	*IFNGR1* associated with severe malaria in Mandinka people of the Gambia
Interleukin-1a and 1b	Pro-inflammatory cytokines	Marginal associations with severe malaria in the Gambia
Interleukin-10	Anti-inflammatory cytokine	Associated with severe malaria in the Gambia
Interleukin-12b subunit	Promotes development of Th1 cells	Associated with severe malaria in Tanzania
Interleukin-4	Promotes antibody-producing B cells	Associated with antimalarial antibody levels in Fulani people of Burkina Faso
Mannose binding protein	Activates classic complement	Associated with severe malaria in Gabon

TABLE 3.2 (Continued)

Protein	Function	Effects on malaria
Inducible NO synthase	Generates nitrous oxide, a free radical	Associated with severe malaria in Gabon, the Gambia, and Tanzania
Tumor necrosis factor	Cytokine with antiparasitic and pro-inflammatory properties	Associated with severe malaria and re-infection risk in the Gambia, Kenya, Gabon, and Sri Lanka
CD40 ligand	T cell–B cell interactions leading to immunoglobulin class switching	Associated with severe malaria in the Gambia

[1] From Kwiatkowski (2005).

What happens in any particular instance when selection operates depends on the details of the genetic control of the phenotype on which selection acts. There are also examples of the repeated appearance of the same mutations in adaptive evolution, a possibility suggested by Haldane (1932, pp. 139–140). These are cases of *parallel evolution* at the level of DNA sequences (**Table 3.3**). From what we have just explained, these must often represent situations in

TABLE 3.3 Some examples of parallel evolution at the level of DNA sequences

Characteristic and species	Source
Drug resistance in HIV virus	Crandall et al. (1999)
Experimental adaptation of the DNA bacteriophage X174 to high temperature and a novel host (*Salmonella typhimurium*)	Wichman et al. (1999)
Experimental adaptation of vesicular stomatitis virus	Novella et al. (2004)
Seasonal changes in influenza A virus hemaglutinin	Wolf et al. (2006)
Multiple origins of sickle cell β-globin in different African and Asian populations of humans	Lapoumiéroulie et al. (1992)
Insecticide resistance genes in insects	Ffrench-Constant et al. (1998)
Resistance to the toxins tetrodotoxin (TTX) and saxitoxin (STX) via parallel amino acid replacements in all eight sodium channels in pufferfishes	Jost et al. (2008)

which there are only very few ways in which the requirements of selection can be met (Orr 2005b), or when the population size is so large that the mutation conferring the highest fitness can arise repeatedly and outcompetes less fit mutations.

3.3. SELECTION ON QUANTITATIVE TRAITS

3.3.i. Introduction

The largesse of the genome applies especially to quantitative traits. As we saw in **Section 1.1.ii.c**, a quantitative trait is usually affected by variants in several different genes, which individually make only small contributions to the total genetic variability in the trait. Understanding the effect of selection on the trait as a whole, as well as at the level of the individual genes involved, is important for understanding the evolution of many traits in natural populations, as well as for animal and plant breeding. We therefore include a brief coverage of this topic, relating it to the results on selection on single loci. The problem of quantitative trait variation and selection on quantitative traits is a major field in its own right, with several full-length books summarizing the major findings (Bulmer 1980; Falconer and Mackay 1996; Lynch and Walsh 1998; Bürger 2000). We will describe only some of the most basic results. Most of the results in this section will not be used in the rest of the book, so that skipping this section (and moving straight to **Section 3.4**, or reading only the text and not the boxes) will not make it difficult to understand the later chapters.

We first describe how to characterize variation for a single trait, and then how to predict the response to directional selection on that trait. We also show how to determine the strength of selection on the variants affecting the trait, and the long-term effects of selection. Finally, we consider selection on multiple traits.

3.3.ii. Quantitative genetic parameters

3.3.ii.a. *Introduction*

The facts discussed in **Section 1.ii.b** imply that we can write the phenotypic value of a quantitative trait for an individual, z, as the sum of its genotypic value, G, and a nongenetic term, E (for simplicity, any interactions between genotypic and environmental effects can be included in E for the present purposes). To describe variation in the trait, we use the fact that the variance of the sum of two independent variables is the sum of the variances of each variable (**Appendix A2.iv**). When G and E are independent of each other, this implies that:

$$V_Z = V_G + V_E \tag{3.12}$$

where V_Z is the *phenotypic variance* in z, V_G is the *genetic variance*, and V_E is the *environmental variance*.

If the trait follows the normal probability distribution, the distribution of z among individuals in the populations is completely described by the trait mean, $\bar{z}$, and the variance V_z (**Appendix A2.v.e**). If z is not normally distributed, it is usually possible to find a scale transformation that converts it to normality (Falconer and Mackay 1996, Chapter 17). We wish to predict the response of $\bar{z}$ to directional selection on the trait, given information on the direction and strength of selection and the amount of genetic variation in the population. The degree of resemblance between parents and their offspring is fundamental to understanding how selection can alter the mean of a population. We will therefore first describe how resemblances between relatives are determined quantitatively.

3.3.ii.b. *Components of genetic variance and covariances between relatives*
We start with the case of a single autosomal nucleotide site, labeled j, which contributes to variability in the trait. Assume that there are two variants at this site, A_{j1} and A_{j2}, with frequencies p_j and q_j. It is convenient to express the values of z as deviations, $\tilde{z}$, from the mean value for the generation under consideration. The deviations from the population mean for each of the three genotypes at site j are $\tilde{z}_{j11}$ (for $A_{j1}A_{j1}$), $\tilde{z}_{j12}$ (for $A_{j1}A_{j2}$), and $\tilde{z}_{j22}$ (for $A_{j2}A_{j2}$). These represent averages over all the possible genotypes at other sites in the genome (the *genetic background*), weighted by the frequencies of these backgrounds in the population (an example of a real case is shown in **Figure 3.5**). We are assuming as usual that the genotypic value of an individual is determined by its own

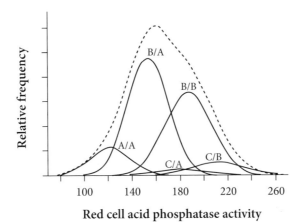

Red cell acid phosphatase activity

FIGURE 3.5 Differences in enzyme activity among genotypes for different red cell acid phosphatase alleles that are polymorphic in human populations. There are three common alleles (A, B, and C), two of which appear as homozygotes (A/A and B/B, as well as the heterozygous genotypes B/A, etc.). The figure shows the distributions of activity of each genotype that is common in human populations, and the dashed line indicates the overall distribution of enzyme activity calculated from the observed genotype frequencies. [Adapted from Figure 12 of Spencer et al. (1964).]

genotype, so that we can ignore the possibility of phenomena such as effects of a mother's genotype on the offspring phenotype (see Falconer and Mackay 1996, pp. 156–157 for a brief discussion of maternal effects).

Section i of **Box 3.5** shows how to use these deviations to describe the contribution of this site to variation in the trait in a randomly mating population. **Section ii** of **Box 3.5** shows that the resemblance between parents and offspring contributed by variation at this site, measured by the *covariance* between parental value and offspring mean (**Appendix A2.iv**), is one-half of the site's contribution V_{Aj} to the *additive genetic variance*, a concept introduced in **Section 1.1.ii.d**. This can be extended to include the joint effects of all relevant sites if we make two further assumptions. Assumption 1 is that variants at different nucleotide sites across the genome are distributed independently of one another in the population, i.e., the chance that a gamete carries a given variant

---

Box 3.5 QUANTITATIVE GENETIC PARAMETERS

B3.5.i. Additive and dominance variances

We can define equivalents of the marginal fitnesses used in selection theory (**Section 2.1.ii.a**) as a_{j1} and a_{j2}, which measure the deviations of z from the population mean for the carriers of variants A_1 and A_2 respectively, at a nucleotide site j. With random mating and autosomal inheritance, these are given by the relations:

$$a_{j1} = p_j \tilde{z}_{j11} + q_j \tilde{z}_{j12}, \quad a_{j2} = p_j \tilde{z}_{j12} + q_j \tilde{z}_{j22} \quad \text{(B3.5.1)}$$

The *breeding value* or *additive genetic value* of genotype $A_{jk}A_{jl}$ (where k and l take values of either 1 or 2) is A_{jkl} (not to be confused with the allele symbols A_{jk} and A_{jl}). This is given by the following expression:

$$A_{jkl} = a_{jk} + a_{jl} \quad \text{(B3.5.2)}$$

From the definition of a, the mean value of the offspring produced by mating an $A_{jk}A_{jl}$ individual to random members of the population, expressed as a deviation from the mean of the parental population, is $A_{jkl}/2$; the population mean value of this is 0. This corresponds to the definition of breeding value in **Section 1.1.ii.d**.

The contribution of variation in breeding values among the three genotypes at site j to the variability in the trait can be expressed by the corresponding *additive variance*, V_{Aj}. With random mating, we have:

$$V_{Aj} = p_j^2 A_{j11}^2 + 2p_j q_j A_{j12}^2 + q_j^2 A_{j22}^2 \qquad \text{(B3.5.3)}$$

Some simple algebra reduces this to:

$$V_{Aj} = 2\left(p_j a_{j1}^2 + q_j a_{j2}^2\right) = 2p_j q_j \alpha_j^2 \qquad \text{(B3.5.4)}$$

where $\alpha_j = a_{j2} - a_{j1}$ measures the effect of the variants at site j on the additive genotypic values. The variable α_j is called the *average effect of the gene substitution* (Fisher 1930b, Chapter 2).

This provides a convenient expression for the contribution of site j to the *additive genetic variance* in the trait, V_A, which is given by the sum over all sites of V_{Aj}, $\Sigma_j V_{Aj}$. V_A is, however, not necessarily the same as the total genetic variance of the trait. Unless the heterozygotes are exactly midway between the two homozygotes at each site, the genotypic values will deviate from the additive genetic values. The deviations are called the *dominance deviations*; for genotype $A_{jk}A_{jl}$, the dominance deviation is $D_{jkl} = \tilde{z}_{jkl} - a_{jk} - a_{jl}$.

This allows us to define the contribution of site j to the *dominance variance* (V_D) as V_{Dj}, which is given by an expression equivalent to **Equation B3.5.3** with D_{jkl} replacing A_{jkl}. The mean of the products, $A_{jkl}D_{jkl}$, can be shown by straightforward algebra to be 0, so that the total genetic variance at site j, V_{Gj} (equal to the mean of the squares of $A_{jkl} + D_{jkl}$), is equal to $V_{Aj} + V_{Dj}$. The dominance variance in the trait is thus $\Sigma_j V_{Dj}$; summing over all sites, we have:

$$V_G = V_A + V_D \qquad \text{(B3.5.5)}$$

B3.5.ii. Resemblances between parent and offspring

The importance of V_A is that it determines the *covariance* (**Appendix A2.iv**) between parents and offspring, C_{OP}. As noted in the previous section of this box, the mean value of the offspring of individuals of genotype $A_{jk}A_{jl}$ is $A_{jkl}/2$, and so the covariance between the value of a parent and the offspring mean is the mean of the product $(A_{jkl} + D_{jkl} + E)A_{jkl}/2$, where E is the nongenetic contribution to the phenotype of the parent. If environmental and genetic effects are independent, the mean of EA_{jk} is 0; the same is true for $D_{jkl}A_{jkl}$ as just mentioned. The only nonzero contribution to C_{OP} is thus the mean of $A_{jkl}^2/2$, which is $V_{Aj}/2$. Summing over all sites, we have $C_{OP} = V_A/2$.

at site j is the same, whatever the state of other sites in the same gamete (see **Section 1.2.v.b**). Assumption 2 is that an individual's genotypic value can be represented by the sum of the contributions from each site, i.e., there are no *interactions* between sites. In **Section 8.4.iv** we relax these assumptions, but for now they can be regarded as approximations that work well in practice (for a detailed discussion, see Turelli and Barton 1994).

Given these assumptions, we can write the total additive genetic variance in the trait as $V_A = \Sigma_j V_{Aj}$. The covariance between parental value and offspring value is $C_{OP} = V_A/2$. This shows that V_A can be estimated by determining C_{OP} from measurements on pairs of parents and offspring, provided that environmental sources of covariance can be excluded. This approach can be extended by similar methods to any desired category of relative (e.g., Nagylaki 1992, Chapter 10; Falconer and Mackay 1996, Chapter 9), as was first done by Weinberg in 1910 and Fisher in 1918 (Provine 1971). **Table 3.4** shows some of the more important formulae for covariances between relatives with autosomal inheritance.

3.3.iii. Response to selection

3.3.iii.a. *Change over one generation*

How can we predict the population mean after one generation of selection? Assume that the population is infinitely large. Directional selection implies that the mean of the parents who contribute to the next generation differ from the overall mean of the population by an amount S, called the *selection differential*. If selection involves only survival differences, S is the difference between the

TABLE 3.4 Covariances between pairs of relatives with autosomal inheritance and random mating

Relationship	Covariance
Parent–offspring	$\frac{1}{2}V_A$
Full siblings	$\frac{1}{2}V_A + \frac{1}{4}V_D$
Half siblings, grandparent–grandchild	$\frac{1}{4}V_A$
Double first cousins	$\frac{1}{4}V_A + \frac{1}{8}V_D$
Single first cousins, great grandparent–great grandchild	$\frac{1}{8}V_A$

mean among the survivors at the time of breeding from the value if all individuals had survived. If there is selection on fertility, individuals are weighted by the number of offspring that they produce. S can be estimated from data on the distribution of z in the population before and after selection (we describe some examples in **Section 3.3.iv**).

It is often convenient to scale S by the standard deviation of the trait, $\sigma_Z = \sqrt{V_Z}$, where σ is the conventional symbol for the standard deviation. This removes dependence on the scale of measurement of the trait. The ratio $i = S/\sigma_Z$ is called the *selection intensity*. In the case of a large population and a normal distribution with *truncation selection* (where only individuals that fall above a threshold value survive or reproduce), there is a known relation between i and the proportion P of the population that fall above the threshold (Haldane 1930b):

$$i = t/P \qquad (3.13)$$

where t is the height (probability density) of the normal distribution at the threshold value (**Problem 3.10***). It is thus easy to find i from tables of the normal distribution. Artificial selection often involves truncation. For example, if 10% of the population are chosen for breeding, $i = 1.8$, but $i = 0.8$ if 50% are chosen.

The *response to selection*, R, is defined as the deviation of the mean of the selected offspring from the mean of the offspring of the population of potential parents in the absence of selection. If the distribution of nongenetic effects does not change between generations, R equals the difference in mean between the offspring and parental generations, $\Delta \bar{z}$. As **Figure 3.6** shows, R is equal to the product of S and the *regression coefficient*, b, of the *regression line* (**Appendix A2.iv**) for the relationship between the mean of each pair of parents (the *midparent* value, M) and that of their offspring.

This result depends on assuming two properties of the inheritance system: (i) the relation between parent and offspring values is linear, and (ii) the scatter around the line is independent of the trait values. These properties hold if the parent and offspring values are both normally distributed (**Appendix A2.v.e**), which will be the case if the trait is the sum of individually small, independent environmental and genetic effects (from the central limit theorem of statistics: **Appendix A2.v.e**). For further details, see Bulmer (1980, Chapters 6 and 8).

The results in **Box 3.6** show that, with these assumptions, the response to selection is given by the *breeder's equation*:

$$R = h^2 S \qquad (3.14)$$

where h^2 is the *heritability*, defined as

$$h^2 = V_A/V_Z \qquad (3.15)$$

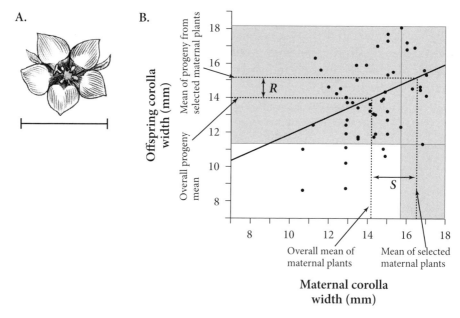

FIGURE 3.6 Selection on a quantitative trait in a natural population. **A.** A flower size character (corolla flare) of a plant, *Polemonium viscosum*. [Adapted from http://www.nps.gov/plants/color/northwest/imgs/.] **B.** Variation in corolla flare among individual plants in an alpine population and the response to natural selection by bumblebees. The means of the maternal and progeny generations are indicated to show the selection differential *S* estimated in the maternal plants and the response to selection *R* in the progeny. [Adapted from Figure 1 of Galen (1997).]

Note that h has nothing to do with the dominance coefficient used in **Section 3.1.iii.b**; it comes from Wright's notation for the decomposition of quantitative trait variability into genetic and nongenetic components (Wright 1920), which was applied to selection theory by Jay Lush in his classic book on animal breeding (Lush 1937).

Given an estimate of V_Z, h^2 can be estimated from measurements of resemblances between relatives (**Table 3.4**). This solves the problem of predicting the response to selection from one generation to the next. Estimates of h^2 are usually between 20% and 80% (see Table 10.1 of Falconer and Mackay 1996), which means that a substantial part of the change induced by selection in the parental generation is passed on to the offspring.

Another way of viewing **Equation 3.14** is that it implies that the change in mean over one generation of selection is equal to the covariance between fitness and the additive genetic value for the trait (**Equation B3.6.2**), where fitness is measured relative to the population mean fitness, i.e., as $w/\overline{w}$. This is a form of the *Price equation* for the effect of selection on a trait (Price 1970), stated by Alan Robertson (1967). If the trait is fitness itself, so that $z = w$, the formula becomes a version of Fisher's *fundamental theorem of natural selection* (Fisher 1930b, Chapter 2): the rate of change of mean fitness is proportional to the additive variance in fitness, $V_A(w)$:

$$\Delta \bar{w} = \frac{V_A(w)}{\bar{w}} \tag{3.16}$$

Fitnesses are here assumed to have constant values. Since, as explained above, the underlying genetic model assumes small effects of individual sites on fitness (i.e., weak selection at individual sites) as well as independence between the effects of genetic differences at different sites, this result is by no means generally true, although it is often a good approximation, even if there is frequency-dependent selection (**Section 8.4.iii.b**).

Box 3.6 PREDICTING THE RESPONSE TO SELECTION

From the argument in the text, we need to obtain an expression for the slope b of the *regression line* (**Appendix A2.iv**) relating mid-parent value, M, and offspring mean, O. From basic statistical theory, b is equal to the covariance between M and the offspring value (C_{OM}), divided by the variance of M (**Appendix A2.iv**). With random mating, the two parental values are independent of each other, so that the variance of M is equal to a quarter of the sum of the variances for each parent, i.e., to $V_Z/2$. In addition, $C_{OM} = C_{OP}$, so that:

$$b = 2C_{OP}/V_Z = V_A/V_Z \tag{B3.6.1}$$

It can also be shown that S is equal to the covariance C_{ZW} between z and the fitness of individuals with trait value z, measured relative to the mean fitness, given by $W = w(z)/\bar{w}$ (**Problem 3.11**). It is also useful to note that b can be interpreted as the regression of breeding value on phenotypic value, since the covariance between breeding value and z is equal to V_A. The expected breeding value, $A(\tilde{z})$, for individuals with deviation $\tilde{z}$ is thus equal to $b\tilde{z}$, and the covariance between $A(\tilde{z})$ and relative fitness is $C_{AW} = bC_{ZW}$. Combining this with **Equation 3.14**, and using the fact that $S = C_{ZW}$, we deduce that:

$$R = C_{AW} \tag{B3.6.2}$$

3.3.iii.b. *Selection coefficients at individual sites*

Equation 3.16 is closely related to Wright's result that variant frequency change at a site depends on the derivative of $\bar{w}$ with respect to variant frequency (**Section 2.1.ii.e**). **Equations B3.7.3** and **B3.7.4** in **Box 3.7*** show that i/σ_Z is approximately equal to the partial derivative of the natural logarithm of $\bar{w}$ with respect to $\bar{z}$ (the ratio of a small change in $\ln(\bar{w})$ to the corresponding small change in $\bar{z}$; see **Appendix A1.i**). This quantity is often referred to as the *selection gradient*, and is denoted by the Greek beta symbol, β (Lande 1979b; Lande and Arnold 1983). β can be used to quantify the strength of selection on quantitative traits (see **Section 3.3.iv** for examples). If z is measured after a logarithmic scale transformation, or is normalized by dividing by $\bar{z}$, β is independent of the scale on which the trait is measured.

The analysis in **Box 3.7*** yields a simple relation between the selection intensity on a trait and the selection coefficient at a given site, a problem first considered by Haldane (1930b). In the case of a semidominant variant that causes a phenotypic difference a_j between the two homozygotes at site j, we have $\alpha_j = a_j/2$, so that **Equation B3.7.6** implies that the selection coefficient s_j for the favored homozygote is:

$$s_j = a_j \beta = i a_j / \sigma_Z \qquad (3.17)$$

Box 3.7* RELATING CHANGE IN MEAN TO CHANGES IN VARIANT FREQUENCIES

We assume that the effects of variants at individual sites on both the trait and fitness are small, so that higher powers in these can be neglected compared with the effects themselves. Using **Equation B2.5.2b** from **Box 2.5*** in **Section 2.1ii.e**, with w replaced by z, we can obtain an approximate expression for the change in the mean of z over one generation, summing over all sites contributing to variation in the trait, as we did in **Box 3.5** of **Section 3.3.ii.b**. This is done by using the first two terms in the Taylor series (**Appendix A1.ii.a**) for $\bar{z}$ as a function of the q_j:

$$\Delta\bar{z} \approx \sum_j \Delta q_j \frac{\partial \bar{z}}{\partial q_j} = 2\sum_j \alpha_j \Delta q_j \qquad (B3.7.1)$$

where α_j is the average effect of the gene substitution at site j, defined in **Box 3.5**.

Equation B2.5.3 gives the following expression for the change in variant frequency at site j (partial derivatives are used since we are considering the effect of this site alone):

$$\Delta q_j = \frac{p_j q_j}{2} \frac{\partial \ln(\overline{w})}{\partial q_j} \qquad \text{(B3.7.2a)}$$

The partial derivative on the right-hand side is equal to twice the average effect of the gene substitution on fitness, α_{jw}, divided by the population mean fitness (see **Box 2.5**). We can thus use $\alpha_{jw}/\overline{w}$ to measure the strength of selection on the variants at site j. By use of the chain rule for differentiation (**Appendix A1.i**), **Equation B3.7.2a** can be written as:

$$\Delta q_j = \frac{p_j q_j}{2} \frac{\partial \overline{z}}{\partial q_j} \frac{\partial \ln(\overline{w})}{\partial \overline{z}} = p_j q_j \alpha_j \frac{\partial \ln(\overline{w})}{\partial \overline{z}} \qquad \text{(B3.7.2b)}$$

so that

$$\Delta \overline{z} = 2 \sum p_j q_j \alpha_j^2 \frac{\partial \ln(\overline{w})}{\partial \overline{z}} = V_A \frac{\partial \ln(\overline{w})}{\partial \overline{z}} \qquad \text{(B3.7.3)}$$

Comparing this with **Equations 3.14** and **3.15**, we find that:

$$\frac{S}{V_z} = \frac{i}{\sigma_z} \approx \frac{\partial \ln(\overline{w})}{\partial \overline{z}} \qquad \text{(B3.7.4)}$$

Equations B3.7.2a and **B3.7.2b** imply that:

$$\Delta q_j = p_j q_j \frac{\alpha_{jw}}{\overline{w}} = p_j q_j \beta \qquad \text{(B3.7.5)}$$

where

$$\beta = \frac{\partial \ln(\overline{w})}{\partial \overline{z}} \qquad \text{(B3.7.6)}$$

The strength of selection at site j is thus given by:

$$\frac{\alpha_{jw}}{\overline{w}} = \alpha_j \beta = \frac{\alpha_j i}{\sigma_z} \qquad \text{(B3.7.7)}$$

This result connects the single-site selection theory of the first part of this chapter with the results of quantitative genetics. It implies that the strength of selection on a variant contributing to variability in a quantitative trait is proportional to the size of its effect on the trait, relative to the amount of variation in the trait. If many sites contribute to the variation, these selection coefficients can be quite small, even with strong selection on the trait itself. For example, with m sites, each with effect $a = 2\alpha$ and intermediate frequencies of the variants, such that $p = q = 1/2$, we have $V_A = 2mpq \, (a^2/4) = ma^2/8$. Writing $\sigma_A = \sqrt{V_A} = (a\sqrt{m/2})/2$ for the additive genetic standard deviation, we have $a = 2\sigma_A\sqrt{2/m}$. Since $h^2 = V_A/V_z$, **Equation 3.17** implies that $s = 2ih\sqrt{2/m}$. With $h^2 = 0.5$ and $i = 0.1$, this give $s = 0.02$ when $m = 100$; with $m = 10$, $s = 0.06$. With such small selection coefficients, a substantial change in trait mean over the course of a selection experiment that lasts only a few generations can be caused simply by shifts in variant frequencies, not fixations.

While we do not know the details of the underlying genetics of any quantitative trait, there is evidence in some cases for considerable variation in the sizes of the effects of different variants, with a few genes or variant sites having large a values, while the majority have small effects (Flint and Mackay 2009). The results above show that sites with the largest a values will undergo the fastest changes in frequencies of the favorable variants, and hence contribute most to the selection response.

3.3.iii.c. *Long-term responses to selection*

An important question for both adaptive evolution and artificial selection is the following: what is the *limit to selection*, R_l, defined as the difference between the final and initial means of a population? We will show that selection can produce a genotype with a combination of variants that creates a phenotype many additive genetic standard deviations beyond the original population mean, especially if the selection response comes primarily from variants that were previously rare in the population. Ignoring new mutations, in an infinite population the selection limit corresponds to the situation in which all the variants with favorable effects on the trait have been fixed by selection. With the above model of semidominant effects, and with variants initially at frequencies of 0.5 at each site, the increase in trait mean at the limit is simply $R_l = \Sigma_j a_j/2$. With m sites, each with the same effect a on trait value, the results in the previous section show that $R_l = ma/2$, so that $R_l/\sigma_A = \sqrt{2/m}$. The larger the number of sites (and the smaller their effects), the larger this ratio; for example, with $m = 10$, it is 4.5, and with $m = 100$, it is 14.1. If many sites have low frequencies of the favorable variants, the ratio can be much larger than this.

Furthermore, this limit can be reached in just a few generations, as can be seen from the theory for rates of change of mean or variant frequencies developed in **Section 3.3.iii.b** above. This is borne out by artificial selection experiments on mice and *Drosophila*. Some results are presented in Table 12.2 of Falconer and Mackay (1996), which shows R_t/σ_A values of between 2 and 14 over 20–30 generations, when the populations had largely ceased responding to selection (see also Bell 2008, Table 6.1). These experiments used small population sizes, which means that the observed selection limits are likely to be much smaller than the theoretical values, because of losses of favorable variants by genetic drift. This effect can be taken into account by using the theory of the joint effects of genetic drift and selection (Robertson 1960), which we develop in **Section 6.2.iii**.

Both theory and experiments thus show that, by combining variants at different sites that are initially present at relatively low frequencies in the population, and fixing the combinations, selection can produce phenotypes far outside the range of variability of the initial population. The greatest responses to selection are produced in large populations and when the intensity of selection is low, since these conditions maximize the effectiveness of selection versus drift; experimental evidence for this is reviewed by Weber and Diggins (1990). Several long-term artificial selection experiments of this kind, in maize, mice, and *Drosophila* (Falconer and Mackay 1996, Chapter 12), show remarkably linear changes in the means of the selected traits over many tens of generations (**Figure 3.7**). These sustained responses imply that new variation is being generated by mutation, balancing the loss of variability caused by drift, so that the additive genetic variance for the trait remains roughly unchanged, as is predicted by models that take account of drift and mutation (Hill 1982).

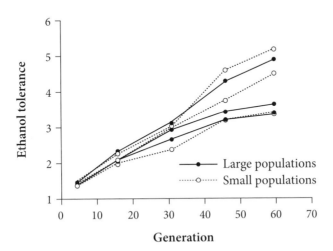

FIGURE 3.7 Responses to selection on ethanol tolerance in *Drosophila melanogaster* in long-term selection on several replicate large and small populations. [Adapted from Figure 1 of Weber and Diggins (1990).]

The magnitudes of the responses observed in these experiments are comparable to some of the most dramatic changes seen over much longer time scales in the fossil record, such as the approximately three-fold increase in cranial capacity in the human lineage over 2.8 million years (about 140,000 generations) between *Australopithecus* and *Homo sapiens*, representing a change of about nine standard deviations (Cronin et al. 1981). There is thus no difficulty in accounting for rapid episodes of evolution in characters subject to novel selection pressures, which are followed by periods of slow or zero change (*stasis*) once an optimum is approached; these are fully consistent with what is known about responses to selection on quantitative variation (Charlesworth et al. 1982; Estes and Arnold 2007). The fluctuating changes observed over several generations in beak characteristics of the Darwin's finch *Geospiza fortis*, in response to changes in the nature and availability of food resources, illustrate this point (Grant and Grant 2002).

However, the prediction that selection can rapidly change the mean of a trait to a value well outside the range of variability in the initial population depends critically on the occurrence of sexual reproduction and genetic recombination between the sites involved, which allow sites to evolve more or less independently of each other (**Section 8.4.iii**). Different favorable variants present in the original population can then eventually be combined together in the same individual, as selection increases their frequencies. In the absence of sex and recombination, this is not possible. Without new mutations, selection can then only fix the fittest genotype available in the population (Fisher 1930b, Chapter 6; Muller 1932). This is very important for the evolution of sex and recombination (as we discuss in **Section 10.2.iv**).

3.3.iv. Multiple traits

Considering a single trait on its own is an abstraction. In reality, many quantitative traits vary simultaneously in populations. If these traits were independent of each other, we could treat each one separately, using the results just outlined. In practice, however, a given quantitative trait is often highly correlated with several others, as measured by estimates of the covariances or correlation coefficients between traits (**Appendix A2.iv**). For example, **Table 3.5** shows the correlations between three linear measurements of body parts of specimens of the pentatomid bug *Euschistus variolarius* (Lande and Arnold 1983). The effects of selection on a set of correlated characters were discussed by Darwin, and the first detailed theoretical treatment of the effect of selection over a single generation (without reproduction) was by Karl Pearson (1903). Changes between generations were first modeled in the context of plant and animal breeding by Smith (1937) and Hazel (1943). Lande (1979b) and Lande and Arnold (1983) later developed a general theory for treating evolutionary change in multiple traits, which has been very widely used (Lande 2000).

TABLE 3.5 Selection on correlated traits of the hemipteran bug
Euschistus variolarius

A. Correlations between traits[1]

	Head	Thorax	Scutellum	Wing
Head width	1	0.72	0.50	0.60
Thorax width		1	0.59	0.71
Scutellum length			1	0.62
Wing length				1

B. Selection differentials (S), selection intensities (i), and selection gradients (β)[2,3]

	S	i	β
Head	−0.004	−0.11	−0.70
Thorax	−0.003	−0.06	11.6**
Scutellum	−0.016*	−0.28*	−2.8
Wing	−0.019*	−0.43**	−16.6**

[1] All correlations were significant at the $p < 0.001$ level.

[2] All traits were measured on a scale of natural logarithms; * and ** indicate significance at the $p < 0.05$ and $p < 0.01$ levels, respectively.

[3] From Lande and Arnold (1983).

This treatment, modified slightly to fit with the theory developed above, is described in **Box 3.8***. We now consider a set of traits, such that z_k is the phenotypic value of the kth trait, forming the kth component of a column vector of trait values, z (see **Appendix A1.iv.a**). We can represent the additive genetic variability in the traits by a variance–covariance matrix, $G = \{g_{kl}\}$, where g_{kl} is the covariance of additive genetic values between traits k and l (if $k = l$, g_{kl} is the additive genetic variance for the trait). The values of the g_{kl} can be estimated by measuring the trait values in sets of related individuals (Falconer and Mackay 1996, Chapter 19), in a similar way to the estimation of V_A for a single trait. The effect of directional selection on the vector of mean values is expressed by the vector β, whose k^{th} component, β_k, is the partial derivative of $\ln(\bar{w})$ with respect to $\bar{z}_k$ (**Equations B3.8.1a** and **B3.8.1b**) — in other words, β_k is the ratio of the change in $\ln(\bar{w})$ to a small change in $\bar{z}_k$, keeping all other traits constant (**Appendix A1.i**).

Equation B3.8.3 shows how β is determined by the set of selection differentials on each trait individually, together with the matrix of phenotypic vari-

Box 3.8* SELECTION ON MULTIPLE TRAITS

We can use the same type of argument as for **Equation B3.5.4** of **Box 3.5** in **Section 3.3.ii.b** to show that $g_{kl} = 2\Sigma_j\, p_j q_j \alpha_{kj} \alpha_{kl}$. Using this equation, and carrying through a similar analysis to that for a single trait in **Box 3.7*** of **Section 3.3.iii.b** (Charlesworth 1993b), we obtain the following result:

$$\Delta\bar{z}_k \approx 2\sum_{jl} p_j q_j \alpha_{kj} \alpha_{lj} \beta_l = \sum_l g_{kl} \beta_l \qquad \text{(B3.8.1a)}$$

where α_{kj} is the average effect of the gene substitution at site j on the kth trait.

In matrix notation:

$$\Delta\bar{z} \approx \mathbf{G}\beta \qquad \text{(B3.8.1b)}$$

The vector of selection gradients, β, can be related as follows to the column vector S, composed of the selection differentials for each trait (Lande 1979b). If the set of traits follows a multivariate normal distribution (**Appendix A2.v.e**), as is expected with small, additive effects of the individual sites affecting its variability, multiple regression theory implies that the set of multivariate regression coefficients of breeding value on z is given by the matrix $\mathbf{GP}^{-1}$, so that:

$$\Delta\bar{z} = \mathbf{GP}^{-1}S \qquad \text{(B3.8.2)}$$

This gives the generalization of **Equation B3.7.4** as:

$$P^{-1}S = \beta \qquad \text{(B3.8.3)}$$

ances and covariances, P. This important result shows that the selection differential or selection intensity observed for a single trait may, at least in part, be the result of directional selection on a trait with which it is correlated phenotypically. Measurements of single traits may thus be misleading as indicators of the true targets of selection. This is illustrated by the data of Lande and Arnold (1983), mentioned above. They compared living and dead specimens of the insect, and calculated the selection differentials with respect to survival for each trait separately. Using their estimate of P, they calculated β from **Equation B3.8.3**, with the results shown in **Table 3.5**. There was no apparent selection differential for thorax width, but a significant selection gradient for increased thorax width was detected. There was no selection differential, because selection to reduce wing length overcame direct selection on thorax width. Conversely, there was no direct selection for reduced scutellum length, but a significantly negative selection differential on scutellum length was detected.

Responses to selection are further complicated because the **G** matrix modulates the transmission of the changes in phenotypic means across generations. **Equations B3.8.1a** and **B3.8.1b** show that a change in one trait depends on the selection gradients for all traits with which it has an additive genetic covariance, and the direction of evolutionary change is not necessarily the same as that of the gradient in mean fitness, although Fisher's fundamental theorem for the change in mean fitness (**Equation 3.16** in **Section 3.3.iii.a**) still holds. It is quite possible for selection on one trait to produce a substantial response in another trait for which the selection gradient is zero, but with which it is genetically correlated. Many artificial selection experiments demonstrate that this can happen (Falconer and Mackay 1996, Chapter 19), and Lande (1979b) used the results of selection experiments, in which brain size in mice changed in response to selection on body size, to argue that much variation in brain size among related species of mammals may be a by-product of selection on body size alone.

It is also possible for selection that maintains a constant mean phenotype for one trait to impede the response to selection for another, slowing down the response to selection. This effect may explain the scarcity of examples of long-sustained, directional changes in the fossil record (Charlesworth 1984). It is therefore difficult to predict evolutionary change in any detail, since one cannot measure all possible traits. However, traits are rarely completely correlated, and selection experiments show that strong selection on a trait nearly always produces a response in the desired direction. An elegant demonstration of this is provided by Weber's experiments on wing shape in *D. melanogaster* (Weber 1992, 1996). It seems likely that, in general, genetic correlations among traits may restrict the responses to selection below what would be expected if the traits were evolving independently, but responses to selection would still be possible in many directions in phenotype space (McGuigan and Blows 2007; Meyer and Kirkpatrick 2008).

3.4. THE EVOLUTIONARY GENETICS OF ADAPTATION OF PHENOTYPES AND PROTEINS

3.4.i. Introduction

The preceding discussion of selection on multiple traits introduces the broader question of how selection theory can explain the evolution of adaptations that involve the mutual adjustment of many different component traits. This question was discussed by Darwin in Chapter 6 of *The Origin of Species*, notably in connection with the vertebrate eye. Darwin's answer was that a complex adaptation arises from a much simpler starting state by successive changes, each of which is advantageous with respect to the current state of the system. He noted the existence of many different types of eyes, from simple groups of light-sensitive cells, through eyes with simple lenses that cannot be focused, to the camera-type eyes of vertebrates and cephalopods. Darwin argued that these intermediates show how a long sequence of simple evolutionary steps can produce a highly complex structure, and this conclusion is strongly supported by later comparative research on eyes (Salvini-Plawen and Mayr 1977; Nilsson and Pelger 1994; Land and Nilsson 2002).

An alternative view, advocated by some Mendelian geneticists in the early twentieth century (Provine (1971), was that a single mutation can produce a coordinated set of changes in a character complex. The fixation of such a *macromutation* might cause an evolutionary jump (*saltation*) to a new structure in one step. Extreme advocates of this idea, notably Richard Goldschmidt (1940), have claimed that macromutations are responsible for new species and higher taxa. An example that initially seemed convincing was provided by the polymorphic mimetic forms of some butterfly species, exemplified by the African swallowtail butterfly *Papilio dardanus* (**Plate 1**, front endpaper). The different mimetic forms appear to be controlled by alleles at a single locus, but differ in multiple aspects of their color and patterns (Punnett 1915; Goldschmidt 1945). Mimicry has, therefore, been important in attempts to distinguish between these two alternative theories of adaptation.

As we will show, the evidence from mimetic butterflies strongly favors the Darwinian view (see the **Case Study** in **Section 3.4.v**), as does the evolution of eyes and many other adaptations. But even accepting this general conclusion, several further questions remain, as with any successful scientific theory that raises new questions in our minds, demanding a more detailed understanding. In the case of adaptation, we would like to know its typical genetic basis. Does it involve many genes of small effect, or a few genes with large effects? Does evolutionary change require new mutations, or can it use variability already present in the population (Orr and Betancourt 2001)? Do regulatory mutations play the major role, or mutations that affect the amino acid sequences of proteins (Hoekstra and Coyne 2007; Wray 2007; Carroll 2008)? Here, we focus on the first of these questions.

3.4.ii. Fisher's geometric model of adaptation

Fisher (1930b, Chapter 2) proposed that evolution of increased adaptation of a complex structure resembles a series of movements in a multi-dimensional space, equivalent to the vector z described in the previous section. He studied the effects of the sizes of the changes that occur during an adaptive change in a phenotype. His model imagined that there is a maximum fitness, associated with a point O in the phenotype space, representing the optimum phenotype. An individual located at another point, A, experiences a mutation that randomly changes its state vector z; this increases fitness only if the change takes the phenotype closer to the optimal point O, i.e., the new phenotype lies within the multi-dimensional sphere of radius r that is centered on O and passes through A (such a sphere is a set of points that are equally distant from a fixed point). A two-dimensional example is given in **Figure 3.8A**, where the "sphere" is simply a circle. A very small change in z evidently has a nearly 50% chance of lying within the sphere, whereas very large changes often lead to a phenotype outside it (i.e., less well adapted). Fisher worked out the probability that a change of size d increases fitness, for an n-dimensional vector with large n. The probability decreases approximately exponentially with the size of the change. Fisher provided no derivation for his result, but proofs have since been given (e.g., Leigh 1987).

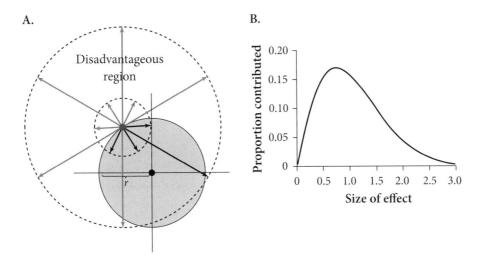

FIGURE 3.8 The effects of mutations of different sizes in a two-dimensional system. **A.** The optimal phenotype is represented by the black dot, and the starting state of a population, with the phenotype at a distance r from the optimum, is shown as a grey dot. Almost half of the small mutations arising in this population (with magnitudes represented by the smaller dotted circle) lead to phenotypes closer to the optimum than the initial state (these mutations are represented by black arrows in the grey circle). Mutations with large effects (represented by the large dotted circle) only rarely lead to phenotypes closer to the optimum than the initial state, and most are disadvantageous (grey arrows). **B.** The predicted relative contributions (the y axis) to adaptation of mutations with different sizes of phenotypic effects (the x axis). [Adapted from Figure 2 of Orr (1998).]

Fisher thus concluded that movement towards an optimum under directional selection is unlikely to involve very large steps, i.e., his model argues against evolution by macromutation. Fisher's result is often thought to imply that mutations that affect multiple characters and are incorporated into the population by selection usually have individually small effects. However, Kimura (1983, pp. 135–137) pointed out that this neglects a second important factor—the chance that an advantageous mutation is incorporated into the population—and that this increases with the size of its effect on fitness; if a mutation falls within the sphere (i.e., is advantageous), the results in **Section 3.2.ii** show that its probability of fixation is approximately $2hs$, where hs is the selective advantage to its carriers. It is reasonable to suppose that the fitness effects of advantageous mutations from point A will be larger, the more closely their phenotype approaches the optimum; Kimura showed that, overall, a mutation of intermediate size effect is the most likely to both fall inside the sphere and become fixed by selection (**Figure 3.8B**), assuming that hs is proportional to the size of the effect of the mutation.

Orr (1998, 2005a) extended this approach to consider the sizes of effects of mutations fixed over an entire sequence of successive steps, in which new mutations are fixed by selection and carry the population closer towards the optimum with each step. The distribution of effect sizes is approximately exponential in form (**Appendix A2.v.c**), with many small-effect mutations and a few larger ones (**Figure 3.9**). This result is nearly independent of the distribution of the sizes of the effects of new mutations. Furthermore, large effect mutations tend to be fixed early in the process. Finally, it is interesting to note that

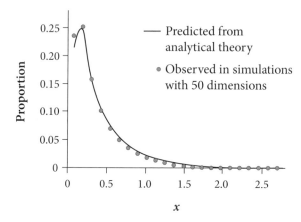

FIGURE 3.9 The size distribution of mutations fixed during adaptation. The line shows the predicted proportion of fixed mutations with different effects, and the open and filled symbols show the results of computer simulations of systems with 50 dimensions. The variable x is a standardized measure of the size of the effect of a mutation. [Adapted from Figure 5 of Orr (1998).]

the mean size of effect of a successful mutation (i.e., one that is fixed during an adaptive sequence) decreases with the number of dimensions of the phenotype space, reflecting the increasing improbability of moving towards the optimum as the number of dimensions increases. An alternative model has recently been proposed by Martin and Lenormand (2006, 2008), based on the multivariate quantitative genetics model discussed in **Section 3.3.iv**. This makes somewhat different predictions about the distribution of mutational effects from the Orr model, but the two give very similar results when the dimensionality of the character space is large (Martin and Lenormand 2008).

3.4.iii. Adaptive evolution of sequences

3.4.iii.a. *Introduction*

An important type of multi-component phenotype is represented by the sequence of a polypeptide, a functional RNA molecule, or a regulatory component of noncoding DNA. Evolutionary change in DNA and protein sequences is the subject of the field of molecular evolution, which we will discuss in **Chapters 6** and **10**. The increasing quantity of information from genome sequences, and the use of sequence data to observe the time courses of evolutionary changes in experimental systems, require models of adaptive sequence evolution, against which the data can be tested. We focus on protein sequences, but similar ideas apply more broadly.

A key idea is the *protein space* (Maynard Smith 1970). Adaptive evolution in a protein must generally occur by the successive fixations of single, favorable mutations, unless the mutation rate is so high that multiple mutations can occur simultaneously in the same gene. Most changes at first and second nucleotide positions in a codon cause a change in the corresponding amino acid, whereas many changes to third positions leave the sequence of the polypeptide unchanged. Overall, in most higher organisms about 70% of mutations to a coding sequence cause a change in amino acid sequence (Kryukov et al. 2007). Since there are three alternatives to the nucleotide present at a site in a coding sequence, a coding sequence of m sites can mutate to approximately $3 \times 0.70m = 2.1m$ alternative amino acid sequences that differ from it by one nucleotide. A subset of these may increase fitness; each of these sequences can mutate to a further set of sequences that again cause increased fitness, and so on. This creates a space of protein sequences connected by single mutational steps, through which an *adaptive walk* occurs.

3.4.iii.b. *The distribution of the fitness effects of mutations fixed in an adaptive walk*

A second important idea is that adaptive evolution is often triggered by environmental changes, so that the starting point for an adaptive walk is probably a sequence that was well adapted to the old environment but is also reasonably fit in the new one (Gillespie 1983, 1984, 1991, Chapter 5). It is reasonable to assume that most sequences that are one mutational step away from the initial sequence will have lower fitnesses than the initial one, but, if the environment has changed, some will have higher fitness. The fittest sequence in the new environment might be i mutational steps away from the initial sequence, where i is small. To study this model, we label the fittest sequence as 0, the next fittest as 1, down to the initial sequence, i. The fitness increase between sequences ranked j and $j - 1$ will be denoted by Δ_j.

The first step in the adaptive walk is a transition from sequence i to a fitter sequence j. If there are no differences in mutation probabilities for different types of sequence, the chance, P_j, that sequence j becomes fixed, among all the alternatives, is simply the fixation probability for a mutation from i to sequence j, Q_j, divided by the sum of the fixation probabilities for all possible mutations with increased fitness:

$$P_j = \frac{Q_j}{\sum\limits_{k=0}^{i-1} Q_k} \tag{3.18}$$

In a large population with small selection coefficients, the Q_j are given by **Equation B3.4.4** of **Box 3.4** in **Section 3.2.ii**, using the value of the selection coefficient given by the difference in fitness between sequence j and i, relative to the fitness of i.

Once this mutation has become fixed, the process starts again with sequence j. Selection coefficients are assumed to be assigned to the new set of sequences that differ from sequence j by single mutations, drawn from the same distribution of fitnesses as in the first step, with sequence j retaining its previously assigned fitness. The rule for selecting the successful mutation is then applied to the set of new sequences, and the process is repeated until no fitter allele can be found. The optimal sequence for the new environment, accessible by successive single mutational steps from the starting sequence, has then been reached and the adaptive walk ends.

It turns out that one does not need to know the distribution of selection coefficients, given the assumption that the initial sequence has a high fitness, and is therefore in the right-hand tail of the distribution of fitnesses. This follows from the statistical theory of the extremes of distributions (Gillespie 1983, 1984; Orr 2002, 2005a; Joyce et al. 2008). In particular, under a wide range of conditions, the Δ_j values at each step are independent, exponentially distributed variables, whose expectations (means) follow the rule $E\{\Delta_j\}/E\{\Delta_{j+1}\} = (j + 1)/j$. Together with **Equation 3.18**, this result can be used to get the simple expression for P_j (Orr 2002) shown in **Equation 3.19**:

$$P_j = \frac{1}{(i-1)}\sum_{k=j}^{i-1}\frac{1}{k} \tag{3.19}$$

For a given value of i, P_j is a roughly exponential function of the value of j, with the smallest values (i.e., the largest jumps in fitness) having the highest probability, although there are some circumstances under which deviations from this can occur (Joyce et al. 2008). In addition, simulations show that the number of steps in the adaptive walk is usually small, and increases only logarithmically with the distance to the optimal sequence; for example, with $i = 10$, it is 2.36, and with $i = 50$, it is 3.36 (Orr 2002). However, the mean selection coefficients associated with successive steps decrease geometrically, such that

$$\frac{E\{s_j\}}{E\{s_{j+1}\}} \approx \frac{E\{s_{j+1}\}}{E\{s_{j+2}\}} \tag{3.20}$$

where the subscripts indicate the number of steps reached in the adaptive walk. This makes intuitive sense: as the system approaches the optimum, it becomes less likely that a large improvement in fitness will occur by mutation.

Finally, the overall probability distribution of selection coefficients for mutations fixed over an entire adaptive walk is approximately exponential, with many more mutations with small effects on fitness becoming fixed than ones with very large effects. An exponential distribution also describes the fitness effects of *all* new beneficial mutations, regardless of whether they are fixed by selection (Orr 2003). The overall expectation is thus that large-effect mutations will be fixed in the initial stages of adaptation, followed by the accumulation of more minor ones (Orr 2002, 2005a; Joyce et al. 2008).

3.4.iv. Testing the theory

3.4.iv.a. *The theoretical predictions*

The model of the evolution of a multi-dimensional phenotype described in **Section 3.4.ii** predicts a wide (exponential) distribution of effects of mutations fixed by selection, with many mutations whose effects are small relative to the net change produced by evolution, and a smaller number with fairly large effects. This prediction is, of course, subject to many qualifications. First, the model assumes that selection only incorporates new mutations. As we have seen (**Section 3.3.iii.c**), a substantial amount of evolutionary change in quantitative traits arises from variants that are already segregating in the population, so that the sizes of effects of variants that become fixed reflect the distribution of effects among the segregating variants (**Equation 3.17**). As we mentioned in **Section 3.3.iii.b**, this distribution is often very wide, resulting in a pattern similar to that predicted by the mutational theory for the effects of variants on a single trait. Second, it assumes that the effects of mutations are randomly directed in pheno-type space. Given the fact that there are often strong correlations between genetic effects on different traits (Falconer and Mackay 1996, Chapter 19; Johnson and Barton 2005), new mutations are likely to have correlated effects on sets of traits (Mackay and Lyman 2005), although few data are currently available about the effects of spontaneous mutations in a multivariate context. This problem can in principle be dealt with by the use of multivariate statistical methods to transform a set of correlated traits into a smaller set of independent variables (e.g., McGuigan and Blows 2007; Meyer and Kirkpatrick 2008).

3.4.iv.b. *Genetics of interpopulation and interspecies crosses*
Genetic analyses of crosses between populations or species, one of which has been selected for a new adaptation, can be used to test the predictions of this theory. In practice, the direction and causes of evolutionary change are often difficult to identify, so we mostly rely on studies of genetic differences between species and populations that differ markedly. Early reviews by Haldane (1932, Chapter 3) and Wright (1940a,b) pointed out that such differences involve a wide spectrum of types of genetic differences, ranging from single Mendelian genes (especially for traits like coloration and color patterns) to quantitative traits controlled by variants at many loci.

This conclusion is confirmed by modern work, in which molecular markers distributed across the genome are used to map *quantitative trait loci* (QTL) associated with differences in the traits of interest. An excellent example, in which the direction of adaptive evolution can be inferred with a high degree of certainty, is that of the monkey flower *Mimulus cardinalis*. This species is polli-nated by humming birds, whereas most species of *Mimulus*, including its close relative *M. lewisii*, are bee-pollinated. *M. cardinalis* has evolved a set of traits that adapt it to humming bird pollination, including red flowers (**Table 3.6** and **Plate 3**, back endpaper). It has long been known that the flower color differ-ence between the two species is controlled by a single locus (Vickery 1978). Mapping of QTL in the F2 cross between the species shows a wide distribution of effects of individual QTL on individual traits (**Table 3.6**), with some con-tributing a substantial proportion of the variability in the F2 generation (Brad-shaw et al. 1998; Schemske and Bradshaw 1999). Studies of the ability of F2 seg-regants to attract the different types of pollinators show that the traits listed have significant effects, with flower color playing a major role (Schemske and Bradshaw 1999; Bradshaw and Schemske 2003). It is clear from this analysis that the complex of traits resulting in adaptation to the new pollinator is due to many different genetic changes, not a *macromutation* of the kind proposed by Goldschmidt (1940), even though some individual genes may have moderately large effects. This is qualitatively in agreement with theoretical expectations.

TABLE 3.6 Floral characteristics of two species of the monkey flower *Mimulus*

Species	*M. lewisii*	*M. cardinalis*
Pollinators	Bee	Hummingbird
Flower size	Small	Large
Flower shape	Wide, with "landing platform"	Narrow, tubular
Flower color	Pink	Red
Nectar	Moderate, high sugar	Abundant, low sugar

3.4.iv.c. *Experimental evolution*

Another source of data relating to these questions is provided by studies of experimental evolution, in which microbial populations are exposed to a novel environment for many generations, allowed to adapt to this environment, and then analyzed genetically (Betancourt and Bollback 2006). Very long-term experiments on the bacterium *Escherichia coli*, in which 12 replicate populations were allowed to evolve in a glucose-limited environment for 10,000 generations (Lenski and Travisano 1994), are especially illuminating. Since bacteria can be deep frozen and later thawed out, it is possible to compare the relative fitnesses of the ancestral and descendant populations by the methods described earlier in **Section 3.1.iii.d**. As shown in **Figure 3.10**, the mean fitnesses of these populations show a rapid initial increase, followed by a gradual

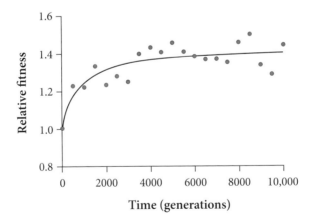

FIGURE 3.10 Mean fitness increase over several thousand generations as the bacterium *Escherichia coli* adapts to a new environment. Fitnesses during this natural selection process were measured by sampling each strain at intervals during the experiment and by growing cells in competition with the ancestor of the strains. The dots show the means over 12 replicate experiments, relative to the fitness of the ancestral strain, and the line is a hyperbolic model fitted to these values. [Adapted from Figure 1 of Lenski and Travisano (1994).]

approach to a plateau. This is consistent with the theoretical predictions explained above.

A more recent study of the result of selection for more rapid replication of a single-stranded bacteriophage, whose 5.6-kb genome can easily be sequenced, identified nonsynonymous mutations responsible for the first step in adaptation (Rokyta et al. 2005). The effects of these mutations on phage fitness, as measured by the rate of increase in phage numbers per hour, were determined, and the distribution of effects for 20 independent mutations was determined. The fit to the prediction of **Equation 3.19** of **Section 3.4.iii.b** was fairly good, but not perfect, with too few mutations with the highest fitness rank, and too many in the intermediate range. This was largely because the different types of change had different mutation rates, requiring a modified version of **Equation 3.19** that takes these differences in mutation rates into account.

In *E. coli* it is possible to determine in detail the effects of mutations at individual nucleotide sites during the evolution of antibiotic resistance, and it has been found that several changes occur in the protein involved before the highest resistance levels are achieved. By making all combinations of the mutations observed at the sites in question and measuring their resistance levels, the possible paths through sequence space were examined to determine the steps that led to increased fitness (in terms of increased resistance to the antibiotic: Weinreich et al. 2006; DePristo et al. 2007). **Figure 3.11** shows the changes inferred in one such experiment (DePristo et al. 2007). This illustrates how a new point in sequence space can be reached by successive steps in an adaptive walk.

3.4.v. Multiple fitness peaks

The theory we have just described assumes the existence of a single optimum or peak of fitness, to which the population evolves by successive fitness increments. This is a reasonable assumption for an evolutionary response to a shift in the environment. However, this is not the only way in which adaptive evolution can occur; the extreme alternative is that the environment stays constant, but there is a "rugged" fitness landscape relating fitness to phenotype, with multiple fitness peaks, such that some mutations permit movement from a lower peak to a higher one (Wright 1932; Wright 1978, Chapter 13). Small mutational steps will then mean that a population starting near a peak with low fitness will evolve towards that peak by the process we have just described. In contrast, a mutation causing a large change in phenotype may cause a jump across a valley of reduced fitness, allowing access to a higher peak by a succession of subsequent smaller changes.

This possibility has been best studied in relation to mimicry in butterflies, as outlined in the **Case Study** below. As mentioned in **Section 3.4.i**, Batesian mimicry, in which the palatable mimetic species gains protection from predators by evolving a close resemblance to a distasteful species, is one of the cases

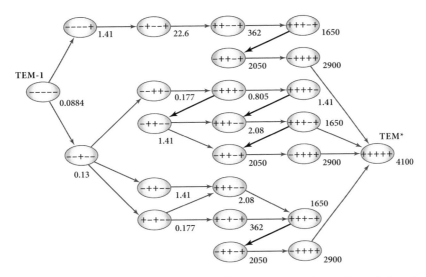

FIGURE 3.11 The changes inferred in the *E. coli* β-lactamase gene during the evolution of resistance to the antibiotic cefotaxime. In the experiment, the bacteria all started with the same allele (TEM-1 with low resistance, indicated by – – – – –), and the most resistant allele (TEM*) had five nucleotide site changes, one in a noncoding region (4205), plus four amino acid changes (at positions 42, 104, 182, and 238 of the protein sequence); the five changes are symbolized by + + + + +. The figure indicates the estimated effects on resistance (positive or negative) of each change, and a measure of the antibiotic amount in which the genotype could grow (high values indicate resistance). Thick arrows indicate mutations from one of the states present in the most resistant allele back to the initial state (reversions), whereas thin arrows indicate forward mutations. [Adapted from Figure 1 of DePristo et al. (2007).]

of adaptation most likely to involve macromutations, since its genetic basis sometimes appears to involve a single locus. As in the case of the *Mimulus* species described above, the detailed genetic studies of mimetic butterfly species described in the **Case Study** below show that this adaptation involves several components evolving by multiple steps, as proposed by Darwin, not by macromutations that affect a whole set of characters, as claimed by Goldschmidt (Turner 1977, 1981).

CASE STUDY
Batesian Mimicry and its Genetic Basis

Edible species usually evolve cryptic colors and patterns, which reduce their conspicuousness to visual predators such as birds. In contrast, dangerous or distasteful species often warn predators off by warning or *aposematic* coloration. For an edible butterfly species to become a *Batesian mimic* (**Section 2.2.i.b**), it must usually change several features of its wing and body patterns in order to achieve the observed near perfect resemblance to the model, and gain the maximum protection from predation. **Plate 1** (front endpaper) illustrates this for *Papilio dardanus*. The nonmimetic form has a yellow ground color with black borders to its wings, and has a tail on its hind wings. With the exception of populations in Ethiopia, at the northern end of the species range, the mimetic forms all lack tails, and have red or white ground colors to the wings, with varying degrees of extension of the black borders into the wings. The model species all lack tails (probably because their absence reduces visibility to predators by disruptive camouflage), so that a high degree of resemblance to the model requires *P. dardanus* to lose its tails (Sheppard 1959b, 1975, Chapter 11).

Increased resemblance to the model species results in an increased conspicuousness to the predators, as well as causing an increased probability of being mistaken for the model. A poor mimetic resemblance is likely to be selectively disadvantageous, whereas a sufficiently large increase in resemblance can be advantageous. This has been shown by modeling the relations between the predation rates on models and nonmimetic forms, conspicuousness, and the probability of being mistaken for the model, as shown in **Figure 3.12** (Charlesworth and Charlesworth 1975a). Thus, only mutations with sufficiently large effects will invade the population; they either spread to fixation, or remain polymorphic as a result of the frequency-dependent selection that operates with Batesian mimicry (**Section 2.2.i.b**). However, even imperfect mimicry can be selectively advantageous, as is shown by the presence of tailed mimics of *P. dardanus* in Ethiopia (Clarke and Sheppard 1960b). Subsequent to the establishment of a mutation causing imperfect mimicry, there is selection for mutations that enhance resemblance to the model, allowing the population to climb towards the peak of perfect mimicry (Nicholson 1927).

In *P. dardanus*, the loss of tails is controlled by a single gene that is unlinked to the locus that controls the wing patterns, as shown by crosses between tailed mimetic forms and tailed forms (Sheppard 1959b; Clarke and Sheppard 1960b). In addition to this major gene that improves mimetic resemblance, there are many *modifier* genes with smaller effects that can improve the mimetic resemblance. This was shown by crosses of mimetic butterflies from mainland Africa to butterflies from Madagascar, where only nonmimetic forms exist (Sheppard 1959b; Clarke and Sheppard 1960a). These crosses lead to a breakdown in mimetic resemblance and blurring of the segregation of the gene controlling mimicry. This implies that the high degree of perfection of mimicry shown in **Plate 1** is the result of several genetic changes involving mutations that enhance the effects of alleles at the mimicry locus. The appearance of single-gene control of mimicry is thus deceptive. We discuss mimicry further in **Section 10.2.iii.b**, where we describe how low recombination may evolve between different genes involved in Batesian mimicry.

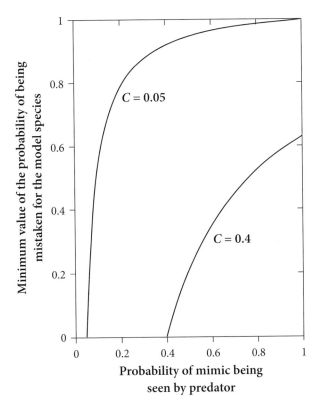

FIGURE 3.12 A case in which mutations can be advantageous only when they have large effects. The example is based on the theory for a palatable butterfly that is fairly inconspicuous, but which can mutate to a more conspicuous form that resembles a distasteful species (the "model") that is more conspicuous. C is the conspicuousness of a nonmimetic form (its probability per unit of time of being seen). The minimum resemblance to the model species (the y axis) that is advantageous depends on the mimic's probability of being seen by the predator (the x axis), and on the probability per unit of time that an individual of the model species is eaten by the predator; in the calculations, this was assumed to be 0.05. If the nonmimetic form is protected by being inconspicuous (C is low), the mimetic form is advantageous only if it has a strong resemblance to the model species, whereas if the mimetic species is rather conspicuous, mimicry may be more likely to evolve and may not require a close resemblance.

PROBLEMS

3.1. **i.** Find an expression for the frequency q_t in generation t of a recessive lethal allele, given an initial frequency q_0, by using the exact expression (**Equation B2.2.2b**) in **Section 2.ii.a**. Find q_1 in terms of q_0, then q_2 in terms of q_0, etc.

  ii. Show how an exact expression similar to **Equation B3.1.2** of **Section 3.1.ii** can be obtained when $w_{11} = 1 + s$, $w_{12} = (1 + s)$ and $w_{22} = (1 + s)^2$.

3.2* i. Use **Equation B3.2.2** in **Section 3.1.iii.b** to obtain formulae for the time needed to change q from initial frequency q_0 to final frequency q_1 for the case of a favorable recessive variant, and for a favorable dominant variant.

ii. Discuss why these formulae imply a very long time for a favorable recessive to increase from a low initial frequency, and for a favorable dominant to spread to fixation.

3.3 Use the general equation for change in allele frequency of an autosomal allele (**Equation B2.2.3** in **Section 2.1.ii.a**) to obtain approximations for the changes in allele frequency of a rare, favorable mutation with arbitrarily strong selection for the cases of recessivity ($h = 0$) and partial or complete dominance ($h > 0$).

3.4* Use the formula for the rate of increase in numbers of a continuously reproducing population consisting of a single genotype i, $dN_i/dt = r_i N_i$, to obtain an expression for the change in frequency of a variant at a biallelic locus in an asexual population, similar in form to the discrete generation **Equation 2.1** in **Section 2.1.i**.

3.5 Use **Equations 3.6** and **3.7** in **Section 3.1.v.b** to compare the conditions for autosomal and X-linked mutations to spread when their selection coefficients are of opposite sign in males and females. Discuss what happens if the mutations are completely recessive ($h = 0$).

3.6 Show how **Equations B3.3.2a** and **B3.3.2b** from **Box 3.3** in **Section 3.1.v.d** give rise to an expression for allele frequency change that is similar to the standard equation for an autosomal locus (**Equation B2.2.3** in **Section 2.1.ii.a**), with inclusive fitnesses replacing individual fitnesses.

3.7 Show how Hamilton's rule (**Section 3.1.v.d**) for the spread of a rare variant promoting altruism can be derived from **Equation 3.11** for inclusive fitness.

3.8* Use **Equation B3.4.1** from **Section 3.2.ii** to show that the extinction probability of a new mutation, E, is *less* than 1 when the mean offspring number per individual for carriers of the mutation is *greater* than 1.

3.9 Assuming that the probability of survival of a favorable mutation with selection coefficient hs in heterozygotes is approximately $2hs$ (**Section 3.2.ii**), derive an equation for the number of mutations that must enter a randomly mating population for there to be a probability greater than 0.95 that at least one mutation survives in the population. Calculate the value of these numbers for $hs = 0.01$ and $hs = 0.05$.

3.10* Derive **Equation 3.13** in **Section 3.3.iii.a** for the selection intensity, i, for a normally distributed trait subject to truncation selection, when only the top fraction, P, of the population is chosen to produce the next generation.

3.11 Show that the selection differential, S, of **Section 3.3.iii.a** is equal to the covariance between trait value, z, and fitness measured relative to the mean fitness of the population, which is defined as $w(z)/\bar{w}$.

CHAPTER FOUR

Migration, Mutation, and Selection

CHAPTER SUMMARY

We now extend the analysis of selection to several biologically important situations where selection interacts with other evolutionary forces. We first consider selection at a locus or nucleotide site when there are differences in fitnesses between two populations connected by migration. This case shows how selection can maintain genetic differences between populations despite gene flow between them. We then describe the theory for three other biologically important situations: (**1**) clines in allele frequencies between populations spread along a continuous geographic gradient, with spatially varying selection, (**2**) hybrid zones, and (**3**) a favorable mutation spreading through the geographic range of a species. We show how data on such situations can be used to estimate the strength of selection in nature, when migration rates are known.

We then examine the interaction between mutation and selection in a single population, for single nucleotide sites, and for both autosomal and sex-linked inheritance. An important conclusion is that most deleterious, autosomal mutations in randomly mating populations are eliminated in the heterozygous state, rather than as homozygotes. We extend the theory to the levels of entire genes and the genome as a whole, and to inbreeding populations. We also show how the theory provides methods of estimating mutation rates at single genes where mutations cause major, deleterious effects.

We next discuss the genetic load—the extent to which the mean fitness of a population is reduced below the maximum possible value. When the population size is large, the load due to deleterious mutations depends only on the mutation rate. The effects of multiple sites across the whole genome that carry mutations can be modeled by assuming multiplicative fitness effects of the mutations, and we discuss whether this mutational load can limit genome sizes. We also describe the "segregational load" caused by variants maintained by heterozygote advantage, and show that a genome cannot tolerate many sites subject to strong balancing selection of this kind. A related question is: how much is the rate of adaptive evolution limited by the fitness cost of the fixation of selectively favorable variants?

In populations with mutational or segregational loads, inbreeding often causes reduced fitness—inbreeding depression. We use data on deleterious mutation rates and dominance coefficients from mutation accumulation experiments to infer the extent to which mutational load is uncovered by inbreeding. Data from experiments on *Drosophila* suggest that deleterious mutations are a major cause of inbreeding depression.

Finally, we model the genetic variance for quantitative traits and fitness components at mutation–selection equilibrium, showing that observed genetic variances for fitness components are higher than predicted, suggesting the involvement of loci under balancing selection. We also discuss stabilizing selection on quantitative traits, and ask how much variation in these traits can be maintained when mutations perturb phenotypes away from their optima. Surprisingly, the equilibrium variance in sexually reproducing populations does not depend on the effects of mutations on the trait, but only on the strength of stabilizing selection and the net mutation rate for the trait. We briefly review evidence on how much quantitative trait variability is maintained by balancing selection versus mutation–selection balance.

> It is at once clear that in equilibrium . . . abnormal genes are wiped out by natural selection at exactly the rate as they are produced by mutation.
>
> Haldane (1937)

INTRODUCTION

In **Chapter 2**, we saw how natural selection can sometimes maintain genetic variation within a population. **Chapter 3** showed, however, that selection often removes variation. Genetic variation can then be maintained only if other forces oppose selection. These forces include the migration of individuals from other populations, and mutations; combinations of these two different evolutionary forces with selection are the main subjects of this chapter. The interplay between migration and selection is especially important in relation to *geographic variation*, the familiar tendency for populations of the same species that live in different areas to diverge with respect to phenotypes of adaptive significance, such as skin pigmentation in humans (see Mayr 1963, Chapter 11). In the first part of this chapter, we examine the strength of selection needed to oppose the homogenizing effects of migration. Situations where there is such opposition allow us to use the theoretical results to estimate the strength of selection in nature.

The interaction between mutation and selection is a major contributor to natural variability in fitness and other quantitative traits. Because most free-living organisms have large genomes (even bacteria such as *Escherichia coli* have over 4,000 genes), deleterious mutations kept at low frequencies by selection can collectively produce a large reservoir of variation within a population, despite the low rate of occurrence of mutations at individual nucleotide sites (**Section 1.3.iii**).

The existence of genetic variants at many sites that affect fitness, scattered throughout the genome, raises the question of the effect of this variation on the fitness of the population as a whole—this is the problem of *genetic load*. Are there limits to the amount of selected variability that a population can tolerate? If so, what determines these limits, and are human and other populations close to such limits? In addition, we saw in **Section 1.1.iii** that inbreeding leads to reduced fitness. Here we ask the important question: how much do deleterious mutations contribute to inbreeding depression, and to variation in fitness in natural populations?

4.1. MIGRATION AND SELECTION

4.1.i. Migration and selection involving two populations

In **Section 2.2.iii**, we discussed what happens if selection acts in different directions in different parts of the environment (Levene's model). For simplicity, we assumed free movement of individuals between the different environments. It seems plausible that restricted migration will make it easier for variation to be maintained. For instance, in a two-allele system with two environmental niches and no migration between them, selection might cause allele A_1 to go to fixation in environment 1, and A_2 to go to fixation in environment 2. This would maintain variation in the species as a whole, but no variation would be present within either environment. Allowing a small amount of migration between the two environments should not alter the fact that both alleles are maintained in the overall population, but would merely introduce variability into the two subpopulations.

Early analyses by Haldane (1930a) and Wright (1931) considered one-way migration of alleles from a "source" population into a "sink" population in which the immigrant alleles are at a selective disadvantage. **Box 4.1** shows a simple, but more general, model based on Moran (1962, Chapter 9), with migration leading to variability within a *metapopulation* consisting of two local populations (*demes*), given local differences in the direction of selection. As one would expect, the extent to which the two demes differ from each other depends on the strength of selection (*s*), and the amount of migration between the demes (*m*).

Box 4.1 MIGRATION AND SELECTION WITH TWO DEMES

A fraction m of individuals in a given deme is assumed to come from the same deme, and a fraction $1 - m$ comes from the other one by migration. Let the frequencies of A_1 and A_2 be p_1 and q_1, respectively, in deme 1 and p_2 and q_2 in deme 2. The corresponding frequencies of A_2 after migration are $(1 - m)q_1 + mq_2$ and $(1 - m)q_2 + mq_1$, so that the changes in frequencies due to migration are:

$$\Delta q_{1m} = -\Delta q_{2m} = m(q_2 - q_1) \qquad (B4.1.1)$$

For the purpose of modeling selection, we assume weak selection and haploidy. The fitness model for the two demes is:

	A_1	A_2
Deme 1	1	$1 - s$
Deme 2	$1 - s$	1

where s is > 0.

Using the haploid equivalent of **Equation 3.2** in **Section 3.1.iii.b** for weak selection, we find that the changes in allele frequencies due to selection are:

$$\Delta q_{1s} \approx -sp_1q_1, \quad \Delta q_{2s} \approx sp_2q_2 \qquad (B4.1.2)$$

If the two demes are to differ in their allele frequencies, migration must not be strong enough to overwhelm the effects of selection. The changes in allele frequencies due to both migration and selection are then sufficiently small that their squared values and their product can be neglected. This means that the approximate net changes for each deme, Δq_1 and Δq_2, can be found simply by adding the terms due to migration and selection; it is then also irrelevant whether individuals migrate before or after selection has occurred within a generation.

The equilibrium can then be found by equating these net changes to 0. By symmetry, it is obvious that at equilibrium we have $q_2 = p_1$; the equilibrium solution is given by the equation for $\Delta q_1 = 0$ obtained from **Equation B4.1.1** and **Equation B4.1.2**:

$$sp_1q_1 + m(q_1 - p_1) \approx 0 \qquad (B4.1.3)$$

It is immediately obvious from this result that the equilibrium is determined solely by the ratio $l = m/s$.

Noting that $p_1 = 1 - q_1$, and solving the resulting quadratic equation we obtain:

$$q_1^* \approx \frac{1 + 2l - \sqrt{1 + 4l^2}}{2} \qquad \text{(B4.1.4)}$$

If $l \ll 1$, then $q_1^* \ll 1$, so that the square of q_1^* can be neglected. This leads to the simple result:

$$q_1^* \approx l \qquad \text{(B4.1.5)}$$

We will not conduct a full analysis of the stability of this equilibrium; note, however, that Δq_2 when q_2 is close to 0 is approximated by $q_2(s - m)$ + $m\,q_1$. A_2 will therefore spread from a low initial frequency in deme 2 if $s > m$. The same applies to the initial spread of A_1 in deme 1. The condition $l < 1$ is therefore sufficient to guarantee the preservation of variability in the whole population.

Equations B4.1.4 and **B4.1.5** show that the local frequency of the variant that is at a disadvantage in a given deme (A_2 in deme 1 and A_1 in deme 2) depends on the ratio $l = m/s$, i.e., the relative strengths of selection and migration control the extent to which the two demes are kept different (**Figure 4.1**). If the m and s values are similar ($l \approx 1$), the two demes will have essentially the same composition, but if $m \ll s$, so that $l \ll 1$, A_1 will be the most frequent allele in deme 1, and A_2 will predominate in deme 2.

As this simple model predicts, many conspicuous phenotypic differences between adjacent populations are maintained by local selection in the face of migration (Jain and Bradshaw 1966; Endler 1977; Barton and Hewitt 1989; Bell 2008, pp. 228–230, 304–307). A classic example of such local adaptation, first studied in the 1920s, is provided by the pocket mouse, *Chaetodipus intermedius*, of the southwestern United States, which lives in semi-desert areas, mainly on light-colored rocks. Mice living on light backgrounds are light colored, whereas mice from places covered by dark lava flows are predominantly dark colored (melanic). The melanic form is controlled by a nearly completely dominant allele of the melanocortin-1 receptor gene (*Mc1r*). This gene codes for the receptor protein that binds the hormone melanocortin, which is involved in regulating pigment formation. There is experimental evidence that the color of the mice provides protection against bird predators on the background with the corresponding color (reviewed by Nachman 2005). Hoekstra et al. (2004) estimated the rate of migration m between lava flow populations and adjacent populations, using DNA sequence differentiation for mitochondrial genes (see **Section 7.2.ii.b**). By substituting m values into an equation similar to **Equation B4.1.4**, together with estimates of the frequencies of the melanic *Mc1r* allele in different populations, they estimated the selection coef-

A.   **B.**

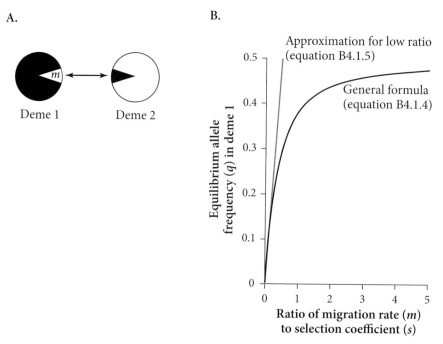

FIGURE 4.1 Allele frequency differences between two demes undergoing different selection, with migration between the demes. **A.** Diagram of the model. The two demes are indicated by circles (deme 1 in black and deme 2 in white), with the proportions of migrant individuals in the other deme's color indicating the migration rate, m. One allele is advantageous in deme 1 and the other in deme 2; selection is symmetrical (i.e., the selection coefficient against the disfavored allele is the same for both demes). **B.** Plot of the relationship between the equilibrium frequency in deme 1 of the variant disfavored in this deme, and the ratio $l = m/s$. **Equation B4.1.4** in **Box 4.1** applies to any value of l, whereas the linear approximation (**Equation B4.1.5**) is applicable only when $l \ll 1$.

ficient (against light-colored mice on a lava flow) to lie between 0.4 and 0.04, depending on the specific assumptions that were made.

Cases such as this contradict the intuition of many biologists that:

> . . . a population cannot change drastically as long as it is exposed to the normalizing effects of gene flow. (Mayr 1963, p. 521)

4.1.ii. Migration and selection in a continuous habitat

4.1.ii.a. Clines

Another important situation where migration and selection interact is in *clines*. In a cline, a trait like body size, or an allele frequency, changes along a continuous geographic gradient, often latitude (**Figure 4.2**). For example, animals in

A. Flower colors

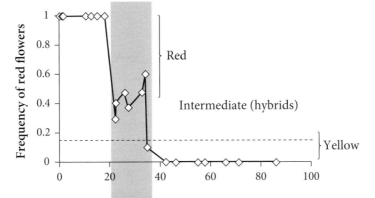

B. Chloroplast DNA haplotype frequencies

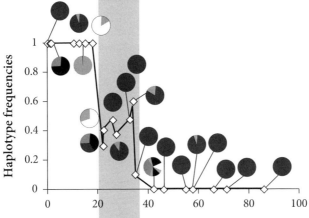

C. Allele frequencies at four loci

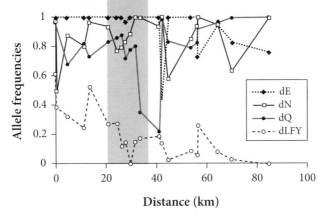

FIGURE 4.2 **A**. Cline in flower color, measured as the frequencies of red versus yellow flowers in several *Mimulus aurantiacus* populations in San Diego County, California. **B**. Cline in the frequencies of chloroplast DNA haplotypes. **C**. Frequencies of single nucleotide polymorphisms at four nuclear loci, showing that these loci do not suggest isolation between the populations; three of the loci (D, E, and Q) have unknown functions, and the fourth locus is the *Leafy* gene. [Adapted from Figure 4 of Streisfeld and Kohn (2005).]

high-latitude populations generally have larger mean body sizes than low-latitude populations (Bergmann's Rule); this applies to many different species, including humans (Mayr 1963, Chapter 11). Clines in the frequencies of amino acid variants in metabolic enzymes are also quite common (Sezgin et al. 2004); for example, the *F* electrophoretic allele at the *Adh* locus of *D. melanogaster* is more frequent at high latitudes in multiple continents and in both hemispheres (**Figure 4.3**; Oakeshott et al. 1982; Umina et al. 2005).

The repeatability of clines for traits such as body size and the frequencies of electrophoretic variants across different species or different geographic regions implies that they are caused by selective responses to local environmental differences (Mayr 1963, Chapter 11; Oakeshott et al. 1982; Umina et al. 2005). Direct evidence on the nature of the selection is usually lacking. However, experiments can help to test whether a putative selective factor is actually involved. In *Drosophila* experiments where flies are kept at different temperatures, large body size evolves at lower temperatures, consistent with the larger sizes found at high latitudes (Anderson 1973); although the physiological basis for this is not well understood, the differences are caused by genetic factors, not merely by a direct effect of the environment.

4.1.ii.b. *Modeling clines*

Clines for a single pair of allelic variants can be modeled in terms of the frequencies of the alternative variants, A_1 and A_2, at a location x on a line, denoted by $p(x)$ and $q(x)$, respectively. The selection coefficient that specifies the advantage or disadvantage of A_2 with respect to the alternative A_1 also depends on x.

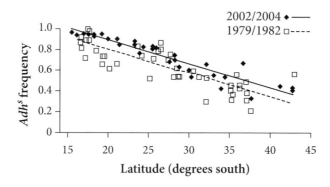

FIGURE 4.3 Cline in the frequencies of the slow (*S*) allele at the *Adh* locus of *D. melanogaster*, showing the higher frequency at low latitudes in the southern hemisphere, and the change between the years 1979/1982 and 2002/2004, with higher temperatures in the more recent years moving the allele frequencies towards values previously found in lower latitudes. [Adapted from Figure 1 of Umina et al. (2005).]

This dependence is modeled by multiplying a constant strength of selection, s, by a quantity that varies with the location, $\gamma(x)$ (assumed to be < 1). For instance, if A_2 is favored at the right-hand end and disfavored at the left-hand end, $\gamma(x)$ will be positive when x is large and negative when it is small.

Migration is represented by a *migration distribution*, $m(y)$: the probability that an individual moves a distance between birth and reproduction of between y and $y + \delta y$ (where δy represents a small change in y) is approximately $m(y)\,\delta y$ (y is negative for movements to the left, and positive for movements to the right). The simplest assumption is that $m(y)$ is symmetrical with respect to y, with mean zero, so that there is no net movement of individuals along the line. The typical distance moved over one generation, disregarding the direction of movement, is conveniently measured by σ, the standard deviation of $m(y)$. If individuals move randomly and independently of each other, $m(y)$ will be a normal distribution, and is thus completely characterized by σ (**Appendix A2.v.e**). Simple techniques for modeling a cline in variant frequencies at a single site are described in **Box 4.2***.

Box 4.2* MODELING MIGRATION AND SELECTION IN A CONTINUOUS HABITAT

For simplicity, we again assume haploidy. The relative fitnesses of A_1 and A_2 at location x are written as 1 and $1 + s\gamma(x)$, respectively, using the notation of Slatkin (1973b). If s is small, we can again use the haploid equivalent of **Equation 3.2** in **Section 3.1.iii.b** to approximate the change in frequency of A_2 due to selection at a point x:

$$\Delta q_s(x) \approx s\gamma(x)p(x)q(x) \qquad \text{(B4.2.1)}$$

We also need to consider the effect of migration on allele frequencies. This can be done by noting that the allele frequency after migration (ignoring selection) is given by:

$$q'(x) = \int q(x - y)\, m(y)\, dy \qquad \text{(B4.2.2)}$$

where the integral is taken over all values of y contributing to a given x.

The integral can be approximated as follows, using a Taylor's series expansion of $q(x - y)$ (**Appendix A1.ii.a**):

$$q'(x) \approx \int \left[q(x) - y\frac{dq(x)}{dx} + \frac{y^2}{2}\frac{d^2 q(x)}{dx^2} \right] m(y)\, dy \qquad \text{(B4.2.3)}$$

where higher-order terms in y have been neglected. This approximation is equivalent to assuming that the distance travelled over one generation is small compared with the overall length of the line, so that the third and higher powers of y are small when compared to this length. Similar reasoning is used in deriving the *diffusion equations* that describe variant frequency changes in finite populations (**Appendix 5A.1**).

Carrying out the integration, and noting that the mean and variance of y over the distribution $m(y)$ are 0 and σ^2, respectively, we obtain the following expression for the change in allele frequency due to migration:

$$\Delta q_m \approx \frac{\sigma^2}{2} \frac{d^2 q(x)}{dx^2} \qquad \text{(B4.2.4)}$$

Since selection and migration are assumed to be weak forces, the net change in allele frequency is approximated by the sum of their effects, $\Delta q_s(x) + \Delta q_m(x)$ (see **Box 4.1**).

Box 4.3* outlines the method for solving for the equilibrium between migration and selection, to obtain a formula for the relation between $q(x)$ and x. What is the relationship between the equilibrium allele frequency differences along a cline and the differences in selection as functions of the location? **Box 4.3*** examines this question for the case when there is an abrupt change in the direction of selection in the middle of the species range, and **Figure 4.4** shows the results for one set of parameter values, showing how migration smoothes out discontinuities in the allele frequencies. In this case, with haploidy and equal intensities of selection at either end of the cline, the maximum slope of $q(x)$ (at $x = 0$) is about $0.6/l_c$, where $l_c = \sigma/\sqrt{s}$ is called the *characteristic length* of the cline (Slatkin 1973b). The same results apply to semidominant selection in a randomly mating diploid population, except that $s/2$ is used instead of s (see **Section 3.1.iii.a**).

With a continuous selection function, so that $\gamma(x)$ changes smoothly with x, the allele frequencies change gradually with position, unless l_c is small compared with the size of the region over which γ changes; in this case, the shape of the cline in that region becomes similar to that in the discontinuous case (Slatkin 1973b). Discontinuous and continuous changes in selection can therefore both produce similar effects, because migration blurs any discontinuities in allele frequencies caused by discontinuities in selection pressures.

Box 4.3* THE PROPERTIES OF AN EQUILIBRIUM CLINE

If the net change in gene frequency at a point x is set to 0, **Equations B4.2.1** and **B4.2.4** yield an approximate expression for equilibrium under migration and selection:

$$\frac{\sigma^2}{2}\frac{d^2 q(x)}{dx^2} \approx -s\gamma(x)p(x)q(x) \qquad (B4.3.1)$$

The effects of the scale of measurement of distance can be removed by dividing x by the quantity $l_c = \sigma/\sqrt{s}$ (Slatkin 1973b), which is the equivalent of the square root of l in **Box 4.1** of **Section 4.1.i**. On the new scale, $z = x/l_c$, we get:

$$\frac{d^2 q(z)}{dz^2} \approx -2\gamma(z)p(z)q(z) \qquad (B4.3.2)$$

The system is therefore completely described by the dependence of q and γ on z.

We can use this to examine the smoothing effect of migration. The steepest change in allele frequency occurs when the selection function is discontinuous at a point in the center of the range, e.g., when there is a line extending from $-\infty$ to $+\infty$, with $\gamma(0) = 0$, $\gamma(z) = -k$ for $z < 0$, and $\gamma(z) = 1$ for $z > 0$ (this allows for the possibility that the strength of selection differs on either side of the center). Clearly, we have $q(-\infty) = 0$ and $q(+\infty) = 1$. The slope of q with respect to z, dq/dz, must approach 0 at either end of the range, and increases as z approaches 0.

If we write $g = dq/dz$, a simple manipulation gives:

$$\int\frac{dg}{dz}dq = \int\frac{dg}{dz}\frac{dq}{dz}dz = \int g\,dg$$

Applying this to **Equation B4.3.2**, and using the boundary conditions above, we obtain:

$$\left(\frac{dq}{dz}\right)^2 = 4k\left(\frac{q^2}{2} - \frac{q^3}{3}\right) \quad (z < 0) \qquad (B4.3.3)$$

A similar expression holds for $z > 0$, replacing q by p and k by 1. At the midpoint $z = 0$, the two expressions must be the same, so that the allele frequency at this point (q_b) can be found by equating them, giving a cubic equation for q_b:

$$2q_b^3 - 3q_b^2 + \frac{1}{(1 + k)} = 0 \qquad \text{(B4.3.4)}$$

These expressions determine the slope of the cline, dq/dz. The maximum value of the slope is reached as z approaches 0 and q approaches q_b; with $k = 1$, it is equal to $1/\sqrt{3} = 0.577$ on the scale of z, or $0.557l_c$ on the original scale.

The complete solution for this case, writing z as a function of q, can be found by inverting **Equation B4.3.3**, giving:

$$\frac{dz}{dq} = \frac{1}{2\sqrt{k\left(\dfrac{q^2}{2} - \dfrac{q3}{3}\right)}} \qquad (z < 0) \qquad \text{(B4.3.5)}$$

There is an equivalent expression for $z > 0$, with q replaced by p and k replaced by 1.

These equations can be integrated, using standard integral formulae and the appropriate boundary conditions (Haldane 1948), giving:

$$z\sqrt{k} = \tanh^{-1}\sqrt{1 - \frac{2q_b}{3}} - \tanh^{-1}\sqrt{1 - \frac{2q}{3}} \qquad (z < 0) \qquad \text{(B4.3.6a)}$$

$$z = \tanh^{-1}\sqrt{\frac{(1 + 2q)}{3}} - \tanh^{-1}\sqrt{\frac{(1 + 2q_b)}{3}} \qquad (z > 0) \qquad \text{(B4.3.6b)}$$

The variable q as a function of z is readily obtained by rearranging these expressions to put the terms involving q on the right-hand side, and taking the tanh of the result (note that $\tanh x = [\exp(x) - \exp(-x)]/[\exp(x) + \exp(-x)]$, and $\tanh^{-1}(x)$ is the function whose tanh is equal to x).

Haldane (1948) also obtained a solution for the case of a dominant allele; the case of heterozygote disadvantage was solved by Bazykin (1969). Most other cases require numerical solutions of the equivalent of **Equation B4.3.2** (e.g., Fisher 1950).

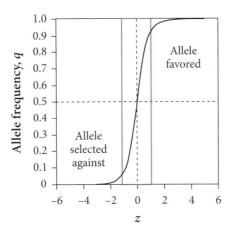

FIGURE 4.4 Expected allele frequencies in a cline, plotted from **Equations B4.3.6a** and **B4.3.6b** in **Box 4.3**. The x axis shows scaled distances, z (see text), from a location where the environment changes, such that the allele whose frequencies are on the y axis is selected against in the region to the left of the place where $z = 0$, and the other allele is selected against in the region where $z > 0$. The diagram shows the steep allele frequency change close to the discontinuity in the selection coefficient. The box shows the region between $z = -1$ and $z = 1$, showing the narrow zone in which most of the allele frequency change occurs. In this case, $k = 1$.

4.1.ii.c. *Using cline widths to measure selection*

It is clear from these results that migration can maintain variability within local populations, although populations at the extremes of a gradient become fixed for one or another type if they are separated by distances that are several times l_c. The inverse of the maximum slope is often used as a measure of the width of the cline. The fact that the width depends inversely on $\sigma/\sqrt{s}$ rather than σ/s means that selection is surprisingly effective in causing an abrupt change in frequency, as first noted by Haldane (1948). If $s = 0.01$, a selection coefficient that is extremely difficult to measure directly in nature (**Section 2.1.ii.d**), the width of the cline for the symmetrical haploid model considered above would only be about 6σ.

Since σ is much easier to measure than s, a comparison of the observed width of a cline with σ provides a useful method for inferring the strength of selection in natural conditions (Endler 1977):

> *It should . . . be possible, without very extensive work, to say whether a given cline suggests selective intensities of the order of 10 or 0.01%.* Haldane (1948)

Haldane applied his theoretical results to a cline in body color of the deer mouse, *Peromyscus polionotus*. In this species, a light-colored form inhabits the sandy beaches and islands of the Gulf of Mexico coast in Florida and Alabama, intergrading with a darker inland form over a belt about 40 miles from the coast, with a change in light phenotype frequency from 75% to 25% over 12 miles (Sumner 1929, 1930), as shown in **Figure 4.5**. This difference involves several genes, but it is now known that there is a major contribution to the light-colored phenotype from a semidominant mutation in the *Mc1r* gene (**Section 4.1.i**) that alters an amino acid in the MC1R protein (Hoekstra et al. 2006). In his pioneering study, Haldane assumed that the light form was controlled by a dominant mutation; he used a value of σ of about half a mile, and estimated the selection coefficient needed to account for the observed shape of the cline as about 0.1% (see Mullen and Hoekstra (2008) for a re-examination of this question).

Endler (1977, Chapter 6) reviewed cases in which direct estimates have been obtained of both the migration distribution and the nature of selection. This allows comparisons of the observed and predicted shapes of clines, and there often appears to be quite good agreement between them. This suggests that the simplifications needed to obtain the theoretical predictions are not too severe.

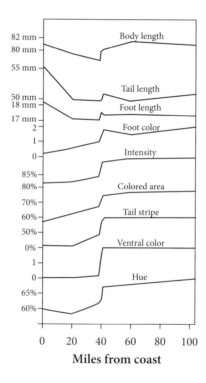

FIGURE 4.5 Cline in body color and various other characteristics of the deer mouse *Peromyscus polionotus*, from the light-colored form on the sandy beaches and islands of the Gulf of Mexico coast in Florida and Alabama, to the darker form found about 40 miles from the coast. Note that tail and foot lengths were measured by standardizing against a body length of 80 mm. [Adapted from Figure 8.1 of Wright (1978).]

An interesting recent extension of the use of this theory is to establish experimental clines with known migration rates in order to estimate selection intensity. This approach has allowed the fitness costs of herbicide resistance genes in the absence of herbicides to be estimated in *Arabidopsis thaliana* (Roux et al. 2006).

4.1.ii.d. *Hybrid zones*

Hybrid zones are closely related to clines. They can arise when two populations of a species, which have been separated for a long time into two isolated geographical areas, come back into *secondary contact* with each other. The populations may have developed partial reproductive isolation, due to fixation for different alleles that reduce fitness when they are brought together again in F1 hybrids between the two populations, as we will discuss in relation to the genetic basis of speciation in **Section 8.4.vi**. This reduces fitness in the region where the populations come into contact (**Figure 4.6**).

Such a hybrid zone can be modeled by assuming that the two populations are fixed for alternative alleles, A_1 and A_2, which show heterozygote fitness disadvantage (**Section 2.1.ii.f**); this ignores the true details of the genetic basis of the low fitness, which probably usually involves more than a single locus. In contrast to the clinal situation, this model has no geographic gradient of selection coefficients, but the model of migration along a continuous line is still useful for determining the location and width of the hybrid zone (Barton and Hewitt 1989). The width depends on the characteristic length defined above, now expressed in terms of the strength of selection against the heterozygotes, in much the same way as in a cline (Slatkin 1973b).

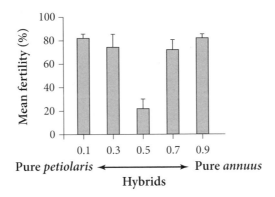

FIGURE 4.6 Reduced fitness in hybrids between two sunflower species. The *x* axis is an index of the extent to which individuals are hybrids, based on allozyme markers, and the fitness measure on the *y* axis is the female fertility of plants in the hybrid zone. [Adapted from Figure 2 of Rieseberg et al. (1998).]

4.1.ii.e. *The wave of advance of a favorable mutation*

The same general framework can be used to ask a very different question: how long does it take for a favorable mutation to spread out from its place of origin? This problem was first studied by Fisher (1937). The details are too complex to be described here, but the model assumes a linear population and haploid selection with selection coefficient s on a newly arisen, favorable mutation. **Equation B4.2.4** in **Box 4.2*** of **Section 4.1.ii.b** gives the allele frequency change due to migration at location x, and **Equation B4.2.1** gives the change due to selection, with $\gamma(x)$ now set equal to 1 because there is no change in selection pressure with location. The equation given by adding these together, to obtain the net change in allele frequency, has a similar general form to the equations for wave motion in physics.

Fisher showed that there is a *wave of advance* of the favorable mutation, with the allele frequency at a location far from the point of origin of the favorable mutation starting out at 0, then increasing as the wave passes over it, eventually reaching fixation at that point. After a good deal of analysis, he concluded that, when enough time has elapsed, the wave front reaches a constant form, and its speed converges to a constant value: $v = \sigma\sqrt{2s}$.

This result for the speed of advance of the mutation along the line implies that the time for a favorable mutation to reach fixation at a location x units from its point of origin is approximately x/v greater than the fixation time for a single population (as given by **Equation B3.1.5** in **Box 3.1** of **Section 3.1.ii**). With a selection coefficient s of 0.01, a location distant 1000σ away from the origin of the mutation would have to wait an additional 7071 generations to become fixed for the mutation. No cases of the spread of a favorable mutation in nature have yet been followed in detail, but an analysis of the spread of the CCR5 mutation that confers resistance to HIV infection in humans has been carried out by fitting the predictions of a two-dimensional version of **Equation B4.2.4** to the spatial pattern of frequencies of this mutation in Europe and the Near East (Novembre et al. 2005).

Similar results apply to the situation when a species is introduced into a new habitat, but with the population growth rate r replacing s in the equation for v. Many studies of such situations have been published (Skellam 1951), and there is often a good fit between the data and the theory, although considerable discrepancies can be caused by departures from the assumptions of the model, such as the occurrence of a non-normal distribution of migration distance (Levin et al. 2003).

The speed of spread of genetic variants subject to selection is important in relation to the problem of the contamination of wild species or organically grown crop plants by populations of genetically modified crops, and detailed models of selection and migration are essential for studying this problem (e.g., Stewart et al. 2003; Kelly et al. 2005).

4.2. MUTATION AND SELECTION

4.2.i. Introduction

As we saw in **Section 1.3.iii**, mutation is an ever-present but usually very weak force, causing changes in DNA sequences to occur at a rate of the order of 10^{-9} to 10^{-8} per nucleotide site per generation in higher eukaryotes. This means that, even if selection has established the fittest possible sequence of a gene under the prevailing environmental circumstances, there will be some rate of occurrence of mutations to less fit variants. For example, the mean rate of mutation per generation for single-nucleotide substitutions in humans is estimated to be about 2×10^{-8} (see **Section 1.3.iii.b**). A typical human nuclear gene has about 500 codons, corresponding to 1500 nucleotides, and about 70% of nucleotide changes lead to a change in the amino acid sequence (Kryukov et al. 2007). This implies a rate of mutation to a changed form of the protein of $0.70 \times 1500 \times 2.0 \times 10^{-8} = 2.1 \times 10^{-5}$ per gene per generation. There is an additional, but substantially smaller, contribution from nonsense mutations, indel mutations, intron splice-site mutations and transposable elements, all of which severely disrupt protein structure (Kryukov et al. 2007).

As we saw in **Section 1.2.iv.a** and will discuss further in **Section 6.4.iv**, there is evidence that most of these amino acid changes are deleterious, but a majority have very small effects on fitness. Some, however, greatly reduce the fitness of their heterozygous carriers, causing Mendelian genetic diseases. These contribute a significant fraction of childhood diseases and mortality; the frequency of single-gene disorders among pediatric hospital admissions in North America is between 4% and 7% (Gelehrter et al. 1998). With at least 20,000 protein-coding genes in the human genome (http://www.ornl.gov/sci/techresources/Human_Genome), each new zygote must contain more than $2 \times 20,000 \times 2.1 \times 10^{-5} = 0.84$ new nonsynonymous mutations, of which at least 70% are probably sufficiently deleterious that selection is virtually certain to eliminate them from the population (see **Section 6.2.iii.a**), i.e., they are subject to *purifying selection*. This yields a net nonsynonymous deleterious mutation rate of more than 0.58 mutations per new zygote per generation.

There is also evidence for widespread purifying selection against mutations in noncoding sequences in the human genome (Keightley et al. 2005; ENCODE Project Consortium 2007). Since noncoding sequences are far more abundant in mammalian genomes than coding sequences, as many as 5% of the nucleotides in the genome may be subject to purifying selection (ENCODE Project Consortium 2007). The net *deleterious mutation rate* in humans may be as high as six mutations per individual per generation. This constant production of deleterious variants by mutation is of great importance for both medical and evolutionary genetics, and requires careful study by the approach initiated by Haldane (1927b, 1937). This involves modeling the balance between the input of new, deleterious mutations and their elimination by selection. We will first consider this at the level of a single nucleotide site, then whole genes, and finally the whole genome. The extent to which deleterious mutations contribute to within-population variability in DNA sequences is discussed in **Section 6.4.iv**.

4.2.ii. Mutation–selection equilibrium

4.2.ii.a. *Autosomes with random mating*

How much variation can mutation maintain? We begin by studying the simple case of mutation at a single autosomal site in a randomly mating, diploid species. Assume a wild-type variant A_1 that mutates at rate u per generation to a deleterious alternative A_2. We also assume that A_1 is sufficiently common in the population that mutation in the reverse direction can be ignored (i.e., selection is so strong that the frequency of A_2, q, is very low). We saw in **Section 1.3.iii.c** that the change in the frequency of A_2 due to mutation is $\Delta q_{mu} = u(1 - q) - q$. If q and u are both small, their product can be neglected, and we have:

$$\Delta q_{mu} \approx u \qquad (4.1)$$

Assume that the population is randomly mating, and that the fitnesses of A_1A_2 and A_2A_2 are $1 - hs$ and $1 - s$, respectively (using the notation introduced in **Section 2.1.ii.a**). Using the results of **Problem 3.3**, we find that the change due to selection is approximately:

$$\Delta q_s \approx -hs\, q \qquad (4.2)$$

(compare this with **Equation B4.1.2** of **Section 4.1.i** for selection with migration.) At equilibrium, the changes due to mutation and selection are equal and opposite, and q^*, the equilibrium frequency of A_2, is thus:

$$q^* \approx \frac{u}{hs} \qquad (4.3)$$

(compare this with **Equation B4.1.5**.) It is easily verified that this is a stable equilibrium (**Problem 4.2**). A similar result applies to haploid organisms, replacing hs with s.

A useful way of understanding this result is to consider the situation when the deleterious mutant allele A_2 is present at a low frequency, so that the frequency of heterozygotes is approximately $2q$. Most selective elimination then involves heterozygotes, giving an average reduction in frequency of $hs/2$ per heterozygous individual per generation. The rate of elimination of A_2 by selection is thus approximately $2qhs/2 = qhs$. This must equal the rate per generation at which new mutations enter the population, which is approximately u when q is small.

The case of completely recessive mutations ($h = 0$) is slightly more complex (**Problem 4.3**). The equilibrium is now given by:

$$q^* \approx \sqrt{\frac{u}{s}} \qquad (4.4)$$

For a given mutation rate and selection coefficient, the equilibrium frequency of a completely recessive variant is much higher than that for a corresponding variant with intermediate or complete dominance, reflecting the ineffectiveness of selection against rare recessive alleles (**Section 3.1.iii.c**).

Recall from **Section 2.2.iv** that most mutations with large effects are recessive with respect to their phenotypic effects. However, it is likely that mutations are rarely fully recessive with respect to their fitness effects. Mutations with recessive lethal effects have an estimated average heterozygous fitness effect of 2–3% in experiments on *Drosophila* (Crow 1993; García-Dorado and Caballero 2000), and mutations with small homozygous fitness effects (of the order of a few per cent) have much higher h values, in the range 0.1–0.4. While direct evidence is lacking for other organisms, there seems to be no reason to assume any radical difference from *Drosophila*. **Equation 4.3** is therefore probably more appropriate than **Equation 4.4** for most cases, even for apparently recessive mutations. The question of *why* mutations are generally or partially recessive with respect to their effects on phenotypes and fitness is one of the oldest in evolutionary genetics (Fisher 1928), and we discuss it in **Section 4.4.v** below.

4.2.ii.b. *Other cases*

Other biologically important situations can be modeled similarly to the autosomal case just explained. With X linkage and random mating, we need to include the possibility that the selection coefficients may differ between the sexes. We can denote the selection coefficients in males and females as s_f and s_m, and use **Equation 3.6** in **Section 3.1.v.b** (with the appropriate change of sign). The equilibrium allele frequency (weighting frequencies in eggs and sperm by 2/3 and 1/3, respectively) is:

$$q^* \approx \frac{3u}{\left(2hs_f + s_m\right)} \qquad (4.5)$$

Under inbreeding with an inbreeding coefficient f, and assuming autosomal inheritance, **Equation 3.10** in **Section 3.1.v.c** gives:

$$q^* \approx \frac{u}{\left[h(1 - f) + f\right]s} \qquad (4.6)$$

As would be expected intuitively, exposing mutations to selection in the hemizygous state (sex linkage) or in the homozygous state (inbreeding) greatly lowers their equilibrium frequencies, provided that the deleterious effects are not completely dominant.

4.2.ii.c. *Mutation and selection at multiple sites*

The equations just derived relate to the situation at a single nucleotide site, such as a mutation that alters an amino acid in a protein sequence. If we assume that selection at each site acts independently of selection at other sites, the equilibrium frequency of mutations at each site can be calculated from the results for each individual site considered in isolation. If we consider a gene whose coding sequence includes m nonsynonymous sites, the overall frequency of mutations in the gene is then simply the sum of the contributions from each site, given by **Equation 4.3** for the case of autosomal inheritance and random mating. If the strength of selection varies across sites, the equilibrium frequency q_g^* of deleterious nonsynonymous mutations present in the gene as a whole is:

$$q_g^* \approx \sum_i \frac{u_i}{h_i s_i} \tag{4.7a}$$

where the subscript i indicates a particular site and runs from 1 to m.

When the strength of selection varies across sites, but the selection coefficients and mutation rates at each site are independent of each other, this simplifies to:

$$q_g^* \approx \frac{u_g}{(hs)_H} \tag{4.7b}$$

where u_g is the mutation rate for the whole gene (i.e., the sum of the mutation rates at each site), and the subscript H denotes the *harmonic mean*, the reciprocal of the mean of the reciprocals.

This can be extended to the whole genome, by summing over all sites capable of mutating to deleterious variants. The mean number, $\bar{n}$, of deleterious mutations carried by a haploid genome is given by:

$$\bar{n} \approx \frac{U}{(hs)_H} \tag{4.7c}$$

where U is the total mutation rate per haploid genome to deleterious alleles for autosomal sites (a different expression must be used for X-linked sites, using the appropriate extension of **Equation 4.5**). If mutations are rare at individual sites, the number of mutations per diploid individual is $2\bar{n}$.

In order to estimate the value of $2\bar{n}$ for a species, we would need to know the abundance of mutations within the genome, and the proportion of those that are sufficiently strongly selected that their frequencies can be well predicted from the infinite-population formulae used here. We discuss this problem in detail in **Section 6.4.iv**, where we show how to combine the theory of selection and genetic drift with genomic data on the frequencies of amino acid variants in populations to obtain estimates of the selection coefficients against deleterious amino acid variants. The results imply values of about 700 deleterious amino acid mutations per individual in human populations, and around 10-fold more in *Drosophila*, mostly with very small selection coefficients (< 0.001). Deleterious mutations in noncoding sequences probably contribute even more than this. Since most of these mutations are rare, a different set of mutations is present in each individual in the population. There is thus an enormous number of slightly deleterious mutations segregating in a natural population of an outbreeding organism.

If the frequencies of variants at one site are independent of those at other sites (**Section 1.2.v.b**), there will be a Poisson distribution of the number of mutations per haploid genome (**Appendix A2.v.d**). This is because we can treat the number of mutations carried in a given haploid genome as a random draw from a binomial distribution with mean $\bar{n}$; if the total number of sites m is large compared with $\bar{n}$, the Poisson approximation to the binomial distribution will apply. The corresponding frequency distribution for diploid individuals with random mating is Poisson with mean $2\bar{n}$. We will use this result frequently later in the book.

An important implication of these results is that individuals completely free of deleterious mutations are very unlikely to be found in natural populations. With $2\bar{n} = 700$, the value of the zero term of the Poisson distribution is $\exp(-700) = 10^{-304}$, so that a population size of 10^{304} would be required for just one mutation-free individual to be produced. Even with $2\bar{n}$ as low as 100, a population size of 10^{43} would be needed.

4.2.ii.d. *Estimating mutation rates from equilibrium frequencies*
If estimates of selection coefficients are available, and the population is at equilibrium, **Equations 4.3**, **4.5**, or **4.7b** can be used to estimate mutation rates (**Problem 4.4**). Such estimates are often called *indirect estimates*. In practice, this method is usually limited to mutations with conspicuous phenotypic effects, which are likely to be only a subset of all mutations that affect a protein sequence, so it substantially underestimates overall mutation rates. The first such estimate was for hemophilia (Haldane 1935), and this approach still forms the basis for many of the published estimates of human mutation rates for genes causing diseases. It is especially useful for recessive X-linked disorders, since these are probably eliminated largely through their effects on male carriers. The results indicate that the net rate of mutation per gene per generation for such mutations is typically around 10^{-5} (Haldane 1949c; Vogel and Motul-

sky 1997, Chapter 9). This is in rough agreement with what is expected from the estimate of the human mutation rate from DNA sequence data (**Section 4.2.i** above).

4.3. GENETIC LOAD

4.3.i. Introduction

What is the overall effect of deleterious mutations on the mean fitness of the population? This question was first asked by Haldane (1937) and later explored in a famous paper by Muller (1950), who coined the term "load" for the reduction in fitness caused by the presence of deleterious mutations in a population. Genetic load was defined by Crow (1958) as the proportional reduction in mean fitness of the population below that for the genotype with the highest possible fitness, the *optimal genotype*:

$$L = \frac{\left(w_M - \bar{w}\right)}{w_M}$$

(4.8)

where w_M is the fitness of the optimal genotype, and $\bar{w}$ is the population mean fitness (note that the optimal genotype, such as an entirely mutation-free individual, may be so rare that it is never observed in the population).

The load effectively measures the fraction of the population that fails to survive or reproduce because of selective differences among its constituent genotypes; this is sometimes referred to as the amount of *selective death*. There must also always be environmental sources of death or reproductive failure, even for the optimal genotype. We will now derive expressions for the genetic load for several different ways in which variation in fitness can arise.

4.3.ii. Mutational load

4.3.ii.a. *Autosomal inheritance with random mating*
The load under mutation at a single nucleotide site in a large, randomly mating population can be found as follows (Haldane 1937). Mutations with heterozygous effects on fitness are mostly eliminated from the population as heterozygotes, as we saw in **Section 4.2.ii.a** above. Since the frequency of heterozygotes is approximately $2q$, the reduction in fitness to the population, measured relative to the fitness of wild-type homozygotes, is $L \approx 2qhs$. If the population is at equilibrium, we can use the value of q from **Equation 4.3** and obtain:

$$L \approx 2u$$

(4.9)

Similarly, for the case of completely recessive mutations (eliminated exclusively as homozygotes), we have (**Problem 4.5.i**):

$$L \approx u \qquad (4.10)$$

The difference between the two cases reflects the fact that a single "selective death" eliminates two mutations in the case of recessivity (when selection acts only on mutant homozygotes), but only one when heterozygotes are the main source of selective elimination, so that selection is twice as efficient with recessivity. The remarkable result that the load is independent of the selection coefficient can be understood as follows. Although weakly selected mutations cause fewer selective deaths among their carriers than strongly selected mutations, they rise to higher equilibrium frequencies. Provided that these frequencies are still sufficiently low that the relevant approximations are valid, the two effects exactly cancel out (Haldane 1937).

4.3.ii.b. *Other cases*

Equation 4.10 applies in the case of haploids. With X-linked inheritance, and equal selection on the two sexes ($s_f = s_m = s$), the load (averaging over males and females) is $3u/2$, exactly intermediate between the two autosomal cases (**Problem 4.5.ii**). If selection acts only on one sex, the load for this sex for nonrecessive mutations is $3u$, and the overall load is half of this value (**Problem 4.5.ii**).

In an inbreeding population with autosomal inheritance, **Equation 4.6** of **Section 4.3.ii.b** gives:

$$L \approx \frac{u\left[2(1 - f) + f\right]}{\left[h(1 - f) + f\right]} \qquad (4.11)$$

This shows that, for nonrecessive mutations, the equilibrium load decreases as the inbreeding coefficient increases, reflecting the more efficient elimination of mutations in the homozygous state. The lower the dominance coefficient h, the faster the rate of decline in the load; if h is small but nonzero, even a relatively small amount of inbreeding greatly reduces the load. If the inbreeding coefficient is close to 1, the load approaches the mutation rate, independently of the dominance coefficient (see **Section 4.3.ii.c** below).

4.3.ii.c. *Multiple sites*

The theory just outlined shows that, in all cases, the mutational load at a site is of the same order as the mutation rate. Since mutation rates per nucleotide site are extremely small in organisms other than RNA viruses and some mitochondrial genomes (**Section 1.3.iii.b**), it might at first sight seem that the mutational load is negligible. However, this ignores the fact that a genome usually includes a very large number of sites capable of producing deleterious variants (**Section**

4.2.i). To assess the genetic load, we thus need to determine its genome-wide value. This can easily be done by applying the assumption already used above, that mutations at different sites affect fitness independently of each other. In addition, we assume that they are distributed independently of each other in the population.

The first assumption is equivalent to the assumption of *multiplicative fitnesses*, as explained in **Box 4.4**. **Box 4.4** also shows how to obtain the mean fitness and the total load on the population. For nonrecessive autosomal muta-

---

Box 4.4 GENETIC LOAD DUE TO VARIATION AT MULTIPLE SITES

If mutations at different sites reduce the probability of survival or reproduction independently by an amount s_i for the ith site, the net fitness of an individual, relative to the fitness of a mutation-free individual, is the product of $(1 - s_i)$. This follows from the rule that the probability of an event caused by the co-occurrence of a set of independent events is given by the product of the probabilities of each event (**Appendix A2.ii**). More intuitively, the overall chance of successful survival or reproduction is like a race with many hurdles. The net probability of getting to the end of the course is the product of the chances of *not* falling at each successive hurdle encountered in the race.

The same principle can be applied to the mean fitness of the population, replacing s_i by the load L_i for the ith site:

$$\bar{w} = \left(1 - L_1\right)\left(1 - L_2\right)\cdots\left(1 - L_m\right) \qquad \text{(B4.4.1a)}$$

so that:

$$\ln \bar{w} = \sum_i \ln\left(1 - L_i\right) \qquad \text{(B4.4.1b)}$$

Since the individual loads are small, $\ln(1 - L_i)$ can be approximated by $-L_i$ (**Appendix A1.ii.c**), so that $\ln \bar{w}$ is approximately $-\Sigma_i L_i$. We therefore have:

$$\bar{w} \approx \exp -\sum_i L_i \qquad \text{(B4.4.2)}$$

The total load, L, is simply $1 - \bar{w}$.

tions in a randomly mating population, we can substitute from **Equation 4.9** into **Equation B4.4.2** to obtain the load:

$$L \approx 1 - \exp(-2U) \qquad (4.12a)$$

where $2U$ is the mean number of new deleterious mutations in a zygote (the genomic diploid deleterious mutation rate).

This approach can be extended to inbreeding populations by using **Equation 4.11** (Lande and Schemske 1985; Charlesworth and Charlesworth 1998):

$$L \approx 1 - \exp\left\{ -\frac{U\left[2(1-f)+f\right]}{\left[h(1-f)+f\right]} \right\} \qquad (4.12b)$$

Figure 4.7 displays the dependence of the load on the dominance coefficient and inbreeding coefficient for a genome-wide deleterious mutation rate of 0.1 per haploid genome.

These theoretical results raise the question of the magnitude of the mutational load in nature. As we saw in **Section 4.2.i**, $2U$ for humans is at least 0.58 per generation. **Equation 4.12a** then implies that the mean fitness of the population is less than 56% of that of a mutation-free individual (i.e., the load is greater than 44%). The mutational load places some limits on the size of the functional portion of the genome that an organism can sustain. The mean fit-

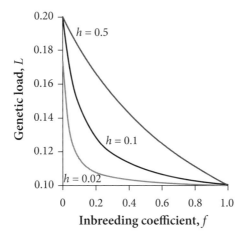

FIGURE 4.7 The predicted equilibrium genetic load, L, due to mutation in populations with different inbreeding coefficients, f, using **Equation 4.12b**. The mutation rate, U, is equal to 0.1, and results for three different values of the dominance coefficient, h, are shown. The loads for a randomly mating population and a completely inbred population are close to 0.2 and 0.1, respectively, independent of h (unless h is very close to 0). The rate of decline in L as f increases from 0 is greatest for mutations with the smallest h values.

ness from **Equation 4.12a** is an upper limit, since it considers only mutational load (and ignores both environmental sources of mortality or sterility, and also contributions from other kinds of genetic load; see below).

A population with two sexes must produce a mean of two offspring per individual for the population size to be stable. The absolute mean fitness of the population (the average number of zygotes contributed to the next generation by a new zygote) must therefore equal or exceed 2 if the population is to remain in existence. A load of 44% then implies that the absolute mean fitness for mutation-free individuals must be 3.6. If mutations in noncoding sequences are taken into account (**Section 4.2.i**), the load is much higher.

With the human reproductive capacity seen in hunter–gatherer societies, which represents the state under which our species evolved, this level of successful offspring production is only just sustainable. A value of 8 for the lifetime expected number of offspring for women at the start of their reproductive life is at the high end of the range (Howell 1976); this must be discounted by early-life mortality. Of course, strictly multiplicative fitness effects are unlikely, especially as selection may often be competitive, so that the proportion of successful individuals depends on the availability of the resources for which they are competing, rather than on their genes. Non-multiplicativity of the fitness effects of different mutations may considerably mitigate the problem of genetic load, as we discuss in **Section 10.2.iv.d**. Nevertheless, the long-term survival of populations of organisms with a functional genome much larger than the human genome seems unlikely, unless their mutation rate per nucleotide is much lower than the estimates above, or their reproductive capacity is much higher than for humans.

4.3.iii. Segregational load

Variability maintained by balancing selection also creates a genetic load, the *segregational load*. Consider the case of heterozygote advantage (**Section 2.1.ii.c**). The formula for the equilibrium variant frequencies (**Equation 2.3**) leads to the following expression for the load (**Problem 4.6**):

$$L = \frac{st}{(s + t)} \tag{4.13}$$

In contrast to the mutational load, the segregational load is of the same order of magnitude as the selection coefficients against the homozygotes (Morton et al. 1956). Unless selection is very weak, the load increases very rapidly with the number of polymorphisms maintained by balancing selection, if multiplicative fitnesses are assumed (**Problem 4.6.ii–iii**). It follows that a genome can have only a few sites subject to the intensity of balancing selection estimated for the malaria resistance polymorphisms discussed in **Section 2.1.ii.d**.

4.3.iv. The cost of selection

The fixation of a variant favored by natural selection also causes a load, called the "cost of selection" (Haldane 1957). At first sight this seems paradoxical; the spread of a favorable mutation increases the fitness of individuals and populations, unless selection is entirely competitive. However, a cost arises because the environment has changed, so that individuals with the initial genotype have low fitness. According to the equations we derived in **Sections 3.1.ii** and **3.1.iii**, adaptation to a novel environment takes a substantial amount of time. During this time, the mean fitness of the population is lower than when the favorable variant has become fixed. Maynard Smith (1976c) called this fitness difference the *lag load*: the loss in fitness due to the population lagging behind the change in environment (for an alternative way of looking at this, see Crow 1970). If the population is regulated by density-dependent mortality or fecundity, this reduction in fitness will imperil the survival of the population if the load becomes so high that the mean number of offspring per parent falls below 2 (see **Section 4.3.ii.c** above).

How large is this load? In the simple case of an autosomal variant with intermediate dominance in a randomly mating population, let the variant that is less fit in the changed environment be at frequency p, and have a homozygous fitness of $1 - s$ relative to a value of 1 for homozygotes for the favored mutation A_2 (frequency $q = 1 - p$). The lag load for this generation is $1 - \bar{w} = sp$. Haldane's *cost of selection*, C, is the sum of the lag loads over all generations while the favorable variant is on its way to fixation from its initial frequency, q_0. With weak selection, C is independent of the selection coefficient, and is approximately equal to $-2 \ln(q_0)$ (Haldane 1957; **Problem 4.7***). This is substantially above 1; for example, with $q_0 = 10^{-6}$, $C = 30$. C for a haploid population is half this value.

Biologically, C means that the number of individuals that suffer selective elimination while a favorable mutation is on its way to fixation is the product of C and the population size, and is thus many times the number of individuals alive in any one generation. If evolution could instantaneously switch the population from being nearly all A_1 to nearly all A_2, selective elimination would not occur. But natural selection is the only way for adaptation to occur, even though it entails a large toll in terms of selective deaths. If multiple favorable variants are in the process of spreading to fixation as the result of steady environmental change, the net lag load is approximated by $1 - \exp(-KC)$, where K is the rate per generation at which a new variant becomes fixed somewhere in the genome, i.e., the *rate of substitution* of favorable variants (Haldane 1957). Haldane suggested that a species could tolerate the substitution of a new favorable mutation about once every 300 generations, i.e., $K \approx 1/300$, on the grounds that a lag load of 10% would be tolerable when combined with other sources of death and sterility (**Problem 4.8**).

The observed value of K inferred from data on the rate of protein sequence evolution is far higher than this (see **Section 6.1.ii.e**), which led to the suggestion that natural selection cannot be the main cause of protein sequence evolution (Kimura 1968). Kimura proposed instead the *neutral theory* that provides the framework for much modern thinking about molecular evolution and variation (**Chapters 5** and **6**). Kimura's assumption of independent fitness effects of mutations was, however, soon criticized (Maynard Smith 1968), and his argument is now mainly of historical interest. Nevertheless, there must be limits to the number of independent traits that can be selected simultaneously, without a risk that the population will fail to reproduce in sufficient numbers to survive. Natural selection is an inherently inefficient process, and populations can go extinct in a rapidly changing environment, because their genetic composition does not track environmental changes sufficiently fast.

4.4. INBREEDING DEPRESSION

4.4.i. Introduction

As described in **Section 1.1.iii**, homozygosity caused by inbreeding is associated with reduced mean values of traits related to fitness, such as survival and fecundity: the phenomenon of *inbreeding depression*. The increased fitness of F1 hybrids between inbred lines compared with the lines themselves is called *heterosis*. Early in the history of genetics, two main alternative genetic theories were proposed to explain these phenomena (Wright 1977, Chapter 1; Charlesworth and Charlesworth 1987). The first possibility is that inbreeding depression and heterosis are caused by deleterious recessive or partially recessive alleles maintained in the population by mutation pressure: the "dominance" hypothesis. If a mutant allele has frequency q, homozygotes are present at frequency q^2 in a randomly mating population, but their expected frequency is q among completely homozygous lines derived from the population, e.g., by using the balancer chromosome technique available in some species of *Drosophila* (**Section 1.1.iii.b**). The ratio of inbred to outbred homozygote frequencies is $1/q$, which is very large if q is very small. Fitness can thus be greatly reduced by inbreeding.

The other possibility is that heterozygotes have higher fitness than homozygotes, so that alternative alleles are maintained in the population, as explained in **Section 2.1.ii.c**. Homozygosity caused by inbreeding will then reduce fitness. This is the "overdominance" hypothesis. While the relative contributions of these two possibilities to inbreeding depression are still uncertain, there must be a major effect of partially recessive, deleterious mutations (Charlesworth and Hughes 2000). In the next section, we describe the relevant theory and data. Readers who are not interested in the causes of genetic variation in fitness and other quantitative traits should skip to **Chapter 5**.

4.4.ii. Mutation and inbreeding depression in a randomly mating population

4.4.ii.a. *The effects of mutation and selection at a single site*
We first consider an autosomal site subject to mutation and selection in a randomly mating population, modeled in **Section 4.2.ii.a** above. Suppose that inbred individuals are experimentally created from individuals sampled randomly from this population. We will ignore terms involving the square of the mutant allele frequency q, and use **Equation 1.3** in **Section 1.3.ii.b** for the genotype frequencies in the inbred population. The mean fitness (relative to the fitness of the optimal genotype, w_M) of individuals with inbreeding coefficient f is then given by the approximate expression:

$$\bar{w}_I \approx 1 - 2qhs(1 - f) - qsf \approx 1 - 2qhs - q(1 - 2h)sf \qquad (4.14)$$

whereas we saw in **Section 4.3.ii.a** above that the corresponding fitness of the outbred population is $\bar{w}_O = 1 - 2qhs$. **Equation 4.14** shows that the inbred population will have reduced fitness if $h < 1/2$, i.e., the fitness of the heterozygote is closer to that of the wild-type homozygote than to the mutant homozygote. More generally, if alleles that reduce the trait value are on average partially recessive, the mean of any quantitative character will be reduced by inbreeding (Falconer and Mackay 1996, Chapter 14), independent of their frequencies in the population (**Problem 4.9**).

Providing that the mutant is not completely recessive in its effect on fitness (i.e., $h > 0$), we can use **Equation 4.3** of **Section 4.2.ii.a** to obtain **Equation 4.15**:

$$\bar{w}_O - \bar{w}_I \approx \frac{u(1 - 2h)f}{h} \qquad (4.15)$$

4.4.ii.b. *Multiple sites*
This result can be generalized to give the effect of inbreeding due to multiple sites across the genome, assuming independence among sites and multiplicative fitness interactions (**Box 4.5**). We obtain the following equations:

$$\frac{\bar{w}_I}{\bar{w}_O} \approx \exp\left(-Bf\right) \qquad (4.16a)$$

with

$$B = \frac{U\left(1 - 2h_H\right)}{h_H} \qquad (4.16b)$$

Box 4.5 INBREEDING EFFECTS DUE TO MUTATIONS AT MULTIPLE SITES IN THE GENOME

We assume independence between sites and multiplicative effects on fitness (**Box 4.4** of **Section 4.3.ii.c**). **Equation 4.15** implies that the mean fitness of inbred individuals (relative to the fitness of the optimal genotype), determined by the effects of mutations at m sites, is given by:

$$\bar{w}_I \approx \prod_{i=1}^{m} \left(1 - 2u_i - \frac{u_i\left(1 - 2h_i\right)f}{h_i} \right) \qquad \text{(B4.5.1)}$$

where the subscript i denotes the value of u or h for a particular site.

Using the same argument as in **Box 4.4**, this can be approximated further by:

$$\bar{w}_I \approx \exp -\sum_{i=1}^{m} u_i \left(2 + \frac{\left(1 - 2h_i\right)f}{h_i} \right) \qquad \text{(B4.5.2)}$$

The first term on the right-hand side, $\exp(-2\Sigma_i u_i)$, gives the mean fitness of the outbred population, $\bar{w}_O$ (see **Equation 4.12a** of **Section 4.3.ii.c**). If u and h are independent of each other, we obtain **Equation 4.16a** and **Equation 4.16b** of the main text.

If we are dealing with a component of fitness, z, a homozygous mutation at site i can be assumed to reduce z relative to the wild-type value z_M by an amount $c_{zi}s_i$, where s_i is the selection coefficient for mutant homozygotes (see text). The equilibrium frequency of the mutation is the same as before, so that the bracketed term in **Equation B4.5.1** is replaced by:

$$1 - 2u_i c_{zi} - \frac{u_i c_{zi}\left(1 - 2h_i\right)}{h_i} \qquad \text{(B4.5.3)}$$

It is possible that only a fraction P_z of deleterious mutations affect z, so that $c_{zi} = 0$ at the remaining $m(1 - P_z)$ sites. We can disregard sites without effects on z, and write c_z for the mean value of c_{zi} for the sites where z is affected by mutations. Provided that c_{zi}, u_i and h_i are independent, we obtain **Equation 4.16d** of the main text, using the argument that yielded **Equation 4.16b**.

where h_H is the harmonic mean of the dominance coefficients across sites and U is the total mutation rate per haploid genome for mutations affecting fitness.

We usually cannot measure fitness directly, but instead work with *components of fitness*, i.e., traits such as juvenile survival or female fecundity, where a decrease in trait value caused by a mutation is likely to be associated with a decrease in fitness. The inbreeding depression for a given component of fitness, represented by a variable z, can be modeled as follows. Let the effect on z of a mutation at a particular site i in the genome be related to its effect on fitness by a coefficient of proportionality c_{zi} (Mukai et al. 1972), i.e., a change in z is related to the change in fitness δw by $\delta z_i = c_{zi} \delta w$ (where $c_{zi} \leq 1$). As shown in **Box 4.5**, we can obtain the equivalent of **Equation 4.16a** for the effect of inbreeding on z:

$$\frac{\bar{z}_I}{\bar{z}_O} \approx \exp{-B_z f} \qquad (4.16c)$$

in which B_z is given by:

$$B_z = \frac{UP_z c_z \left(1 - 2h_H\right)}{h_H} \qquad (4.16d)$$

where P_z is the proportion of sites at which deleterious mutations affect z (i.e., for which $c_{zi} > 0$), and c_z is the mean of c_{zi} over such sites; as before, U is the genome-wide deleterious mutation rate).

For a given value of f, the inbreeding depression for a trait is often measured by $(\bar{z}_O - \bar{z}_I) / \bar{z}_O$, usually represented by the symbol δ. **Equations 4.16c** and **4.16d** suggest that the absolute value of the natural logarithm of $\bar{z}_I / \bar{z}_O$ is also a useful measure of the effect of inbreeding. This is equal to $B_z f$, where B_z is the slope of $-\ln(\bar{z}_I)$ with respect to f (Morton et al. 1956). The intercept, A_z, of this line at $f = 0$ is $-\ln(\bar{z}_O)$, which is equal to $2UP_z c_z + C_z$, where the term C_z is added to represent the reduction of $\ln(\bar{z}_O)$ below its maximum value due to the combined effects of environmental factors and any genetic effects that are not included in the model.

4.4.ii.c. *The inbreeding load*

B_z is often called the *inbreeding load*, and is probably the most useful measure of the effect of inbreeding, since it is independent of the values of f used for estimating inbreeding effects. Some examples are given in **Table 4.1**. In some species of *Drosophila*, balancer chromosomes can be used to assess the relative egg-to-adult viabilities of balancer heterozygotes and homozygotes for chromosomes extracted from natural populations, by the method described in **Section 1.1.iii.b**. This allows the inbreeding load due to homozygous lethal chromosomes to be distinguished from that due to nonlethal *detrimental* chromosomes, by comparing the mean viability of heterozygotes for pairs of independently isolated chromosomes with the mean viability of homozygotes carrying nonlethal chromosomes. The *B* values for lethals and for the viability effects of detrimental autosomes are approximately equal. We can also use the fitness estimation method described in **Section 2.1.ii.c** to estimate *B* for the net fitness effects of homozygous nonlethal chromosomes under laboratory conditions, and it is much larger than the value for viability alone (**Table 4.1** in **Section 4.4.iv**). It is currently unknown whether this is true for other organisms, where such genetic manipulations are impossible.

When dealing with survival probabilities, the quantities *A*, *B*, and *C* defined in the previous section are often used to express the load as the number of genetic deaths that would be caused by the mutations carried by a typical gamete when dispersed among separate individuals; this is given by $A + B - C$ (Morton et al. 1956). For a trait involving survival, this quantity is often referred to as the *number of lethal equivalents* carried by a typical gamete (Morton et al. 1956). Given the difficulties in estimating $A - C$ (see below), *B* is frequently used as a lower bound estimate of the number of lethal equivalents. The utility of these measures depends on the assumption of multiplicative effects; there is conflicting evidence on the validity of this assumption, as we discuss in **Section 10.2.vi.b**. As with genetic load, the inbreeding depression and inbreeding load are independent of the selection coefficients. They are strongly affected by the dominance coefficients, however, as well by the strength of the relationship between the fitness component and net fitness.

Finally, **Equations 4.5** and **4.6** in **Section 4.2.ii.b** lead straightforwardly to expressions for the inbreeding depression for X-linked mutations and in inbreeding populations. The last topic will be discussed further in **Section 9.2.ii.d** in relation to the evolution of mating systems. We will not discuss the details here, but simply note that inbreeding depression should usually be lower for the X chromosome, reflecting the reduced equilibrium frequencies of deleterious mutations compared with autosomes (Crow 1970). Studies on *D. melanogaster* have indeed found much smaller values of *B* for the X chromosome than the autosomes, especially for egg-to-adult viability (Wilton and Sved 1979; Eanes et al. 1985), consistent with this expectation.

4.4.iii. Heterozygote advantage and inbreeding depression

It is simple to derive the equivalent of **Equation 4.14** in **Section 4.4.ii.a** for the case of heterozygote advantage (**Problem 4.10**). By combining this with **Equation 4.13** in **Section 4.3.iii** for the segregational load, we obtain the following expression for *A* and *B* for a trait *z* when there are multiple sites with pairs of alleles maintained by heterozygote advantage, assuming that the components of fitness themselves show heterozygote advantage:

$$A_z - C_z = B_z = \sum_i \frac{s_i t_i c_{zi}}{\left(s_i + t_i\right)} \qquad (4.17)$$

This equation shows that the genetic load in an outbred population caused by heterozygote advantage is equal to the inbreeding load (Morton et al. 1956).

4.4.iv. Testing alternative explanations of inbreeding depression

In this section and the following one, we discuss how to discriminate among the possible causes of inbreeding depression and genetic variation in quantitative traits. There are as yet no conclusive answers to these questions (probably because both types of genetic load exist), and we shall only briefly review the reasoning, and describe some of the evidence, which mainly comes from data on *Drosophila*.

4.4.iv.a. *Use of B/A ratios*

The results derived in **Sections 4.4.ii.b** and **4.4.iii** suggest that the value of the ratio $B/(A - C)$ might distinguish between the dominance and heterozygote advantage theories of inbreeding depression. The basis for this is that, if there is heterozygote advantage for a fitness component, values above 1 are unlikely, but such values are quite possible with a mutational source of inbreeding depression, provided that h is sufficiently small (Morton et al. 1956). It is, however, not easy to estimate the reduction, $A - C$, in the log value of a trait for an outbred population below its maximum, since C is usually unknown. The upper limit for survival probability, however, is 1. An upper bound on $A - C$ for a survival trait is thus obtained from the natural logarithm of the observed survival probability of outbred individuals.

Several long-term studies of natural populations of birds and mammals have yielded $A - C$ and B estimates using this approach. High B values are often found for survival early in life, and A values are frequently much smaller than B (Keller 1998; Keller and Waller 2002). One of the largest such data sets (on a Swedish population of collared flycatchers) yielded a B value of 7.5 for the chance of survival to breeding age (Kruuk et al. 2002). The survival probability of non-inbred individuals was 0.125, giving $A = \ln(0.125) = 2.1$. Given that this must include a substantial contribution from environmental sources of mortality, this suggests that $B/(A - C) \gg 1$ for this trait.

These examples show that $B/(A - C)$ for natural populations can greatly exceed 1, as was originally proposed for humans on the basis of rather limited

TABLE 4.1 Inbreeding loads in a number of laboratory and field studies

A. Animals

Species	Natural mating system	Method	Character(s) studied	Inbreeding coefficients	Outbred measure	Inbred measure	B	Source
Drosophila melanogaster	Outcrossing	Balancer chromosomes	Egg-to-adult viability for chromosome 2 detrimental mutations	0, 1	1	0.79	0.23	Simmons and Crow (1977)
			Egg-to-adult viability for chromosome 2 lethal mutations	0, 1	1	0.78	0.25	
			Net fitness for chromosome 2	0, 1	1	0.16	1.8	Sved (1971)
		Sib mating	Productivity of a mating	0, 0.25	108	62.7	2.18	Dahlgaard and Hoffmann (2000)
House mouse (*Mus musculus*)	Outcrossing	Sib mating	Survival of pups sired in semi-natural enclosures	0, 0.25	0.75	0.64	0.63	Meagher et al. (2000)
Collared flycatcher (*Ficedula albicollis*)	Outcrossing	Sib mating, parent–offspring	Recruits to the breeding population per nest	0, 0.25	0.60	0.12	7.5	Kruuk et al. (2002)

B. Plants

Species	Natural mating system	Method	Character(s) studied	Inbreeding coefficients	Outbred measure	Inbred measure	B	Source
Scots pine (*Pinus sylvestris*) (Conifer)	Outcrossing	Open-pollination versus hand self-fertilization	Seed set	0, 0.5	0.71	0.18	2.74	Koelewijn et al. (1999)
Arenaria uniflora (Flowering plant)	Outcrossing (selfing-rate 0.3)	Self-fertilization and cross-fertilization of flowers on the same plants	Fecundity (total flowers)	0, 0.5	47	38	0.42	Fishman (2001)
Arenaria uniflora	Inbreeding (selfing-rate ~ 1)	As above	Fecundity	0, 0.5	24	23	0.08	Fishman (2001)

data (Morton et al. 1956). For human populations, most data on the effects of inbreeding come from studies of first-cousin marriages, where the inbreeding coefficient f is only 1/16, so that the estimates are noisy. There are also biases due to social differences between groups with different levels of inbreeding, which are hard to eliminate. A meta-analysis of 38 studies of early life survival (Bittles and Neel 1994) suggested that inbred individuals tend to survive worse than outbred individuals, and that the difference in mortality rates is roughly independent of the outbred mortality (**Figure 4.8**). The overall estimate of B from this mortality difference is 0.70 (Bittles and Neel 1994). Using life tables for the 1964 population of the United States of America (Keyfitz 1968), in order to infer mortality rates when most environmental sources of mortality have been removed, A for survival to age 20 is about 0.04, giving $B/(A - C) \geq 17.5$.

However, such high $B/(A - C)$ ratios, while definitively ruling out heterozygote advantage at the level of the fitness components themselves, do not necessarily imply that variation contributing to inbreeding depression is maintained by mutation alone. As we discussed in **Section 2.2.iv**, one way in which heterozygote advantage for fitness can arise is when alternative variants at a site

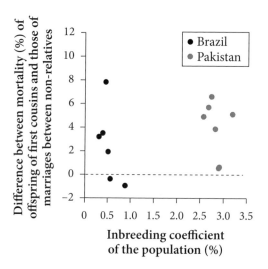

FIGURE 4.8 The increased mortality caused by inbreeding in humans. The figure shows the effect of inbreeding on percentage mortality values of individuals from populations in two countries. In the set from each country, offspring from parents who were first cousins were compared with those whose parents were not known to be related. The populations span a range of inbreeding coefficients. For the data from Brazil, mortality for the first 21 years was recorded, and for the data from Pakistan, mortality was for the first 10 years of life. Despite the different inbreeding coefficients, the more inbred offspring suffered higher mortality in both countries. [From data in Bittles and Neel (1994).]

have opposite effects on two different fitness components, with their deleterious fitness effects being at least partially recessive. This type of model can produce values of $B/(A - C)$ much higher than 1, for a given fitness component, although not for fitness itself or for all the fitness components affected by a set of antagonistically pleiotropic loci. This is because the assumption of antagonistically pleiotropic effects implies that the high frequency allele at a locus is favorable for some subset of traits, but is deleterious for others. Consistently high B/A ratios across fitness components therefore suggest a predominance of mutational contributions to inbreeding depression.

4.4.iv.b. *Alternative approaches*

The best way to resolve this question would be to compare the observed value of B with that predicted from **Equations 4.16a–d** of **Section 4.4.ii.b** above, using estimates of the mutation rate and the harmonic mean dominance coefficient. To date, the closest approximation to this comes from work on *D. melanogaster*. There is little difficulty in accounting for the lethal contribution to the inbred load in *Drosophila* purely by mutations, since lethal mutations are present only at low frequencies (**Section 1.iii.b**; Crow and Temin 1964). The evidence concerning detrimental mutations is less clear cut, as we show next.

Case Study
Estimates of Genome-wide Detrimental Mutation Rates and Dominance Coefficients

For detrimental mutations, mutation rates and selection coefficients have been estimated from mutation accumulation (MA) experiments in a number of different organisms, using the breeding scheme described in **Section 1.3.iii.b**, starting with Mukai (1964) and Mukai et al. (1972); see Charlesworth et al. (2004) and Avila et al. (2006) for overviews of work on *D. melanogaster*, and Keightley and Eyre-Walker (1999) for a more general review. The principle of the method is described in **Box 4.6**. Most work on *Drosophila* has concentrated on the viability from egg to adult of homozygotes for mildly detrimental chromosomes (those with at least 50% of normal viability when homozygous), assayed by using balancer chromosomes to clone chromosomes from the MA lines (**Section 1.1.iii.b**).

The results from different MA studies with *D. melanogaster* have unfortunately yielded very disparate values of U for viability, ranging from a few percent to close to 0.5. On the basis of recent experiments, however, estimates of U percent in the range 0.06–0.15 for the genome as a whole were proposed by Charlesworth et al. (2004). This corresponds to values of 0.022–0.055 for chromosome 2, which represents about 37% of the *D. melanogaster* euchromatic genome (Misra et al. 2002).

These experiments also provide estimates of the arithmetic mean of h, obtained from the regression coefficient of the viabilities of heterozygotes for two different chromosomes on the means of the viabilities of the corresponding homozygotes (Mukai 1969; Crow 1993; García-Dorado and Caballero 2000). García-Dorado and Caballero (2000) estimated a value of 0.2 for the mean h for mildly detrimental chromosomes. Using the above U estimates for the second chromosome, this yields $B = 0.07–0.16$, which is reasonably consistent with the observed B value of 0.15 for mildly detrimental second chromosomes (Crow 1993), especially as the harmonic mean of h is likely to be considerably lower than 0.2 if there is a wide distribution of h values.

Mutation accumulation experiments are likely to miss contributions from mutations with very small effects. The best method for estimating U for all mutations affecting fitness is to use DNA sequence comparisons between species, which allow estimates of the proportion of mutations that are sufficiently deleterious for selection to prevent them from becoming fixed by genetic drift (Kondrashov and Crow 1993), based on the principles we describe in **Section 6.2.iii.a**. Combined with direct estimates of the mutation rate in DNA sequences from mutation accumulation lines, we can then estimate the genome-wide mutation rate for mutations with evolutionarily significant deleterious effects. This approach has yielded a genome-wide deleterious mutation rate estimate for *D. melanogaster* of about 0.6 (Haag-Liautard et al. 2007), implying a value of 0.22 for chromosome 2. Using $h_H = 0.2$, we find that the predicted B value for net fitness is 0.66, while the observed value is 1.8 (**Table 4.1**). As already noted, the harmonic mean of h is likely to be much smaller than 0.2, so that the discrepancy may not mean much (with $h_H = 0.1$, the predicted B value is 1.98). Overall, these data suggest that variability caused by deleterious mutations contributes much, but possibly not all, of the inbreeding load in *Drosophila*.

Box 4.6 CHANGES IN MEAN AND VARIANCE IN MUTATION ACCUMULATION EXPERIMENTS

These experiments involve either maintaining a large number of sib-mated lines, started from an initial, highly inbred base stock (**Section 1.3.iii.b**), or the use of balancer chromosomes to create sets of lines with *D. melanogaster* X, 2nd or 3rd chromosomes derived from an initial single wild-type chromosome (**Figure 1.7** of **Section 1.1.iii.b**). With the second method, lines are maintained by backcrossing males heterozygous for a wild-type chromosome and the balancer to females from the balancer stock, with a single male used to create each generation of a given line. In either type of experiment, mutations rapidly become fixed within lines; a given line thus carries mutations derived from a single haploid genome, whose effects are measured in the homozygous state. In sets of sib-mated lines, mutations that occur in the whole genome contribute to changes in the mean and variance across lines, whereas with a chromosome extracted by a balancer, only mutations on the extracted chromosome contribute.

In general, we expect a wide distribution of magnitudes of effects for different mutations, and so the variance in these effects must be taken into account. For a fitness component z, we can write the reduction in trait value due to a mutation at site i as a_{zi}. In general, there must be a probability distribution of effects across different sites, with mean $\bar{a}_z$ and variance V_{az}. The change in mean per generation over lines is:

$$DM_z = \sum_i u_i a_{zi} = U_z \bar{a}_z \qquad (B4.6.1)$$

where U_z is the net mutation rate for fitness component z.

The rate of increase in the variance of line means is:

$$DV_z = \sum_i u_i a_{zi}^2 = U_z\left(\bar{a}_z^2 + V_{az}\right) \qquad (B4.6.2)$$

where U_z is the mean number of new mutations per haploid genome per generation that affect z (Mukai et al. 1972). Combining these equations and rearranging, we obtain the Bateman–Mukai relations:

$$U_z \geq \frac{DM_z^2}{DV_z} \ , \ \bar{a}_z \leq \frac{DV_z}{DM_z} \qquad (B4.6.3)$$

By using estimates of the rates of change in mean and variance in mutation–accumulation experiments, a lower bound to the mutation rate U_z and an upper bound to the mean reduction in trait value $\bar{a}_z$ can thus be estimated. More complex methods of estimation have been proposed, but this is the one most widely used: see García-Dorado and Gallego (2003) and Keightley (2004) for discussions of the merits of different methods.

4.4.v. Does dominance evolve?

As we saw in **Sections 4.4.ii** and **4.4.iv**, the experimental evidence from *Drosophila* shows that most deleterious mutations have partially recessive effects on fitness, with an average fitness reduction for heterozygotes for lethal mutations and wild-type alleles of 2–3% (i.e., $h = 0.02$–0.03), and a much larger value (about 0.2) for detrimental mutations. In **Section 2.2.iv**, we also mentioned the nearly complete recessivity of mutations with major phenotypic effects. Fisher (1928, 1930b, Chapter 3, 1931) first suggested that the prevalence of dominance of wild-type alleles over most mutant alleles requires an evolutionary explanation.

He proposed that natural selection could act on the level of dominance of deleterious mutations at a locus in the following way. With random mating, deleterious mutations will mostly be heterozygous with wild-type alleles, and the relative effects of the wild-type and mutant alleles in heterozygotes could potentially be modified by other loci, i.e., by variants at *modifier* loci that change the dominance coefficient, h, at a gene undergoing mutation and selection (these variants may otherwise be neutral). Since reducing h increases the fitness of the mutant heterozygotes, which is equal to $1 - hs$, modifier variants that increase the dominance of wild-type alleles over mutant ones will be advantageous. Fisher provided empirical evidence for the existence of genetic variation that affects dominance (Fisher 1938); other evidence is reviewed by Sved and Mayo (1970) and Mayo and Bürger (1997).

Fisher's reasoning is unquestionably correct, but overlooks an important difficulty. Wright (1929, 1934) pointed out that the frequency of heterozygous carriers of rare mutations at a locus is approximately $2q^* = 2u/(hs)$ (from **Equation 4.3** in **Section 4.2.ii.a**). The fitness advantage of a modifier that makes the fitness effects of a heterozygous mutation completely recessive is hs. If modifiers are inherited independently of the locus under selection, which is a reasonable assumption, their fitness advantage will thus equal the product of the two quantities above: $2q^*hs = 2u$. In other words, the selective advantage is only about twice the mutation rate. If the modifiers act on a gene-by-gene basis, as Fisher proposed, then u will usually be around 10^{-5} (the estimated deleterious mutation rate per gene: see **Section 4.2.i** above). Wright suggested that such a small selective advantage would be unlikely to overcome other fitness effects of the modifier; even in the absence of such effects, the modifier would behave nearly neutrally unless the population size is bigger than 10^5 (see **Sections 5.3.iii.e** and **6.2.iii.a** for the basis for this).

This argument is not conclusive, however, since species-wide population sizes are often large enough for such small selection coefficients to be effective (see **Section 5.2.iii.e**). But two other arguments cast doubt on Fisher's theory. First, as just mentioned, the evidence from *Drosophila* shows that detrimental mutations are less recessive (have larger h values) than lethals, which is not predicted by Fisher's model (Charlesworth 1979c). More strikingly, Orr (1991) compiled data on the phenotypic effects of mutations in the single-celled green alga, *Chlamydomonas reinhardtii*. This species has only a transient diploid resting stage in its life cycle, and many of the mutants listed by Orr are expressed only in the haploid stage. Nevertheless, when made heterozygous with wild-type alleles in diploids, most mutations were recessive.

This evidence strongly suggests that recessive phenotypic effects of rare mutations do not result from selection on dominance modifiers. How, then, can they be explained? As first proposed by Wright (1934), and elaborated by Kacser and Burns (1981), dominance may reflect nonlinear relations between the output of developmental or biochemical pathways and reductions in the level of expression or activity of the proteins involved in individual steps of the pathways, as shown in **Figure 2.6** of **Section 2.2.iv**. As can be seen from this figure, a 50% reduction in the activity of one of the components, corresponding to loss of activity of one of the two alleles at a locus, may reduce the amount of final product by only a small amount, whereas the homozygotes for a complete loss-of-function mutation have drastically reduced values. This implies nearly complete recessivity of loss-of-function mutations. In contrast, heterozygotes for a mutation that reduces activity only slightly are much closer to being intermediate between the wild-type and mutant homozygotes, consistent with what is observed for detrimental mutations. There is thus no reason to appeal to selection on dominance modifiers to explain the dominance of wild-type over loss-of-function mutations, although the properties of the pathways themselves may be the outcome of selection (reviewed by Keightley 1996).

4.5. GENETIC VARIATION IN FITNESS COMPONENTS AND OTHER QUANTITATIVE TRAITS

Another way of thinking about the population consequences of mutation and other processes that maintain variation is to determine the amount of genetic variance that they create in fitness components and other quantitative traits. Theoretical predictions can then be compared with experimental estimates.

4.5.i. Mutational variation

4.5.i.a. *Direct effects of mutations on fitness*

Mutation–selection balance may be an important mechanism maintaining genetic variability in all types of quantitative traits. **Box 4.7** shows how to calculate the equilibrium genetic variance in a trait z in a large, randomly mating population, contributed by a single autosomal site subject to mutation and

Box 4.7 GENETIC VARIANCE DUE TO DELETERIOUS MUTATIONS

Under the model described in the text, the only two genotypes that need to be considered are A_1A_1 (fitness 1, frequency $1 - q_i^*$) and A_1A_2 (fitness $1 - hs$, frequency $2q_i^*$), since the frequency of A_2A_2 is negligible under random mating. The variance in fitness is the same as the variance in the fitness deviations from 1. Using **Equation 4.3** of **Section 4.2.ii.a** for q_i^*, the variance is thus approximated by:

$$\frac{2u_i\left(h_is_ic_{zi}\right)^2}{h_is_i} - (2u_i)^2 \approx 2u_ih_is_ic_{zi}^2 \qquad \text{(B4.7.1)}$$

where terms in u_i^2 have been neglected in the approximation.

Using the definition of average effect α_i in **Box 3.5** of **Section 3.3.ii.b**, we have:

$$\alpha_i = [q_i^*h_is_i - (1 - q_i^*)h_is_i - q_i^*s_i]c_{zi} \approx -h_is_ic_{zi}$$

From **Equation B3.5.4** in **Section 3.3.ii.b**, after neglecting second-order terms in q_i^*, the additive variance contributed by site i is given by:

$$V_{Ai} \approx 2p_i^*q_i^*\alpha_i^2 \approx 2u_ih_is_ic_{zi}^2 \qquad \text{(B4.7.2)}$$

The genetic variance and additive genetic variance are therefore approximately the same, showing that the dominance variance is negligible in this case (Mukai et al. 1972).

selection. This uses the simple model introduced in **Section 4.4.ii.b** for relating the effect of a mutation on a fitness component to its effect on total fitness. As shown in **Box 4.7**, the dominance variance is negligible, even if $h \ll 0.5$, and so the genetic variance is approximately equal to the additive genetic variance (**Section 3.3.ii.b**). Perhaps counterintuitively, the effective absence of dominance variance is quite consistent with the existence of inbreeding depression contributed by rare, recessive or partially deleterious mutations, derived in **Section 4.4.ii.b**. This reflects the fact that the dominance variance contributed by a segregating site is proportional to the square of the product of the variant frequencies (i.e, to p^2q^2) (Falconer and Mackay 1996, p. 126), whereas the additive variance is proportional to pq. Rare variants thus contribute little to dominance variance relative to additive variance, even with complete recessivity.

If variants at different sites act additively or multiplicatively, and the variance at each site is small, the genetic variance is approximately:

$$V_{Az} \approx 2\sum_i u_i h_i s_i c_{zi}^2 \qquad (4.18)$$

This result is quite general. It applies both to components of fitness (as defined in **Section 4.4.ii.b**) and also to any trait for which mutations affect fitness, sometimes increasing it, and sometimes decreasing it. In the second case, there is no consistent relation between trait value and fitness. Individuals with extreme trait values are, however, likely to carry larger than average numbers of deleterious mutations, and hence suffer reduced fitness, giving the appearance of stabilizing selection (Kondrashov and Turelli 1992; Johnson and Barton 2005). This is discussed further in **Section 4.5.i.b** below.

This model predicts that most genetic variance will be additive, even for fitness components, since the underlying variants are at low frequencies. This prediction seems often to be verified (Houle 1992; Hill et al. 2008), although there are examples of fitness components with nonadditive variance (Crnokrak and Roff 1995; Charlesworth and Hughes 2000), suggesting that there are nonmutational contributions to variability in some cases.

Another prediction is that the additive genetic variance for a fitness component should be related to the rate of decline in trait mean under mutation pressure, DM_z (Burt 1995). This follows from the fact that, at equilibrium between mutation and selection, the mutational decline must balance the increase due to selection, which is equal to the additive genetic covariance between the trait and fitness (**Section 3.3.iii.a**). With the model used above, the covariance is given by:

$$C_{Az} \approx 2\sum_i u_i h_i s_i c_{zi} \qquad (4.19)$$

Given that the c_{zi} values are always less than 1, C_{Az} is greater than the additive genetic variance for the trait, given by **Equation 4.18**. Both DM_z and the additive genetic variance can be estimated experimentally, providing a way of testing the hypothesis that genetic variance in fitness traits is maintained by mutation pressure. For egg-to-adult viability in *D. melanogaster*, DM_z for chromosome 2 homozygotes is less than 0.005 per generation (Charlesworth et al. 2004). From the information on h reviewed in **Section 4.4.iv.b** above, the corresponding value for heterozygous carriers must therefore be about 0.001.

The additive genetic variance for viability should thus be substantially less than 0.001. Extensive measurements have been made of the additive genetic variance for egg-to-adult viability among heterozygotes for pairs of nonlethal second chromosomes extracted from natural *D. melanogaster* populations

(Mukai 1988); the estimates are always at least 0.002, and are much higher for some populations. This suggests that more than 50% of the additive genetic variance in fitness components such as viability is derived from sources other than mutation, contrasting with what was found for inbreeding depression. There is no contradiction between these conclusions, however, since we are comparing different kinds of variables. Inbreeding depression requires h to be less than 0.5 on average for variants that reduce trait values, or for there to be heterozygote advantage. There is no such requirement for variation maintained by other forms of balancing selection, e.g., with frequency-dependent selection, variation can be maintained in the complete absence of dominance (**Section 2.2.i**). Loci that are under balancing selection, and that contribute to additive variance in a fitness component, may thus contribute little or nothing to inbreeding depression in that trait.

4.5.i.b. *Mutation and stabilizing selection*

Another way that quantitative trait variation can be maintained under mutation pressure is by *stabilizing selection* on the trait. Stabilizing selection means that individuals with intermediate trait values have the highest fitness, with fitness falling off on either side of the optimal value, conventionally denoted by θ_z. A convenient and plausible way of representing the fitness w_z of individuals with trait value z is by a formula similar to a normal distribution function, the *nor-optimal* fitness model:

$$w_z = \exp- \frac{\left(z - \theta_z\right)^2}{2V_S} \qquad (4.20)$$

where V_S is an inverse measure of the intensity of stabilizing selection (i.e., a large value of V_S causes the fitness function to be very wide: **Figure 4.9**). This model has the useful property that a trait that is normally distributed before selection is also normally distributed after selection (**Box 4.8***).

If the trait is normally distributed, with mean $\bar{z}$ in a given generation, the results in **Box 4.8*** show that the change in mean over one generation is given by:

$$\Delta\bar{z} = \frac{\left(\theta_z - \bar{z}\right)V_z h^2}{\left(V_z + V_S\right)} \qquad (4.21)$$

where V_z is the phenotypic variance of the trait and h^2 is its heritability (**Section 3.3.iii.a**).

Equation 4.21 shows that, if h^2 is in the typical range of 0.2–0.8 (**Section 1.1.ii**), $\bar{z}$ will converge rapidly towards θ_z, provided that the environment is

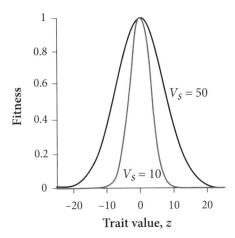

FIGURE 4.9 The nor-optimal fitness model. The figure plots the equation for fitness (w_z in **Equation 4.20**) for two values of the intensity of stabilizing selection, V_S; selection is assumed to be symmetrical, i.e., the optimum trait value, θ_z, is equal to 0.

constant so that θ_z remains unchanged. This reinforces the conclusion reached in **Section 3.3.iii.c**, that selection on quantitative traits is very effective in causing rapid evolutionary change.

Selection within a generation reduces the value of V_z (**Box 4.8***). Haldane (1954) used this result to estimate the genetic load $(1 - \bar{w})$ caused by selection on a quantitative trait, applying the equivalent of **Equation B4.8.4** to data on the means and variances of traits before and after selection, obtained in the pioneering studies of natural selection on quantitative traits by Weldon (1901), di Cesnola (1907), and Rendel (1943). He concluded that these gave a load of approximately 0.10.

This raises the question of how this reduction in variance can be balanced by mutation. This can be examined by assuming that the trait mean has converged to θ_z, so that only stabilizing selection is acting. At first sight, one might assume that the equilibrium genetic variance could be obtained by equating the change in variance to the mutational variance, DV_z. However, this overlooks the fact that the extent to which the change in variance caused by selection is transmitted to the next generation depends on both the genetic basis of the trait and the population's mating system. Very different results apply to asexual populations (in which the whole genome is inherited as a unit), and to sexual populations with recombination among sites that affect the trait (Lande 1976; Turelli 1984; Johnson and Barton 2005).

Box 4.8* CHANGES IN MEAN AND VARIANCE UNDER STABILIZING SELECTION

We use the nor-optimal model of **Equation 4.20** and assume that z is normally distributed with mean $\bar{z}$ and variance V_z (**Appendix A2.v.e**). We will determine the probability density of z after selection within a generation. For this purpose, it is convenient to measure z as its deviation from the optimum value, θ_z (which is equivalent to setting $\theta_z = 0$). The post-selection probability density of z is proportional to the product of the fitness function and the probability density before selection; this product is:

$$\phi'(z) = \frac{1}{\sqrt{2\pi V_z}} \exp\left(-\frac{z^2}{2V_S} - \frac{(z - \bar{z})^2}{2V_z}\right) \quad \text{(B4.8.1)}$$

Writing $y = (z - \bar{z})\sqrt{(V_S + V_z) / V_S V_z}$, we have:

$$\phi'(z) = \frac{1}{\sqrt{2\pi V_z}} \exp\left(-\frac{\bar{z}^2}{2V_S} - \frac{y\bar{z}\sqrt{V_z}}{\sqrt{(V_S + V_z)}} - \frac{1}{2}y^2\right) \quad \text{(B4.8.2)}$$

The post-selection probability density is obtained by dividing this expression by its integral over the entire range of z, which gives the mean fitness of the population. Noting that $dz = dy = \sqrt{V_S V_z}/\sqrt{(V_S + V_z)}$, we have:

$$\bar{w} = \sqrt{\frac{V_S}{(2\pi)(V_S + V_z)}} \exp\left(-\frac{\bar{z}^2}{2V_S}\right) \int_{-\infty}^{\infty} \exp\left(-\frac{y\bar{z}\sqrt{V_z}}{\sqrt{V_S(V_S + V_z)}} - \frac{1}{2}y^2\right) dy \quad \text{(B4.8.3)}$$

The integral can be simplified by writing $u = y - \bar{z}\sqrt{V_z} / \sqrt{V_S(V_S + V_z)}$ in the first term in the exponent in the integral. This yields:

$$\bar{w} = \sqrt{\frac{V_S}{(2\pi)(V_S + V_z)}} \left[\exp-\left(-\frac{\bar{z}^2}{2V_S} - \frac{\bar{z}^2 V_z}{2V_S(V_S + V_z)}\right)\right] \int_{-\infty}^{\infty} \exp\left(\frac{1}{2}u^2\right) du$$

The integral divided by $\sqrt{2\pi}$ is the integral of the standardized normal distribution (with mean 0 and variance 1), which is equal to 1. Simplifying the remaining terms and returning to the original scale with arbitrary optimal value θ_z, we obtain:

$$\bar{w} = \sqrt{\frac{V_S}{\left(V_S + V_Z\right)}} \, \exp\left(-\frac{\left(\bar{z} - \theta_z\right)^2}{2\left(V_S + V_Z\right)} \right) \qquad \text{(B4.8.4)}$$

This can be combined with **Equation B4.8.1** to obtain the post-selection probability density of z. This in turn can be used to derive expressions for the mean and higher moments of the post-selection distribution, by use of its moment-generating function (**Appendix A2.iii**):

$$M(\xi) = \bar{w}^{-1} \int_{-\infty}^{+\infty} \exp(\xi z)\phi_z' \, dz \qquad \text{(B4.8.5)}$$

where ξ is an indicator variable whose powers in the expansion of $M(\xi)$ give the moments of the distribution.

After some manipulations similar to those leading to **Equation B4.8.4**, we obtain (again using the scale with $\theta_z = 0$):

$$M(\xi) = \exp\left(\xi\bar{z}\left(1 - \frac{V_Z}{\left(V_S + V_Z\right)}\right) + \frac{\xi^2 V_S V_Z}{2\left(V_S + V_Z\right)} \right) \qquad \text{(B4.8.6)}$$

This is the moment-generating function of a normal distribution, so that the post-selection distribution is normal. The coefficient of ξ corresponds to the mean of the distribution; on transformation back to the scale with arbitrary θ_z, this gives the post-selection mean as:

$$\bar{z}' = \bar{z} - \frac{\left(\bar{z} - \theta_z\right)V_Z}{\left(V_S + V_Z\right)} \qquad \text{(B4.8.7)}$$

Multiplying the change in mean given by this expression by the heritability (**Section 3.3.iii.a**) yields **Equation 4.21** for the change in mean between two successive generations.

The coefficient of $\xi^2/2$ in the expansion of $M(\xi)$ is the second moment about 0 of the distribution, from which the post-selection variance can be obtained by subtracting the square of the mean, giving:

$$V_Z' = \frac{V_S V_Z}{\left(V_S + V_Z\right)} \qquad \text{(B4.8.8)}$$

This implies that the variance is always reduced by nor-optimal selection, independently of the location of the mean with respect to the optimum.

If the trait is controlled by a set of independent autosomal sites, and the effects of mutations on z are not completely recessive, it can be shown that stabilizing selection will drive variants at all sites to extreme frequencies (**Box 4.9***). With weak stabilizing selection, the additive genetic variance for z is the same as the total variance and is approximately:

$$V_A \approx 4U_z V_S \qquad (4.22)$$

where U_z is the total mutation rate for the trait (Bulmer 1980, Chapter 10; Turelli 1984).

Box 4.9* EQUILIBRIUM GENETIC VARIANCE UNDER STABILIZING SELECTION

We assume that the mean has converged to the optimum, as will happen in a constant environment (**Equation B4.8.7**). If we consider a given site i in the genome that is fixed for allele A_{1i}, the effect of a mutation on fitness, relative to the mean fitness of the population, can be approximated by its effect on the natural logarithm of fitness (**Appendix 1.ii.c**). If a mutation from A_{1i} to A_{2i} changes the trait value of $A_{1i}A_{2i}$ individuals by a_{zi}, its effect on log fitness can in turn be approximated using the first three terms of the Taylor series expansion of $\ln(w)$ with respect to z (this generalizes the approach used in **Box 3.7*** of **Section 3.3.iii.b**):

$$\frac{\left(w_{12} - \bar{w}\right)}{\bar{w}} \approx \ln\left(\frac{w_{12}}{\bar{w}}\right) \approx a_{zi}\left(\frac{d\ln(w)}{dz}\right)_{\bar{z}} + \frac{a_{zi}^2}{2}\left(\frac{d^2\ln(w)}{dz^2}\right)_{\bar{z}} \qquad (B4.9.1)$$

where w_{12} is the fitness of $A_{1i}A_{2i}$.

Using the expression for fitness under the nor-optimal model (**Equation 4.20**), it follows that $d\ln(w)/dz$ vanishes at $z = \theta_z = \bar{z}$, so that the first term on the right-hand side is 0. Evaluating the second derivative of $\ln(w)$ using **Equation 4.20**, we obtain:

$$\frac{\left(w_{12} - \bar{w}\right)}{\bar{w}} \approx -\frac{a_{zi}^2}{2\left(V_S + V_z\right)} \qquad (B4.9.2)$$

This shows that selection tends to eliminate mutations that change the trait, regardless of the direction of their effects. From the theory developed for mutation–selection balance (**Section 4.2.ii.a** above), if the mutation rate from A_{1i} to A_{2i} is u_i, and the frequency of A_{2i} is q_i, then we have:

$$\Delta q_i \approx u_i - \frac{a_{zi}^2 q_i}{2(V_S + V_Z)} \qquad \text{(B4.9.3a)}$$

Setting this to 0 gives the equilibrium frequency of A_{2i} as:

$$q_i^* \approx \frac{2(V_S + V_Z)u_i}{a_{zi}^2} \qquad \text{(B4.9.3b)}$$

From the argument in **Box 4.7** of **Section 4.5.i.a**, the genetic variance contributed by this site is approximately $2q_i^* a_{zi}^2$. If stabilizing selection is weak, the bracketed term is dominated by V_S, and contributions of non-random associations between variants to the genetic variance (**Section 8.4.iv.a**) can be neglected. The total genetic variance in the trait is then given by the sum of $2q_i^* a_{zi}^2$ over all sites, yielding **Equation 4.22**.

Counterintuitively, the variance is independent of the magnitude of the effects of mutations on the trait, and depends only on the strength of stabilizing selection on the overall trait and the net mutation rate for the trait. It is difficult to test this prediction rigorously, because it is hard to estimate U_z for traits subject to stabilizing selection, since the method described in **Box 4.6** of **Section 4.4.iv.b** is only useful if all mutations affect the trait in the same direction. Based on a literature survey of empirical estimates of the effects of selection on quantitative traits in natural populations, Kingsolver et al. (2001) suggested a median value of V_S/V_Z of 5, corresponding to a value of 0.8 for the ratio of post-selection variance to pre-selection variance (**Equation B4.8.8**), very similar to Haldane's (1954) results. To account for a heritability value of 0.5 under **Equation 4.22**, a mutation rate of about 0.05 would be required for the trait. This seems excessive, given the large number of quantitative traits that characterize an organism. This discrepancy probably arises from mutations having pleiotropic effects on several traits (Johnson and Barton 2005).

4.5.ii. Variability maintained by selection

As we discussed in **Chapter 2**, there are many ways in which variation can be maintained by selection. This makes it extremely difficult to develop tests that can distinguish between the different types of selection that may, in theory, contribute to variability in quantitative traits. The evidence discussed in **Section 4.5.i.a** above suggests that balancing selection, as well as mutation, must contribute to variation in fitness components. Some examples of DNA sequence polymorphisms maintained by balancing selection are discussed in **Section 8.3.ii.a**. Heterozygote advantage for the traits themselves seems to be ruled out in most cases, given the evidence for additive variation in many fitness components (Houle 1992; Charlesworth and Hughes 2000). At equilibrium under balancing selection alone, the additive covariance between a trait and fitness must be 0 (from **Equation B3.6.2** in **Section 3.3.iii.a**), and this in turn is proportional to the additive variance in the trait, if the effects of segregating variants on fitness are proportional to their effects on the traits. It follows that the latter must be 0.

However, with antagonistically pleiotropic effects of fitness components on overall fitness (**Section 2.2.iv**), there can be additive variance for fitness components even if it is absent for fitness itself; additive genetic variance in fitness traits, or other quantitative traits, is therefore compatible with variability maintained by heterozygote advantage or other forms of balancing selection (Charlesworth and Hughes 2000). At present, we lack firm knowledge of how quantitative trait variation is maintained; various possibilities and relevant data are discussed by Johnson and Barton (2005). Further progress will probably come from molecular characterization of the genes controlling quantitative trait variability. This will permit tests for the type of selection acting on them, of the kinds discussed in **Sections 6.4** and **8.3.ii**.

PROBLEMS

4.1.   **i.** In an example of the two-population model of migration and selection of **Section 4.1.i**, use the fact that the allele frequencies of A_2 in populations 1 and 2 are 0.1 and 0.9, respectively, and the frequency of migration, m, is 0.01, to estimate the selection coefficient, s. What is the value of s when $m = 0.05$?

   ii. A mutation to a favorable semidominant, autosomal mutation occurs in a randomly mating species of moth with a generation time of one year, which confers resistance to a pesticide. It is observed to have spread about 7 kilometers since its origin 55 years ago. Knowing that the standard deviation of its dispersal distance, σ, is 0.5 km, estimate the selection coefficient s using Fisher's expression for the speed of the wave of advance of a favorable mutation (**Section 4.1.ii.e**).

4.2   Show that the equilibrium under mutation and selection for a nonrecessive deleterious allele (**Section 4.2.ii.a**) is stable, by finding an expression for the rate of change in the difference between current allele frequency, q, and the equilibrium value q^*. Discuss how long it takes to approach equilibrium, in terms of the half-life of the process.

4.3. **i.** Derive **Equation 4.4** of **Section 4.2.ii.a** for the equilibrium frequency of a deleterious recessive allele with mutation–selection balance. Make use of the answer to **Problem 3.3**.

ii. Compare the equilibrium frequencies of mutations that are homozygous lethal when the selection coefficients for the heterozygous carriers are 0.01 and 0, respectively, for a mutation rate of 10^{-6}.

4.4 A study of achondroplastic dwarfism in humans, which is inherited as a dominant allele, found that 10 out of 94,075 births had the trait. 108 affected individuals had a total of 27 children, whereas 457 of their normal siblings produced 582 offspring (Haldane 1949c). Estimate the strength of selection against carriers of the mutation, and use the appropriate equation for equilibrium under mutation and selection (**Section 4.2.ii.a**) to estimate the mutation rate at this locus to alleles causing dwarfism.

4.5. **i.** Show that the genetic load (**Section 4.3.ii.a**) for a population at mutation–selection equilibrium for recessive mutations at a locus is approximately equal to the mutation rate.

ii. Determine the mutational load for the case of an X-linked locus, when selection acts either equally on both sexes or only on one sex (use **Equation 4.5** of **Section 4.2.ii.b**).

4.6. **i.** Derive **Equation 4.13** of **Section 4.4.iii** for the genetic load with heterozygote advantage.

ii. Obtain an expression for the genetic load caused by m independent biallelic sites with heterozygote advantage and selection coefficients s against the homozygotes at each site.

iii. Use this result to determine the maximum numbers of independent sites maintained by heterozygote advantage (with either $s = 0.1$ or $s = 0.001$) that would be compatible with the survival of a *Drosophila* population, assuming that females can produce up to 100 eggs per generation.

4.7* Derive the cost of selection, C, (**Section 4.3.iv**) for the case of a semidominant favorable mutation with initial frequency, q_0. (Approximate the sum of the loads over each generation by an integral, and use **Equation B3.2.1** in **Section 3.1.iii.b** to change the variable of integration from dt to dq.)

4.8 Derive the lag load for a model in which an average of K sites in the genome become fixed each generation for a favorable variant, when the cost of selection for each variant is C (**Section 4.3.iv**). If $C = 30$, what value of K gives a lag load of 0.10?

4.9 Obtain an expression for the mean of a population, derived by inbreeding from a randomly mating population, for arbitrary allele frequencies at a biallclic locus. Use a notation similar to that for fitness, expressing trait values relative to those for a standard genotype. Show that some degree of dominance is required for the mean to differ from that of the original population (see **Section 4.4.ii.a**), and determine the conditions for an increase versus a decrease in the mean following inbreeding.

4.10 Derive an equation for the mean fitness of an inbred population with inbreeding coefficient f, derived from a randomly mating population at equilibrium under heterozygote advantage at a single site, without any change in allele frequencies (see **Section 4.4.iii**). Use this to obtain **Equation 4.17** for a fitness component.

CHAPTER FIVE

The Evolutionary Effects of Finite Population Size: Basic Theory

CHAPTER SUMMARY

We now introduce genetic drift—the random changes in variant frequencies caused by the sampling of the genetic composition of populations that occurs when each new generation is formed. Variants at many sites in the genome probably have little or no effect on any phenotype, and affect fitness only slightly or not at all. Over long time periods, genetic drift has large, cumulative effects on the frequencies of such variants. It is therefore a process that is central to understanding the evolution of DNA sequences. This chapter considers drift in populations with no spatial structure.

We first study the behavior of selectively neutral autosomal variants in a diploid, randomly mating hermaphrodite population of constant size, in which all individuals have the same chance of contributing to the next generation (the Wright–Fisher model). All the allelic copies of a given nucleotide site that are present in the population at a given time eventually trace back to a single allele present in an ancestral population, so that they are identical by descent from this allele. Over time, unless mutations occur, the population therefore approaches genetic uniformity. This is described by the population's inbreeding coefficient (the probability of identity by descent of a pair of randomly sampled alleles from the population). Among a set of isolated populations, the variance in allele frequencies between populations increases in parallel with the increase in the inbreeding coefficient.

With mutation, a statistical equilibrium is eventually reached between the loss of variability caused by drift and the input of new mutations. We derive the expected equilibrium values of measures of variability, especially for the infinite sites model. This model assumes that mutation rates are low enough that new mutations occur only at nonvariable nucleotide sites.

We also examine the effects of drift on genetic variability using coalescence theory, which considers the history of a sample of alleles by tracing the alleles' ancestry backwards in time, generating a gene tree. This approach yields expressions for the probability distribution of the times to common ancestry of the lineages leading to the sample, and provides formulae for the means and variances of measures of DNA sequence variability. This procedure is extremely useful for simulating DNA sequence or microsatellite variability. By distributing mutations over gene trees, we can study the statistical properties of variability measures.

The assumptions of the Wright–Fisher model are biologically unrealistic. Its limitations can be overcome with the use of the concept of the effective population size (N_e). We describe this concept using the coalescent approach, and outline a general, approximate method for calculating theoretical values of N_e. This approach allows for various modes of inheritance, nonrandom variation in reproductive success, and inbreeding. We describe how the results can be extended to several situations of biological interest, such as variable population size and overlapping generations. We show how N_e can be estimated from data on genetic variability.

A full description of a population or sample involves the frequencies of variants at segregating sites, which can provide information about the action of selection, especially when using the neutral expectation as a null hypothesis. We show how the equilibrium probability distribution of variant frequencies can be derived using diffusion equations, which assume that both genetic drift and deterministic forces (mutation and selection) are weak. Specific results are given for some simple situations, both for the neutral case and for the joint action of selection and mutation. If the product of N_e and the selection coefficient is greater than 1, the population behaves in much the same way as an infinitely large population, supporting the conclusion that most phenotypic evolution is under the control of selection.

> I returned and saw under the sun that the race is not to the swift,
> nor the battle to the strong . . . but time and chance happeneth to
> them all.
>
> Ecclesiastes 9.11

> The very small range of selective intensity in which a factor may be
> regarded as effectively neutral suggests that such a condition must
> in general be extremely transient.
>
> R. A. Fisher, *The Genetical Theory of Natural Selection*, Chapter 4

INTRODUCTION

In the last three chapters, we studied the processes of evolution while largely ignoring the fact that populations are finite in size. Even when considering the chance extinction of new mutations (**Section 3.2**), we assumed that the population was so large that an advantageous mutation is certain to become fixed if it succeeds in spreading to a sufficiently high frequency. Similarly, variants subject to balancing selection were assumed to be maintained indefinitely in the population. In other words, although we did not define precisely what we meant by a large population, we treated evolutionary change within species as a completely predictable, deterministic process (apart from the possibility of the chance loss of new mutations). We therefore disregarded an important evolutionary process—the sampling from the parental population that occurs when each new generation is formed, causing random changes in variant frequencies (**Figure 5.1**). This is called *genetic drift* (Wright 1931).

Since the number of individuals in a species is usually very large, other than for species on the verge of extinction, disregarding drift may seem a good approximation to the truth. It was established early in the history of population genetics that changes caused by genetic drift decrease in importance as the number of breeding individuals increases (Fisher 1922, 1930a,b; Wright 1931). In large populations, drift therefore produces only very slow evolutionary change compared with natural selection. Most traits with functional significance to organisms must therefore evolve by selection rather than drift; we give quantitative versions of this argument in **Sections 5.3.iii.e** and **6.2.iii**.

It is nevertheless very important to study the random fluctuations in variant frequencies produced by finite population size. A major reason for taking genetic drift seriously is that there are many sites in the genome where mutations probably have little or no effect on any phenotype, and hence do not affect fitness. Much evidence from studies of molecular evolution and variation strongly suggests that many DNA sequence variants are subject to little or no selection. Over a sufficiently long time period, genetic drift may have large cumulative effects on the frequencies of *selectively neutral* variants, even in very large populations, and we shall see that drift also affects non-neutral variants if selection is sufficiently weak. Many evolutionary changes in DNA sequences are likely to involve random changes caused by drift, including the fixation of variants with slightly deleterious fitness effects (Kimura 1968, 1983; King and Jukes 1969). This is a major topic of **Chapter 6**.

Furthermore, as repeatedly emphasized by Sewall Wright (1931, 1969, 1977), species are often broken up into fairly small local populations, with limited migration among them. Drift may then cause local populations to diverge from each other, even when selection is fairly strong. The spatial subdivision of populations may greatly affect evolutionary processes, and this complicates the interpretation of data from natural and human populations, as we discuss in detail in **Chapter 7**.

The present chapter deals with drift in populations that lack any spatial structure. It includes more mathematical theory in relation to biological illustrations than most chapters, particularly after **Section 5.3.ii**. The results are,

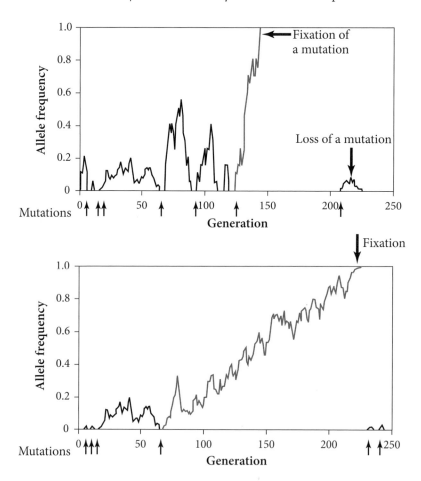

FIGURE 5.1 Genetic drift. The figure shows the allele frequencies of neutral mutations in a simulated population over 250 generations, showing the occurrence of several mutations (indicated by the arrows below the *x* axis). Most mutations are quickly lost from the population (black lines); when this happens, a new mutation is (biologically unrealistically) introduced into the population in the next generation. Occasionally, however, a mutation becomes fixed in the population (grey lines). The upper graph shows a small population (10 diploid individuals), while the lower graph shows a population of size 100. This illustrates the slower rate of genetic drift (the larger fixation time for the mutation that becomes fixed) in the larger population.

however, fundamental to an understanding of molecular variation and evolution, the subject of **Chapter 6**, as well as providing the general principles governing evolutionary change in populations of finite size. The main conclusions are stated in the text, and can be understood without following all of the mathematical details.

5.1. MODELING GENETIC DRIFT

5.1.i. Drift without mutation

5.1.i.a. *The Wright–Fisher model*

We first consider the effects of genetic drift on selectively neutral variants, assuming the population to be closed (i.e., there is no migration from elsewhere), and *panmictic* (i.e., there is no subdivision according to spatial location or genotype, so that all parental genotypes potentially contribute to the same pool of offspring). The simplest situation of this kind is a randomly mating population consisting of N diploid, hermaphroditic individuals, with discrete generations, each generation being counted at the time of breeding. New individuals are formed each generation by random sampling (with replacement) of gametes produced by the parents. Each parent thus has an equal probability of contributing a gamete to an individual that survives to breed in the next generation. This is called a *Wright–Fisher* population. A population of hermaphroditic marine organisms that shed a very large number of gametes into the water to form zygotes by random encounters of male and female gametes, of which only a small fraction survive to produce the new generation of N breeding individuals, comes closest to this model. We show in **Section 5.2** that it is surprisingly easy to apply the most important results to situations that are more realistic for other types of organism.

As usual, we assume two alternative allelic variants at an autosomal site, A_1 and A_2, with frequencies p_0 and $q_0 = 1 - p_0$ in an initial generation. The state of the population in the next generation can then be described by the probability that the new frequency of A_2 is $i/2N$, where i can take any value between 0 and $2N$. The reason for using $2N$ is simply that, with diploid inheritance, there are $2N$ allele copies in N individuals; if the species were haploid, we would use N. The Wright–Fisher model is identical to the classic problem in probability theory of determining the chance of i successes out of a specified number ($2N$) of trials (a success being the choice of A_2 rather than A_1), when the chance of success on a single trial is q. Tossing an unbiased coin $2N$ times corresponds to the case where $q = 0.5$.

Standard probability theory tells us that the probability of obtaining i copies of A_2 in the next generation, corresponding to a frequency of $q = i/2N$, is given by the *binomial distribution* (**Box 5.1**). It is obvious that the new mean frequency of A_2 is simply q_0, because drift cannot affect the mean frequency taken over a large set of replicated populations. But the frequency in any given population will probably change somewhat, becoming $q_0 + \delta q$, where the change δq has variance $V_{\delta q}$ given by:

$$V_{\delta q} = \frac{p_0 q_0}{2N} \tag{5.1}$$

Box 5.1 THE BINOMIAL DISTRIBUTION APPLIED
TO GENETIC DRIFT

If the frequency of A_2 is q, the chance that a set of $2N$ alleles contains i
copies of A_2 and $2N - i$ copies of A_1 in a particular order (e.g., the first i
alleles in the set are all A_2 and the second set are all A_1) is simply $p^{(2N-i)}q^i$.
The number of possible orderings is the number of combinations of i
objects of a given type among a total of $2N$, which is given by:

$$\binom{2N}{i} = \frac{(2N)!}{i!(2N-i)!} \qquad \text{(B5.1.1)}$$

where $i! = i(i-1)(i-2)\cdots 1$ is the number of permutations of i objects
(i.e., the number of ways in which they can be ordered).

The probability that there are i A_2 alleles out of $2N$ is thus:

$$\binom{2N}{i}p^{2N-i}q^i \qquad \text{(B5.1.2)}$$

This is the *binomial distribution* (**Appendix A2.v.a**). The mean of i is
$2Nq$, so that the mean value of the new frequency of A_2 is simply q, i.e.,
the mean change in q is 0. The variance of the new frequency of q is
$pq/2N$; this is the same as the variance of the change in frequency.

(See **Appendix A2.v.a**.) After a further generation, the new frequency will be q_0
$+ \delta q + \delta q'$, where $\delta q'$ has a mean of 0 and a variance of $(p_0 - \delta q)(q_0 + \delta q)/2N$,
and so on. As we will see, **Equation 5.1** plays a fundamental role in the theory
of genetic drift.

If we follow a single population, there will be a succession of random changes
in q, until eventually A_2 either becomes fixed in the population ($q = 1$) or is lost
($q = 0$) (see **Figure 5.1** in the **Introduction**). Either of these events terminates
the process, so that the states $q = 0$ and $q = 1$ are *absorbing boundaries*. Ideally,
we would like to be able to specify the probability of a given value of q in any
generation, conditional on an initial value, but this is a very hard problem to
solve (see **Section 5.3.ii** below). It is possible, however, to obtain simple but very
useful results by noting that the process of genetic drift has two aspects.

(1) The tendency for a population of finite size to become genetically uniform, owing to the fact that all the allelic copies of a site present in the popu-

lation at a given time trace back to a single allele in the initial generation (**Figure 5.1**).

(**2**) The tendency of the frequencies of variants in isolated populations to diverge over time, because independent replicates of a population with the same initial state arrive at different variant frequencies by chance (**Figure 5.2**).

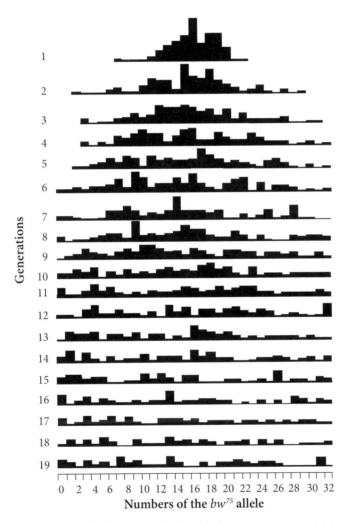

FIGURE 5.2 Frequencies of the brown eye (bw^{75}) allele in an experiment with laboratory populations of *Drosophila melanogaster* with 16 breeding individuals in each generation. The starting populations (generation 0) had a frequency of 0.5 (16 copies of bw^{75}), and the figure shows the distribution of the number of copies of the bw^{75} allele in each replicate population over 19 successive generations. [Adapted from Figure 7 of Buri (1956).]

5.1.i.b. *Approach to genetic uniformity of a population*

The first process is closely related to the increase in *homozygosity* that accompanies the inbreeding of close relatives, which we studied using the concept of *identity by descent* (*i.b.d.*) in **Section 1.3.ii.** Two different allelic copies of a given nucleotide site drawn from a population are identical by descent if they trace their ancestry back to a single ancestral copy. The progress of a population towards genetic uniformity is measured by the probability that a pair of randomly sampled alleles are i.b.d. (the *inbreeding coefficient*) relative to an initial generation. At this stage, we ignore the states of the nucleotides present at the site in question.

Box 5.2 CHANGE IN IDENTITY IN A FINITE POPULATION

Let the inbreeding coefficient in generation t be F_t. Consider a pair of distinct alleles sampled from the population. The chance that one of the two alleles is a copy of a given allele in the preceding generation is $1/2N$, since there are $2N$ distinct alleles in the breeding population. The chance that both alleles are derived from this allele is $(1/2N)^2$. There are $2N$ alleles present in the population in generation $t - 1$, so the net chance that the two chosen alleles are copies of the same allele from generation $t - 1$, and are thus identical by descent, is $2N(1/2N)^2 = 1/2N$.

There is a probability $1 - 1/2N$ that the pair are copies of two different alleles in generation $t - 1$, in which case their probability of identity by descent is F_{t-1}. We therefore obtain the recurrence relation:

$$F_t = \frac{1}{2N} + \left(1 - \frac{1}{2N}\right)F_{t-1} \qquad (B5.2.1a)$$

so that:

$$1 - F_t = \left(1 - \frac{1}{2N}\right)\left(1 - F_{t-1}\right) \qquad (B5.2.1b)$$

Iterating this back to generation 0 gives **Equation 5.2** of the main text.

With diploid inheritance (i.e., with $2N$ allelic copies among the breeding adults), the results in **Box 5.2** show that F_t, the inbreeding coefficient in generation t, is given by:

$$1 - F_t = \left(1 - F_0\right)\left(1 - \frac{1}{2N}\right)^t \tag{5.2a}$$

We use F rather than f to represent the inbreeding coefficient in order to distinguish the increase in i.b.d caused by genetic drift from that due to an excess of matings between relatives, which was studied in **Section 1.3.ii.a**. If N is fairly large, we can approximate this discrete generation process by treating time as a continuous variable t, as was done in **Section 3.1.iii.b**. This means that we can also approximate $(1 - 1/2N)^t$ by $\exp(-t/2N)$ (see **Box 4.4** of **Section 4.3.ii.c**), so that:

$$1 - F_t \approx \left(1 - F_0\right)\exp\left(-\frac{t}{2N}\right) \tag{5.2b}$$

This shows that there is an approximately exponential decline in the probability that two randomly chosen alleles are nonidentical by descent, $1 - F_t$ (**Figure 5.3**).

The population thus tends towards complete identity by descent, reflecting the fact that only one of the alleles in a closed population at a given time is the ultimate ancestor of the alleles in the future population (**Figure 5.1**). Because

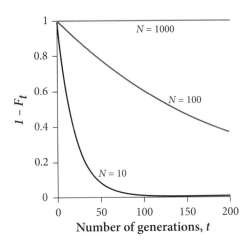

FIGURE 5.3 Decline with time in the probability of nonidentity by descent, $1 - F$, of a pair of randomly chosen alleles, as given by the approximate **Equation 5.2b**.

we assume that all alleles are equivalent in fitness, each of the $2N$ alleles present in the ancestral population has the same probability, $1/2N$, of being this ancestor; there is a much larger probability, $1 - 1/2N$, that a given allele ultimately fails to leave any descendants. This has important implications for the evolution of DNA sequences, which we examine in **Section 6.2.iii.**

5.1.i.c. *Increase in variance of allele frequencies*

What about the increase in variability among different populations? Imagine a large set of completely isolated populations, all starting, as before, with variants A_1 and A_2 at frequencies p_0 and q_0. Ultimately, each of these populations will become fixed for one of these two variants (**Problem 5.1**).

Box 5.3 derives a recursion relation for the variance in frequency among populations, V_q. A comparison with the corresponding formula for F (**Equation B5.2.1a** in **Section 5.1.i.a**) shows that the two are related by:

$$F = \frac{V_q}{p_0 q_0} \tag{5.3}$$

Box 5.3 CHANGE IN VARIANCE OF ALLELE FREQUENCY

Let the variance in frequency of A_2 in the new generation be V_q'. This is equal to the variance of $q + \delta q$ over all values of q, where δq is the change in q conditional on the value of q. The expectation of δq for a given q is 0, so that the expectation of $q\delta q$ is also 0. Using **Equation 5.1**, we therefore have:

$$V_q' = V_q + E\left\{\frac{pq}{2N}\right\} \tag{B5.3.1}$$

where E{} denotes the expectation over the probability distribution of the frequency of A_2.

We have:

$$E\{pq\} = E\{q - q^2\} = q_0 - q_0^2 - V_q$$
$$= p_0 q_0 - V_q \tag{B5.3.2}$$

Substituting into **Equation B5.3.1**, we obtain:

$$V_q' = \frac{p_0 q_0}{2N} + \left(1 - \frac{1}{2N}\right) V_q \tag{B5.3.3}$$

The dynamics of the loss of variation by drift within a single Wright–Fisher population with constant size is thus the same as for the increase in variability in allele frequencies among isolated populations. This can also be seen from the following argument. The probability that a pair of alleles drawn from the same population are both A_2 is given by the expectation of q^2 over the distribution of allele frequencies; this is equal to $V_q + q_0^2$ (**Equation B5.3.2**). But this probability is also equal to $(1 - F) q_0^2 + Fq_0 = Fp_0q_0 + q_0^2$, using the argument in **Box 1.5** of **Section 1.3.ii.b**, so that $V_q = Fp_0q_0$. **Equation 5.3** shows that V_q ultimately approaches p_0q_0, the value when all populations are fixed (**Problem 5.1**). F can therefore be viewed as measuring the between-population variance, relative to this maximum possible value. This concept is frequently used in the analysis of data on differences between populations of the same species (**Section 7.1.i.a**).

Because the populations are assumed to be randomly mating, $V_q + q_0^2$ corresponds to the frequency of A_2A_2 homozygotes. A similar argument shows that the frequency of A_2A_2 homozygotes is $V_q + p_0^2$, and the frequency of A_1A_2 heterozygotes is $2p_0q_0 - 2V_q$. Differentiation of allele frequencies among a set of randomly mating populations thus leads to an excess of homozygotes over the random mating expectation when genotype frequencies are averaged over the populations, regardless of the cause. This is the *Wahlund effect* (Wahlund 1928).

5.1.ii. Drift and mutation

5.1.ii.a. *Identity in state of a pair of alleles and the infinite alleles model*
The use of identity by descent to look at the loss of variability caused by drift merely traces lines of descent from the alleles present in an initial generation, ignoring the states of the nucleotides in the sequence or sites. Information on the "state" of a nucleotide site (or non-recombining sequence of nucleotides) can then be added, allowing one to take into account mutations that counteract the loss of variability. For simplicity, we ignore differences in the probabilities of mutation for the four possible nucleotides (which may, in reality, be substantial; see **Section 6.1.ii.e**), and simply assume a probability u of mutating to a different state (for a sequence, rather than a single site, this is the net probability that a change occurs somewhere in the sequence). We also assume a low enough mutation rate, u, that a variant arising by mutation is not already present in the population. This is often referred to as the *infinite alleles* assumption, because it is strictly valid only if an indefinitely large number of possible variants can be generated by mutation (Malécot 1948, 1969; Kimura and Crow 1964). This is a good approximation to the properties of a long sequence of nucleotides.

The probability that two alleles randomly chosen in generation t are different in state is the theoretical value of the *gene diversity*, H_t, whose empirical counterpart was introduced in **Section 1.2.i.b.** The results of **Box 5.4** show that if u is the neutral mutation rate per site, H_t approaches the approximate equilibrium value given by:

$$H^* \approx \frac{\theta}{1 + \theta} \qquad (5.4)$$

where $\theta = 4Nu$. Theta (θ) is often referred to as the *scaled mutation rate*, and can be thought of as a measure of the rate of mutation for the whole population.

Box 5.4 THE EQUILIBRIUM LEVEL OF VARIABILITY UNDER DRIFT AND MUTATION

Assume that we have a Wright–Fisher population of size N, with a set of non-recombining sequences (loci), each with the same neutral mutation rate u. Let H_t be the probability that two randomly sampled alleles at a locus are nonidentical in allelic state in generation t. Using the same argument as for the inbreeding coefficient in **Section 5.1.i.b** of the main text, there is a chance of $1/2N$ that the pair are descended from the same allele in the previous generation; however, in this case they are identical in state if *neither* member of the pair has mutated during the passage from generation $t-1$ to t. The probability of this is $(1 - u)^2$. There is a similar chance of $1 - 1/2N$ that the pair are descended from two distinct alleles in the previous generation. In this case, there is a probability $1 - H_{t-1}$ that they were identical in state (in the absence of mutation), so that their chance of identity in generation t is $(1 - H_{t-1})(1 - u)^2$. We therefore obtain the relation:

$$1 - H_t = (1 - u)^2 \left[\frac{1}{2N} + \left(1 - \frac{1}{2N}\right)\left(1 - H_{t-1}\right) \right] \qquad \text{(B5.4.1)}$$

Since u is very small and N is usually large in most cases of biological interest, we can ignore terms in u^2 and u/N, so that this simplifies to:

$$1 - H_t \approx \frac{1}{2N} + \left(1 - 2u - \frac{1}{2N}\right)\left(1 - H_{t-1}\right) \qquad \text{(B5.4.2)}$$

At equilibrium, $H_t = H_{t-1} = H^*$. Substituting this into the above equation and rearranging, we obtain **Equation 5.4** of the main text (**Problem 5.2**).

5.1.ii.b. *The infinite sites model*

For individual nucleotide sites, as opposed to an entire sequence of non-recombining nucleotides, it is often realistic to assume that the mutation rate is so low that $\theta \ll 1$. **Equation 5.4** can then be approximated further (**Problem 5.2**) to give the equilibrium probability, π^*, that two randomly chosen sequences differ at a given site (the *nucleotide site diversity*):

$$\pi^* \approx \theta \qquad (5.5)$$

A sufficient condition for this to be a good approximation is that at most two alternative variants are ever present in the population at any given site, an assumption that is often called the *infinite sites* model (Kimura 1971). This is often found to be true in surveys of DNA sequence variation (**Section 1.2.iii.b**), so that **Equation 5.5** is frequently regarded as the more useful theoretical formula for interpreting data on DNA sequence variability, because it enables θ to be estimated by equating it to an observed value of nucleotide site diversity as measured by π (see **Box 1.3**). The theoretical value of π given by **Equation 5.5** can be usefully thought of as being equivalent to the mean nucleotide site diversity over a large number of sites subject to an average mutation rate u (hence the term "infinite sites" model). Under neutrality, this mean is unaffected by the lack of independence due to linkage among the sites under study, whereas the variances of estimates of nucleotide site diversity are strongly affected by correlations between sites (**Problem 5.3**).

5.1.iii. The coalescent process

5.1.iii.a. *Basic properties*

Another very useful way of describing a population subject to genetic drift for neutral variants is to look backwards in time (**Figure 5.4**), rather than forwards, and to concentrate attention on the time scale of the process, rather than on identity probabilities. The importance of this way of thinking, known as *coalescence theory* (Griffiths 1980; Kingman 1982; Hudson 1983a; Tajima 1983), is that it shows how the properties of samples of alleles from populations depend on the genealogies connecting them. Because we have empirical information only about samples, not whole populations, understanding the properties of samples is crucial. Coalescence theory is thus central to inferences from data on molecular variation; excellent reviews of varying length are provided by Hudson (1990), Donnelly and Tavaré (1995), Hein et al. (2005), and Wakeley (2008). It involves the properties of two distinct processes: the formation of a genealogy for the alleles in a sample and the imposition of mutations on the tree. The latter process generates the variability observed in the sample.

5.1.iii.b. *The genealogy of two alleles*

We first examine the properties of the simplest possible genealogy, that of two alleles, sampled from a Wright–Fisher population of size N in a certain genera-

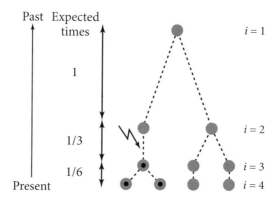

FIGURE 5.4 Coalescent events in the ancestry of a sample of four alleles (grey circles). The two alleles on the left trace their ancestry back to the same ancestral allele (i.e., they coalesce) at a time when the other two alleles have distinct ancestors. Going further back in time, the pair on the right then coalesce into an ancestral allele that is distinct from the ancestor of the other pair; finally, the two ancestral alleles themselves coalesce into a common ancestor of the whole set. The lines showing the ancestry of the alleles are represented by dashes, to indicate that the times to coalescence involve many generations. The mean times between successive coalescent events (obtained from the results in the text) are shown on the left of the diagram, in units of $2N$ generations. The flash symbol indicates a mutation in the sequence whose ancestry is depicted, and the presence of this mutation is also shown (by black dots) in the descendants of the allele in which it occurred.

tion. The approach is very similar to that used for identity by descent in **Section 5.1.i.b**, using essentially the same concepts. The results apply both to a single nucleotide site and to a non-recombining sequence, such as a mitochondrial genome or a Y chromosome. From **Box 5.2**, there is a probability of $1/2N$ that the two alleles descend from the same ancestral allele in the previous generation. Two such alleles are said to *coalesce* (or "coalesce into this allele," using the term introduced in **Section 1.3.ii.a**). In general, the probability that two alleles coalesce t generations back in time is:

$$Pr(t) = \left(1 - \frac{1}{2N}\right)^{t-1} \frac{1}{2N} \qquad (5.6a)$$

This is the probability that no coalescent events happened in the preceding $t - 1$ generations, and that a coalescent event occurred in generation t.

This is a *geometric distribution* (**Appendix A2.v.c**). If N is moderately large, we can again treat time as continuous, and replace the geometric distribution by the more convenient *exponential distribution*:

$$P(t) \approx \frac{1}{2N}\exp\left(-\frac{t}{2N}\right) \tag{5.6b}$$

where $P(t)$ is now the *probability density* of coalescence at time t, i.e., the chance of coalescence over a small interval of time, t to $t + \delta t$, is $P(t)\delta t$ (**Appendix A2.ii**). From the properties of the exponential distribution, the mean time to coalescence of a pair of alleles is $2N$ generations, and the variance of the coalescent time is $(2N)^2$; these results can be obtained directly, using simple calculus (**Problem 5.4***). The standard deviation is therefore $2N$, so that the scatter in the coalescence time is of the same magnitude as its mean.

5.1.iii.c. *Gene genealogies in general*

We can build on these results to model a sample of k alleles, assuming that k is much smaller than the total number of alleles in the population, $2N$. If $2N$ is moderately large, it is then reasonable to assume that at most one coalescent event occurs in a given generation among a set of alleles. The genealogy connecting the alleles is then a bifurcating tree (a *gene tree*), with a steadily decreasing number of nodes (**Figure 5.4**). With i lineages present at a given time, there are $j_i = i(i-1)/2$ ways of choosing a pair of different lineages to coalesce with each other. It is a matter of chance which alleles actually coalesce during the sample's ancestry, so that the net probability of a coalescent event in a given generation is $j_i/2N$. The probability density of the time t_i to the next coalescent event (going back in time from the present) is given by substituting $j_i/2N$ for $1/2N$ into **Equation 5.6b**, so that t_i is exponentially distributed, with mean and standard deviation $2N/j_i$.

Starting with $i = k$ lineages (representing a sample at the present time), this process can be repeated for $i = k - 1, k - 2$, until $i = 1$, which represents coalescence of all the alleles in the sample into their *most recent common ancestor* (**Figure 5.4**). Coalescent events occur fast at first (when i is largest), and then more and more slowly as the number of distinct alleles in the ancestry of the sample decreases (**Figure 5.4**). The probability densities for the different t_i values are independent of each other. Their joint distribution can thus be obtained by multiplying the densities for each t_i. The mean time back to the most recent common ancestor of a sample of k alleles can be shown to be $4N(k-1)/k$ (**Problem 5.5**), which is close to $4N$ when k is large. For a large sample of alleles, about half the total average time to the common ancestor is thus taken up with the coalescence of the last remaining pair of alleles.

5.1.iii.d. *Mutations, the coalescent process, and sequence diversity*

This description of a gene tree is purely theoretical, because gene trees cannot be observed directly. However, the results are relevant to data on population samples, because variation in a sample of k allelic sequences reflects mutations that have arisen in different branches of the tree since the most recent common ancestor (**Figure 5.4**). To model a sample, we simply allow mutations to occur on the lineages in the gene tree. The simplest model to use is the infinite sites model, introduced above (the mutation rate per generation per site is u, and u is assumed to be low, so that at most one mutation arises per site in the tree). We will briefly consider other mutational models in **Section 5.1.iii.f** below.

We can then derive the formula for the expected nucleotide site diversity very simply. Consider a given pair of alleles taken randomly from the sample. There is a time t connecting each of them to their common ancestor. They will be identical in state at a site if no mutation has arisen over the time separating them from each other, which is $2t$. The probability that a mutation *has* arisen at that site, and caused them to differ in state, is $2tu$. From the considerations given above, t has an expected value of $2N$, so that the net probability of a difference in state at a given nucleotide site is $4Nu$. Averaging over all pairs of alleles in the sample, and over a large number of sites, gives the expected value of the nucleotide site diversity for the sample, as in **Equation 5.5**.

A very similar argument can be used to derive the expected total number of segregating sites in the sample. **Box 5.5** derives the expected value of the sum L of the lengths of all the branches in a gene tree (the size of the tree, in numbers of generations):

$$E\{L\} = 4N\sum_{i=1}^{k-1}\frac{1}{i} = 4Na_k \qquad (5.7)$$

Box 5.5 THE EXPECTED SIZE OF A GENE TREE

The contribution to L from the portion of the tree involving the transition from i to $i-1$ alleles is the sum of i branches of length t_i, i.e., it_i (**Figure 5.4**). From the results derived in the text, the expectation of t_i is $2Nj_i = 4N/i(i-1)$. The net contribution of this section of the tree to the expected total length of the tree is thus:

$$E\{it_i\} = \frac{4Ni}{i(i-1)} = \frac{4N}{(i-1)} \qquad (B5.5.1)$$

The expected total length of the tree is obtained by summing $E\{it_i\}$ over all values of i, from the initial value, k, down to the final value, $i = 2$. This is equal to $4Na_k$, where a_k is the sum of $1/j$ from $j = 1$ to $k-1$.

where a_k is the *Watterson's correction factor* (Watterson 1975) described in **Box 1.3** of **Section 1.2.iv.b**. The probability that a given site has mutated in the gene tree connecting the k sampled alleles is thus Lu. For a sequence of m nucleotides, the expectation of Lmu (using **Equation 5.7**) gives the expected value of the number of segregating sites S_k in the sequence, in a sample of size k:

$$E\{L\}mu = \theta m a_k \qquad (5.8)$$

Equating the observed S_k value to the expected value on the left of **Equation 5.8**, we can thus estimate θ from the following relation:

$$\theta_w = \frac{S_k}{m a_k} \qquad (5.9)$$

This shows why nucleotide site diversity can be estimated by the θ_w statistic described in **Box 1.3**.

We can also obtain the variances of these estimates of θ by using the coalescent approach for two extreme cases: independence among the variants at different sites in the DNA sequence used in the study and complete linkage. **Box 5.6** shows this for θ_w. The derivation of the variance of the pairwise diversity estimator is more complicated (Tajima 1983), and only the final result is given in **Box 5.6**. Note that there are two components to these variances (e.g., **Equation B5.6.2**); the first component arises from the fact that a small sample is used to obtain the data, and the other results from the highly stochastic evolutionary process that generates the state of the population from which the sample is obtained. This causes great variation in the branch lengths, contributing what is called the *evolutionary stochastic variance*.

The variances are large and do not get much smaller even with large sample sizes. They are especially large when the sites are completely linked, i.e., the evolutionary histories of the sites are not independent of each other (**Problem 5.7**). In order to obtain a reliable estimate of the mean of θ across the genome, the best strategy is to use a large number of relatively short sequences that recombine freely, rather than a few long sequences within which recombination is infrequent. It is now common practice to include multiple sequences of around 500 bp in surveys of DNA sequence variability in both human populations (e.g., Yu et al. 2002) and natural populations (e.g., Glinka et al. 2003; Nordborg et al. 2005).

5.1.iii.e. *Simulating the coalescent process*

In many applications of coalescent theory, such as asking the important question of whether the properties of a sample deviate significantly from the expectation under neutral equilibrium, we need to determine the probability distribution of a statistic estimated from the data, such as the pairwise diversity

Box 5.6 VARIANCES OF TREE LENGTHS AND DIVERSITY ESTIMATES

On the infinite sites model, the number of segregating sites in a sample, S_k, is given by the number of mutations that occurred over the entire tree connecting the alleles in the sample (**Figure 5.4**). The probability of a given number of mutations conditional on the size of the tree follows a Poisson distribution, whose mean is the product of the probability of the mutation rate for the sequence (mu) and the length of the tree (L). The variance of S_k for a non-recombining sequence is equal to the sum of two terms (**Problem 5.6**):

(1) The variance of the number of mutations for a tree with length $E\{L\}$; from **Equation 5.8**, this is $mu(4Na_k)$.

(2) The product of the variance of the length of the tree (V_L) and the square of the mutation rate.

V_L can be obtained by considering the contribution from the part of the tree connecting i to $i - 1$ alleles, with length t_i (**Figure 5.4**). From the properties of the exponential distribution, the variance of t_i is $[4N/i(i - 1)]^2$. The contribution from all i branches to L is it_i, so the net contribution to V_L is $i^2[4N/i(i - 1)]^2 = [4N/(i - 1)]^2$. Summing over all branches of the tree:

$$V_L = \sum_{i=2}^{k} \frac{(4N)^2}{(i - 1)^2} \qquad \text{(B5.6.1)}$$

The squared mutation rate for the sequence is $(mu)^2$, so the term in V_L gives a contribution $(mu)^2 V_L$ to the variance in the number of mutations in the sample, S_k, and the term in $E\{L\}$ gives a contribution of $muE\{L\}$. **Equation 5.9** for θ_w implies that $V(\theta_w) = V(S_k)/(ma_k)^2$, so that:

$$V_{\theta_w} = \frac{m\theta a_k}{\left(ma_k\right)^2} + \frac{(m\theta)^2}{\left(ma_k\right)^2} \sum_{i=2}^{k} \frac{1}{(i - 1)^2}$$

This simplifies to:

$$V_{\theta_w} = \frac{\theta}{ma_k} + \frac{\theta^2}{a_k^2} \sum_{i=1}^{k-1} \frac{1}{i^2} \qquad \text{(B5.6.2)}$$

An estimate of the variance can be obtained by substituting θ_w, the estimate of θ, into this expression. Only the first term depends inversely on the sequence length; the second term decreases with a_k, which increases approximately as the logarithm of the sequence length, so that the standard deviation of θ_w decreases only slowly with sequence length.

If the sites are completely independent of each other, rather than non-recombining, we can regard θ_w as the mean of m independent estimates of θ, and use the result that the variance of a mean of m independent variables with the same variance is equal to the variance divided by m (**Appendix A2.iv**). **Equation B5.6.2** is then replaced by a similar expression in which the last term is divided by m; the variance in this case declines with sequence length.

The derivation of the variance of the pairwise diversity estimate, π, is more complex, due to the non-independence of the terms involved (Tajima 1983). For the case of two alleles in the sample, it is the same as the formula for θ_w, because the two quantities are the same. The general formula for m non-recombining sites is:

$$V_\pi = \frac{(k + 1)\theta}{3(k - 1)m} + \frac{2(k^2 + k + 3)\theta^2}{9k(k - 1)} \qquad \text{(B5.6.3)}$$

For m independent sites, the last term is divided by m.

estimator, π, or the number of segregating sites, S_k. It is usually very hard or impossible to obtain analytic expressions for the distributions of such statistics, although this has been done for S_k in the case of a non-recombining sequence (Tavaré 1984).

It is, however, relatively simple to generate random gene trees for a non-recombining sequence by simulating the exponential distribution of branch sizes described in **Section 5.1.iii.b** above (**Figure 5.4**). Starting with $i = k$, we can draw a sequence of $k - 1$ random numbers that represent the times between successive coalescent events for a random gene tree like that in **Figure 5.4**. It is convenient to measure time on the *coalescent time scale* in which one time unit is equal to $2N$ generations, i.e., if the time to coalescence of i alleles is T, we work with $T_c = T/2N$ instead of T. The details of how to obtain random numbers describing the exponential distribution of T_c are described in **Box 5.7***. The next step in such a simulation is to add mutations onto the trees, and extract the sample statistics of interest. The procedure depends on the muta-

Box 5.7* SIMULATING A GENE TREE

We need to be able to obtain a random number generated by a computer program, which represents the exponential distribution. This can be done as follows, in the context of the coalescent process. At the stage when a gene tree still has i extant alleles, the cumulative probability function (**Appendix A2.ii**), representing the probability that the time until the next coalescence event is less than or equal to T, is the integral of **Equation 5.6b** from 0 to T, replacing $1/2N$ with $j_i/2N$. This is given by $1 - \exp(-j_i T/2N)$. For convenience, we can use the coalescent time scale in which one time unit is equal to $2N$ generations and determine a value of $T_c = T/2N$ for the simulation by the following method. It is known from probability theory that the cumulative probability function for a continuous variable is distributed as a uniform variate, with equally probable values between 0 and 1. To choose a value of T_c, we thus can draw a uniform random number with value X; if $X = 1 - \exp(-j_i T_c)$, we obtain T_c as $-\ln[(1 - X)/j_i]$. Methods for generating uniform random numbers are available as standard computer packages (e.g., *Numerical Recipes* at http://www.nr.com/), and can readily be used in a program for simulating a gene tree.

tion process that is to be modeled. We have already described the infinite sites model; with a mutation rate of u per site per generation, the number of mutations on a branch of length t_i (on the *coalescent time scale*, in units of $2N$) follows a Poisson distribution with mean $t_i mu$, for a sequence of length m. It is easy to simulate random draws from a Poisson distribution using standard computer algorithms (available from sources such as *Numerical Recipes*: see **Box 5.7***). By repeating this procedure independently for every branch of the gene tree, appropriate numbers of mutations can be assigned to each branch of the tree.

This simulation procedure yields a complete description of the properties of the simulated sample, because it generates the lengths of the branches of the gene tree, and these lengths determine the properties of the sequences that evolve in the tree. For instance, under the infinite sites model, each mutation is assumed to occur at a different site in the sequence. The frequency of a mutation in the sample is determined by where it occurred on a tree, with mutations that occurred early in the sample's ancestry tending to be at higher frequencies than ones that arose in the most recent branches of the tree (see **Figure 5.4**). In particular, mutations on the external branches of the tree occur only once in the sample.

The simulation procedure can then be repeated many times. Making many gene trees using the same assumptions yields accurate approximations to the probability distribution of the lengths of the branches of the tree, and therefore of the probability distribution of the diversity statistics in samples. This is extremely useful for testing for departures from the assumptions of the standard neutral model, as we discuss in more detail in **Section 6.4.iii**. Computer packages that run coalescent simulations and generate statistics that describe variability in a simulated sample are publicly available, e.g., DnaSP (http://www.ub.es/dnasp/), Genetree (http://www.stats.ox.ac.uk/~griff/research.htm), and ms (http://home.uchicago.edu/~rhudson1/).

It is important to note that this procedure assumes complete genetic linkage between the different sites along the sequence; this is realistic for most organelle and Y chromosomal sequences, but not for the nuclear genes of eukaryotes and even for many prokaryotes (Maynard Smith et al. 1993). In **Section 8.2.iii.b**, we will describe how to introduce genetic recombination into the simulation procedure, which is implemented in the programs just mentioned.

5.1.iii.f. *Microsatellite loci*

Microsatellite loci (**Section 1.2.iii.c**) represent quite a different type of genetic polymorphism, but coalescent simulations are again very useful for obtaining predictions about their properties (e.g., Valdes et al. 1993). Detailed reviews of the population genetics of microsatellites are provided in Goldstein and Schlötterer (1999), and we only sketch the basic ideas here. The simplest realistic model for microsatellites is the "two-phase" or "stepwise" mutation model (Slatkin 1995a,b), in which a variant with a given number of repeats has a constant probability of either gaining or losing one or more repeats. With this model, the variance in the difference in repeat numbers between two alleles drawn from a population is equal to the product of the rate at which the variance in repeat numbers increases as a result of mutation and the genealogical time separating them (Slatkin 1995a,b), i.e. this variance behaves similarly to pairwise diversity under the infinite sites model and is thus a useful diversity measure.

However, it is difficult to obtain analytical results for other aspects of microsatellite variability, and so coalescent simulations are usually used. Gene trees are constructed just as before, but then mutations at a single microsatellite locus are modeled. This is done by assigning mutations to each branch of a sampled gene tree, using a Poisson random variate, just as above. The change that occurs at each mutation is then decided by draws from the probability distribution of mutations assumed in the model being simulated (e.g., a probability of one-half of adding a repeat unit, and one-half of losing one), and the state of the locus is recorded. For microsatellite mutations, it is necessary to start at the base of the tree and work up towards the tips of the branches, because the allelic state changes each time a new mutation is added. By comparing the distributions of copy numbers in real samples of alleles with the results of simulating different mutation models, it is possible to discriminate among different mutational processes, and to use this knowledge to increase the power of hypothesis testing (e.g., Di Rienzo et al. 1998).

5.2. EFFECTIVE POPULATION SIZE

5.2.i. Introduction

Our discussion of genetic drift so far has been based on the Wright–Fisher model, and so we have ignored sex differences between members of the population and other complications. It is possible to predict the effects of genetic drift in more realistic situations without having to do elaborate calculations for each individual case. Early in the history of population genetics, Wright (1931) introduced the concept of the *effective population size*, conventionally denoted by N_e. This is a quantity that appears in the equations for genetic drift, replacing the number of breeding individuals N in the Wright–Fisher model. For a given situation, N_e is no longer the census number of breeding individuals, but depends on the mode of inheritance, the level of inbreeding in the population, the numbers of males and females, the variance in reproductive success, changes in population size over time, age structure, and any geographic and genetic structuring of the population. In general, the use of N_e only gives an approximation to the rate of drift and is often valid only asymptotically, i.e., after sufficient time has elapsed since the start of the process. Exact calculations are, therefore, often needed in applications where the population size is very small or the time scale is short, as in animal and plant improvement or in conservation breeding programs (Crow and Kimura 1970, Chapter 7; Caballero 1994; Wang and Caballero 1999; Vitalis 2002). Difficulties also arise when the distribution of offspring number among individuals is highly skewed, so that multiple coalescent events can occur in the same generation, even with a large population size; ways of dealing with this problem are discussed by Wakeley and Sargsyan (2009).

There is a large literature on effective population size (reviewed by Wang 2005); we consider only the most basic aspects here, assuming a panmictic population in which there is no subdivision according to spatial location or genotype (**Section 5.1.i.a**). Spatially and genetically structured populations are considered in **Chapter 7** and **Section 8.3**, respectively. We use the framework of coalescent theory, rather than classical methods that investigate changes in the inbreeding coefficient or the variance of allele frequencies. These methods are described by Crow and Kimura (1970, Chapters 7 and 8), Nagylaki (1992,

Chapter 9), and Ewens (2004, Chapter 3). The coalescent approach is essentially the same as the inbreeding approach, as mentioned in **Section 5.1.iii.b**, but has the advantage that the formula for N_e can be inserted directly in place of N into the results on the properties of DNA sequence diversity arising from coalescent theory. This is very useful for interpreting data on natural and human populations. We continue to assume discrete generations for most of the discussion, but briefly discuss populations structured by age or stage at the end of this section.

5.2.ii. Effective population size and the coalescent process

5.2.ii.a. *Defining effective population size*

Under the Wright–Fisher model, the probability of coalescence in one generation for a pair of autosomal alleles is $P_c = 1/2N$; alternatively, the mean time to coalescence is $T_c = 1/P_c = 2N$ generations (**Section 5.1.iii.b**). Can one define an effective number N_e that yields the correct T_c for other situations? We will show that this is possible by illustrating the approach with the case of a population with two sexes and discrete generations, allowing for nonrandom variation in reproductive success and various modes of inheritance. This involves a *structured coalescent* process in which there are several "compartments" in the population from which alleles can be sampled (Hey 1991). Alleles are initially sampled from one or more of these compartments, and the probabilities of allele movements to the other compartments, as we go back in time, are determined by the rules of inheritance. A useful simplification is to assume that alleles flow among the different compartments on a much faster time scale than the coalescent process. This assumption means that we can treat the sampled alleles as coming from the equilibrium state of the process (Nagylaki 1980; Nordborg 1997; Rousset 1999; Laporte and Charlesworth 2002; Nordborg and Krone 2002; Wakeley 2008, Chapter 6): the *fast time scale approximation*.

To give a specific example, with X-linked inheritance we might sample alleles from both males and females. Going back one generation, an X-linked allele from a male must come from his mother, but has an equal probability of coming from his grandfather or his grandmother, and so on back down the generations. During the many generations before coalescence is likely to occur, the ancestors of this allele are present in males 1/3 of the time, on average, and 2/3 of the time in females. Unless the population size is very small, these averages should accurately represent the times spent in the male and female compartments. The same averages apply to an X-linked allele sampled from a female (even though initially it has an equal chance of coming from the female's father or mother). We can therefore approximate the process by treating all X-linked alleles in a sample as having a probability of 1/3 of coming from a male and 2/3 from a female.

5.2.ii.b. *Calculating N_e with constant population size*

Consider two alleles that are sampled from distinct individuals to ensure their independence. In general, let α_r and α_s be the probabilities that they come from individuals of classes (compartments) r and s, respectively. In the case of the two sexes, r and s can either have the value f (denoting a female) or m (male). We want to find the probability, P_c, that two alleles coalesce in the previous generation. Clearly, this can happen only if both alleles came from the same individual in this generation. Let β_{rsu} be the probability that the pair of alleles with origins r and s both come from a parent of sex u (u also takes values of either f or m). Then write Θ_{rsu} for the probability that both alleles come from the *same* individual, given that an rs pair of alleles was derived from a parent of sex u. Finally, let γ_{rsu} be the probability that the alleles coalesce within that individual. Putting all these events together, we have:

$$P_c \approx \frac{1}{2N_e} = \sum_{rsu} \alpha_r \alpha_s \beta_{rsu} \Theta_{rsu} \gamma_{rsu} \qquad (5.10)$$

This is a general formula for the coalescent probability, not restricted to our example of two sexes; the approximately equals sign indicates that (as explained above) we use the equilibrium values for the probabilities of origin of the sampled alleles (the α values). To get useful results about N_e for different biologically important situations, we need to obtain expressions for the individual components of this expression. The values of the α, β, and γ coefficients are derived in **Box 5.8.i**; an outline of how to find the Θ coefficients is given in **Box 5.8.ii**, and expressions for the Θ coefficients in specific cases are given in **Table 5.1**. The reader who is not interested in the algebraic details can skip over this material.

> **Box 5.8** PROBABILITIES OF ORIGINS OF PAIRS OF ALLELES
>
> **B5.8.i.** The α, β and γ coefficients
>
> The α values in **Equation 5.10** are determined by the genetic system. With autosomal inheritance, $\alpha_f = \alpha_m = 1/2$. With X-linked inheritance, the argument given in the text shows that $\alpha_f = 2/3$ and $\alpha_m = 1/3$ (the reverse is true for Z-linked inheritance with female heterogamety). With Y-linked inheritance, alleles are inherited only from their father, so that $\alpha_f = 0$ and $\alpha_m = 1$; the reverse is true for maternally inherited cytoplasmic factors and for W-linked genes with female heterogamety.

The β coefficients are determined in a slightly different way. For autosomal inheritance, they are all 1/4, because each sex has an equal probability of contributing an autosomal allele to the progeny, so there is a probability of $(1/2)^2$ that any pair of alleles originates from a parent of a given sex. For X-linked inheritance, this is also true for alleles from females, so that $\beta_{fff} = \beta_{ffm} = 1/4$. But males inherit their X chromosome from females only, so that $\beta_{fmf} = 1/2$, $\beta_{fmm} = 0$, $\beta_{mmf} = 1$, and $\beta_{mmm} = 0$ (males and females are interchanged for Z-linked inheritance). For Y-linked inheritance, we need consider only male transmission, so that the only nonzero value is $\beta_{mmm} = 1$ (females are substituted for males under W-linked and cytoplasmic inheritance).

It is also easy to derive expressions for the γ coefficients. With autosomal inheritance, the chance that two alleles sampled from the same individual coalesce is $(1 + f_{IS})/2$, where f_{IS} is the inbreeding coefficient due to departures from random mating (**Section 1.3.ii**). The subscript IS was introduced by Wright (1951) to indicate that f_{IS} is the inbreeding coefficient for individuals (I) caused by nonrandom mating within a subpopulation of the species (S), as distinct from the inbreeding coefficient that describes the effects of genetic drift (see **Section 5.1.i.a**).

For autosomal inheritance, γ is independent of sex. With X-linkage, the same result holds for females, so that $\gamma_{rsf} = (1 + f_{IS})/2$. Males are effectively haploid, so that $\gamma_{ffm} = 1$ and no other terms need be considered. Males and females are exchanged for Z-linked inheritance. With Y-linkage, there is only male transmission through a haploid lineage, so that the only relevant term is $\gamma_{mmm} = 1$ (for W-linkage and cytoplasmic inheritance, males and females are exchanged).

B5.8.ii. The Θ coefficients

Consider a pair of alleles that both originated from female parents. We will derive an expression for Θ_{ffp} the chance that that they both came from the mothers of the individuals sampled. Assume that there are N_f females of breeding age in the parental generation, and that the kth female produces d_{fk} daughters that survive to breeding age. Let the mean of d_{fk} over all mothers be $\bar{d}_f$. The chance that a given female member of the progeny generation is a daughter of the kth female of the parental generation is $d_{fk}/(N_f\bar{d}_f)$. If the population size is sufficiently large, the chance that a second female member of the progeny generation is the sister of this individual is approximately $(d_{fk}-1)/(N_f \bar{d}_f)$. Summing over all individuals, we get:

$$\Theta_{fff} \approx \frac{\sum_k d_{fk}(d_{fk}-1)}{(N_f \bar{d}_f)^2} \qquad (B5.8.1)$$

This can usefully be simplified by noting that, with the same expected number of offspring for each individual (as assumed in the Wright–Fisher model), so that there is purely random variation in offspring number, the distribution of offspring number is a binomial distribution, which can be approximated by a Poisson distribution (see **Box 3.4** of **Section 3.2.ii**). If N_f is large enough, the variance of the number of daughters among the individuals in the population will also approach that given by a Poisson distribution, i.e., it will be approximately equal to $\bar{d}_f$. We can then write ΔV_{ff}, the deviation from the Poisson value of the variance of numbers of daughters per mother, as:

$$\Delta V_{ff} \approx V_{ff} - \bar{d}_f \qquad (B5.8.2)$$

and so we obtain:

$$\Theta_{fff} \approx \frac{(\Delta V_{ff}/\bar{d}_f^2) + 1}{N_f} \qquad (B5.8.3)$$

This can be simplified further when the population is stationary in size (so that each female produces an average of one daughter that survives to maturity), and there is a binomial distribution of the proportion of males and females among breeding individuals. Writing c for the expected proportion of sons, the expected number of daughters of a mother with a total of x offspring is $(1 - c)x$. Using the same type of argument as in **Problem 5.6**, it can be shown that:

$$\Delta V_{ff} = (1 - c)[(1 - c)V_f - 1] = (1 - c)^2 \Delta V_f \qquad (B5.8.4)$$

where ΔV_f is the deviation of the variance in total offspring number per female from that expected under purely random variation (Laporte and Charlesworth 2002). Similar expressions can be obtained for other offspring–parent pairs, but we omit the details. These results can be inserted into the expressions for the Θ values to obtain the equations for N_e shown in the text.

TABLE 5.1 Probabilities of origins of alleles from different classes of parents

Sexes of sampled individuals	Sex of Parent	Probability of common parent of given sex[1]
Female, female	Female	$\Theta_{fff} \approx \dfrac{\left(\Delta V_{ff}/\bar{d}_f^2\right)+1}{N_f}$
Male, male	Female	$\Theta_{mmf} \approx \dfrac{\left(\Delta V_{fm}/\bar{s}_f^2\right)+1}{N_f}$
Male, female	Female	$\Theta_{fmf} \approx \dfrac{\left(C_{ffm}/\bar{d}_f\,\bar{s}_f\right)+1}{N_f}$
Female, female	Male	$\Theta_{ffm} \approx \dfrac{\left(\Delta V_{mf}/\bar{d}_m^2\right)+1}{N_m}$
Male, male	Male	$\Theta_{mmm} \approx \dfrac{\left(\Delta V_{mm}/\bar{s}_m^2\right)+1}{N_m}$
Male, female	Male	$\Theta_{fmm} \approx \dfrac{\left(C_{mfm}/\bar{d}_m\,\bar{s}_m\right)+1}{N_m}$

[1] ΔV_{rs} is the excess over the Poisson value of the variance in number of offspring of sex s produced by breeding individuals of sex r; C_{rfm} is the covariance between the numbers of sons and daughters produced by breeding individuals of sex r; $\bar{d}_r$ and $\bar{s}_r$ are the mean numbers of daughters and sons produced by breeding individuals of sex r. (The numbers of offspring refer to numbers alive at the time of reproduction.)

5.2.iii. Effective population sizes for some biologically interesting cases

By substituting values of the coefficients described above into **Equation 5.10**, we can obtain expressions for N_e for specific cases of biological interest. The effective population size derived in this way is essentially a measure of the rate of coalescence of a pair of alleles, such that their expected time to coalescence is equal to $2N_e$. For consistency, we will apply this definition even to haploid organisms, such as bacteria, or haploid components of the genome, such as Y chromosomes.

5.2.iii.a. *Different modes of inheritance (constant population size)*
We start with autosomal inheritance and a constant population size, and allow for nonrandom variation in numbers of offspring per individual, with ΔV_f and ΔV_m representing excesses over the Poisson values of the variances in total off-spring numbers for males and females, respectively. As described in **Box 3.4** of **Section 3.2.ii**, a Poisson distribution applies when there is purely random variation in offspring numbers among individuals of a given sex, so that this can be treated as the null model.

To obtain a specific formula, we combine **Equation 5.10** with the results derived above, and those in **Box 5.8** of **Section 5.2.ii.b** and **Table 5.1**. If the breeding population is made up of N_f females of breeding age and N_m males, so that the sex ratio among adults is $c = N_m/(N_f + N_m)$, the effective population size with autosomal inheritance, N_{eA}, is given by:

$$\frac{1}{N_{eA}} \approx \frac{\left(1 + f_{IS}\right)}{4}\left\{ \frac{1}{N_f} + \frac{1}{N_m} + \frac{(1-c)^2\Delta V_f}{N_f} + \frac{c^2\Delta V_m}{N_m} \right\} \qquad (5.11)$$

where f_{IS} is the inbreeding coefficient of an individual due to an excess of inbred matings within the population over the expectation with random mating (Wright 1951).

Equation 5.11 brings out several important points. First, with no inbreeding and a Poisson distribution of offspring number for both sexes, $1/N_{eA}$ is simply:

$$\frac{1}{N_{eA}} \approx \frac{1}{4N_f} + \frac{1}{4N_m} \qquad (5.12a)$$

This is the classic result of Wright (1931). More exact calculations show that this is quite accurate even for a population size as small as 10 (Kimura and Crow 1963; Crow and Kimura 1970, Chapter 7). With a 1:1 sex ratio among breeding individuals, the effective size in this case is equal to the total popula-tion size ($N = 2N_f = 2N_m$), so that the population has the same properties as the Wright–Fisher model. If the numbers of females and males are not the same, the effective size is much closer to the smaller of N_f and N_m. For example, if there is only a small number of breeding males, the reciprocal of N_m dominates in **Equation 5.12a** (**Problem 5.8**).

With nonrandom variation in offspring numbers, but the same variance in offspring number for the two sexes, we get the following equation for the case of a 1:1 sex ratio:

$$\frac{1}{N_{eA}} \approx \frac{(2 + \Delta V)}{2N} \qquad (5.12b)$$

An excess variance in offspring numbers compared with random expectation thus reduces N_e below N. Conversely, if there is less than random variation, N_e can be greater than N; it equals $2N$ in the limit when all individuals have equal reproductive success. This is important for conservation breeding programs, where it is desired to maximize N_e so as to slow down the approach to homozygosity (Frankham et al. 2002, Chapter 17).

In animals, a major cause of a nonrandom distribution of reproductive success is *sexual selection*, when males compete with each other for access to mates (Andersson 1994). In the most extreme case, only one male succeeds in breeding, producing $2N_f$ offspring if the population size is stationary. The variance in offspring number among males is then $(2N_f)^2/(N_m) - (2N_f/N_m)^2 \approx 4N_f^2/N_m$ in a large population. With a 1:1 sex ratio and a random distribution of offspring among females, $\Delta V_m \approx 2N$, so that **Equation 5.11** gives $N_e \approx 4$. This situation is approached in practice in populations of domestic animals, where artificial insemination is used in selective breeding, causing serious problems with inbreeding (Brotherstone and Goddard 2005). The situation with such small N_e values, however, violates our assumption of at most one coalescent event per generation; Eldon and Wakeley (2006) and Wakeley and Sargsyan (2009) give a more rigorous treatment of such situations. The reduction in N_e in less extreme cases depends on the details of the mating system (Nunney 1991, 1993).

Equation 5.11 also shows that an excess of matings between relatives (relative to random mating) reduces N_e by a factor of $1/(1 + f_{IS})$. This is because inbreeding causes faster coalescence of an individual's maternal and paternal alleles than under random mating (Nordborg and Donnelly 1997); for instance, with self-fertilization, the chance that an individual's alleles coalesce in the preceding generation is 1/2. With partial self-fertilization at rate S in an hermaphrodite population, we saw in **Section 1.3.ii.c** that the equilibrium inbreeding coefficient is $S/(2 - S)$. Selfing causes N_e to be multiplied by a factor of $(2 - S)/2$ when there is a Poisson distribution of offspring number; this approaches 1/2 for 100% selfing (Pollak 1987; Laporte and Charlesworth 2002).

This result suggests that neutral variability within populations of highly self-fertilizing species, such as *Arabidopsis thaliana* and *Caenorhabditis elegans*, should be reduced to about half the value for randomly mating populations of similar size. These species indeed have low variability (Nordborg et al. 2005; Cutter 2005) compared with their outcrossing relatives (Wright et al. 2006; Cutter et al. 2006). Additional possible reasons for this low variability are discussed in **Sections 8.3.iii.a** and **8.3.v.a**.

5.2.iii.b. *Comparisons of different modes of inheritance*

The mode of inheritance, and factors such as the intensity of sexual selection, can greatly alter effective population sizes, and hence expected levels of neutral diversity. With X-linked inheritance, the results in **Table 5.1** yield:

$$\frac{1}{N_{eX}} \approx \frac{1}{9}\left\{\frac{4(1 + f_{IS})}{N_f} + \frac{2}{N_m} + \frac{4(1 + f_{IS})(1 - c)^2 \Delta V_f}{N_f} + \frac{2c^2 \Delta V_m}{N_m}\right\} \qquad (5.13)$$

Similarly, with Y-linked inheritance we obtain:

$$\frac{1}{N_{eY}} \approx \frac{2(1 + c^2 \Delta V_m)}{N_m} \qquad (5.14)$$

and for maternally transmitted organelle genomes or W-linkage:

$$\frac{1}{N_{eC}} \approx \frac{2\left[1 + (1 - c)^2 \Delta V_f\right]}{N_f} \qquad (5.15)$$

With random mating, a 1:1 sex ratio, and Poisson distributions of offspring numbers, **Equation 5.13** for X-linked inheritance gives $N_{eX} = 3N/4$, versus N in the autosomal case, consistent with the fact that there are only 3/4 as many X chromosomes as autosomes in the population. It is therefore often stated that X-linked nucleotide site diversity should be three-quarters of that of autosomal loci, and it is a common practice to adjust diversity estimates for X-linked loci by multiplying by 4/3, when comparing them with data for autosomal genes (e.g., Moriyama and Powell 1996).

But **Equation 5.13** shows that this is an oversimplification. The intensity of sexual selection can greatly alter effective population sizes. If there is strong sexual selection among males, such that $\Delta V_m \gg \Delta V_f$, the effective size for X-linked loci may approach or even exceed that for autosomal loci. N_{eX}/N_{eA} has an upper limit of 1.125; the reason this ratio can exceed 1 is that autosomes are transmitted through males more often than X chromosomes, and the males' effective population size is small. Surveys of SNP variability in the putatively ancestral African populations of *Drosophila melanogaster* show that mean silent-site diversity for X-linked loci is indeed slightly higher than for autosomal loci, consistent with the operation of very strong sexual selection, although other factors may be involved as well (Andolfatto 2001; Hutter et al. 2007; Vicoso and Charlesworth 2009). For ZW sex determination systems, the predicted difference between males and females is reversed. For Z-linked inheritance, N_{eZ}/N_{eA} with strong sexual selection can be as low as 9/16 = 0.5625. Data on DNA sequence variability in introns in domestic chickens gave a ratio of Z-linked to autosomal variability of 0.24, even lower than expected under strong sexual selection (Sundström et al. 2004).

Similarly, sexual selection can greatly reduce N_{eY} below the value of $N/4$ expected with a random distribution of offspring numbers and a 1:1 sex ratio, down to a minimum of $N_{eY}/N_{eA} = 0.125$. The opposite applies to maternally transmitted organelle genomes; with strong sexual selection, the ratio N_{eC}/N_{eA} can be much larger than 1, rather than the value of 1/4 expected with random offspring distributions for both sexes.

The relative levels of diversity for different modes of inheritance may also be affected by mutation rate differences between the genomes involved. For example, between-species comparisons (Lynch et al. 2006), using the methods described in **Section 6.1.ii**, and some direct estimates of mutation rates (Denver et al. 2000; Howell et al. 2003; Haag-Liautard et al. 2008), show that animal mitochondria often have much higher mutation rates than nuclear genes, while plant mitochondria and chloroplast genomes have lower mutation rates. In birds and mammals (but not *Drosophila*), the mutation rate is higher in the male germ line than in females (Miyata et al. 1987; Axelsson et al. 2004; Vicoso and Charlesworth 2006). The equations for diversity in such cases must use mutation rates obtained by weighting the male and female mutation rates by the appropriate values of α_m and α_f respectively, which are given in **Box 5.8** (Laporte and Charlesworth 2002).

5.2.iii.c. *Varying population size*

It is possible to generalize **Equations 5.6a** and **5.6b** for the Wright–Fisher model (**Section 5.1.iii.b**) to allow the population size N to change over time. We denote the population size by N_0 for the generation in which the alleles are sampled, and N_t for t generations back in time. Summing from $t = 0$ to $t = T - 1$, we see that the probability that the time to coalescence is greater than or equal to T is thus:

$$G(T) = \left(1 - \frac{1}{2N_0}\right)\left(1 - \frac{1}{2N_1}\right)\cdots\left(1 - \frac{1}{2N_{T-1}}\right) \qquad (5.16a)$$

If the N_t are sufficiently large, **Equation 5.16a** can be approximated by:

$$G(T) \approx \exp\left(-\frac{T}{2N_{HT}}\right) \qquad (5.16b)$$

where N_{HT} is the harmonic mean of N over T generations, such that $N_{HT} = T/(\Sigma_t 1/N_t)$ (Slatkin and Hudson 1991). A population that has recently grown from a much smaller size will thus have a much lower effective size than one that has always remained at its present size.

The result just derived suggests that the effective size of a population whose size varies can be determined by equating N_e and N_{HT}, at least if the variation in population size follows a process such that N_{HT} becomes independent of T over a time scale that is short compared with that of the coalescent process. This would be the case, for example, for regular cycles of population size with a period much shorter than the minimum population size. In this case, the expected coalescence time is similar to that with constant population size, i.e., approximately $2N_H$, where N_H is the harmonic mean defined above. This allows use of N_H instead of N in the results we derived above for expected neutral diversity.

More generally, the mean coalescence time is given by the sum or integral of $G(T)$ from 0 to infinity, and there is not necessarily any analytical formula for this. But, given a model of population growth, it is straightforward to simulate the coalescent process by using numerically generated values of $G(T)$ in conjunction with random numbers to generate gene trees, in a similar way to the constant population size case (**Section 5.1.iii.e**) (Slatkin and Hudson 1991). This method is frequently used in computational studies that fit data on microsatellite variants or SNPs to models of population history (e.g., Beaumont 1999, 2003; Beaumont et al. 2003).

Finally, for more complex population structures, we can replace the N_t in **Equation 5.15** by N_e values for each t, providing that the flow between different compartments equilibrates over a short time scale compared with changes in population size.

5.2.iii.d. *Age- and stage-structured populations and the Moran model*

There has been a good deal of work on how to determine effective population sizes for populations where reproductive individuals have a range of ages (see **Section 9.5** for a discussion of selection in such populations). This has been reviewed by Charlesworth (1994b, Chapter 2) and Charlesworth (2001a). Nordborg and Krone (2002) have studied the related problem of populations divided into different developmental stages, often distinguished by their mean body sizes.

The fast time scale approximation used for the two-sex case with discrete generations can be applied to these cases, where alleles flow between ages or stages, as well as sexes (Rousset 1999; Charlesworth 2001a; Nordborg and Krone 2002). Formulae for age-structured populations that are very similar to **Equations 5.11** and **5.13–5.15** can be obtained for the case of an age-structured population reproducing at discrete time intervals, such as annually breeding species of birds or mammals. Here individuals breed over several seasons, allowing one to assume that reproductive events in different years are independent of each other. There is, however, no completely satisfactory treatment of populations in which individuals reproduce more or less continuously, as in humans and many tropical species, since these cases violate the assumption of independence between reproductive events of the same individual at successive times (Charlesworth 2001a).

The simplest type of age-structured population model was proposed by Moran (1958) as an alternative to the Wright–Fisher model, and is often called the *Moran model*. This assumes a haploid population of N individuals. In each time unit, one individual is randomly chosen to die, and is replaced by an offspring produced by another randomly chosen individual. The expected lifespan of an individual is thus N time units, which corresponds to a single generation in the Wright–Fisher model. This causes the offspring distribution to depart from a Poisson distribution (it is close to a geometric distribution: Moran 1958). The rate of coalescence in this case is approximately $2/N$ per generation, twice the value for the corresponding haploid Wright–Fisher model. Mathematically, this model is often simpler to deal with than the Wright–Fisher model (Ewens 2004, Section 3.4), and has been used in some coalescent modeling (e.g., Neuhauser and Krone 1997; Ewens 2004, Section 10.6).

5.2.iii.e. *Estimating effective population sizes*

For very small populations, N_e can be estimated from observations on changes in frequencies of variants between generations, on the assumption that these are neutral (Crow and Morton 1955; Wang 2005; Waples and Yokota 2007; Jorde and Ryman 2007). Effective population sizes for large natural populations can be estimated from neutral diversity estimates, because diversity depends on effective sizes and neutral mutation rates, as shown by replacing N by N_e in **Equation 5.5** of **Section 5.1.ii.b**. If the mutation rate is known, either from a direct experimental estimate (**Section 1.3.iii.b**) or from data on DNA sequence divergence between species with known dates of separation (**Section 6.2.ii.a**), this equation can be used to estimate N_e (**Problem 5.9**). Some examples of estimates of N_e obtained in this way are shown in **Table 5.2**.

As might be expected from the results of **Sections 5.2.iii.a–5.2.iii.c**, effective population sizes estimated in these ways, or by using estimates of the relevant demographic variables together with population censuses, are often found to be much lower than the observed numbers of breeding individuals in natural or laboratory populations (Crow and Morton 1955; Frankham 1995). The human population, for example, is estimated to have an N_e of about 10,000–20,000, presumably because of its long past history of small numbers of individuals and relatively recent expansion in size (in terms of numbers of generations relative to N_e: see **Section 5.2.iii.c** above), e.g., Wall and Przeworski (2000) and Voight et al. (2005). Larger population sizes in the past than for extant populations have, however, sometimes been inferred from diversity estimates. This is illustrated by the case of Atlantic whales, which probably reflects the devastating effects of whaling on their population sizes (Roman and Palumbi 2003). The two methods of estimating N_e may therefore yield very different results if there have been large changes in population sizes, because the first approach relates to the present-day population size, and the second to the harmonic mean value of population size over the long period of time required for diversity levels to equilibrate.

TABLE 5.2 Effective population size estimates from standing levels of sequence variability.

Species[1]	N_e	Genes used	References
Species with good mutation rate estimates			
Humans	10,400	50 nuclear sequences	Yu et al. (2002)
Drosophila melanogaster (African population)	1,150,000	252 nuclear genes[2]	Shapiro et al. (2007)
Caenorhabditis elegans	80,000	6 nuclear genes	Cutter (2005)
Escherichia coli	25,000,000	410 genes	Charlesworth and Eyre-Walker (2006)
Species with mutation rate estimates based on sequence divergence from related species (see **Section 6.2.ii**) or mutation rates for related species			
Bonobo	12,300	50 nuclear sequences	Yu et al. (2004)
Chimpanzee	21,300	50 nuclear sequences	Yu et al. (2004)
Gorilla	25,200	50 nuclear sequences	Yu et al. (2004)
Gray whale	34,410	9 nuclear gene introns	Alter et al. (2007)
Caenorhabditis remanei	1,600,000	6 nuclear genes	Cutter et al. (2006)
Plasmodium falciparum	210,000–300,000	204 nuclear genes	Mu et al. (2002)

[1] For species such as *C. elegans*, with high levels of departures from panmixia, the estimates are for species-wide samples of alleles (see **Section 7.2.ii.d** for a discussion of the problem of defining N_e in such a situation).

[2] For data from genes, synonymous site diversity was used as the basis for the calculations, unless otherwise stated.

5.3. PROBABILITY DISTRIBUTIONS OF ALLELE FREQUENCIES

5.3.i. Introduction

We have concentrated so far on the effects of genetic drift on convenient quantities (such as DNA sequence diversity) that are empirically measurable in populations, rather than complete descriptions of variability; such quantities are often called *summary statistics*. However, we sometimes need a fuller description of the effects of drift. For example, it is possible to infer the action of selection on a gene from the frequencies of nucleotide variants in a sample taken from a population by comparing the numbers of sites at which variants are found at different frequencies with the numbers predicted by models of selection and drift (**Section 6.4.iv**). It is sometimes possible to predict the distribution of variant frequencies in a sample using coalescent theory, but this becomes very difficult when selection is acting, even using simulation methods that incorporate the effects of selection on genealogies (Neuhauser and Krone 1997; Coop and Griffiths 2004; Fearnhead 2006). In practice, other methods are usually employed. These date back to the early days of population genetics (Fisher 1930a,b; Wright 1931, 1937, 1938). Remarkably, many of the classical results are very useful for modern research on DNA sequence variation and evolution, as we discuss in **Section 6.4**. However, this final section of **Chapter 5** is more advanced than the earlier parts, and can be skipped in a first reading.

5.3.ii. The Wright–Fisher model with mutation and selection

5.3.ii.a. *The transition from one generation to the next*

The Wright–Fisher model introduced in **Section 5.1.i.a** can be generalized to include both mutation and selection at a single site with alternative variants A_1 and A_2. Let the frequency of A_2 among the N breeding individuals of the parental generation be q, and let the changes in one generation caused by the deterministic forces of mutation and selection be Δq_m and Δq_s, respectively. Formulae for these frequency changes are given in **Sections 1.3.iii.c** and **2.1.ii**. If the changes are small, the net change in q caused by the deterministic forces is well approximated by $M_{\delta q} = \Delta q_m + \Delta q_s$. The symbol $M_{\delta q}$ emphasizes that we are now dealing with a finite population and that we are interested in the mean value of the change in q over all the values generated by drift, conditional on the current value of q.

Under the Wright–Fisher model, with its assumption of a very large pool of gametes produced by the parents, it is reasonable to assume that the change due to mutation occurs at the time of gamete production, and that the new frequency of q produced by mutation ($q + \Delta q_m$) applies to zygotes formed in the next generation. We assume that selection operates on a large pool of the diploid offspring, after which population size is reduced to N breeding individuals by density-dependent mortality that regulates the population size. This assumption allows us to use a single value for the population size for every generation.

The joint effects of mutation and selection result in a new frequency $q' = q + M_{\delta q}$. The effect of drift reflects the random sampling of individuals from this large pool to form the N breeding individuals of the new generation. It seems reasonable to assume that we can replace the binomial distribution for the neutral case of **Section 5.1.i.a** by a binomial distribution with q' and variance $p'q'/2N$, where $p' = 1 - q'$. This is exact for a haploid model (replacing $2N$ by N). In the diploid case, however, there are three genotypes, A_1A_1, A_1A_2, and A_2A_2; the sample of N individuals is therefore really a *multinomial* distribution

(**Appendix A2.v.b**), generated by sampling N times from the three genotype frequencies after selection but before the population size is reduced to N by density-dependent mortality. This generates a binomial distribution of variant frequencies only if the two alleles of each individual are sampled independently of each other, i.e., when there are Hardy–Weinberg genotype frequencies. Selection, however, causes deviations from Hardy–Weinberg frequencies, whose magnitude is similar to the selection coefficient (s). The resulting departure from the binomial variance is of the order s/N, suggesting that the binomial distribution is a good approximation if selection is weak and the population is large, so that the departures from it are second order. A rigorous proof has been given by Ethier and Nagylaki (1980) and Nagylaki 1990), summarized by Nagylaki (1992, Section 9.9).

With weak selection, the expression for the variance can be simplified further by noting that $p'q'/2N$ can be replaced by $pq/2N$, neglecting terms of order s/N. Thus, the frequencies after mutation, selection, and drift are approximated by a binomial distribution with mean $q^* = q + M_{\delta q}$ and variance $V_{\delta q} = q(1 - q)/2N$. This is, however, just an approximation, valid only for weak selection and large population size, which ensure that second-order terms in s and $1/N$ are small compared with the main terms. Models of strong selection or small population size require computer simulations of the selection and sampling processes over successive generations (so-called *forward simulations*).

5.3.ii.b. *The probability distribution of variant frequencies*

We can now model the state of a population subject to a specified set of mutation, selection, and drift parameters. This is done by determining the probability $P(q, t, q_0)$ that the frequency of A_2 is q at time t, given an initial value q_0. For an autosomal locus, q can clearly take any one of the $2N + 1$ values 0, $1/2N$, $2/2N$, …, $1 - 1/2N$, 1. An example of such a distribution for a set of laboratory populations of *Drosophila melanogaster* was shown in **Figure 5.2** of **Section 5.1.i.a**.

With a frequency q in generation $t - 1$, the reasoning in the previous section implies that the probability of frequency $i/2N$ in generation t is given by the binomial probability distribution with mean $q' = q_{t-1} + M_{\delta q}$. Summing over all possible states in generation $t - 1$, we find that the formula for the binomial distribution in **Box 5.1** of **Section 5.1.i.a** gives the probability of a given value of q in generation t as:

$$P\left(\frac{i}{2N}, t, q_0\right) = \sum_{q=0}^{1} \binom{2N}{i} p'^{2N-i} q'^{i} P(q, t-1, q_0) \qquad (5.17)$$

By iterating this equation back over previous generations, we can express this probability as a function of the initial state, q_0.

Equation 5.17 is simple in form, and is in fact the standard representation of a multi-dimensional stochastic process by a *stochastic matrix* (**Appendix A1.iv.a**), whose components are the binomial probabilities on the right-hand side of the equation. It is, however, impossible to write down a simple algebraic expression for *P*, even without selection and mutation. **Equation 5.17** is useful for obtaining numerical results for relatively small populations, but becomes computationally demanding when *N* becomes very large; see Ewens (2004, Chapter 3) for a detailed discussion.

5.3.ii.c. *Diffusion equations*

The way out of this difficulty is to approximate the process further, by treating *q* as a continuous rather than discrete variable. We replace the probability *P* by a probability density $\phi(q, t, q_0)$, such that the probability that the frequency of A_2 lies in the small interval *q* to $q + \delta q$ is approximately $\phi(q, t, q_0) \, \delta q$ (this raises some questions about how to deal with the probabilities of frequencies 0 and 1; these are considered in **Section 5.3.iii.c** below). If the deterministic forces are weak and population size is small, we can also replace discrete time by a continuous time approximation, as in **Section 5.1.iii.b**. Using these approximations, **Equation 5.17** is replaced by an equation that describes the rate of change of ϕ over time (see **Section 1** of the **Appendix** to this chapter). The argument is very similar to that used to obtain the expression for the change in allele frequency in a cline (**Section 4.1.ii.b**).

Equations of this type are known as *diffusion equations*, because they were originally used in physics to describe the behavior of molecules diffusing by random motion. They were first introduced into population genetics by Fisher (1922, 1930a, 1930b, Chapter 4). Diffusion equations allow the change in probability density to be determined simply by the mean and variance of the change in allele frequency per generation—$M_{\delta q}$ and $V_{\delta q}$, respectively. There is thus no need to assume the binomial distribution used to obtain **Equation 5.17**, and so we can use diffusion equations in situations other than the Wright–Fisher model. Before discussing specific results, we will therefore consider a general representation of the effects of drift on variant frequencies.

5.3.iii. Allele frequency distributions: beyond the Wright–Fisher model

5.3.iii.a. *Variance effective population size*

In dealing with the coalescence of neutral alleles, we found that it is often possible to replace the population size *N* of the Wright–Fisher model by an effective population size, N_e (see **Sections 5.2.ii** and **5.2.iii** above). This suggests that we should also be able to replace the denominator *N* in the expression for the variance of change in allele frequency, $V_{\delta q} = pq/2N$, by an effective population size, often called the *variance effective population size*. Intuitively, it might seem that we can just use the expressions for N_e derived for the neutral coalescent process.

However, there are situations where this is not correct, as was first noticed by James Crow (Crow 1954; Kimura and Crow 1963). One way to see this is to compare the recursion relation for the probability of identity by descent, f (**Box 5.2** of **Section 5.1.i.b**), under the Wright–Fisher model (which is the same equation as for the coalescent process) with the recursion for the variance of allele frequency, V_q (**Box 5.3** of **Section 5.1.i.c**). If the population size changes between generations, the recursions for f and V_q differ; the change in f depends on the population size in the parental generation, whereas the change in V_q depends on that in the offspring generation. Calculating the variance population size can then become very complicated (Crow 1954; Kimura and Crow 1963; Crow and Kimura 1970, pp. 352–361; Crow and Denniston 1988; Nagylaki 1992, Section 9.13).

Most treatments using diffusion equations assume constant population size, so this is not necessarily a difficulty. Dealing with the complications that arise with two sexes, and when selection and mutation are acting, is a more intricate problem (Ethier and Nagylaki 1980; Nagylaki 1990). However, the argument given in **Section 5.3.ii.a** concerning departures from the binomial distribution caused by deterministic forces implies that the neutral coalescent expressions for N_e should provide good approximations when the evolutionary forces are all weak.

5.3.iii.b. *Solving the forward diffusion equation*

While the use of the diffusion equation approach reduces the problem to a single dimension (i.e., the current frequency of a variant or allele), determining the probability distribution of frequency as a function of time is still a very difficult problem. Some special cases, including neutral drift, have been solved (Kimura 1955, 1964; Crow and Kimura 1970, Chapter 8; Ewens 2004, Chapter 5), and methods have recently been developed for obtaining numerical solutions to the diffusion equation when the population size is changing over time (Evans et al. 2006). Most attention, however, has been given to the situation in which an equilibrium between the forces of drift, mutation, and selection has been reached, so that ϕ is stationary in form over time. A nontrivial equilibrium of this type is possible only if there is reversible mutation between two variants A_1 and A_2, allowing a population that has become fixed for one variant to regain variability. In the absence of mutations, fixation eventually occurs, as we saw above in **Section 5.1.i.b**.

Wright (1937) determined the equilibrium form of ϕ by setting the changes in the mean and variance of allele frequency to 0. The approximations that Wright used are equivalent to those used in deriving the diffusion equation. **Section 5A.2** of the **Appendix** to this chapter shows how to use the diffusion equation to derive **Equation 5A.7**, Wright's general formula for the equilibrium frequency distribution with mutation at rate u from A_1 to A_2 and v from A_2 to A_1. Assuming autosomal inheritance, constant fitnesses, and random mating, we can write the equilibrium distribution in the simple form:

$$\phi(q) \approx C\bar{w}^{2N_e} p^{4N_e v-1} q^{4N_e u-1} \tag{5.18}$$

where C is a constant that ensures that the integral of ϕ from 0 to 1 is 1 (**Problem 5.10***).

Formulae of this kind allow one to study the relative importance of selection, mutation, and drift, a question that is clearly fundamental for understanding evolution. Wright studied many other biologically important cases, e.g., Wright (1931, 1937, 1969, Chapters 13 and 14). A useful way of applying **Equation 5.18** is to think of it as representing the distribution of the state of the population over a large number of independent nucleotide sites scattered across the genome, all subject to the same mutational process and selection regime.

5.3.iii.c. *The equilibrium distribution with mutation between neutral variants*
We first consider mutation alone, i.e., the neutral model studied above. As shown in **Box 5.9***, with no selection **Equation 5.18** simplifies to:

$$\phi(q) \approx Cp^{4N_e v-1} q^{4N_e u-1} \tag{5.19}$$

This is a well-known distribution in statistics, the *beta distribution* (**Box 5.9***). From its properties, the means of p and q are $v/(u+v)$ and $u/(u+v)$, respectively (**Problem 5.11***), the same as the equilibrium frequencies, p^* and q^*, under mutation pressure in an infinite population, which were derived in **Section 1.3.iii.c**.

This model can be applied when mutation from A_1 and A_2 (and vice versa) corresponds to mutations between GC and AT base pairs in a DNA sequence (see **Section 5.3.iii.c** below). As discussed in **Section 10.1.ii**, there is considerable interest in what determines the base composition of genomes (the frequencies of GC versus AT base pairs). This simple model of reversible mutation between two states provides a basis for interpreting data on base composition. In this example, q^* corresponds to the mean frequency of AT pairs over many different sites sampled from a random individual from the population. The results above show that a long sequence of neutral nucleotide sites is expected to reach the equilibrium base composition produced by mutation pressure in an infinite population (Sueoka 1962, 1988).

In reality, there are, of course, four possible states at a given nucleotide site, and a complete model should thus consider all possible mutational transitions among them, not just GC to AT. An analytic expression for the multiple-state case has so far been derived only when all variants mutate at the same rate to a

Box 5.9* AN EXAMPLE OF AN EQUILIBRIUM PROBABILITY DISTRIBUTION

With constant fitnesses, the contribution of selection to $M_{\delta q}$ in **Equations 5A.4a** and **5A.4b** of the **Appendix** to this chapter can be written as $pq(\mathrm{d}\ln(\bar{w})/\mathrm{d}q)/2$ (see **Section 2.1.ii.e**), and the mutation term is $up - vq$ (**Section 1.3.iii.c**). Since $V_{\delta q} = pq/2N_e$, this gives:

$$\frac{M_{\delta q}}{V_{\delta q}} = 2N_e\left[\frac{\mathrm{d}\ln\bar{w}}{\mathrm{d}q} + \left(\frac{u}{q} - \frac{v}{p}\right)\right]\qquad(\text{B5.9.1})$$

Substituting into **Equation 5A.7**, and carrying out the integration, we obtain **Equation 5.18** of the main text.

If there is no selection, this yields **Equation 5.19** of the main text. The constant C in that equation is equal to the reciprocal of the *beta function*, defined as:

$$\mathrm{B}(\alpha,\beta) = \int_0^1 x^{\alpha-1}(1 - x)^{\beta-1}\mathrm{d}x = \frac{\Gamma(\alpha)\Gamma(\beta)}{\Gamma(\alpha + \beta)}\qquad(\text{B5.9.2})$$

(In this case, $\alpha = 4N_e v$ and $\beta = 4N_e u$.) Here, Γ is the *gamma function*:

$$\Gamma(\alpha) = \int_0^\infty x^{\alpha-1}\exp(-x)\mathrm{d}x\qquad(\text{B5.9.3})$$

which has the useful property that $\Gamma(\alpha + 1) = \alpha\Gamma(\alpha)$, implying that $\Gamma(\alpha + 1) = \alpha!$ for integral values of α. This result is used in **Box 5.11*** below.

given state (Wright 1969, Chapter 14; Ewens 2004, Section 5.10); there is, however, no reason to doubt the general conclusion that the equilibrium base composition with neutral mutation and drift is the same as the deterministic equilibrium.

Although drift does not affect the mean frequencies of variants, it produces a scatter of variant frequencies at individual sites, centered around the deterministic equilibrium value q^* (**Figure 5.5**). From the known properties of the beta distribution (**Problem 5.10***), the variance of the distribution is:

$$V_q \approx \frac{p^* q^*}{1 + 4N_e(u + v)}\qquad(5.20)$$

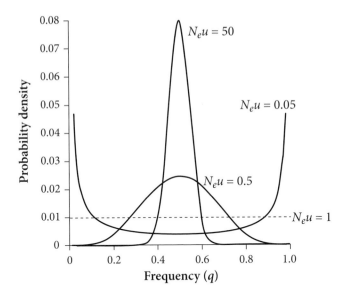

FIGURE 5.5 The equilibrium probability distribution of frequencies of neutral variants, from **Equation 5.19**, showing that genetic drift produces a scatter around the deterministic equilibrium value q^*, but does not affect the expected mean variant frequency. In the example, the forward and backward mutation rates are the same ($u = v$), so that $q^* = 0.5$. Lower values of $N_e u$ values lead to greater scatter.

We can express the scatter of the distribution in terms of an inbreeding coefficient, F, using the following argument. Normalizing the variance in allele frequency by its maximal value $p^* q^*$ (**Section 5.1.i.c**), we can define the inbreeding coefficient as:

$$F \approx \frac{1}{1 + 4N_e(u + v)} \tag{5.21}$$

As might be expected intuitively, F is close to 1 if $4N_e(u + v)$ is $<< 1$, indicating a wide scatter around the mean, with most sites fixed for one variant or the other (**Box 5.10*** shows how to calculate the proportion of fixed sites). Because mutation rates at single nucleotide sites are usually of the order of 10^{-9} to 10^{-8} (**Section 1.3.iii.b**) and effective population sizes for multicellular organisms are not usually more than a few millions (**Section 5.2.iii.e**), this condition is likely to hold for many cases of interest. Thus, under mutation pressure alone, variant frequencies will depart from mutational equilibrium with a wide distribution of frequencies at different polymorphic sites, and many sites will be fixed

Box 5.10* THE EQUILIBRIUM FREQUENCIES OF THE FIXED CLASSES

One approach to determining the approximate equilibrium frequencies of sites fixed for A_1 or A_2 (denoted by f_0 and f_1) is to equate these to the integrals of ϕ between $q = 0$ and $1/2N$, and between $q = 1 - 1/2N$ and 1, respectively. This is justified heuristically by the requirement for the integral of ϕ over its whole range to equal 1. The distribution between $q = 1/2N$ and $1 - 1/2N$ describes the frequencies at sites that are segregating; the probability, f_i, that a site takes frequency $i/2N$ is approximated by $\phi(i/2N)/2N$, because the variant frequencies are in steps of $1/2N$. Thus, the remainder of the integral of ϕ, after integrating between $1/2N$ and $1 - 1/2N$, approximates the probabilities of the fixed classes.

An alternative method was used by Wright (1931), who determined the conditions for the mutational flow of probability out of the fixed classes to equal the flow in the opposite direction from the adjacent classes. This method can be shown to give the same results as the first method (Crow and Kimura 1970, p. 441). These are only approximations, however, and may well be inaccurate in individual cases (Ewens 2004, Section 5.7).

For example, in the case of mutation alone, the first method gives the frequency of the class fixed for A_1 as:

$$\int_0^{1/2N} \phi(q)\mathrm{d}q \approx C \int_0^{1/2N} q^{4N_e u-1} \mathrm{d}q = \frac{C/(2N)^{4N_e u}}{4N_e u} \tag{B5.10.1}$$

The frequency of the class fixed for A_2 is obtained by substituting v for u in this equation. If $4N_e u$ and $4N_e v$ are both $\ll 1$, the infinite sites model is approached. The term involving $1/2N$ raised to the power of $4N_e u$ and $4N_e v$ in this equation is then approximately 1, so that the ratio of the frequencies of the classes fixed for the A_2 and A_1 classes is approximately u/v. A fraction $u/(u + v)$ of fixed sites is thus in the state A_2, just as for the deterministic equilibrium (**Section 1.3.iii.c**).

for one variant or the other. This corresponds well with what is seen in surveys of DNA sequence variation (**Section 1.2.iii.b**).

The mean diversity, H^*, in the population is equal to the expectation of $2pq$, which is equal to $2p^*q^* - 2V_q$. Using **Equation 5.21**, some simple algebra shows that:

$$H^\star \approx \frac{2(4N_e u)(4N_e v)}{4N_e(u + v)[1 + 4N_e(u + v)]}$$ (5.22a)

If both $4N_e u$ and $4N_e v$ are $\ll 1$, we obtain the following approximation:

$$H^\star \approx 4N_e \tilde{u}$$ (5.22b)

where $\tilde{u} = 2uv/(u + v)$.

Equation 5.22b has the same form as the infinite sites result derived earlier (**Equation 5.5** of **Section 5.1.ii.b**), indicating that $\tilde{u}$ can be equated to the neutral mutation rate in the infinite sites model. This is because, as shown in **Box 5.10***, a fraction $u/(u + v)$ of the fixed sites are in the state A_2 (yielding $vu/(u + v)$ mutations to A_1 each generation), and a fraction $v/(u + v)$ are A_1 (yielding $uv/(u + v)$ mutations to A_2 each generation). The overall rate of input of mutations from fixed sites is thus $\tilde{u}$; under the assumptions of the infinite sites model, nearly all sites are fixed, so this is close to the overall mutation rate. The identity probability, coalescent theory, and diffusion equation methods all yield

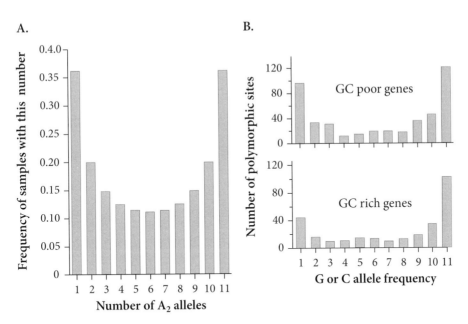

FIGURE 5.6 The expected distribution of frequencies of variants at segregating sites in a sample from a population at equilibrium under drift and mutation. The prediction (from **Equation B5.11.2** in **Box 5.11***) applies to a site where mutation occurs between two neutral variants, A_1 and A_2 (e.g., GC to AT mutations). In the example, the sample size, k, is 12. **B.** An example of data on GC to AT mutations in a population of *Drosophila melanogaster*. [Adapted from Figure 1 of Galtier et al. (2006).]

the same approximation for the expected diversity under the infinite sites assumption, indicating their underlying unity.

We can also use **Equation 5.19** to determine the equilibrium frequency distribution of variants in a sample of k alleles from a population (**Box 5.11***), as shown in **Figure 5.6A**. As would be expected intuitively, in the absence of selection this distribution is symmetrical: in a sample of size k, for example, configurations with i A_2 alleles, or with $k - i$ A_1 alleles, are equally probable. In the case of GC to AT mutations, a survey of segregating sites should yield a distribution of this type, if the population is at equilibrium and no forces other than drift or mutation are acting. This means that we can test for the action of such forces, or for departure from equilibrium, by testing for a departure of the distribution from this expectation. **Figure 5.6B** shows an example, using data on the frequencies of GC versus AT SNPs from a large set of noncoding

---

Box 5.11* THE DISTRIBUTION OF VARIANT FREQUENCIES IN A SAMPLE

The probability that we observe i copies of A_2 and $k - i$ copies of A_1 in a sample of k alleles, assuming a frequency q of A_2 in the population as a whole, is given by the binomial distribution with parameter q. The overall probability is obtained by integrating over the probability distribution of q, i.e., by:

$$C\binom{k}{i}\int_0^1 p^{4N_e v + k - i - 1} q^{4N_e u + i - 1}\, dq = \frac{Ck!\,B(4N_e v + k - i, 4N_e u + i)}{i!(k - i)!} \tag{B5.11.1}$$

where B is the beta function (**Box 5.9*** in **Section 5.3.iii.c** above).

If $4N_e u$ and $4N_e v$ are < 1, B in **Equation B.5.11.1** is approximated by $B(k - i, i)$, provided that we restrict ourselves to segregating sites ($1 \le i \le k - 1$). From the properties of the beta function (see **Box 5.9***), $B(k - i, i) = (k - i - 1)!(i - 1)!/(k - 1)!$. The right-hand side of **Equation B.5.11.1** therefore simplifies to $Ck/i(k - i)$.

The probability of observing a segregating site with i copies of A_2, among samples with segregating sites, is thus equal to:

$$P_i = \frac{1}{b_k}\left(\frac{1}{i} + \frac{1}{k - 1}\right) \tag{B5.11.2}$$

where $b_k = \Sigma_i[1/i(k - i)] = \Sigma_i[1/i + 1/(k - i)] = 2a_k$, and a_k is Watterson's correction factor (**Box 1.3** of **Section 1.2.iii.b**).

sequences in *Drosophila melanogaster* (Galtier et al. 2006). There is a clear bias towards higher frequencies than 0.5 of GC variants, which means that the assumption of an equilibrium distribution under neutrality is incorrect in this case. The possible forces acting on SNPs in cases like this will be discussed in **Section 10.1.iv.a**.

5.3.iii.d. *The equilibrium distribution with mutation and selection*
We can use **Equation 5.18** in **Section 5.3.iii.b** to examine the importance of selection relative to mutation and drift. A general insight is provided by the fact that $\bar{w}$ is raised to the power of $2N_e$ in this expression. Given the fact that N_e for a species is likely to be at least several thousand (**Section 5.2.iii.e**), this immediately suggests that even weak selection will overcome the effects of drift, producing a large peak in the allele frequency distribution around the point where $\bar{w}$ is at its maximum.

This conclusion is confirmed by examining specific cases. Consider, for example, selection against A_2 with semidominance (i.e., the fitnesses of A_1A_1, A_1A_2, and A_2A_2 are 1, $1 - s/2$, and $1 - s$, respectively). The mean fitness is $1 - sq$, and so the term involving $\bar{w}$ is equal to $(1 - sq)^{2N_e}$. This can be approximated by $\exp(-2N_e sq)$, and so we obtain:

$$\phi(q) \approx C \exp(-2N_e sq) p^{4N_e v - 1} q^{4N_e u - 1} \tag{5.23}$$

If $N_e s \gg 1$, the exponential term will be very small for large q, so that the probability of a high value of q is small (**Figure 5.7**). Conversely, if $N_e s \ll 1$, the

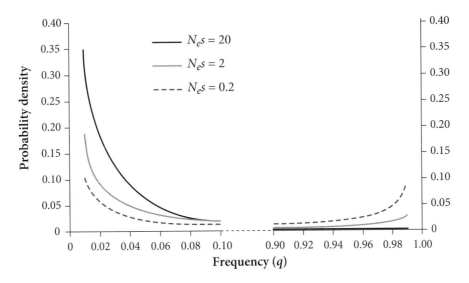

FIGURE 5.7 The equilibrium distribution with mutation and selection with different $N_e s$ values, from **Equation 5.23**. This shows that strong selection prevents high q values, while weaker selection leads to an outcome similar to the neutral one shown in **Figure 5.5**.

exponent decreases only slightly for high values of q, and the distribution is close to that under drift and mutation, with a U-shaped distribution of frequencies and a large fraction of sites fixed for A_2. For intermediate values of $N_e s$, the effect of drift is non-negligible, but the distribution has a mean of $q^* = 2u/s$, the equilibrium value for a large population.

It is straightforward to extend this approach to other situations, such as heterozygote advantage, by inserting the appropriate expression for $\bar{w}$ into **Equation 5.18**. Wright (1969, Chapter 13) describes examples for several different types of selection.

5.3.iii.e. *Biological implications*

With mutation and selection, we have seen that it is the product of N_e and the strength of the deterministic force that controls the outcome of evolution, not either factor on its own. If N_e is very large, as seems to be the case for most species, this suggests that even a very small selection pressure (e.g., $s = 0.0001$) will control the outcome of evolution and bring the population close to the equilibrium state under selection. In the next chapter, we follow up on this result by examining the effect of weak selection on the probabilities of fixation of mutant alleles in a finite population (**Section 6.2.iii.a**), thus extending the results of **Section 3.2**.

This type of result famously led R. A. Fisher (1930b, Chapter 4) to the conclusion that drift is an unimportant factor in evolution, and that natural selection acting on genetic differences among individuals within a population predominates over other forces (see the quotation at the beginning of this chapter). While this conclusion is almost certainly correct for phenotypic traits affected by selection (**Chapters 2–4),** the considerations mentioned in the **Introduction** to this chapter mean that it is important to take drift seriously as a factor in evolution. We therefore continue to study it in **Chapter 6** by examining the variation and evolution of DNA sequences when mutations are close to being selectively neutral. In **Chapter 7**, we examine the effects of limited migration among populations of a single species, allowing us to relax the assumption of panmixis that has been made in this chapter.

APPENDIX TO CHAPTER 5

Diffusion equations

5A.1. GENERAL FORMULATION

Let $\phi(q, t)$ be the probability density of frequency q in generation t (for convenience, we omit the dependence on the initial frequency, q_0), and write $g(\delta q, q - \delta q)$ for the probability density of a change from $q - \delta q$ to q over one generation. From first principles, we have:

$$\phi(q, t) = \int \phi(q - \delta q, t - 1)g(\delta q, q - \delta q)\mathrm{d}(\delta q) \tag{5A.1}$$

where the integration is taken over all possible values of δq.

Using Taylor's theorem (**Appendix A1.ii.a**), we can express the integrand as:

$$\phi(q - \delta q, t - 1)g(\delta q, q - \delta q) = \phi(q, t - 1)g(\delta q, q)$$
$$-\delta q\frac{\partial(\phi g)}{\partial q} + \frac{(\delta q)^2}{2}\frac{\partial^2(\phi g)}{\partial q^2} + o\left[(\delta q)^2\right] \tag{5A.2}$$

where ϕ and g in the derivatives are evaluated at time $t - 1$ and allele frequency q, and $o(x)$ indicates a quantity whose ratio to x tends to 0 as x approaches 0.

If the deterministic forces are weak, and the population size is sufficiently large, δq will be small, so that the terms in $o[(\delta q)^2]$ can be neglected, to a high level of accuracy. We can therefore approximate **Equation 5A.1** by:

$$\phi(q,t) \approx \int\left[\phi(q,t - 1)g(\delta q,q) - \delta q\frac{\partial(\phi g)}{\partial q} + \frac{(\delta q)^2}{2}\frac{\partial^2(\phi g)}{\partial q^2}\right]\mathrm{d}(\delta q) \tag{5A.3}$$

This can be simplified by noting that:

$$\int \phi(q,t - 1)g(\delta q,q)\mathrm{d}(\delta q) = \phi(q,t - 1)\int g(\delta q,q)\mathrm{d}(\delta q) = \phi(q,t - 1)$$

(using the fact that the integral of a probability density over its range is 1).

Similarly,

$$\int (\delta q)\frac{\partial(\phi g)}{\partial q}\mathrm{d}(\delta q) = \frac{\partial}{\partial q}\left[\phi\int (\delta q)g(\delta q)\mathrm{d}(\delta q)\right] = \frac{\partial(\phi M_{\delta q})}{\partial q}$$

where $M_{\delta q}$ is the expected change in allele frequency at frequency q.

In addition, we have:

$$\int (\delta q)^2 \frac{\partial^2 (\phi g)}{\partial q^2} \mathrm{d}(\delta q) = \frac{\partial^2}{\partial q^2} \left[\phi \int (\delta q)^2 g(\delta q) \mathrm{d}(\delta q) \right] = \frac{\partial^2 \left[\phi \left(V_{\delta q} + M_{\delta q}^2 \right) \right]}{\partial q^2}$$

where $V_{\delta q}$ is the variance of the change in allele frequency at frequency q.

Since $M_{\delta q}$ is assumed to be small, the term $M_{\delta q}^2$ in this equation can be neglected, yielding a further simplification. Putting all of these terms together, we therefore obtain the following expression:

$$\phi(q,t) - \phi(q,t-1) \approx \frac{\partial(\phi M_{\delta q})}{\partial q} + \frac{1}{2} \frac{\partial^2 \left(\phi V_{\delta q} \right)}{\partial q^2} \tag{5A.4a}$$

If we replace discrete time by continuous time, with one time unit corresponding to a generation (replacing $\approx$ by $=$ for convenience), this can be written as:

$$\frac{\partial \phi(q,t)}{\partial t} = -\frac{\partial(\phi M_{\delta q})}{\partial q} + \frac{1}{2} \frac{\partial^2 \left(\phi V_{\delta q} \right)}{\partial q^2} \tag{5A.4b}$$

This is a second-order, linear partial differential equation with many applications outside population genetics. It is known as the *Kolmogorov forward equation* or *Fokker–Planck* equation. For a more rigorous treatment, see Ewens (2004, Chapter 4).

5A.2. DETERMINING THE EQUILIBRIUM DISTRIBUTION OF ALLELE FREQUENCIES

At equilibrium between drift and the deterministic forces, the rate of change of ϕ given by **Equations 5A.4a** and **5A.4b** must be 0; i.e., we have:

$$\frac{\partial(\phi M_{\delta q})}{\partial q} - \frac{1}{2} \frac{\partial^2 \left(\phi V_{\delta q} \right)}{\partial q^2} = 0 \tag{5A.5}$$

Integrating this equation, we have:

$$P(q) = \phi M_{\delta q} - \frac{1}{2} \frac{\partial \phi V_{\delta q}}{\partial q} = k \tag{5A.6a}$$

where k is a constant of integration (**Appendix A1.iii**).

A lengthy argument, which we omit, shows that $P(q)$ is the rate at which probability flows across the point q, where $P(q)$ is positive for flow in the direction of increasing q (Wright 1945; Kimura 1964), i.e., $P(q)$ is the difference between the number of jumps per generation across q from the left and right, respectively. A stationary distribution, with no net transfer of probability across the classes, requires P to be 0 over all values of q between 0 and 1 (i.e., $k = 0$), giving:

$$\phi M_{\delta q} = \frac{1}{2} \frac{\partial \left(\phi V_{\delta q} \right)}{\partial q} = 0 \qquad (5A.6b)$$

Dividing both sides of **Equation 5A.6a** by $\phi\, V_{\delta q}$, taking the integral of the resulting expression and rearranging, we obtain:

$$\ln \left(\phi V_{\delta q} \right) = A + 2 \int \frac{M_{\delta q}}{V_{\delta q}} \mathrm{d}q$$

where A is another constant of integration.

This can be rearranged to give the final expression:

$$\phi(q) = \frac{C}{V_{\delta q}} \exp 2 \int \frac{M_{\delta q}}{V_{\delta q}} \mathrm{d}q \qquad (5A.7)$$

where C is such that the integral of ϕ between 0 and 1 is equal to 1, since ϕ is a probability density.

PROBLEMS

5.1. Consider a polymorphic site with neutral variants A_1 and A_2 at frequencies p_0 and $q_0 = 1 - p_0$, respectively.

    i. Using the results in **Section 5.1.i.b**, show that a large set of completely isolated populations with initial frequencies of neutral variants A_1 and A_2 at a site of p_0 and $q_0 = 1 - p_0$, respectively, will eventually become fixed, with probabilities p_0 and q_0 of fixation for A_1 and A_2.

    ii. Show that the variance in allele frequency among this set of fixed populations is $p_0 q_0$.

5.2.   **i.** Show that **Equation B5.4.2** of **Box 5.4** in **Section 5.1.ii.a** leads to **Equation 5.4**, and that this gives **Equation 5.5** of **Section 5.1.ii.b** when Nu is small.

    ii. Calculate the value of H^* from **Equation 5.4** when $Nu = 0.05$ and 0.005, respectively, and compare these with the corresponding values of π in **Equation 5.5**.

5.3 Show that the expected value of the mean nucleotide site diversity over a set of m sites is unaffected by a lack of independence among the values at each site, and that the variance of this mean *is* affected by non-independence. Do we expect the variance to be increased or decreased by correlations between sites?

5.4* Determine the mean and variance of the exponential distribution with parameter λ, such that the probability density of x is $\lambda \exp(-\lambda x)$ (**Appendix A2.v.c**). Apply these results to the mean and variance of the time to coalescence of a pair of alleles in a Wright–Fisher population of size N (**Section 5.1.iii.b**).

5.5 Using the results in **Sections 5.1.iii.b** and **5.1.iii.c**, derive an expression for the time to coalescence of the most recent common ancestor of a set of k alleles in a Wright–Fisher population. (Use the time taken for i alleles to coalesce to $i - 1$, for all possible values of i.)

5.6 Show why the variance of the number of segregating sites, S, for a non-recombining sequence is the sum of the variance of the number of mutations that occur on a tree with length $E\{L\}$, and the product of the variance of the length of the tree (V_L) with the square of the mutation rate for the whole sequence (see **Box 5.6** of **Section 5.1.iii.e**).

5.7 Use **Equation B5.6.2** of **Box 5.6** in **Section 5.1.iii.e** to determine the variance and standard deviations of θ_w for the cases of no recombination among sites with complete independence between them, respectively. Assume that the sequence length is either 100 or 1000 bp, with a sample size of 5 alleles, and that the true value of θ is 0.01.

5.8 Use the results of **Section 5.2.iii.a** to calculate the effective sizes for autosomal loci in randomly mating populations with Poisson distributions of offspring numbers, when there are 50 breeding males and 50 breeding females, and when there are 5 males and 95 females. Compare the results with the case when there are 50 individuals of each sex of breeding age, but the variance in the offspring number of males is 10 rather than 2. Comment briefly on the relation between N and N_e in each of these cases.

5.9 A survey of a large number of autosomal genes in an African and a U.S. population of *D. melanogaster* gave mean π values for synonymous sites of 0.023 and 0.011, respectively. The estimated mutation rate per nucleotide site per generation is 5×10^{-9} (**Section 1.3.iii.b**). Use **Equation 5.5** of **Section 5.1.ii.b** to estimate the effective sizes for these two populations. Suggest reasons why the two estimates are very different, in light of the fact that the closest relatives of *D. melanogaster* live in Africa.

5.10* Derive **Equation 5.18** of **Section 5.3.iii.b** from the general equation for the stationary allele frequency distribution, **Equation 5A.7** of the **Appendix** to this chapter, by using **Equation B5.9.1** in **Section 5.3.iii.c**.

5.11* Determine the mean and variance of the beta distribution by using **Equations B5.9.2** and **B5.9.3**. Use **Equation 5.19** to show how this gives the mean of the allele frequency distribution and **Equation 5.20**.

CHAPTER SIX

Molecular Evolution and Variation

CHAPTER SUMMARY

The study of evolution at the molecular level requires comparisons of DNA and protein sequences between different species and within species. We start by showing how divergence between homologous sequences can be measured. We then relate sequence divergence over time to models of the underlying rates of change at individual nucleotide sites, emphasizing the highly stochastic nature of sequence changes. Rates of evolution differ greatly between nonsynonymous and synonymous or silent nucleotide sites, so we also discuss how divergence can be estimated for different classes of sites.

The basic population genetics theory needed to interpret data on rates of sequence evolution starts with the model of selectively neutral evolution. For comparisons between sufficiently divergent species, the rate of neutral evolution is equal to the mutation rate, consistent with the "molecular clock" (the observed approximate constancy of the rate of sequence evolution). On the neutral theory, the number of differences accumulated in a given length of sequence over a period of time should follow a Poisson distribution, and this provides one means of testing for neutrality.

Other tests that can detect departures from neutral expectations are based on the population genetics theory of how mutation, genetic drift, and selection jointly determine polymorphism levels and rates of sequence divergence between species. We therefore incorporate selection into the models, showing how this affects the probability of fixation of weakly selected mutations, and relating the results to the expected rate of sequence evolution. Lower levels of nonsynonymous than synonymous divergence are generally observed when coding sequences are compared between species; this observation implies selection against deleterious amino acid sequence mutations (purifying selection). For a small proportion of genes, the rate of nonsynonymous evolution is significantly higher than for synonymous sites, implying the fixation of selectively favorable amino acid sequence mutations (positive selection).

Models of polymorphism and between-species divergence show how comparisons of nonsynonymous and silent polymorphism levels with the corresponding values for divergence can be used to detect both purifying and positive selection on amino acid sequences, using methods such as the McDonald–Kreitman test. Selection acting on noncoding sequences can be studied with similar methods. This approach can also provide estimates of the proportion of differences between related species in amino acid sequences or noncoding sequences caused by the fixation of positively selected mutations. We describe how the observed frequencies of polymorphic nucleotide variants can be compared with theoretical predictions, and used to detect selection and estimate its strength. In describing these tests and estimation methods, we consider some of their difficulties and biases, including the effects of factors such as population size changes.

> In many respects the ideal measurement of evolutionary rate would
> be the amount of genetic change in continuous (ancestral and
> descendant) populations per year . . . .
>
> G. G. Simpson, *The Major Features of Evolution*, 1953, p. 4

INTRODUCTION

To understand DNA sequence evolution and variation, a fundamental first step is to be able to interpret data on the rate of evolution of DNA sequences. The first section of this chapter therefore describes how to estimate the numbers of DNA sequence differences between species and rates of DNA sequence evolution. Such sequence differences are now widely used to construct phylogenies, an important part of evolutionary biology, and the methodologies for this present some formidable problems, as reviewed by Felsenstein (2004) and more briefly by Graur and Li (2000, Chapter 5). Here, we deal with a different set of problems—the evolutionary processes of molecular evolution and variation.

A fundamental question, which has occupied the attention of evolutionary biologists since the beginnings of the study of molecular evolution in the 1960s, concerns the extent to which the evolution of DNA and protein sequences involves the random fixation by genetic drift of neutral or nearly neutral mutations, compared with the spread of favorable mutations under the control of natural selection. The neutralist view was eloquently argued in the classic book by Motoo Kimura (1983), whereas John Gillespie (1991) presented the contrary case for selection.

To resolve this controversy (or rather, to evaluate the relative contributions of the various processes involved in sequence evolution), it is necessary to employ the population genetics theory of molecular evolution and variation

(**Sections 6.2** and **6.3**). We show how this theory can be used to test for the action of various forms of selection and to estimate the strength of selection acting on DNA sequence variants, using data on DNA sequence polymorphism and between-species divergence (**Section 6.4**). Data of this kind are becoming available on a very large scale with the advent of new DNA sequencing technologies and genome-wide scans of variability. The methods that we describe are an essential foundation for their analysis and interpretation.

6.1. MEASURING RATES OF DNA SEQUENCE EVOLUTION

6.1.i. Comparing a pair of sequences

Most of our information about the rate of evolution of DNA sequences comes from comparisons between different species. Rates of sequence evolution can also be studied directly with sequences from rapidly evolving viruses, such as influenza and HIV viruses, for which samples can be collected from populations in different years; even the time course of sequence evolution over single infections can be followed (Drummond et al. 2003). Microbial evolution can also be observed by direct sequencing during experiments on evolving populations of organisms, such as bacteriophage viruses (Wichman et al. 2005). Evolutionary rates can then be estimated directly, using principles similar to those that we will describe for inter-species comparisons. Sometimes "ancient" DNA can be sequenced from long dead organisms that lived at known dates (e.g., Shapiro et al. 2004). With recent advances in sequencing technology, increasingly larger portions of the nuclear genome from samples of ancient DNA can be sequenced, including genes from Neanderthal humans (Green et al. 2006; Krause et al. 2007).

We shall focus here on comparisons between different living species. The simplest situation arises when we have aligned a pair of homologous sequences from two related species and can determine the number of differences between them. For simplicity, we assume that these involve only nucleotide substitutions, ignoring (or assuming an absence of) differences caused by insertions or deletions (indels). When coding sequences are compared, indels are usually rare or absent (see **Figure 1.7** of **Section 1.2.ii** for an example), indicating that they are strongly selected against. However, indels are commonly found in comparisons of noncoding regions. This causes two problems: first, how to align the sequences correctly, and second, how to quantify the sequence divergence due to indels. We will not discuss these important problems, which are well reviewed by Graur and Li (2000, pp. 86–97) and Wang et al. (2006).

6.1.ii. Estimating rates of divergence

6.1.ii.a. *Basic principles*

The most basic data for estimating rates of sequence evolution come from comparing two homologous sequences and calculating the proportion of nucleotide sites that are different (for the moment, we will ignore any distinctions between different classes of nucleotide changes, such as synonymous versus nonsynonymous mutations). We denote this proportion by D. If T, the amount of time separating the two species (usually expressed in years), is known from the fossil record or from biogeographic information, this proportion gives an estimate of the evolutionary rate per site: $\lambda = D/2T$. The factor of 2 arises because each sequence is connected by an amount of time T back to the common ancestor of the two species. A two-species comparison gives no information about which of the two branches of the tree experienced the mutations we observe (**Figure 6.1**).

In the alignment shown in **Figure 1.8** of **Section 1.2.iii.b**, for example, 59 bases differ between the human and chimpanzee sequences, giving $D = 59/5022 = 0.0117$. The time since the separation of the human and chimpanzee lineages (T) is thought to be about five million years. The *rate of substitution* per nucleotide site per year is then estimated as:

$$\lambda = 0.0117/(2 \times 5 \times 10^6) = 1.17 \times 10^{-9}$$

There are two problems with this type of estimate. First, information on the date of separation of the two species is often either unavailable or uncertain in accuracy, as is the case in this example. A more fundamental problem is that we are assuming that the time of separation of the two sequences corresponds to the date of separation of the two species, i.e., that speciation was instantaneous, after which the sequences of each species started accumulating fixed differ-

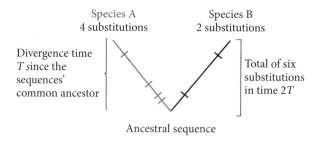

FIGURE 6.1 The time T back to the common ancestor of two sequences, with mutations that have occurred in the two sequences since divergence from their ancestor. The diagram shows time in the direction that is conventional for phylogenetic trees, with the ancestor at the bottom.

ences. In reality, T is made up of two components (**Figure 6.2**): T_1, the time back to the final split of the two species from a common ancestral population, and T_2, the time to the coalescence of the two sequences within the common ancestral species. When the species are well separated in time, most of the evo-

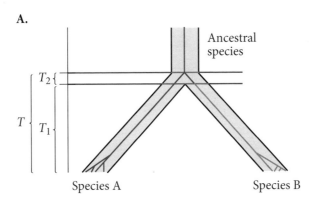

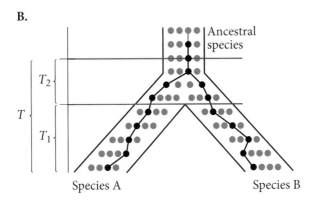

FIGURE 6.2 The time T separating two sequences sampled from two different species. In this diagram, time is shown in the direction that is used when thinking of the coalescence of alleles into a common ancestor, with the ancestor at the top. **A.** The case when the time to the split of the ancestor of the two species into two separate species, T_1, is large relative to the coalescence time within each species, so that the variants in a sample from a species arose from a common ancestor within that species. **B.** The case with a small T_1 relative to the within-species coalescence time. In this case, a pair of alleles from the two species may have a lineage (shown in black) that fails to coalesce with any alleles from their own species, so that their first coalescent event is with each other, within the ancestral species. This means that polymorphic variants may be shared between the two species (as is the case for the mutation indicated by the black circle; the grey circles indicate alleles with different ancestries from the black ones).

lutionary time separating the sequences does indeed correspond to the time since the species last shared any common ancestry (**Figure 6.2.A**). However, as shown in **Figure 6.2.B**, there may be difficulties when comparing close relatives, where T_2 and T_1 are comparable in magnitude (see **Section 5.1.iii.b** for the time for coalescence within a population). For these reasons, comparisons are often expressed simply as sequence divergence values, without translating these into rates (but researchers often confusingly refer to D values as rates).

6.1.ii.b. *Correcting for multiple changes*

A further important problem is that the method of estimating D by simply counting the number of differences is equivalent to assuming that each site that differs between the species has undergone only one substitution of a nucleotide base in the history of the sequences since their common ancestor. In fact, two or more changes may have occurred at the same nucleotide site. For example, a G versus A difference at a site could have involved the sequence of changes G $\rightarrow$ C $\rightarrow$ A. Similarly, two sites which are the same in two species (e.g., both are G) may actually have changed and subsequently reversed (e.g., G $\rightarrow$ C $\rightarrow$ G); reversal is less likely than a change to another base, because each nucleotide can mutate to three different ones. Multiple changes become increasingly likely if a long time separates the two species (especially in genome regions with high mutation rates).

The simplest method of correction, which is useful for dealing with multiple changes that do not restore the original state, is to use the Poisson distribution (**Appendix A2.v.d**); reversals are dealt with in **Section 6.1.ii.d** below. If changes occur at a constant rate, λ, per site per time unit, the expected number of changes at a site over a time interval of $2T$ is:

$$K = 2T\lambda \tag{6.1}$$

Probability theory tells us that, in the biologically realistic case when the rate λ is small, the actual number of changes that occur at the site follows a Poisson distribution, i.e., the probability of i differences is given by:

$$P_i = \frac{K^i}{i!}\exp(-K) \tag{6.2}$$

This means that the true number of changes (K) that have occurred can be estimated by the method described in **Box 6.1**. Applying **Equation B6.1.3** to the example above gives an estimated λ only marginally different from the earlier value of 1.17×10^{-9} (**Problem 6.1**). With higher divergence, the difference would have been much bigger. With 2000 observed differences between the two sequences, for example, the uncorrected λ is close to 4.0×10^{-8}. The corrected value is then 5.1×10^{-8}, about 28% larger.

Box 6.1 CORRECTING FOR MULTIPLE SUBSTITUTIONS

We assume that all sites are evolving independently of each other, at the same constant rate. The probability of getting no changes at a site is the first term of the Poisson distribution, i.e., $P_0 = \exp(-K)$, where K is given by **Equation 6.1**. If the length of the sequence is m, the expected number of sites that have not changed is mP_0. This can be equated to the observed number of unchanged sites, i.e., to $n = (1 - D) m$:

$$n = m \exp(-K) \tag{B6.1.1}$$

Taking natural logarithms on both sides, we obtain:

$$\ln(n/m) = -K \tag{B6.1.2}$$

That is,

$$K = -\ln(1 - D) \tag{B6.1.3}$$

6.1.ii.c. *Stochasticity of the substitution process*

The Poisson formula can be applied to calculate the chance that j changes occur over a sequence of length m, rather than for an individual site. With the simple Poisson model of **Equations 6.1** and **6.2**, the expected number of changes in a sequence of m bases is $2Tm\lambda$ (this is valid even if different sites evolve at different rates). This model is the simplest example of a *Hidden Markov Model* of molecular evolution; this name refers to the fact that the observed sequence differences reflect an underlying model with a constant rate of changes per unit time. It is very important to understand that, even with an unchanging evolutionary process (producing a constant rate of evolution, λ, in terms of numbers of substitutions per site per unit time), wide variation is to be expected in the actual number of differences between sequences of the same length from two species. This can be seen as follows.

A Poisson distribution with a fairly large number of expected changes (more than 10 or so) is close to a normal distribution, so that approximately 95% of the distribution falls within plus or minus 1.96 standard deviations of the mean; if we use the estimated standard deviation and mean, we obtain the 95% *confidence interval* for the mean. The standard deviation of a Poisson distribution is the square root of its mean (**Appendix A2.v.d**), so the number of changes observed can thus easily deviate from the expected value of $2Tm\lambda$ by as much as $\pm 1.96\sqrt{2Tm\lambda}$, i.e., for short sequences, there is considerable random noise in estimates of divergence and evolutionary rates. Using the observed number of differences to estimate the mean, we obtain the 95% confidence interval for the mean. The error relative to the estimate of K decreases in proportion to $1/\sqrt{m}$ (**Problem 6.1**). Accurate divergence estimates therefore require hundreds or thousands of bases to be sequenced.

6.1.ii.d. *Correcting for reverse and multiple changes*

The Poisson correction explained above assumes that each mutation is a unique event, and also ignores the possibility that a substitution reverses itself. A number of methods for correcting for both multiple and reverse changes are reviewed by Graur and Li (2000, Chapter 3) and Ewens (2004, Chapter 12). The simplest is the method of Jukes and Cantor (1969), which assumes that each change involves substitution by one of the other three bases with equal probability. If the net rate of change is λ, the rate of change to one of the three alternatives to a given base is $\lambda/3$. This yields a simple equation for the true divergence, D, expressed as the proportion of sites in either of the two sequences that have experienced changes (**Box 6.2***):

$$K = -\frac{3}{4}\ln\left(1 - \frac{4D}{3}\right) \tag{6.3}$$

Box 6.2* CORRECTING FOR REVERSE AND MULTIPLE SUBSTITUTIONS

Rates of nucleotide substitution, measured per site per year, are very low. We can therefore treat the process of sequence divergence as though time is a continuous variable, with changes taking place at a rate λ per nucleotide site per year. The chance of a substitution of one nucleotide for another between the two sequences at a particular site over a small interval of time, δt, is $2\lambda\delta t$, because changes may take place on either branch leading back to the common ancestor. (The value of $\lambda\delta t$ is assumed to be so small that multiple changes in this interval are negligibly rare.)

If the two sites were different at the start of this interval (probability D), the chance that they become the same is $2\lambda\delta t/3$, because each has a chance of 1/3 of mutating to the state as the other one. If they are the same (probability $1 - D$), the chance that they become different is $2\lambda\delta t$. The change in $1 - D$ is thus $(2D\lambda\delta t/3) - 2(1 - D)\lambda\delta t = -[8(1 - D)/3 - 2/3]\lambda\delta t$. Dividing by $2\delta t$, and taking the limit as $\delta t \to 0$, we obtain:

$$\frac{d(1 - D)}{d(2T)} = -\frac{4\lambda}{3}(1 - D) + \frac{\lambda}{3} \qquad \text{(B6.2.1)}$$

This equation can be solved by noting that the equilibrium value of D, obtained by setting the derivative to 0, is 3/4. We can then work with deviations from this equilibrium, $\tilde{D} = D - 3/4$:

$$\frac{d(1 - \tilde{D})}{d(2T)} = -\frac{d\tilde{D}}{d(2T)} = \frac{4\lambda\tilde{D}}{3} \qquad \text{(B6.2.2)}$$

The solution is:

$$\tilde{D} = \tilde{D}_0 \exp\left(-\frac{8\lambda T}{3}\right) \qquad \text{(B6.2.3)}$$

where $\tilde{D}_0$ is the initial deviation from equilibrium, –3/4. The true level of sequence divergence, in terms of the mean number of events that have occurred at a site, is $K = 2\lambda T$, so that the exponential term can be replaced by $-4K/3$. Rearranging, and taking natural logarithms on both sides, gives **Equation 6.3** of the main text.

Since K becomes indefinitely large as time increases, this implies that the bracketed term on the right-hand side of **Equation 6.3** approaches 0, so that the observed divergence D tends to 3/4, i.e., one-quarter of the sites compared are identical. This reflects the fact that (assuming equal mutation rates among all four nucleotides) the ultimate state of a sequence is a random collection of nucleotides with frequency of one-quarter of each, so that each site has a chance of one quarter of having the same state in two sequences separated by a long evolutionary time. The sequence differences are then said to be *saturated*. Corrections for multiple and reverse changes thus become increasingly meaningless as the divergence time increases.

Assuming equal mutation rates for each type of change is, of course, unrealistic, because there is evidence both from DNA sequence comparisons and studies of spontaneous mutations that the rates of mutation between the four different nucleotides are not the same. In particular, transitions (mutations between purine and purine or pyrimidine and pyrimidine) are usually more common than transversions (mutations between purine and pyrimidine) (Graur and Li 2000, pp. 125–126). Kimura (1980) proposed extending the sequence evolution model above to include separate rates for transitions and transversions; this "Kimura two-parameter method" is very widely used. Several even more general models have subsequently been proposed, but we will not consider these in detail. The important point is that a model of the underlying process of sequence evolution is necessary for interpreting raw sequence divergence data (Graur and Li 2000, Chapter 3; Ewens 2004, Chapter 12; Yang 2006).

We have so far assumed that all parts of a sequence evolve according to the same rate model. Recent modeling work has incorporated the fact that the rates and nature of mutations at a given site may be influenced by the nature of the adjoining bases, i.e., there is "context dependence" of mutational processes (Siepel and Haussler 2004; Arndt and Hwa 2005; Kryukov et al. 2007). The best-known such effect is the high mutability in mammals of methylated C bases in positions 5′ to a G base in a sequence (CpG sites), causing an elevated frequency of mutations from CG dinucleotides to TG (Graur and Li 2000, p. 126; Kim et al. 2006; Subramanian and Kumar 2006).

6.1.ii.e. *Nonsynonymous and synonymous changes*

Differences in mutation rates are only one cause of divergence rate variation. Another very important source of such variation is that most changes affecting the amino acid sequence of a protein impair its function, at least mildly. Thus, in the protein coding sequences of genes, most mutations that change an amino acid are prevented by selection from spreading through the population and replacing the existing amino acid, as already mentioned in **Section 1.2.iii.b**. Because of the *degeneracy* of the genetic code, some mutations do not change the amino acid encoded by a codon, i.e., these mutations are *synonymous*. In the standard genetic code of most genomes (Knight et al. 2001), only two of the 20 amino acids are encoded by single codons, i.e., they are nondegenerate (AUG for methionine and UGG for tryptophan). The codons for nine amino acids are *two-fold degenerate* (i.e., two codons for the same amino acid; e.g., cysteine with UGU and UGC), isoleucine is *three-fold degenerate* (AUU, AUC, and AUA), five amino acids are *four-fold degenerate* (e.g., glycine with GGN, where N can be any nucleotide), and three are *six-fold degenerate* (e.g., serine with UAN, AGU, and AGC). In addition, there are three stop codons, UAA, UAG, and UGA, which terminate translation of the message.

This means that some changes in the first positions of codons, and a substantial proportion of changes to third codon positions, are synonymous and hence less likely to significantly affect fitness than the nonsynonymous mutations that alter the amino acid sequence of the protein. These variants may thus readily become fixed by genetic drift, giving a higher divergence rate than for nonsynonymous mutations. The chance that a mutation changes an amino acid depends on the base composition of the gene encoding the amino acid sequence. In mammalian genomes, roughly 70% of random single-nucleotide site changes to a coding sequence produce a change in an amino acid (Kryukov et al. 2007).

The expectation that nonsynonymous nucleotide changes will accumulate more slowly than synonymous or silent changes (except at sites where the amino acid changes have little or no effect on fitness) is confirmed by sequence comparisons between species. These use methods that can estimate numbers of nonsynonymous and synonymous changes between two homologous coding sequences, relative to the number of sites that are potentially capable of generating these two types of change. For some coding sequence sites, only one type of change is possible (for instance, four-fold degenerate sites can undergo only synonymous changes, and nondegenerate sites only nonsynonymous ones); these sites can thus be counted directly. It is not obvious how to define the two categories for sites of other degeneracy classes.

Several approaches have been developed, based on different models for sequence evolution. For instance, each two-fold degenerate site can mutate to three different nucleotides (two transversions are possible and one transition). In the nuclear genetic code of most organisms, transitions at these sites are synonymous changes and transversions are nonsynonymous; thus, one crude model might assume equal probabilities of transitions and transversions, and, for each two-fold degenerate site, we could thus count it as one-third of a synonymous site and two-thirds of a nonsynonymous one (Nei and Gojobori 1986; Graur and Li 2000; pp. 80–81). More sophisticated methods for comparing nonsynonymous and synonymous changes use observed substitution patterns in supposedly neutral sequences, such as small introns (ignoring splice-site sequences), to calculate the expected numbers of nonsynonymous and synonymous changes in a gene of interest for comparison with the observed numbers (Eyre-Walker and Keightley 1999; Subramanian and Kumar 2006; Kryukov et al. 2007).

Estimates of the numbers of nonsynonymous changes per nucleotide site capable of generating nonsynonymous changes are usually denoted by K_A or d_N. The corresponding quantity for synonymous changes is K_S or d_S. For the vast majority of genes that have been examined, K_A is much less than K_S. **Figure 6.3** shows some examples. In comparisons of *Drosophila* species, the mean K_A across large sets of genes is typically less than 10% of the mean K_S value (Larracuente et al. 2008), whereas for hominids it is 23% (The Chimpanzee Sequencing and Analysis Consortium 2005).

A.

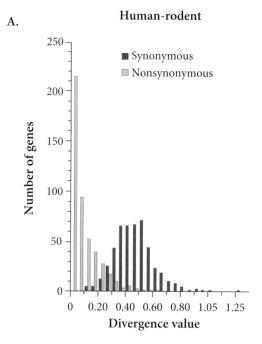

B.

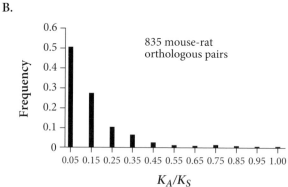

FIGURE 6.3 Some K_A and K_S values indicating purifying selection on nonsynonymous mutations. **A.** Human–rodent nucleotide site divergence values, plotted from data in Makalowski and Boguski (1998). **B.** Values of the ratio K_A/K_S in comparisons of mouse and rat gene sequences. [Adapted from Figure 1 of Hurst (2002).]

6.1.ii.f. *Approximate rate constancy of molecular sequence evolution*
An important observation, made very early in the history of the study of molecular evolution, is that the rate of molecular evolution of a given gene's sequence is approximately constant over time, at least within a defined set of lineages

(Zuckerkandl and Pauling 1965). This contrasts strikingly with the highly variable evolutionary rates of morphological (and other) phenotypes (see the **Introduction** to **Chapter 3**). We discuss explanations for this *molecular clock* below (**Section 6.2.ii**), but here we regard it as an empirical finding. Although the molecular clock model is only approximately valid (Graur and Li 2000, pp. 150–155), it is nonetheless useful for estimating divergence dates between species for which there is no adequate fossil dating, e.g. humans and chimpanzees. This, of course, requires adequate rate calibration, using known divergence times for species that are not too distant from those of interest. Methods for estimating divergence times that relax the assumption of a molecular clock are becoming available (e.g., Drummond et al. 2006).

6.2. THE CAUSES OF DNA SEQUENCE EVOLUTION

6.2.i. Introduction

Up to now, we have been considering sequence evolution purely in terms of the accumulation of nucleotide differences along lineages. But, as we have discussed in previous chapters, evolutionary divergence between a pair of reproductively isolated species requires mutations to become fixed, or to spread to high frequencies, within each species. Evolutionary rates therefore depend on the processes that cause mutations to spread. There are two extreme possibilities. First, mutations may be favorable, and driven to fixation by selection (including processes of intra-genomic selection, such as biased gene conversion, mentioned in **Section 1.3.i.b**). Alternatively, mutations could be neutral with respect to their effects on fitness, and spread through the population by genetic drift. More generally, selection, mutation, and drift jointly determine the fates of new mutations (**Section 5.3.iii.d**). The dynamics of frequency changes under selection at single nucleotide sites in infinite populations has already been described in **Section 3.1**, so we focus attention here on finite populations, and examine both neutral and weakly selected mutations.

6.2.ii. Neutral sequence evolution

6.2.ii.a. *Expectations for sequence divergence under neutrality*
We first consider strict neutrality, i.e., the case when fixations occur purely by genetic drift, without any action of natural selection. The simplest way to model neutral sequence evolution is to consider one sequence from each of two species. For neutral sites with a mutation rate u per nucleotide site per generation, u new mutations are expected to accumulate per site per generation. Thus, the rate λ in the sequence evolution models of **Section 6.1.ii** is u/t_g per site per

year, where t_g is the species' generation time in years. If the time separating the two sequences is $2T$ years, as in **Section 6.1.ii.a** above, the expected number of mutations that will accumulate is:

$$K = 2Tu/t_g \qquad (6.4)$$

6.2.ii.b. *Fixation of neutral mutations and divergence of sequences from closely related species*

A different perspective, whose usefulness will appear when we include selection, is to follow the fate of mutations that arise in one of the two species under consideration. Using this approach, **Box 6.3** shows that the expected rate of fixation of neutral mutations in this species is u per generation. This result implies that, if two related species are compared, and the time back to their divergence from a common ancestor is T_1, the number of fixations of mutations that are expected to accumulate between them over this period of time is $2T_1$ times the mutation rate per year, as pointed out by Kimura (1968). Division of the number of mutations by $2T_1$ gives the *neutral rate of substitution*. This result applies to a single-nucleotide site, an entire gene, or the genome as a whole, if we define the λ and u values appropriately.

If T_1 is much larger than the time to coalescence within each species ($T_2 \ll T_1$, as in **Figure 6.2.A** of **Section 6.1.ii.a**), polymorphisms that existed in the ancestral population will contribute little to the overall divergence between the two species as compared to mutations that arose subsequent to the split, and the within-species level of polymorphism will be small compared to the between-species divergence. To a good approximation, the total divergence time, T, is then the same as the time since the two species became reproductively isolated, as is commonly assumed in studies of molecular evolution. We can then equate T to T_1 and use the overall sequence divergence between the two species as a measure of the number of fixations that occurred since the species separated from each other.

This is inadequate for closely related species, for which T_2 and T_1 may be similar in magnitude. In this case, if the effective size (**Section 5.2**) of the ancestral population was N_{ea}, the expected number of years separating a pair of alleles is their mean coalescence time in generations multiplied by the generation time in years, i.e., $4N_{ea}t_g$ (**Section 5.1.iii.b**). This implies that the expected number of differences per site is given by:

$$E\{K\} = 2u(2N_{ea} + T_1/t_g) \qquad (6.5)$$

N_{ea}, however, is usually unknown. If DNA sequence diversity estimates are available from the species being compared, we can assume that N_e has remained unchanged, and roughly estimate $4N_{ea}u$ using the mean of the nucleotide site diversity estimates for the two extant species, or the value for one of them (**Sections 5.1.ii.b** and **5.1.iii.b**). Subtracting this estimate from the

Box 6.3 THE RATE OF SUBSTITUTION OF NEUTRAL MUTATIONS

A new mutation at a nucleotide site will initially be present in a single individual, or in a few individuals if the mutation arose as a pre-meiotic event that caused a cluster of mutations, as is frequently the case in multicellular organisms (**Section 1.3.iii.b**; Woodruff and Thompson 1992, 2005). With diploid autosomal inheritance, the initial frequency of a single new mutation in a population of N breeding individuals is $1/2N$, independent of the sex of the individual in which it appeared (**Problem 6.2**). This result can easily be extended to other situations, such as X-linkage, with slight modifications (**Problem 6.2**).

Similarly, if the initial number of carriers of the mutation is j, the initial frequency, q_0, of the mutation in a diploid population of size N is $j/2N$ (**Problem 6.2**). From the argument given in **Section 5.1.i.b**, the probability of fixation of a copy of the mutation is q_0, because all $2N$ alleles present in the population have an equal probability of eventually becoming fixed. If u_j is the expected number of new mutations per site per generation for clusters of size j, the expected number of clusters of size j arising each generation is $2Nu_j/j$. Because only one member of a cluster can become fixed, use of the rule for the probability of mutually exclusive events (**Appendix A2.ii**) implies that the expected number of mutations that eventually become fixed is $2Nu_j/j \times j/2N = u_j$. Averaging over the distribution of j, the expected number of new mutations that enter the population each generation that will eventually become fixed is u, the overall mutation rate (each member of a cluster is counted as a separate mutation).

If the process is in equilibrium, the rate of origination of mutations that eventually become fixed must correspond to the expected number of mutations that become fixed each generation, so that the rate of fixation of mutations is equal to the mutation rate. The results do not depend on the population size N being constant over time. The approach used here is originally due to Kimura (1968), but the result was first obtained by Wright (1938), using a more complex argument.

right-hand side of **Equation 6.5** provides an expression for the contribution to K from mutations accumulated since the two species diverged (or, equivalently, the number of substitutions between the species). More sophisticated methods for estimating ancestral effective population sizes have been developed (Wakeley and Hey 1997; Wall 2003; Hey and Nielsen 2004).

6.2.ii.c. *Testing the neutrality of sequence evolution*

As Kimura (1968) and King and Jukes (1969) pointed out, the theoretical equality between the neutral substitution rate and the neutral mutation rate provides a simple explanation for the molecular clock: provided that u is constant for the sequence investigated, across the different lineages compared, the rate of evolution is invariant (even if the population size fluctuates). Sequences free of selective constraints should thus have a rate of nucleotide substitution equal to the mutation rate. To estimate divergence times or mutation rates, it is therefore important to identify sequences that can plausibly be assumed to behave neutrally. We shall see that they are also needed for testing for selection. Examples of probable neutral sequences are pseudogenes and portions of polypeptides that serve only as packaging, and are removed during the production of the mature protein (Graur and Li 2000, pp. 111–113). As we described in **Section 1.3.iii.b**, estimates of the human mutation rate obtained by comparing homologous human and chimpanzee pseudogenes (Nachman and Crowell 2000) agree with a direct estimate from disease-causing mutations (Kondrashov 2003).

In contrast, it has been argued that most nonsynonymous mutations in functionally significant DNA sequences, such as coding sequences, are deleterious and likely to be eliminated by selection (**Section 6.2.iii.a** below), and that the only mutations that contribute to sequence divergence are those that are neutral or nearly neutral (Kimura 1968; King and Jukes 1969). Nonfunctional sequences should then have the highest rates of DNA sequence evolution, whereas functionally very significant sequences, such as the active centers of enzymes, should be much more conserved in evolution. Kimura (1983, pp. 156–168) presented a wide range of evidence supporting this prediction, and this result has been confirmed by later work.

In the previous section, we saw that, with a constant sequence evolutionary rate, the numbers of neutral substitutions (i.e., fixations) that actually occur along an evolving lineage are expected to follow a Poisson distribution (see **Section 6.1.ii.c**, where T_2 was neglected). For a pair of closely related species, however, the divergence between sequences taken from each of them will have an added contribution due to polymorphisms within species. Different sequences will have different T_2 values, due to the stochasticity of the time back to their ancestors within the ancestral population, and this makes the variance in the number of differences between such species greater than the expectation under the Poisson distribution (Gillespie and Langley 1979).

A constant rate of sequence divergence for lineages at a suitable divergence distance is testable by comparing the variance of numbers of substitutions along different branches of a phylogeny with the mean (Kimura 1983, pp. 76–80; Gillespie 1986, 1991, Chapter 3; Graur and Li 2000, pp. 142–145). An excess variance in the rate of substitution of nonsynonymous mutations among lineages is often interpreted as evidence against neutrality, providing evidence for adaptive evolution of protein sequences (Gillespie 1986, 1991, Chapter 3). However, other processes, such as the fixation of slightly deleterious amino-acid mutations, can also increase the variance (Cutler 2000). The most recent analyses using genome sequence comparisons provide mixed evidence for an excess variance in the numbers of nonsynonymous substitutions, with uncertainty concerning whether this reflects positive selection (Kim and Yi 2008; Bedford et al. 2008).

6.2.iii. Selection in finite populations

6.2.iii.a. *Rates of substitution for unique mutations*
We now extend the approach above to deal with mutations under selection in a finite population. We assume that mutations enter the population as unique events, and are either lost or fixed, as in the neutral model in **Section 6.2.ii.b** above. The direction and strength of selection on a particular type of nucleotide site variant are assumed to be constant over the whole time period under consideration. Following the argument of Kimura and Ohta (1971, p. 11), we can obtain the expected rate of substitution as follows. Instead of the chance of fixation of a neutral mutation with initial frequency q_0 (which, as we saw, is simply q_0), we determine the probability of fixation of a new selected mutation, denoted by $Q(q_0)$, and multiply this by the expected number of such mutations that enter the population each generation ($2Nu$), so that we have:

$$\lambda = 2NuQ(q_0) \tag{6.6a}$$

Equation B6.4.1 of **Box 6.4*** gives an expression for $Q(q_0)$; this assumes a randomly mating population with a large effective population size and weak selection, so that second-order terms in s and $1/N_e$ can be neglected. The equation also assumes diploid, semidominant autosomal inheritance, with selection coefficients s and $s/2$ for homozygous and heterozygous carriers, respectively (here, s is negative for a deleterious mutation and positive for a favorable one). For a new mutation, $q_0 = 1/2N$, where N is the number of breeding adults, and the expression for Q simplifies to **Equations B6.4.2a** and **B6.4.2b**. It is often useful to express the substitution rate relative to the neutral value, i.e., as $\tilde{\lambda} = \lambda/u$. This gives:

$$\tilde{\lambda} = \lambda/u \tag{6.6b}$$

Equation B6.4.2b gives an explicit expression for Q. It is convenient to write $\gamma = 2N_e s$; using **Equation B6.4.2b**, we then have

$$\tilde{\lambda} \approx \frac{\gamma}{[1 - \exp(-\gamma)]} \tag{6.7}$$

Box 6.4* FIXATION PROBABILITY WITH SEMIDOMINANCE

The fixation probability for semidominant, autosomal mutations can be found from **Equations 6A.6** and **6A.7** in the **Appendix** to this chapter (as usual, analogous results apply to haploid populations by doubling the selection coefficient). In this case, we have $M_{\delta q} = spq/2$ and $V_{\delta q} = pq/2N_e$ (See **Section 5.3.ii.a**), so that $\psi(q)$ in these equations is equal to $\exp(-\gamma q)$, where $\gamma = 2N_e s$. From **Equation 6A.7**, a simple integration shows that

$$Q(q_0) = \frac{1 - \exp(-\gamma q_0)}{1 - \exp(-\gamma)} \tag{B6.4.1}$$

As in **Box 6.3** of **Section 6.2.ii.b**, the initial frequency of a single new mutation is independent of the sex of the individual in which it appeared, so that $q_0 = 1/2N$. **Equation B6.4.1** then simplifies to:

$$Q(1/2N) \approx \frac{1 - \exp(-N_e s/N)}{1 - \exp(-2N_e s)} \tag{B6.4.2a}$$

For small s, as assumed in the derivation of the diffusion equation, this is accurately approximated by:

$$Q(1/2N) \approx \frac{\gamma}{(2N)\left[1 - \exp(-\gamma)\right]} \tag{B6.4.2b}$$

(see **Problem 6.3.i*.**)

This can be generalized to account for pre-meiotic germ line clusters of mutations, as in **Box 6.3**. For a cluster of size j, each member of the cluster has an initial frequency of $1/2N$. As in the neutral case, only one member of the cluster can become fixed in the population, so $Q(q_0) = jQ(1/2N)$. Averaging over clusters of different sizes (using the method used for the neutral case in **Box 6.3**), we have $\lambda = 2NuQ(1/2N)$.

For favorable mutations ($s > 0$), when $N_e s > 1$, the second term in the denominator of **Equation 6.4.2b** is close to 1, so $Q(1/2N)$ reduces to sN_e/N. This is similar to **Equation B3.4.4** in **Section 3.2.ii** for a favorable mutation in an infinite population, except for the factor of N_e/N. For a Wright–Fisher model, which implies a Poisson distribution of offspring numbers, the two results are identical.

For deleterious mutations, the denominator is large when $N_e|s| \gg 1$, and the fixation probability becomes very small. Conversely, for $N_e|s| \ll 1$, the denominator approaches the value for the neutral case, $1/2N$ (**Problem 6.3.ii***).

The effect of selection on the substitution rate thus depends only on γ. Selection against deleterious mutations is extremely effective when $N_e|s| > 1$, where $|s|$ denotes the absolute value of s. The substitution rate rapidly approaches 0 as $N_e|s|$ increases above this value (**Figure 6.4**). Conversely, for $N_e|s| \ll 1$, rates of substitution for both deleterious and favorable mutations are close to the neutral value. The rate for favorable mutations becomes much larger than the neutral value when $N_e s > 1$, and rapidly approaches that for an infinite population (**Figure 6.4**).

These results are consistent with the conclusions from the properties of the stationary distribution of variant frequencies, derived in **Section 5.3.iii.d**. Even

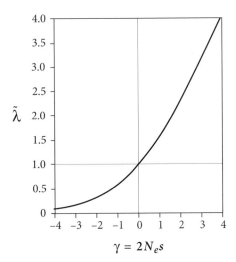

$$\gamma = 2N_e s$$

FIGURE 6.4 Plot of the predicted substitution rate of semidominant mutations (**Equation B6.7.2b** in **Box 6.7**), relative to the neutral rate (the horizontal line indicates the value of one for the neutral case). The figure shows that selection against deleterious mutations (negative $N_e s$) is extremely effective, with the rate of substitution rapidly approaching 0 as $N_e|s|$ increases above 1, and that the rate for favorable mutations rapidly rises above the neutral value when $N_e|s| > 1$. The figure also shows that mutations, whether deleterious or favorable, have a rate of substitution that is close to the neutral value when $N_e|s| \ll 1$.

in populations with effective sizes as low as those of the human population (about 10,000, as discussed in **Section 5.2.iii.e**), selection coefficients greater in magnitude than the very small value of around 1 in 10,000 are sufficient for the fate of mutations to be controlled by selection, rather than being subject to genetic drift. Such selection coefficients are far below the resolution of even the most powerful experimental estimation methods described in **Section 3.1.iii.b**. This conclusion applies much more broadly than under the specific assumptions we have made above. In particular, it is likely to apply to species divided into partially isolated local populations (**Section 7.3.ii**).

6.2.iii.b. *Rates of substitution for recurring mutations*

It should be pointed out that the assumption that selection acts in a constant direction on unique, newly arisen mutations may not be correct for mutations that are favored by selection. For example, if there is a change in the environment, a previously neutral or deleterious amino acid at a given site in a polypeptide might become advantageous, as in the case of insecticide or antibiotic resistance mentioned in **Section 3.1.i**. The mutation's initial frequency can be denoted by q_0, where (instead of being $1/2N$, as for a new mutation) q_0 is drawn from a probability distribution for segregating variants, such as that described in **Section 5.3.iii.d**. If the environment remains in its new state, and the population is large, many copies of the favorable mutation are available for selection to act on, rather than just one, and each of them has a chance of becoming fixed. The ultimate probability of fixation of one of these thus approaches 1.

In such cases, therefore, fixation probabilities are almost irrelevant to the rate of the evolutionary process (Maynard Smith 1976c; Gillespie 1984, 1991, Chapter 5). They may affect the order in which different types of mutations are likely to be established, as we saw in **Section 3.4.iii.b**, but the overall rate of substitution per nucleotide site for a given sequence is simply driven by the number of sites where environmental changes have altered the direction of selection. The pattern of evolution in genes subject to such selection may thus be *episodic*, with bursts of change at multiple nonsynonymous sites associated with shifts in the selective environment (Gillespie 1986; 1991, Chapters 5 and 7). Sites that have been affected by positive selection will, of course, show higher rates of substitution than other sites, so that the principle still holds that selection for favorable mutations increases the rate of substitution over the neutral rate.

6.2.iv. Implications for molecular evolution and for studying selection

The result that even very weak selection can prevent deleterious mutations from contributing to sequence evolution provides a natural explanation for the much slower rate of nonsynonymous substitutions than of synonymous or silent ones (**Section 6.1.ii.e** above), if the latter are either neutral or subject to much weaker selection than nonsynonymous sites (this may not always be correct—see Parmley and Hurst 2007). Similarly, very weak selection on favorable mutations produces a much higher rate of fixation than with neutrality. If we can estimate the rate of evolution for sites that can confidently be identified as close to neutral, we can compare these with the rates for other sites to test for such *positive selection*.

A widely used method is to test for a higher rate of nonsynonymous substitutions than the neutral standard, which suggests that a protein sequence is evolving under positive selection. Such situations are infrequent (e.g., Yang and Bielawski 2000); for example, a comparison of the human and chimpanzee genome sequences estimated that only 585 out of 13,454 human–chimpanzee pairs of homologous genes have $K_A > K_S$, and many of these may be cases of type I error (apparent statistical significance arising by chance, or "false positives"; see The Chimpanzee Sequencing and Analysis Consortium 2005).

Several well-documented examples that do provide evidence for positive selection involve genes in situations that lead to "evolutionary arms races," such as sequences of pathogens evolving to escape from the recognition systems of their hosts. For instance, orthologous chitinase I genes (which encode proteins involved in defenses against plant pathogens) have been sequenced in multiple related species in the Brassica family, and sometimes have significantly higher nonsynonymous substitution rates than synonymous rates in interspecies comparisons (Bishop et al. 2000). Other examples, also probably involving arms races, have been found in proteins associated with the fusion of eggs and sperm. Competition is likely between males to fertilize eggs, while eggs are selected to avoid multiple fertilizations; in one such sperm protein, nonsynonymous divergence is higher than synonymous divergence only in the parts involved in binding activities (Panhuis et al. 2006).

As already mentioned in **Section 6.1.ii.e**, slower divergence than the neutral standard suggests *purifying selection* against deleterious mutation. Some important applications of this principle will be discussed later in **Section 6.4.ii**. An interesting pattern, also mentioned in **Section 6.1.ii.e**, is that K_A/K_S tends to be lower for sequences from *Drosophila* (10%), compared with those from primates (> 20%). This may be a consequence of lower effective sizes in primate populations, whose values are probably around 10^4, whereas those of *Drosophila* species are estimated to be around 10^6. This is consistent with **Equation 6.7**, which, for sequences under predominantly purifying selection, predicts a higher ratio of nonsynonymous to synonymous substitutions with smaller N_e (**Figure 6.4**). Similarly, there appears to be a higher rate of protein sequence evolution for island taxa compared with related mainland ones, as would be expected if island populations have substantially lower population sizes (Woolfit and Bromham 2005). However, the relationship between substitution rate and $N_e|s|$ for deleterious mutations is highly nonlinear (**Figure**

6.4), and substitution rates decline rapidly to near 0 as $N_e|s|$ becomes large. Thus, to explain the observed K_A/K_S difference between primates versus *Drosophila*, deleterious nonsynonymous variants must mostly be in just the right range of *s* values, such that a change of N_e from 10^6 to 10^4 halves the fixation probability. Other possibilities, such as selection intensities differing between very different groups of organisms or environments, cannot be ruled out.

Observing that overall K_A/K_S values are nearly always much less than 1 does not imply that all the amino acid substitutions that occurred during the evolution of these genes are either deleterious or neutral. Kimura and Ohta have proposed that this is indeed the predominant mode of protein sequence evolution (Kimura 1983; Ohta 1992), and it is certainly likely that most nonsynonymous mutations are strongly selected against and have no chance of fixation by drift. This hypothesis can certainly explain why the molecular clock applies to protein sequences as well as silent DNA sequences (Zuckerkandl and Pauling 1965)—most substitutions are from the subset of mutations that are close to neutral. However, it is possible that a small minority of nonsynonymous mutations are advantageous and become fixed by selection, even if purifying selection is also acting to reduce substitution rates at most amino acid sites, leading to an overall value of K_A less than K_S. Is there evidence for such positively selected mutations from analyses of patterns of sequence evolution?

6.2.v. Detecting positive selection when purifying selection also acts: codon-based models

In the sperm protein example cited in **Section 6.2.iv** above, where only part of the coding sequence shows $K_A > K_S$, and K_A in most of the sequence is less than K_S, the adaptively evolving region was identified through prior knowledge of its function—the region of the protein involved in binding the sperm to the egg. In some cases of this kind, the functional sites are not organized into domains of the protein, but are sets of nonadjacent amino acids, as in the case of the antigen recognition sites of MHC proteins, which are encoded largely by the second exon of the MHC genes, although this exon also encodes stretches of amino acids not involved in recognition (Hughes et al. 1990; Garrigan and Hedrick 2003)

Methods have recently been developed that can detect individual codons that are under positive selection, even without functional information, using data on homologous sequences from a range of taxa (Muse and Gaut 1994; Yang 1997, 2006; Suzuki and Gojobori 1999; Yang and Bielawski 2000; Yang et al. 2000; Kosakovsky Pond et al. 2005; Zhang et al. 2005). The most sophisticated of these approaches, including PAML (http://abacus.gene.ucl.ac.uk/software/paml.html) and HYPHY (http://www.hyphy.org/), use *codon-based* models of sequence evolution, which specify transition probabilities between all possible codons, as shown in **Table 6.1**, instead of modeling nucleotides as we have done so far. The model takes into account the possibility that mutation frequencies are affected by the nucleotide sequence, by allowing transitions and transversions to occur at different rates (the parameter κ in **Table 6.1** quantifies the mutational bias in favor of transitions), as well as selective constraints on nonsynonymous substitutions (the parameter ω is analogous to K_A/K_S). The parameters of the sequence evolution model are estimated by maximum likelihood (**Appendix A2.vi.b**), using knowledge of the phylogenetic tree over which the set of sequences evolved. Nonsynonymous sites that are evolving unusually rapidly can then be identified.

Likelihood ratio tests can be used to test alternative hypotheses (**Appendix A2.vi.b**). The simplest hypothesis is that all codons in the protein have the same ω value in all branches of the tree. This hypothesis can be extended to allow some sites to be under purifying selection ($\omega < 1$), while the rest are neutral ($\omega = 1$); the fraction of these two categories of sites is estimated as a parameter of the model. A model with a third category of sites evolving under positive selection ($\omega > 1$) can then be fitted, and a likelihood ratio test used to ask whether this fits the data significantly better than without this category. If so, it suggests that some sites have been positively selected.

The case of the Y-linked *DAZ* gene of higher primates (Yu et al. 2008) illustrates the inability to infer the nature of selection pressure using the average K_A/K_S ratio over all sites in the sequence. *DAZ* is closely related to the autosomal gene *DAZL1*, which is present in all vertebrates. The formation of *DAZ* probably involved a translocation of a *DAZL1* copy to the Y chromosome after the divergence of New World monkeys and apes (Yu et al. 2008). *DAZ* was then amplified several times, forming a gene cluster on the Y chromosome. *DAZ* is thought to have an important role in human spermatogenesis, since a small deletion containing these genes is the commonest deletion in infertile men. Surprisingly, a K_A/K_S analysis found similar rates of nonsynonymous, synony-

TABLE 6.1 Probabilities of change from codon *i* to codon *j*

Category of change between the codons	Probabilities[1]	
	Transversions	Transitions
Differences in more than one position	0	0
Synonymous changes with one mutational difference	x_i	κx_i
Nonsynonymous changes with one mutational difference	ωx_i	$\omega \kappa x_i$

[1] The probabilities are relative values for different types of change; x_i is the equilibrium frequency of codon *i* in the sequence in question.

mous, and intron site substitutions, so that the *DAZ* coding sequence appeared to be experiencing no selective constraints. However, this turned out to be misleading. An analysis of the evolution of the *DAZ* gene family, using sequences from several primate species, suggested that most amino acids are under strong selective constraint, but a few sites are under positive selection (with $\omega > 1$), resulting in an overall ω close to 1 (Bielawski and Yang 2001).

Similarly, these codon-based methods have recently been applied to the genome sequences of *Drosophila* species in the *melanogaster* group. They suggest that, while most nonsynonymous sites are under purifying selection, about 10% of genes show evidence that some of their nonsynonymous divergence has been driven by positive selection (Larracuente et al. 2008).

Further tests can be done with these models, including comparing a fixed rate model with models in which ω differs among lineages. Finally, if enough data are available, it is possible to identify individual amino acid sites that are consistently under positive selection in a set of sequences, using a Bayesian statistical procedure (see **Appendix A2.vi.c**). Examples include the analysis of sequences of plant self-incompatibility genes (Takebayashi et al. 2003) and sperm lysin genes (Yang and Bielawski 2000). In both these examples, parts of the protein with known important functions were shown to have evolved under positive selection.

However, a potentially doubtful assumption of this general approach is the constancy of the frequencies of all of the possible types of codons in the sequences under consideration. This assumption is not necessarily correct, and phylogenetic analyses of coding and noncoding sequences often show that different lineages are evolving differences in base composition (see **Section 10.1.ii.d**). This may call into question apparent inferences of positive selection from purely phylogenetic evidence. Evidence of this kind should therefore be supplemented by other data. In **Section 6.4**, we describe methods that combine data on sequence polymorphism with data on divergence among species. These methods are based on theories of the effects of selection, mutation, and drift on nucleotide site polymorphisms, which we describe in the next section. These also lead to a more realistic way of looking at sequence evolution than the widely used ones described in **Section 6.2.iii.a** above. Before describing these tests and analyses, we next describe the theory predicting variant frequencies within a population.

6.3. THE DISTRIBUTION OF VARIANT FREQUENCIES UNDER MUTATION, SELECTION, AND DRIFT

6.3.i. Introduction

We previously derived the probability distribution of the frequencies of two alternative variants at a site by using the stationary distribution with mutation in both directions (**Section 5.3.iii**). A difficulty in applying this distribution to data on variants under selection in natural populations is that we would need to know in advance which of the two alternative types of variant observed at a site is the one favored by selection. As discussed in **Section 5.3.iii.c**, relevant information is sometimes available; for example, in the case of GC versus AT variants at silent sites, there is evidence that biased gene conversion in heterozygotes generally favors GC over AT (Marais 2003). Similarly, in some species we can determine that a particular codon for a certain amino acid is used much more frequently than alternative codons, and so we infer that this codon is probably favored by some type of selection pressure (Graur and Li 2000, pp. 132–139; **Section 10.1.iii**).

When prior information of this kind is lacking, it is nevertheless possible with suitable data to infer which of the two alternative variants at a segregating site is the ancestral state, and which has been derived by mutation, by comparing a sequence or set of sequences with that from a related species. **Figure 6.5** shows an example, based on the data shown in **Figure 1.8** of **Section 1.2.iii.b**. This displays portions of the aligned sequences of human and chimpanzee G6PD homologues, showing nucleotide sites with polymorphisms among the human sequences, and the inferences of the changes that occurred in humans at these sites (sites that are the same in both species, or different between the single chimpanzee sequence and all those from humans, are omitted).

Information about whether a variant at a given site is ancestral or derived does not tell us whether it is favored or disfavored by selection, but is nevertheless very helpful in analyzing sequences for the purpose of detecting selection, as we will show below. As more genomes are sequenced from sets of related species, such information is becoming increasingly available. We can then compare the distribution over a large number of different polymorphic nucleotide sites of the within-population frequency of the mutant variant at each site (often called the *frequency spectrum*) to the probability distribution expected under a given set of assumptions, such as neutrality. This provides ways of testing hypotheses and estimating parameters, as we describe in **Section 6.4** below. Even when information on ancestral and derived states of variants is not available, the theory can be used for inferences about the nature and strength of selection.

The following section describes how to predict the relevant distribution for a simple scenario, similar to that already studied in **Sections 5.3.iii.c** and **5.3.iii.d**. As usual, for the purpose of obtaining specific results we assume diploid, autosomal inheritance and random mating, although it is not difficult to generalize to other cases. This section deals with topics that are more advanced than the earlier parts of this chapter, and it is possible to skip it on a first reading and go straight to **Section 6.4**.

A.

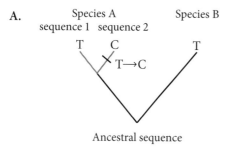

Ancestral sequence

B.

Number in Homo sapiens sample	29	340	1073	1656	1799	2015	2178	2495	2906	2942	3006	3055	3617	3916	4062	4141	4712	4974
24	A	G	A	C	C	G	C	C	C	C	C	G	G	C	T	C	A	C
4	A	G	A	C	C	G	C	C	T	C	C	G	G	C	T	C	A	T
2	A	G	A	C	C	A	T	C	C	C	T	G	A	C	C	C	G	C
2	A	G	A	C	C	A	T	C	C	C	C	G	G	C	C	C	G	C
2	A	G	A	C	C	G	C	C	C	C	C	G	G	C	C	C	G	C
4	A	G	A	C	C	G	C	C	C	C	C	G	G	T	C	C	G	C
2	A	G	A	C	C	G	C	C	C	C	C	A	G	C	C	T	G	C
1	A	G	A	T	C	G	C	C	C	C	C	G	G	C	C	T	G	C
1	A	G	A	C	C	G	C	C	C	C	C	G	G	C	C	T	G	C
7	G	A	G	C	G	G	C	T	C	T	C	G	G	C	C	C	G	C
2	G	G	G	C	G	G	C	C	C	C	T	C	G	G	C	C	G	C
Chimpanzee	A	G	A	C	C	G	C	C	C	C	C	G	G	C	C	C	G	C
Inferred changes	A→G	G→A	A→G	C→T	C→G	G→A	C→T	C→T	C→T	C→T	C→T	G→A	G→A	C→T	C→T	C→T	G→A	C→T

FIGURE 6.5 Determining the ancestry of a polymorphic variant when multiple sequences are available from at least one of a pair of closely related species, A and B. **A.** The method of inference for a case with only two sequences from one species. In this example, the variant C is found at a polymorphic nucleotide site that is not present in the related species, and is thus inferred to be derived by mutation within A, whereas the other variant (T) is inferred to be ancestral to both species' sequences. **B.** An example of this analysis for the partial human and chimpanzee G6PD sequences described in **Figure 1.8** of **Section 1.2.iii.b**. The figure shows the nucleotide sites inferred to have changed in humans, for the sites in which some humans differ from others. The changes that occurred in humans can be distinguished using a chimpanzee sequence, ignoring the possibility that chimpanzees may also be polymorphic at some of these sites (this has a relatively low probability). The bottom row of the figure shows the inferred changes at each such site.

6.3.ii. Statistical equilibrium under mutation, selection, and drift

6.3.ii.a. The general model

Suppose that, over a long stretch of m nucleotide sites, each site has two alternative types, A_1 and A_2, with mutation rates u and v from A_1 to A_2 and vice versa. A_1 and A_2 might correspond to GC and AT base pairs, or to selectively favored and disfavored amino acid variants. At any given time, some sites are fixed for the A_1 type, some for A_2, and others (usually a small minority) segregate for both. Let the number of sites fixed for A_1 and A_2 be m_1 and m_2, respectively. If the population is at statistical equilibrium, the mean number of sites in each state is constant over time, despite continual changes at individual sites. Each generation, $2Num_1$ sites fixed for A_1 and $2Nvm_2$ sites fixed for A_2 are expected to experience a mutation and start to segregate. At equilibrium, these changes are balanced by segregating sites that become fixed for A_1 or A_2. If the mutation rate is low enough relative to the population size, we can use the infinite sites assumption (**Section 5.1.ii.b**). Each site then segregates for at most one mutation: either an A_2 mutation at a site formerly fixed for A_1, or A_1 derived from a site fixed for A_2. The details of how to obtain the relevant distributions of variant frequencies are given in **Box 6.5***.

Box 6.5* DETERMINING THE STATISTICAL EQUILIBRIUM FOR VARIANTS DERIVED BY MUTATION

We focus initially on the frequency spectrum for A_2 variants that arose at sites fixed for A_1. We cannot use the stationary distribution formulae derived in **Section 5.3.iii**, because these do not discriminate between ancestral variants and variants derived by mutation. The method used to deal with the case studied here is described in **Section 6A.2** of the **Appendix** to this chapter. Under the infinite sites model, **Equation 6A.13** gives a general formula for $\phi(q)$, the probability density of finding a frequency q of A_2 at a site whose ancestral state was fixation for A_1. This involves the assumption that sites fixed for A_1 have a constant rate λ_1 per generation of becoming fixed for A_2. The expression for λ_1 is given by **Equation 6A.11**, which is equivalent to **Equation 6.6a** of **Section 6.2.iii.a**.

Exactly the same argument can be carried through for A_1 mutations derived from sites fixed for A_2; we simply exchange the fitnesses of A_1A_1 and A_2A_2, and substitute v for u, m_2 for m_1, and λ_2 for λ_1 (where the fixation probability in the formula for λ_2 is that for a new A_1 mutation at site fixed for A_2).

An alternative method is often used; instead of determining the probability distribution of variant frequencies, we work with the expected time that a mutation from A_1 to A_2 at a site spends in the frequency range q to $q + \delta q$ before it is fixed or lost from the population. For small δq, this is given by an expression of the form $t(q)\delta q$, where $t(q)$ is the *sojourn time density*. Determining $t(q)$ requires methods that are more advanced than those used in this book (Ewens 1964; Ewens 2004, Chapter 5; Kimura 1983, Chapter 8). Under statistical equilibrium, it can be shown that the probability that a segregating site has frequency q is proportional to $t(q)$, the principle of *ergodicity* (**Problem 6.4***). The functions $\phi(q)$ and $t(q)$ are thus closely related.

To describe the system fully, we also need to determine the proportions of sites that are fixed for A_1 and A_2. The expected number of sites fixed for A_1 that acquire mutations that eventually become fixed for A_2 is $m_1\lambda_1$, and the number of fixations in the opposite direction is $m_2\lambda_2$. At statistical equilibrium, these events must be equal in frequency, so that $m_1\lambda_1 = m_2\lambda_2$ (Bulmer 1991; Gillespie 1994). This implies that, among fixed sites, the equilibrium frequency of sites fixed for A_2 is:

$$x^* = \frac{m_2}{\left(m_1 + m_2\right)} = \frac{\lambda_1}{\left(\lambda_1 + \lambda_2\right)} \tag{6.8}$$

Under the infinite sites assumption, most sites are fixed for one variant or the other, so that x^* is close to the overall proportion of sites that are fixed for A_2, and also to the proportion of sites that are in state A_2 in a sequence taken at random from the population (McVean and Charlesworth 1999). If we write $\phi_1(q)$ for the probability density of frequency q of an A_2 variant derived from a site fixed for A_1, and $\phi_2(q)$ for the density when the derived variant is derived from a site fixed for A_2, the overall probability density function for a derived variant with frequency q is then given by $\phi_0(q) \approx x^*\phi_1(q) + (1 - x^*)\phi_2(q)$.

We can also use this approximation to determine the overall equilibrium rate of nucleotide substitutions, averaging over changes in both directions:

$$\lambda^* = (1 - x^*)\lambda_1 + x^*\lambda_2 = \frac{2\lambda_1\lambda_2}{\left(\lambda_1 + \lambda_2\right)} \tag{6.9}$$

We show below how these general results can be applied to some specific cases of biological interest.

6.3.ii.b. *The equilibrium frequency distribution under neutrality*
First, we describe the equilibrium results for neutral mutations, which provide the null hypothesis for testing for selection or departures from equilibrium; the details are given in **Box 6.6***. **Equation B6.6.2b** states the extremely useful

Box 6.6* THE EQUILIBRIUM DISTRIBUTION OF VARIANT FREQUENCIES WITH ONE-WAY MUTATION: NEUTRALITY

The distribution for neutral mutations can be found by setting $M_{\delta q} = 0$ in **Equations 6A.6.b** and **6A.13** of **Section 6A.2** of the **Appendix** to this chapter, giving:

$$\phi(q) \approx \frac{4N_e u}{q} \tag{B6.6.1}$$

where $1/2N \leq q \leq 1 - 1/2N$.

This equation yields two important results. First, the probability of a mutation being present at frequency q is inversely proportional to q. This can be used to obtain the frequency spectrum for a sample of k alleles, as was done for the stationary distribution in **Box 5.9*** of **Section 5.3.iii.c**. Approximating the integration limits $1/2N$ and $1 - 1/2N$ by 0 and 1, the probability of observing i copies of A_2 (where $1 \leq i \leq k - 1$ for a segregating sample) is given by:

$$\frac{4N_e uk!}{i!(k-i)!} \int_0^1 p^{k-i} q^{i-1} \, \mathrm{d}q = \frac{4N_e uk!}{i!(k-i)!} \mathrm{B}(k-i,i) = \frac{4N_e u}{i} \tag{B6.6.2a}$$

where B is the beta function (see **Box 5.9***).

The probability that a site is segregating in the sample is $4N_e u a_k$ (**Problem 6.5.i**), where a_k is Watterson's correction factor (**Box 1.3 of Section 1.2.iv.b**). The probability of observing i copies of a mutant variant at a segregating site in a sample of size k is thus:

$$P_i = \frac{1}{i a_k} \tag{B6.6.2b}$$

Exactly the same result applies when A_2 is the ancestral variant, because the distinction between A_1 and A_2 is arbitrary in this case. This result thus provides a general description of the frequency spectrum for a variant derived by mutation.

Second, the expected nucleotide site diversity for sites where A_2 is the mutant variant is:

$$\pi_1^* \approx 4N_e u \int_0^1 \frac{2pq}{q} dq = 8N_e u \int_0^1 (1-q) dq = 4N_e u \qquad \text{(B6.6.3)}$$

This is the same as the expression derived by the identity probability and coalescent methods in **Sections 5.1.ii.b** and **5.1.iii.d**, respectively.

result that the probability of finding i copies of a derived variant in a sample of k alleles is proportional to $1/i$, and is thus independent of the mutation rate (Watterson 1975). This result can also be derived from coalescent theory (Fu and Li 1993; Fu 1995). Perhaps counterintuitively, it implies that, among polymorphic sites, those with variants present only once in the sample (*singletons*) are expected to be the most frequent class. This formula provides a basis for tests of departure from neutral equilibrium, using comparisons between the observed and expected spectra of variant frequencies in a sample (see **Section 6.4.iii**).

We can also obtain the expected equilibrium nucleotide site diversity, π_1^*, at sites where the ancestral state was fixation for A_1 (**Equation B6.6.3**). The diversity at sites where A_2 is the ancestral state, π_2^*, is given by a similar formula, with u replaced by v. To obtain the approximate overall expected nucleotide site diversity, we average over these two classes of sites, weighting π_1^* by $1 - x^*$ and π_2^* by x^*. With neutrality, $\lambda_1 = u$ and $\lambda_2 = v$, so that **Equation 6.8** gives $x^* = u/(u + v)$; the mean diversity π^* is then the product of $4N_e$ and the net rate of mutation (in both directions), $\tilde{u} = 2uv/(u + v)$. This is the same as **Equation 5.2.2.b** of **Section 5.3.iii.c**. This approach can also be used to show that the expected number of segregating sites in a sample of k alleles is $4N_e a_k \tilde{u} m$ (**Problem 6.5.ii**), as expected from the results in **Section 5.3.iii.c**. From **Equation 6.9**, $\tilde{u}$ is also the overall equilibrium rate of nucleotide substitutions, λ.

With equal mutation rates in each direction, $x^* = 1/2$ and $\tilde{u} = u$. In general, the two mutation rates may well be unequal. Let $\kappa = v/u$ be the *mutational bias* parameter, which measures the relative rates of mutation away from and to A_2. Substituting this into **Equations 6.8** and **6.9** for the neutral case, $x^* = 1/(1 + \kappa)$ and $\lambda = \tilde{u} = 2\kappa u/(1 + \kappa)$. For a given u, $\tilde{u}$ increases with κ, from a value of u for $\kappa = 1$ to close to $2u$ for large κ. An important implication is that, if the mutation rates at a site are very asymmetrical, its equilibrium diversity and substitution rate will be close to twice the value for the less mutable type, because the base composition of the genome will be heavily biased towards nucleotides with this state. GC base pairs are usually more mutable than AT pairs (Graur and Li 2000, pp. 125–126). Regions of the genome with an unusually high mutational bias in the direction of GC to AT mutations will thus have a higher frequency of AT base pairs relative to GC pairs at neutral noncoding or synonymous sites, leading to an increased rate of DNA sequence evolution and level of variability (assuming that u is the same for all sites); see **Section 10.1.ii** for further discussion.

6.3.ii.c. *Properties of the statistical equilibrium with selection: rates of evolution*

The approach we have just used also provides a way of estimating the intensity of selection on nucleotide site variants. These applications usually assume a semidominant, autosomal selection model in order to obtain explicit analytical formulae; other cases have to be studied by numerical integrations or approximations. With a selection coefficient s in favor of A_2A_2 homozygotes, the frequency of sites fixed for A_2 (see **Section 6.3.ii.a** above) can be approximated by x^* in **Equation 6.8**. In this case, we have:

$$x^* = \frac{1}{1 + \kappa \exp(-\gamma)} \tag{6.10}$$

where $\gamma = 2N_e s$ (**Problem 6.6**).

This is known as the Li–Bulmer equation (Li 1987; Bulmer 1991) and is very useful for interpreting data on base composition and codon usage bias (**Section 10.1.iv**). It shows that x^* is determined jointly by the level of mutational bias, κ, together with the selection intensity scaled by N_e. If $N_e s$ is $<< 1$, x^* is close to the neutral value, but if $N_e s > 2$, x^* is close to 1. There is thus only a small range of values of $N_e s$ for which x^* is intermediate (**Figure 6.6**), unless mutational bias is extremely high.

For this selection model, using **Equation 6.9** to average over mutations in both directions gives the equilibrium rate of nucleotide substitution as a function of γ:

$$\lambda^* = \frac{2\gamma v}{\left[1 + \kappa \exp(-\gamma)\right]\left[\exp(\gamma) - 1\right]} \tag{6.11}$$

An interesting effect was first noticed by Eyre-Walker (1992), but largely overlooked until recently (McVean and Charlesworth 1999; Takano-Shimizu 1999; Kondrashov et al. 2006; Charlesworth and Eyre-Walker 2008). With a sufficiently high mutational bias against the selectively favored variant, and with low γ (< 0.25), the rate of substitution may *increase* as γ increases, contrary to the prediction of **Equation 6.7** of **Section 6.2.ii.a** for purifying selection (and to what is often believed). The reason for this is as follows. If γ is very low, x^* is close to the equilibrium under mutation pressure alone; sequences then mostly

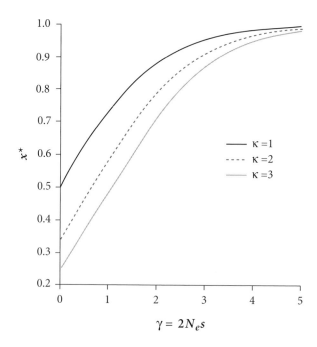

FIGURE 6.6 The equilibrium frequency, x^*, of sites fixed for the preferred variant under the model of reversible mutation and selection, showing the effect of different $\gamma = 2N_e s$ and κ values (from **Equation 6.10**). This shows that there is only a small range of values of γ for which x^* is intermediate, unless the mutational bias towards deleterious variants is extremely high.

contain nucleotides with low mutability. With an increased effective strength of selection (as measured by γ) against the more mutable nucleotides, sequences in an equilibrium population will contain fewer of the less mutable nucleotides. Although these produce mutations that are each less likely to become fixed than their alternatives, their larger mutation rate may dominate λ (this occurs if $\kappa > 2$), so that the rate of substitution is increased compared with the neutral situation. This will continue as γ is increased, until it becomes large enough for selection to control the substitution process (**Problem 6.7***). Furthermore, over a broader range of γ values, an increase in N_e can result in a temporary increase in the rate of substitution as favorable mutations at weakly selected sites enter the population (Takano-Shimizu 1999); there is some evidence for this at the level of amino acid sequence evolution in species that have shifted from islands to the mainland (Charlesworth and Eyre-Walker 2008).

It is also important to note that equilibrium implies that the rate of substitution of favored variants from sites fixed for the disfavored variant is the same as that in the opposite direction. The model of sequence evolution we are examining here, which ignores any adaptive evolution that may occur in response to a changing environment (see **Section 6.2.iii.b**), is often thought of as involving only the fixation of slightly deleterious mutations by drift (Ohta 1992). These must, however, eventually be balanced by mutations in the reverse direction (Bulmer 1991; Gillespie 1994). At statistical equilibrium, about half the nucleotides that have become fixed in a species since its divergence from a close relative will thus be selectively superior to their alternatives, regardless of the strength of selection, and half will be inferior. A lack of equality indicates a departure from equilibrium (see **Section 10.1.ii.e**).

6.3.ii.d. *Properties of the statistical equilibrium with selection: variability*
Box 6.7* derives the frequency distribution for segregating A_2 mutations derived from ancestral A_1 sites (**Equation B6.7.1**), as well as the expected equilibrium nucleotide site diversity (**Equation B6.7.3**). Similar expressions can be obtained for the frequency spectrum of A_1 when A_1 is the variant derived from A_2 by mutation. We simply change the sign of s in the terms involving selection, and interchange u and v. The two expressions for $\phi(q)$ can be averaged over the two cases using **Equation 6.8**, as in **Section 6.3.ii.a**, to give a single probability density for derived variants, $\phi_0(q)$, and a corresponding frequency spectrum for a sample. The resulting formulae predict an excess of variants at low frequencies, compared with the neutral case. This reflects the fact that the majority of derived variants under this process are deleterious, so that selection on polymorphic variants is predominantly purifying in nature.

Again using **Equation 6.8** to average over both types of site, we can derive the equilibrium nucleotide site diversity (McVean and Charlesworth 1999):

$$\pi^* \approx \frac{4v\left[1 - \exp(-\gamma)\right]}{s\left[1 + \kappa\exp(-\gamma)\right]} \tag{6.12}$$

If N_e is fixed, π^* relative to its neutral value depends on γ in a way that is qualitatively similarly to the substitution rate. The effect of increasing γ away from 0, however, is larger than for the substitution rate. With a high mutational bias, π^* starts to decline again for $\gamma > 2$, a larger value than for the substitution rate.

For $\gamma > 2$, π^* becomes largely independent of N_e, and approaches the equilibrium value expected under mutation–selection balance in an infinite population, $4v/s$ (this is twice the equilibrium frequency of a deleterious mutation given by **Equation 4.3** of **Section 4.2.ii.a**, since this frequency is very low with large $N_e s$: note that here we are assuming that A_1 is the variant disfavored by selection). Weak selection on synonymous or noncoding variants may therefore help to explain why the range of their diversity values across species seems to be much smaller than their range of population sizes (**Figure 1.10** of **Section 1.2.iii.b**).

Box 6.7* THE EQUILIBRIUM DISTRIBUTION OF VARIANT FREQUENCIES WITH ONE-WAY MUTATION: SELECTION WITH SEMIDOMINANCE

The distribution for semidominant, autosomal mutations A_2 derived from sites fixed for A_1, with a selection coefficient s in favor of A_2, can be found by setting $M_{\delta q} = spq/2$ in **Equation 6A.13** of **Section 6A.2** of the **Appendix** to this chapter. As in the derivation of **Equation B6.4.1** in **Box 6.4** of **Section 6.2.iii.a**, we have $\psi(q) = \exp(-\gamma q)$, where $\gamma = 2N_e s$. Carrying through the integrations involved in **Equation 6A.13** (**Problem 6.8.i***), we obtain:

$$\phi(q) \approx \frac{4N_e u \left[1 - \exp(-\gamma p)\right]}{pq\left[1 - \exp(-\gamma)\right]} \tag{B6.7.1}$$

This equation can be used to derive expressions similar to those for the neutral case. The probability of observing i copies of A_2 and $k - i$ copies of A_1 in a sample of k alleles is given by an approximate expression similar to **Equation B6.6.2a**:

$$\frac{k!}{i!(k-i)!} \int_0^1 p^{k-1} q^i \phi(q) \mathrm{d}q \tag{B6.7.2}$$

where $\phi(q)$ is given by **Equation B6.7.1**.

Similarly to **Equation B.6.6.2b**, a formula for P_i, the probability of observing i copies of A_2 in a segregating sample, is obtained by dividing this expression by its sum over all values of i for which $1 \leq i \leq k$. The integral cannot be expressed in simple algebraic terms, but must be evaluated numerically.

The approximate expected equilibrium nucleotide site diversity is given by:

$$\pi_1^* \approx \frac{8N_e u \int_0^1 \left[1 - \exp(-\gamma p)\right] \mathrm{d}q}{\left[1 - \exp(-\gamma)\right]} = \frac{4u\left[\gamma - 1 + \exp(-\gamma)\right]}{s\left[1 - \exp(-\gamma)\right]} \tag{B6.7.3}$$

It is easily verified that, as γ approaches 0, **Equations B6.7.1** and **B6.7.3** yield the corresponding neutral expressions (**Problem 6.8.ii***).

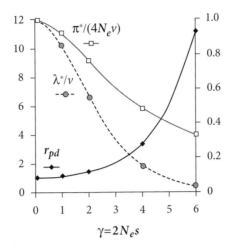

FIGURE 6.7 The equilibrium rates of substitution (λ^*), nucleotide site diversity (π^*), and ratio of polymorphism to divergence (r_{pd}) as a function of γ, all expressed relative to their values under neutrality, based on **Equations 6.11** and **6.12**. The left-hand y axis gives the values of r_{pd}; the right-hand axis is for λ^* and π^*. Symmetrical mutation ($u = v$) is assumed.

Diversity within populations is less affected by selection than divergence between species, as measured by the product of divergence time and λ. This makes sense on the theory above, which shows that selection is less effective at preventing a deleterious mutation from entering the population and becoming polymorphic than at preventing it spreading to fixation. To illustrate this, **Figure 6.7** plots the equilibrium ratio of polymorphism to divergence (r_{pd}) against γ; this ratio is always higher than the value for neutrality (i.e., the relative r_{pd} value in the figure always exceeds the neutral value of 1). Selection of this kind is thus indicated if r_{pd} for sites of interest is higher than for sites that can be assumed to be neutral. The same applies when the level of polymorphism is measured by the number of segregating variants in the sample, rather than the diversity value, but the calculation is more complex (Sawyer and Hartl 1992). As we show in **Section 6.4.ii**, this property of r_{pd} provides a way of examining the nature of selection on amino acid mutations.

6.4. TESTING FOR SELECTION FROM POLYMORPHISM AND DIVERGENCE DATA

6.4.i. Introduction

One of the most important advances to have come out of the study of variation and evolution at the DNA sequence level is that it provides the opportunity to ask, and potentially answer, such basic questions as: how many nucleotides in the genome are neutral, what fraction are subject to selection, and what is the nature and intensity of such selection? In this section of the chapter, we examine what may be called "direct" tests for selection, i.e., tests that detect selection from its effects on the sites that are themselves the targets of selection. In **Section 8.3**, we will discuss the effects of selection on sites in the surrounding genomic region, e.g., on other regions of the same gene, or on nearby genes or noncoding sequences, and show how these effects can be used in indirect tests for selection.

Here we outline several tests that use the direct effects of selection on DNA sequence variation within populations and divergence between species. The McDonald–Kreitman (MK) test, which we describe first, can detect the presence of too many nonsynonymous site substitutions between species to be accounted for without positive selection, and can also reveal the operation of purifying selection. Other tests detect patterns of variation within species that are unlikely on the null hypothesis of neutrality and demographic equilibrium. Another direct indicator of selection within a species, involving the level of divergence between local populations, is discussed in **Section 7.3.iii**, when we describe the population genetic theory of subdivided populations.

6.4.ii. The McDonald–Kreitman (MK) test

6.4.ii.a. *Introduction*

We saw in **Section 6.2.iv** above that K_A is generally much less than K_S, and that this is evidence for selective constraints acting on the coding sequence in question, i.e., for selection removing amino acid mutations that are detrimental to the protein's function. We also saw how adaptive evolution of a protein sequence can be inferred when K_A is larger than K_S, but that this situation is unusual. Even when positive selection occurs, there is often only a limited region within a protein's coding sequence in which K_A is larger than K_S, with the rest of the sequence being under purifying selection. When we observe an unusually high value of K_A relative to K_S for a particular species pair or lineage, but K_A is not significantly greater than K_S, this might indicate either a higher rate of adaptive protein sequence evolution for some of the sites in the protein sequence, or weak selective constraints on some or all of them. It is important to be able to distinguish these alternative possibilities, both of which are biologically plausible.

The difficulties of interpreting substitution patterns are even greater for noncoding sequences, which are being increasingly studied by means of sequence comparisons. Sequence differences between species are often significantly lower for some types of noncoding sequence than for others, and lower than synonymous sites: some intergenic sequences and intronic sequences are remarkably conserved between different species (Bergman and Kreitman 2001; Haddrill et al. 2005b; Keightley et al. 2005; ENCODE Project Consortium 2007). In some cases, sequence conservation extends over a very wide taxonomic range, such as primates versus rodents, and experiments on their effects on gene expression indicate that the sequences involved have important regulatory functions, such as acting as enhancers of gene expression (Visel et al. 2008). While sequence conservation is usually interpreted as evidence for greater selective constraints on the more slowly evolving sequences, the possibility of different mutation rates, or simply smaller mutational bias (**Section 6.3.ii.b**), cannot be excluded without further information. In the next section, we show how polymorphism data can help to solve these problems.

6.4.ii.b. *Principle of the MK test*

The McDonald–Kreitman (MK) test can detect selection on amino acid sequences by combining data on sequence divergence between two species with polymorphism data from at least one of them (McDonald and Kreitman 1991). Ideally, the same length of sequence is used for both polymorphism and divergence estimates, so that the total number of mutable sites is the same in both cases. Under the null hypothesis that all types of sequence variants are neutral, the results in **Sections 6.2.ii** and **6.3.ii.b** imply that the total number of synonymous and nonsynonymous differences in a coding sequence are expected to be proportional to the mutation rates for the two classes of variants: between-species differences and within-species polymorphisms. Selection on the protein sequence produces a departure from this proportionality, which can be tested for by a simple 2 × 2 contingency table (**Figure 6.8**).

Although this simple view is helpful in understanding the principle of the MK test, it is illuminating to relate it more quantitatively to the theory, espe-

A.

	Fixed	Polymorphic
Synonymous	D_s	P_s
Nonsynonymous	D_n Excess suggests directional selection	P_n

B.

	Fixed	Polymorphic
Synonymous	D_s	P_s
Nonsynonymous	D_n Deficit suggests purifying selection	P_n Excess suggests balancing selection

FIGURE 6.8 The 2 × 2 contingency table used in the McDonald–Kreitman test. **A.** A case when the ratio of the numbers of nonsynonymous variants to synonymous variants for differences between species (D_n and D_s, respectively) exceeds the ratio for within-species variation (P_n and P_s, respectively), suggesting positive selection. **B.** A case with a deficit of nonsynonymous substitutions between species relative to nonsynonymous polymorphisms, suggesting either purifying selection or balancing selection.

cially as this leads to an approach that can measure the strength of selection, not just test for it (see **Section 6.4.iv.c**). With no selection, we have seen that the ratio of the number of nonsynonymous polymorphic variants in a gene to the number of nonsynonymous fixed differences between the species (nonsynonymous r_{pd}) is expected to be the same as the synonymous r_{pd}, the ratio of synonymous polymorphisms to synonymous fixed differences. But, as shown in **Section 6.3.ii.d**, mutation–selection–drift equilibrium (with reversible mutations between favored and disfavored variants) means that r_{pd} at selected sites will be greater than the neutral value. If synonymous (or noncoding) mutations are close to neutral, or at least more weakly selected than nonsynonymous mutations in the same gene, their r_{pd} is thus expected to be lower than the nonsynonymous r_{pd}. The difference in these ratios thus forms the basis for a test for purifying selection on nonsynonymous sites.

Dividing the r_{pd} value for sites that are potentially under selection by the r_{pd} for putatively neutral or nearly neutral sites gives a measure that is often called the "neutrality index," *NI* (Rand and Kann 1996; Weinreich and Rand 2000); sometimes the reciprocal of this measure is used (Charlesworth 1994a; Presgraves 2005; Shapiro et al. 2007). Under neutrality, we expect this ratio to equal 1. The statistical significance of a departure of *NI* from 1 for a given gene can be assessed by a 2×2 contingency table test (**Figure 6.8**). Results for different genes can be combined by standard statistical procedures, such as the Mantel–Haenszel test (Sokal and Rohlf 1995, p. 766) or by permutation tests (Shapiro et al. 2007).

NI values significantly greater than 1 (an excess of polymorphisms in the category being tested for selection) thus suggest purifying selection. *NI* > 1 is, indeed, frequently observed when tests are pooled over many different genes (Rand and Kann 1996; Weinreich and Rand 2000; Charlesworth and Eyre-Walker 2006; Shapiro et al. 2007). However, *NI* > 1 can also result from balancing selection, causing one or more amino acid variants to be present at intermediate frequencies; this possibility can usually be tested using information on variant frequencies, as we describe in **Sections 6.4.iii** and **8.3.ii.d**.

In contrast, some amino acid substitutions that distinguish a pair of species may have been fixed by relatively strong directional selection, for instance because of a recent change in the environment (**Section 6.2.iii.b**), or an evolutionary arms race (**Section 6.2.iv**). Such adaptive variants contribute little or nothing to variation within a species, so they will be associated with a low r_{pd}, creating an excess of nonsynonymous fixed differences over the neutral expectation. *NI* is thus less than 1 for such sites. The original paper on the MK test described such a situation for the *Adh* gene of *Drosophila yakuba* and its relatives, *D. melanogaster* and *D. simulans* (McDonald and Kreitman 1991), and many other examples have since been reported (Eyre-Walker 2006). **Table 6.2** shows MK test results on a gene in some *Drosophila* species, which provides evidence for positive selection.

TABLE 6.2 An example of the use of the McDonald–Kreitman test[1]

Species compared	Diversity data from	Polymorphisms			Divergence			G test value	P value
		N[2]	S	N/S	N	S	N/S		
D. melanogaster and D. simulans	Pooled data from both species	27	108	0.250	27	34	0.794	11.89	0.0006
D. melanogaster and D. yakuba	D. melanogaster only	5	43	0.116	69	152	0.454	9.98	0.0016
D. simulans and D. yakuba	D. simulans only	21	68	0.309	60	139	0.432	1.33	0.24 ns
D. melanogaster and D. mauritiana	D. melanogaster only	5	43	0.116	32	51	0.627	13.20	0.0003
D. simulans and D. mauritiana	D. simulans only	22	69	0.319	3	13	0.231	0.233	0.63 ns
D. melanogaster lineage	D. melanogaster only	5	43	0.116	16	21	0.762	12.35	0.0012
D. simulans lineage	D. simulans only	22	69	0.319	10	8	1.25	6.57	0.010

[1]The study tested whether a gene (*nup96*) identified as being important in hybrid inviability in crosses between *Drosophila* species has evolved differences under positive selection. Data were analyzed in two ways. The first five rows show analyses of polymorphisms within species (or pooled data from the two species being compared, as indicated in column 2), and divergence between the species pair, with the significant MK test results highlighted in grey. The two lines at the bottom, below the dividing line, show results when fixations are not estimated from the numbers of fixed differences between the species, but inferred in the two separate lineages, using sequences from an outgroup species. This analysis suggests that both the *nup96* gene is evolving adaptively in both *D. melanogaster* and *D. simulans* lineages (whereas, without an outgroup, it appeared that only the former was under positive selection). From Presgraves et al. (2003).

[2]*N* values are numbers of nonsynonymous differences. The symbol R was used in the original paper, and many subsequent ones, for replacement changes, another term for nonsynonymous changes. *S* values are numbers of synonymous differences.

This study illustrates the fact that the test can be used to compare numbers of synonymous and nonsynonymous substitutions, not just for pairs of species, but also in specified lineages of interest. It is also not confined to comparing nonsynonymous versus synonymous variants, but can also be used to compare other types of variants. For example, the test can be used to test for selectively driven substitutions at noncoding sites by comparing r_{pd} for noncoding sites that are candidates for selection, using comparisons with r_{pd} values for supposedly neutral sites, such as putatively neutral noncoding sites or synonymous sites (Andolfatto 2005; ENCODE Project Consortium 2007).

6.4.ii.c. *Potential difficulties with the MK test*

One problem is that the use of the 2×2 test of significance is based on the assumption that the frequencies of observations that fall into the different categories follow a multinomial distribution (**Appendix A2.v.b**). This is appropriate if there is no recombination in the gene of interest, so that all mutations occur on the same genealogy, or if there is sufficiently frequent recombination, so that mutations at different sites evolve independently of each other. For the more realistic case when there is occasional recombination within a gene, the evolution of different sites is not fully independent, as is explained more fully in **Section 8.1.iii.c**. A more stringent significance level than usual should thus probably be used, and procedures for doing this are discussed by Andolfatto (2008).

We have already mentioned that positive selection causing an excess of nonsynonymous fixed differences may be obscured by the effect of purifying selection on nonsynonymous mutations, which elevates the *NI*. Because it is likely that only a small minority of nonsynonymous mutations are positively selected, and most are deleterious, this may be a serious problem. Because deleterious variants will mostly be present at lower frequencies than neutral or nearly neutral variants (as we discuss further in **Section 6.4.iii**), it has been suggested that this problem can be dealt with by removing low-frequency variants (both nonsynonymous and synonymous) from the polymorphism data (Fay et al. 2001). This procedure can change the overall picture from one of neutrality or predominantly purifying selection to one with evidence for positive selection (Charlesworth and Eyre-Walker 2008; Shapiro et al. 2007). The effect of excluding rare variants does not, however, completely remove the bias caused by purifying selection, so that estimates of the frequency of adaptive evolution (see the next section) may often be underestimated (Charlesworth and Eyre-Walker 2008). It also suffers from the difficulty that the cut-off frequency for variants is arbitrary, and that the conclusions are affected by the cut-off chosen.

Another problem is that genetic drift in small populations allows slightly deleterious variants to become fixed (**Section 6.2.iii.a**). If drift has been occurring over a prolonged period of reduced population size, and the population then expands to a large effective size, so that new nonsynonymous, deleterious mutations now become unlikely to get fixed, but can still accumulate as polymorphisms, the older fixed variants will be interpreted as adaptive substitutions (Eyre-Walker 2002). This bias is worsened if low-frequency variants are removed. On the other hand, the bias is reduced if the synonymous variants used as the neutral standard of comparison are not neutral, but are subject to weak selection for codon usage bias (Eyre-Walker 2002); as explained in **Section 10.1.iii**, this is the case for many species used in population genetic studies. The problems posed by population size changes are, however, less severe for MK tests than for tests for selection based on variant frequencies, as we shall discuss below in **Section 6.4.iii.e**.

6.4.ii.d. *Estimating the fraction of amino acid fixations caused by positive selection*

The MK test provides a way to estimate the proportion of fixed differences between species at nonsynonymous sites that were caused by positive selection, as opposed to genetic drift. This topic has been hotly debated since the beginnings of the study of molecular evolution (Kimura 1983; Gillespie 1991). This proportion is often denoted by α. **Box 6.8** outlines a simple way of estimating this, based on the neutrality index introduced in **Section 6.4.ii.b**. The number of fixed substitutions depends both on the rate of origin of new mutations and their fixation probabilities relative to the neutral value (with favorable mutations implying greater fixation probabilities; see **Section 6.2.iii.a**); thus it is important to distinguish between α and the (much lower) fraction of newly arisen nonsynonymous mutations that are advantageous.

Box 6.8 ESTIMATING THE FRACTION OF AMINO ACID MUTATIONS FIXED BY POSITIVE SELECTION

This is based on the method proposed by Smith and Eyre-Walker (2002) and Fay et al. (2002). Let the observed number of nonsynonymous and synonymous or silent fixed differences between the two species be NF_A and NF_S, respectively (these may need to be corrected for multiple and reverse substitutions, by the methods described in **Section 6.1.ii.d**). Similarly, let the number of nonsynonymous and synonymous or silent segregating sites in the within-species sample of sequences be NP_A and NP_S.

We assume that all of the synonymous/silent mutations and a fraction f of the nonsynonymous mutations are effectively neutral, the rest being eliminated by selection. The number of nonsynonymous differences fixed by drift is thus equal to $NF_S f$. We also assume that there is an additional number, NF_a, of nonsynonymous differences that are fixed by positive selection, so that $NF_A = NF_S f + NF_a$. The fraction of nonsynonymous differences fixed by positive selection is thus $\alpha = NF_a/NF_A$. These mutations are assumed to make no contribution to within-species variation, so that $NP_A = NP_S f$. This means that we can estimate f from the ratio NP_A/NP_S. We then obtain the theoretical value of NF_A as $(NP_A/NP_S)NF_S + NF_a$. This can be rearranged to give:

$$\alpha = 1 - \frac{NF_S NP_A}{NF_A NP_S} = 1 - NI \qquad \text{(B6.8.1)}$$

where NI is the neutrality index (**Section 6.4.ii.b**).

As discussed in **Section 6.4.ii.c**, with purifying selection acting on most amino-acid variants, NP_A/NP_S will be greater than f, so that α is likely to be underestimated by this method.

This is the expression for a single gene. It is not entirely straightforward to combine information from different genes; methods for doing so, and for obtaining confidence intervals on the resulting estimate of α, are discussed by Bierne and Eyre-Walker (2004), Welch (2006), Shapiro et al. (2007), and Andolfatto (2008).

Several different multi-locus surveys of DNA sequence variability in *D. melanogaster* and *D. simulans*, combined with estimates of divergence between the two species, have consistently suggested an α value of about 0.25; i.e., about 25% of nonsynonymous fixed differences are the result of positive selection (Eyre-Walker 2006); the latest study included data on about 400 autosomal genes (Shapiro et al. 2007). Similarly, data from multiple genome sequences of *Escherichia coli* and *Salmonella typhimurium/enterica* suggested an α of about 50%, once low-frequency variants have been removed (Charlesworth and Eyre-Walker 2006). In contrast, an analysis of human polymorphism data and divergence from Old World monkeys for 149 genes yielded little evidence for positive selection (Zhang and Li 2005). In *Drosophila*, there is evidence that certain categories of genes, especially those involved in male reproductive functions, may have unusually high rates of protein sequence evolution, and high α values (Pröschel et al. 2006; Baines et al. 2008). Future large-scale studies of genome-wide variability should lead to more reliable estimates of α for different categories of genes. In **Section 6.4.iv.c** below, we describe an alternative approach to estimating α in the presence of purifying selection, which does not require removal of low-frequency variants.

6.4.iii. Tests based on variant frequencies

6.4.iii.a. *Introduction*

We next describe a different type of test, also using polymorphism data at a large number of nucleotide sites in a sample of alleles from a population, but now testing observed numbers of segregating variants against the neutral frequency distribution described in **Section 6.3.ii**. As shown there, variants subject to purifying selection should show an excess of sites with rare variants compared with the neutral expectation.

We first note that there is often no outgroup from which to infer which variants are ancestral, or there may be concerns about the reliability of such inferences. A way of avoiding these problems is to use a "reflected" variant frequency spectrum, adapting the theory to predict the frequencies of the rarer variant at each polymorphic site, regardless of which is the ancestral state; this is similar to the method used in **Section 5.3.iii.c**, although the underlying logic is different. For a sample of size k, we use as data the minority variants at the sites surveyed (variants whose number of copies at the sites in question are $\leq k/2$ if k is an even number, or $< k/2$ for odd k). Instead of the probability of observing i copies of a mutant, we sum the probabilities of either i or of $k - i$ variants (**Equation B6.6.2b** in **Section 6.3.ii.b**), to obtain the neutral expectation for the probability of observing a minority variant present as i copies at a site. The observed distribution across all segregating sites can then be compared with this expectation, as shown in **Figure 6.9**.

A simplified way of testing for selection on nonsynonymous variants is to compare the relative number of nonsynonymous versus synonymous or noncoding sequence variants in different broad frequency categories, such as "rare" and "common." If, as is usually expected, there is stronger purifying selection on nonsynonymous than on silent variants, we expect to see synonymous or silent derived variants at higher frequencies than nonsynonymous variants; this is indeed often found in the case for both human and natural populations (e.g., Fay et al. 2001; Shapiro et al. 2007).

At first sight, it might seem easy to use a simple statistical test for heterogeneity, such as χ^2, to compare the observed and expected numbers of variants in the different frequency categories. Such tests are, however, legitimate only when variants at different sites are distributed independently of each other in the population, which is often not the case, especially for non-recombining portions of the genome or sites within the same gene, which recombine rarely (**Section 8.1.iii**). Such correlations do not affect the expected numbers in the different frequency categories, but cause the true sampling distribution for the

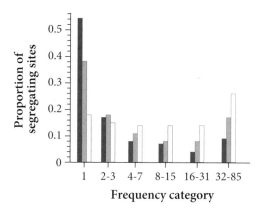

FIGURE 6.9 The observed allele frequency distribution of a set of SNPs in a sample of genes from the human population, compared to the values expected for neutral mutations in an equilibrium population (open bars). The black bars show nonsynonymous variants, and the grey bars show intron SNPs. Alleles are grouped into frequency classes (frequency category 1 represents singletons, and the other classes represent SNPs present in five non-singleton categories). [Adapted from Figure 1 of Eyre-Walker et al. (2006).]

observed numbers to be more variable than with independence. Unless most of the segregating sites come from different, independent loci, the conclusions from these tests may be questionable unless a highly significant result is found.

To deal with situations with little recombination, such as data on different sites in the same gene, tests have been developed that use coalescent theory for non-recombining sequences to generate test statistics measuring the departure of the data from neutral expectation, and to determine their distribution under the null hypothesis of neutral equilibrium (usually by coalescent simulations). The assumption of no recombination is the most conservative basis for testing for departures from the null hypothesis, because, as we saw in **Section 5.1iii.c**, recombination reduces the sampling variances of the statistics by providing multiple independent estimates. Extensions to the tests have been developed, which estimate recombination parameters from the DNA sequence polymorphism data, and employ these in coalescent simulations with recombination (**Section 8.2.iii.b**) to determine the distribution of the test statistics under the null hypothesis. These are available in computer packages such as DnaSP (Rozas et al. 2003).

6.4.iii.b. *Tajima's D*

At statistical equilibrium with neutrality, the two measures of sequence diversity described in **Section 1.2.iv.b**, π and θ_w, have the same expectation (see **Section 6.3.ii.b**); the expected value of both measures is equal to the scaled neutral mutation parameter, $\theta = 4N_e\tilde{u}$. Common variants contribute more than rare variants to diversity measures such as π, whereas θ_w is simply based on counting the number of variants, regardless of their frequencies. For a given number of variants in a set of sequences, an excess of rare variants will lead to a lower π value than if the variants were at higher frequencies, and thus π will be less than θ_w. Tajima's test (Tajima 1989b) is based on the same principle as the t test familiar in statistics, and uses the ratio of the difference ($\pi - \theta_w$) to its standard deviation on the assumption of no recombination; differences much larger than the standard deviation are unlikely if the null hypothesis is correct (we discuss below the types of deviations from neutrality that may be detected). This statistic is called Tajima's D or D_T. A large magnitude, negative D_T suggests an excess of rare variants (and a positive value suggests the reverse). The distribution of D_T, as determined from coalescent simulations, is fairly insensitive to the true value of the scaled mutation parameter, which of course is unknown.

Since we expect selection against deleterious amino acid variants to cause them to be present at lower frequencies than neutral variants, we would, in general, expect to find significantly negative D_T values for nonsynonymous variants, and nonsignificant (or at least much smaller) D_T values for silent or synonymous variants. This is sometimes observed. For instance, the human *Toll-like receptor 4* gene (*TLR4*), which is involved in the immune response, shows significantly negative D_T for nonsynonymous variants, in contrast to noncoding sites in the same gene (Smirnova et al. 2001). In a study prompted by this evidence for selection, rare *TLR4* amino acid variants were indeed found to be associated with meningococcal infections (Smirnova et al. 2003). However, a large survey of genes implicated in several other human diseases failed to detect an overall significant D_T for nonsynonymous variants (Cargill et al. 1999). Large sample sizes are needed for clear patterns to emerge, and the population's recent history may affect variant frequencies and cause departures from the null hypothesis, even for neutral sites (see **Section 6.4.iii.e** below).

Positive D_T values might be expected for variants maintained by balancing selection (**Sections 2.1.ii.c** and **2.2**), because these are maintained at intermediate frequencies for much longer than neutral variants. Except for unusual highly polymorphic loci, such as the MHC locus and self-incompatibility loci of plants (**Section 2.2.i.b**), few amino acid mutations within a gene are likely to be subject to balancing selection. Thus, D_T has little power to detect the effects of balancing selection on nonsynonymous sites. The influence of such selected sites on closely linked neutral or nearly neutral variants may, however, leave a detectable "footprint" of selection, as we will discuss in **Section 8.3.ii**.

6.4.iii.c. *Fu's and Li's test statistics*

A related test for selection is based on the idea that some parts of the ancestry of a sample of alleles from a population are relatively old compared with others (Fu and Li 1993). The low frequencies of deleterious mutations mean that they will tend to be seen only on the more recent, "external" branches of the genealogy (**Figure 5.4** of **Section 5.1.iii.a**). In contrast, variants under balancing selection will tend to be on internal branches, since they persist for long evolutionary times (see **Section 6.4.iii.b**). These differences can be used to test for departure from neutral equilibrium expectations, by asking whether singleton variants (variants restricted to the outermost branches of a gene genealogy) are more abundant (or less abundant) in a sample of sequences than is consistent with neutral expectation.

If we can identify variants derived by mutation by using an outgroup, the expected number of singleton-derived variants can be used to test for selection. Use of **Equation B6.6.2a** in **Section 6.3.ii.b** with $i = 1$ shows that, in a sequence of length m, this expected number is $m\theta$, where θ is the scaled mutation rate. In contrast, the expected total number of segregating variants in a sample of size k is $ma_k\theta$, where a_k is Watterson's correction factor (**Box 6.6**). This suggests that we can base a test on the difference between the observed total number of segregating sites and the number of singletons for derived variants multiplied by a_k. As in Tajima's test, this difference is divided by the value of its standard deviation for a non-recombining sequence, giving the statistic D_F (Fu and Li 1993), whose significance can be evaluated by coalescent simulation, as explained for D_T. If an outgroup is unavailable for inferring ancestry, the difference between the product of the number of segregating sites multiplied by $k/(k - 1)$, and the number of singletons multiplied by a_k, with a modified expression for the standard deviation in the denominator, generates a test statistic D_F^* (Fu and Li 1993). In general, these tests yield similar results to D_T, and the two tests are not fully independent.

6.4.iii.d. *Other test statistics*

Several other direct methods for testing for departures from neutral equilibrium expectation have been proposed, using principles similar to those we have just described. Some of the relevant papers are Slatkin and Hudson (1991), Fu (1997), Mizawa and Tajima (1997), and Sunyaev et al. (2000).

6.4.iii.e. *Problems of interpretation of tests using variant frequencies*

Although these effects of selection on the frequencies of selected variants are simple to understand, it is important to realize that selection is not the only factor that influences variant frequencies. The only firm conclusion from a significant test result is that the model assumed in the null hypothesis is incorrect for the data in question. The constant population size assumed in the theoretical models is particularly troubling, since population expansion leads to an excess of rare variants, because the larger size of more recent populations means that a relatively high proportion of mutations are of recent origin, and hence are present at low frequencies (Tajima 1989c), mimicking selection against deleterious mutations. Similarly, complete loss of variability due to an extreme or prolonged bottleneck of population size leads to a situation in which all diversity must come from new mutations, which will initially be at much lower frequencies than those in an equilibrium population (this effect is discussed in detail in **Section 8.3.v.b**, in relation to the effects of "selective sweeps" on variability).

Conversely, if only a proportion of variation is lost in a bottleneck (as is most likely in nature), the remaining variants will include an excess of high frequency ones, since rare variants have a higher chance of being lost (Tajima 1989c). Subdivision of a population into local, partially isolated populations can also cause departures from neutral expectation (Tajima 1989a) by causing an excess of rare variants (in samples pooled across the populations) or a deficiency (in samples taken from individual populations) (see **Section 7.2.v.a**).

As already mentioned, selection against deleterious nonsynonymous variants can be detected by virtue of a difference in the magnitude of a test statistic for nonsynonymous variability compared with synonymous or noncoding sites. For example, a survey of 18 genes of *D. miranda* found no significant difference between the mean π and mean θ_w values for silent variants, but there was a highly significant difference for nonsynonymous ones (Bartolomé et al. 2005).

One might be tempted to assume that departures from expectation at silent sites are effects of the population's demographic history, rather than being caused by selection, especially if similar departures are found in many sequences distributed across the genome. We cannot, however, ignore the possibility of selection at silent sites, or even of positive selection on some types of noncoding sequences, as discussed in **Section 6.4.ii.a** above. In addition, as we discuss in **Section 8.2.iii.e**, population size changes affect the patterns of associations among variants at different sites in the genome, and information on such associations may thus help us test for their operation, as opposed to, or in addition to, selection on silent sites. Another approach that is increasingly being used is to simulate alternative demographic histories, and to ask whether they can explain multiple aspects of data from many loci (e.g., Wall and Przeworski 2000; Haddrill et al. 2005a; Jensen et al. 2005; Ometto et al. 2005; Thornton and Andolfatto 2006; Thornton et al. 2007).

6.4.iv. Estimating selection intensities

6.4.iv.a. *Introduction*

In this section, we describe how the theory of variability and divergence under the joint effects of selection, mutation, and drift can be used to estimate the intensity of selection on nonsynonymous variants; we consider the related problem of selection on synonymous and noncoding variants in **Section 10.1**. Here, we examine two approaches in detail, which illustrate both the utility of methods based on the population genetics theory that we have described, and also their potential pitfalls.

6.4.iv.b. *Fitting the distribution of variants to a one-way mutation model*
Perhaps the most obvious way of estimating the intensity of selection on non-synonymous variants is to fit their observed frequency spectra to the distributions described in **Section 6.3.ii** (Akashi 1999). We examine how this can be done, using a specific example from a study of the human population.

CASE STUDY
Estimating the Distribution of Selection Coefficients from Human Polymorphism Data

Eyre-Walker et al. (2006) analyzed a set of about 170 allelic DNA sequences for 320 different human autosomal genes (illustrated in **Figure 6.9** of **Section 6.4.iii.a** above). The details of their method are given in **Box 6.9***. It assumes that all new nonsynonymous mutations are deleterious and semi-dominant, with a continuous probability distribution of the selection coefficient s against homozygotes for new mutations (a gamma distribution). Variants in introns were used as presumptively neutral controls.

The study of the fitness effects of variants yielded maximum likelihood estimates of the parameters of the gamma distribution, corresponding to a mean and standard deviation for $N_e s$ of 666 and 1315, respectively. With N_e for the human population of about 10,000 (**Section 5.2.iii.e**), this implies a mean heterozygous selection coefficient (half the homozygous value) of 0.033 against a newly arisen nonsynonymous mutation, with a standard deviation of 0.066 (giving a coefficient of variation of 2). There thus appears to be a wide distribution of selection coefficients, with a large fraction of very low values. About 80% of the distribution has $N_e s > 2$, corresponding to strongly deleterious mutations that are under the control of selection rather than drift. Mutations detected as polymorphisms within the population are enriched for the least severely deleterious variants, and the mean $N_e s$ for such segregating variants was estimated to be about 15.2, about 2% of the mean for new mutations. This corresponds to a mean selection coefficient against heterozygotes of 7.6×10^{-4}.

The method makes two doubtful assumptions—that the population size is constant and that intron mutations are neutral. In fact, the frequency spectrum for the intron variants did not fit the equilibrium expectation for neutral mutations, but showed too many rare variants (**Figure 6.9**). This may reflect either population size expansion or some weak selection on intron variants. Simulations showed that population size expansion may cause the mean selection coefficient to be overestimated by a factor of four or so (Eyre-Walker et al. 2006). Selection on mutations in introns would cause a bias in the opposite direction.

Box 6.9* FITTING VARIANT FREQUENCY SPECTRA TO A PURIFYING SELECTION MODEL

B6.9.i. The gamma distribution

The gamma distribution has been widely used in models of molecular evolution, because it is mathematically convenient and can take a wide range of shapes, from nearly symmetrical to highly skewed to the right. It has two parameters, a and b. The mean and variance of the gamma distribution are equal to ab and ab^2, respectively (a is sometimes called the "shape parameter" and b the "location parameter"). The probability density for a new nucleotide site mutation with scaled selection coefficient, $\gamma = 2N_e s$, where s is the homozygous selection coefficient against a deleterious mutation, is given by:

$$\psi(\gamma) = \frac{\gamma^{a-1}\exp(-\gamma/b)}{b^a\Gamma(a)} \qquad (B6.9.1)$$

where $\Gamma(a)$ is the gamma function (see **Box 5.9*** of **Section 5.3.iii.c**). A small value of a (< 1) indicates a highly skewed, peaked distribution, with a large coefficient of variation ($1/\sqrt{a}$).

B6.9.ii. Predicting the frequency spectra

For a given value of γ, the probability density of finding i deleterious non-synonymous mutations at a site in a sample of size k is the product of the number of combinations of i items chosen from k, times the integral of $q^i(1-q)^{k-i}\phi_A(q)$ between 0 and 1, where $\phi_A(q)$ is given by the equivalent of **Equation B6.7.1** in **Section 6.3.ii.d** for a deleterious mutation (the γ term in that equation is equivalent to $-\gamma$ here). The relevant mutation rate is the per nucleotide site rate of origin of nonsynonymous mutations for the gene. For variants in introns, neutrality is assumed, so that $\phi_A(q)$ is replaced by $\phi_I(q)$, given by **Equation B6.6.1** in **Section 6.3.ii.b**, where the same mutation rate per site in the gene is assumed for intron and nonsynonymous sites.

The expected number of nonsynonymous and intron sites with mutations at frequencies i and j, respectively, among k allelic sequences of a given gene is then given by:

$$E_A\{i\} = m_A\binom{k}{i}\int_0^\infty\int_0^1 q^i p^{k-i}\phi_A(q)\psi(\gamma)\,dq\,d\gamma \qquad (B6.9.2)$$

and

$$E_I\{j\} = m_I \frac{4N_e\tilde{u}}{j} \tag{B6.9.3}$$

where m_A and m_I are the numbers of nonsynonymous and intron sites, respectively; $\tilde{u}$ is the net mutation rate per site for the gene in question (see **Section 6.3.ii.b**). The intron sites thus provide a means of estimating the mutational parameters.

B6.9.iii. Estimating the parameters

If variants at different sites in the same gene segregate independently of each other in the population, the number of variants in each category for a given gene follow Poisson distributions with the above expectations, so that the probabilities of observing I nonsynonymous and J synonymous sites with variants at frequencies i and j are given by the Ith and Jth terms of Poisson distributions with means $E_A\{i\}$ and $E_I\{j\}$, respectively. This is called the *Poissson random field* assumption (Sawyer and Hartl 1992). In general, mutation rates may vary among genes, so that different expectations apply to different genes. The logarithms of the probabilities for each gene can be summed over all observed values of I and J for each gene surveyed, to give a log-likelihood function (**Appendix A2.vi.b**) for the data. If an outgroup is not used to distinguish between ancestral and derived variants, as was the case in the analysis of Eyre-Walker et al. (2006), the expressions for the frequencies of minority variants (**Section 6.4.iii.a**) should be used, but the principle is unchanged. The estimates of the parameters of the gamma distribution can then be found by maximum likelihood or Bayesian computational methods.

6.4.iv.c. *Other methods and implications of the results*

Other approaches to analysing human sequence polymorphism data have been proposed. These yield general conclusions similar to those just described, while differing in the numerical details (Yampolsky et al. 2005; Kryukov et al. 2007; Keightley and Eyre-Walker 2007; Boyko et al. 2008). All the results indicate a wide distribution of selection coefficients around a mean whose current value is uncertain, but likely to be on the order of a few per cent. This also seems to apply to *Drosophila* (Loewe et al. 2006; Keightley and Eyre-Walker 2007), although the mean selection coefficient may be an order of magnitude smaller than in humans. Some of these methods have the advantage that they include the possibility of past population size changes, using putatively neutral variants to estimate their effects (Keightley and Eyre-Walker 2007; Boyko et al. 2008); this removes at least some of the biases discussed in **Section 6.4.iii.e** above.

These results suggest that each individual in the human population carries many heterozygous mutations, which are individually very rare in the population, and mostly have very small effects on fitness. By how much do these mutations affect the mean fitness of the population? For the more strongly selected mutations (i.e., those with $N_e s > 2$), the expected frequency of nonsynonymous variants is close to the infinite population equilibrium value, $2u/s$ (see **Section 6.3.ii.d**). Using **Equation 4.7c** of **Section 4.2.ii.c**, the mean number of such mutations per individual is given by the total mutation rate per individual for mutations in this category, divided by the harmonic mean of the selection coefficient against heterozygotes, which was estimated as 8.5×10^{-4} by Eyre-Walker et al. (2006). In **Section 4.2.i**, we estimated the rate of occurrence of new nonsynonymous mutations as 0.84 per individual per generation. Since 80% of the distribution was estimated by Eyre-Walker et al. (2006) to have $N_e s > 2$, the net rate for these more strongly selected mutations is $0.84 \times 0.8 = 0.67$. The predicted equilibrium mean number of mutations per individual is thus $0.67/(8.5 \times 10^{-4}) = 784$. On the assumption of multiplicative fitnesses, the mean fitness of the human population, relative to the fitness of individuals that are free of this type of mutation, is approximately $\exp -(784 \times 7.6 \times 10^{-4}) = 0.55$, similar to the value inferred by a slightly different method in **Section 4.3.ii.c**.

As several authors have pointed out (Pritchard 2001; Wright et al. 2003; Eyre-Walker et al. 2006; Kryukov et al. 2007), and as discussed in **Section 4.5.i**, the presence of such a large number of low-frequency deleterious mutations in populations creates a substantial variance in fitness (**Problem 6.9**). This suggests that much human genetic susceptibility to disease may be the effects of rare mutations with minor phenotypic effects, consistent with the evidence on the *TLR4* gene described in **Section 6.4.iii.b** above. This implies that even very large scale searches for associations between genetic markers and diseases (e.g., Wellcome Trust Case Control Consortium 2007; Donnelly 2008) may uncover only the (possibly small) portion of the total variability caused by common major effect variants (Kryukov et al. 2007).

Of course, many assumptions have been made in these estimation procedures, so the exact values of the parameters may be questioned. In the first place, the inferences are based on the properties of mutations found to be segregating in samples of moderate size (i.e., ones that are not at extremely low frequencies). These are necessarily the least deleterious variants, and so the estimated distribution for new mutations depends heavily on information about the portion of the distribution relating to the most weakly selected amino acid variants. Furthermore, as mentioned in **Section 6.3.ii.c**, the assumption of one-way mutation is an oversimplification (a few of the low-frequency variants may be favorable variants that have recently entered the population, rather than deleterious mutations). In addition, the likelihood formula assumes independence between sites within the same gene, whereas the empirical evidence from human and natural populations suggests that this is far from the truth (see **Section 8.1.iii.c**). This introduces some bias into the estimates, and certainly means that the confidence intervals on the estimates are too low. The assumption of semidominance is also implausible, since most deleterious mutations appear to be partially recessive, although this may not apply to mutations with very small effects, as discussed in **Section 4.4.v**. For $N_e s > 2$, however, most deleterious mutations will not reach high frequencies, so that only their heterozygous fitness effects matter.

Finally, given an estimate of the distribution of $N_e s$ values for new nonsynonymous mutations, $\psi(N_e s)$, the probability of fixation of a nonsynonymous variant, relative to the neutral value, can be estimated by integrating the product of $\psi(N_e s)$ and $\tilde{\lambda}$ in **Equation 6.7** of **Section 6.2.iii.a**. This yields the ratio of expected K_A and K_S values for divergence from a related species, on the hypothesis that mutations like these are the only cause of divergence, i.e., there is no separate category of mutations driven to fixation by positive selection. If this is subtracted from the observed value of K_A/K_S, we can estimate the fraction of nonsynonymous differences between the two species that are driven by positive selection, defined as α in **Section 6.4.ii.d** (Loewe et al. 2006). This approach has been applied to polymorphism data on over 20,000 human genes and the divergence of these genes from their chimpanzee counterparts, suggesting an α value of about 10% (Boyko et al. 2008). In addition, Eyre-Walker and Keightley (2009) have applied a related approach to human and *D. melanogaster* datasets.

6.4.iv.d. *Using polymorphism and divergence statistics*

A less direct method (Sawyer and Hartl 1992; Akashi 1999; Bustamante et al. 2002) fits a model based on the equations in **Boxes 6.6*** and **6.7*** of **Sections 6.3.ii.b** and **6.3.ii.d** to data on the number of segregating nonsynonymous and putatively neutral synonymous or silent variants within a species, together with divergence from a related species, assuming a single selection coefficient for all mutations in a given gene. This approach uses information on the variant frequencies only indirectly.

However, as we saw for the human example described in **Section 6.4.iv.b** above, it is unrealistic to assume the same selection coefficient for all nonsynonymous mutations that arise within a gene, and this assumption may bias the s estimates. A modified approach assumed a normal distribution of s for new mutations, with a fixed variance across sites within each gene, but allows different means for different genes (Sawyer et al. 2003, 2007). When applied to a large data set on polymorphism in an East African population of *Drosophila* *melanogaster*, and divergence from its relative *D. simulans*, this yielded the startling conclusion that $\alpha = 0.95$, contrasting with the value of approximately 0.25 from the methods described in **Section 6.4.ii.d**. The problem with this method is that the assumption of a normal distribution may bias the conclusions about *s*, especially since the fact that there are equal numbers of fixations of deleterious and favorable mutations at equilibrium implies a bimodal distribution of *s* for newly arising mutations. This probably explains the very high fraction of positively selected mutations found by this method.

Piganeau and Eyre-Walker (2003) dealt with these difficulties in a different way, by assuming a gamma distribution for $N_e s$, but using **Equation 6.10** of **Section 6.3.ii.c** to determine the fraction, $x^\star$, of nonsynonymous sites that are fixed for variants favored by selection for a given value of $N_e s$. These generate deleterious mutations with selection coefficient *s*, whereas the sites fixed for deleterious variants (fraction $1 - x^\star$) generate favorable mutations with the same magnitude of selection coefficient. The net expectations for divergence and the number of segregating sites in a gene were found by integrating over the gamma distribution. A maximum likelihood method was then applied to polymorphism and divergence data on mitochondrial genes in 18 different animal taxa. This gave clear evidence for a wide distribution of $N_e s$ values, but the shape parameter was mostly much larger than in the human example in **Section 6.4.iv.b**, suggesting a much tighter distribution of selection coefficients.

It is likely that there will be many extensions and enhancements to these approaches, which will eventually be applied to the large data sets on natural and human variability that will become available in the future, as new sequencing technologies are applied to polymorphism studies (e.g., http://www.1000 genomes.org/, and http://1001genomes.org/).

APPENDIX TO CHAPTER 6

More diffusion equation results

6A.1. FIXATION PROBABILITIES

Let $Q(q_0, t)$ be the probability that variant A_2 has become fixed by time *t*, given that its initial frequency was q_0. The following method yields a recursion relation for *Q*. Using the notation of **Chapter 5: Appendix 5A.1**, we can write $g(\delta q, q_0)$ for the probability density for a change in the frequency of A_2 from q_0 to $q_0 + \delta q$ in the first generation of the process. The value of *Q* in generation *t* is thus given by integrating over all possible values of $Q(q_0 + \delta q, t - 1)g(\delta q, q_0)$; this gives the probability of fixation after t generations have elapsed for a population that started at q_0:

$$Q(q_0, t) \approx \int Q(q_0 + \delta q, t - 1)g(\delta q, q_0)d(\delta q) \qquad (6A.1)$$

We can approximate the integrand using the same approach as in **Chapter 5: Appendix 5A.1**, i.e., by:

$$\left[Q(q_0, t-1) + (\delta q)\frac{\partial Q}{\partial q_0} + \frac{(\delta q)^2}{2}\frac{\partial^2 Q}{\partial q_0^2} \right] g(\delta q, q_0) + o\left[(\delta q)^2 \right] \qquad (6A.2)$$

where the derivatives are evaluated at $q = q_0$ and time $t - 1$.

Integrating over all possible values of δq, and simplifying in the same way as in the derivation of **Equations 5A.4a** and **5A.4b**, we get:

$$\frac{\partial Q}{\partial t} \approx M_{\delta q_0}\frac{\partial Q}{\partial q_0} + \frac{V_{\delta q_0}}{2}\frac{\partial^2 Q}{\partial q_0^2} \qquad (6A.3)$$

where $M_{\delta q}$ and $V_{\delta q}$ are the mean and variance, respectively, of the change in frequency of A_2 at an allele frequency q.

This is an example of the *backward Kolmogorov equation*, which has played an important role in many aspects of theoretical population genetics, especially in the hands of Motoo Kimura and Tomoko Ohta (Kimura and Ohta 1971). Using this equation, a solution for Q as a function of time can be obtained in the case of neutrality (Kimura 1964), but usually we must be content with finding the probability of ultimate fixation of A_2, i.e., the value of Q as t tends to infinity (we drop the dependence on t in this case). This is equivalent to the probability of fixation that we found for an infinite population in **Section 3.2.ii**. The derivative of Q with respect to time is then 0, and the partial derivatives on the right-hand side of **Equation 6A.3** can thus be replaced by ordinary derivatives, giving the following ordinary differential equation (for convenience, the approximately equals symbol has been replaced by an equals symbol):

$$M_{\delta q_0}\frac{dQ}{dq_0} + \frac{V_{\delta q_0}}{2}\frac{d^2Q}{dq_0^2} = 0 \qquad (6A.4)$$

This can be rearranged to give:

$$\frac{1}{(dQ/dq_0)}\frac{d^2Q}{dq_0^2} = -2\frac{M_{\delta q_0}}{V_{\delta q_0}} \qquad (6A.5)$$

Integrating, we get:

$$\frac{dQ}{dq_0} = \psi \tag{6A.6a}$$

where

$$\psi(q) = \exp\left(-2\int \frac{M_{\delta q}}{V_{\delta q}} dq\right) \tag{6A.6b}$$

$Q(q_0)$ is equal to the integral of ψ between $q = 0$ and q_0, because $Q(0) = 0$. We have $Q(1) = 1$, so we can remove any constants of integration by writing:

$$Q(q_0) = \frac{\displaystyle\int_0^{q_0} \psi(q)dq}{\displaystyle\int_0^1 \psi(q)dq} \tag{6A.7}$$

This is the general expression for fixation probability, derived in this form by Kimura (1962), although a formula for the special case of semidominance and a Wright–Fisher population was first obtained by Fisher (1930a, 1930b, Chapter 4).

6A.2. THE EQUILIBRIUM DISTRIBUTION OF VARIANT FREQUENCIES WITH ONE-WAY MUTATIONS

This material is based on Wright (1945), who extended earlier work by Fisher (1930a; 1930b, Chapter 4) and Wright (1938). Kimura (1964) provides a more detailed derivation. We consider sites where the initial state was fixation for A_2, and where A_2 mutations are now segregating. The probability density over all sites for the frequency of A_2, $\phi(q)$, is assumed to remain constant over time. From **Equation 5A.6a** of **Appendix 5A.2**, this is equivalent to the following condition:

$$\frac{\partial \phi}{\partial t} = 0 = -\frac{\partial P}{\partial q} \tag{6A.8}$$

where P is the probability flow across the point q.

The relation $\partial P/\partial q = 0$ implies that P is independent of q and is equal to the flow of probability across each class of the distribution. If the process has reached stationarity, this flow must equal λ_1, where λ_1 is the rate at which sites fixed for A_1 eventually become fixed for A_2. We can thus convert **Equation 6A.8** into an ordinary differential equation:

$$\frac{1}{2}\frac{d\,V_{\delta q}\phi(q)}{d\,q} - M_{\delta q}\phi(q) + \lambda_1 = 0 \qquad (6A.9)$$

As shown by Wright (1945), this has the following solution:

$$\phi(q) = \left(C - 2\lambda_1 \int_0^q \psi(x)\,dx\right) \bigg/ \left(V_{\delta q}\psi(x)\right) \qquad (6A.10)$$

where $\psi(q)$ is given by **Equation 6A.6b** and C is a constant of integration (this can be verified by differentiating **Equation 6A.10** with respect to q).

For autosomal sites, the approximate values of the constants λ_1 and C can be found as follows (Wright 1938, 1945; Kimura 1964). From **Equations 6.6** of **Section 6.2.iii.a**, λ_1 is the expected number of mutations that enter the population per site per generation from sites fixed for A_1, multiplied by the fixation probability for each mutation, i.e., $\lambda_1 = 2NuQ(1/2N)$, where $Q(1/2N)$ is given by **Equation 6A.7** with $q_0 = 1/2N$. Neglecting second-order terms in the strength of selection and $1/2N$, we obtain:

$$\lambda_1 \approx u \bigg/ \int_0^1 \psi(q)\,dq \qquad (6A.11)$$

Most mutations will be lost before they reach high frequencies, so that $\phi(1 - 1/2N)$ is close to 0; using this in **Equation 6A.10** and rearranging, we get:

$$C = 2\lambda_1 \int_0^1 \psi(q)\,dq \qquad (6A.12a)$$

The formula for λ_1 and **Equation 6A.6b** imply that:

$$C \approx 2u \qquad (6A.12b)$$

Substituting into **Equation 6A.10** and simplifying, we obtain:

$$\phi(q) \approx \frac{2u}{V_{\delta q}\psi(q)} \frac{\displaystyle\int_q^1 \psi(x)\,dx}{\displaystyle\int_0^1 \psi(x)\,dx} \qquad (6A.13)$$

Note that we have calculated this probability density function using the rate of flow (λ_1) of probability out of sites fixed for A_1, ultimately into sites fixed for A_2. The integral of $\phi(q)$ over the whole range of q is constant and equal to 1. The integral of $\phi(q)$ over the range $1/2N$ to $1 - 1/2N$ thus approximates the equilibrium value of the probability that sites whose original state was fixation for A_1 are segregating.

PROBLEMS

6.1.   **i.**   The human and chimpanzee sequences of a gene shown in **Figure 1.8** of **Section 1.2.iii.b** showed 59 differences among a total of 5022. Use **Equation B6.1.3** in **Box 6.1** of **Section 6.1.ii.b** to estimate the rate of change per site, λ, assuming that the human and chimpanzee lineages separated five million years ago.

     ii.   Use the fact that the standard deviation of the number of differences between species in a sequence of length m evolving at rate λ over a time T is equal to $\sqrt{(2m\lambda T)}$ (**Section 6.1.ii.c**) to estimate the 95% confidence interval for the estimate of λ obtained in **Problem 6.1.i**.

6.2.   **i.**   Extend the argument in **Box 6.3** of **Section 6.2.ii.b** to show that, for a discrete generation population with N_f breeding females and N_m breeding males, the probability of fixation of a new neutral mutation at an autosomal site is $1/2N$, where $N = N_f + N_m$.

     ii.   Obtain the corresponding result for a mutation at an X-linked site.

     iii.   Show that, if new mutations at a site appears as clusters of size j among the breeding individuals in the population, the above probabilities are multiplied by j.

6.3.*   i.   Show that **Equation B6.4.2a** of **Section 6.2.iii.a** can be approximated by **Equation B6.4.2b** when s is small.

     ii.   Show that **Equation B6.4.2b** reduces to the neutral value as $N_e s$ tends to 0.

6.4* Show that, if $t(q)$ is the sojourn time density, the equilibrium probability density that a segregating mutation has frequency q is proportional to $t(q)$ under the infinite sites assumption.

6.5.   **i.**   Use **Equation B6.6.2a** of **Section 6.3.ii.b** to derive an expression for the expected number of segregating neutral mutations in a sample of k alleles, given a sequence of length m.

     ii.   Derive a corresponding expression for the overall expected number of segregating sites when the forward and backward mutation rates at each site are u and v, respectively.

6.6   Using the results in **Sections 6.2.iii.a** and **6.3.ii.a**, derive the Li–Bulmer equation, i.e., **Equation 6.10** of **Section 6.3.ii.c**.

6.7.*   i. Use the Taylor series expansions of $\exp(x)$ and $1/(1 + x)$ (**Appendix A1.ii.c**) to obtain an approximation for **Equation 6.11** of **Section 6.3.ii.c** that is accurate to terms in γ^2 (it is easiest to approximate each of the two terms in the denominator of **Equation 6.11** separately and then combine the results).

 ii. Use the resulting expression to find an approximate condition for λ to increase when γ is small, and to find an approximation for the value of γ that maximizes λ. Determine this value when $\kappa = 2$ and $\kappa = 4$.

6.8.*   i. Use **Equation 6A.13** of the **Appendix** to this chapter to derive **Equation B6.7.1** of **Section 6.3.ii.d** for the case of semidominant selection in favor of A_2.

 ii. Show that, as γ approaches 0, **Equations B6.7.1** and **B6.7.3** approach their respective neutral values.

6.9 Use **Equation 4.18** of **Section 4.5.i.a**, together with the estimates of the fitness effects of amino acid mutations obtained in **Section 6.4.iv.b**, to estimate the genetic variance in fitness contributed by slightly deleterious amino acid mutations in the human population. Use this to estimate the upper and lower 2.5 percentiles of the distribution of fitness among individuals.

CHAPTER SEVEN

Genetic Effects of Spatial Structure

CHAPTER SUMMARY

This chapter deals with the complexities of spatially structured populations. Such structure is the reality for many species of evolutionary interest. It arises when a species is divided into separate sub-populations (demes) with limited amounts of migration, or when individuals are distributed continuously over the species range, but disperse only over short distances.

We describe methods for measuring the genetic consequences of spatial subdivision by partitioning allele frequencies or diversity measures among different hierarchical levels, e.g., into differences between demes versus differences among individuals within them. These generate measures of divergence among populations relative to total variability, such as F_{ST}.

We explain how coalescent theory can predict the properties of neutral variation in structured populations. For the classic island and stepping-stone models, the mean coalescence time for a pair of alleles sampled from the same deme (and hence the expected within-deme diversity value under the infinite sites model) is the same as in a panmictic population, with an effective population size given by the sum of the N_e values for each deme. This property also applies to some more general migration models, including continuous populations (specifically, when migration is conservative, an important concept that we define). For alleles from different demes, between-deme diversity values are always higher than those within demes, but decrease with the migration rate. We describe situations under which a high proportion of total diversity is between demes (leading to high F_{ST} values), including the effects of deme extinction and re-colonization.

When there is a large number of demes, some general properties emerge. In particular, the place where an allele is sampled may be largely unrelated to where its ancestors were located. This can be understood in terms of the "scattering" and "collecting" phases of the coalescence of alleles sampled from a set of demes. Details of the migration process can then have surprisingly little importance in determining divergence patterns, explaining why effects of isolation by distance are often weak. With large deme numbers, population structure has little effect on the expected values of many of the properties of samples of sequences, provided that each allele is sampled from a different deme, even if variants are subject to selection as well as drift.

In real organisms, population ranges may change, involving bottlenecks in population sizes, and species may increase or decrease in size over historical time. It is essential to take these possibilities into account when testing for selection, so we briefly discuss their effects.

When selection pressures are uniform in space, the details of migration patterns do not strongly affect the fixation probabilities of new, weakly selected semidominant mutations, if deme numbers are large or migration is conservative. The fixation probability of recessive or partially recessive deleterious mutations is, however, reduced by population structure, and the fixation of recessive favorable mutations is promoted. With weak selection and genetic drift, individual recessive and weakly deleterious mutations tend to be locally distributed, contributing to between-population heterosis.

Spatially uniform balancing selection opposes the effects of drift within demes, reducing F_{ST} relative to the neutral value, whereas local selective differences may increase F_{ST}. Comparisons between populations can thus potentially detect outlier loci that may be affected by selection. Comparisons between F_{ST} estimates for quantitative traits and neutral or nearly neutral molecular markers are also useful for detecting selection on phenotypes in structured populations.

Finally, differences in mean fitness among demes may allow inter-deme selection, which is sometimes proposed to be a major factor in adaptive evolution. However, it requires restricted conditions to be effective in opposing selection on individuals within populations.

> The breeding structure of natural populations thus is likely to be intermediate between the model of subdivision into partially isolated territories and that of local inbreeding in a continuous population.
>
> Sewall Wright (1940a)

INTRODUCTION

Apart from describing the interaction between migration and local differences in selection pressures in **Section 4.1**, we have so far assumed that populations are panmictic, i.e., there is no spatial subdivision, so that all individuals contribute equally to the entire population in the next generation, apart from any fitness differences or accidents of genetic drift. In reality, species are usually spread out over a wide area, or along a linear habitat, so that individuals from the same locality are more likely to contribute offspring to that locality than individuals from elsewhere. The extent to which individuals move from one place to another between birth and reproduction must affect the extent to which genetic differences will be maintained between different local populations.

This may have important consequences for the amounts and patterns of both neutral and selected genetic variants in the species, producing genetic differences between populations and affecting the total level of variability present in the species as a whole. Importantly, tests for selection based on patterns of DNA sequence polymorphism, of the type described in **Section 6.4.iii**, may be biased by departures from the predictions for a panmictic population, caused by spatial structure. In addition, local populations may differ genetically as a result of genetic drift. If variants are affected by selection as well as drift, the populations' mean fitnesses may differ, potentially allowing selection to operate at the level of populations (rather than on individuals, as we have so far assumed); it is important to examine this possibility, because of its implications for the mechanism of adaptive evolution.

In order to understand observed patterns of genetic variability in spatially subdivided populations, to see how methods for detecting and estimating selection are affected by subdivision, and to evaluate the evolutionary consequences of interactions between selection, migration, and drift, we need to extend our theoretical models and relate them to the data. This is the purpose of this chapter.

7.1. DESCRIBING POPULATION DIFFERENTIATION

7.1.i. Use of genetic markers to describe differences among populations

Here we provide an overview of the methods for quantifying genetic differentiation among populations of the same species. More detailed accounts of the statistical properties of the various estimators of the parameters described here can be found in Weir and Hill (2002) and Excoffier (2003), and reviews of their theoretical bases are given by Nagylaki (1998a) and Rousset (2003). Software packages such as Genepop (http://genepop.curtin.edu.au/) and DnaSP (http://www.ub.es/dnasp/) are available to implement these methods.

7.1.i.a. *Variances in allele frequencies and F_{ST}*

Differences between local populations of the same species can be described in many different ways, depending on the types of genetic variants being used. For variants such as allozymes (**Section 1.2.i**), a natural measure of the differences between a set of populations at a given locus is the variance in allele frequencies among them, at least when there are just two alleles per locus. Populations are often organized into hierarchical sets with increasing levels of subdivision, such as subspecies within a species, geographic races within subspecies, and local populations within races (**Figure 7.1**). Variances between groups at any level of such a hierarchy can then be estimated.

The use of variances suffers from the problem that their values depend on the mean allele frequencies at the loci. As discussed in **Section 5.1.i.c**, this problem can be overcome for a biallelic locus by dividing the variance in allele frequency by its maximum possible value, the product $\bar{p}(1 - \bar{p})$, where $\bar{p}$ is the mean allele frequency over the set of populations. When used to measure differentiation among a set of populations, this quantity is usually denoted by the symbol F_{ST}. Its mean over a set of loci is a good measure of differentiation between populations.

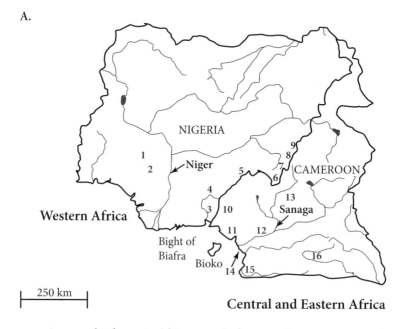

FIGURE 7.1 An example of a species (chimpanzees) whose populations are organized into hierarchical sets with increasing levels of subdivision. **A.** Locations from which individuals were sampled. **B.** Two possible interpretations (**i** and **ii**) of chimpanzee subspecies. **C.** Phylogenetic tree using a part of the mitochondrial genome, showing two major lineages (corresponding to interpretation **i**), and possible sub-lineages, suggesting interpretation **ii**. The two major lineages are estimated to have separated more than 500,000 years ago, and their present geographic ranges are in central and eastern Africa, and western Africa, respectively. Evidence for migration between these regions, across the Sanaga River, is limited. The numerical values on the nodes indicate the frequency with which these are supported by bootstrap resampling of the data (see Felsenstein 2004, Chapter 20). [Adapted from Figures 2, 4, and 6 of Gonder et al. (2006).]

B.

(i)

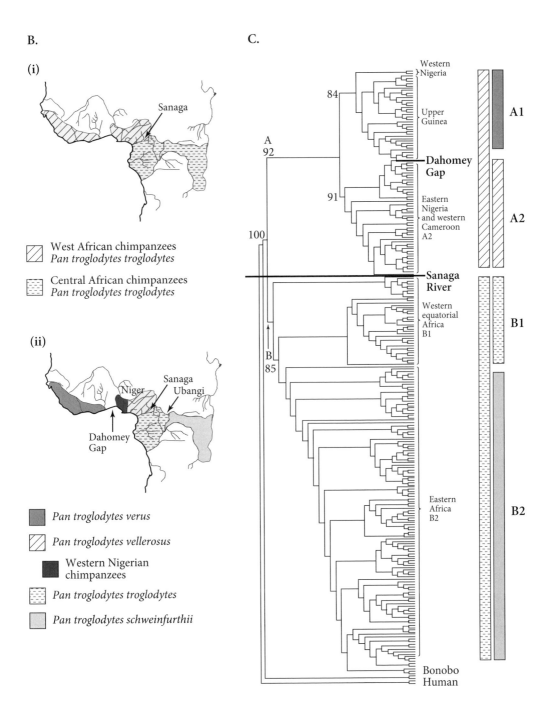

West African chimpanzees
Pan troglodytes troglodytes

Central African chimpanzees
Pan troglodytes troglodytes

(ii)

Pan troglodytes verus

Pan troglodytes vellerosus

Western Nigerian
chimpanzees

Pan troglodytes troglodytes

Pan troglodytes schweinfurthii

C.

FIGURE 7.1 (Continued)

F_{ST} statistics can be estimated for all levels of a hierarchy of samples from successively finer geographic regions (Wright 1951), using a *nested* or *hierarchical* analysis of variance to estimate components of variance in allele frequency at a locus at successive levels of the hierarchy (Excoffier 2003). The lowest level of the hierarchy represents individuals within local populations or demes (**Section 4.1.i**). The departure of genotype frequencies from Hardy–Weinberg expectation at a biallelic locus within a deme can also be quantified by a measure that is similar to the inbreeding coefficient (**Section 1.3.ii**). This is denoted here by F_{IS}, where the subscripts indicate that this F statistic measures the departure of individual genotype frequencies within a local population from random mating expectation. As in **Section 5.1.i.b**, F is used here instead of f for the statistics describing all levels of a hierarchy, in order to distinguish these measures from the inbreeding coefficient based on identity by descent caused by matings between related individuals within a single population; F_{IS} is often referred to as the "fixation index" (Wright 1951). Unlike the inbreeding coefficient, F_{IS} can be negative if a population has an excess of heterozygotes (caused, for example, by selection or assortative mating).

7.1.i.b. *Gene diversities for allelic variants*

The method just described does not allow for the possibility of multiple alleles at the loci of interest, unless we adopt the expedient of simply considering one variant versus all alternatives, thereby throwing away information (which is undesirable). Microsatellite loci, for example, usually have large numbers of alleles (**Section 1.2.iii.c**), and are often used to study population differentiation. A commonly used method that avoids this problem is to use the gene diversities introduced in **Section 1.2.i.b**. Pooling over all samples from a species (i.e., using the frequencies of all alleles in the entire sample), we can estimate the species-wide diversity, H_T, for a set of loci, using the standard formula (**Box 1.2**). For the next level of the hierarchy, such as subspecies, we can calculate diversity for each subspecies and determine the mean over subspecies. Denoting this by H_S, it seems reasonable to define an analog of F_{ST} by:

$$G_{ST} = \frac{\left(H_T - H_S\right)}{H_T} \tag{7.1}$$

G_{ST} describes the difference between the level of diversity in the species as a whole and the mean within-subspecies diversity, relative to the overall diversity. It is thus a natural measure of the extent to which subspecies are genetically differentiated (Nei 1973). This can be extended to successive levels of a hierarchy by computing the values of the mean diversities within successive levels and obtaining the corresponding G_{ST} statistics (Nei 1973; Excoffier 2003). For loci with two alleles, it can be shown that the theoretical values of F_{ST} and G_{ST} are equivalent (**Problem 7.1**).

For microsatellite loci, some information is lost by using allelic diversity, because this ignores the information about differences in alleles' repeat numbers, which one should take into account, because they tend to increase more linearly than diversity with genealogical divergence; the variance of repeat lengths among alleles is therefore often used to measure variability (**Section 5.1.iii.f**). This can be extended to differentiation between populations by estimating the relevant mean squared differences in repeat lengths among pairs of alleles, sampled from the same and different populations. This provides an analog of F_{ST} and G_{ST}, denoted by R_{ST} (Goldstein et al. 1995; Slatkin 1995b).

7.1.i.c. *Nucleotide site diversity*
If DNA sequence information is used, we could use the gene diversity method just described to characterize population differentiation, replacing allele frequencies by haplotype frequencies (**Section 1.2.v.a**). This is often done, especially in the extensive literature on mitochondrial gene variation (Excoffier et al. 1992). However, this approach loses information about the extent of sequence differences, and it is better to treat each nucleotide site as an observational unit and to define an analog of G_{ST}, K_{ST}, at this level (Hudson et al. 1992). The values of π_T and π_S are estimated for the set of populations in question, where the nucleotide site diversity π for a sequence or set of sequences (**Section 1.2.iii.c**) replaces gene diversity in **Equation 7.1**. This gives:

$$K_{ST} = \frac{\left(\pi_T - \pi_S\right)}{\pi_T} \tag{7.2}$$

The statistical significance of K_{ST} can be assessed by resampling. To do this, variants at individual sites are reassigned randomly to populations, so as to compare the observed K_{ST} value with the distribution of K_{ST} over many randomizations that assume no subdivision of the species (Hudson et al. 1992). A similar procedure can be performed with an alternative measure, S_{nn}, which measures the extent to which related sequences are found in the same population (Hudson 2000).

7.1.i.d. *Other methods*
In addition to these widely used methods, many other procedures have been devised for quantifying genetic differentiation between populations (e.g., Holsinger 1999; Song et al. 2006). One approach is to assign individuals to distinct groups, based on their multi-locus genotypes, and to use Bayesian statistical inference to find the best assignment (Pritchard et al. 2000; Dawson and Belkhir 2001; Falush et al. 2003; Corander et al. 2004; Guillot et al. 2005); see http://pritch.bsd.uchicago.edu/structure.html, http://www.genetix. univ-montp2.fr/partition/partition.html, and http://www.abo.fi/fak/mnf/mate/ jc/smack_software_eng.html. This does not require prior knowledge of the

individuals' spatial origins, but estimates the number of distinct groups into which individuals appear to fall. Unlike the methods described above, however, the results cannot easily be related to the underlying population genetic processes.

7.1.ii. Empirical patterns

A very large amount of data has been accumulated on genetic variation among populations of animals and plants. The results range from low levels of differentiation in mobile organisms, such as the fruitfly *Drosophila pseudoobscura* (Schaeffer and Miller 1992) or humans (Lewontin 1972a; Akey et al. 2002), which have mean G_{ST} or K_{ST} values of around 0.10, to levels approaching 1 in highly self-fertilizing species of plants, where gene flow between populations due to pollen dispersal is likely to be very limited, because ovules receive little pollen from other individuals (Hamrick and Godt 1996; Charlesworth 2003). **Figure 7.2** shows this effect in a plant species.

It is often assumed that statistics like K_{ST} and G_{ST} reflect the amount of migration between populations, but, as we will see in **Section 7.2** below, they are also strongly affected by the effective sizes of local populations, reflected in the level of variability within populations. For example, **Equation 7.2** shows that, if $\pi_S = 0$, $K_{ST} = 1$. High levels of the divergence statistics are indeed often associated with low diversity within local populations, which suggests low effective population sizes (Jarne and Staedler 1995; Hamrick and Godt 1996; Charlesworth 2003), as can be seen in **Figures 7.2** and **7.3**. **Figure 7.3** shows a comparison between an outcrossing and a highly selfing species of the nematode worm *Caenorhabditis*. Here, migration depends on movements of individuals, not gametes, and so self-fertilization should not directly affect migration. In this case, the much higher K_{ST} value in the inbreeding species is probably due to the reduced effective population size caused by selfing (see **Sections 5.2.iii.a** and **8.3.i**).

7.2. THE THEORY OF NEUTRAL VARIATION WITH SPATIAL STRUCTURE

Now that the relevant measures have been defined and illustrated with examples from nature, we next explain how they are expected to behave. Much of this theory deals with F_{ST} and related statistics, but in order to understand their properties it is necessary to examine the behavior of diversity measures at the different hierarchical levels. We begin with neutral variation, which must be understood before dealing with variants that are under selection, which we consider in **Section 7.3**.

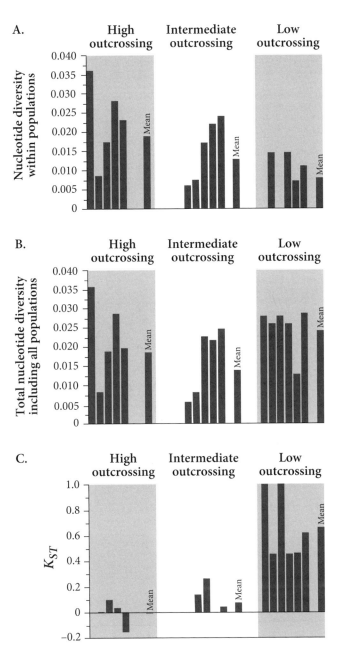

FIGURE 7.2 The effect of inbreeding on patterns of DNA sequence diversity at six loci, and on their K_{ST} values, in a plant species, *Leavenworthia crassa*, whose populations differ in their outcrossing rates, from highly outcrossing (on the left of each panel) to highly inbreeding (on the right of each panel). **A.** Mean within-population silent nucleotide site diversities, π_S, for each gene. **B.** Total silent site diversities, π_T. **C.** The corresponding K_{ST} values. [Adapted from Figure 4 of Charlesworth (2003).]

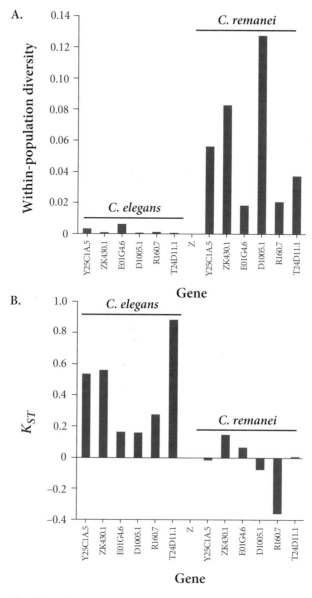

FIGURE 7.3 The effect of inbreeding on patterns of diversity at six loci and on K_{ST} in two species of the nematode worm *Caenorhabditis* with very different breeding systems; *C. elegans* is a highly self-fertilizing hermaphrodite, whereas *C. remanei* has males and females. **A.** Mean within-population silent site diversities, π_S, for each gene. **B.** The corresponding K_{ST} values. [Plotted using data from Cutter (2005) and Cutter et al. (2006).]

7.2.i. General principles

7.2.i.a. *Introduction*

The equilibrium pattern of neutral genetic variability for a stable structured population is determined by the interaction of genetic drift, migration, extinction/colonization events involving local populations, and mutation (nonequilibrium situations will be discussed briefly in **Section 7.2.vi**). As we saw in **Section 5.1**, there are several different ways to quantify genetic drift: through its effects on the rates of change in probability of identity by descent and variance in allele frequencies, on equilibrium genetic diversity measures, and on the coalescence times of alleles sampled from a population. To avoid duplication of material, we use the last method as our principal conceptual tool, but refer to the other approaches when appropriate. As discussed in **Section 5.1.iii.d**, under the infinite sites model, DNA sequence diversity is directly related to coalescence time, so it is straightforward to move between the different approaches. The same is true for variances of differences in microsatellite repeat numbers between pairs of alleles, at least under the simplest realistic mutational models (**Section 5.1.iii.f**).

With spatial structure, it is essential to specify the places from which alleles are sampled. This was not needed in **Section 5.2.ii**, when we previously dealt with structured coalescent processes with different sex or age classes, because we assumed that alleles move so fast between classes that there are soon no effects of their initial class—the *fast time scale approximation*. When spatial structure has significant evolutionary consequences, however, this assumption is not necessarily valid, although it can still be used to simplify the effects of demographic factors, as shown in the next section.

In the next few sections, we describe a general way to treat structured coalescent processes. Before doing this, however, it may be helpful for advanced readers to relate the coalescent process to the infinite alleles model (**Section 5.1.ii.a**), which has been widely used for modeling the effects of geographic structure on neutral variability. **Box 7.1*** shows how the probability of identity by descent of a pair of alleles under the infinite alleles model is related to the distribution of their time to coalescence. Although the infinite alleles model is relevant only to certain types of genetic data, these results show that formulae derived for the infinite alleles model allow one to find the mean coalescence times (and other moments) of pairs of sampled alleles (e.g., Maruyama 1970b; Maruyama 1977; Nagylaki 1982; Wilkinson-Herbots 1998). These times are independent of the mutational process.

7.2.i.b. *Fast time scales and migration models*

We first consider a *metapopulation* consisting of a set of *d* discrete local populations (i.e., *d* demes of constant size, interconnected by migration). If migration rates and coalescence probabilities are both small, we can treat coalescence

Box 7.1* THE COALESCENT PROCESS AND
IDENTITY PROBABILITIES

Under the *infinite alleles* mutational model (Kimura and Crow 1964), each allele is assumed to have a probability u of mutating to a different allele, where u is assumed to be very small (**Section 5.1.ii.a**). We can then determine the probability of identity in state of two alleles, $1 - H$, given $\psi(t)$, the probability distribution of their time to coalescence. If they coalesced at time t, they can be identical only if neither allele has mutated since time t, which has probability $(1 - u)^{2t}$ (Hudson 1990; Wilkinson-Herbots 1998). This implies that:

$$1 - H = \sum_{t=1}^{\infty} (1 - u)^{2t} \psi(t) \qquad \text{(B7.1.1a)}$$

As shown in **Sections 5.1.iii.b** and **5.2.ii**, for a *panmictic* population (i.e., one that lacks any spatial structure), $\psi(t) \approx 2N_e \exp(-2N_e t)$. In general, the distribution of t for a pair of alleles will differ from this in the presence of spatial structure and will be affected by the demes from which the alleles originated.

Since $u \ll 1$, $(1 - u)^{2t}$ can be accurately approximated by $\exp(-2ut)$ (**Appendix A1.ii.c**), giving:

$$H \approx 1 - \sum_{t=1}^{\infty} \exp(-2ut) \psi(t) \qquad \text{(B7.1.1b)}$$

The summation term on the right-hand side is the *moment generating function* of $\psi(t)$, with argument $-2u$ (**Appendix A2.iii**). From the properties of moment-generating functions, the mean time to coalescence is given by the derivative of H with respect to $-2u$, evaluated at $u = 0$; higher-order moments around 0 can be determined from the corresponding higher-order derivatives.

and migration as mutually exclusive events, i.e., we can neglect the possibility that two or more such events occur in a single generation. We will use this approximation repeatedly. We can also use the approximations developed in **Section 5.2.ii** for modeling the effects of sex, life history stage, or age, assuming that the flow of alleles between these "compartments" is fast, compared with either the flow between demes or coalescence. Sampled alleles can then be treated as drawn randomly from the equilibrium probability distribution of being present in a given compartment. This implies that the rate of drift within deme i is determined by its effective population size, N_{ei}, where $P_{ci} = 1/2N_{ei}$ is the probability that two alleles sampled from deme i coalesced within this deme in the previous generation, conditional on their remaining within deme i; the corresponding conditional mean time to coalescence within deme i is $2N_{ei}$.

Migration can be treated in a similar way by obtaining a net migration probability m_{ij} for demes i and j, i.e., the probability that an allele sampled from deme i originated in deme j in the previous generation. To do this, we specify the probabilities of migration events for individuals in different demographic classes (e.g., sexes); we write $m_{ir\,js}$ for the probability that a gene sampled from an individual of class r in deme i originated from an individual of class s in deme j in the previous generation. Using the fast time scale approximation for flow of lineages between different compartments (**Section 5.2.ii**), we average these over all possible classes to obtain m_{ij}, the required net migration probability per generation for demes i and j.

Formulae for different situations, including cytoplasmic, X-linked, Y-linked, or autosomal loci, can be derived in this manner, and these show that sex-specific differences in migration rates can cause very different patterns of between-population differentiation for different modes of inheritance (Laporte and Charlesworth 2002; Hedrick 2007). As mentioned in **Section 7.1.ii**, reproduction predominantly by self-fertilization also has a major effect in plants, because it restricts migration of genes through pollen, leading to increased isolation (Wright 1946; Ingvarsson 2002; Laporte and Charlesworth 2002). However, the assumptions of weak migration and slow coalescence within a deme do not always apply to real-life situations; more complex, exact models are then required (Chesser et al. 1993; Wang and Caballero 1999; Vitalis 2002; Vitalis et al. 2004; Eldon and Wakeley 2009).

7.2.ii. Mean coalescence times for migration models

7.2.ii.a. *Introduction*

Alleles can move by migration of individuals or gametes (e.g., pollen flow in plants or male gamete movement in the sea in an animal with external fertilization), or they can move in space when demes go extinct and colonists from other demes re-establish extinct demes. The consequences of local extinction and the formation of new demes by colonization from surviving demes will be examined in **Section 7.2.iv**. Here, we assume that the flow of genes among a set of d demes is caused purely by migration, and that there is no extinction of demes (which remain constant in size over the generations). **Box 7.2** derives equations for the mean coalescence time T_{ij} of a pair of alleles sampled from two different demes (labelled i and j); T_{ij} is the time for a pair of alleles from the same deme, i. It is possible, in principle, to solve **Equations B7.2.1a** and **B7.2.1b** for all of the $d + d(d - 1)/2$ mean coalescence times, if the migration rates and N_e values are specified; this is easy to do in the case of a pair of populations, for example (Nordborg 1997). However, most theoretical work has focused on simpler types of population structure.

Box 7.2 MEAN COALESCENT TIMES IN A SPATIALLY STRUCTURED POPULATION

In this box, we follow the approach of Nagylaki (1998b). Consider a pair of alleles sampled from demes i and j. If $i = j$, the two alleles are from the same deme. In this case, we assume that they are sampled from separate individuals, so that their probabilities of origin by migration are independent. The probability that both alleles came from deme i in the previous generation is m_{ii}^2, in which case they have probability P_{ci} of coalescing in the previous generation. This contributes a net amount $m_{ii}^2 P_{ci}$ to their mean coalescent time. If they fail to coalesce, there is a net contribution of $m_{ii}^2(1 + T_{ii})(1 - P_{ci})$ to the mean coalescent time, because one generation has elapsed, and the coalescent time for alleles in deme i is T_{ii}. When alleles are sampled from different demes, they have no chance of coalescence in the previous generation, so that an additional generation is always added to their coalescence time. The probability that a pair of alleles sampled from demes i and j originated in demes k and l, respectively, is $m_{ik} m_{jl}$, giving a net contribution of $m_{ik} m_{jl}(1 + T_{kl})$ to the mean coalescent time.

The expressions involving P_{ci} can be simplified by noting that $m_{ii} = 1 - \Sigma_k m_{ik}$ ($i \neq k$), so that $m_{ii}^2 P_{ci} \approx P_{ci}$. Hence, $m_{ii}^2 P_{ci} + m_{ii}^2(1 + T_{ii})(1 - P_{ci}) \approx m_{ii}^2(1 + T_{ii}) - P_{ci} T_{ii}$.

The final expression for the mean coalescent time can then be written as:

$$T_{ij} \approx \sum_{kl} m_{ik} m_{jl} \left(1 + T_{kl}\right) - \delta_{ij} P_{ci} T_{ii} \qquad \text{(B7.2.1a)}$$

where $\delta_{ij} = 1$ for $i = j$ and is otherwise 0.

This can be simplified by noting that $\Sigma_k m_{ik} = \Sigma_l m_{il} = 1$, because an allele must have come from somewhere in the previous generation, so that:

$$T_{ij} \approx 1 + \sum_{kl} m_{ik} m_{jl} T_{kl} - \delta_{ij} P_{ci} T_{ii} \qquad \text{(B7.2.1b)}$$

This expression can be further simplified by neglecting the second-order terms in the migration probabilities (**Problem 7.2**).

7.2.ii.b. *The island model*

The simplest type of population structure is the "island model" (Wright 1943, 1951), in which all populations (demes) are identical in size and have equal probabilities of exchanging migrants with each other. Wright modeled a metapopulation with an infinite number of demes, but we consider a set of d demes, all with the same effective population size, N_e, in which there is a constant probability, m, that an allele in a given deme is derived by migration from any other deme in each generation.

The demes are all equivalent (*exchangeable*) in their properties, so there is a probability $1/(d-1)$ that an allele migrating into a given deme comes from any specified other deme (**Figure 7.4.A**). The system can then be described by two variables: T_S, the mean time to coalescence of a pair of alleles sampled from the same deme (equivalent to T_{ii} above), and T_B, the mean time to coalescence of alleles from different demes (equivalent to T_{ij}). It is then easy to use **Equations B7.2.1a** and **B7.2.1b** to obtain expressions for T_S and T_B (**Problem 7.3**):

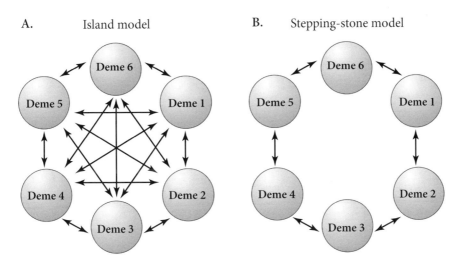

FIGURE 7.4 **A.** The island model of population structure, for which all pairs of demes have the same probabilities of exchanging migrants. **B.** The circular stepping-stone model. Here, migration occurs only between adjacent demes, with the same migration rate for each pair of demes, and the demes at the end are also assumed to exchange migrants with each other.

$$T_S \approx 2dN_e \tag{7.3a}$$

$$T_B \approx 2dN_e + \frac{(d-1)}{(2m)} \tag{7.3b}$$

To quantify genetic differentiation among demes, it is helpful to introduce an additional quantity, T_T, the mean time to coalescence of a pair of alleles drawn randomly from the population as a whole (with probabilities $1/d$ and $1 - 1/d$ of coming from the same deme and different demes, respectively). Using **Equations 7.3a** and **7.3b** for the corresponding mean coalescence times, we find that:

$$T_T \approx 2dN_e + \frac{(d-1)^2}{(2dm)} \tag{7.3c}$$

With a large number of demes, T_B and T_T are very similar.

One definition of the effective size of the metapopulation is one-half of T_T, by analogy with the definition for a single panmictic population (**Section 5.2.ii.a**). We call this the *total effective population size* for the metapopulation, N_{eM}. **Equation 7.3.c** shows that this is higher than N_e for a panmictic population with the same total size by $(d-1)^2/(dm) \approx d/2m$. This reflects the fact that two alleles sampled from two different demes must go back $d/2m$ generations on average before they are found in the same deme (Slatkin 1991), in which case they have a mean coalescence time of $T_S \approx 2dN_e$ generations.

These times give us the expected diversity values. Under the infinite sites model, mean nucleotide site diversities corresponding to the different origins of alleles are equal to twice the product of the mutation rate and the mean coalescence times (**Section 5.1.iii.d**). **Equation 7.3a** thus implies that the mean diversity within a deme, π_S, is given by:

$$\pi_S = 4dN_e u \tag{7.4}$$

This is the same value as in a panmictic population of effective size dN_e, and is independent of the migration rate (provided that this is nonzero). This *invariance* with respect to population structure was first noticed by Takeo Maruyama (Maruyama 1971b; 1974; 1977, Chapter 10), and analyzed further by Nagylaki (1982, 1998b) and Strobeck (1987); such invariance arises for some other important genetic parameters, as we will discuss later in **Section 7.3.ii**.

In contrast, **Equation 7.3.b** shows that the mean diversity for alleles sampled from *different* populations is inversely related to the product of the migration rate and the local effective population size and can be much higher than π_S if $N_e m$ is small. It is therefore incorrect to ignore the effects of population subdivision and say simply that *high levels of heterozygosity . . . , according to the Neutral Theory, should be caused by elevated mutation rates or large effective population sizes. . . .* (Small et al. 2007)

or

In general, large and continuously distributed habitats facilitate the maintenance of high levels of genetic diversity in plants. (Nassar et al. 2003)

A natural definition of the degree of population differentiation is given by replacing the values for π in **Equation 7.2** of **Section 7.1.i.c** by the corresponding coalescence times (Slatkin 1991; Takahata 1991), i.e., we can define a coalescent version of F_{ST}, corresponding to the empirically observable quantity K_{ST} in **Equation 7.2**:

$$F_{ST} = \frac{\left(T_T - T_S\right)}{T_T} \tag{7.5a}$$

Substituting from **Equations 7.3a–7.3c**,

$$F_{ST} \approx \frac{1}{1 + \dfrac{4N_e md^2}{(d-1)^2}} \tag{7.5b}$$

If the number of demes is large, this becomes:

$$F_{ST} \approx \frac{1}{1 + 4N_e m} \tag{7.6}$$

An important implication is that, if $N_e m$ is equal to 1 or less, there will be at least moderate differentiation among populations ($F_{ST} > 0.2$), whereas when $N_e m > 2$, the population behaves as nearly panmictic ($F_{ST} < 0.1$).

A similar formula was originally derived by Wright (1943, 1951), using the variance in allele frequencies among populations and the cognate expression for the inbreeding coefficient under the reversible mutation model of **Section 5.3.iii.c**. If the mutation rates are much less than m, as is usually biologically realistic, Wright's approach yields **Equation 7.6**.

This formula is frequently used to estimate the scaled migration rate, $N_e m$, even when the population structure of the species under consideration does not correspond to the island model with many demes (Slatkin 1993; Whitlock and McCauley 1999); the extent to which this is justified is discussed in **Sections 7.2.ii.c** and **7.2.v**. Since $N_e m$ represents the contribution of migrants to the effective deme size each generation, this has led to the often-repeated statement that one or more migrants per generation are sufficient to prevent substantial differentiation among demes.

For mutational models other than the infinite-sites model, allelic differences are not linearly related to genealogical divergence, so estimates of F_{ST} and G_{ST} from data on allele frequencies are not necessarily comparable with the coalescent formulae. For microsatellite data, the stepwise mutational model (**Section 5.1.iii.f**) is often assumed. Pairwise differences in repeat numbers can be used instead of pairwise diversities, and F_{ST} should then be replaced with the R_{ST} statistic described in **Section 7.1.i.b** above.

Clearly, the definition of the theoretical value of F_{ST} in terms of mean coalescence times can, in principle, be applied to any type of population structure; we can always define a mean time to coalescence for a pair of alleles sampled from the same deme, and from the population as a whole, by taking appropriate averages over demes. However, it is often difficult or impossible to obtain explicit analytical expressions for T_S and T_T. Therefore, although empirical estimates of the corresponding π values can always be obtained using the methods described in **Section 7.1.i.c**, they cannot always be easily interpreted.

7.2.ii.c. *Stepping-stone models*

While the island model has the advantage of simplicity, the assumption that each population has an equal probability of exchanging migrants with every other one is unrealistic for most species. A more realistic spatial structure can be introduced, while retaining the convenient assumption of equal effective population sizes for each deme, by allowing populations to exchange migrants only with their immediate neighbors. These *stepping-stone* models were first studied by Kimura (1953). Later work is reviewed by Wilkinson-Herbots (1998).

The one-dimensional case with d demes, each with effective size N_e, is shown in **Figure 7.4B**. In order to avoid the complications of edge effects, the demes are arranged in a circle, with probability m that an allele in a given deme is derived by migration, assigning an equal probability of its coming from the left-hand or right-hand neighbor (Maruyama 1970a). As the number of demes increases, the difference from a straight line becomes less and less important. Based on the symmetry of the model, the mean time to coalescence of a pair of alleles depends purely on the distance l between the demes from which they are sampled, where T_S is the value for $l = 0$. T_S is given by **Equation 7.3a**, as can be shown using the results for general migration models developed in **Section 7.2.ii.d**.

Equations B7.2.1a and **B7.2.1b** in **Section 7.2.ii.a** provide a set of equations for T_l $(d > l > 0)$ (Slatkin 1991, 1993). Following Slatkin (1993), we can define an F_{ST}-like measure of divergence between demes at a given distance l as:

$$F_{ST}(l) = \frac{\left(T_l - T_S\right)}{\left(T_l + T_S\right)} \tag{7.7}$$

The equations for T_l given by the results in **Box 7.2** then yield the following relation:

$$F_{ST}(l) \approx \cfrac{1}{1 + \cfrac{4dN_em}{l(d-l)}} \qquad (7.8\text{a})$$

It is also possible to determine the overall F_{ST} value by using **Equation 7.5a**:

$$F_{ST} \approx \frac{d^2 - 1}{d^2 - 1 + 12dN_em} \qquad (7.8\text{b})$$

The corresponding expression for the mean coalescence time for a random pair of alleles can be determined from **Equations 7.3a** and **7.8b** as:

$$T_T = \frac{T_S}{\left(1 - F_{ST}\right)} \qquad (7.8\text{c})$$

These results show that the expected sequence divergence between alleles sampled from different demes increases with the distance between the demes; the rate of increase is inversely related to the scaled migration parameter, N_em. If the number of demes d is large compared with N_em, **Equation 7.8b** shows that the overall F_{ST} will always be close to 1, i.e., we expect very high divergence between distant demes in a linear stepping-stone model with a large number of demes, even with N_em values high enough to produce effective panmixia in the island model.

The two-dimensional stepping-stone model is similar, except that the demes are now spread out over a plane not a line. Edge effects can be avoided by wrapping the plane into a *torus* (doughnut shape). The d_1 demes are arranged along the edge of the plane in one dimension, and d_2 demes along the other, giving a total of $d = d_1 d_2$ demes. The net migration rates in the two dimensions may differ (m_1 and m_2, respectively). The coalescence times for pairs of demes are determined by the number of steps separating them in the two dimensions, l_1 and l_2. Once again, the mean coalescence time for alleles sampled within a deme is given by **Equation 7.3a**.

The mean coalescence time for two demes separated by distances l_1 and l_2 in the two dimensions is given by a rather complex formula (Maruyama 1971a):

$$T_{l_1,l_2} \approx 2dN_e + \sum_{j=1}^{[d_1/2]} \sum_{k=1}^{[d_2/2]} \frac{1 - \cos\left(2\pi jl_1/d_1\right)\cos\left(2\pi kl_2/d_2\right)}{m - m_1\cos\left(2\pi jl_1/d_1\right) - m_2\cos\left(2\pi kl_2/d_2\right)} \qquad (7.9)$$

where m_1 and m_2 are the migration probabilities for the two dimensions, $m = m_1 + m_2$, and $[d_i/2]$ is the largest integer less than or equal to $d_i/2$. This allows

F_{ST} for a pair of demes to be calculated, as well as the overall F_{ST}, from the equivalent of **Equations 7.8a–7.8c**.

Although it is not obvious from **Equation 7.9**, the two-dimensional case shows far less differentiation among demes than the one-dimensional stepping-stone model for a given value of $N_e m$, unless the plane is very narrow in proportion to its length; the condition for approximate panmixis is thus similar to that for the island model. Unless $N_e m$ is quite small, differences between demes increase rather fast over small separations, and then become nearly independent of distance. The reasons for this feature of two-dimensional populations are discussed in **Section 7.2.v.c**.

This is a very useful conclusion, because it suggests that results from the island model may provide a good approximation to more realistic cases (see the example in **Figure 7.5**). An absence of evidence for a relationship between genetic differentiation and between-deme distances thus does not imply that the metapopulation obeys the island model; the migration probabilities may well depend strongly on spatial separation. Slatkin (1993) suggested that an index of genetic similarity between pairs of demes can be based on the estimate of $N_e m$ obtained from the demes' F_{ST} value, using **Equation 7.6** for the island model. A significant decline in this index with distance suggests an effect of distance on genetic differentiation, i.e., a departure from the island model. **Figure 7.6.A** shows the theoretical relationship between the distance between demes from which pairs of alleles are sampled and the estimate of $N_e m$ obtained by substituting the estimated F_{ST} values into **Equation 7.6**, when the true $N_e m$ value is 1. For a moderate degree of separation, $N_e m$ is quite accurately estimated for most shapes of the metapopulation, unless it approaches a narrow strip. A real-life example of the strong increase of genetic differentiation with distance in a linear habitat is shown in **Figure 7.6.B**.

7.2.ii.d. *Mean coalescence times for general migration models*

Despite the complexities of migration processes when less restrictive assumptions than the island model are made, a very general result for a quantity related to T_S in the island and stepping-stone models can be derived (**Box 7.3***). This result is especially simple in the case of "conservative" migration (Nagylaki 1980), i.e., migration among demes that leaves their relative sizes unchanged (the mean allele frequency across demes is also unchanged by such migration; see **Problem 7.5**). One situation where this condition is met is when the number of migrants that move from population i to j is the same as the number moving in the opposite direction (Nagylaki 1992, p. 136); this includes the island and stepping-stone models as special cases.

With conservative migration, when each deme is a Wright–Fisher population, the mean time to within-deme coalescence, weighting demes by their relative sizes, is equal to the total number of breeding individuals in the metapopu-

A.

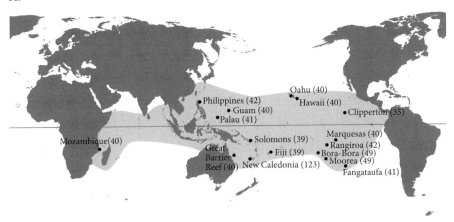

B.

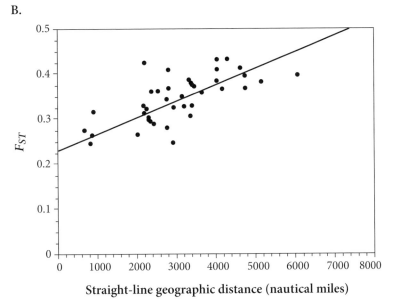

FIGURE 7.5 Genetic differentiation between populations of fish, *Acanthurus triostegus*, across large geographic distances in the Indo-Pacific region, measured from allozyme loci. **A.** Map of the locations of the populations sampled. **B.** F_{ST} values between pairs of populations as a function of the distance between them. (A nautical mile is 1852 meters, somewhat larger than a statute mile.) This shows a relatively small change in F_{ST} with increasing separation between populations, despite the large distances involved. [Adapted from Figure 3 of Planes and Fauvelot (2002).]

lation, N_T (Nagylaki 1982; Nagylaki 1998b). More generally, this result applies to the sum of the effective population sizes over all demes, N_{eT} (**Equations B7.3.5**), provided that all demes have the same ratio of the effective size to the number of breeding individuals in the deme. In other words, under any migration model that satisfies these assumptions, the expected mean within-population nucleotide site variability is the same as for a panmictic population with effective population size N_{eT} (see **Box 7.3***). This is the *migration effective population size*, N_{eS}, defined more generally by **Equation B7.3.3**.

If migration is non-conservative, as may often be the case in reality, the migration effective population size is less than N_{eT} (Whitlock and Barton 1997; Nagylaki 1998b; Wakeley 2001). This happens if some populations send out migrants disproportionately to their population sizes. It is then difficult to

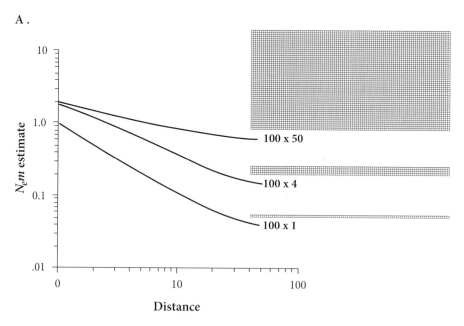

A.

$N_e m$ estimate

Distance

100 x 50

100 x 4

100 x 1

FIGURE 7.6 **A.** The theoretical relationship between the distance between demes from which pairs of alleles are sampled, and the $N_e m$ estimate obtained by substituting the estimated F_{ST} values into **Equation 7.6 of Section 7.2.ii.b**, when the true $N_e m$ value is 1. The different shapes of the metapopulations modeled are shown by the rectangles on the right. For a moderate degree of separation between demes, $N_e m$ is quite accurately estimated for most shapes of the metapopulation, unless it approaches a narrow strip. [Adapted from Figure 1 of Slatkin (1993).] **B.** A real-life example of F_{ST} values in a linear habitat for populations of guppies in a river in Trinidad, showing high values over short geographic distances. The F_{ST} values are based on six microsatellite loci, after removing one locus with null alleles, and were estimated as R_{ST}. Only populations not separated by waterfalls are included. [Adapted from Figure 3 of Crispo et al. (2006), whom we thank for their help in preparing the figure.]

B.

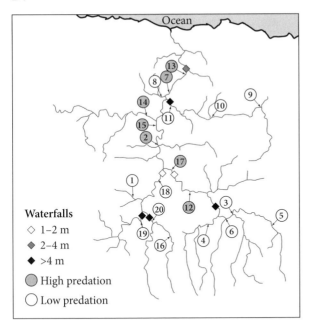

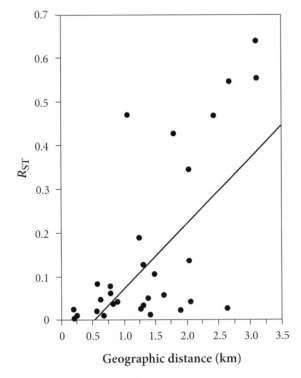

FIGURE 7.6 (Continued)

Box 7.3* THE MIGRATION EFFECTIVE POPULATION SIZE OF A STRUCTURED POPULATION

B7.3.i. The general case

The migration matrix $M = \{m_{ij}\}$ for a general migration model describes the set of probabilities m_{ij} that an allele sampled from deme i was in deme j in the previous generation, where i and j range from 1 to d. As noted in **Box 7.2**, we have $\Sigma_j m_{ij} = 1$. M is thus a *stochastic matrix* (**Appendix A1.iv.a**). It thus has a leading eigenvalue of 1, if at least one of the m_{ii} is nonzero (Nagylaki 1998a). The left eigenvector of M that corresponds to the eigenvalue of 1 is denoted here by $\mathbf{v}$; this consists of elements v_i such that v_i is proportional to the equilibrium probability that an ancestor of a given allele was present in deme i at some specific time in the past. This follows from the fact that, by definition, we have $\mathbf{v}M = \mathbf{v}$ (**Problem 7.4.i***).

Following Nagylaki (1998b), we define the following weighted mean coalescence times:

$$\tilde{T}_T = \sum_{ij} v_i v_j T_{ij} \tag{B7.3.1a}$$

$$\tilde{T}_S = \frac{\sum_i v_i^2 P_{ci} T_{ii}}{\sum_i v_i^2 P_{ci}} \tag{B7.3.1b}$$

$\tilde{T}_S$ is the mean within-deme coalescent time, which weights the coalescent time for each deme by the product of the equilibrium probability that a pair of alleles are found in the deme (v_i^2), and the reciprocal of twice its effective population size, i.e., $P_{ci} = 1/2N_{ei}$. Nagyylaki (1998b) showed that:

$$\sum_i v_i^2 P_{ci} T_{ii} \approx 1 \tag{B7.3.2}$$

(See **Problem 7.4.ii***.) Substituting this into **Equation B7.3.1b**, we have:

$$\tilde{T}_S \approx \frac{1}{\sum_i v_i^2 P_{ci}} \tag{B7.3.3}$$

When divided by 2, this expression defines an effective population size for the structured population, the *migration effective population* size, N_{eS} (Nagylaki 1980, 1998b). This is equal to the reciprocal of the sum of the reciprocals of the effective sizes for each deme, weighting each deme by the equilibrium probability that a pair of alleles is present in deme i, and so has an intuitively reasonable biological meaning.

B7.3.ii. Some special cases

$\tilde{T}_S$ reduces to a simpler form in some special cases. First of all, assume a Wright–Fisher model for each deme, with population size N_i for the ith deme, so that $P_{ci} = 1/2N_i$. Let the total size of the species be $N_T = \Sigma_i N_i$ and write $c_i = N_i/N_T$ for the proportion of the total population contributed by the ith deme. Under these conditions, **Equation B7.3.3** becomes:

$$\tilde{T}_S \approx \frac{2N_T}{\displaystyle\sum_i \frac{v_i^2}{c_i}} \tag{B7.3.4}$$

For conservative migration models, we have $v_i = c_i$ (**Problem 7.5.iii***); the intuitive interpretation of this is that the probability that an allele is found in deme i is simply the proportion of the total population contributed by this deme. In this case, we simply use weights of c_i when calculating $\tilde{T}_S$, and so obtain:

$$\tilde{T}_S \approx 2N_T \tag{B7.3.5a}$$

When $v_i \neq c_i$ for at least some i, the denominator of **Equation B7.3.4** exceeds unity, and so $\tilde{T}_S < 2N_T$, i.e., the migration effective population size is less than the total number of breeding individuals.

For more general demographic models, for which the effective population size of a deme is not the same as the number of breeding individuals, the situation is more complex. We can write $\tilde{c}_i$ for the ratio N_{ei}/N_{eT}, where N_{eT} is the sum of the effective population sizes over demes, and substitute into **Equation B7.3.3**. But $\tilde{c}_i$ lacks a direct relation to the migration process, unless we make the additional, fairly plausible, assumption that the N_{ei} value for a deme is proportional to its number of breeding individuals, N_i. If the coefficient of proportionality has the same value for all demes, $\tilde{c}_i = c_i = v_i$, and **Equation B7.3.2** for the case of conservative migration gives the following result:

$$\tilde{T}_S \approx \sum_i c_i T_{ii} = 2N_{eT} \tag{B7.3.5b}$$

interpret diversity data, especially as relative deme sizes are usually poorly known. The usual procedure for estimating total and within-deme diversity is to weight samples from different localities by their sample sizes, and trust to luck.

Overall, it is clear from the theory that measures of mean nucleotide site diversity for alleles sampled within the same deme are less influenced by details of population structure than the total pairwise diversity measure, π_T, which is very sensitive to the amount of migration, as shown by **Equations 7.3a–7.3c** for the island model. It is thus difficult to assess differences in effective population sizes between species on the basis of species-wide genetic diversity values; the mean within-deme diversity is likely to be more suitable for such comparisons (see **Figures 7.2** and **7.3**).

Computationally intensive methods using maximum likelihood or Bayesian inference from simulations of the coalescent process with migration have recently been developed to estimate deme sizes and migration rates (scaled by the mutation rate) from DNA sequence variability data for sets of local populations (Bahlo and Griffiths 2000; Beerli and Felsenstein 2001; Nielsen and Wakeley 2001; Beerli 2004; Hey and Nielsen 2004). Given sufficient data on all populations in the metapopulation, and an estimate of the mutation rate, all parameters of interest can be estimated, thereby removing many of the difficulties in interpreting pairwise diversity data. If only some demes are sampled, however, as is usually the case in practice, there can be considerable biases in these estimates (Slatkin 2005), so that these methods must be used with caution.

7.2.ii.e. *The strong migration limit*

If the products of migration rates and effective population sizes are sufficiently large, the metapopulation approaches the *strong migration limit*, with little differentiation among demes. This is because migration moves alleles around the metapopulation faster than they coalesce within demes. We can then use a fast time scale approximation, and assume that alleles are drawn at random from the equilibrium distribution of the deme of origin generated by the migration process (**Box 7.4**), just as we treated the flow of alleles among different demographic classes in **Section 7.2.i.b**. The coalescence time between a pair of alleles from different individuals, regardless of place of origin, then obeys **Equation B7.3.3** of **Section 7.2.ii.d** (Nagylaki 1980, 1998b).

Under the strong migration limit, a set of k alleles can be modeled using the same coalescent process as for a panmictic population (**Section 5.1.iii.c**), since coalescence probabilities are now independent of the demes from which alleles are sampled. The migration effective population size of **Section 7.2.ii.d** is used, instead of the single population effective size. The properties of genetic diversity in species with little genetic differentiation among populations can thus be

Box 7.4 THE STRONG MIGRATION LIMIT

In this case, the flow of alleles among demes by migration occurs much faster than the rate of coalescence within demes (Nagylaki 1980, 1998b) and there is effectively no genetic differentiation among demes. This enables us to use a fast time scale approximation, as with the flow of alleles among different demographic classes (**Section 5.2.ii.b**). The identities of the initial demes from which two alleles are sampled can thus be disregarded, and an allele can be treated as having probability v_i of being in deme i at any given time.

Since two alleles can coalesce only if they originate from the same deme (with probability P_{ci} for deme i), the net probability of coalescence, P_c, of a randomly chosen pair of alleles is given by:

$$P_c \approx \sum_i v_i^2 P_{ci} \qquad (B7.4.1)$$

The mean time to coalescence of any pair of alleles, T_T, is the reciprocal of P_c; in this case, it is identical to the within-deme mean coalescence time given by **Equation B7.3.3**, as expected from the lack of differentiation between demes.

interpreted relatively straightforwardly, although the relevant effective population size depends on the properties of the migration process in a way that is often overlooked.

Equation B7.4.1, which describes the inverse of the mean coalescent time under the strong migration limit, applies to any structured coalescent process where a fast time scale approximation can be used, provided that the probability of coalescence within a given class is independent of the class from which alleles were originally sampled. This result appears in a variety of contexts, including age-structured and stage-structured populations (Woolliams et al. 1993; Nordborg and Krone 2002), and will be used again in **Sections 8.3.ii–8.3.iv** for dealing with some aspects of the genetic structuring of populations.

7.2.iii. Continuous populations

7.2.iii.a. *General considerations*

Biological populations are often distributed over more or less continuous geographic ranges, rather than being split into easily identifiable discrete demes.

We can then treat individuals as occupying points along a line or plane. A good deal of attention has been paid to such *isolation by distance* models of spatial continua, first introduced by Wright (1943, 1946, 1951), and subsequently studied by Malécot (1948, 1969, 1975); see also Nagylaki (1989). Migration in continuous habitats is measured by the distribution of distances between the locations of birth and reproduction, as in **Section 4.1.ii.b**. Breeding individuals are assumed to be distributed with a uniform density over a line or plane, and to disperse randomly and independently of each other.

Under these assumptions, migration is conservative. The equilibrium coalescent time for a pair of alleles sampled from points arbitrarily close to each other is thus given by the equivalent of **Equation 7.3a** in **Section 7.2.ii.b.**:

$$T_S \approx 2S\rho_e \qquad (7.10)$$

where S is the size of the entire population (in terms of length or area) and ρ_e is the effective density of individuals per unit length or area, defined in **Box 7.5** (Charlesworth et al. 2003). The product $S\rho_e$ is thus analogous to the migration effective population size of **Section 7.2.ii.d**. Multiplying T_S by twice the mutation rate per nucleotide site, we get the expected equilibrium nucleotide site diversity under the infinite sites model for alleles sampled from the same location.

Box 7.5* THE NEIGHBORHOOD SIZE FOR A CONTINUOUS POPULATION

B7.5.i. A linear population

Assume that the population is distributed along an indefinitely long line with a uniform density of breeding individuals per unit length. Let the net probability density of dispersal of distance y between birth and reproduction be $m(y)$. As in **Section 4.1.ii.b**, $m(y)$ has mean 0 and variance σ^2. The probability that two alleles sampled from distinct individuals, located close to a given point x, are derived from points located between distances y and $y + dy$ from x is $[m(y)\,dy]^2$, where dy is arbitrarily small.

Conditioning on this, the chance that they coalesce is $1/2\rho_e\,dy$, where ρ_e is the effective density along the line, such that the effective number of individuals between points x and $y + dy$ is $\rho_e\,dy$. The effective density is related to the density of breeding individuals in the same way that the effective population size in the discrete deme model is related to the number of breeding individuals (**Section 5.2.ii**).

The net probability of coalescence of the original pair of alleles is thus $[m(y)dy]^2/2\rho_e dy$. Integrating this over the entire range of y, the net probability of coalescence is:

$$\frac{1}{2N_b} = \frac{1}{2\rho_e} \int_{-\infty}^{+\infty} m(y)^2 dy \qquad \text{(B7.5.1)}$$

where N_b is Wright's (1946) "neighborhood size."

It is usually assumed that dispersal follows a random walk, implying that $m(y)$ is a normal distribution. In this case, it can be shown from **Equation B7.5.1** that we have:

$$N_b = 2\sigma\rho_e\sqrt{\pi} \qquad \text{(B7.5.2)}$$

i.e. the neighborhood size is the number of individuals within a distance $\pm \sigma\sqrt{\pi}$ of the place where the alleles were sampled (**Problem 7.6***).

However, there are often considerable departures from normality for migration functions, with a long tail of low probabilities of dispersal over long distances, so that skewed distributions, such as the exponential distribution, provide a better fit, in which case different formulae for N_b can be derived (Wright 1978, pp. 303–307).

B7.5.ii. An area continuum

It is straightforward to generalize this approach to a two-dimensional plane if we assume that dispersal occurs independently in the two dimensions, with the same variance σ^2 in each direction, y and z. The effective number of individuals in a small region of area $dydz$ is then $\rho_e dydz$, and the chance of migration of two alleles into this region from a point at distances y and z away is $[m(y)m(z)dydz]^2$. The same argument as before leads to the following result:

$$\frac{1}{N_b} = \frac{1}{\rho_e} \int_{-\infty}^{+\infty}\int_{-\infty}^{+\infty} m(y)^2 m(z)^2 dydz \qquad \text{(B7.5.3)}$$

If the dispersal distance is normally distributed, this gives:

$$N_b = 4\pi\sigma^2\rho_e \qquad \text{(B5.7.4)}$$

The neighborhood size is thus the number of individuals within a circle of radius 2σ (**Problem 7.6***).

7.2.iii.b. *Genetic differentiation in continuous populations*

The extent of genetic differentiation between locations in continuous populations is mainly determined by the *neighborhood size*, N_b (Wright 1946), defined in **Box 7.5***. For two-dimensional populations, the theoretical results of Wright and Malécot mean that genetic differences between different localities for neutral variants will be large for neighborhood sizes of a few tens of individuals, moderate for a few hundred or so, and negligible for sizes of 1000 or more (Wright 1946, 1951). As expected from the results for the stepping-stone model (**Section 7.2.ii.c**), linear populations are most differentiated. Neighborhood sizes can be estimated for natural populations from data on dispersal and population density, reviewed by Wright (1978, pp. 61–76) and Crawford (1984). As might be expected, estimates range from values in the thousands for highly mobile animals like *Drosophila* and birds, down to values of less than 10 for some plant populations, broadly consistent with the wide range of F_{ST} values described in **Section 7.1.ii** above.

The problem of predicting the pattern of genetic differentiation as a function of distance is formidable. A linear population can be treated as the limit of a one-dimensional stepping-stone model, noting that the variance of distance moved in that case is *m*. Nagylaki (1978) showed that **Equations 7.8a–7.8c** can then be applied to the linear continuum around a circle, replacing *m* by the variance σ^2 of the migration distribution from **Equation B7.5.i**, dN_e by $S\rho_e$, and separation *l* by distance *x*. The case of a straight line has been treated by Wilkins and Wakeley (2002).

It is more complicated to deal with a two-dimensional population by this approach, because formally there is a zero probability that two alleles sampled at different points will ever move to exactly the same location, where they can coalesce (Nagylaki 1978). Following Wright (1943, 1946, 1951) and Malécot (1948, 1969, 1975), this can be side-stepped by assuming that two alleles sampled from individuals that are very close to each other coalesce with a probability equal to one-half of the reciprocal of the neighborhood size, N_b (Barton and Wilson 1995, 1996; Barton and Wilson 1996; Barton et al. 2002). The equations derived by Malécot for a two-dimensional, continuous population of indefinite size then approximate the results for differentiation among populations in a finite-sized, two-dimensional stepping-stone model with $m = \sigma^2$ (Barton and Wilson 1995, 1996; Barton et al. 2002), at least for distances substantially greater than σ. The results of **Section 7.2.ii.c** for such distances describe many of the properties of an area continuum. As with discrete deme models, the two-dimensional continuum often shows little effect of distance on genetic differentiation between samples, except for alleles sampled within a few σ of each other, reflecting the fact that fluctuations in variant frequencies occur only over short spatial scales (Charlesworth et al. 2003). Indeed, Wilkins (2004) showed that the mean coalescence time of two alleles is little affected by the distance separating the locations where they are sampled, unless N_b is < 12 (see also **Section 7.2.v**).

It is, however, biologically unrealistic to assume a uniform density of breeding individuals. Interactions between neighboring individuals, which are required to regulate population density, violate the assumption of uniform density (Felsenstein 1975). Barton et al. (2002) studied this problem in an ecological model of density regulation involving competitive interactions between nearby individuals. The results suggest that local fluctuations in population density are a serious difficulty only for predicting genetic differentiation over distances that are smaller than a few multiples of σ, providing that some adjustments are made to the classical formulae—notably an increase in neighborhood size.

7.2.iv. Colonization and extinction models

7.2.iv.a. *Introduction*

In many species, local populations are ephemeral and are liable to extinction and replacement by re-colonization from other populations (Hanski and Gilpin 1997). This was first discussed in the context of population genetics by Wright (1940a), who pointed out that it implies that genetic drift in the metapopulation as a whole may be controlled by the size of the single population that was the ancestor of the current population (**Figure 7.7**), rather than by the size of the set of populations as a whole.

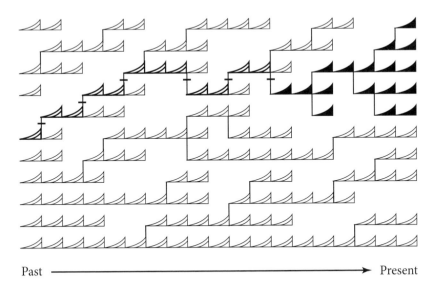

Past ⟶ Present

FIGURE 7.7 A metapopulation with frequent extinction and re-colonization of local populations, showing the descent of several demes from a single ancestor population. The group of demes symbolized in black represents a currently large set of populations, which are the sole surviving populations of this ancestor and have descended by a history involving multiple re-colonization events with small numbers of founders (these events are symbolized by the six small horizontal ticks). [Adapted from Figure 2 of Wright (1940a).]

7.2.iv.b. *The exchangeable deme model*

Slatkin (1977) introduced a simple model of extinction and colonization, based on the island model of **Section 7.2.ii.b**, which has formed the basis for most subsequent theoretical studies (reviewed by Pannell and Charlesworth 2000). It assumes d exchangeable, Wright–Fisher demes, each of size N, with migration obeying the island model. In addition, there is a constant probability, e, that a deme goes extinct in a given generation, in which case it is immediately replaced by n colonists drawn from one or more other demes, and the population instantaneously grows back to size N. Under the Wright–Fisher assumption, there is a probability $1/2n$ that two alleles that are sampled from the deme coalesce in the generation after such a colonization event. Other demographic models can be studied similarly by defining an effective number of colonists, n_e, and an effective deme size, N_e.

There are two extreme alternative possibilities for the origin of the colonists (Slatkin 1977): the *propagule-pool model*, where all colonists come from the same deme, and the *migrant-pool model*, in which each colonist comes from a random deme sampled from the entire set of demes (**Figure 7.8**). In the migrant-pool model, extinction and colonization mix alleles from different demes, whereas there is no mixing in the propagule-pool model. Mean coalescent times for the two extreme cases are given by Pannell and Charlesworth (1999). A more general model is to assign a probability ϕ that all colonists come from the same deme, and $1 - \phi$ that they are of independent origin (Whitlock and McCauley 1990).

If we make the standard assumptions of the coalescent process, that e, m and $1/n_e$, and $1/N_e$ are all small enough that their second-order terms can be neglected, and also assume a large number of demes, a simple approximation for F_{ST} can be derived (Whitlock and McCauley 1990; see **Section 7.2.v.d**):

$$F_{ST} \approx \frac{1 + \left(N_e e/n_e\right)}{1 + 4N_e m + 2N_e e\left[1 - \phi\left(1 - 1/2n_e\right)\right]} \tag{7.11}$$

This expression implies that an increase in the extinction rate e causes an increase in F_{ST} if:

$$n_e < \frac{2N_e m}{(1 - \phi)} + \frac{1}{2} \tag{7.12}$$

(see **Problem 7.7***.)

As expected, under the propagule-pool model ($\phi = 1$), an increase in the extinction rate e always increases F_{ST}, although the effect is slight unless $e > m$.

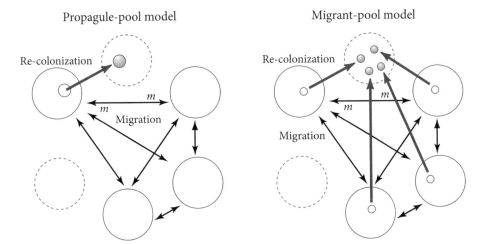

FIGURE 7.8 Slatkin's alternative possibilities for the origin of colonists of unoccupied sites in a metapopulation (Slatkin 1977). In both models, migration into unoccupied sites (symbolized by dashed circles) follows the island model, and can also occur between all occupied sites, as indicated by the black arrows. In the propagule-pool model, all colonists of a given unoccupied site come from the same deme (so that mixing of alleles from different demes occurs only through migration between occupied demes). The migrant-pool model allows colonization by individuals from random demes in the entire set, so that alleles from different demes are mixed during colonization of unoccupied sites; in the next generation after colonization (not shown in the diagram), newly occupied demes are assumed to exchange migrants again with other demes, at the same rate, m.

With $e \gg m$ and $N_e e \gg 1$, F_{ST} approaches 1, even when migration rates are high enough to cause an approach to panmixia in the absence of extinction. In contrast, under the migrant-pool model, increasing e increases F_{ST} only if $n_e < 2N_e m + 0.5$ (see **Figure 7.9**).

Under the propagule-pool model, both total and within-population coalescent times and diversities can be greatly reduced, when compared with the corresponding island model. Under the same assumptions as above, we obtain:

$$T_S \approx \frac{2dN_e n_e m}{(m + e)(n_e + N_e e)} \tag{7.13a}$$

Using the relation $T_T = T_S / (1 - F_{ST})$, **Equations 7.12** and **7.13a** yield:

$$T_T \approx \frac{d(n_e + 4N_e n_e m + N_e e)}{2(m + e)(n_e + N_e e)} \tag{7.13b}$$

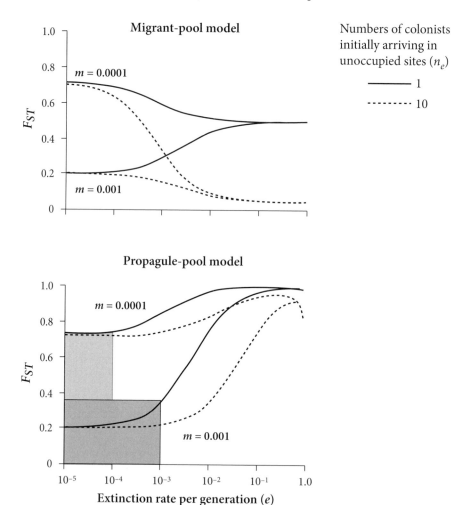

FIGURE 7.9 Example of the behavior of F_{ST} under the two colonization models of **Figure 7.6**, each with high and low migration rates per generation ($m = 10^{-3}$ and 10^{-4}, respectively), and high and low effective numbers of colonists (n_e). In each case, there were $d = 500$ demes, each with $N = 1000$ individuals. The figure shows that the migrant-pool model does not always lead to increased F_{ST} as the extinction rate increases (see the text). In contrast, increased extinction always increases F_{ST} under the propagule-pool model, except for very high values of e; the effect is slight unless the extinction rate is high ($e > m$, as indicated by the grey boxes, whose width on the x axis corresponds to $e = m$). [Adapted from Figure 2 of Pannell and Charlesworth (1999).]

These expressions reduce to the standard island model results of **Equations 7.3a–7.3c** when $e = 0$. But when extinction dominates over migration and coalescence within demes, we have $T_S \ll 2dN_e$ while T_T approaches $d/2e$, i.e., T_T depends only on the total number of demes and the extinction rate (Maruyama and Kimura 1980).

As might be expected, the effects of extinction are less extreme for migrant-pool colonization, unless the effective number of colonists is small (**Figure 7.9**), although both total coalescent times and within-deme coalescent times are again shortened compared with the standard island model. Under the usual coalescence assumptions (**Section 7.2.i**), we have:

$$T_s \approx \frac{d\left[4N_e n_e m + N_e e(2n_e - 1)\right]}{2(m + e)(n_e + N_e e)} \qquad (7.14a)$$

$$T_T \approx \frac{d\left(n_e + 4N_e n_e m + 2N_e n_e e\right)}{2(m + e)(n_e + N_e e)} \qquad (7.14b)$$

Overall, extinction and colonization can reduce mean genetic diversity both within local populations and across the metapopulation as a whole, compared with the corresponding pure migration case. In the case of the propagule-pool model, these reductions are accompanied by high values of F_{ST}. Unless the extinction rate is higher than the migration rate, however, the differences from the pure migration case are slight. In self-fertilizing species of plants, even stronger effects can occur, because selfing reduces the availability of ovules to pollen from other plants (**Section 7.1.ii**). This means that a newly founded colony is more likely to have reduced genetic diversity than with outcrossing. With a single population, within-population diversity is at most halved, which requires complete selfing (**Section 5.2.iii.a**). A moderate extinction rate per generation, however, can halve the within-population diversity with a much lower selfing rate, particularly under the propagule-pool model, where the effects of population turnover are generally greatest (Ingvarsson 2002). A highly skewed distribution of offspring number can also lead to low within-deme diversity and high F_{ST}, even in the absence of extinction and re-colonization events (Eldon and Wakeley 2009).

In a few cases, it has been possible to compare empirical estimates of the migration and extinction parameters with the theoretical predictions (Giles 1997). An excellent example is the work of Ingvarsson et al. (1997) on the beetle, *Phalacrus substriatus*, living on tussocks of sedges on an island in northern Sweden. The tussocks were censused for four years during the beetle's breeding season. If no beetles were found on a tussock in two successive years, that colony was scored as extinct. If beetles were found in two successive years on a tussock whose population was previously extinct, a re-colonization event was recorded. Migration rates were also estimated by mark–release–recapture experiments, giving a value of m of 0.37. Extinctions and re-colonizations occurred at approximately equal frequencies (0.24 and 0.28 per year, respectively), with a harmonic mean number of colonists per empty tussock of about

2. The harmonic mean population size was about 11 beetles per tussock. Assuming that ϕ in **Equation 7.11** is 0.5, these values predict an F_{ST} of about 0.070, which compares well with the value of 0.077 obtained from a survey of seven allozyme loci. Ingvarsson et al. (1997) estimated that F_{ST} was approximately 40% greater than would have been expected with a pure migration process, and that extinctions reduced the total effective population size for the metapopulation by about 55%.

7.2.v. Sampling of alleles from spatially structured populations

7.2.v.a. *Introduction*

Up to now, we have mainly discussed the expected values of pairwise diversity statistics. As we saw in **Chapters 5** and **6**, however, there is likely to be a wide distribution around these expected values when we consider samples of alleles from a real population. Because many tests for selection use the frequency distributions of nucleotide site variants (**Sections 6.4.iii** and **6.4.iv**), we need to understand their properties in spatially structured populations. For a sample from a single deme, drift may have caused variants to reach locally high frequencies, compared with their frequencies in the population as a whole, creating an excess of relatively high frequency variants compared with the panmictic case. In terms of the coalescent process, this corresponds to rapid coalescence within the local deme, so that there are fewer external branches of the gene tree. Samples taken from more than one deme may show an excess of intermediate frequency variants, due to the long time needed for the coalescence of alleles sampled from different demes.

The extent of such departures from the panmictic case can first be assessed in the following way. Expressions for the expectation of pairwise diversities (π values) and numbers of segregating sites for the case of two demes were obtained by Tajima (1989a). The latter can be converted into expectations of the θ_w estimator of variability by using Watterson's correction factor (**Section 1.2.iv.b**). As discussed in **Section 6.4.iii.b**, the expectations of π and θ_w differ if the frequency spectrum of variants departs from that expected under panmixia and neutralilty: an excess of intermediate frequency variants causes π to exceed θ_w. As can be seen from the numerical examples in **Table 7.1**, these differences can be large when there is restricted migration between demes, especially when many alleles are sampled across the two demes. Such departures from panmictic expectations clearly create difficulties for the tests of selection of the type discussed in **Sections 6.4.iii** and **8.3.ii**, at least when there is substantial genetic differentiation among local populations, as is often the case (**Section 7.1.ii**).

TABLE 7.1 Effects of population structure on the expected values of π and θ_w

$4N_e m$	Sample size	All alleles from a single deme	Equal samples from both demes	Random samples across demes
0.01	2	1.00	51.0	26.0
	10	1.00	2.72	1.39
	50	1.00	4.17	2.13
0.1	2	1.00	6.00	3.50
	10	1.01	2.11	1.24
	50	1.03	2.76	1.66
1.0	2	1.00	1.50	1.25
	10	1.03	1.23	1.04
	50	1.10	1.47	1.22
10	2	1.00	1.05	1.02
	10	1.01	1.02	1.00
	50	1.05	1.03	1.00

[1]This assumes two demes of equal effective size N_e, with symmetrical migration between them. The results for sample size 2 show the expected values of π for a pair of alleles, relative to the value for a panmictic population of effective size $2N_e$. The entries for sample sizes of 10 and 50 are the ratios of the expectations of π and θ_w for the mode of sampling in question (adapted from Tajima 1989a).

For more complex types of population structure, this problem has mostly been studied by computer simulations of the coalescent process, using one or more of the population structure models described in **Sections 7.2.ii** and **7.2.iv**—e.g., Wall (1999), Ptak and Przeworski (2002), Pannell (2003), and De and Durrett (2007). However, the results are limited to the specific models assumed. It might seem impossible to develop general results, but it has recently been found that populations with large numbers of demes can have surprisingly general properties (Wakeley 1998, 1999, 2001, 2008, Chapter 6; Wakeley and Aliacar 2001; Matsen and Wakeley 2006). They behave in a way that suggests that the problems created by population structure for testing for selection can be minimized by adopting the appropriate strategy for sampling alleles across demes. In the next section we examine these properties, treating both migration and extinction/colonization models by the same method.

7.2.v.b. *The large deme number approximation*

The general problem is to describe the coalescence of a sample of k lineages from a specified set of demes (which we will label 1 to j), with k_i alleles in deme i ($\Sigma_i k_i = k$). The two extreme possible sampling schemes are shown in **Figure 7.10**. As we shall see, it is simplest to study the first case, where each of the k alleles is sampled from a different deme ($k = j$), but the theory can be extended to include other sampling schemes.

The key to simplifying the problem is to note that a large number of demes means that, if we trace the history of the alleles in our sample back to a given ancestral generation, most demes will not contain any alleles ancestral to those in the sample; these are called "unoccupied" demes. Thus, (looking backwards in time) when migration and extinction/colonization events cause an allele to move from an occupied deme to another deme, it only rarely moves to a deme that contains another lineage in the ancestry of the sample—instead, most such events will move alleles into unoccupied demes (indicated by demes containing only grey dots in **Figure 7.11**). It thus takes a long time for two alleles from separate demes to come into the same deme.

In this situation, coalescence within demes can be regarded as effectively instantaneous, relative to the time it takes for a pair of alleles from different demes to move into the same one. Furthermore, for many types of population structure, the ancestor of an allele sampled from a deme rapidly reaches a fixed probability of being found in a given deme i. This probability is determined by the parameters of the migration and extinction/colonization processes (see **Box 7.6*** for details), but it does not depend on which deme the allele was sampled

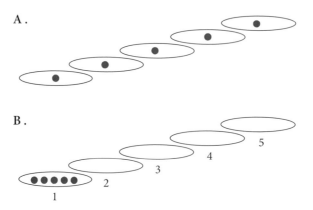

FIGURE 7.10 Two of the possible different ways in which samples of k lineages can be taken from a set of $j = 5$ demes. **A.** Each of the k alleles is sampled from a different deme ($j = k$). **B.** At the other extreme, all of the alleles are sampled from a single deme ($j = 1$).

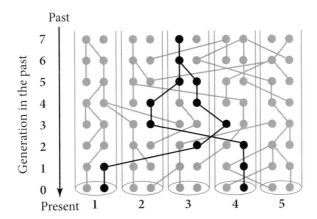

FIGURE 7.11 The history of the lineages (alleles) in a sample of two alleles traced back in time going upwards (black dots and lines), showing migration and coalescence events. Grey dots and lines indicate alleles present in the populations and their ancestral alleles, which are not represented among the alleles in the sample.

from. This means that alleles from different demes can be treated as exchangeable, even if demes differ in their sizes and probabilities of migration or extinction.

The time to coalescence is then determined largely by the time that it takes two alleles from separate demes to trace their ancestry back to the same deme. This implies that we can use a fast time scale approximation, similar to that used to derive the effective population size of a single population (**Section 5.2.ii**), or the strong migration limit of **Section 7.2.ii.e**, and neglect the contribution from the time to coalescence within a deme to the total coalescence time (Wakeley 2008, Chapter 6). This is evidently not always true, because otherwise there could never be any effects of distance on the extent of genetic differentiation. It requires the migration and extinction/colonization processes to be such that an allele moves a long way before coalescing with an allele that was initially in a different deme.

7.2.v.c. *Scattering and collecting phases*

These considerations lead to a distinction between two phases of coalescence of a sample of alleles from a structured population—the "scattering" and "collecting" phases (Wakeley 1999). Again looking backwards in time (**Figure 7.12**), during the relatively brief scattering phase, alleles in the same deme either coalesce within this deme or they disperse to unoccupied demes, where they remain as single lineages for long periods of time. This phase ends when all the sample's remaining (non-coalesced) ancestral alleles are dispersed among

Box 7.6* COALESCENCE IN THE COLLECTING PHASE

Consider a pair of alleles located in separate demes. We note that, given the assumptions concerning the magnitudes of the migration and extinction parameters needed for the coalescent method to be valid, we can treat both migration and extinction/colonization (**Section 7.2.iv**) as mutually exclusive processes by which an allele is derived from a different deme in the preceding generation (Wakeley and Aliacar 2001). We write $b_{ij} = m_{ij} + e_{ij}$ for the probability that an allele found in deme i was derived from deme j in the previous generation, so that the stochastic matrix $B = \{b_{ij}\}$ replaces the migration matrix M used in **Box 7.3*** of **Section 7.2.ii.d**. This applies to both the migrant-pool and propagule-pool models, if we assume that demes are labelled by location.

As in **Box 7.3***, we assume that B is such that an allele from any deme will eventually be found in any other deme, with an equilibrium probability v_i that this is deme i, where v_i is the ith element of the left eigenvector $\mathbf{v}$ of B that corresponds to its leading eigenvalue of 1 (**Appendix A1.iv.c**). Conditioning on a given allele being in deme i, we need to know the probability that another allele currently located in a different deme will have been in deme i in the previous generation. This is found by summing $v_j b_{ji}$ over all $j \neq i$ and dividing by the corresponding sum of the v_j values. Using the fact that $\mathbf{v}B = B$ (**Appendix A1.iv.c**), we have:

$$\sum_{j \neq i} v_j b_{ji} = \sum_j v_j b_{ji} - v_i b_{ii} = v_i \left(1 - b_{ii}\right) \qquad \text{(B7.6.1)}$$

where b_{ii} is the probability that an allele in deme i came from the same deme in the previous generation. The net probability per generation that an allele found in deme i meets an allele from another deme is thus $v_i (1 - b_{ii})/(1 - v_i)$. With a large number of demes, v_i will be small, so that the denominator can be neglected and the encounter probability is approximated by $v_i (1 - b_{ii})$. We can write m_i and e_i for the sums of m_{ij} and e_{ij} over $j \neq i$, respectively, so that $1 - b_{ii} = m_i + e_i$.

Conditioning on the occurrence of an encounter between two alleles from different demes, what is the probability that the two alleles coalesce within this deme? This can easily be determined for the cases of migration and migration-pool extinction/colonization, where the two alleles must have come from distinct demes if a migration or colonization event took place. The probability of coalescence by colonization from another deme is $e_i/2n_{ie}$ (on the assumption that an effective number n_{ie} of propagules colonizes deme i if it goes extinct), so the net probability of coalescence per generation is:

$$P_{ci} = \frac{1}{2N_{ei}} + \frac{e_i}{2n_{ei}} \qquad (B7.6.2)$$

The probability that one of the two alleles came from another deme by migration is $2m_i$ in each generation, and the probability of extinction and colonization without coalescence is $e_i(1 - 1/2n_{ie})$.

The times to each of these events are independently and exponentially distributed, on the usual assumptions of the coalescent process, and the history of the alleles in the deme must terminate by one of them taking place. From the properties of competing exponential distributions (**Appendix A2.v.c**), the probability that the terminal event involves coalescence is given by:

$$F_i \approx \frac{P_{ci}}{\left[P_{ci} + 2m_i + e_i\left(1 - 1/2n_{ei}\right)\right]} = \frac{1}{\left\{1 + \left[2m_i + e_i\left(1 - 1/2n_{ei}\right)\right]/P_{ci}\right\}} \qquad (B7.6.3)$$

The net probability per generation of ultimate coalescence for an allele currently in deme i, due to entry of an allele from a different deme, is thus approximated by $v_i(m_i + e_i)F_i$. The net probability of coalescence, P_{Cc}, for a pair of alleles sampled randomly from the entire set of demes is obtained by averaging this expression across demes, using weights of v_i (the chance that two alleles are sampled from the same deme is negligible, given the assumption of a large number of demes). We also need to multiply by 2, to take into account the fact that either of the two sampled alleles could have been the migrant. This gives the final expression:

$$P_{Cc} \approx 2\sum_i v_i^2\left(m_i + e_i\right)F_i \qquad (B7.6.4)$$

The inverse of P_{Cc} is the expected coalescence time for a pair of randomly sampled alleles, T_T. This can be equated to twice the effective population size of the metapopulation as a whole, N_{eM}.

Because the probability of coalescence of alleles is independent of the deme from which they were originally sampled, the probability of coalescence of a set of k_c alleles present at the start of the collecting phase can be treated as in **Section 5.1.iii.c**, multiplying P_{Cc} by $k_c(k_c - 1)/2$; in this way, we can apply the standard results of the panmictic coalescent process to the collecting phase.

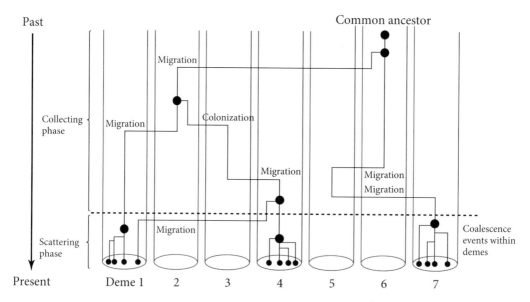

FIGURE 7.12 The "scattering" and "collecting" phases of the coalescence of a sample of alleles from a structured population. In the example, four alleles (small black dots) were sampled from each of three demes (demes 1, 4, and 7) of a metapopulation. Coalescence events in the ancestry of the sample are indicated by large black dots. Going backwards in time, during the scattering phase the lineages either coalesce within the demes or move between demes. Initially, within-deme coalescence events are more likely than migration if migration is restricted, but one case where migration precedes coalescence is shown (involving demes 1 and 4). When all alleles in the initial sample have at most one ancestral allele in a given deme, the much longer collecting phase starts, and the remaining lineages (three out of the original 12 in the example) eventually come together in the same deme and coalesce (sometimes moving between demes before this happens).

separate demes, representing the start of the collecting phase. This is a much slower process, in which pairs of alleles from distinct demes eventually come together in the same deme, and either coalesce or move to other demes.

When each allele in the initial sample is taken from a different deme (**Figure 7.10**), there is only a collecting phase. More generally, the scattering phase has only a minor effect on the gene genealogy of the sample, except for determining the number of alleles in separate demes, k_c, when the collecting phase begins. Formulae for the probability distribution of k_c are given by Wakeley (1998, 1999, 2001) and Wakeley and Aliacar (2001). The collecting phase must obey a standard coalescent process, because it involves an initial set of k_c distinct alleles, for which there is a constant probability per unit time that a pair of alleles enter the same deme and then coalesce, which we will write as P_{Cc} (the probability of collecting phase coalescence). The mean time to coalescence of a pair of alleles sampled from two different demes is $T_B = 1/P_{Cc}$, which is nearly the same as the mean time to coalescence for a random pair of alleles from the metapopulation (T_T), given the assumption of large d. All the standard coalescent results can thus be applied once this probability is known, which depends on the details of the migration and extinction model (**Box 7.6***). If each allele is sampled from a different location, it is safe to apply standard coalescent theory to the properties of the sample. The only difference is that we no longer have a simple theoretical relationship between mean coalescent time and the size of the population. A more rigorous treatment of this problem is given by Wakeley (2001; 2008, Chapter 6), Wakeley and Aliacar (2001), and Matsen and Wakeley (2006).

This method provides a quick way of simulating gene trees in spatially structured populations, and of examining the frequency distributions of nucleotide site variants by placing mutations onto the genealogies according to the infinite sites model, as described in **Section 5.1.iii.e** for a panmictic population (Wakeley 2001; 2008, Chapter 6; Wakeley and Aliacar 2001; Matsen and Wakeley 2006). Some examples are shown in **Figure 7.13**, for the case of a sample from a single deme, showing the departures from panmictic expectation that can arise in a structured population. It also provides a framework for using maximum likelihood or Bayesian statistical methods (**Appendix A2.vi.b–c**) to estimate population parameters in a way that is less affected by the details of the migration process than those mentioned in **Section 7.2.ii.d**.

Conditions under which the assumptions needed for this approach are likely to apply are discussed by Matsen and Wakeley (2006) for discrete populations, and by Wilkins (2004) for the area continuum. Insight into these conditions is provided by the two-dimensional stepping-stone model. For a square of d demes with $\sqrt{d}$ demes along each edge, so that there are approximately $\sqrt{2d}$ demes along the diagonal, it will take an allele sampled from one corner about $\sqrt{2d}/(m/2)$ generations to migrate to the opposite corner along the diagonal. The mean time to coalescence within a deme is $2dN_e$, which gives a lower bound to the mean time to coalescence of a pair of alleles sampled from different demes. A sufficient condition for the mean coalescent time of a pair of alleles to exceed the maximum time for their diffusion across the metapopulation is $2dN_e > (2\sqrt{2d})/m$, i.e., $N_e m\sqrt{2d} > 1$. This is easily satisfied for large d, unless $N_e m \ll 1$. In contrast, in a one-dimensional stepping-stone model, both the mean within-deme coalescence time and the migration time increase in proportion to d, so that an increase in deme number does not remove the dependence of coalescence time on distance between locations of alleles (see **Equations 7.8a** and **7.8b** of **Section 7.2.ii.c**). A simulation study shows that the large deme number approximation is valid, even for a two-dimensional stepping-stone model with a square of only 100 demes (De and Durrett 2007).

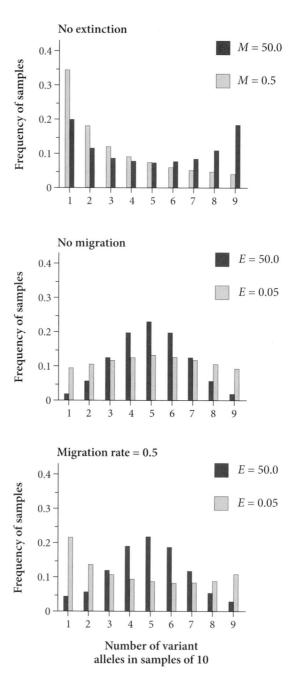

FIGURE 7.13 The distribution of variant frequencies at a single site is expected in a sample of 10 sequences from a single population, using simulated data (no distinction is made between ancestral and derived states, so this is comparable to the panmictic results in **Figure 5.6** of **Section 5.3.iii.c**). The results of the simulations are shown for several migration rates (expressed in terms of $M = 4N_e m$) and extinction/re-colonization rates (expressed in terms of $E = 4N_e e$). The case with $M = 50$ and no extinction is closest to the panmictic case. The distortions of the frequency distribution caused by population structure are visible. [Adapted from Figure 4 of Wakeley and Aliacar (2001).]

7.2.v.d. *Within-population coalescence times*

Another useful feature of this way of looking at spatial structure is that it provides a simple way of obtaining within-deme mean coalescence times (the T_{ii} in **Box 7.2** of **Section 7.2.ii.a**). The assumption of negligibly small coalescence times when conditioning on alleles remaining in a deme, relative to coalescence times for alleles from separate demes, implies that T_{ii} is approximated by the probability that two alleles from deme i fail to coalesce, multiplied by the expected time to coalescence of a pair of alleles from different demes, T_T:

$$T_{ii} \approx T_T(1 - F_i) \tag{7.15}$$

where F_i is the probability that the history of the pair of alleles terminates in a coalescent event (**Equation B7.6.3**). Averaging this over all demes gives the mean within-deme coalescence time as:

$$T_S \approx T_T(1 - F_{ST}) \tag{7.16}$$

where F_{ST} is the mean of the F_i over all demes (compare **Equation 7.5a** of **Section 7.2.ii.a**).

We can use any method we choose for weighting the contributions of individual demes to T_S, because T_T is independent of the choice of weights under the large deme number approximation. We can define a corresponding effective population size as $T_S/2$. For the infinite sites model, this predicts the mean within-deme nucleotide site diversity for the chosen weighting system. For example, if we use weights of $v_i^2 P_{ci}$, as in **Box 7.3*** of **Section 7.2.ii.d**, we recover **Equation B7.3.2** for the case of migration alone (**Problem 7.8***), so that $T_S/2 = N_{eS}$.

Equations B7.6.3–B7.6.4 are especially easy to apply to the case of d exchangeable demes, all with the same effective size, which satisfy the assumptions of the migrant-pool model of **Section 7.2.iv.b**. In this case, there is no dependence of migration and colonization probabilities on deme identity. Substituting the relevant parameters into these equations yields **Equations 7.11–7.14** (**Problems 7.9** and **7.10**). This method can also be used to study the propagule-pool model of extinction and colonization when demes are exchangeable, by noting that a pair of non-migrant alleles in the same deme must both have been present in the deme from which the colonists that founded the deme came. The probability of a switch to a different deme without coalescence is thus $e(1 - 1/2n_e)$, in which case the alleles remain in the collecting phase and can be treated as though they stayed in the same deme. It can be shown that this leads to **Equations 7.13a** and **7.13b** of **Section 7.2.iv.b** (**Problems 7.9** and **7.10**).

7.2.vi. The effects of population history

7.2.vi.a. *Introduction*

The models reviewed in the last section all made the rather strong assumption that deme sizes, and migration rates and patterns, remain constant over a sufficiently long period for statistics such as F_{ST} to reach equilibrium. Even in the models of extinction and colonization, where individual populations turn over, there is statistical equilibrium for the metapopulation as a whole. The models can be extended by introducing fluctuations in local demes sizes and/or allowing numbers of migrants to be random variables rather than fixed quantitities (Rannala and Hartigan 1995; Hudson 1998), but this still assumes an overall stationary state.

In all but the simplest cases, this assumption is essential, if results from explicit mathematical models are to be obtained (see **Section 7.2.vi.b** below). It is to some extent justified by the fact that F_{ST} equilibrates over a much shorter time scale than the measures of genetic diversity itself (Crow and Aoki 1984) and should therefore recover relatively quickly from a major perturbation to population structure. In addition, if population size undergoes a sharp reduction, equilibrium diversity levels can be reached relatively quickly; however, with a population size expansion, new mutations are required before equilibrium can be restored, so that the approach to the equilibrium diversity level is very slow (Pannell and Charlesworth 1999).

There are, however, many observed patterns of genetic variation that suggest relatively recent perturbations to population size and geographic distribution. For example, the general decrease with latitude in Europe of genetic diversity in local populations of many animal and plant species suggests re-colonization after the last ice age in northern Europe (Hewitt 2001), with more northerly populations having not yet had time to reach their equilibrium levels of variability. Similarly, the lower levels of genetic diversity of non-African compared with African populations of humans, and the fact that the nucleotide variants found in non-African populations are mostly a subset of those found in Africa, are consistent with the non-African populations having originated fairly recently from relatively small founding populations of African origin (Yu et al. 2002; Voight et al. 2005; Handley et al. 2007).

7.2.vi.b. *Distinguishing historical changes from stable structure*

Even in these cases, though, it is surprisingly difficult to distinguish between the very different situations of equilibrium with smaller population sizes of the supposed founder populations, leading to asymmetrical gene flow from larger ancestral to smaller descendant populations, and populations that are completely isolated from each other, but related by a succession of founding events (e.g., Takahata et al. 2001). In either case, substantial genetic differentiation between the populations may coexist with shared polymorphisms.

Many phylogeographic studies infer population history from spatial patterns of variability of a single non-recombining locus or genome, such as the mitochondrion or Y chromosome, because these allow reconstructions of genealogies that can be mapped onto geographical locations (Avise 2000). However, the history of a single non-recombining sequence represents only one realization of the highly stochastic coalescent process, and there is no guarantee that it is representative of the genome as a whole (Knowles and Maddison 2002). For example, an apparently clear division of a non-recombining genome into two geographically distinct clades (which might be taken to suggest a long-established isolating barrier) can occur quite frequently by chance in a subdivided population if migration rates are sufficiently low (Irwin 2002).

For two populations recently separated from a common ancestor, such as a pair of related species, Wakeley and Hey (1997) have developed models predicting the numbers of fixed nucleotide differences, and of polymorphisms shared between the two populations, and also the expected numbers of polymorphisms that are unique to each of the populations. The expectations of these summary statistics across a set of loci (assuming neutrality) depend on the rate of gene flow, the time since divergence, and the relative sizes of the populations and their common ancestor. Simulations of the distribution of the summary statistics can therefore be used to test for gene exchange between populations (Wakeley and Hey 1997; Wang et al. 1997; Kliman et al. 2000). An alternative approach is to calculate the likelihood of a set of nucleotide sequence data for a single, non-recombining locus (Nielsen and Wakeley 2001).

For more complex cases, simulations of individual scenarios are necessary to compare predictions from assumed population histories with the observed patterns (Austerlitz et al. 2000; Estoup et al. 2004; Ray et al. 2005; Klopfstein et al. 2006; Burton and Travis 2008). Wakeley (1999) has shown how to use the fast time scale approximation for a large number of demes to infer a historical change in human population migration structure; this approach removes some of the need to include many different parameters in the models.

There are still many difficulties, even with data on multiple, independently inherited sets of markers. As we will discuss in **Section 8.3.iv**, in geographical regions where the environment changes, selection acting on some loci may affect markers in linked genome regions, even if the marker alleles are neutral, and inconsistent marker behavior is indeed often observed in ecological or geographical boundary regions (Martinsen et al. 2001; Shaw 2002). Even in the absence of such effects, the results of **Sections 7.2.ii–7.2.v** show that much of the neutral diversity in an outcrossing species is present within any local population, unless migration is very restricted. In other words, most of the ancestry of a sample of genes is in the distant past, and is spread across the species' entire ancestral range. Phylogenetic inferences about outcrossing populations can therefore be extremely difficult, since shared polymorphisms are common, and few markers have patterns that reflect the populations' history, if indeed there is any clear meaning to this concept.

7.3. THE THEORY OF SELECTION WITH SPATIAL STRUCTURE

7.3.i. Introduction

So far in this chapter, we have assumed neutral variants. We previously assumed panmixis when examining the joint effects of selection and genetic drift, for the purpose of interpreting data on molecular evolution and variation (**Chapters 5** and **6**). We are now ready to put together models of selection, drift, and population structure, and to ask questions about how population structure affects variants under selection in several biologically important contexts. For instance, how are the properties of populations (such as mean fitness and the level of inbreeding depression, discussed in **Section 4.4**) affected by population structure? Can differences in the mean fitnesses of different local populations be an effective evolutionary force, driving a process of inter-deme selection?

In contrast to **Sections 2.2.iii** and **4.1**, we will mostly assume here that selection pressures are constant across the metapopulation, because we are interested in how to deal with the consequences of departures from panmixis, not in how selection leads to local adaptation (this topic is considered further in **Section 8.3.iv**). The problem of selection in spatially structured populations is comprehensively reviewed by Rousset (2004).

7.3.ii. Fixation probabilities in structured populations

Maruyama (1970c, 1974) and Nagylaki (1982) showed that, with semidominant or haploid selection, the fixation probability of a new mutation in a structured population consisting of a set of Wright–Fisher populations connected by conservative migration (**Section 7.2.ii.d**) is invariant with respect to migration patterns and rates. In this case, the product of the selection coefficient and the migration effective population size, N_{eS}, determines the fixation probability in the same way as the selection coefficient and effective population size in a single, panmictic population (**Equations B6.4.2a** and **B6.4.2b** in **Section 6.2.iii.a**). In this case, N_{eS} is the same as the effective size for a single population with the same number of breeding individuals as the metapopulation.

How applicable is this result when the relevant assumptions do not apply? Recent work suggests that an approximate diffusion equation can be derived for fairly general selection and migration models, using the fast time scale approximation for large deme numbers discussed in **Section 7.2.v** (see **Box 7.7***). This has been rigorously established only for semidominant and haploid selection for the island model (Cherry and Wakeley 2003), and for a haploid extinction/colonization model with a propagule number of 1 (Cherry 2003b), but simulations suggest that it applies more generally.

Box 7. 7* DRIFT AND SELECTION WITH A LARGE DEME NUMBER

The approximation for the effects of drift and selection in a metapopulation with a large deme number depends on the fact that, under the conditions discussed in **Section 7.2.v**, the time scale of drift within a deme is negligible compared with that for drift in the metapopulation as a whole. The latter is given by the reciprocal of the coalescence probability, P_{Cc}, for a pair of alleles sampled from two different demes, i.e., $2N_{eM}$, where N_{eM} is the effective size of the metapopulation (**Box 7.6***). We also assume that the selection coefficient s is of the same order as P_{Cc}, so that the time scale of the change under selection is very long.

The effects of drift and selection on the state of the metapopulation with respect to the frequencies of two alternative variants at a site, A_1 and A_2, are represented by the expected change in $\bar{q}$, the mean over demes of the frequency of A_2, and the variance of the change in $\bar{q}$ (Wakeley 2003; Roze and Rousset 2003). (As described in **Section 7.2.v.c**, any convenient method of weighting the contributions of individual demes to the mean over demes can be used in this context.) An argument similar to that used in **Section 5.1.i.c** implies that the variance of the change in $\bar{q}$ is equal to $\bar{q}(1 - \bar{q})/2N_{eM}$ (Whitlock and Barton 1997). We can thus make use of the standard diffusion equations for a panmictic population (**Sections 5.3.ii.c** and **6.2.iii.a**), with $\bar{q}$ replacing the single population frequency q.

We can now write the following equation:

$$\sum_i \omega_i q_i (1 - q_i) = \bar{q} - (\bar{q}^2 + V_q) = \bar{q}(1 - \bar{q})(1 - F_{ST}) \qquad (B7.7.1)$$

where q_i is the variant frequency for deme i and and ω_i is the weight given to deme i when taking the mean of the q_i (where $\Sigma_i \omega_i = 1$). The mean of the selection terms for each deme with semidominant selection, $sq_i(1 - q_i)/2$, is thus equal to $s\bar{q}(1 - \bar{q})(1 - F_{ST})/2$. The assumption of weak selection implies that F_{ST} is given by the same equation as in the neutral case (**Equation 7.16** of **Section 7.2.v.d**), to an accuracy of the order of s^2.

In order to remove any contribution to the expected change in $\bar{q}$ from migration or extinction, we should use the weighting $\omega_i = v_i$, where v_i is the probability that an ancestral allele is present in the ith deme (**Section 7.2.v.b**). Using the value of F_{ST} with this weighting of deme contributions, we then have $N_{eM} = N_{eS}/(1 - F_{ST})$ under the large deme number approximation of **Section 7.2.v.d**.

Only the ratio of the selection and variance terms appears in the diffusion equation formula for fixation probability (see **Box 6.4*** of **Section 6.2.iii.a**), so that the common factor $\bar{q}(1 - \bar{q})(1 - F_{ST})$ cancels, implying that the fixation probability is determined by $N_{eS}s$ and the initial value of $\bar{q}$. It is easily shown that this leads to the same formula for the fixation probability as **Equation B6.4.2b**. If deme sizes do not differ much, the exact choice of weight will have only a second-order effect on the fixation probability.

This result means that the effective population size that determines the mean within-deme nucleotide site diversity under the infinite sites model can be used to determine fixation probabilities in general migration or extinction/colonization models with semidominant or haploid selection, at least as a first-order approximation. This is very useful, because it implies that spatial structure may not have much effect on the fixation probabilities of weakly selected mutations, so that the models used in **Chapter 6** to interpret data on molecular evolution apply even to highly subdivided populations. Tests of predictions of the effects of differences in species' effective population sizes on rates of sequence evolution for species (e.g., Woolfit and Bromham 2005) should therefore use estimates of N_e based on mean within-population diversities.

With dominance, however, population structure can cause important departures from the panmictic results for selection (Slatkin 1981; Barton 1993; Whitlock 2002, 2003; Cherry 2003a; Roze and Rousset 2003). In particular, the increased local frequencies of homozygotes caused by drift within demes enhance the effectiveness of selection, similar to the effect of inbreeding within a population (**Section 3.1.v.c**). Thus, even with conservative migration, fixation probabilities are reduced for recessive or partly recessive deleterious mutations, and increased for recessive or partly recessive advantageous mutations, relative to the value for a panmictic population with an effective size of N_{eT}, as

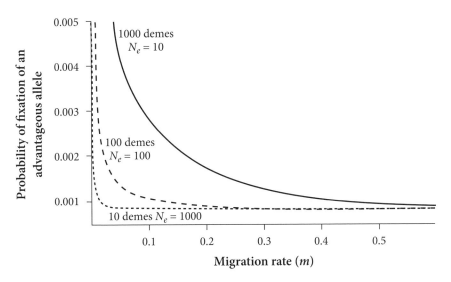

FIGURE 7.14 Effect of population structure on the probability of fixation of a recessive advantageous mutation (dominance coefficient $h = 0$) that arises as a single copy in a population with N adult individuals in each of the stated numbers of demes. In the example, the selection coefficient (s) was set equal to 0.01. [Adapted from Figure 4 of Roze and Rousset (2003).]

shown in **Figure 7.14**. The reverse is true for dominant or partially dominant mutations. The overall effect of population subdivision on the rate of DNA sequence evolution thus depends on both the level of dominance of new mutations and the extent to which advantageous or deleterious mutations contribute to sequence evolution. Few data are available to determine whether such effects occur in nature.

7.3.iii. Variant frequency distributions with selection and spatial structure

We next examine the distribution of variant frequencies at selected sites in a subdivided population, which is important for the tests of neutrality described in **Sections 6.4.iii** and **6.4.iv**.

7.3.iii.a. *The infinite-deme island model*
The problem of determining the distribution of variant frequencies was solved by Wright (1940a) for an island model with an infinitely large number of demes with equal effective population sizes. **Box 7.8*** derives a general formula for the probability density of allele frequency q in a given deme, with different demes being treated as independent draws from this distribution. It is assumed that selection is strong enough to hold allele frequencies close to their equilibrium values under mutation–selection balance or balancing selection.

Box 7.8* APPROXIMATIONS FOR THE DISTRIBUTION OF ALLELE FREQUENCIES UNDER SELECTION IN THE ISLAND MODEL

B7.8.i. General treatment

We will show that the probability density of allele frequency q for a single deme, $\phi(q)$, can be approximated by a beta distribution (**Section 5.3.iii.c**) when the expected change in q, $M_{\delta q}$, is a linear function of the departure of q from an equilibrium value, q^*. This is the case if q is always close to its equilibrium value, so that we can use a Taylor's series (**Appendix A1.ii.a**) to approximate $M_{\delta q}$ by $\kappa(q - q^*)$, where κ is the derivative of $M_{\delta q}$ with respect to q at q^* (**Section 2.1.ii.c**). If the equilibrium is stable, $\kappa < 0$. We can then write:

$$M_{\delta q} \approx \kappa(q - q^*) = \kappa(1 - q^*)q - \kappa q^*(1 - q)$$

This can be substituted into **Equation B5.9.1** in **Box 5.9** of **Section 5.3.iii.c** to give the exponential term in the expression for $\phi(q)$ as:

$$\exp -4N_e\kappa[(1 - q^*)\ln(1 - q) + q^*\ln q] \qquad (B7.8.1)$$

Evaluating this and substituting the result into **Equation B5.9.1**, we get:

$$\phi(q) \approx Cp^{-(4N_e\kappa p^*+1)}q^{-(4N_e\kappa q^*+1)} \qquad (B7.8.2)$$

This has the same form as **Equation 5.19**, and is thus also a beta distribution, with parameters $\alpha = -4N_e\kappa p^*$ and $\beta = -4N_e\kappa q^*$.

From the properties of this distribution, the mean of q is equal to $q^* = \beta/(\alpha + \beta)$, and the variance is $V_q = p^*q^*/(1 + \alpha + \beta) = p^*q^*/(1 - 4N_e\kappa)$.

B7.8.ii. Large deme number approximation

Under the large deme number approximation used in **Section 7.2.v**, the above result can be applied to the mean of q across demes, $\bar{q}$, using the argument in **Box 7.7*** of **Section 7.3.ii**, so that the distribution of $\bar{q}$ also approaches a beta distribution with mean q^*. With an indefinitely large number of demes, $\bar{q}$ can be approximated by q^*. This implies that, for a deme with frequency q, we can write the expected change in frequency due to migration as:

$$(1 - m)q + mq^* - q = -m(q - q^*) = m(1 - q)q^* - mq(1 - q^*)$$

This has the same form as the linearized selection term, so that κ can be replaced by $\kappa - m$ in **Equations B7.8.1** and **B7.8.2**, thus giving the final expression for the distribution of q among demes as:

$$\phi(q) \approx Cp^{4N_e(m-\kappa)(1-q^*)-1}q^{4N_e(m-\kappa)q^*-1} \qquad (B7.8.3)$$

Use of **Equation (B7.8.3)** and the formula for the variance of a beta distribution gives:

$$F_{ST} \approx \frac{1}{1 + 4N_e(m - \kappa)} \qquad (B7.8.4)$$

Since κ is negative, a comparison with the neutral formula (**Equation 7.6** of **Section 7.2.ii.b**) shows that F_{ST} is reduced below the neutral value; this is because selection tends to restore allele frequencies to their equilibrium, in opposition to the effects of drift.

B7.8.iii. Mutation, selection, and migration

Here, we assume that mutations occur from A_1 to A_2 at rate u, with selective disadvantage hs to A_2 in heterozygotes. In this case, use of **Equation 4.2** of **Section 4.2.ii.a** gives the approximate change in frequency of A_2 as $-(qhs - u)$. The equilibrium value of q is $q^* = u/hs$ (**Equation 4.3**), so that $(qhs - u) = hs(q - q^*)$, giving $\kappa = -hs$.

B7.8.iv. Heterozygote advantage with drift

Using the notation of **Section 2.1.ii.c**, the equilibrium frequency q^* is $s/(s + t)$. From **Equation B2.4.3** of **Section 2.1.ii.c**, we have $\kappa \approx -st/(s + t)$, assuming that selection is weak.

The box also gives explicit formulae for F_{ST} for the cases of mutation–selection balance and heterozygote advantage. In these cases, selection always reduces F_{ST} below the neutral value, a result that holds more widely than for this particular model of population structure (e.g., Malécot 1969, Section 3.3). Differences in F_{ST} among loci may therefore provide evidence for the action of selection on some of them, and methods for testing whether there is more variation among loci in F_{ST} than expected under neutrality have been devised (Lewontin and Krakauer 1973; Beaumont and Nichols 1996; Beaumont and

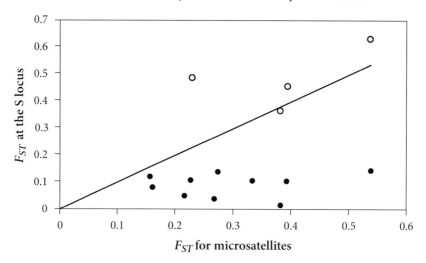

FIGURE 7.15 Comparison between F_{ST} estimated from microsatellite alleles in populations of the plant *Brassica insularis*, and from alleles at the self-incompatibility locus, which is under strong balancing selection (see text). Open circles are F_{ST} values estimated from microsatellite markers, whereas black ones are estimates from the self-incompatibility alleles. [Adapted from Figure 5 of Glémin et al. (2005).]

Balding 2004). In **Section 8.3.iv** we discuss how F_{ST} for presumptively neutral sites can be used to detect the effects of selection at linked sites.

There is evidence that some polymorphisms maintained by balancing selection have lower F_{ST} values than other loci. A study of self-incompatibility alleles (see **Section 2.2.i.b**) in a wild *Brassica* species found that alleles are only mildly differentiated between five populations in Corsica (Glémin et al. 2005). These populations are strongly ecologically isolated, and **Figure 7.15** shows that pairwise F_{ST} values for microsatellite markers are much higher than for the SI locus (mean pairwise $F_{ST} = 0.28$).

CASE STUDY

Lethal mutations in *Drosophila pseudoobscura*

A classic example of how this theory can be applied to real populations, to ask interesting questions about selection and migration, is the extraordinarily thorough investigation of the spatial distribution of lethal mutations on chromosome 3 of *Drosophila pseudoobscura* in southern California (Dobzhansky and Wright 1941; Wright et al. 1942). The data were collected using crosses with balancer chromosomes, as described in **Section 1.1.iii.b**. Two different localities separated by 200 miles, Death Valley and Mt. San Jacinto, were sampled, with sites separated by several miles within each of the two localities, and with several different nearby collecting stations at each site.

The results were summarized in terms of the frequencies of lethal-bearing chromosomes and the frequency of allelism between different lethal chromosomes found in a sample, i.e., the frequency with which a cross between carriers of two different lethal chromosomes yields a lethal combination (**Section 1.1.iii.b**). Chromosome 3 is homologous to the right arm of chromosome 2 of *D. melanogaster*. Accordingly, the mean frequency of lethal chromosomes was about half that for *D. melanogaster* second chromosomes (a mean of about 15.5% over the two localities). For lethal chromosomes collected at the same station, the mean allelism rate (with standard error) was 2.6% ± 0.5%, versus 0.41% ± 0.08% for chromosomes from different sites. Different stations from the same site had intermediate allelism rates of about 0.88% ± 0.20%. These results show that individual lethal mutations differ in their frequencies between the different locations, despite the fact that the flies can migrate over several kilometers in their lifetime (Dobzhansky and Wright 1947; Coyne et al. 1982).

As shown in **Box 7.9**, several parameters of interest can be estimated by combining the data on natural populations with an estimate of the rate at which lethal mutations arise on chromosome 3 (0.31% ± 0.04% per generation; Dobzhansky and Wright 1941). First, from **Equation B7.9.2b**, F_{ST} values among stations at San Jacinto and Death Valley are 2.39×10^{-3} and 5.01×10^{-3}, respectively. **Box 7.9** also outlines the reasons for inferring that lethals are removed mainly through their heterozygous effects on fitness (h): the mean h value was estimated to be 0.019 over the two localities. This inference was confirmed by later experiments that showed heterozygous fitness effects of lethals in *Drosophila* (**Section 4.2.ii.a**).

Second, the data allow N_e for a deme to be estimated, despite the possibility of migration. Using **Equation B7.8.4** and setting $\kappa = -h$, the estimated values of $N_e(m + h)$ for San Jacinto and Death Valley are 104 and 50, respectively. This is much higher than the $N_e m$ value of about 2 estimated from DNA sequence variation at the *Adh* locus for North American populations of this species (Schaeffer and Miller 1992). The term involving h is therefore the major contributor to $N_e(m + h)$. After subtracting the $N_e m$ value (estimated from the *Adh* data), and using the h estimate of 0.019 derived above, N_e can be estimated. The N_e values are 5358 for San Jacinto and 2400 for Death Valley. Further discussions of these results are given by Wright et al. (1942) and Wright (1978, pp. 157–170).

Box 7.9 BEHAVIOR OF LETHAL MUTATIONS IN SUBDIVIDED POPULATIONS

Let the frequency of lethal mutations for the ith gene on the chromosome be q_i. Using an approach similar to that of **Section 4.2.ii.c**, the expected number of lethal mutations per chromosome is $\Sigma_i q_i$, where the sum is taken over all genes capable of mutating to recessive lethal mutations. If the mean frequency of lethal mutations per gene is $\bar{q}$, and there are n genes capable of mutating to lethals per chromosomes, $\Sigma_i q_i = n\bar{q}$. If mutations in different genes are distributed independently of each other, the number of lethal mutations per chromosome follows a Poisson distribution with mean $n\bar{q}$, so that the observed frequency of lethal chromosomes is $P = 1 - \exp(-n\bar{q})$. This enables $n\bar{q}$ to be estimated from $-\ln(1 - P)$. For small P, $n\bar{q}$ is close to P.

The frequency with which two lethal chromosomes carry a mutation in gene i is equal to $p_i = (q_i/n\bar{q})^2$, because the frequency of lethals at locus i, relative to the overall frequency of lethals, is $q_i/n\bar{q}$. The frequency with which two lethal chromosomes carry a mutation in the same gene is thus $p_a = \Sigma_i p_i^2$. From the relation between p_i and q_i, we obtain:

$$p_a = \frac{\left(\bar{q}^2 + V_q\right)}{\left(n\bar{q}^2\right)} \tag{B7.9.1}$$

where V_q is the variance in q among loci.

V_q has two components—the variance in lethal allele frequencies due to differences in selection coefficients and mutation rates among different loci, and the variance due to the effect of genetic drift discussed in **Box 7.8***. For chromosomes collected from remote localities, with allelism rate p_{ar}, only the first component needs to be considered, and we can write $V_q = V_{qd}$, where d indicates the variance among loci caused by the variance in deterministic effects. For chromosomes from the same collecting station, the allelism rate is p_{as} and $V_q = V_{qd} + V_{qc}$, where c indicates the stochastic component. Substituting these into the versions of **Equation B7.9.1** for remote populations and the same station, we obtain:

$$V_{qc} = \left(p_{as} - p_{ar}\right)n\bar{q}^2 \tag{B7.9.2a}$$

For a structured population, V_{qc} can be equated to the variance in allele frequencies among demes due to restricted migration and finite population size. Since $\bar{q}$ is expected to be << 1, this gives:

$$F_{ST} = \left(p_{as} - p_{ar}\right)n\bar{q} \qquad \text{(B7.9.2b)}$$

As shown above, $n\bar{q}$ can be estimated from the frequency of lethal chromosomes, and the other two variables are obtainable from the relevant allelism frequencies, allowing F_{ST} for lethal chromosomes to be estimated.

Knowing the rate of mutation to lethals allows further conclusions to be drawn. Let this rate be U_l per chromosome per generation; for chromosome 3 of *D. pseudoobscura*, this is 0.0031. For allelism between remote populations, **Equation B7.9.1** gives:

$$n = (1 + V_{qd}/\bar{q}^2)/p_{ar} \qquad \text{(B7.9.3)}$$

A lower bound estimate of n is thus provided by $1/p_{ar}$. This is equal to $1/0.0035 = 285$ in the case of the California populations of *D. pseudoobscura* discussed in the main text. If lethals are completely recessive in their fitness effects and mating is random, **Equation 4.4** of **Section 4.2.ii.a** implies that $\bar{q} = \sqrt{U_l/n}$, i.e., $n\bar{q} = \sqrt{nU_l}$. For the *D. pseudoobscura* data, the predicted value of $n\bar{q}$ is thus greater than 0.939, more than five times the observed value. This implies that lethals are either eliminated mainly as a result of their heterozygous fitness effects (h, since $s = 1$), or that populations are partially inbred. The latter is very unlikely, however, given the breeding biology of this species. We can therefore infer that this discrepancy is due to selection against the heterozygous effects of lethals. On this basis, the expected value of $n\bar{q} = U_l/h$, so that $h = U_l/n\bar{q}$. This gives an average h for the two localities of about 0.019.

7.3.iii.b. *Quantitative traits and Q_{ST} values*

Variability in quantitative traits can also be partitioned into total and within-population components of genetic variance. The difference between these components, relative to the total genetic variance, provides a way of partitioning the variance, similar to F_{ST} (Prout and Barker 1993), which is often called Q_{ST} (Lynch et al. 1999). Under neutrality, F_{ST} and Q_{ST} should be similar, but if stabilizing selection is acting on the quantitative trait in the same way in each deme, $Q_{ST} < F_{ST}$. If different trait values are favored in different locations, we expect the reverse. Strictly speaking, this result assumes additive inheritance of the trait (**Section 3.3.ii.b**), and dominance effects are likely to lead to $Q_{ST} < F_{ST}$ for a subdivided populations, but the bias is small if many loci affect the trait (Goudet and Martin 2007; Santure and Wang 2009). Methods have been developed to test the statistical significance of differences between Q_{ST} and F_{ST} (Prout and Barker 1993; Lynch et al., 1999).

The amount of differentiation for molecular markers, such as SNPs or microsatellites, is commonly much lower than that estimated for phenotypic characters, suggesting that phenotypic differences between populations have often evolved by adaptation to local environments (**Section 4.1**), despite the flow of neutral or weakly selected variants between the populations. The first such test was used to detect selection in natural populations of *Drosophila buzzatii* (Prout and Barker 1993), and some other animal studies are reviewed by Lynch et al. (1999) and McKay and Latta (2002). Most uses of this test so far are in plants, where there is already much evidence for local adaptation from classic approaches, such as the reciprocal transplantation experiments reviewed in Linhart and Grant (1996) (see **Figure 7.16**). However, Whitlock (2008) has pointed out that the statistical tests used in these studies overlook the fact that F_{ST} for neutral loci has a probability distribution generated by the coalescent process in a structured population, and describes a method for overcoming this problem.

7.3.iii.c. *The sampling properties of alleles under the large deme number approximation*

The approach outlined in **Section 7.3.iii.a** above ignored the fact that the mean allele frequency for the metapopulation is not fixed, but itself evolves under

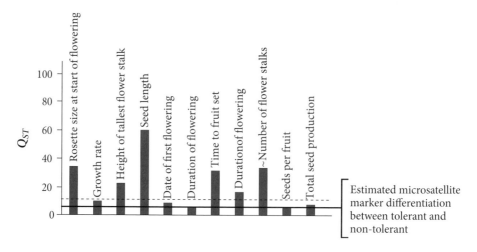

FIGURE 7.16 Comparison between F_{ST} estimated from microsatellite alleles in populations of the plant *Thlaspi caerulescens*, and Q_{ST} values for a number of phenotypic characters. The figure compares metal-tolerant and non-tolerant plants, showing that there is differentiation for several phenotypic characters, but not for the microsatellite markers (as can be seen from their mean and upper 95% confidence interval, indicated by the black line and the dashed line, respectively; the lower confidence interval coincides with the *x* axis). [Adapted from Figure 3 of Jimenez-Ambriz et al. (2007).]

selection, mutation, and drift. This complicates the analysis of the distribution of variant frequencies. In general, it is necessary to use simulations to determine the probability distribution of variant frequencies for a sample of alleles from a set of populations. A powerful approximate method for doing this for a pair of selected variants at a single site has recently been devised (Fearnhead 2006).

If the large deme number approximation is valid, however, approximate analytical results can be derived for the island model (Wakeley 2003). These results are particularly simple when each allele in the sample comes from a different deme. The large deme number approximation then implies that the distribution of the mean frequency of A_2 for the metapopulation, $\phi(\bar{q})$, is approximately the same as that for a panmictic population, with the effective population size for the metapopulation, N_{eM} (**Section 7.2.ii.b**), replacing N_e for a single population (Wakeley 2003). If k alleles are sampled, each from a different deme, these can be considered as k independent draws from a population with frequency $\bar{q}$. The distribution of variant frequencies discussed in **Section 6.3.ii.d** can then be used, with N_{eM} instead of N_e for a single population. These results suggest that tests for selection using frequencies of variants should use samples of this kind when there is evidence for significant effects of population structure.

7.3.iv. Effects of population structure on mean fitness and selection among demes

7.3.iv.a. *The effects of drift on population mean fitness*

The effect on population mean fitness of the random differentiation of allele frequencies caused by population structure is similar to the effects of inbreeding modeled in **Section 4.4.ii**. To the level of accuracy of the approximations used in **Box 7.8*** of **Section 7.3.iii.a.** we can simply replace f in **Equations 4.14–4.16** by F_{ST} to obtain the mean over all demes of the mean fitness of a deme. The reduction in this mean fitness caused by a set of loci with deleterious alleles whose fitness effects combine multiplicatively, relative to that for a panmictic population, is thus BF_{ST}, where B is given by **Equation 4.16.b**. This is true for any type of population structure, but simple expressions incorporating the effect of selection on F_{ST} have been derived only for the island model (**Box 7.8***). If selection is weak relative to migration, the neutral value of F_{ST} is a good approximation, but this will considerably overestimate F_{ST} for strongly deleterious variants. If there is a large reservoir of very weakly selected deleterious variants in natural populations, as suggested by the analyses of data on amino acid polymorphisms described in **Section 6.4.iv**, the neutral value may be a good approximation.

Populations formed by intercrossing local populations should thus experience an increase in mean fitness of approximately BF_{ST}, and this prediction is supported by more rigorous theoretical analyses (Whitlock 2002; Roze and Rousset 2003, 2004; Glémin 2005). Interpopulation hybrids do indeed often exhibit heterosis, which can be substantial (Escobar et al. 2008), although it is unclear whether it can be explained by this simple mechanism alone. One of the first such studies demonstrated heterosis for survival and fecundity in crosses between different *D. pseudoobscura* populations (Vetukhiv 1953, 1956). Heterosis was also found in a water flea (*Daphnia magna*) metapopulation living in small pools of water in Finland, in an experiment in which pools were emptied and then seeded with equal numbers of animals previously collected from the pool, and animals collected from a different pool. In most pools, hybrid genotypes (detected using markers) rapidly appeared and increased in frequency in successive samples, while the previously resident genotypes became rare (4 pools) or disappeared (11 pools); in only 2 pools were the initial genotypes still common at the end of the experiment (Ebert et al. 2002).

Interpopulation heterosis has also been observed in many experiments with plants (Van Treuren et al. 1993; Ouborg and Van Treuren 1994; Willi and Fischer 2005). In the plant *Silene latifolia*, for example, seed germination rates were significantly higher in interpopulation crosses than those generated on the same maternal plants by within-population outcrossing (Richards 2000).

7.3.iv.b. *Inter-deme selection*

Many biologists who have not been trained in evolutionary biology, and even some who have, assume that a trait will evolve if it enhances the fitness of the population or the species as a whole, without asking whether it can spread through a population by the type of selection that we have considered up to now (including kin selection, as discussed in **Section 3.1.v.d**). A famous example is the claim by V. C. Wynne-Edwards (1962) that animals have evolved mechanisms to reduce their reproductive rate, because this prevents them from exhausting their resources, thereby causing local populations to become extinct. Appeals to group- or species-level selection have also been made in relation to the evolution of sexual reproduction (**Section 10.2**) and the mating systems of sexual populations (**Section 9.2**).

If such arguments are to be made, we need to consider carefully how selection at higher levels than that of the individual can be effective, especially when it opposes selection involving fitness differences between individuals (if there is no such opposition, group selection might accelerate the rate of evolution, but adds nothing essentially new). For example, altruistic behavior increases the fitness of other members of the population, at the expense of the fitness of the individuals who exhibit the behavior (**Section 3.1.v.d**).

We therefore focus on how selection might act at the level of demes in a subdivided population of the type we have been considering in this chapter. There are two ways in which such *inter-deme selection* can act. First, a deme that acquires a higher mean fitness can contribute more to the pool of migrants or propagules than demes with lower fitness, and so genotypes that contribute to this effect have a higher chance of spreading from deme to deme. This effect is often called *hard selection* and has been explicitly included in models of migration and selection by Whitlock (2002, 2003) and Roze and Rousset (2003, 2004). This contrasts with the assumption that we made when treating selection in a spatially structured population—that there is strict density-dependent population regulation, so that the mean fitness of a deme does not influence its contribution to the migrant pool. However, as pointed out by Nagylaki (1982), under the diffusion approximations used to develop explicit results on selection in subdivided populations, the products of migration rates and selection strength are second order, and hence negligible in the context of the assumptions used in the derivations. The results of **Sections 7.3.ii** and **7.3.iii** above are therefore valid for this case as well. Significant effects of hard selection require stronger selection than can be modeled by diffusion equations.

The second process is the differential extinction of demes as a function of their genetic composition, which is involved in most models in which group selection opposes individual selection. For such opposition to be effective, both types of processes require demes to diverge genetically, so that there are heritable differences in mean fitness among them. From the discussions above, this requires local N_e values and migration rates to be small enough that a variant that reduces fitness can rise to a high frequency within a deme, and hence affect the deme's contribution to the migrant pool, or its rate of extinction. This has been modeled in a number of different ways. Some examples are described in the following publications: Boorman and Levitt (1973), Levin and Kilmer (1974), and Gilpin (1975); for a general review, see Wade (1978).

But, as pointed out by Maynard Smith (1976a), there is a simple general rule that can be applied to all of these cases, which is analogous to the condition in epidemiology for the maintenance of a population of pathogens: the pathogen's net reproduction rate must exceed 1 (Anderson and May 1991). This means that a single pathogen individual that infects a host produces at least one descendant that successfully infects another host before the original host dies. We can classify demes as being "uninfected" by the selfish genotype (U), or as "infected" (I), in the sense of containing at least one copy of the variant conferring selfish behavior, which we assume to have a selective advantage within a deme. If the mean number M of successful emigrants from an I deme (from the time that it becomes infected to the time it goes extinct) exceeds 1, the frequency of I demes will increase within the metapopulation and the altruistic variant will disappear. By successful, we mean that the migrants contribute selfish progeny to a previously uninfected deme or establish a new I deme. Conversely, with $M < 1$, altruism will be maintained in the metapopulation, if not completely fixed.

While it is certainly possible for inter-deme selection to promote the evolution of traits that are beneficial to the group but harmful to the individual, the conditions on deme sizes and migration rates must be quite restrictive, as is confirmed by the specific models cited above. Restricted migration among demes means that members of the same deme are more closely related than individuals chosen randomly from the population. The evolution of altruism in a structured population can therefore be viewed as a form of kin selection, and the formalism of kin selection can be adopted for modeling purposes (Frank 1998; Rousset 2004, Chapter 7; West et al. 2006). The distinction between kin selection and group selection in this situation is, therefore, largely a semantic one (West et al. 2006; Wilson 2007; Wilson and Wilson 2007), but the kin selection perspective offers the considerable advantage of providing a set of powerful mathematical tools for predicting the outcome of selection (West et al. 2006, Wild et al. 2009).

It has been proposed that, contrary to what we have just said, group selection can operate when there is no isolation between demes, and is hence much more widespread and effective than this argument suggests. This is exemplified by the *trait-group* model (Wilson 1975, 1980), which postulates that groups of individuals are formed each generation by random sampling from the population, undergo behavioral interactions, and then mate randomly among demes in the same way as in Levene's model (**Section 2.2.iii**). Provided that the between-group variance in the frequency of an altruistic variant is greater than the value for a binomial distribution (**Appendix A2.v.a**), so that altruistic genotypes are associated with each other within groups, and hence receive benefits from each other, the altruistic variant can spread through the population. It is unclear, however, what process could plausibly generate such an excess variance, other than relatedness among the individuals within the group, in which case this process is not distinct from kin selection (Maynard Smith 1976a). If this is not the case, it is hard to envisage how more than one altruist could become present in a given deme when a variant is very rare in the metapopulation, in which case its altruism cannot benefit others with the same genotype (Charlesworth 1979a).

PROBLEMS

7.1. **i.** For the case of a biallelic site or locus, show that F_{ST} and G_{ST} of **Section 7.1.i** are equivalent. (Use the results in **Section 5.1.c.**)

ii. Four populations of *Arabidopsis thaliana* had frequencies 0.1, 0.15, 0.20, and 0.3 of a SNP variant. Calculate the values of F_{ST} and G_{ST}.

7.2 Simplify **Equations B7.2.1a** and **B7.2.1b** in **Box 7.2** of **Section 7.2.ii.a** by neglecting second-order terms in the probabilities of migration between demes. Treat T_{ii}

and T_{ij} ($i \neq j$) separately. Note that $m_{ii}^2 = (1 - \Sigma_{k \neq i} m_{ik})^2 \approx 1 - 2\Sigma_{k \neq i} m_{ik}$, and $m_{ik} m_{jl} \approx$ 0 for $i \neq k$ except for $l = j$, when $m_{ik} m_{jj} \approx m_{ik}$ (similarly, $m_{ii} m_{jl} \approx m_{jl}$).

7.3   Use the simplified version of **Equations B7.2.1a** and **B7.2.1b** derived in **Problem 7.2** to obtain **Equations 7.3a–7.3c** of **Section 7.2.ii.b** for the island model.

7.4.*   i. Show that the definition of the leading left eigenvector, $\mathbf{v}$, of the migration matrix, M, (**Box 7.3*** of **Section 7.2.ii.d**) implies that $\mathbf{v}$ is proportional to the equilibrium probability that an ancestor of a given allele was present in deme i at some time in the past.

    ii. Use this result together with **Equation B7.2.1a** to show that **Equation B7.3.1b** leads to **Equation B7.3.2**.

7.5.*   i. Use the results in **Box 7.3** of **Section 7.2.ii.d** to show that the mean allele frequency across demes is unchanged by migration, if the allele frequency for a deme is weighted by the corresponding component of the leading left eigenvector, $\mathbf{v}$, of the migration matrix, M.

    ii. Show that conservative migration (where the number of individuals in each deme is unchanged by migration events) implies that v_i is proportional to the size of deme i. (To do this, it is necessary to relate the components of the migration matrix M to the probabilities that individuals in deme j move to deme i.)

    iii. Use the results of **Problems 7.5.i** and **7.5.ii** to show that conservative migration implies constancy after migration of the mean allele frequency, weighting the contribution from each deme by its size relative to the total number of individuals in the metapopulation, N_T.

7.6* Use the properties of the standardized normal distribution (**Appendix A2.v.e**) to derive **Equations B.7.5.2** and **B7.5.4** of **Section 7.2.iii.a** for the neighborhood sizes in one- and two-dimensional habitats.

7.7* Determine the condition for F_{ST} in **Equation 7.11** of **Section 7.2.iv.b** to increase with the extinction rate, by examining its derivative with respect to e.

7.8* Use **Equations 7.15** and **7.16** of **Section 7.2.v.d**, together with **Equation B7.6.4** of **Section 7.2.v.c**, to obtain **Equation B7.3.2** of **Section 7.2.ii.d**.

7.9   Use **Equations 7.15** and **7.16** of **Section 7.2.v.d** to obtain **Equation 7.11** of **Section 7.2.iv.b** for the migrant-pool and propagule-pool models of extinction and re-colonization, when all demes are exchangeable with each other.

7.10 Use **Equation B7.6.4** of **Section 7.2.v.b**, and the results of **Problem 7.9**, to derive **Equations 7.13** and **7.14** of **Section 7.2.iv.b** for the total and within-deme coalescence times for the migrant-pool and propagule-pool models with large deme numbers.

CHAPTER EIGHT

Multiple Sites and Loci

CHAPTER SUMMARY

We first show how to detect and measure nonrandom associations between variants at different sites in the genome (linkage disequilibrium or LD). In outcrossing species, such associations are generally found only for relatively closely linked genomic sites, within a few hundred bases in organisms like *Drosophila* with a large effective population size, N_e. They extend over much greater distances in organisms like humans with smaller N_e, and in predominantly asexually reproducing or inbreeding species. These observations are consistent with the expectation of a rapid breakdown of LD between polymorphisms in randomly mating populations, at a rate determined by the recombination frequency, c, between the sites in question. Strong inbreeding leads to a low frequency of heterozygotes, which reduces the effectiveness of recombination. In this case, the relatively weak forces that produce LD can overcome the effects of recombination; these forces are completely unopposed in non-recombining, asexual populations.

One important way in which LD can arise is by genetic drift, and we show how to predict the equilibrium level of LD in a finite population. The expected LD between a pair of sites is inversely related to $N_e c$, broadly consistent with the observed differences in LD between species with different N_e values. Population structure and population growth also affect LD, as do hot and cold spots of recombination activity. In human populations, LD is often much higher than expected from estimates of $N_e c$, probably because of recent bottlenecks in population size.

LD can lead to neutral sites being affected by linked, selected variants. Both the amount and pattern of variability at the neutral sites are affected. Balancing selection causes high variability at sites near the target of selection, and an excess of common variants, consistent with observations on highly polymorphic genes such as the MHC loci. Conversely, the selective elimination of deleterious mutations reduces variability at linked sites (background selection), as does the spread of favorable mutations to fixation (selective sweeps). These processes probably explain the low variability and excess of rare variants seen in regions of the genome with low recombination, and in highly inbreeding species.

Another cause of LD is selection acting on two or more sites in the genome when there is epistasis (i.e., the joint effects on fitness of the selected variants are nonadditive). This effect is important for understanding the evolution and maintenance of recombination and sexual reproduction. However, high LD between a pair of polymorphic sites at equilibrium under epistatic selection requires strong selection relative to the recombination frequency, and is therefore likely to be important only for closely linked variants. If recombination between the polymorphisms is frequent, the rate of change in the population's mean fitness is approximated by the additive genetic variance in fitness, and the rate of change in the mean of a quantitative trait by the additive genetic covariance between trait value and fitness, justifying the approximations used in **Chapter 3** when treating selection on quantitative traits.

Epistatic fitness interactions between loci are also important in the evolution of isolated or partially isolated populations. Some types of epistatic selection can lead to multiple, alternative stable equilibria, which may cause partially isolated populations to come to differ in mean fitness. Wright proposed that this allows an alternative to within-population selection as a mode of adaptive evolution (the shifting balance theory). He proposed that random genetic drift in a small, local population moves the population to a new equilibrium with higher mean fitness, and the resulting good genotypes can then spread through the species, increasing its level of adaptation. It is still debated whether this process is important in evolution.

Isolated or partially isolated populations can also arrive at different multilocus equilibria, purely because different mutations become fixed independently in the different populations. New mutations are selected to produce high fitness on the genetic background of the population in which they arise, and may interact harmfully with genotypes in other populations. This leads to hybrid fitness loss when individuals from different populations are intercrossed with each other; this plays a major role in the evolution of reproductive isolation between species.

> In panmictic populations, two factors are mainly responsible for linkage disequilibrium or non-random association of genes between different loci. One is epistatic interaction in fitness. . . . Another is random frequency drift caused by sampling of gametes in a finite population.
>
> Motoo Kimura and Tomoko Ohta (1971, p. 105)

INTRODUCTION

Up to now, we have concentrated on thinking about variation and evolution in terms of the alternative states of an individual nucleotide site, or other types of genetic variants at a single locus in the genome. This is a very useful first approximation. It would be completely valid if there were no associations between variants at different sites, i.e., if the genetic constitutions of individuals at one position in the genome were independent of those at other positions. In practice, studies of molecular variation demonstrate that associations do exist, especially among variants at closely linked sites.

In addition, our treatment of quantitative trait variability and selection assumed that variants at different sites have independent effects on trait values or fitness, so that the effects of multiple loci on variation in quantitative characters or fitness could be described by adding (**Section 3.3.ii**) or multiplying (**Section 4.3.ii.c**) their individual effects. This is at best an approximation to reality, and there is abundant evidence for interactions between the phenotypic effects of variants at different sites in the genome.

For a fuller understanding of variation and evolution, we therefore need to answer three major, interrelated questions:

1. How do we describe associations and interactions between variants at different sites?
2. How can we include such effects in models of evolutionary processes?
3. When are they important for the interpretation of data on variation and evolution?

These questions are the subject of this chapter and are further explored in **Section 10.2**, where we discuss the evolutionary significance of sex and recombination. We show there that nonrandom associations between variants at different sites can have many important effects on observable features of the genome. Nevertheless, most of the results of earlier chapters concerning the effects of selection and other evolutionary factors are still true, at least as good approximations in cases when evolutionary forces are weak.

8.1. DESCRIBING THE STATE OF A POPULATION

8.1.i. Haplotype frequencies

Data from surveys of DNA sequence variation for a particular region of the genome yield sets of *haplotypes*. As described in **Section 1.2.v.a**, a haplotype is simply the set of states of the variable nucleotide sites in a region of interest (e.g., a chromosome or a gene) for a single haploid genome of an individual.

For a haploid species or genomic region, such as the mitochondrial genome, or the X and Y chromosomes in males in a species with male heterogamety, the haplotype of an individual is its genotype.

In outbred organisms, with distinct maternal and paternal contributions, it may be difficult to establish an individual's two haplotypes for an autosomal region and to describe haplotype frequencies in a sample from a population. This is because the PCR amplifications used in sequencing amplify both alleles, and we often cannot distinguish the alleles in which variants occur when the variants are detected by direct sequencing of PCR products. In species of *Drosophila* with balancer chromosomes, this problem can be overcome by extracting chromosomes and making them homozygous, using the procedure shown in **Figure 1.5** of **Section 1.1.iii.b**. In other cases, such as humans, we must either use information from families to infer the maternal and paternal haplotypes (**Figure 8.1**), or else employ statistical methods to infer haplotype frequencies in a sample (e.g., Excoffier and Slatkin 1995; Ding et al. 2006; Scheet and Stephens 2006). Alternatively, one can use the laborious procedure of making multiple PCR clones from each individual, with sets of primers overlapping across the region of interest. For the rest of this chapter, we will assume that haplotypes can reliably be established. A sample from the population can then be described in terms of the frequencies of the different haplotypes present. If the population is random mating, this provides a full description of its state.

With a large number of variable sites, however, the list of haplotypes becomes very large. As with the description of variability at single sites (**Sec-

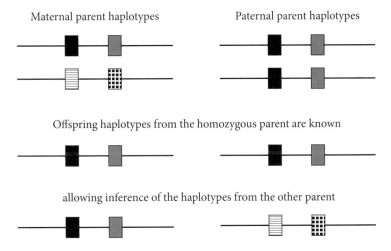

FIGURE 8.1 The diagram shows how families can be informative about the phase of variants in naturally occurring haplotypes, for the simple case when only two offspring haplotypes are expected because the sites are closely linked. Variants at the two sites are indicated by different symbols.

tions **1.2.i.b** and **1.2.iv.b**), simple summary statistics that describe *linkage dise-quilibrium* (LD) among pairs of polymorphic sites in samples or populations can help us understand the extent of nonrandom associations among variants.

8.1.ii. Measures of linkage disequilibrium

8.1.ii.a. *Pairs of sites*

Box 8.1 shows how the state of a population can be represented in the simplest possible case: two sites, each with two variants. The departure of the haplotype frequencies from random combinations of variants at the two sites is measured by the quantity D, given by **Equations B8.1.1** and **B8.1.2**. For a single pair of sites, this is easily calculated from the haplotype frequencies (**Problem 8.1**). Although D measures the departure from randomness, it also depends strongly on the variant frequencies at the two sites. If any of the frequencies are low, D is also low. For this reason, other measures derived from D, which are less sensitive to the variant frequencies, are often used, including the correlation coefficient r (**Box 8.2**). However, these methods do not completely remove the dependence of LD measures on variant frequencies (Hedrick 1987; Lewontin 1988).

Box 8.1 MEASURING NONRANDOM ASSOCIATIONS BETWEEN PAIRS OF SITES

Assume that we have two variants at each of two sites: A_1 and A_2 at the first site (A), and B_1 and B_2 at the second site (B). There are four possible haplotypes or gamete types that can be present in the population: A_1B_1, A_1B_2, A_2B_1, and A_2B_2, with frequencies $x_1, ..., x_4$, respectively. The frequencies of the variants at each site can be obtained by summing over the relevant gamete frequencies. For A_1, we have $p_A = x_1 + x_2$; for A_2, $q_A = x_3 + x_4$; for B_1, $p_B = x_1 + x_3$; for B_2, $q_B = x_2 + x_4$. The four haplotype frequencies sum to 1, so only three variables are needed to describe the state of the population (i.e., there are three *degrees of freedom*); similarly, the two variant frequencies at each site add up to 1, so that two of the degrees of freedom are used up by these. This leaves one to be accounted for, implying that a single variable describes the extent of nonrandom association between the two sites.

This is the *coefficient of linkage disequilibrium*, usually denoted by the symbol D (not to be confused with the measure of divergence between two sequences used in **Section 6.1.ii.a**). We can find an expression for D by the following argument. First, consider the haplotype frequencies expected if there is no departure from randomness. This means that the chance of a haplotype carrying a given variant at site B is independent of its state at site A (and the other way round). The frequencies of the haplotypes are then just the products of the frequencies of the alleles that they contain (**Appendix A2.ii**). If there are departures from randomness, the frequencies of the four haplotypes can be written as follows:

Haplotype	Frequency	Haplotype	Frequency
A_1B_1	$x_1 = p_A p_B + \delta_1$	A_1B_2	$x_2 = p_A q_B + \delta_2$
A_2B_1	$x_3 = q_A p_B + \delta_3$	A_2B_2	$x_4 = q_A q_B + \delta_4$

where δ_i measures the departure from the random frequency for haplotype i.

It is easy to show that only one variable is needed to measure the departure from randomness for all four haplotypes. We begin by considering A_1B_1 and A_1B_2. The sum of their frequencies is:

$$p_A = x_1 + x_2 = p_A p_B + \delta_1 + p_A q_B + \delta_2 = p_A(p_B + q_B) + \delta_1 + \delta_2$$

But $p_B + q_B = 1$, so that $p_A = p_A + \delta_1 + \delta_2$, implying that $\delta_1 = -\delta_2$.

A similar argument can be applied to each haplotype. Writing D for δ_1, it is easily shown that we have:

Haplotype	Frequency	
A_1B_1	$x_1 = p_A p_B + D$	(B8.1.1a)
A_1B_2	$x_2 = p_A q_B - D$	(B8.1.1b)
A_2B_1	$x_3 = q_A p_B - D$	(B8.1.1c)
A_2B_2	$x_4 = q_A q_B + D$	(B8.1.1d)

D is positive if the haplotypes A_1B_1 and A_2B_2 are in excess of their random frequencies, and negative if they are in deficiency.

A frequently used formula is:

$$D = x_1 x_4 - x_2 x_3 \qquad \text{(B8.1.2)}$$

This can be verified by substituting the expressions for the haplotype frequencies $x_1 = p_A p_B + D$, etc., into this expression, and then canceling suitable terms of the form $p_A p_B$ (**Problem 8.1**).

Box 8.2 OTHER MEASURES OF NONRANDOM ASSOCIATIONS BETWEEN SITES

B8.2.i. The D' statistic

One widely used measure for reducing the dependence of measures of nonrandom associations on variant frequencies is to use the D' statistic (Lewontin 1964). This is simply the ratio of D to its maximum absolute value (i.e., the value after changing a negative value to a positive one) that can be attained for the given variant frequencies at each site, i.e.,

$$D' = D/\left|D\right|_{\text{MAX}} \tag{B8.2.1}$$

If $D > 0$, then $\left|D\right|_{\text{MAX}}$ is the smaller of $p_A q_B$ and $q_A p_B$ (because **Equations B8.1.1b** and **B8.1.1c** imply that $D \leq p_A q_B$ and $q_A p_B$). If $D < 0$, then $\left|D\right|_{\text{MAX}}$ is the smaller of $p_A p_B$ and $q_A q_B$.

B8.2.ii. The correlation coefficient

Another method is to use the fact that D can be regarded as a measure of the covariance between the states of the variants at each site. We can define "indicator" variables X and Y for sites A and B, respectively. X is 1 if a haplotype carries A_1 and 0 if it carries A_2; Y is 1 for a haplotype carrying B_1 and 0 for a haplotype carrying B_2. The means of X and Y are simply p_A and p_B, respectively. The covariance, C, between X and Y is thus the mean of $(X - p_A)(Y - p_B)$ over the set of four haplotypes, weighting each value by the relevant haplotype frequency (**Appendix A2.iv**). Some simple algebra shows that $C = D$.

This suggests that we can use the correlation coefficient, r, between X and Y as a measure of the association between the two sites. This is because a correlation coefficient is bounded between -1 and 1, and is independent of the scale on which the variables are measured (**Appendix A2.iv**). The general definition of a correlation coefficient is the ratio of the covariance to the square root of the product of the variances. In our case, the variances of X and Y, V_X and V_Y, are the means of the squares of $(X - p_A)$ and $(Y - p_B)$, respectively. Some further algebra shows that:

$$V_X = p_A(1 - p_A) \text{ and } V_Y = p_B(1 - p_B) \qquad \text{(B8.2.2)}$$

It follows that we can define r for a pair of sites by the following expression:

$$r = \frac{D}{\sqrt{p_A q_A p_B q_B}} \qquad \text{(B8.2.3)}$$

It is easily verified that D' and r are far less sensitive to the frequencies of the variants at the two sites than D, the primary measure of LD (**Problem 8.2**). Of course, the three quantities all have the same sign. In many applications, the sign is of no interest, because it is arbitrary which variant is assigned subscript 1 or 2, and so either the squares or absolute values of these statistics are used.

B8.2.iii. Other measures

Other statistics that have been proposed are reviewed by Hedrick (1987). Extensions of D' to systems with more than two variants per site are described by Weir (1996, pp. 112–117) and Zapata (2000).

Statistics that describe pairwise associations are not unique to genetics; they are widely used in statistical analyses of pairs of discrete variates, each of which can have either of two alternative states, i.e., so-called 2×2 contingency tables. The statistical significance of departures from randomness for pairs of sites or loci can be assessed by the procedures developed for 2×2 tables (**Box 8.3**).

8.1.ii.b. *Multiple sites*

With the increasing availability of data on DNA sequence variation, it is now almost routine to have data sets with variants at numerous, closely linked nucleotide sites. Computer software packages (e.g., DnaSP: Rozas et al. 2003) can compute the LD statistics just described, commonly in terms of the values of r or D' for all pairs of sites, or the probabilities obtained from the 2×2 table tests described in **Box 8.3** (**Problem 8.2**).

LD measures can be generalized to multiple sites by the approach used in **Box 8.2.ii** (Bennett 1954; Slatkin 1973b). Alternative approaches are described by Smouse (1974) and Gorelick and Laublicher (2004), but are little used in practice.

Box 8.3 PROPERTIES OF 2 × 2 TABLES

As mentioned in the text, the data on haplotypes at two sites with two variants can be conveniently described by a 2×2 table, where the entries (k_{ij}) indicate the numbers of observations of combinations $A_i B_j$:

	B_1	B_2
A_1	k_{11}	k_{12}
A_2	k_{21}	k_{22}

The total number of observations is k (the sum of the entries in the table).

There is a standard formula for the chi-squared statistic that describes the departure from randomness in a 2×2 table. This gives the following relation:

$$\chi^2 = kr^2 \tag{B8.3.1}$$

where r is the correlation coefficient defined by **Equation B8.2.3**.

This has one degree of freedom and can be used to test for significant departures from the null hypothesis of random combinations (i.e., for significant LD), unless the expected cell numbers under the null hypothesis are very small (≤ 5). More generally, we can use Fisher's Exact Test, which computes the probabilities of each possible configuration of a 2×2 table, given fixed variant frequencies at each site (for further details, see Weir 1996, pp. 114–116).

8.1.iii. Empirical results on linkage disequilibrium

8.1.iii.a. *Observations based on classical genetics*

Before the introduction of molecular methods for studying genetic variability, few examples of LD in natural populations were known. One example involves the shell color and pattern polymorphisms of the snail *Cepaea nemoralis* (**Section 1.1.i; Plate 2**, front endpaper). The allele conferring a brown background color is usually associated with the allele at a separate but closely linked locus that confers an unbanded shell (Sheppard 1975, p. 90). Indirect evidence suggests that the polymorphisms for Batesian mimicry in *Papilio* butterflies, which appear to segregate as alleles at a single locus (**Section 1.1.i; Plate 1**, front endpaper), are in fact controlled by two or more genes. Different elements of the mimetic patterns are probably controlled by alleles at distinct loci that are in

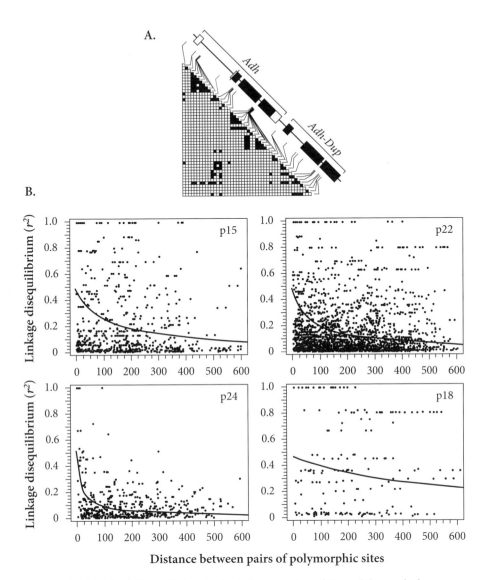

FIGURE 8.2 A Linkage disequilibrium in the *Adh* gene region of *Drosophila pseudoobscura*, showing that LD is infrequent except for polymorphic sites that are close together. The size of the region is less than 3.1 kilobases. Comparisons between neighboring sites are near the diagonal and between more distant sites are further from the diagonal. The black cells indicate statistically significant LD values ($P < 0.05$ in Fisher's exact test), and it can be seen that significant LD falls off rapidly over a few hundred base pairs. [Adapted from Figure 5 of Schaeffer and Miller (1993).] **B.** LD in *C. remanei*, an outcrossing *Caenorhabditis* species with a large effective population size. The figure shows the results for several loci, with about 600 bp of sequence from each locus. Pairwise r^2 values for all pairs of polymorphic sites are plotted against the distance in base pairs between each pair. The curves are the theoretical predictions under drift, mutation, and recombination (see **Section 8.2.iii.a**), using the estimated recombination parameter for each locus. [Adapted from Figure 3 of Cutter et al. (2006).]

strong LD and recombine so rarely that the entire complex of genes appears to be transmitted as a single locus with several alleles (Clarke and Sheppard 1960a; Turner 1977). This type of system is referred to as a *supergene*. We discuss supergenes in more detail in **Section 10.2.iii.b**.

8.1.iii.b. *Results from electrophoretic surveys*

After the introduction of gel electrophoresis of proteins as a tool for studying variation in natural populations (**Section 1.2.i**), many loci were studied in samples from populations of *Drosophila melanogaster,* and little or no LD was found among loci in natural populations, although LD was sometimes observed in laboratory cage populations (Langley et al. 1978). In contrast, LD is often observed between inversions and electrophoretic loci in *Drosophila* species (Krimbas and Powell 1992a), in predominantly asexually reproducing species like many bacteria (Whittam et al. 1983; Maynard Smith et al. 1993), and in plant populations with high levels of self-fertilization, such as wild oats, *Avena barbata* (Allard et al. 1972).

8.1.iii.c. *Results from surveys of DNA sequence variability*

DNA sequencing and SNP detection methods can now provide information on haplotypes in population samples for any degree of physical and genetic distance between variants. As a result, we now have very detailed information on LD in both natural and human populations, with the most extensive data coming from the systematic genome-wide survey of previously ascertained human SNP variants—the "HapMap project" (McVean et al. 2006). We briefly describe some patterns that have been observed in a representative set of organisms, ordered according to the rate at which LD falls off with distance between pairs of variable sites. In **Section 8.2**, we discuss the causes of these patterns, as well as those for the electrophoretic variants mentioned in **Section 8.1.iii.b** above.

First, outbreeding organisms with large effective population sizes, such as *Drosophila,* typically show high levels of LD for closely adjacent SNPs, but this falls off rapidly with distance in genomic regions with normal levels of crossing over, with little evidence for significant LD for sites separated by more than a few hundred base pairs (**Figure 8.2A**; Schaeffer and Miller 1993). The same is true in the outbreeding *Caenorhabditis* species, *C. remanei,* whose effective population size is estimated to be very large (**Figure 8.2B**; Cutter et al. 2006).

In outbreeding species with small effective population sizes, however, such as human populations, LD decays much more slowly with distance, and often extends over a hundred or more kilobases (**Figure 8.3A**). In humans, *haplotype blocks* are found; LD is virtually complete between all pairs of SNPs within blocks and often breaks down abruptly at their ends (Reich et al. 2001).

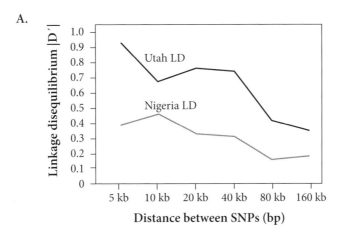

A.

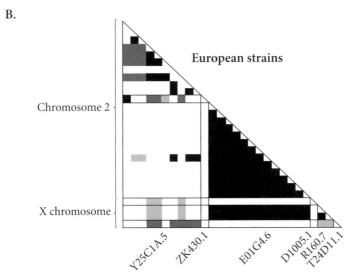

B.

FIGURE 8.3 **A.** High LD over long distances between sites in human populations. LD is higher in European and North American populations of humans than in samples from sub-Saharan Africa (the Utah sample consists only of individuals of European descent). [Adapted from Figure 1 of Reich et al. (2001).] **B.** High LD in strains from European natural populations of the self-fertilizing species, *C. elegans*, illustrating the differences from the outbreeding species in **Figure 8.2**. The figure shows LD between genes shown in **Figure 8.2** (although the names of the genes differ in the two species). Light shading indicates a significance level of $P < 0.05$ in Fisher's exact test, and dark shading is $P < 0.01$. Three of the genes shown here are on the X chromosome, as indicated in the figure, and three on chromosome 2, an autosome; results for the different loci are separated by lines. (The X-linked genes have low diversity, so that the regions of the LD plots for these loci are small.) LD decays on a scale of many megabases (tens of centiMorgans) and can even exist between genes on different chromosomes. [Adapted from Figure 2 of Cutter (2005).]

When reproduction is effectively asexual, as in genomic regions such as the mitochondrial genome, high levels of LD are seen among SNPs (Piganeau et al. 2004). The same is found in predominantly self-fertilizing species like *C. elegans* (Koch et al. 2000; Cutter 2005) (**Figure 8.3B**), in sharp contrast with the outcrossing species *C. remanei* (**Figure 8.2B**).

8.1.iii.d. *Mapping disease genes and quantitative trait loci*

Intense efforts are currently underway to survey human populations for associations between diseases and SNP markers distributed across the entire genome, with the aim of detecting previously unknown genes or regulatory sequences that affect susceptibility to diseases. These methods involve comparisons of the genotypes of markers among individuals diagnosed as having the disease (case individuals), and disease-free individuals from the populations (control individuals), thereby maximizing the chance of detecting variants at any sites that are associated with the disease (Wellcome Trust Case Control Consortium 2007; Easton et al. 2007; Donnelly 2008). The same approach was attempted 50 years ago using blood group markers (Sheppard 1959a), but now that hundreds of thousands of markers are available, it has enormously greater power. There are, however, formidable statistical problems, because it is important to correct for false positive results, due to the large numbers of statistical tests involved, and to avoid associations caused by immigration from populations with different variant and disease frequencies (see **Section 8.2.iii.d**). Nonetheless, these methods offer great power for characterizing the genetic causes of disease, compared with conventional mapping approaches (using just one or a few generations, followed in pedigrees), because they use information from thousands of generations of opportunities for recombination between markers and variants affecting disease susceptibility and other quantitative traits (Wellcome Trust Case Control Consortium 2007; Donnelly 2008).

Similarly, associations between molecular marker variants and quantitative trait values can be used to try to identify variants responsible for quantitative trait variation. The largest studies of this kind outside humans have been carried out in *Drosophila melanogaster* (Gruber et al. 2007; Flint and Mackay 2008).

8.2. CAUSES OF LINKAGE DISEQUILIBRIUM: NEUTRAL SITES

8.2.i. Introduction

In order to understand the patterns just described, it is necessary to investigate the theoretical population genetic processes that influence LD. **Table 8.1** summarizes the main processes that create linkage disequilibrium.

TABLE 8.1 Summary of processes creating linkage disequilibrium

Process	Chapter section
Genetic drift in haplotype frequencies	**8.2.iii.a**
Spatial structure (population subdivision)	**8.2.iii.d**
Balancing selection	**8.3.ii.a**
Selective sweeps	**8.3.v**
Selection at multiple sites with epistasis in the genome	**8.4.ii–iv**

In this section of the chapter, we initially ignore these causes of LD, and simply ask what happens to LD among neutral variants in an infinitely large population, first considered by Robbins (1918). We then go on to examine the effects of finite population size and population structure on LD for purely neutral variants. The effects of selection are studied in **Sections 8.3** and **8.4**.

8.2.ii. Neutral sites at two loci in an infinite population

8.2.ii.a. *Random mating*

For two polymorphic sites, the state of the population in a given generation is represented by the frequencies of the four haplotypes, or by the variant frequencies and LD parameter, D, as described in **Box 8.1** of **Section 8.1.ii.a**. In order to determine the state in the next generation, we must specify the recombination frequency for the pair of sites in question. We designate this by c. This includes the frequencies of two biologically distinct processes: reciprocal crossovers between the two sites and gene conversion events that affect only one of the two sites. For polymorphisms that are only a few hundred base pairs or so apart, gene conversion may be the predominant mode of recombination (Hilliker and Chovnick 1981). However, only the overall c value is needed to predict the population process for a pair of sites.

Box 8.4.i shows how D changes between generations with random mating; because the variant frequencies at each site remain constant, the same result applies to the other measures of LD in **Box 8.2** of **Section 8.1.ii.a**. Thus, **Equation B8.4.4** fully describes the dynamics of the population. D is reduced by a factor of $(1 - c)$ each generation. For loci on separate chromosomes, or far apart on the same chromosome (in eukaryotes), c is 0.5, so that D is halved each

Box 8.4 CHANGES IN LD WITH NEUTRALITY IN INFINITE POPULATIONS

B8.4.i. Randomly mating populations

The following argument is due to Malécot (1948, 1969, p. 17). Consider the haplotype A_1B_1. In a new generation, a gamete has probability $1 - c$ of being derived from a non-recombinant gamete of the parental generation; if the frequency of A_1B_1 in this generation is x_1, the net contribution of this class of gametes to the new frequency of A_1B_1, x_1', is $(1 - c)x_1$. There is a probability c that a gamete is a recombinant, with the A_1 and B_1 variants coming from two distinct gametes of the parental generation. In this case, because random mating is assumed, there is a probability $p_A p_B$ that these parental gametes carried A_1 and B_1. The net contribution to x_1' is $cp_A p_B$. This gives:

$$x_1' = (1 - c)\, x_1 + cp_A p_B \tag{B8.4.1}$$

Subtracting $p_A p_B$ from both sides, and noting that $D = x_1 - p_A p_B$ (**Equation B8.1.2** in **Section 8.1.ii.a**), we obtain:

$$x_1' = x_1 - cD \tag{B8.4.2a}$$

Similar arguments can be applied to the other haplotypes, yielding:

$$x_2' = x_2 + cD \tag{B8.4.2b}$$
$$x_3' = x_3 + cD \tag{B8.4.2c}$$
$$x_4' = x_4 - cD \tag{B8.4.2d}$$

We also obtain the simple recursion formula for D (**Problem 8.3.i**):

$$D_t = (1 - c)D_{t-1} \tag{B8.4.3}$$

If this is applied recursively to successive generations, the value of D in generation t can be written in terms of its initial value, D_0, as:

$$D_t = (1 - c)^t D_0 \tag{B8.4.4}$$

This shows that D declines approximately exponentially towards 0 (**Problem 8.3.ii**).

B8.4.ii. Inbreeding populations

The argument can be extended as follows to an inbreeding population with inbreeding coefficient f. An A_1B_1 gamete still has a probability $1 - c$

of being derived from a non-recombinant gamete of the parental generation, and c of being derived from a recombinant. As before, the first case contributes $(1 - c)x_1$ to the frequency of A_1B_1. In the second case, a gamete derives its A site from a gamete formed by a recombination event with the B site. There is a probability f that the two alleles at the B site are identical by descent. In this case, the probability that the recombinant gamete is A_1B_1 is the same as for a non-recombinant, i.e., x_1. There is a probability $1 - f$ that the alleles at the B site are non-i.b.d., in which case the probability that the recombinant is A_1B_1 is $p_A p_B$. The net contribution to the new frequency of A_1B_1 from all of these types of event is thus:

$$x_1' = (1 - c + cf)x_1 + c(1 - f)p_A p_B \qquad \text{(B8.4.5)}$$

Using the relation $D = x_1 - p_A p_B$, this gives:

$$D' = [1 - c(1 - f)]D \qquad \text{(B8.4.6)}$$

The recombination rate is therefore effectively reduced by a factor of $1 - f$. Alternative derivations of this result are given by Dye and Williams (1997) and Nordborg (1997).

generation. For close linkage, the decay of LD is close to an exponential process, and the time taken to reduce D to one-half of its initial value is approximately $0.69/c$ (**Problem 8.3.ii**).

If there are sex differences in recombination rates, as is often the case, the value of c used in the population genetic equations for dealing with autosomal sites should be the mean of the male and female values. For X-linked variants, the female value is weighted by 2/3 and the male value by 1/3, using the same argument as in **Section 5.2.ii.b** (the converse is true for Z-linked sites, with female heterogamety). Sex differences in recombination rates can be considerable. For example, there is sometimes no crossing over or gene conversion in the heterogametic sex (such as *Drosophila* males and Lepidopteran females; see White 1973); in human males, crossover rates are lower than in females, except near the tips of the chromosomes (Morton 1991; International Human Genome Sequencing Consortium 2001).

What level of LD is expected for genes sampled from the genome as a whole? Even for loci on the same chromosome, c is normally high enough that D for a pair of loci at an average distance apart will be reduced to 0 within a few tens of generations. To show this, we need to know the mean value of c for a pair of loci chosen randomly from a chromosome. The distribution of recombination frequencies among pairs of loci on the same chromosome of a typical map length of about 100 centiMorgans is approximately uniform between 0 and 0.5, with a mean of close to 0.25 (Morton 1955). We can also ask about genes sampled from the genome as a whole. For a species with many chromosomes, such as humans (with 23 pairs), the probability that a random pair of loci are on separate chromosomes is very high (approximately 0.95), so that the overall mean c for a random pair of loci is about 0.49.

The results just described imply that rather strong evolutionary forces are needed to overcome the breakdown of LD by recombination for most pairs of sites (for a more rigorous analysis, see **Section 8.4.ii.c** below). This is the reason why one can often use models that ignore LD, as was done earlier in this book. It also explains why variants, such as electrophoretic loci, which are thinly dispersed over the genomes of the organisms in which they have been studied, show little LD in randomly mating populations; very few pairs of electrophoretic loci are tightly linked (with recombination frequencies of 0.01 or lower), even in *Drosophila*. However, if c is very small for all pairs of sites, as for bacteria that are nearly completely asexual, weak forces can produce LD, and strong LD is indeed observed among electrophoretic loci in bacterial species such as *E. coli*, as described in **Section 8.1.iii.b**.

8.2.ii.b. *Inbreeding populations*

As we saw in **Section 1.3.ii.b** (**Equation 1.3**), an inbreeding coefficient of f is associated with a reduction in the frequency of heterozygotes by a factor of $1 - f$, compared with the value for a randomly mating population with the same variant frequencies. **Box 8.4.ii** shows that the ability of recombination to break down LD is also reduced by a factor of $1 - f$, reflecting the fact that crossing over or gene conversion in double homozygotes is ineffective (because the production of recombinant genotypes requires heterozygotes). We can therefore replace c by an "effective recombination frequency," $\tilde{c} = c(1 - f)$. In highly self-fertilizing populations, f is very close to 1 (**Equation 1.4** of **Section 1.3.ii.c**), so that effective recombination rates are always small, even for loci with $c = 0.5$. The observations of LD among distant markers in highly selfing species (**Sections 8.1.iii.b** and **8.1.iii.c** above) can thus be explained by low recombination that cannot rapidly break down LD.

8.2.iii. The effects of finite population size on linkage disequilibrium

For neutral or nearly neutral variants, LD in a single finite population is likely to be mostly produced by genetic drift (**Table 8.1** of **Section 8.2.ii** above), and we consider this process before examining any effects of selection.

8.2.iii.a. *Linkage disequilibrium and drift in a panmictic population*
With the model of two sites with two variants described in **Section 8.2.ii** above, we can think of the effect of genetic drift as though the four haplotype frequencies are the frequencies of four alleles at a single locus. Drift has no effect on mean frequencies, but only increases the variance in frequencies (**Section 5.1.i**), so that it cannot cause a trend towards negative or positive D, and so the expected (mean) value of D for neutral variants must tend to 0. One way of characterizing the effect of drift on the state of a population is therefore to determine the change in the expectation of D^2 under recombination and drift (Hill and Robertson 1968). Because the expectation of D is 0, this change in $E\{D^2\}$ is the same as the variance in D.

If mutations occur at both sites, as is biologically realistic, there will be some variability (without mutation, variant frequencies tend to 0 or 1; see **Section 5.1.i.c**). An expression for the equilibrium $E\{D^2\}$ at segregating sites has been derived under the infinite sites model of mutation, using a modification of the backward diffusion equation method introduced in the **Appendix** to **Chapter 6** (Ohta and Kimura 1971). In order to compare theoretical predictions with data on LD, we would also like to know the expected value of the square of the correlation coefficient r (**Box 8.2** of **Section 8.1.ii.a**), but it is difficult to calculate this explicitly. Instead, the quantity σ_d^2 is used; this is the ratio of $E\{D^2\}$ to one-quarter of the expectation of the product of the nucleotide site diversities at the two sites (this expectation is equivalent to the expectation of the square of the denominator of **Equation B8.2.3**; see Ohta and Kimura 1971). If sites with rare variants (frequencies < 10%) are excluded, $E\{r^2\}$ and σ_d^2 are very close to each other (Hudson and Kaplan 1985; McVean 2002).

The derivation of the equilibrium value of σ_d^2 for a randomly mating population is sketched in **Section 8A.2** of the **Appendix** to this chapter. The final expression is:

$$\sigma_d^2 = \frac{10 + \rho}{22 + 13\rho + \rho^2} \tag{8.1a}$$

where $\rho = 4N_e c$, the *scaled recombination rate*.

For large ρ, this expression gives the following approximation, which is often used in the literature:

$$\sigma_d^2 \approx \frac{1}{\rho} \tag{8.1b}$$

This shows that there is little LD if $N_e c > 1$, as would be expected. Conversely, for completely linked sites ($c = 0$), **Equation 8.1a** gives:

$$\sigma_d^2 = 0.45 \qquad (8.1c)$$

Even with complete linkage, the stochastic element in the process by which mutations are generated thus leads to a less than complete correlation between the two sites.

These results are very useful for interpreting data on neutral or nearly neutral molecular variation. First, they predict that LD depends jointly on both the recombination frequency and on the effective population size. This means that we expect much higher LD for a given recombination rate in small populations than in large ones. Thus, in humans, with their N_e of 10^4, we expect much more LD than in *Drosophila melanogaster* or outcrossing plants, with N_e values in the millions (**Section 5.2.iii.e**). Second, it predicts a strong decline of LD with the genetic distance between markers, even in humans. In *Drosophila* and humans, the average effective recombination frequency between adjacent nucleotides in a gene is about 10^{-8} (Andolfatto and Przeworski 2000; International Human Genome Sequencing Consortium 2001). With N_e of 10^6 and 10^4, respectively, we expect similar σ_d^2 values of 0.43 and 0.45 between adjacent nucleotide sites in the two species. For sites 1 kilobase apart, however, there is a large difference: the respective values drop to 0.02 versus 0.31, while for sites 10 kb apart, the value for *D. melanogaster* becomes tiny (0.0025), but it is still substantial (0.09) for humans.

These predictions are in fairly good agreement with what is seen in *D. melanogaster*, although there is a tendency for there to be more LD than predicted from recombination and N_e estimates (Andolfatto and Przeworski 2000). But in humans, as mentioned in **Section 8.1.iii.c**, strong LD often extends over 100 kb or more (Reich et al. 2001). This is a far greater distance than is predicted by these simple calculations. Possible reasons for this discrepancy are discussed in **Sections 8.2.iii.c** and **8.2.iii.d** below.

8.2.iii.b. *Linkage disequilibrium and the coalescent process*

An alternative way to think about LD in a finite population is in terms of the coalescent process (**Section 5.1.iii**). Up to now, we have considered this in terms of either a single nucleotide site or a non-recombining DNA sequence in which all sites share the same genealogy. However, coalescent theory can be applied to a pair of recombining sites with recombination rate c, and can be extended to sequences of nucleotides of any length within which recombination occurs. This makes it possible to estimate recombination rates from data on LD, using maximum likelihood approaches. In addition, genealogies with recombination can be simulated in situations where the assumption of a panmictic population of constant size is violated. This is a complex topic, with many recently developed methods for applying the theory to data, and we only outline the basic approach (detailed accounts are given by Hein et al. 2005, Chapter 5, and Wakeley 2008, Chapter 7).

Simulating the process of coalescence with recombination between a pair of closely linked sites is straightforward (Hudson 1983a; Hudson 1990). Assume that a set of k alleles is sampled from a genome region from a population. The problem is to describe the genealogical history of both sites. We assume that the recombination rate between the sites is sufficiently small that we can neglect the chance that more than one of the alleles (haplotypes) have been derived by a recombination event in the preceding generation, and that we can also neglect the chance of simultaneous recombination and coalescent events. If there are k distinct alleles in the initial sample, the probability of a recombination event in a given generation is then kc, because any one of the k alleles may have been derived by a recombination event. We saw in **Section 5.1.iii.c** that the probability of a coalescent event, such that two of the alleles are derived from a single ancestral allele, is $j_k = k(k-1)/4N_e$. These are independent events with constant probabilities, so the probability that a recombination event occurs first is:

$$P_r(k) = \frac{kc}{\left(kc + j_k\right)} = \frac{\rho}{\left(\rho + k - 1\right)} \tag{8.2}$$

where ρ is the scaled recombination rate, defined in **Section 8.2.iii.a** above. The probability that a coalescent event happens first is $1 - P_r(k)$.

In simulations, the time to this event in units of $2N_e$ generations can be determined by drawing an exponentially distributed random number with rate constant equal to the sum of the rates of coalescence and recombination, given by $k[\rho + (k - 1)]/2$. A uniform random number is then generated, and compared with $P_r(k)$ to determine which of these two possibilities occurs in a generation in which an event occurs. If a recombination event occurs, the number of alleles is increased from k to $k + 1$ (a new haplotype is created by the recombinants produced); if a coalescent event occurs, the number of alleles is decreased to $k - 1$ and the pair of alleles that coalesce is chosen at random from the set of k.

This process is repeated until only one allele remains; the most recent common ancestor of all the alleles in the sample has then been found. This method can be used to draw the history of the alleles, known as the *ancestral recombination graph* (**Figure 8.4**). By placing mutations onto the lineages in this graph (just as in the case of coalescence without recombination; see **Section 5.1.iii.e**), observable properties of samples can be simulated. A sequence of m nucleotides can be simulated in a similar way, with recombination events occurring at random points within the sequence (Hudson 1983a; Hudson 1990).

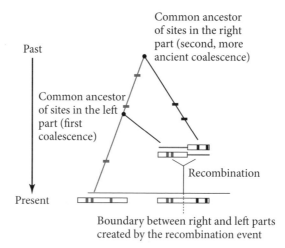

FIGURE 8.4 The ancestral recombination graph. The bottom of the figure shows two alleles with several different mutations at sites indicated by the vertical grey and black lines. The allele on the right was produced by a recombination event from two different alleles, creating three distinct ancestral alleles. The recombinant allele with the left-hand portion of the original allele coalesces with a common ancestor of the allele on the left (with whom it shares the leftmost pair of mutations). The resulting allele then coalesces with the recombinant allele carrying the right-hand portion of the original allele on the right.

To evaluate expected levels of LD for a pair of neutral sites using the coalescent approach, it is helpful to consider a pair of haplotypes in the sample and to think about the covariance or correlation between the lengths of the genealogies connecting each of the two sites to their most recent common ancestor. Without recombination, the lengths of the genealogies at the two sites must be identical; with free recombination, they are completely independent of each other. This suggests that the correlation between the lengths of the genealogies at the two sites is related to the extent of LD. McVean (2002) showed how **Equation 8.1.a** can be obtained from the properties of the genealogy of a sample of alleles (for further explanation, see Wakeley 2008, Section 7.2.4).

8.2.iii.c. *Estimating recombination rates*

The use of coalescent theory has led to the development of methods for estimating rates of recombination from data on haplotype frequencies and LD in natural and human populations (Hein et al. 2005, Chapter 6; Wakeley 2008, Chapter 7). Ideally, we should use the principle of maximum likelihood to find the set of values of ρ for all pairs of variable sites that gives the highest probability of producing the set of haplotypes observed in the sample. If a mutation rate estimate is available, N_e can be simultaneously estimated from diversity values and the ρ estimates can then be translated into estimates of recombination rates (e.g., Kuhner et al. 2000 and Fearnhead and Donnelly 2001). This is, however, computationally very demanding, and is difficult to apply to large datasets.

An alternative method is to calculate individual likelihood functions for each pair of segregating sites in the genomic region of interest, and to obtain an overall approximate or *composite* likelihood of the sample by summing the log likelihoods. Although this is only an approximation, because it ignores the non-independence of different pairs of sites, simulations show that it works well and can provide reliable recombination rate estimates. Several different methods for obtaining the likelihoods have been proposed (Hudson 2001; McVean et al. 2002; Padhukasahasram et al. 2006).

This approach has been applied to genome-wide surveys of SNPs in humans, to provide a very detailed picture of recombination rate variation across the genome ("hotspots" and "coldspots"; see Myers et al. 2005). Genetic mapping also detects such differences in various organisms, including yeast (Gerton et al. 2000) and other mammalian species (Kauppi et al. 2004). The borders of the human haplotype blocks described in **Section 8.1.iii.c** are often recombination hotspots (Myers et al. 2005), and similar evidence is starting to emerge in other species (Kim et al. 2007).

8.2.iii.d. *The effects of spatial structure*

Spatial structure is another important cause of LD. The coalescent approach is very useful in giving insights into the effects of population spatial structure on LD (Wakeley and Aliacar 2001; Wakeley and Lessard 2003). An exact treatment exists for the island model (Wakeley and Lessard 2003), but useful results can be obtained more easily using the large deme number approximation, dividing the evolutionary process into scattering and collecting phases, as we described in **Section 7.2.v**.

In a subdivided population, recombination occurs at an effective rate, $\tilde{c}$, analogous to that derived in **Section 8.2.ii.b** above for inbreeding populations. This rate can be determined using the large deme number approximation, in which an allele has a probability v_i of being present in deme i in a given generation (where v_i is the ith component of the leading left eigenvector of the migration matrix; see **Box 7.6*** of **Section 7.2.v.c**). If it experiences a recombination event while in this deme (probability c), there is a probability F_i that the two products will coalesce before one of them leaves the deme (**Equation 7.15** of **Section 7.2.v.d**), leaving the genealogy of the alleles in the sample unchanged, so that the recombination event can be disregarded. The net rate of recombination for this deme is thus $c(1 - F_i)$. Averaging over all d demes gives:

$$\tilde{c} = c\sum_{i=1}^{d} v_i \left(1 - F_i\right) \tag{8.3}$$

The rate of coalescence per generation during the collecting phase, P_{Cc}, is given by **Equation B7.6.4**. Recombination and coalescent events during this phase can be treated as competing exponential processes as above, with respective rates $\tilde{c}$ and $P_{Cc} = 1/2N_{eM}$, where N_{eM} is the effective size of the metapopulation as a whole, as defined in **Section 7.2.ii.b**. N_{eM} and $\tilde{c}$ thus can be used in **Equation 8.2** of **Section 8.2.iii.b** above, in place of N_e and c. Simulations of the ancestral recombination graph can then be carried out in the same way as for a panmictic population.

When each allele in the initial sample comes from a separate deme, the whole process follows these rules. The expected level of LD can then be predicted from **Equation 8.1.a**, replacing the scaled recombination parameter, ρ, by $\tilde{\rho} = 4N_{eM}\tilde{c}$, using the argument in **Box 7.6***. In the case of conservative migration and proportionality between the deme size and effective size of each deme (**Section 7.2.ii.d**), it can be shown that $\tilde{\rho}$ is equal to $4N_{eT}c$, where N_{eT} is the sum of the effective population sizes of each deme (**Problem 8.4***). Thus, there is no difference between the panmictic and metapopulation expectations in this case, because the effect of subdivision in inflating the metapopulation effective size over the panmictic value is exactly cancelled by its effect in reducing the effectiveness of recombination, by causing individual demes to become somewhat inbred.

The main situation in which there will be a large departure from panmictic expectation is when a sample contains multiple alleles contributed by one or more demes, rather than one allele per deme. In this case, the scattering phase causes variants at different sites to be more strongly associated than expected under panmixia, since they have a reduced opportunity of recombining before coalescence. This effect has been studied quantitatively for the island model by Ohta (1982), Wakeley and Lessard (2003), and De and Durrett (2007). This increase in LD may explain some of the discrepancies between observed levels of LD and predictions assuming panmixis that we mentioned in **Section 8.2.iii.a** above. **Figure 8.5** shows an example for human data, in which the relation between mean r^2 for a large collection of genome-wide SNPs is plotted against the distance between pairs of sites, together with the corresponding predicted values of σ_d^2 for a single deme sample in an island model.

High levels of LD can arise when two or more formerly isolated populations hybridize. If the populations have markedly different variant frequencies at many sites, the resulting hybrid population will exhibit LD. If hybrids rarely reproduce successfully, even variants at unlinked loci will be associated, because only newly formed hybrid individuals will carry immigrant alleles. Hybrid zones, where there is a narrow region of hybridization between two geographically disjunct populations (**Section 4.1.ii.d**), characteristically show LD for loci throughout the genome (e.g., Szymura and Barton 1991).

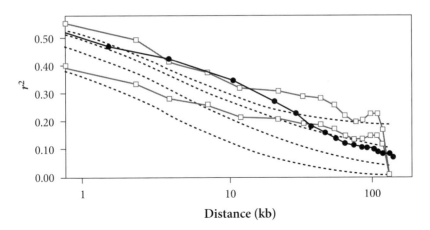

F I G U R E 8.5 Mean r^2 for a large collection of genome-wide human SNPs plotted against the distance between pairs of sites, together with the corresponding predicted values of σ_d^2 for a single deme sample in an island model. The black line with dots shows the estimated LD values for the Utah sample, similar to that shown in **Figure 8.3A**, but including the entire sample, whether of European descent or not. Because the variants scored were ascertained as SNPs in a small initial sample, they are biased towards higher frequency variants, biasing the LD estimates. The grey lines with open squares indicate upper and lower bounds that take these estimation problems into account. The predicted values, assuming $4N_e m = 4$ (**Section 7.2.ii.b**), are shown as dotted lines; four lines are shown, assuming different numbers of demes in the model (parameter $d = 1, 4, 16,$ and infinity, from bottom to top). [Adapted from Figure 2 of Wakeley and Lessard (2003).]

8.2.iii.e. *Other factors affecting linkage disequilibrium*

Population growth also causes departures from panmictic equilibrium expecta-tions. A sudden reduction in population size (e.g., because of a bottleneck asso-ciated with the foundation of a new population) will increase the magnitude of LD, because entire low-frequency haplotypes are preferentially lost. Con-versely, population growth tends to reduce LD, because it causes a star-shaped ancestry, with variants arising on the different independent lineages (Slatkin 1994; McVean 2002). This means that mutations at different sites are more likely to occur in separate lineages, and hence in different alleles in a sample taken from the population, than with constant population size and a standard coalescent process.

These effects are seen in haplotype data from natural populations. For exam-ple, European and North American populations of *Drosophila melanogaster* and *D. simulans* originated relatively recently from ancestral populations in Africa, almost certainly in association with human activities. Compared with African populations, these show low variability and high levels of LD (Glinka et al. 2003; Haddrill et al. 2005a). Similarly, non-African human populations have lower SNP diversity (Yu et al. 2002; Boyko et al. 2008) and much longer haplotype blocks than African populations (Reich et al. 2001), as expected if human populations went through bottlenecks as they spread out of Africa (see **Figure 8.3A**) in **Section 8.1.iii.c.**

8.3. NEUTRAL AND NEARLY NEUTRAL VARIANTS LINKED TO SELECTED SITES

8.3.i. Introduction

Investigations of DNA sequence variability have shown that diversity is not constant across the genome. For example, silent-site DNA sequence variability is elevated in the neighborhoods of the highly polymorphic major histocompatibility (MHC) loci of mammals (Shiina et al. 2006) and the self-incompatibility (SI) loci of plants (Richman et al. 1996; Kamau et al. 2007). This suggests that balancing selection that maintains variation over long evolutionary time periods has led to increased variability at closely linked sites that are not themselves the targets of selection (reviewed by Charlesworth 2006).

In *D. melanogaster*, crossing over is essentially absent in the gene-poor heterochromatin (consisting largely of repetitive DNA), which surrounds the centromeres (Smith et al. 2007). It is also highly suppressed in the euchromatin adjacent to the heterochromatin (Lindsley and Sandler 1977; Ashburner et al. 2005, p. 458), and near the telomeres, especially on the X chromosome (Lindsley and Sandler 1977; Ashburner et al. 2005, p. 449). The small "dot" chromosome also lacks detectable crossing over under normal conditions (Ashburner et al. 2005, p. 449), possibly because the whole chromosome is adjacent to the centromere. The reduction in crossing over near centromeres appears to be an effect of the centromere itself, and is intensified by deletions of the heterochromatin that bring euchromatin close to the centromere (Yamamoto and Miklos 1978). Reduced crossing over near centromeres is widespread, as is seen in budding yeast, with its simple centromere structure (Lambie and Roeder 1986; Gerton et al. 2000), and in mammalian genomes with their very complex centromeres (International Human Genome Sequencing Consortium 2001). In flowering plants, large parts of the chromosomes around the centromeres have very little recombination (Sherman and Stack 1995; Copenhaver et al. 1998, **Figure 8.6**).

These features of the genome provide important opportunities for studying the influence of rates of genetic recombination on variation and evolution. For example, if long-term balancing selection were common, we might expect to

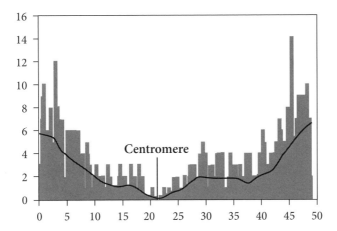

FIGURE 8.6 Low recombination frequencies in the centromere region of maize chromosome 1 (center of the *x* axis), compared with the chromosome arm regions. The estimates were made using the density of synaptonemal complex recombination nodules observed in electron microscopic images of meiotic cells (the *y* axis). The *x* axis shows 0.2 μm intervals along the image of the chromosome. The line superimposed over each distribution shows the general trend of the distribution of numbers of recombination nodules. [Adapted from Figure 6 of Anderson et al. (2003).]

see higher variability in regions of low recombination, because there would be more opportunity for close linkage between neutral sites and sites maintained polymorphic by selection. But the reverse pattern is found in *Drosophila melanogaster* (Begun and Aquadro 1992; Presgraves 2005; Shapiro et al. 2007), humans (Hellmann et al. 2003; Cai et al 2009), and some plant species (Roselius et al. 2005): silent-site variability correlates positively with the local rate of genetic recombination, and is extremely low in regions where there is little or no recombination (**Figure 8.7**). In addition, species or populations with high levels of inbreeding often exhibit very reduced levels of variation compared with outcrossing relatives (**Section 5.2.iii.a**).

In *Drosophila*, the higher diversity of regions with higher recombination cannot result from an association of recombination with higher mutation rates because there is no significant relation between recombination and silent-site divergence between related species (Begun and Aquadro 1992; Presgraves 2005), but in primates silent-site divergence increases weakly with the estimated local recombination rate (Hellmann et al. 2003; Spencer et al. 2006); nevertheless, a relation between recombination rate and diversity remains after taking this effect into account (Cai et al. 2009). As we describe later in this chapter, a plausible explanation for these patterns is that neutral variability is

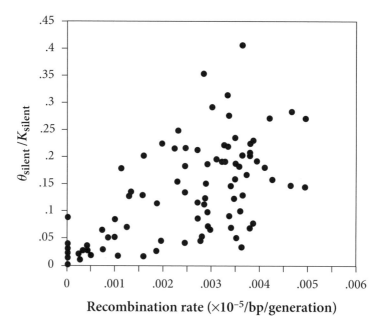

FIGURE 8.7 The relationship between the local rate of genetic recombination of a gene in the *D. melanogaster* genome and its silent-site diversity. Each dot is the result for a different gene. The diversity estimates for the genes (values of θ_w for silent sites) are standardized by dividing by the silent-site divergence from the closely related species, *D. simulans* (K_{silent} values), in order to avoid possible effects of difference in mutation rates. The y axis is thus θ_w/K_{silent}. [Adapted from Figure 1 of Presgraves (2005).]

reduced by selection occurring at closely linked sites (or, in inbreeding populations, sites that rarely recombine with selected ones).

Before explaining these effects, the following two sections show how the concepts and methods we have already used to study the effects of spatial structuring of populations (**Section 7.2**) can also be used to understand stable *genetic* structure, where different genotypes are maintained in the population, either by balancing selection (which we consider first) or by recurrent mutation to deleterious alleles. In either case, selection creates different compartments in the population, between which neutral variants at other sites can move by recombination. As in **Section 8.2** above, we use coalescent theory to study the effects of genetic structure on neutral variants. After showing that stable genetic structure due to selection can affect both neutral diversity at sites linked to the variants under selection and the frequency spectra of neutral variants, we consider transient changes in frequencies of neutral variants caused by selection on variants at linked sites.

8.3.ii. The effects of balancing selection

8.3.ii.a. *Effects of balancing selection on neutral variability*
Consider an autosomal site with two variants, A_1 and A_2, maintained by balancing selection in a randomly mating population with effective population size N_e. We assume that selection is sufficiently strong and constant in direction that intermediate frequencies of A_1 and A_2 are maintained over a much longer time period than $2N_e$, the mean time to coalescence for neutral variants. A neutral site recombines with the A site at rate c. By how much will haplotypes carrying A_1 and A_2 differ at the neutral site? The flow of neutral variants by recombination between the haplotypes carrying A_1 and A_2 is similar to conservative migration between demes (**Sections 7.2.ii.b–7.2.ii.d**) (Hudson and Kaplan 1988; Hudson 1990; Nordborg 1997; Wakeley 2008, Section 5.3). High equilibrium levels of differentiation between the A_1 and A_2 haplotypes would thus be predicted only at closely linked neutral sites (i.e., in the situation equivalent to low migration). The relevant theory is outlined in **Box 8.5.i** for the simple case of equal frequencies of A_1 and A_2.

Box 8.5 SITES LINKED TO VARIANTS MAINTAINED BY BALANCING SELECTION

B8.5.i. Mean coalescent times

Recombination between a neutral site and the selected site has an effect only in heterozygotes for A_1 and A_2 at the selected site. If these have the same frequency, an A_1 gamete has a probability of one-half of encountering A_2. We use the standard coalescent process assumptions, i.e., c and $1/2N_e$ are both sufficiently small that second-order terms can be neglected. If this also applies to the strength of selection, the probability that a neutral allele sampled from a gamete carrying A_1 came from an A_2 background in the previous generation is approximated by $c/2$. This is equivalent to the migration rate m between two populations, each of size $N_e/2$, under the island model with two demes (**Section 7.2.ii.b**).

Using **Equation 7.3a** for the island model with $d = 2$, the mean coalescent time, T_A, at the neutral site for two alleles with the same genotype at the selected site is simply $2N_e$, i.e., the same as if the population were not partitioned among the two selected variants. Using **Equation 7.3b**, the mean coalescence time for an A_1 allele and an A_2 allele (T_B) is increased by approximately $1/c$, or by $1/2N_e c$ relative to T_A. From **Equation 7.3c**, the mean coalescent time for a pair of alleles sampled randomly from the population is $T_T \approx T_A + 1/2c$. We can thus write the equivalent of **Equation 7.5a** as:

$$F_{AT} = \frac{(T_T - T_A)}{T_T} = \frac{1}{1 + \rho} \qquad (B8.5.1)$$

B8.5.ii. Associative overdominance

We have $T_T = (T_A + T_B)/2$, so that $T_B = 2T_T - T_A$. Because diversity values are proportional to mean coalescent times, the probability that a heterozygote at the neutral site occurs in an $A_1 A_2$ individual (frequency 1/2 and fitness 1) is $(2T_T - T_A)/2T_T$. The probability that it occurs in a homozygote (fitness $1 - s$, with symmetrical overdominance) is 1 minus this expression: $T_A/2T_T$. The expected fitness of heterozygotes at the neutral site is thus:

$$\frac{\left[2T_T - T_A + T_A(1 - s)\right]}{2T_T} = 1 - \frac{T_A s}{2T_T} = 1 - \frac{1}{2}\left(1 - F_{AT}\right)s \quad (B8.5.2)$$

From the results of **Section 4.3.iii**, the mean fitness of the population as a whole is $\bar{w} = 1 - s/2$. If s is sufficiently small that s^2 is negligible, the difference in fitness between heterozygotes at the neutral site and $\bar{w}$, expressed relative to $\bar{w}$, is approximated by $F_{AT}s/2$. The equivalent expression for homozygotes at the neutral locus is $-F_{AT}s/2$, so that the net apparent fitness advantage to heterozygotes is $F_{AT}s$. This is the same as the expression derived by Ohta and Kimura (1970), using the diffusion equation method described in **Section 8A.1** of the **Appendix** to this chapter.

Just as F_{ST} measures the level of differentiation between demes in a spatially structured population, we can measure genetic differentiation between the A_1 and A_2 haplotypes (or alleles), using the similar quantity F_{AT} defined by **Equation B8.5.1**. Another way of looking at this situation is in terms of linkage disequilibrium between variants at the neutral and selected sites. It turns out that F_{AT} is equivalent to σ_d^2 described in **Section 8.2.iii.a** above (Nei and Li 1973;

Charlesworth et al. 1997). F_{AT} approaches its maximum value of 1 for neutral sites close to the selected site, and declines to 0 for sites for sites that are so far from the target of selection that $N_e c > 1$. Similarly, there will be high LD among neutral variants at sites close to the target of selection (Charlesworth et al. 1997).

From this model, it is evident that balancing selection produces local peaks of diversity and LD in the genome, and that diversity and LD decline over a genetic distance on the order of $N_e c$. The phenomena of local peaks of variability can be seen in the classic cases of the mammalian MHC genes (Shiina et al. 2006, see **Figure 8.8**) and the self-incompatibility (SI) genes of plants (Richman et al. 1996; Kamau et al. 2007) described in **Section 2.2.i.b**. For sites that are extremely closely linked to a target of long-term balancing selection, the coalescent times can be much greater than the time during which the species has existed, and the alleles can differ by more than closely related species do at other loci (Innan and Nordborg 2003). Neutral variants that distinguish the selected alleles may also persist across the species boundaries. This is called *trans-specific polymorphism*, and is seen, for example, in the SI polymorphisms of plants and in genes closely linked to the S-loci (Charlesworth et al. 2006).

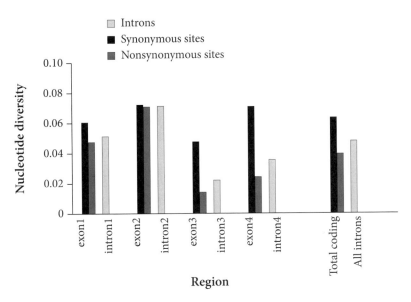

FIGURE 8.8 High diversity in different regions of DRB1, a human MHC gene. The vertical bars show that diversity is high at all classes of sites, compared with the mean value for silent sites across the genome (0.001). [Data on the DRB1 gene come from sequences analyzed in Raymond et al. (2005).]

These simple predictions suggest that polymorphisms maintained by long-term balancing selection could be discovered by scanning the genome for local peaks of silent-site diversity and/or polymorphisms shared between species. The stochasticity of the coalescent process, however, implies considerable random fluctuation around these expectations, so that peaks of diversity caused by balancing selection, or the trans-specific polymorphisms it causes, may be difficult to detect in practice (Nordborg 2000; Wiuf et al. 2004). (**Sections 8.3.ii.c** and **8.3.ii.d** below describe other tests used to detect balanced polymorphisms.) Such scans using the human and chimpanzee genomes have so far been largely negative, suggesting that there are rather few cases of long-term balancing selection (Asthana et al. 2005; Bubb et al. 2006), although a few convincing examples have been discovered (e.g., Baysal et al. 2007).

In addition, the assumption that variants are maintained indefinitely at the selected site is a crude one. In reality, the selected variants themselves arose at a definite time in the past. The origin of these variants by mutation can be included in the models to give more realistic predictions of the diversity expected surrounding the selected locus (Takahata and Satta 1998; Kamau et al. 2007). Incorporating fluctuations in the frequencies of the selected variants requires more complex calculations, which show that diversity is lower than predicted by the simple models, unless the product of N_e and the selection coefficient is > 1 (Barton and Etheridge 2004). Extensions to the case of balancing selection acting at several linked sites, such as the pollen and stigma genes involved in self-incompatibility, have been made by Navarro and Barton (2002).

8.3.ii.b. *Associative overdominance caused by heterozygote advantage*

The LD between variants at neutral sites closely linked to sites under balancing selection causes apparent fitness differences at the neutral sites. If there is heterozygote advantage at the selected site, heterozygotes for different *neutral* variants at closely linked polymorphic sites appear to have higher fitnesses than the homozygotes, because heterozygotes for neutral variants have a higher than average probability of being heterozygous at the selected locus (associative overdominance—see Sved 1968, 1972; Ohta and Kimura 1970). The magnitude of the apparent fitness differences can be calculated from the expression for F_{AT} derived in **Section 8.3.ii.a** above (**Equation B8.5.1**), using the fact that expected diversity under the infinite sites model is proportional to the mean coalescent times. The argument is outlined in **Box 8.5.ii**.

Equations B8.5.1 and **B8.5.2** show that the effect will be small except for sites so close to the target of selection that $\rho \ll 1$. Therefore, unless variants maintained by heterozygote advantage are widely distributed throughout the entire genome, neutral variants in a population will show apparent heterozygote advantage only in small populations. Associative overdominance can also be caused by deleterious mutations (discussed in **Section 8.3.iii.b** below), and again the effect is small except at sites very close to those under selection. This phenomenon is thus most likely to be important in laboratory populations or domesticated animals and plants. In such populations, one should therefore be very cautious about inferring selection from observations of apparent fitness differences among genotypes at a given site. Associative overdominance also explains why the rate of decline in the within-population variability in small populations is often slower than is predicted by neutral theory; if a locus is maintained heterozygous by selection, the segment of genome around it will also tend to remain heterozygous (Sved 1968, 1972; Gilligan et al. 2005; Barrière et al. 2009).

8.3.ii.c. *Testing for the signature of balancing selection: the HKA test*
The results in **Section 8.3.ii.a** deal with expectations (i.e., predicted average values); in practice, the randomness of the coalescent process means that samples from a finite population will often depart greatly from the expectations. It is therefore important to have ways of asking whether a set of data that suggest balancing selection could be generated under the null hypothesis of no selection. Various methods have been proposed.

One approach is the Hudson–Kreitman–Aguadé (HKA) test (Hudson et al. 1987). This tests whether a locus of interest has unexpectedly high diversity at putatively neutral silent or synonymous sites, by comparing it with sites of the same kind in a reference locus (or set of loci). In the test, the observed numbers of variable sites for a locus, or set of loci, in population samples from one or a pair of species are compared with expected numbers. The expected numbers are calculated on the null hypothesis of the standard neutral model, under which diversity and neutral divergence are both proportional to the mutation rate (this principle was used in the McDonald–Kreitman test for comparing nonsynonymous and synonymous polymorphism and divergence levels—see **Section 6.4.ii**). By including the sequence divergence from a related species, the test allows for possible mutation rate differences at the different loci, which would affect their diversity even if neither locus was experiencing balancing selection. The test sums the squared deviations of observed from expected numbers of variable sites, divided by the expected values, and the significance of the departure from the null model is tested using the estimated sampling variances given by coalescent theory with no recombination within loci (this is conservative, because this case has the maximum variance, as described in **Section 5.1.iii.d**). The statistic obtained in this way is distributed approximately as chi-squared (Hudson et al. 1987).

A difficulty in using this test to compare more than two loci is how to determine the significance of the deviation for a candidate outlier locus from the behavior of the reference loci. A maximum likelihood version of the HKA test has therefore been developed. This uses a formula for the probability of finding a given number of segregating sites at a locus in a sample of k alleles from a population, given a value of $4N_e u$ (Hudson 1990). The test compares the joint maximum log-likelihood of the number of segregating and divergent sites, on the null hypothesis of equal levels of variability for each locus surveyed (Wright and Charlesworth 2004). This can be compared with the maximum log-likelihood of an alternative hypothesis, e.g., that a candidate locus has an unusual diversity value. This was used to show that the human succinate dehydrogenase A gene (*SDHA*), which was suspected to be under balancing selection, indeed had unusually high silent diversity compared with four other genes (Baysal et al. 2007); tests that we describe in **Section 8.3.ii.d** below gave further support for balancing selection acting on this gene.

8.3.ii.d. *Testing for balancing selection: D_T, D_F, and haplotype tests*
Another consequence of the elevated variability caused by linkage to a balanced polymorphism is that the genealogy is distorted in a way that is similar to the distortion caused by population subdivision (**Section 7.2.v.a**). If we sample randomly from the population, the tips of the genealogy at linked neutral sites will be shorter relative to the total length of the tree than for a standard panmictic population, because of the expected rapid coalescence of those alleles in the sample that have the same allele at the selected locus. As in a subdivided population, we thus expect an excess of common variants over the standard neutral expectation, giving positive values of Tajima's D_T statistic and Fu and Li's D_F statistics, which were described in **Section 6.4.iii**. These methods can thus be adapted to test for such effects (Nordborg and Innan 2003). An example is provided by the *SDHA* gene of humans, mentioned in **Section 8.3.ii.c** above, for which D_T is close to 2, higher than for most other human genes (Baysal et al. 2007, **Figure 8.9A**). The large, positive value of D_T reflects the division of the variants into two, strongly differentiated major groups of haplotypes (**Figure 8.9B**).

We can also test for the elevated linkage disequilibrium expected in the neighborhood of a balanced polymorphism by asking whether the observed number of haplotypes in a sample is consistent with the expectation with random mating (Watterson 1978; Depaulis and Veuille 1998; Wall 1999; Depaulis et al. 2001). Balancing selection will usually lead to a smaller number of distinctively different haplotypes than expected under neutrality.

8.3.iii. Background selection

8.3.iii.a. *Reduced neutral variability caused by linked deleterious mutations*
Another important type of genetic structuring in populations is caused by deleterious alleles that are present because of recurrent mutation (Charlesworth et al. 1993). These reduce neutral diversity at linked sites, because the elimination

A.

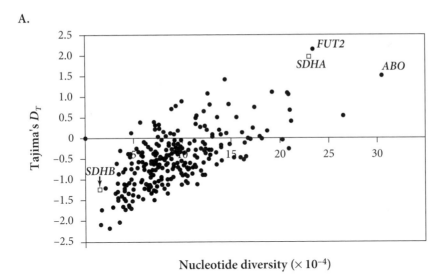

B.

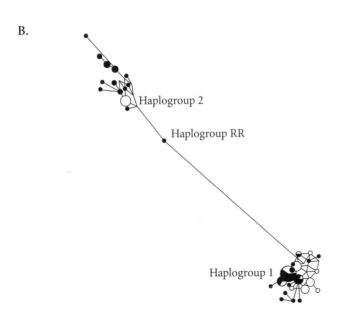

FIGURE 8.9 Sequence-based evidence for balancing selection acting on SDHA, a human succinate dehydrogenase gene. **A.** The values of Tajima's D_T plotted against nucleotide site diversities for a large collection of human genes. The values of both the nucleotide diversity and Tajima's D_T are unusually high for the SDHA gene and three other loci, including the ABO blood group gene. **B.** The two divergent haplotypes that cause these effects for SDHAs are called "haplogroups" in the figure. [Adapted from Figure 2 of Baysal et al. (2007).]

of a deleterious mutation carried on a particular chromosome also reduces the frequencies of any associated neutral or nearly neutral variants (**Figure 8.10**). This effect is known as *background selection*. The effect of deleterious mutations at an individual nucleotide site is, of course, very weak, because mutations are rare events and sites have the wild-type state most of the time. The cumulative effect on a genome region may, however, be substantial, especially if recombination is infrequent in a region containing many functional loci, because hundreds of thousands of sites within the region may be subject to deleterious mutations. A recent study suggests that there may even be a slight effect of background selection in *Drosophila* in genes in regions with normal levels of recombination, because of the abundance of slightly deleterious amino acid mutations in natural populations (Loewe and Charlesworth 2007).

Background selection can be modeled as follows. If selection against deleterious mutations is sufficiently strong compared with $1/N_e$, the equilibrium frequencies of mutations in a randomly mating population are given by standard deterministic theory, assuming no fitness interactions between mutations at different sites (**Section 4.2.ii**). For a given selected site, the equilibrium frequency, q, of the mutant variants at an autosomal site in a randomly mating population is then u/t, where $t = hs$ is the selective disadvantage of heterozygotes for the mutation and u is the mutation rate from the wild-type to the mutant state ($A_1 \rightarrow A_2$; see **Equation 4.3 in Section 4.2.ii.a**). In a finite population, we can still use u/t for the calculations (Nordborg et al. 1996), because the expected value of q is close to u/t if $N_e t > 1$ (**Section 6.3.ii.d**).

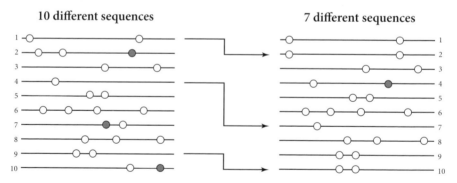

FIGURE 8.10 Background selection. The diagram shows that the elimination of strongly deleterious mutations (grey dots) causes the elimination of any associated neutral or nearly neutral variants (circles) at linked sites, and that the greater success of mutant-free haplotypes leads to their replacing haplotypes carrying deleterious mutations (haplotypes 2, 7, and 10), thus reducing the overall diversity (or, equivalently, lowering the effective population size). New deleterious mutations constantly arise in the population, as indicated by haplotype 4 on the right.

We assume that recombination between a neutral site and the selected site occurs at rate c. As before, gametes carrying the A_1 and A_2 alleles can be regarded as two classes; the rate of gene flow between them depends on the mutation and recombination rates. **Box 8.6.i*** derives the mean time to coalescence of a pair of alleles at the neutral site, when there is selection against deleterious mutations at the linked selected site.

Box 8.6* BACKGROUND SELECTION

B8.6.i. Effect of a single site subject to mutation and selection

We can treat the effect on a linked neutral site of a single site subject to selection and mutation as though the wild-type and mutant variants represent two classes, where the two-dimensional matrix B represents the flow of alleles between classes caused by both mutation and recombination (Nordborg 1997). This corresponds to the generalized migration matrix in **Box 7.6*** of **Section 7.2.v.c**. The probability that a neutral variant present in the wild-type class A_1 is derived from the mutant class A_2 in the previous generation is:

$$b_{12} = qc(1 - t) = uc(1 - t)/t \qquad \text{(B8.6.1a)}$$

Similarly, the probability that a neutral allele present in the class A_2 is derived from the class A_1 in the previous generation is given by:

$$b_{21} = pc(1 - t) + pu/q = p[c(1 - t) + t] \qquad \text{(B8.6.1b)}$$

because a fraction pu of the population at the start of a generation consists of newly arisen A_2 mutations. In order for the expected frequency of A_2 to correspond to the deterministic equilibrium value, we need $N_e t > 1$, so that b_{21} is necessarily greater than the rate of coalescence. This means that a fast time scale approximation, equivalent to that used in **Section 7.2.ii.e**, can be used.

The other coefficients of the B matrix are $b_{11} = 1 - b_{12}$ and $b_{22} = 1 - b_{21}$. Using the fact that the left eigenvector corresponding to the leading eigenvalue of 1 is given by $vB = B$, the probability that a neutral allele is in class 1 is $v_1 = b_{21}/(b_{12} + b_{21})$, and the probability that it is in class 2 is $v_2 = b_{12}/(b_{12} + b_{21})$. The probabilities of coalescence conditional on being in class 1 and class 2 are $P_{c1} = 1/2pN_e$ and $P_{c2} = 1/2qN_e$, respectively.

We can substitute these expressions into **Equation B7.4.1** in **Section 7.2.ii.e** for the rate of coalescence P_{Cc} under the fast-time-scale approximation. After some algebra (**Problem 8.5***), we obtain the following expression for the mean coalescence time, $T_T = 1/P_{Cc}$:

$$T_T \approx T_0 \left\{ 1 - \frac{q}{\left[1 + c(1 - t)/t\right]^2} \right\}$$

(B8.6.2)

where $T_0 = 2N_e$ is the mean coalescence time in the absence of background selection.

B8.6.ii. Effects of multiple selected sites

Equation B8.6.2 implies that a single selected site will have only a small background selection effect, because $q \approx u/t$ is expected to be very small (**Section 4.2.ii.a**). The effects of many sites can be predicted by the following heuristic argument (for alternative derivations, see Hudson and Kaplan 1995; Nordborg et al. 1996; Santiago and Caballero 1998). Suppose that the first selected site to be considered causes the mean coalescence time to be reduced to $T_0 f_1$, where f_1 is given by the bracketed term in **Equation B8.6.2**. If the genotypes at each site are distributed independently of each other, and the second selected site is associated with a reduction factor f_2, this factor will apply to a mean coalescent time for the neutral site under consideration of $T_0 f_1$, not T_0, and so on. In general, the effect of a set of m sites is to cause the mean coalescent time for the neutral site to be reduced from T_0 to:

$$T_T \approx T_0 \prod_{i=1}^{m} \left\{ 1 - \frac{u_i}{t_i \left[1 + r_i(1 - t_i)/t_i\right]^2} \right\} \approx T_0 \exp - \sum_{i=1}^{m} \frac{u_i}{t_i \left[1 + r_i(1 - t_i)/t_i\right]^2}$$

(B8.6.3)

where the subscript i indicates the parameter value for the ith selected site.

This theory can be extended to the effects on a neutral site of purifying selection at multiple, linked selected sites, assuming that variants at these selected sites are independently distributed in the population, i.e., there is no LD between them (**Box 8.6.ii***). **Equation B8.6.3** predicts the net effect on variability at a given neutral site due to background selection at an array of linked selected sites (this can be extended to inbreeding populations by using the appropriate changes in the recombination and selection parameters; see Nordborg 1997). With plausible parameter values for the mutation rate, selection coefficient, and recombination rate, this mechanism can account for much of the observed relationship between recombination rate and variability in *Drosophila melanogaster* (Hudson and Kaplan 1995; Charlesworth 1996). Background selection is unlikely to be the sole cause of this relationship (see **Section 8.3.v** below), and a full explanation of the relationship between recombination rates and levels of genetic diversity cannot yet be given.

In addition, the structured coalescent approach can be used to predict the extent to which genealogies with background selection depart from neutral expectation. If selection is strong, the departures are small, but when $N_e t < 1$, the effect can be large enough that a significant excess of rare variants can be detected using Tajima's D_T or Fu and Li's D_F statistics (Charlesworth et al. 1995; Gordo et al. 2002).

8.3.iii.b. *Associative overdominance due to linkage disequilibrium with deleterious mutations*

In **Section 8.3.ii.b**, we saw how associative overdominance (i.e., apparent heterozygote advantage at a site that is in reality neutral) can be caused by linkage to a locus with variants maintained by true heterozygote advantage. We now show how effects of this kind can be caused by partially recessive deleterious mutations, without any variants with heterozygote advantage. Consider a neutral site with two intermediate frequency variants, A_1 and A_2, linked to a gene or region of genome, B, in which deleterious mutations can occur. In a finite population, it is quite possible for a deleterious mutation in B (B_1, say) to arise in an A_1 gamete, and an independent one (B_2, at a different site in the same region) to arise in an A_2 gamete. If linkage is tight, as is likely for sites within the same gene, the associations between the neutral and selected variants will persist for some time (until recombination separates them). Homozygotes at the A site thus have a higher probability than $A_1 A_2$ heterozygotes of being homozygous for mutations at the other sites: most A_1/A_1 will be B_1/B_1 and most A_2/A_2 will be B_2/B_2, whereas A_1/A_2 will mostly be B_1/B_2. This creates an apparent fitness advantage to $A_1 A_2$ relative to the other two A genotypes.

A simple model of this process can be made, based on the results of **Section 7.3.iv.a** on mutation and selection in a subdivided population; a more exact treatment is given by Ohta (1971), using the diffusion equation method described in **Section 8A.1** of the **Appendix** to this chapter. We assume two neutral variants, A_1 and A_2, with nearly constant frequencies, p_A and q_A, dividing the population into two compartments. Deleterious mutations at a linked locus, B, have short persistence times in the population and are subject to selection and drift independently in the two compartments (Ohta 1971).

Recombination between the A and B loci can be treated like migration between demes, as in **Section 8.3.ii.a** above. The method of **Section 7.3.iv.a** can thus be used and shows that the effect on the B locus of the partitioning of the population into A_1 and A_2 compartments depends on F_{AT}, the analog of F_{ST} (**Section 8.3.ii.a**). The excess of the fitness of A_1A_2 over the mean fitness of the population is given by the product $B_B F_{AT}$, where B_B is the inbreeding load caused by deleterious mutations at the selected locus (see **Equation 4.15** of **Section 4.4.ii.a**). When selection is weak compared with the recombination frequency, F_{AT} can be approximated by the neutral value. With equal frequencies of A_1 and A_2, F_{AT} is given by **Equation B8.5.1**; because half of the population is A_1A_2, the excess fitness of A_1A_2 over the homozygotes' fitness is $2B_B F_{AT}$.

Unless the deleterious mutations at the selected locus are nearly fully recessive, B_B is on the order of the mutation rate for the gene, and so the effect is very small. The cumulative effect of mutations on the entire chromosome is on the order of $1/2N_e$ times B for the whole chromosome, multiplied by the length of the genetic map in Morgans (Ohta 1971; Sved 1972). This is still likely to be quite small, unless the population size is small or recombination is severely reduced, in which case the apparent selective advantage to heterozygotes may be considerable, because deleterious mutations make a substantial contribution to inbreeding depression (**Section 4.4.iv**). Deleterious mutations can, however, have a large effect on the rate of loss of heterozygosity at linked neutral sites in small laboratory populations (Wang and Hill 1999).

It may seem paradoxical that background selection both reduces the level of neutral diversity and yet induces apparent heterozygote advantage, but these are two very different aspects of the same process. The effect on diversity is seen across all neutral sites, whereas the apparent heterozygote advantage is observed for a pair of variants segregating at intermediate frequencies at a particular site; its cause is similar to that of the heterosis observed when two isolated populations (fixed for different recessive deleterious mutations) are hybridized (**Section 7.3.iv.a**).

8.3.iii.c. *Associative overdominance due to identity disequilibrium*
In populations that reproduce partly by random mating and partly by inbreeding, individuals will differ in their level of inbreeding. This causes *identity disequilibrium*, such that the probability that an individual is homozygous at one site is greater when they are homozygous at another site. The statistics describing the associations between sites with respect to their levels of heterozygosity are known as coefficients of identity disequilibrium (Cockerham and Weir 1968). For example, when a population is partially self-fertilizing, some progeny result from outcrosses and have Hardy–Weinberg proportions of genotypes, whereas selfed progenies will have excesses of homozygotes at all variable sites. The identity disequilibrium caused in such situations can occur without any LD,

and can even involve unlinked variants (Haldane 1949a; Cockerham and Weir 1968). In partially selfing populations, this effect can produce substantial associative overdominance, if homozygosity is associated with reduced fitness (**Figure 8.11**). Identity disequilibrium is probably the most common cause of correlations, often strong, that are observed in many natural populations between fitness and the number of marker loci for which individuals are heterozygous (David 1998; Tsitrone et al. 2001; see **Figure 8.12**). These associations are

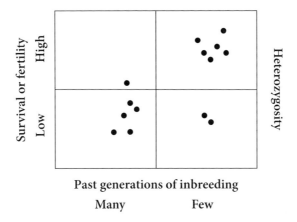

FIGURE 8.11 Heterozygosity–fitness correlations. The *y* axes show the frequencies of heterozygotes and expected fitnesses of individuals, divided by a horizontal line separating high from low values. The *x* axis shows the mean level of heterozygosity in the genome for individuals with a recent history of high inbreeding (on the left) and low inbreeding (on the right).

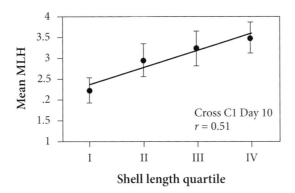

FIGURE 8.12 An example of an observed correlation between a component of fitness (i.e., size, on the *x* axis) and the mean multi-locus heterozygosity (MLH) of individuals at a set of microsatellite loci (*y* axis) in larvae of the oyster, *Ostrea edulis*. [Adapted from Figure 2 of Bierne et al. (2000).]

important for the evolution of the frequency of inbreeding versus outbreeding, as we will discuss in **Section 9.2**.

For linked loci, correlations of this kind will exist even if all individuals have the same level of inbreeding (Haldane 1949a), because inbreeding causes whole segments of chromosome to become homozygous simultaneously (rather than individual nucleotide sites). The distribution of such genome segments along chromosomes can be described by a complex mathematical model, called the "theory of junctions" (Fisher 1954; Stam 1980; MacCleod et al. 2005).

8.3.iv. The joint effects of spatial and genetic structure

The processes we have been discussing must usually act jointly to influence levels of neutral genetic variability within and between demes. For example, because the elimination of deleterious alleles by selection is a process that occurs largely within demes, background selection reduces expected coalescent times within demes (T_S) without altering the coalescence time for alleles sampled from different demes (Charlesworth et al. 1997; Nordborg 1997). From the expression $K_{ST} = 1 - \pi_S/\pi_T$ (**Equation 7.2** of **Section 7.1.i.c**), it is evident that reducing π_S without changing π_T increases K_{ST}. Indeed, almost any factor that reduces local N_e values increases K_{ST} (or any F_{ST}-like measure of population differentiation) for a given migration rate.

In contrast, balancing selection causes reduced differentiation between demes compared with neutral sites (see **Box 7.8*** of **Section 7.3.iii.a**). This is because an incoming migrant allele at the selected locus that is not already present in a deme has an increased chance of establishment in that deme, so that its effective migration rate is increased. Neutral sites linked to a target of balancing selection may thus also have reduced between-population differentiation compared to the genome as a whole (Schierup et al. 2000). For example, in the fungus *Schizophyllum commune*, the incompatibility loci are much less differentiated than a reference locus (James et al. 1999). Such differences in diversity patterns may allow balancing selection to be detected, particularly when polymorphism data are available from multiple loci in samples from multiple populations (Baer 1999; Akey et al. 2002; Akey 2009).

A different situation exists when different alleles at a locus are favored in different demes (**Section 4.1.i**). This reduces the effective rate of migration at neutral sites linked to the selected locus, because migrants into a population carrying a locally deleterious allele are selected against. For the simple case of a biallelic locus and two demes, with symmetrical migration and strong selection against locally maladapted alleles, the effective migration rate at a neutral site is approximately $m\tilde{c}/(s + \tilde{c})$, where m is the migration rate, s is the selection coefficient against an allele in the "wrong" deme, and $N_e\tilde{c}$ is the effective rate of recombination between the selected and neutral sites (Charlesworth et al.

1997). This can be substituted into **Equations 7.5a** and **7.5b** of **Section 7.2.ii.b** with $d = 2$, to find the expected equilibrium level of differentiation among demes at equilibrium under mutation, migration, and drift. A similar result applies when there is a selective disadvantage to heterozygotes at a selected locus, as may occur with interbreeding between partially reproductively isolated populations (Barton and Bengtsson 1986), or immediately after the spread of a variant that is favored in only a subset of the local populations of a metapopulation (Slatkin and Wiehe 1998), as discussed in **Section 8.3.v** below for the case of a panmictic population.

The reduced effective migration rate due to these effects leads to increased values of F_{ST} and other measures of population differentiation. The size of the increase depends on the relative values of the recombination rates and selection coefficients. Increased variability can therefore extend over a much wider genome region than with balancing selection, whose effect depends on $N_e \tilde{c}$ (Charlesworth et al. 1997). Via and West (2008) review evidence that quantitative trait loci causing adaptive differences between the alfalfa and red clover host races of the pea aphid are associated with elevated F_{ST} values for linked neutral markers, as expected on these models.

This raises the question of how to determine whether an unusually high or low value of F_{ST} is truly a product of selection or simply represents a chance effect of the coalescent process. As shown by Lewontin and Krakauer (1973), the distribution of F_{ST} values across independent loci under neutrality is approximately distributed as chi-squared (**Appendix A2.v.f**). This result seems to be fairly robust to the details of population structure (Beaumont and Nichols 1996; Beaumont and Balding 2004; Whitlock 2008), so that tests for outliers that indicate the effects of local selection (higher F_{ST} than expected) or balancing selection (lower F_{ST} than expected) can be carried out (e.g., Akey et al. 2002; Grahame et al. 2006; Teschke et al. 2008; Akey 2009; Coop et al. 2009).

8.3.v. Hitchhiking by favorable mutations (selective sweeps)

8.3.v.a. *Effects on levels of neutral variability at linked sites*
We now consider the effect on neutral variability of the spread of a selectively favorable mutation that arises at a nearby site in the genome (**Figure 8.13**). This was the first hitchhiking situation to be modeled, and the term "hitchhiking" was initially applied to such situations by Maynard Smith and Haigh (1974), who suggested that the effects of recurrent hitchhiking might explain the relative weak relation between levels of allozyme diversity and population size. Gillespie (2000) has recently made a similar proposal in connection with DNA sequence diversity, which he called the "genetic draft" hypothesis.

"Hitchhiking" is now used for any situation in which changes in genotype frequencies caused by selection affect the frequencies of neutral variants at

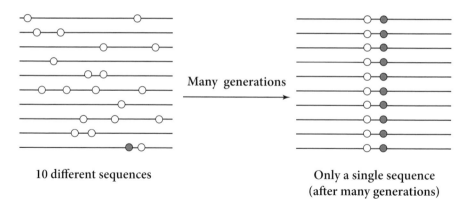

10 different sequences

**Only a single sequence
(after many generations)**

FIGURE 8.13 A selective sweep. As the advantageous mutation (grey dot) increases in frequency in a population, the diversity of neutral or nearly neutral variants (circles) is lost in a closely linked region.

other genetically linked sites in the genome. Background selection (**Section 8.3.iii.a**) thus represents a type of hitchhiking, where deleterious variants at low frequencies are constantly being eliminated by selection. The term *selective sweep* is now often used to denote a hitchhiking event caused by the spread of a favorable mutation in a population (Berry et al. 1991).

The easiest way to visualize this process is to consider a non-recombining genome segment. Imagine a new, selectively favorable mutation that arises in a haplotype with a given set of nucleotide variants at a set of variable sites. These variants will then spread to fixation along with the favored mutation and other variants in the region will disappear from the population (**Figure 8.13**). This process was first observed in experimental populations of the bacterium, *Escherichia coli*, which had been set up to be polymorphic for easily scored, neutral marker variants. Whenever a favorable variant originates in association with a given marker variant, the marker variant rises in frequency; the rise ceases only when mutations back to the other variant break the association down (**Figure 8.14**) This process is called *periodic selection* (Atwood et al. 1951), and it provides a means of estimating the rate at which favorable mutations become fixed in a non-recombining population (Dykhuizen 1990; Perfeito et al. 2007).

The algebraic description of a selective sweep, including the effect of recombination, is given in **Box 8.7*** for semidominant autosomal variants in a diploid, randomly mating population. As before, c represents the frequency of recombination between the neutral and selected sites, and s is the selection

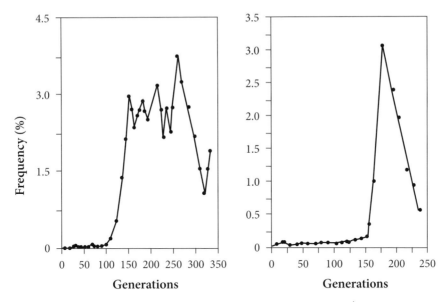

FIGURE 8.14 Periodic selection in bacteria. Selection events in response to drug treatments or host evolution cause selective sweeps reducing diversity. The example in the figure shows a chemostat experiment in which a genetic marker was followed. The two panels show replicate experiments that followed the frequencies of a marker (an allele for resistance to phage T5), in which the strain with the marker also carried a mutator gene. In the experiments, the competing genotypes of cells were introduced into a novel glucose-limited environment, allowing the mutator gene to cause beneficial mutations and adapt to this environment faster than nonmutator cells. Although the phage resistance allele was not under selection in the experiment, its frequency initially increased rapidly. After the initial increase due to advantageous mutations caused by the mutator, however, the marked lineages later decreased in frequency as deleterious mutations occurred (caused by the mutator), and beneficial mutations arose on the other genetic background in the populations of cells, or the marker reverted to wild-type. [Adapted from Figure 1 of Notley-McRobb et al. (2002).]

coefficient in favor of homozygotes for the favorable allele. The selection coefficient is assumed to be sufficiently large that $N_e s > 1$, so that the spread of the favorable mutation can be treated deterministically, except for an initial period in which it is vulnerable to chance loss (**Section 3.2.i**). The derivation in **Box 8.7*** follows Barton (2000).

The expected neutral diversity, π, at a nucleotide site, immediately after a selective sweep, is changed by the following amount:

$$\Delta\pi \approx -\pi_0(2N_e s)^{-4c/s} \tag{8.4}$$

where π_0 is the diversity before the sweep.

Box 8.7* HITCHHIKING BY A SELECTIVELY FAVORABLE MUTATION

At a given nucleotide site, A, there are two neutral variants, A_1 and A_2, with initial frequencies p_A and q_A, respectively. A nearby site, B, experiences a selectively favorable mutation from B_1 to B_2. We arbitrarily assume that this occurs on an A_2 background. At a given time, the frequencies of B_1 and B_2 are p_B and q_B. The selective advantage of B_2 in homozygotes is s, which is small, so that s^2 is negligible. The recombination frequency for A and B is c.

It is convenient to describe the system in terms of the frequencies of A_2 within B_1 and B_2, the two different backgrounds with respect to B. These conditional frequencies are denoted by q_{A1} and q_{A2}, respectively. It is easy to show that the coefficient of linkage disequilibrium for sites A and B (**Box 8.1** of **Section 8.1.ii.a**) is given by:

$$D = p_B q_B (q_{A2} - q_{A1})$$ (B8.7.1)

(See **Problem 8.6.i***.)

The initial values of q_{A1} and q_{A2} are q_{A0} and 1, respectively, where q_{A0} is the initial frequency of A_2, so that $q_{A2} - q_{A1}$ is initially equal to $1 - q_{A0} = p_{A0}$. Recombination changes the frequency of $A_2 B_2$ by $-cD$ (**Equation B.8.4.2a** in **Section 8.2.ii.a**), and selection has no effect on the frequency of A_2 within B_1 or B_2 gametes. The change in q_{A2} caused by recombination is thus equal to $-cD/q_B$. Using **Equation B8.7.1**, this is equal to $cp_B(q_{A1} - q_{A2})$. Similarly, q_{A1} changes by $cq_B(q_{A2} - q_{A1})$. It follows that $(q_{A2} - q_{A1})$ is changed by a factor of $(1 - c)$ each generation. At time t, we thus have $(q_{A2} - q_{A1}) \approx p_{A0} \exp(-ct)$, provided that c is small (see **Problem 8.3.ii**).

At any given time, the change in frequency of A_2 caused by hitchhiking with B_2 is equal to $\Delta q_B(q_{A2} - q_{A1}) \approx p_{A0} \Delta q_B \exp(-ct)$, where Δq_B is the change in frequency of B_2 due to selection. If Δq_B is small, we can approximate it by dq_B/dt (see **Box 3.2*** of **Section 3.1.iii.b**), and write the total change in q_A between time 0 and time t as Δq_A, which is given by:

$$\Delta q_A \approx p_{A0} \int_0^\infty \exp(-ct)\frac{dq_B}{dt}dt = p_{A0} \int_{q_{B0}}^1 \exp(-ct)dq_B$$ (B8.7.2)

The term $\exp(-ct)$ can be eliminated by the following trick. We know from **Equation 3.3** of **Section 3.1.iii.b** that in generation t we have q_B/p_B

$\approx (q_{B0}/p_{B0}) \exp(st/2)$, where the subscript 0 denotes the generation when the mutation occurred. Furthermore, $p_{B0} \approx 1$. Hence, $\exp(-st/2) \approx p_B q_{B0}/q_B$. In addition, $\exp(-ct) = \exp(-sct/s) = [\exp(-st/2)]^{2c/s}$. We thus have:

$$\Delta q_A \approx p_{A0} q_{B0}^{2c/s} \int_{q_{B0}}^{1} \left(\frac{p_B}{q_B}\right)^{2c/s} dq_B \qquad \text{(B8.7.3)}$$

The integral does not have a simple algebraic form. However, for $c/s < 1$, we can expand $p_B^{2c/s} = (1 - q_B)^{2c/s}$ as a binomial series (**Appendix A2.v.a**) in powers of q_B, and integrate its product with $(1/q_B)^{2c/s} = q_B^{-2c/s}$ term by term. For sufficiently small c/s, the resultant integral is close to 1 (**Problem 8.6.ii***). If the mutation to B_2 is present as a single copy, $q_{B0} = 1/2N$ and so the final result is:

$$\Delta q_A \approx p_{A0} \left(\frac{1}{2N}\right)^{2c/s} \qquad \text{(B8.7.4a)}$$

For greater accuracy, we must include the fact that most favorable mutations are lost by chance soon after their occurrence, and that their probability of survival is approximately sN_e/N (**Box 6.4*** of **Section 6.3.iii.a**), provided that $N_e s > 1$, where N_e is the effective population size. This initial process of random fluctuations in frequencies does not affect the expected frequency of B_2, taking both lost and surviving mutations into account, which remains at $1/2N$. The expected frequency of a surviving mutation is thus $(1/2N)/(sN_e/N) = 1/2N_e s$. The final expression is thus:

$$\Delta q_A \approx p_{A0}(2N_e s)^{-2c/s} \qquad \text{(B8.7.4b)}$$

This gives the expected change in the frequency of A_2 caused by hitchhiking, given that the favorable mutation arose in association with A_2, which has probability q_{A0}. A similar calculation can be performed for the case when the favorable mutation occurred in an A_1 gamete, which has probability q_{A0}. The change in frequency of A_1 is then given by **Equation B8.7.4b**, with p replacing q. The expected effect on diversity of a hitchhiking event is obtained by averaging $2pq$ over both possibilities, giving **Equation 8.4** of the main text.

As shown in **Figure 8.15A**, the expected effect of a selective sweep is very sensitive to the ratio c/s, and is small unless this is substantially less than 1. This

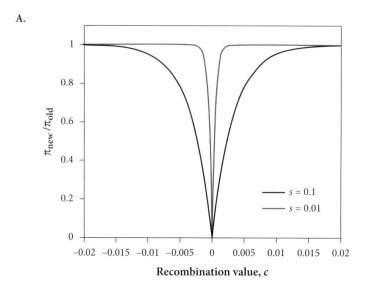

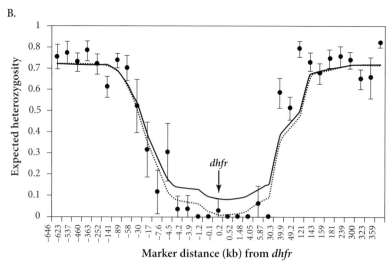

FIGURE 8.15 Effect of a selective sweep on diversity in a genome region. **A.** The expected neutral diversity immediately after a selective sweep (**Equation 8.4**). **B.** A selective sweep in malaria parasites due to pyrimethamine treatment in Southeast Asia. The y axis shows microsatellite diversity, measured by the expected heterozygosity (**Box 1.2** of **Section 1.2.i.b**), for loci in the region of the genome around the dihydrofolate reductase (*dhfr*) gene, which was previously known to be important for resistance. The x axis of the diagram indicates the markers on either side of this "candidate" gene (the distances represent the true physical distances on the chromosome), showing that *dhfr* is in the center of a low diversity region. [Adapted from Figure 1 of Nair et al. (2003).]

is because a rate of recombination that is high compared with the selection coefficient allows more opportunities for neutral variants to recombine away from the favorable variant with which they were initially associated. Similarly, for a given selection coefficient and recombination distance, the effect of a sweep is smaller the larger N_e, because it takes longer for a new mutation to reach a high frequency in a large population than in a smaller one, giving more opportunities for recombination to separate the neutral and selected variants. The figure also shows the wider window of reduced diversity with stronger selection, which takes less time to fix the beneficial allele than when selection is weak. The HKA test (**Section 8.3.ii.c** above) is useful for testing the significance of reduced variability associated with a putative selected sweep. **Figure 8.15B** shows an example of a selective sweep in the malaria parasite genome, caused by a strongly selected drug resistance mutation.

The selective sweep model can be extended to allow a steady rate of substitution of favorable variants at sites scattered randomly over a genomic region. An application of the diffusion equation method described in the **Appendix** to this chapter (Stephan et al. 1992; Stephan 1995) yields the following approximate formula for the expected equilibrium neutral nucleotide site diversity under the infinite sites model, relative to the value in an equilibrium panmictic population without any hitchhiking effects:

$$\frac{\pi}{\pi_0} \approx \frac{1}{c_n + \varepsilon\gamma\lambda_a} \tag{8.5}$$

where c_n is the mean recombination rate per nucleotide site in the genomic region in question, ε is a constant (approximately 0.15), $\gamma = 2N_e s$, and λ_a is the rate of substitution of selectively favorable mutations per nucleotide site. This equation was fitted to data on DNA sequence variability across the third chromosome of *D. melanogaster*, and fits with a value of $\varepsilon\gamma = 4.6 \times 10^{-8}$ (Stephan 1995). However, as noted in **Section 8.3.iii.a** above, a similarly good fit is generated by the background selection model. Andolfatto (2007) has recently shown that this equation provides a good fit to data on X-linked variability in *D. melanogaster* by using empirical estimates of the proportion of amino acid divergence between species due to positive selection from the methods discussed in **Section 6.4.ii.d**. His analysis suggests that neutral diversity is reduced by about 10% of the level expected in the absence of selective sweeps.

8.3.v.b. *Effects on variant frequency spectra*

With a selective sweep, once a new favorable mutation (and its associated haplotype) has reached fixation, variability will start to increase again through mutations producing new neutral variants. For sites very close to the selected site (or in a completely non-recombining region), where variability has been completely lost, the process resembles the decrease in variability after a severe population bottleneck. The new variants are initially much less frequent than expected at equilibrium, and hence the frequency spectrum of neutral variants (**Section 6.3.ii.b**) is skewed towards a high frequency of rare, derived variants.

This effect of a simple selective sweep, like that in **Figure 8.15**, can be modeled quantitatively using coalescent theory (Hudson and Kaplan 1988; Kaplan et al. 1989; Barton 2000). If we sample k alleles from the population at a time T_c generations after completion of the sweep, coalescent events of the lineages will follow the normal rules for a panmictic population until generation T_c in the past. At T_c generations ago, we can assume as a good approximation that all lineages simultaneously coalesce (i.e., they all descend from the common ancestor with the favorable mutation), because the time to fixation of the favorable variant will be very short compared with the standard neutral coalescent time, $2N_e$. If $T_c \ll 2N_e$, so that variability is still much lower than before the sweep, the genealogy will thus resemble a "star phylogeny" with nearly all k branches tracing back to a single ancestor derived from the pre-sweep population (**Figure 8.16**). New mutations must arise independently on these branches, and hence are represented as singleton variants in the sample. Such a genealogy predicts negative Tajima's D_T and Fu and Li's D_F statistics (Braverman et al. 1995; Simonsen et al. 1995). This potentially allows such events to be detected from data on sequences from a population, if a selective sweep occurred long enough

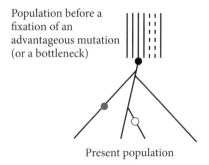

Population before a
fixation of an
advantageous mutation
(or a bottleneck)

Present population

F I G U R E 8.16 The similarity between a selective sweep and a bottleneck in population size, both of which produce a genealogy that resembles a "star phylogeny," because there is little time for coalescent events to occur between the time of sampling and the fixation event or bottleneck. The upper part of the figure shows a population before a fixation of an advantageous mutation with two different haplotypes in a genome region (indicated by black and dotted lines). An advantageous mutation (black dot) in a black haplotype (or a severe bottleneck) leads to a population with only this haplotype. Thus, all samples from this population have a common ancestor carrying the black haplotype, even if there has been enough time for new neutral mutations (grey dot, open circle) to occur. Unless a large amount of time has elapsed since the event, however, there will not have been enough time for these new variants to reach equilibrium; their frequencies will therefore tend to be lower than predicted at equilibrium, and so Tajima's D values will be negative.

ago for diversity in the region affected to have partially recovered, but not so long ago that variants have reached their neutral equilibrium frequencies.

A caveat concerning these results is that they assume that a unique selectively favorable mutation spreads to fixation. It is possible, however, that several different mutations of independent origin were present before the initiation of the sweep, or arise during the sweep—a so-called *soft sweep*. Such an event only mildly reduces the variability at linked sites, and only slightly affects the shape of the sequences' genealogy, because several copies of the mutation with independent genetic backgrounds can coexist at the end of the fixation process (Hermisson and Pennings 2005; Pennings and Hermisson 2006). The frequency of occurrence of this type of sweep is not yet known (Orr and Betancourt 2001).

When low diversity is found, it should be possible to discriminate between the effects of background selection and selective sweeps using the variant frequency spectra. Unfortunately, the predictions of the two models are similar for regions with low frequencies of recombination, even though, for a given reduction in diversity, sweeps will generally cause a greater excess of low-frequency variants compared with background selection (Braverman et al. 1995; Simonsen et al. 1995; Charlesworth et al. 1995; Gordo et al. 2002). There are thus few cases in which this approach has definitely ruled out one or the other process; one example is the dot chromosome (**Section 8.3.i**) of *Drosophila americana*, which lacks crossing over and has a low level of sequence variability. An analysis of patterns of variation in a gene on this chromosome gave results that were inconsistent with a selective sweep (Betancourt et al. 2009).

8.3.v.c. *Selective sweeps with recombination*

One might think that selective sweeps should also be detectable because they would produce a high frequency of the variant that is the target of selection. Thus, a high frequency of a nonsynonymous variant might suggest that selection had acted. However, selected variants generally spread rapidly and it is unlikely that one will be observed in the brief period before fixation. Real selective sweeps are not always as simple as the one in the figure, however, and sweeps with recombination can create a situation in which high variant frequencies can be informative.

Suppose that, before fixation of the haplotype in which the selected allele arose, a recombination event occurs in an ancestor of a sampled allele. The difference between this case and the one in the previous section is that not all neutral sites in the allele sampled will trace their ancestry from the haplotype carrying the selected allele. Instead, some variants will be present that come from other allelic descendants from the pre-sweep population. If non-recombinants are present among the k sampled alleles, the allele that is their ancestor will be a single random allele from the pre-sweep population (i.e., the haplotype in which the selected allele arose is modeled by randomly sampling an allele from the pre-sweep population; Barton 2000; Durrett and Schweinsberg 2004). When the advantageous variant has reached a high frequency, any variants at linked sites carried by this random haplotype will thus have much higher frequencies in post-sweep samples than in the ancestral population. If some recombination has occurred, a pattern of unexpectedly high variant frequencies across a genome region can therefore provide another signature by which a selective sweep may be detectable (Fay and Wu 2000; Zeng et al. 2006). This approach has the advantage that the prediction of the sweep model is quite different from that of background selection; on the other hand, it has little statistical power to detect a sweep unless it is fairly recent, as mutations that arose after the sweep contribute little to variability (Przeworski 2002).

For regions with high recombination frequencies, it can be difficult to distinguish between a recent selective sweep at a single locus versus a demographic event, such as an extreme population bottleneck or population expansion (Thornton et al. 2007). More sophisticated methods that incorporate both types of effect have been developed (Jensen et al. 2005; Nielsen et al. 2005; Voight et al. 2006; Thornton et al. 2007; Williamson et al. 2007). Use of these methods has provided evidence for the occurrence of numerous recent selective sweeps in humans, mostly confined to populations in one of the major geographical areas, suggesting the importance of local conditions for selective pressures (reviewed by Nielsen et al. 2007 and Akey 2009)

8.3.v.d. *Partial sweeps*

Another possible outcome of a hitchhiking event is that the new variant (A_2) at the selected locus does not go to fixation, but instead reaches a stable equilibrium frequency as a result of balancing selection. Immediately after such an event, sites closely linked to the A_2 site will have only the neutral variants present in the haplotype in which it arose, so that (similar to the example without recombination just outlined) A_2 haplotypes will be non-recombinant and will have little or no genetic diversity. Variants in the population will thus show strong LD with the A_1/A_2 site (**Figure 8.17**). This distinction between A_1 and A_2 will gradually be eroded by recombination. If recombination rates are known, the association between A_2 and linked variants provides a way of estimating the age of recently selected alleles (Slatkin 2002; Innan and Nordborg 2003; Toomajian et al. 2003). Coalescent-based tests for the extent of increased LD associated with a candidate selected variant can be used to test the statistical significance of the evidence for selection of this kind (Hudson et al. 1994; Sabeti et al. 2002; Schaffner et al. 2005; Voight et al. 2006).

Such situations have two possible long-term outcomes. First, the balanced polymorphism may persist indefinitely, with new mutations arising independently at closely linked sites on the A_1 and A_2 backgrounds, i.e., to long-term balancing selection (**Section 8.3.ii.a** above). The second alternative is that conditions may change, and one or other of the selected variants may disappear from

Observed numbers	Polymorphic site								
	−989	−780	−710	−703	−551	−543	−521	−491	(AT)x
1	G	A	T	C	T	C	C	A	8–6
1	G	A	T	T	T	C	C	A	8–6
4	G	A	T	T	T	C	C	A	8–5
4	G	A	T	T	T	T	C	C	8–4
1	G	A	T	C	T	C	C	C	7-7
1	G	A	T	T	T	C	C	A	7-7
1	G	A	T	C	T	T	C	A	7–6
22	C	A	G	T	T	C	T	A	8–4
4	C	A	T	T	T	C	C	A	9–4
2	C	A	T	C	T	C	C	A	8–5
3	C	A	T	T	T	C	C	A	8–5
1	C	A	G	T	T	C	T	A	8–4
3	C	T	T	T	T	T	C	C	8–4
1	C	T	T	T	T	C	C	C	8–4
2	C	A	G	T	T	C	C	A	7–6
1	C	A	G	T	T	C	C	A	7–7
2	C	A	T	C	T	C	C	A	7–7
1	C	A	T	C	T	C	C	C	7–7
2	C	A	T	C	T	T	C	C	7–7
1	C	A	T	T	T	C	C	A	7–7
4	C	A	T	C	T	T	C	C	7–6
1	C	A	G	T	T	C	T	A	7–6
3	C	A	T	T	T	T	C	C	6–8
1	C	A	T	T	T	T	C	C	6–8
1	C	A	T	C	T	T	C	C	5–9
1	G	A	T	C	C	C	C	A	9–5
1	G	A	T	C	C	C	C	A	8–5
7	G	A	T	T	C	C	C	A	7–7
3	G	A	T	T	C	C	C	A	6–9
1	C	A	T	C	C	C	C	A	8–5
3	C	A	T	T	C	C	C	A	7–7
6	C	A	T	C	C	C	C	A	7–7
1	C	A	G	T	C	C	T	A	7–6
2	C	A	T	T	C	C	C	A	6–9
1	C	A	T	C	C	C	C	A	6–7

FIGURE 8.17 A partial selective sweep at the human β-globin locus. The figure shows the variants at polymorphic sites in the region 5′ to the coding sequence (most of these variants are SNPs, but the one on the right-hand side is a microsatellite). The first column shows the number of each haplotype found in the sample, which was taken from a single African population. The Hb[S] allele (boxed) has evidently increased in frequency recently in this population, because it is found in many individuals as a single, uniform haplotype. This has not caused loss of other haplotypes, because the strong disadvantage of this allele in homozygotes (who suffer from sickle cell anemia) prevents its fixation (**Section 2.1.ii.c**). [Adapted from Table 3 of Currat et al. (2002).]

the population or be replaced by yet another variant. Currently, the data on variability in human populations (Bubb et al. 2006; Voight et al. 2006) and in *Drosophila* (Eanes 1999) suggest that a majority of recently established balanced polymorphisms display the signatures just described, with few long-term balanced polymorphisms persisting in the population. This predominance of the second type of outcome is consistent with data on diversity in genomic regions with low recombination. If balancing selection were common and persisted for long evolutionary times, these regions would be predicted to contain loci with high diversity. From the theory outlined in **Section 8.3.ii.a**, they should have higher diversity than genome regions with more recombination. As mentioned in **Section 8.3.i**, however, the data show that variability is generally reduced in these regions.

8.4. SELECTION AT MULTIPLE SITES IN THE GENOME

8.4.i. Introduction

We now consider the problem of analyzing the effects of selection at multiple sites in the genome, relaxing our previous assumption that the selective effects of variants at different sites are independent of each other, or that recombination is sufficiently powerful in relation to selection that LD between pairs of sites is negligible. We first examine the general properties of two-site systems with fitness interactions, then show how some useful general features emerge when recombination is strong relative to selection, and finally discuss the dynamics of selection on quantitative traits. These are some of the most complicated problems in the population genetics of infinite populations. Detailed reviews can be found in the books by Christiansen (2000), Bürger (2000), and Ewens (2004, Chapters 6 and 7), and we give only a brief survey. Readers may wish to skip forward to **Section 8.4.v–vi**, where we discuss Wright's shifting balance model of evolution, the fixation of chromosomal rearrangements, and hybrid fitness loss.

8.4.ii. Two-site systems

8.4.ii.a. *General description*

It is straightforward to include selection in the equations for changes in autosomal haplotype frequencies in a diploid, randomly mating population, assuming two variants at each of two linked sites, as in **Box 8.4** of **Section 8.2.ii.a**. We can assign a fitness w_{ij} to the diploid genotype formed by haplotypes i and j with frequencies x_i and x_j, respectively (see **Box 8.1** of **Section 8.1.ii.a**). It is usually assumed that the fitnesses of the two possible doubly heterozygous genotypes, A_1B_1/A_2B_2 and A_1B_2/A_2B_1, are the same, i.e., $w_{14} = w_{23}$. It is then possible to extend the model of selection of **Section 2.1.ii.a** (**Box 2.2**) to the two-site case, as shown in **Box 8.8**. Because the haplotype frequencies sum to 1, only three of the four members of the set of **Equations B8.8.1** need be used.

Box 8.8 SELECTION WITH TWO LINKED SITES

B8.8.i. General recursion equations

As in the case of a single site with two variants (**Section 2.1.ii.a**), we can assign marginal fitnesses to each haplotype, such that $w_{i\cdot} = \Sigma_j x_j w_{ij}$ ($i = 1, 4$). The mean fitness of the population can then be written as $\bar{w} = \Sigma_i x_i w_{i\cdot}$. The marginal fitnesses can be used to represent the effect of selection on the haplotype frequencies, just as in the single-site case. Recombination is effective only in the double heterozygotes, which have fitnesses $w_{14} = w_{23}$, so that the neutral expressions for the effects of recombination on haplotype frequencies (**Equations 8.4.2a–8.4.2d** of **Section 8.2.ii.a**) are modified by multiplying the terms in cD by $w_{14}/\bar{w}$ to correct for the fact that only individuals who have survived the effects of selection contribute to the next generation. This leads to the following equations:

$$x_i' = \frac{x_i w_{i\cdot} \pm cD w_{14}}{\bar{w}} \qquad (i = 1,..,4) \qquad \text{(B8.8.1)}$$

where haplotypes 1 and 4 (A_1B_1 and A_2B_2) are associated with $-cDw_{14}$ in the numerator, and haplotypes 2 and 3 (A_1B_2 and A_2B_1) are associated with $+cDw_{14}$.

Another way of representing the system is in terms of the changes in variant frequencies at the two sites and the change in D. As in **Box 8.1** of **Section 8.1.ii.a**, we have $p_A = x_1 + x_2$ and $p_B = x_1 + x_3$. **Equations B8.8.1** imply that:

$$p_A' = \frac{x_1 w_{1\cdot} + x_2 w_{2\cdot}}{\bar{w}} = \frac{p_A w_{1\cdot A}}{\bar{w}} \qquad \text{(B8.8.2a)}$$

and

$$p_B' = \frac{x_1 w_{1\cdot} + x_3 w_{3\cdot}}{\bar{w}} = \frac{p_B w_{1\cdot B}}{\bar{w}} \qquad \text{(B8.8.2b)}$$

where $w_{1\cdot A} = (x_1 w_{1\cdot} + x_2 w_{2\cdot})/p_A$ and $w_{1\cdot B} = (x_1 w_{1\cdot} + x_3 w_{3\cdot})/p_B$ are the marginal fitness of variants A_1 and B_1, respectively. Similar recursion relations and marginal fitnesses can be written down for A_2 and B_2 by simply changing the labels. These are identical in form to the equations for single segregating sites (**Section 2.1.ii.a**).

Using the definition of D in **Equation B8.1.2**, and substituting into **Equation B8.8.1**, we obtain (**Problem 8.8**):

$$D' = \frac{\left(x_1 x_4 w_{1\cdot} w_{4\cdot} - x_2 x_3 w_{2\cdot} w_{3\cdot}\right)}{\bar{w}^2} - \frac{cDw_{14}}{\bar{w}} \tag{B8.8.2c}$$

This shows that D does not necessarily decrease with time, as it does in the neutral case.

B8.8.ii. Equilibrium conditions

At equilibrium, $x_i' = x_i$. Using $*$ to denote equilibrium values, **Equations B8.8.1** imply that:

$$x_i^*\left(w_{i\cdot}^* - \bar{w}^*\right) = \pm cD^* w_{14} \quad (i = 1,..,4) \tag{B8.8.3}$$

where the sign of $cD^* w_{14}$ is positive for $i = 1$ or 4 and negative for $i = 2$ or 3.

The deviations of the marginal fitnesses from the mean fitness are therefore related to D^*; in particular, they take the same sign as D^* for haplotypes 1 and 4, and the opposite sign for haplotypes 2 and 3. By dividing both sides of each of these equations by x_i^*, changing the signs associated with haplotypes 2 and 3, and summing over all haplotypes, we obtain the following relation:

$$cD^* w_{14} = \frac{\bar{e}^*}{\sum_i \frac{1}{x_i}} \tag{B8.8.4}$$

where $\bar{e} = w_{1\cdot} + w_{4\cdot} - w_{2\cdot} - w_{3\cdot}$ is a measure of the strength of epistasis.

It can be shown that $\bar{e} = 0$ if the marginal fitness of each haplotype is the sum of the additive effects on fitness of the variants at sites A and B that it contains (**Problem 8.9**), so that D^* is then necessarily 0. This means that interactions between the additive fitness effects at each site are required for nonzero D to be maintained at equilibrium, because dominance deviations cannot contribute to the marginal values of haplotypes (**Section 3.3.ii.b**). This type of epistasis is known as *additive* × *additive* epistasis (Falconer and Mackay 1996, pp. 129–130; Bürger 2002, pp. 58–69; Ewens 2004, pp. 74–78). The converse clearly also applies, i.e., $\bar{e} = 0$ at equilibrium if $D^* = 0$, and there is then no contribution from additive × additive epistasis to the variance in fitness (see **Equation B8.11.5** in **Section 8.4.ii.d** for an explicit expression for the additive × additive component of the genetic variance).

The apparent simplicity of the equations in **Box 8.8** is deceptive, because we have three nonlinear equations for the haplotype frequencies, and informative expressions for equilibrium frequencies and dynamics can be obtained only by making simplifying assumptions (see **Section 8.4.ii.b** below). However, if there is no recombination ($c = 0$), the system is equivalent to a single locus with four alleles, because the terms in D in **Equations B8.8.1** disappear. For equilibrium, the marginal fitnesses of any haplotypes with nonzero frequencies must thus be equal. Expressions for the equilibrium frequencies for this case are known (Mandel 1959; Crow and Kimura 1970, p. 275).

In addition, with constant fitnesses and $c = 0$, stable equilibria for multi-allelic systems correspond to local maxima in mean fitness (**Section 2.ii.e**). The stable equilibria with small c values must be close to those with $c = 0$ (Karlin 1975), so that we can gain insight into the properties of multisite systems by examining the corresponding multi-allelic systems; this applies equally well to systems with any number of variable sites. Since nonzero c values change the haplotype frequencies away from those with $c = 0$, the mean fitness is always reduced below the $c = 0$ value, unless $D = 0$. Stable equilibria, therefore, correspond to fitness maxima only when either $c = 0$ or $D = 0$. When linkage disequilibrium is maintained at equilibrium, recombination thus reduces the mean fitness of the population (Lewontin 1972b).

In the two-site model, an equilibrium with $D > 0$ implies that the frequencies of A_1B_1 and A_2B_2 are in excess of their frequencies with random combinations of variants at the two sites, and that there is a deficit of A_1B_2 and A_2B_1. With recombination, the equilibrium **Equations B8.8.2a–B8.8.2b** imply that the marginal fitnesses of A_1B_1 and A_2B_2 are higher than the mean fitness, and those of A_1B_2 and A_2B_1 are lower (and the opposite with $D > 0$). This suggests that maintenance by selection of an equilibrium state with LD requires that some *combinations* of variants at different sites must have higher fitnesses than other combinations (Fisher 1930b, p. 102). This is made more explicit in **Section 8.4.ii.b** below.

8.4.ii.b. *Two-site fitness matrices*

We are mostly concerned here with cases where polymorphisms are maintained at both sites by balancing selection. With constant fitnesses and a uniform environment, this requires heterozygote advantage for the variants at both sites (**Section 2.ii.c**). The fitnesses of the nine possible genotypes with distinct fitnesses can be represented as in **Table 8.2** below. If the fitness effects combine additively, as we assumed for genetic effects on quantitative traits in **Section 3.3.ii**, the effect of simultaneous homozygosity at both sites is simply the sum of the effects of each site. In general, however, this will not be the case, and we need to take into account the interactions between the fitness effects of variants at the two sites (*epistasis*). This is done by introducing four *epistatic parameters*, the e_{ij} ($i, j = 1, 2$), as shown in the table, which express the extent to

TABLE 8.2 The general fitness matrix for variants at two sites

	A_1A_1[1]	A_1A_2	A_2A_2
B_1B_1	$1 - s_A - s_B - e_{11}$	$1 - s_B$	$1 - t_A - s_B - e_{12}$
B_1B_2	$1 - s_A$	1	$1 - t_A$
B_2B_2	$1 - s_A - t_B - e_{21}$	$1 - s_B$	$1 - t_A - t_B - e_{22}$

[1]The parameters s_A and t_A are the selection coefficients against the two types of homozygotes at site A, and s_B and t_B are the corresponding selection coefficients at site B. The e parameters describe the fitness interactions between the variants at the two sites.

which simultaneous homozygosity for two sites has a fitness effect that differs from the additive value.

Box 8.8.ii shows that when fitness effects are completely additive, there can be no LD at equilibrium with $c > 0$; this result also holds for arbitrary numbers of variable sites (Karlin and Liberman 1979). The equilibrium frequencies of the variants at sites A and B are then identical to those in the single-site case (**Section 2.1.ii.c**). However, there can be transient LD during the approach to equilibrium (Felsenstein 1965).

LD at equilibrium with multiple variable sites thus requires nonzero epistatic parameters (nonadditivity). In **Section 4.3.ii.c**, we stated that independence of the fitness effects of variants at different sites means *multiplicative* fitness effects, i.e., a departure from additivity, as shown in **Table 8.3** below. The e_{ij} values shown in parentheses in each of the cells in the four corners of the matrix (found by expanding the products and using the expressions in **Table 8.2**) show that the fitness disadvantages of the double homozygotes in this case are smaller than with additivity. This suggests that multiplicative fitnesses

TABLE 8.3 The fitness matrix with multiplicative fitnesses

	A_1A_1	A_1A_2	A_2A_2
B_1B_1	$(1 - s_A)(1 - s_B)$ $(e_{11} = -s_A s_B)$	$1 - s_B$	$(1 - t_A)(1 - s_B)$ $(e_{12} = -t_A s_B)$
B_1B_2	$1 - s_A$	1	$1 - t_A$
B_2B_2	$(1 - s_A)(1 - t_B)$ $(e_{21} = -s_A t_B)$	$1 - s_B$	$(1 - t_A)(1 - t_B)$ $(e_{22} = -t_A t_B)$

allow nonzero LD to exist at equilibrium, which has been confirmed by detailed analyses (Bodmer and Felsenstein 1967; Roux 1974). More extreme departures from additivity should make it easier for LD to exist. Some properties of equilibria with epistatic selection are considered next. These shed light on the relations between the strength of epistatic selection, the recombination frequency, and the amount of LD maintained at equilibrium.

8.4.ii.c. *Properties of stable equilibria*

We first consider how to evaluate the stability of equilibria. The method for testing local stability of equilibria introduced in **Section 2.1.ii.c** can be extended to a multi-dimensional system (**Box 8.9***). In the neighborhood of a known equilibrium, we can usually reduce the nonlinear **Equations B8.8.1** of **Section 8.4.ii.a** to a set of linear equations by considering the deviations of the haplotype frequencies from the equilibrium values, and by neglecting second-order terms in these deviations (**Box 8.9***). The criterion for stability is described in the box.

We can now discuss the properties of stable equilibria, first looking at the simple case of multiplicative fitnesses. The equilibrium version of **Equation B8.8.2c** shows that there is always an equilibrium with $D = 0$, with the allele frequencies given by the single-site equations (**Problem 8.9.ii**). This is not always stable, however. A necessary and sufficient condition for such an equilibrium to be stable (Bodmer and Felsenstein 1967; Roux 1974) is:

$$c > \frac{s_A t_A s_B t_B}{\left(s_A + t_A\right)\left(s_B + t_B\right)} \tag{8.6}$$

The term on the right-hand side is the product of the genetic loads (**Section 4.3.iii**) at the two sites.

If this condition is violated (i.e., when c is small compared with the products of the loads), there can be two stable equilibria with polymorphisms at both sites, with D values of opposite sign (Ewens 2004, p. 213). This can be understood by considering the extreme case of $c = 0$; there are two stable equilibria, corresponding to single-site systems segregating for haplotypes A_1B_1 and A_2B_2, or A_1B_2 and A_2B_1. The introduction of rare A_1B_2 and A_2B_1 haplotypes into an A_1B_1/A_2B_2 population is resisted by selection under some selection regimes, as is the introduction of A_1B_1 and A_2B_2 into an A_1B_2/A_2B_1 population (**Problem 8.10**). By continuity, a low frequency of recombination will create low frequencies of these deleterious haplotypes, but the system will remain stable (Karlin 1975). In general, the values of p_A and p_B will differ between the two equilibria, which have positive and negative values of D, respectively. From the general result derived in **Section 8.4.ii.a** above, the mean fitness of the population at these stable equilibria must be lower than that of populations consisting

Box 8.9* THE STABILITY OF EQUILIBRIA

We write the set of deviations of the haplotype frequencies x_i from their equilibrium values ($\delta x_i = x_i - x_i^*$) as a column vector $\boldsymbol{\delta x}$. We need consider only three of the δx_i, because they sum to 0; arbitrarily, we use only the first three. (Alternatively, we can use the deviations of p_A, p_B and D from their equilibrium values.)

Proceeding as in **Box 2.4*** of **Section 2.1.ii.c**, we can use the first terms of the Taylor's series (**Appendix A1.ii.a**) representation of the expressions for the x_i' in **Equations B8.8.1** to obtain:

$$\delta x_i' \approx \sum_{j=1}^{3} \delta x_j \left(\frac{\partial f_i}{\partial x_j'} \right)_{x^*} \tag{B8.9.1a}$$

where f_i is the expression for x_i', and x^* is the equilibrium value of the vector of haplotype frequencies, x (for details of the derivation of this and the following results, see Otto and Day 2007, Chapter 8).

This can be expressed compactly in matrix notation as:

$$\delta x' \approx J \delta x \tag{B8.9.1b}$$

where J is the *Jacobian* matrix of partial derivatives of the f_i, evaluated at x^*. From standard matrix theory, we know that, under rather general conditions, δx will approach 0 if the leading eigenvalue, λ_1, of J is less than 1 in absolute value (**Appendix A1.iv.d**). The value of λ_1 can be found by solving the following characteristic equation:

$$\det(J - \lambda I) = 0 \tag{B8.9.2}$$

where $\det(A)$ is the determinant of the matrix A and I is the unit matrix (**Appendix A1.iv.b**). It can be difficult to obtain explicit formulae for λ_1, even using mathematical computer programs such as Mathematica (http://www.wolfram.com/products/mathematica/) or Matlab (http://www.mathworks.com/), but it is always possible to obtain numerical values for specific cases.

If λ_1 is real and < 1, the approach to equilibrium is smooth; if λ_1 is real and > 1, the haplotype frequencies will diverge away from equilibrium. If there are two complex conjugate leading eigenvalues (**Appendix A1.iv.c**) with an absolute value < 1, the haplotype frequencies oscillate around the equilibrium, but converge towards equilibrium. If there are two complex conjugate leading eigenvalues with an absolute value > 1, there is a divergent series of oscillations around the equilibrium. If selection maintains variation at both sites, this indicates the existence of a limit cycle around the equilibrium or chaos, as in the case of frequency-dependent selection (**Box 2.6** of **Section 2.2.i.a**). These situations seem to be comparatively rare in multisite systems (but see Hastings 1981 and Akin 1982 for some examples).

entirely of A_1B_1 and A_2B_2 or A_1B_2 and A_2B_1. It follows that mean fitness decreases during the approach to equilibrium from either of these initial starting points for systems with $c > 0$, at least over part of the trajectory (Moran 1964).

This analysis of the multiplicative fitness model brings out some important features of two-site systems with epistatic selection.

1. Linkage disequilibrium can be maintained in populations that are at stable equilibrium, especially when the frequency of recombination is low in relation to the strength of selection.
2. The magnitude of D is 0 or small when c is sufficiently large relative to the magnitude of the epistatic parameters. In general, therefore, large values of LD require tight linkage or large epistatic fitness effects.
3. There may be two or more different stable equilibria, which can differ in the sign and magnitude of D, as well as in the variant frequencies.
4. Mean fitness does not always increase as haplotype frequencies change during the approach to an equilibrium, even with constant fitnesses.
5. The mean fitness at a given equilibrium point usually (but not always) decreases steadily with c.

These conclusions apply to much more general fitness models, as has been established by a large body of theoretical work; for overviews, see Karlin (1975), Christiansen (2000, Chapters 7 and 8), Bürger (2002, Chapter 2), and Ewens (2004, pp. 208–221).

8.4.ii.d. *Weak selection and loose linkage*

These complexities of the possible behavior of even the simplest two-site systems might lead to despair about the possibility of obtaining any useful general conclusions. However, the fact that strong epistatic effects are required to generate LD, unless recombination is infrequent, suggests that the results on selection described in **Chapters 2–4**, which ignored these complications, may be fairly widely applicable. We will now show that, with sufficiently weak selection in relation to the frequency of recombination, D tends to settle down to a near-constant value, and the population moves along a trajectory that is close to one in which the mean fitness increases each generation. This is the state of *quasi-linkage equilibrium* (QLE), first noticed by Kimura (1965a).

Box 8.10* outlines a rigorous method for establishing this result, invented by Nagylaki (1976a, 1992, pp. 176–184). The final result shows that D approaches a value of the order of s, where s is a positive constant that measures the strength of selection (i.e., the greatest difference in fitness between genotypes is less than s), provided that $c > s$. We can then write $D = O(s)$, where $O(s)$ indicates a quantity whose ratio to s is bounded by a constant of order 1 as s

---

Box 8.10* CHANGE IN D WITH WEAK SELECTION: QUASI-LINKAGE EQUILIBRIUM

For convenience, we work with the value of D measured *after* selection in a given generation, denoted by $\tilde{D} = (x_1 x_4 w_{1.} w_{4.} - x_2 x_3 w_{2.} w_{3.})/\bar{w} = D + O(s)$. Using **Equation B8.8.2c** from **Section 8.4.ii.a**, we can write:

$$\tilde{D}(t) = (1 - c)\tilde{D}(t - 1) + sg(x[t - 1]) \qquad (B8.10.1)$$

where $g(x)$ is a function of the vector of haplotype frequencies, x. We can assume that $|g| < g_{MAX}$, where g_{MAX} is of order 1 so that $sg(x)$ tends to 0 as s approaches 0. The detailed nature of g is irrelevant in what follows.

By iterating this equation, we obtain:

$$\tilde{D}(t) = (1 - c)^t \tilde{D}(0) + s(1 - c)^t \sum_{u=1}^{t} (1 - c)^{-u} g(x[u]) \qquad (B8.10.2)$$

From the formula for the sum of a geometric series (**Appendix A1.ii.c**), the absolute value of the second term on the right-hand side is less than sg_{max}/c; it follows that, for a sufficiently large value of t, $\tilde{D}$ approaches a value of $O(s)$ if c is large compared with s. With loose linkage, this time will be a few generations (Nagylaki 1992, p. 178). Because $\tilde{D} = D + O(s)$, D also approaches $O(s)$ over this time period. It follows from **Equations B8.8.1** that the change in x across generations also becomes $O(s)$.

An extension of this argument leads to the conclusion that $\Delta\tilde{D}(t)$, and hence $\Delta D(t)$, approaches $O(s^2)$ over a similar time scale (Nagylaki 1976, 1992, pp. 178–179).

approaches 0. An extension of this argument shows that the variant frequencies at the two sites both approach values that correspond to those in the absence of linkage disequilibrium, with deviations of magnitude $O(s)$ (Nagylaki 1976a, 1992, pp. 179–181). Less intuitively, it is also possible to show that D asymptotically changes at a rate $O(s^2)$. These results also hold when there is frequency-dependent selection, provided that changes in genotypic fitnesses each generation are $O(s^2)$, as would generally be expected if selection is weak (Nagylaki 1976a).

These results yield a simple expression for the change in mean fitness under selection. The argument is described in **Box 8.11**⋆. The final result can be written as:

$$\Delta \bar{w} = \frac{V_{Aw}}{\bar{w}^2} + O(s^3) \qquad (8.7)$$

where V_{Aw} is the sum of the additive genetic variances in fitness contributed by each site. This is often called the *genic variance*. Again, this expression is valid with frequency-dependent selection under the condition just mentioned.

The implication of this result is that, when selection is sufficiently weak in relation to the frequency of recombination, for most of the time the population will follow a trajectory that is close to one in which mean fitness increases according to Fisher's Fundamental Theorem (**Section 3.3.iii.a**), although there may be limited periods in which mean fitness decreases. Stable equilibria are located close to points with zero D values, with departures of the order of s. All of this suggests that the conclusions from single-site theory do indeed provide good approximations to the behavior of two-site systems, when the assumptions that lead to QLE are satisfied. This will be generalized to arbitrary numbers of sites in **Section 8.4.iii** below.

In addition, the additive genetic variance in fitness is 0 at equilibrium, even without the QLE approximation, because the marginal fitnesses of the two variants at each site must be equal at equilibrium (**Section 2.1.ii.c**). Thus, the only contribution to the variance of the marginal fitnesses of the haplotypes comes from the interactions between the additive effects of variants at the two sites (additive × additive interactions: see **Box 8.8**), i.e., from V_{AA} as defined in **Box 8.11**⋆. This is nonzero at equilibrium only if D is nonzero. The maintenance of nonzero LD by selection is thus associated with the existence of additive × additive epistatic variance in fitness, suggesting that experimental detection of such variance in fitness components might indicate the existence of underlying sets of loci with LD maintained by selection. In practice, however, it is difficult to measure V_{AA}, because it is partially confounded with V_A in the expressions for correlations between relatives (Falconer and Mackay 1996, p. 154; Ewens 2004, pp. 77–78). It also makes a similar contribution to the estimator of the "domi-

Box 8.11* CHANGE IN MEAN FITNESS WITH QUASI-LINKAGE EQUILIBRIUM

If we neglect higher-order terms in fitness differences, we can write:

$$\bar{w}\Delta\bar{w} = \Delta\left\{\sum_i x_i w_{i\cdot}\right\} = \sum_i \left\{w_{i\cdot}\Delta x_i + x_i\Delta w_{i\cdot}\right\} \tag{B8.11.1}$$

where the approximation is accurate to terms of order s^2 under QLE, because Δx_i is $O(s)$ and $\Delta w_{i\cdot}$ is $O(s^2)$.

Using **Equations B8.8.1** in **Section 8.4.ii.a** and simplifying (**Problem 8.11***), we obtain:

$$\bar{w}^2\Delta\bar{w} \approx 2\sum_i x_i\left(w_{i\cdot} - \bar{w}\right)^2 - 2\bar{e}cD + O\left(s^3\right) \tag{B8.11.2}$$

where $\bar{e} = w_{1\cdot} + w_{4\cdot} - w_{2\cdot} - w_{3\cdot}$ is the measure of epistasis introduced in **Box 8.8.ii**.

The first term on the right-hand side of this equation is reminiscent of the expression for the additive genetic variance in **Section 3.3.ii.b**, and is often known as the *gametic variance*; we will denote it by V_g. If there is no epistasis and no LD, it is equal to the sum of the additive genetic variances contributed by each site (see **Equation B3.5.4**), denoted by V_A.

A further approximation to **Equation B8.11.2** with QLE can be obtained as follows. Let $Z = x_1 x_4 / x_2 x_3$. We then have:

$$\Delta\ln(Z) \approx \frac{\Delta x_1}{x_1} + \frac{\Delta x_4}{x_4} - \frac{\Delta x_2}{x_2} - \frac{\Delta x_3}{x_3} = \frac{1}{\bar{w}}\left(\bar{e} - cD\sum_i \frac{1}{x_i}\right) + O\left(s^2\right) \tag{B8.11.3}$$

Comparison with the right-hand side of **Equation B8.8.4** suggests that the term in parentheses must be small if D is small. This can be confirmed as follows. We have $Z = 1 + D/(x_2 x_3)$, so that ΔZ and $\Delta\ln(Z)$ are $O(s^2)$ under QLE, from the results in **Box 8.10***. It then follows from **Equation B8.11.3** that:

$$cD = \left(\bar{e}\bigg/\sum_i \frac{1}{x_i}\right) + O\left(s^2\right) \tag{B8.11.4}$$

and so the right-hand side of **Equation B8.11.2** can be written as:

$$\bar{w}^2\Delta\bar{w} = V_g - 2\left(\bar{e}^2\bigg/\sum_i \frac{1}{x_i}\right) + O\left(s^3\right) \tag{B8.11.5}$$

The fact that the second term on the right-hand side of this equation involves $\bar{e}^2$ suggests that it is related to the epistatic component of V_g. Detailed analysis indeed shows that this term represents the component of V_g that remains after the additive effects of the two sites are removed, i.e., it is the additive × additive component of genetic variance, V_{AA} (Kojima and Kelleher 1961; Kimura 1965a; Ewens 2004, p. 74).

As a result, only the additive genetic variance term remains. Under QLE, this is approximated by the sum of the additive variances contributed by each site under QLE, because D is $O(s)$ and any terms in the variance in fitness contributed by the covariances between the two sites involve the product of D and terms of order s^2. We can therefore write **Equation B8.11.5** in the same form as **Equation 8.7**.

nance" variance component, V_D. The fact that many fitness components fail to show evidence for significant "dominance" variance when fitting models that ignore epistatic variances (**Section 4.5.i.a**) suggests that epistatic variance associated with LD maintained by selection does not often contribute much to variation in these traits (Charlesworth and Hughes 2000).

8.4.iii. Multisite systems

8.4.iii.a. *Properties of equilibria*

It is far harder to obtain the properties and stability of equilibria with an arbitrary number of segregating sites than for the two-site model, and only limited results are available (reviewed by Christiansen 2000, Chapters 7–9). Epistasis, in the form of departures from additive fitness effects, is always required for the existence of equilibria with nonzero LD in randomly mating populations (**Section 8.4.ii.b**; Karlin and Liberman 1979), i.e., equilibria where haplotype frequencies depart from the values expected with random combinations of variants at each site.

With multiplicative fitnesses, a stable equilibrium with zero LD exists if **Equation 8.6** is obeyed for all pairs of adjacent sites (Roux 1974). There can, however, also be equilibria with nonzero LD, as has been established by simulations of finite populations with large numbers of linked sites subject to selection with multiplicative fitnesses (Franklin and Lewontin 1970). In addition, the pairwise D values tend to be larger than predicted from two-site expectations. When selection is sufficiently strong in relation to the recombination frequencies among adjacent sites, the population consists largely of two "complementary" haplotypes with all variants in LD, so that each haplotype's variants form a set excluding those in the other, and heterozygotes for the haplotypes are heterozygous at nearly all selected sites. Franklin and Lewontin (1970) therefore proposed that the maintenance of variation by selection should cause genome-wide departures from linkage equilibrium, even in randomly mating populations. In surveys of LD in natural populations, however, even loci that are as closely linked as one centiMorgan usually fail to show LD (**Section 8.1.iii.b**), implying that the conditions for this effect to occur are not met.

8.4.iii.b. *Dynamics under QLE*

In the absence of selection, it is obvious from the results for pairs of sites (derived in **Section 8.2.ii.a**) that eventually all pairwise D values will approach 0, given nonzero recombination. Multisite systems will thus eventually approach linkage equilibrium (Bennett 1954). Thus, we should be able to generalize the two-site results for weak selection and loose linkage to multisite systems, because LD measures should asymptotically approach values of $O(s)$, provided that recombination is sufficiently strong compared with selection. We might also guess that the trajectory of the population will approach that expected with linkage equilibrium, with deviations of order s, and that the rate of change of mean fitness will obey the multisite equivalent of **Equation 8.7**. These conjectures have been confirmed for randomly mating populations (Nagylaki 1993; Nagylaki et al. 1999), but the details are too complex to be given here.

8.4.iv. The evolutionary dynamics of quantitative traits

In this section, we relax some of the simplifying assumptions made in **Sections 3.3.ii** and **4.5.i** for modeling the effects of selection on a quantitative trait.

8.4.iv.a. *The infinitesimal model*

It is possible obtain an expression for the change in a trait mean under QLE that is similar to **Equation B3.6.2** of **Section 3.3.iii.a**, which involves the additive genetic covariance between trait value and fitness (Nagylaki 1993). This shows that the response of the mean to selection can be predicted, to a good approximation, under much more general conditions than those assumed in **Section 3.3.iii**. In particular, the epistatic component of the genetic variance of a trait is irrelevant with respect to a permanent change in mean under selection (Griffing 1960; Bulmer 1980, pp. 160–162). This approach has been extended to inclusive fitness models by Gardner et al. (2007).

This, however, leaves unsolved the problem of predicting the changes in the components of genetic variance of the trait distribution. In general, these cannot be predicted simply from changes in frequencies of variants at individual sites, even in the absence of LD (e.g., Barton and Turelli 1987). One solution to this problem is to assume that the trait is affected by a large number of sites, each of which contributes a small, similar amount to the additive genetic variance, i.e., the *infinitesimal model*. If there are m variable sites, for a given value of the additive variance, V_A, the results following **Equation 3.17** in **Section 3.3.iii.b** show that the effect on the trait caused by a variant at a given site is on the order of $\sqrt{V_A/m}$. In addition, **Equation 3.17** implies that the rate of change of variant frequency at each site is also of this order. From the formula for V_A (see **Section 3.3.ii.b**, **Box 3.5**), the contribution to the change in V_A from the change in variant frequency at a given site is on the order of the site's contribution to V_A, multiplied by the change in variant frequency at the site. Summing over all sites, the net change in V_A must therefore be on the order of V_A times $\sqrt{V_A/m}$. It follows that, when m is sufficiently large, this effect is negligible relative to V_A.

It is also known, however, that the variance in a quantitative trait is usually changed by selection within a single generation. For example, in **Section 4.5.i.b** (**Box 4.8***), we showed how the variance of a normally distributed trait is reduced by nor-optimal selection. Stabilizing selection is usually detected through such a variance reduction, when there is no change in mean (Haldane 1954; Kingsolver et al. 2001). Truncation selection has a similar effect, but also changes the mean (Bulmer 1974, 1980, p. 153), whereas disruptive selection (favoring the extremes of the distribution) increases the variance. More generally, if fitness, w, is a continuous function of a trait value, z, the variance of a normally distributed trait is reduced by selection when the second derivative of $\ln w(z)$ with respect to z is negative; conversely, if $\ln w(z)$ has a positive second derivative, the variance increases, and there is no change in variance if $\ln w(z)$ is linear (Shnol and Kondrashov 1993).

What are the genetic implications of such changes in variance? This question can be answered fairly easily when the genetic effects are purely additive, so that both dominance and epistasis are absent. The only effect of selection on the variance under the infinitesimal model then comes from LD among sites, and the departures from Hardy–Weinberg genotype frequencies at individual sites (*Hardy–Weinberg disequilibrium*). The trait value of an individual, expressed as the deviation from the population mean, can be represented by the sum of the values contributed by the maternal and paternal alleles at each site, i.e.,

$$z - \bar{z} = \sum_i \left(\alpha_{im} + \alpha_{ip} \right) \tag{8.8}$$

where α_{im} and α_{ip} are the additive effects of the maternal and paternal variants at the ith site, respectively.

Using the standard formula for the variance of a sum (**Appendix A2.iv**), the total additive genetic variance at the start of a generation in a randomly mating population is:

$$V_A = \sum_i V_{Ai} + 4 \sum_{i<j} C_{ij} \tag{8.9}$$

where V_{Ai} is the additive variance contributed by the ith site, and C_{ij} denotes the covariance between α_{im} and α_{jm} or α_{ip} and α_{jp} at sites i and j (with random mating, the covariance between the maternal and paternal contributions is 0).

The first term on the right-hand side of this expression is the genic variance in the trait. **Box 8.2** of **Section 8.1.ii.a** implies that the second term is directly related to the D values for all pairs of sites, multiplied by the products of the effects of the variants at each site. Under the infinitesimal model, the change caused by selection in the genic variance is negligible, because it reflects changes in variant frequencies at individual sites (see above). Similarly, the argument used for the genic variance implies that any covariances between maternal and paternal effects at individual sites that are associated with Hardy–Weinberg disequilibrium are negligible, since they involve differences in variant frequencies created by selection over one generation. This leaves the contribution of the covariances between values at different sites that are within and between the maternal and paternal gametes of each individual; under random mating, only the within-gamete component can be transmitted.

Any heritable change in the additive genetic variance thus reflects the component associated with LD (Bulmer 1971, 1974). This can be negative or positive, depending on the effect of selection on the variance; a reduction in variance corresponds to an overall negative covariance term, whereas an increase corresponds to an overall positive term. These changes in variance are opposed by recombination; we might therefore expect an equilibrium to be reached under a constant selection regime, provided that recombination is occurring. This is verified by direct calculations (**Box 8.12**) and multi-locus computer simulations (Bulmer 1974). Even though individual D values are very small, as indicated by the QLE analysis, the fact that there are $m(m-1)/2$ pairs of sites means that the net contribution of LD to the genetic variance in the trait can be significant, especially under the strong selection applied in artificial selection.

This effect is especially important for species with small numbers of chromosomes, because the effect decreases with the harmonic mean, c_H, of the recombination frequencies among the sites involved (**Box 8.12**). For *Drosophila melanogaster*, which has only three pairs of major chromosomes and no crossing over in males, Bulmer (1974) estimated a value of c_H of 0.1 for sites drawn randomly from the genetic map. But for species like mice or humans, with more than 20 chromosomes with average map lengths of over 100 centiMorgans, c_H is close to 0.4, only slightly less than with completely free recombination. If selection causes a change in variance of 10% of the pre-selection value, and the heritability of the trait is 0.5, **Equation B8.12.3** shows that the LD component of the genetic variance represents a proportion of $0.0125/c_H$ of the genetic variance; this is equal to about 12% for $c_H = 0.1$, but

Box 8.12 CHANGE IN ADDITIVE GENETIC VARIANCE UNDER THE INFINITESIMAL MODEL

Selection within a generation t changes the covariances within the maternal and paternal gametes of individuals from the pre-selection value $C_{ij}(t)$ for a pair of sites i and j, and also induces a covariance between maternal and paternal values at these sites. Let the post-selection values of these covariances be $\tilde{C}_{ij}(t)$ and $\tilde{C}'_{ij}(t)$, respectively. It is reasonable to assume that the change in covariance caused by selection is the same for both terms, i.e., we have $\tilde{C}_{ij}(t) - C_{ij}(t) = \tilde{C}'_{ij}(t) = \Delta C_{ij}(t)$. If c_{ij} is the recombination frequency for the pair of sites, by using the same argument as in **Box 8.4** of **Section 8.2.ii.a**, we obtain the recursion relation:

$$C_{ij}(t) = (1 - c_{ij})C_{ij}(t-1) + \Delta C_{ij}(t-1) \qquad \text{(B8.12.1)}$$

In order to obtain an expression for ΔC_{ij}, we assume that selection changes the variance of z by an amount $\Delta V_z(t)$ in generation t. Under the normality assumption, the regression of genotypic value on z is linear with slope h^2, where h^2 is the proportion of the total variance attributed to additive genetic effects, including any contributions from LD (**Section 3.3.iii.a**). The change in the genetic variance caused by selection is thus $h^4 \Delta V_z(t)$.

If we write $C_D(t)$ for the net contribution to the genetic variance from LD (i.e., the second term in **Equation 8.9**), it must change as a result of selection within generation t by an amount $\Delta C_D(t) = 8\Sigma_{ij}\Delta C_{ij}(t) = h^4 \Delta V_z(t)$. If all m sites contribute equally to variability, at equilibrium we thus have:

$$C^*_{ij} = \frac{h^{*4}\Delta V^*_z}{4m(m-1)c_{ij}} \qquad \text{(B8.12.2)}$$

Summing over all pairs of sites, we obtain:

$$C^*_D = \frac{h^{*4}\Delta V^*_z}{2c_H} \qquad \text{(B8.12.3)}$$

where c_H is the harmonic mean of the recombination fractions among all pairs of sites that contribute to variability in the trait. If an explicit expression for ΔV_z is available, this equation can be solved to yield the value of C_D at equilibrium, given an expression for the genic component of the additive genetic variance (Bulmer 1974, 1980, pp. 158–160).

only 3% for $c_H = 0.4$. This might suggest that these effects are of little biological significance, but in **Section 10.2.iv** we will show that they are of great importance for the evolution of sexual reproduction and recombination.

8.4.iv.b. *Extending the infinitesimal model*

We have assumed that a normal phenotypic distribution is preserved by selection, although this is only exactly true for a nor-optimal fitness function of the type used in **Section 4.5.i.b** (Bulmer 1980, p. 151). Even if the infinitesimal model holds, there is thus some question about this assumption for many types of selection. Normality turns out to be a good approximation under most circumstances, even when the assumptions of the infinitesimal model are not strictly met (Turelli and Barton 1994). The proof uses equations describing changes in all the moments of the trait distribution, derived under the assumption of QLE (Barton and Turelli 1991). This approach has been extended to a variety of genetic systems (Kirkpatrick et al. 2002), although a formal proof of QLE is available only for random mating. This important topic is too complex to be described here, but is reviewed in Bürger (2002, Chapter 5).

The infinitesimal model can be extended to partially self-fertilizing populations, where dominance contributions to the genetic variance become important (Kelly 1999a,b; Kelly and Willamson 2000). With high levels of selfing, the low effective rate of recombination due to high homozygosity (**Section 8.2.ii.b**) means that the reduction in genetic variance caused by negative LD can greatly slow the response to selection, compared to a randomly mating population with the same level of genetic variance. Genetic variances tend in any case to be lower for selfing populations of plants than for comparable outcrossers (Charlesworth and Charlesworth 1995), so that overall we may expect reduced rates of adaptive evolution in inbreeding populations. Asexual populations are an even more extreme case. Here, the approximations used for QLE and the infinitesimal model fail, and a different set of approximations must be used (Kimura 1965b; Lande 1976).

8.4.v. Wright's shifting balance model of evolution

8.4.v.a. *Epistasis and multiple peaks of mean fitness*

Departures from random mating may also affect the dynamics of evolutionary change when there is population subdivision, such that drift, selection, and migration are operating simultaneously. We have already discussed single-site models in **Section 7.3**; the behavior of additively inherited quantitative traits is closely related to the properties of these models. However, the effects of drift in producing departures from random mating expectation for the metapopulation as a whole may allow nonadditive effects to become more important than with random mating. We saw an example of this in the case of the effect of dominance on the fixation probabilities of favorable mutations (**Section 7.3.ii**).

The most important evolutionary theory that involves a major role for non-additive effects in structured populations is Wright's *shifting balance* theory (Wright 1932, 1940a,b, 1977, Chapter 13; see also Haldane 1932, pp. 102–104). This theory is intended to solve the following problem. Selection in a large population drives the population to a local equilibrium. As we have seen (**Section 8.4.iii.b**), to a good approximation this will be close to a maximum of mean fitness if selection is weak relative to recombination, even with frequency dependence. But there may well be more than one such local optimum, if the surface of mean fitness as a function of genotype frequences is "rugged." Multiple, stable equilibria often arise in multisite systems with epistatic interactions between variants at different sites (**Section 8.4.ii.c**). Evolution by natural selection alone may therefore cause a population to arrive at a local equilibrium that has a lower mean fitness than other equilibrium points, so that the population is in some sense sub-optimal.

Genetic drift in a metapopulation may cause individual populations to drift away from the current local equilibrium. Occasionally, a population might cross a low fitness region (an *adaptive valley*), separating two nearby stable equilibria, and move into the zone of attraction of the other equilibrium (this was called a *peak shift* by Wright). If this new equilibrium is associated with a higher mean fitness than the old one, which is still occupied by the rest of the metapopulation, then interdeme selection of the type discussed in **Section 7.3.iv.b** might cause this "better" equilibrium state to spread through the entire metapopulation, so that the species as a whole would become better adapted. In contrast to group selection involving an altruistic genotype disfavored by selection within a deme, this process produces a population that is stable under the existing selection regime, and so is mechanistically more plausible.

Wright's theory has generated controversy over the many years since it was first proposed (for recent critiques of the shifting balance theory, see Coyne et al. 1997, 2000; for defenses, see Wade and Goodnight 1998, 2000). A major difficulty compared with selection acting purely within populations is the one referred to when we discussed group selection (**Section 7.3.iv.b**), i.e., there must not only be sufficient isolation between demes that adaptive valley crossing is possible by drift, but also enough movement between demes that a new equilibrium configuration can spread through the metapopulation. There is abundant evidence for epistatic effects among genetic variants affecting quantitative traits (Wolf et al. 2000; Mackay and Lyman 2005), but we do not yet know enough about the relationships between genotypes and fitness in most systems to determine how frequently multiple, stable equilibria, associated with significant differences in mean fitness, exist in nature. Another difficulty is that there are few predictions of the shifting balance theory that can be used to distinguish it from selection within populations.

8.4.v.b. *The fixation of chromosomal rearrangements*

A process that is closely related to the shifting balance model of evolution is the fixation by genetic drift of an allele in a biallelic system with heterozygote disadvantage, such as a chromosomal rearrangement that causes fertility loss when heterozygous with wild-type, which we mentioned in **Section 2.1.ii.f.** This is in fact the simplest case of an unstable equilibrium in population genetics; there are two fitness peaks, at allele frequencies 0 and 1, with a minimum in mean fitness at the unstable polymorphic equilibrium point. In a finite population, there is thus a certain probability that a new rearrangement becomes fixed by drift, which can be calculated by methods similar to those described in **Section 6.2.iii.a.** As might be expected from general principles, this probability is very small unless the product N_e and the extent of heterozygote fitness loss is on the order of 1 or less (Wright 1941; Lande 1979a; Walsh 1982; Charlesworth et al. 1987; Charlesworth 1992).

In a subdivided population, this product could be sufficiently small that a rearrangement could rise to a high frequency within a local population, even with gene flow from nearby populations. There are many examples of small, isolated populations that have become fixed for rearrangements, such as domestic mice populations in mountainous and peripheral regions of Europe, which have accumulated one or more centric fusions between different chromosomes, creating metacentric chromosomes from pairs of acrocentrics (White 1978, Chapter 6; Pialek et al. 2005). Laboratory experiments have shown that hybrids between members of these populations and wild-type mice suffer fertility losses; these are greater, the larger the number of chromosomal differences involved (Pialek et al. 2005).

Rearrangements will not contribute to the evolution of the species as a whole unless they are able to spread from the population in which they arose. The shifting balance process will thus only favor the spread of a new rearrangement if it confers a fitness advantage when homozygous. In the absence of such an advantage, however, extinction and re-colonization of local populations can result in the chance spread of a rearrangement throughout a species by the processes discussed in **Section 7.2.iv.b**, because a single local population is ultimately the ancestor of the whole species (Wright 1941). Lande (1979a, 1984) has made detailed calculations of the circumstances under which this process can contribute to chromosomal evolution; however, it is still unclear to exactly what extent this mechanism of fixation of rearrangements is involved, as opposed to the selective mechanisms that we discuss in **Section 10.2.iii.a**.

8.4.vi. The evolution of hybrid fitness loss

Loss of fitness in interspecies hybrids is one situation in which there is no doubt that epistatic interactions in fitness effects play the main causal role. This was early recognized (Bateson 1909, p. 98), and later discussed in a population genetics framework by Dobzhansky and by Muller (Dobzhansky and Tan 1936; Muller 1940). Crosses among individuals within species generate progeny with high fitness, whereas crosses between species often result in F1 or F2 progeny with greatly reduced viability or fertility. Genetic analyses of hybrid fitness loss can be made when fertility loss is incomplete, or if one sex is viable and fertile (this is usually the heterogametic sex, a pattern that is called Haldane's rule; Haldane 1922). Such analyses reveal that the hybrid fitness loss is usually caused by interactions between alleles at different loci (Coyne and Orr 2004, Chapter 8). In the simplest case, only two distinct genes, A and B, are involved, with species 1 fixed for the allelic combination A_1B_1 and species 2 for A_2B_2. The hybrid combination A_1B_1/A_2B_2 can have reduced fitness if A_1 and B_2 interact badly. Severe hybrid fitness loss can arise if there are either interactions between a few pairs of genes with major effects or interactions between many genes with small effects. Genetic studies have provided examples of both situations (Coyne and Orr 2004, Chapter 8).

This type of situation could arise because of a peak shift as a result of the shifting balance process described above (**Figure 8.18A**), such that an ancestral species (e.g., one fixed for A_1B_1) has generated two, geographically isolated descendant species (Wright 1940a,b). One species remains as A_1B_1, while the other has become fixed for A_2B_2, despite the fact that individuals with A_1 and B_2 have reduced fitness. An alternative is a peak shift in a species whose population size is small, and that is isolated from the ancestral population as a result of a founder event (Mayr 1954, 1963, Chapter 17; Gavrilets 2004). The idea of "founder effect speciation" has been studied theoretically in a variety of contexts, but there is only limited support for its plausibility (Barton and Charlesworth 1984; Barton 1996; Coyne and Orr 2004, Chapter 11). The problem is that the probability of a peak shift is severely reduced by even a slight dip in mean fitness associated with the adaptive valley to be crossed in a single founder event (Barton and Charlesworth 1984; Barton 1996). Many separate events are therefore required to generate severe hybrid fitness reduction and this will result in a severe loss of variability, for which there is little empirical evidence (Barton and Charlesworth 1984; Barton 1996; Coyne and Orr 2004, Chapter 11). For a different view, see Carson and Templeton (1984).

Dobzhansky and Tan (1936) and Muller (1940) proposed the alternative hypothesis that hybrid fitness is reduced due to different adaptations evolving in two related species, or by neutral divergence (**Figure 8.18B**). One population descended from an A_1B_1 ancestor becomes fixed for a variant B_2 that is

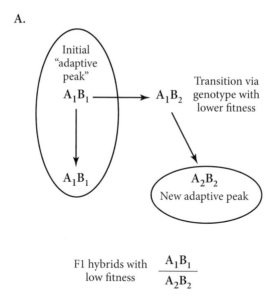

A.

Initial "adaptive peak"

A_1B_1 → A_1B_2 Transition via genotype with lower fitness

A_1B_1

A_2B_2
New adaptive peak

F1 hybrids with low fitness
$$\frac{A_1B_1}{A_2B_2}$$

B.

Advantageous mutation in one population

A_1B_1

Advantageous mutation in other population

A_1B_2 A_2B_1

F1 hybrids with low fitness
$$\frac{A_1B_2}{A_2B_1}$$

FIGURE 8.18 Models of speciation. **A.** The peak shift model. An ancestral population (fixed for A_1B_1) has generated two geographically isolated, descendant populations. One population remains as A_1B_1, whereas the other has shifted to a different "adaptive peak" by becoming fixed for two different alleles, one at each locus, to become A_2B_2. In the model, interactions between the ancestral and the derived alleles (e.g., A_1 and B_2) cause reduced fitness. **B.** The Dobzhansky–Muller model, where two geographically isolated populations descended from a common ancestor are initially fixed for the combination A_1B_1. One descendant population has become fixed for a variant A_2 that is neutral or advantageous on the background of B_1. Similarly, a variant B_2 became fixed in the other descendant population on an A_1 background. If the populations are isolated, there is no selection to ensure that B_2 will have high fitness in the A_2 background of the other population. Hybrid fitness may thus be low when the two derived alleles (A_2 and B_2) are brought together.

advantageous or neutral on the background of A_1. Similarly, a variant A_2 that is neutral or advantageous on the background of B_1 is fixed in the other descendant population (alternatively, two changes could occur in one population, and none in the other; e.g., Fishman and Willis 2006). No adaptive valley need be crossed under this model; the hybrid fitness loss is simply the result of independent evolution in two isolated populations, so that there is no selection to preserve high fitnesses of the new allelic combinations (unlike new variants that spread within each population, which must have high fitness on the genetic background that exists at the time they arise). As we saw in **Section 3.2.iii**, even populations exposed to identical selection pressures will tend to diverge in their genetic constitution, as a result of the chance accumulation of different favorable mutations, so that this phenomenon is an almost inevitable outcome of adaptive evolution.

The evolution of *Dobzhansky–Muller incompatibilities*, as this mechanism is called, has been extensively modeled (Orr 1995; Turelli and Orr 1995, 2000; Orr and Turelli 2001), and it has been shown that a range of phenomena, including Haldane's rule, are consistent with the predictions of these models (Turelli and Orr 2000; Coyne and Orr 2004, Chapter 8). In accord with the Dobzhansky–Muller model, there are now several studies of genes that have been shown to cause hybrid fitness loss, showing convincing evidence of adaptive evolution from their patterns of DNA sequence evolution and variation (**Table 6.2** of **Section 6.4.ii.b**; Orr et al. 2004; Brideau et al. 2006). A recent study of experimental evolution in budding yeast showed that partial Dobzhansky–Muller effects evolved after only a few hundred generations of adaptation to new environments (Dettman et al. 2007).

APPENDIX TO CHAPTER 8

Linkage disequilibrium at neutral sites in finite populations

8A.1. THE GENERAL METHOD

The initial aim is to obtain an expression for the expectation of a function or set of functions of several variables subject to evolutionary forces under similar assumptions to those used for the diffusion equations of **Chapter 5 (Appendix)** and **Chapter 6 (Appendix)**. We consider a vector y of n variables y_i ($i = 1, \ldots, n$), such that the mean change in y_i over one generation is $M_{\delta y_i}$, and the covariance between the change in y_i and y_j is $C_{\delta y_i, \delta y_j}$ (the variance of the change in y_i, $V_{\delta y_i}$, is the special case $i = j$). We define a function of interest as $f(\mathbf{y})$. Using the multi-dimensional version of Taylor's theorem (**Appendix A1.ii.a**), the value of $f(\mathbf{y})$ in generation $t + 1$ can be approximated by:

$$f\left[\mathbf{y}(t+1)\right] \approx f\left[\mathbf{y}(t)\right] + \sum_{i=1}^{n}(\delta y_i)\left(\frac{\partial f}{\partial y_i}\right) + \sum_{i=1}^{n}\sum_{j=1}^{n}\frac{(\delta y_i)(\delta y_j)}{2}\left(\frac{\partial^2 f}{\partial y_i \partial y_j}\right)$$

(8A.1)

where $\mathbf{y}(t)$ is the value of the vector $\mathbf{y}$ in generation t, and δy_i is the change in its ith component between generations t and $t + 1$. The partial derivatives are evaluated at $\mathbf{y}(t)$.

Using $E\{f\}$ to denote the expectation of f over the distribution of $\mathbf{x}(t)$, this gives:

$$\Delta E\{f\} \approx E\left\{\sum_{i=1}^{n}\left[M_{\delta y_i}\left(\frac{\partial f}{\partial y_i}\right) + \frac{1}{2}V_{\delta y_i}\left(\frac{\partial^2 f}{\partial y_i^2}\right)\right] + \sum_{i<j}^{n}C_{\delta y_i, \delta y_j}\left(\frac{\partial^2 f}{\partial y_i \partial y_j}\right)\right\}$$

(8A.2)

where Δ denotes the change between generation t and $t + 1$.

8A.2. APPLICATION TO LINKAGE DISEQUILIBRIUM

In the case of two sites, A and B, each with two alternative variants, we can write $q_A = y_1$, $q_B = y_2$, and $D = y_3$. For neutral sites under the infinite-sites model, there is no contribution from drift to $M_{\delta x_1}$ and $M_{\delta x_2}$. There is a deterministic contribution to the change in D, equal to $-cD$, from the results in **Box 8.4** of **Section 8.2.ii.a**. The expected changes in each of the four haplotype frequencies (x_i, where $i = 1, \ldots, 4$) are 0. Generalizing the argument used in **Section 5.3.iii.a**, the variances and covariances of the changes in haplotype frequencies due to drift are approximated by the values for the *multinomial distribution* (**Appendix A2.v.b**), i.e., we have:

$$V_{\delta y_i} \approx \frac{x_i(1 - x_i)}{2N_e}, \quad C_{\delta x_i, \delta x_j} \approx -\frac{x_i x_j}{2N_e}$$

(8A.3)

We have $y_3 = x_1 x_4 - x_2 x_3$, so that:

$$\delta y_3 = x_1 \delta x_4 + x_4 \delta x_1 - x_2 \delta x_3 - x_3 \delta x_2 + (\delta x_1)(\delta x_4) - (\delta x_2)(\delta x_3)$$

where δx_i is the change in frequency of haplotype i due to drift.

Taking expectations over all values generated by drift, and using **Equation 8A.3**, the expected change in y_3 due to drift is easily seen to be approximately

$-D/2N_e$. This reflects the reduction in D due to the loss of variability. The final expression for $M_{\delta x_3}$ is thus:

$$M_{\delta y_3} \approx -\left(c + \frac{1}{2N_e}\right)D \tag{8A.4}$$

A similar approach can be used for determining the variance and covariance terms, which can then be substituted into **Equation 8A.3** (Ohta and Kimura 1971). We obtain the following:

$$V_{\delta y_1} \approx \frac{y_1(1 - y_1)}{2N_e}, \; V_{\delta y_2} \approx \frac{y_2(1 - y_2)}{2N_e}, \; C_{\delta y_1, \delta y_2} \approx \frac{y_3}{2N_e} \tag{8A.5a}$$

$$C_{\delta y_1, \delta y_3} \approx \frac{y_3(1 - 2y_1)}{2N_e}, \; C_{\delta y_2, \delta y_3} \approx \frac{y_3(1 - 2y_2)}{2N_e} \tag{8A5b}$$

$$V_{y_3} \approx \frac{\left[y_1(1 - y_1)y_2(1 - y_2) + y_3(1 - 2y_1)(1 - 2y_2) - y_3^2\right]}{2N_e} \tag{8A.5c}$$

This gives us a complete set of equations for describing the effects of drift and recombination. The next step is to define a suitable set of functions whose expectations are to be determined. These are suggested by the terms in **Equation 8A.5c**; we write $f = y_1(1 - y_1)y_2(1 - y_2)$, $g = y_3(1 - 2y_1)(1 - 2y_2)$, and $h = y_3^2$. Writing the expectations of these three functions as X, Y, and Z, respectively, and substituting into **Equations 8A.2–8A.5**, we obtain:

$$\Delta X \approx \frac{(Y - 2X)}{2N_e}, \; \Delta Y \approx \frac{\left[4Z - (5 + 2N_e c)Y\right]}{2N_e},$$

$$\Delta Z \approx \frac{\left[X + Y - (3 + 4N_e c)Z\right]}{2N_e} \tag{8A.6}$$

These terms give only the changes caused by drift and recombination, conditional on both sites segregating for a pair of variants. Under the infinite sites model, we also need to consider the effects of mutations entering the population from fixed sites. If we assume that the chance that two fixed sites simultaneously acquire mutations is negligible, and that new mutations are initially present at initial frequencies of $1/2N$, the mutational term for Y is close to 0, and the values for X and Z are equal to K and $K/2N$, respectively, where K is a constant that involves the mutation rate (Ohta and Kimura 1971).

The conditions for equilibrium under mutation, drift, and recombination are obtained by adding the mutational terms to **Equation 8A.6** and setting the results to 0. Because the resulting equations are linear in X, Y, and Z, they are completely soluble. It turns out that K can be eliminated from the ratio Z/X, giving **Equation 8.1** of the main text after neglecting small terms in $1/N$ (Ohta and Kimura 1971).

PROBLEMS

8.1   Show that **Equation B8.1.2** from **Box 8.1** in **Section 8.1.ii.a** follows from **Equations B8.1.1**.

8.2.   **i.** Use the formulae in **Boxes 8.1** and **8.2** of **Section 8.1.ii.a** to calculate the values of the three major descriptors of LD when the frequencies of the four haplotypes at a pair of sites are 0.6 (A_1B_1), 0.2 (A_1B_2), 0.1 (A_2B_1), and 0.1 (A_2B_2), respectively. Compare the results with those when the frequencies are 0.8, 0.1, 0.05, and 0.05, respectively.

 ii. Sixty alleles from a natural population gave the following results for the haplotype frequencies with respect to a pair of SNPs at the *Adh* locus of *Drosophila pseudoobscura*: $x_1 = 0.55$, $x_2 = 0.16$, $x_3 = 0.12$, and $x_4 = 0.17$. Determine whether there is evidence for statistically significant LD in these data (see **Box 8.3** of **Section 8.1.ii.a**).

8.3.   **i.** Use **Equations B8.4.2a–B8.4.2d** in **Section 8.2.ii.a** to derive **Equation B8.4.3**.

 ii. Assuming that c is $\ll 1$, obtain an expression for the half-life of the process of decay of LD for a pair of neutral sites in a randomly mating population.

8.4*   Use **Equation 7.16** of **Section 7.2.v.d** and **Equation 8.3** of **Section 8.2.iii.d** to derive the result that $\tilde{\rho} = 4N_{eT}c$ can be used to predict LD for alleles sampled from separate demes under conservative migration. Assume proportionality between deme size and the effective size of each deme, as described in **Box 7.3*** of **Section 7.2.ii.d**.

8.5*   Use **Equations B8.6.1a** and **B8.6.1b** in **Section 8.3.iii.a** and **Equation B7.4.1** of **Section 7.2.ii.e** to derive **Equation B8.6.2** for the mean coalescence time with background selection.

8.6*   **i.** Derive **Equation B8.7.1** of **Section 8.3.v.a**.

 ii. Show that the integral of $(1 - q_B)^{2c/s}(1/q_B)^{2c/s}$ in **Equation B8.7.3** is approximately 1.

8.7   A population is segregating for alleles A_1 and A_2 at a neutral locus with frequencies 0.2 and 0.8, respectively. A mutation occurs to a dominant allele B_2 at a closely linked locus (the recombination frequency is 0.001) such that B_2 has a frequency-dependent selective advantage with respect to the alternative allele B_1. B_2 occurs in an A_1 gamete and rises quickly to its equilibrium frequency of 0.25 without any crossovers having occurred.

   i. Calculate the initial coefficient of linkage disequilibrium between the two loci at this point in time.

   ii. Calculate the coefficient of linkage disequilibrium, D, when the frequency of A_2 has become equal to 0.5 among B_2 gametes as a result of crossing over.

   iii. How many generations will it take for the system to reach this state?

8.8   Use the definition of D, together with **Equations B8.8.1** of **Section 8.4.ii.a**, to derive **Equation B8.8.2c**.

8.9   **i.** Show that $\bar{e}$ in **Equation B8.8.4** of **Section 8.4.ii.a** is 0 if fitness effects are additive across loci.

   ii. Use the equilibrium versions of **Equations B8.8.1** to show that there is always an equilibrium with $D = 0$ with multiplicative fitnesses. (Use the fact that all the marginal fitnesses are equal to each other at an equilibrium with $D = 0$.)

8.10   Show that, with multiplicative fitness interactions between a pair of non-recombining loci with heterozygote advantage and symmetrical selection coefficients against each homozygote, new haplotypes are always eliminated by selection when introduced into a population at equilibrium for A_1B_1 and A_2B_2 alone, as discussed in **Section 8.4.ii.c**.

8.11*   Use **Equations B8.8.1** of **Section 8.4.ii.a**, together with **Equation B8.11.1** in **Section 8.4.ii.d**, to obtain **Equation B8.11.2**.

CHAPTER NINE

The Evolution of Breeding Systems, Sex Ratios, and Life Histories

CHAPTER SUMMARY

This chapter considers the evolution of three important types of traits that have great importance for the ecology of natural populations and for which the effects of natural selection are strongly affected by the rules of genetics:

1. The evolution of breeding systems (sexual vs. asexual reproduction, hermaphroditism vs. separate sexes, and the frequency of the production of offspring by inbreeding).
2. The evolution of the sex ratio in species with separate sexes.
3. The evolution of life history traits in populations with age structure.

We start by showing that an asexual variant introduced into a population with two sexes has an automatic two-fold transmission advantage (assuming no other effects on fitness), raising the question of why most multicellular organisms are sexual. Mutations causing increased self-fertilization (selfing) in hermaphrodites have a similar, but smaller, advantage. In order to deal with more complex problems, we describe a simple and general method for studying the evolution of the phenotypes of individuals that affect the breeding system of a population. This helps us to understand why many species have evolved mechanisms to prevent selfing, such as self-incompatibility. The results show that inbreeding depression is a major factor affecting the evolution of inbreeding avoidance mechanisms.

We also discuss the invasion of hermaphroditic populations by unisexual variants, leading to polymorphisms for females and hermaphrodites (gynodioecy), or (much more rarely) males and hermaphrodites (androdioecy). Gynodioecy may eventually evolve into a population with separate sexes (dioecy). A major factor in the evolution of unisexuality, in addition to inbreeding depression, is the allocation of resources between the two sex functions, and its reallocation by male or female sterility mutations. For dioecy to evolve, close linkage between the genes involved is required, and we explain how this helps to understand the initial evolution of sex chromosomes.

Many species with genetic sex determination have a sex ratio of close to 1:1 males and females among the zygotes. This may often simply be due to Mendelian segregation at a sex-determining locus or chromosome, but a 1:1 sex ratio is also predicted to evolve under natural selection under many circumstances. To study this topic, we introduce the concept of the Evolutionarily Stable Strategy (ESS). The ESS is the state of a population with respect to a phenotype like sex ratio, which is such that the population resists invasion by rare variants with altered phenotypes. For a panmictic dioecious population, the ESS is to expend equal resources on males and females, when the sex ratio is controlled by nuclear genes. In contrast, local competition among males for mates can select for a female-biased sex ratio. The predicted ESS sex ratios in haplodiploid species, whose females can directly control their offspring sex ratio, agree well with empirical data. Similarly, the proportion of resources invested in female versus male reproduction in hermaphrodites is affected by factors such as the level of self-fertilization.

Our third topic is the evolution of life histories, i.e., age-specific patterns of survival, fecundity, and growth. To understand life history evolution, it is necessary to model natural selection in populations with age structure, where the breeding individuals are of different ages. We briefly describe some basic concepts of demography, and then show how selection at a single locus can be modeled and fitness can be defined and measured. We then model life history evolution. We illustrate the concept of an optimal life history, using the example of a trade-off between age-specific survival and reproduction, and then discuss the two main theories of the evolution of aging.

> . . . [I]t may be doubted if it would be possible to point to any . . .
> character, with the possible exception . . . of sexuality itself,
> which could be interpreted as evolved for the specific
> rather than for the individual advantage.
>
> R. A. Fisher, *The Genetical Theory of Natural Selection,*
> 2nd edition, Chapter 2

> There is much difficulty in understanding why hermaphrodite
> plants should ever have been rendered dioecious.
>
> Charles Darwin, *The Different Forms of Flowers
> on Plants of the Same Species,* Chapter 7

INTRODUCTION

The aim of this chapter is to show how to use ideas and methods from evolutionary genetics to approach questions in evolutionary ecology, the branch of biology concerned with understanding how the major features of organisms reflect the selection pressures imposed by the environment. Many aspects of phenotypic evolution can be understood without detailed knowledge of the genetics of the traits concerned; provided that there is some additive genetic variance, it is usually reasonable to assume that the mean value of a trait will approach its optimum under natural selection (for some significant caveats, see **Section 3.3.iv**). Many problems in evolutionary ecology are, therefore, studied by calculating the selectively optimal values of traits, without reference to genetics (Maynard Smith 1982; Bulmer 1994; Krebs and Davies 1997). In this chapter, we describe some important examples of phenotypic evolution in which the constraints imposed by genetics play a crucial role.

The first topic is the evolution of mating systems. We ask: what factors control the evolution of sexual versus asexual reproduction, inbreeding versus outcrossing, and separate sexes versus hermaphroditism? Next, we study the evolutionary control of sex ratios in species with separate sexes and the allocation of resources to male and female functions in hermaphrodites. Both topics have long been studied by evolutionary biologists, and fundamental contributions to our understanding were made by Darwin (Darwin 1862, 1871, 1876, 1877), but their full development requires models based on population genetics principles. These models yield clear predictions that can be tested against comparative and experimental data.

The third topic concerns the action of selection in populations with age structure. This is crucial for understanding the evolution of species' life histories, including the duration of the juvenile stage, the pattern of change in body size with age, and age- and sex-specific patterns of mortality and reproductive success. Many of these characteristics vary greatly among species; for example, species of mammals have maximum life-spans ranging from a few months to over 100 years. In contrast, aging, the decline in nearly all aspects of bodily performance with age that results in an increase in mortality and decline in reproductive success, is an almost universal feature of the life histories of multicellular organisms, and even of some microbes. In addition to answering important questions about these topics, the results we discuss illustrate some general questions and principles relating to the evolution of phenotypes. An especially important example is the fact that selection can sometimes act in indirect ways, not just through individuals' survival or fertility, or that of close relatives (kin selection: see **Section 3.1.v.d**), but through the fitnesses of individuals' mates.

In addition, the evolutionary questions in this chapter often involve models in which one characteristic trades off against another, i.e., an increase in one function reduces performance with respect to another. Such situations may have the interesting property that characteristics can evolve that are apparently disadvantageous for the population as a whole. An example of this is a 1:1 sex ratio; this contrasts with the fact that a mostly female population produces more offspring, provided that enough males exist to fertilize the females' eggs or ovules. We shall see several examples in which a phenotype can evolve, even if it reduces the population's mean fitness in the short term. Evolution in a direction that is inimical to the long-term mean fitness of the population can also occur, such as the transmission advantage to asexual reproduction that we discuss below.

9.1. SEXUAL VERSUS ASEXUAL REPRODUCTION; INBREEDING VERSUS OUTCROSSING

9.1.i. Introduction

In this section, we consider three important aspects of breeding systems.

1. Is the population sexually reproducing or asexual?
2. If it is sexually reproducing, what sexual phenotypes are present in the population, i.e., what is the *sexual system*?
3. What determines whether individuals mate preferentially with relatives (i.e., they *inbreed*) or cross with randomly chosen members of the population (i.e., they *outcross*)?

9.1.ii. Asexual versus sexual reproduction

9.1.ii.a. *Introduction*

Sex is the most prevalent mode of eukaryote reproduction, whereas a regular cycle of sexual reproduction is absent from prokaryotes (Maynard Smith 1978; Bell 1982). Regular sexual reproduction (with a life cycle involving the fusion of haploid gametes to produce a zygote, with a subsequent two-stage meiotic division at another life cycle stage, resulting in four haploid products) probably evolved very early in eukaryote history. Most, if not all, asexually reproducing eukaryotes are thus the result of the secondary evolution of asexuality. They nearly all have sexual relatives, and have rarely proliferated into diverse forms, suggesting that they evolved recently (Maynard Smith 1978; Bell 1982; Normark et al. 2003). There are, however, some exceptions. In particular, the Bdelloid rotifers have evolved into several hundred "species," with substantial anatomical and molecular differences. No males have been found and a recent analysis of their genomes suggests that they have been asexual for many millions of years (Mark Welch et al. 2008).

9.1.ii.b. *The cost of sex*

A fundamental problem for understanding the prevalence of sexual reproduction among multicellular organisms is the *cost of sex*, first brought to the attention of biologists by John Maynard Smith (1971a). Consider a large sexual population with separate sexes and equal numbers of males and females. Imagine a mutation that causes females to reproduce asexually, with all-female broods, simply replacing males with females while leaving the average number of offspring unchanged. Mutant females thus produce twice as many daughters as their competitors, and so the frequency of the mutants within the female population initially doubles each generation. They therefore spread rapidly through the population, replacing the sexual females and causing the extinction of males. There is thus a two-fold "cost" of sex, which is an example of a *transmission advantage* to a variant, due solely to its effect on the transmission of genes from one generation to the next. **Box 9.1** models and generalizes this scenario.

Box 9.1 SPREAD OF AN ASEXUAL VARIANT INTO A POPULATION WITH SEPARATE SEXES

Assume that the sexually reproducing component of the population contains adult males and females with frequencies c and $1 - c$, respectively (c should not be confused with the c used for the frequency of recombination in **Chapter 8**). An asexual variant is present at frequency q among the females of a given generation; it causes females to produce only daughters that are genetically identical to their mothers, whereas a proportion $1 - c$ of the offspring of sexual females are female. Let the fitness of asexual females relative to sexual females be $1 - s$, where s may be > 0, e.g., because of defects in egg cell production associated with the replacement of regular meiosis by a novel mechanism of cell division. The new frequency of the asexual variant in the female population is thus:

$$q' = \frac{q(1 - s)}{p(1 - c) + q(1 - s)} \tag{B9.1.1}$$

When the asexual variant is rare, this is approximated by:

$$q' \approx \frac{q(1 - s)}{(1 - c)} \tag{B9.1.2}$$

The asexual variant will therefore spread if and only if $s < c$. This condition is hardest to satisfy if the population is heavily female biased. With a 1:1 sex ratio, which is approximately the case for most species with separate sexes (**Section 9.4.i**), s would have to be > 0.5 to prevent the spread of the variant; if asexuals escape a fitness penalty, so that $s = 0$, a rare variant will double its frequency each generation.

A cost of sex also exists with other types of sexually reproducing populations. In outcrossing hermaphrodite plants, the initial advantage of a mutant producing asexual eggs (but that still produces pollen by meiosis) is close to 1.5, still a substantial advantage (Charlesworth 1980a; Lloyd 1980) (**Problem 9.1**). A well-studied example is the dandelion, *Taraxacum officinalis*, where plants in most parts of Europe are triploid and propagate through asexually produced seeds (Van Dijk 2003).

9.1.ii.c. *The maintenance of sexual reproduction*

Sexual populations of multicellular organisms are thus highly vulnerable to invasion by asexual variants, unless the asexuals have lower fertility, e.g., if their mechanism for producing functional egg cells and embryos is imperfect (Lamb and Willey 1979; Bierzychudek 1987). The prevalence of sexual reproduction would thus seem to require a very large fitness advantage of sexually reproducing individuals relative to asexual ones. However, several population processes can reduce the long-term mean fitness of an asexual lineage, compared with its sexual relatives. We defer a detailed discussion of these processes until **Sections 10.2.iv–10.2.vi**, because they are closely connected with the advantages and disadvantages of genetic recombination. It is easy to see, however, that if the extinction rate of asexual lineages relative to sexuals is higher than the rate of origin of new asexual lineages, the frequency of asexual lineages is determined by a similar process to that of mutation and selection for deleterious variants within a population, as described in **Section 4.2.ii.a** (Van Valen 1975; Nunney 1989). A low origination rate, compared with the extinction rate, can thus explain the sporadic distribution of asexual taxa among predominantly sexual lineages, without involving any large selective advantages to sexual reproduction.

This may well be a major factor in the long-term evolutionary maintenance of sexual reproduction, though it cannot explain the initial origins of sex and recombination. However, primitive unicellular eukaryotes in which cells with mating competence are the same size (*isogamy*) probably lack a two-fold cost of sex. Asexual variants may have little or no advantage in such situations; there is no division into males and females, and so there is no analogue to the advantage of producing all-female progeny (Maynard Smith 1978, p. 3; Charlesworth 1980a). In ancestral eukaryotes, a relatively slight selective advantage to sex at the within-population level could have driven its initial evolution.

We will return to the possible selective advantages of sexual reproduction in **Section 10.2**; in the rest of this chapter, we will discuss only the much simpler, and better understood, problem of the advantages and disadvantages of outcrossing.

9.1.iii. Sexual systems and mating systems

9.1.iii.a. *Sexual systems*

Before discussing the processes involved in the evolution of sexual systems, we first briefly survey the main types of system that exist. **Table 9.1** shows some of the commonest sexual systems of plants and animals, and the terminology used to describe them. It is helpful to distinguish between systems in which all individuals are essentially alike (i.e., sexually *monomorphic*, as in *hermaphroditic* organisms), and sexually *polymorphic* species, in which there are different kinds of individuals (as in species with separate sexes: *dioecy*). Many animals are hermaphrodites (e.g., some slugs and snails, and even fish: Jarne and Charlesworth 1993). Flowering plants have a large diversity of sexual systems (Barrett 2002), requiring some complex terminology; unfortunately, some of the terms differ between animals and plants. In plants, unlike most animals, the reproductive structures are repeated modules, i.e., there are usually multiple flowers on an individual. These often all have both male and female functions, in which case the plant is a true hermaphrodite. In some species, some flowers are hermaphrodites and others have only male or female parts. In other cases, all flowers are unisexual, with male and female flowers carried on the same individual (*monoecy*). The term *cosexual* is often used by botanists to describe individuals with both sex functions, and we use this term below. In animals, with just testes and ovaries, there is no need to distinguish between hermaphrodite and monoecious species, and zoologists use these terms interchangeably.

For many plants and animals, the sexual system is easily established by morphological examination. However, morphology can be misleading in dioecious plants, because female plants often have rudimentary anthers and males may have quite well-developed ovaries, as Darwin noticed for spindle trees (*Euonymus europaeus*) (Darwin 1877, pp. 287–293); this suggests that dioecy in these cases evolved recently from an hermaphrodite ancestor. Populations of the same species may also differ; for instance, in gynodioecious species, the frequencies of females often differs widely between different populations (Manicacci et al. 1997; Nilsson and Agren 2006), and some populations may have no females, but only cosexuals (Ganders 1978; Ramsey et al. 2006).

9.1.iii.b. *Mating systems*

A cosexual organism runs the risk of self-fertilization, with the associated reduced fitness of the resulting progeny (*inbreeding depression*) that we discussed in **Sections 1.1.iii** and **4.4**. As Darwin first pointed out (Darwin 1876, 1877), this should result in selection for avoiding inbreeding. Indeed, this applies even to dioecious species, if there is a possibility of matings between relatives. Many non-hermaphrodite animals have behaviors that lead to avoidance of inbreeding, including incest taboos and dispersal (Morgan 2002). It is difficult to determine mating systems in such populations without very laborious

TABLE 9.1 Some plant and animal sex systems, with some examples, and rough estimates of their frequencies

Term	Definition and examples in plants	Occurrence and terminology	Animal cases	Inbreeding
ASEXUAL				
Apomictic	Seeds have the same genotype as maternal parent (male gamete sometimes fertilizes endosperm: pseudogamy)	Several onion species	Aphids, Daphnia (often cyclical) A few species of col-blooded vertebrates	No
SEXUALLY MONOMORPHIC				
Cosexual				
Hermaphrodite	Both male and female functions; in plants, both functions within each flower	~ 90% of flowering plants (Cleistogamous flowers self-fertilize before flowers open; they must be hermaphrodite; often underground flowers, e.g., peanuts)	Less common; many slugs, snails, earthworms, some fish, some parasites	Possible
Monoecious	Separate sex flowers on the same individuals	~ 5% of flowering plants, often those with catkins (e.g., hazel) and many gymnosperms (e.g., pines)	No corresponding situation	Possible between flowers
SEXUALLY POLYMORPHIC				
Dioecious (unisexual)	Separate sex individuals (male and female plants)	~ 5% of flowering plants (e.g., holly) and some gymnosperms (e.g., yew, cycads)	Many animals (Gonochoristic)	No
Gynodioecious	Cosexual and female individuals	~ 5% of flowering plants	A few animals	Possible
Androdioecious	Cosexual and male individuals	Very rare	A few animals (e.g., C. elegans)	Possible

observations, and it is also difficult to test whether any particular behavior has evolved due to selection to avoid inbreeding, or for other reasons (Waser et al. 1986).

Plants and, to a lesser extent, hermaphrodite animals such as slugs and snails are therefore used for much work that is directed towards understanding breeding system evolution, and most such work deals with the frequencies of outcrossing versus self-fertilization (selfing) in cosexual populations. As Darwin documented in detail (Darwin 1862, 1876, 1877), many such species have mechanisms that prevent selfing, such as separate timing of male and female reproduction (*dichogamy*). Plants with *protandrous* flowers, which release pollen before their stigmas are receptive, are especially common (Barrett 2002). Spatial separation of male and female reproduction is also frequent. Flowers of many plants have anthers far from the receptive surface of the stigma, which lowers the chance of self-fertilization (Brunet and Eckert 1998; Barrett 2002), and hermaphrodite animals often have separate ducts for the male and female gametes.

9.1.iii.c. *Self-incompatibility*

A particularly interesting inbreeding avoidance mechanism is represented by genetically controlled chemical self-recognition, or *self-incompatibility* (SI), which we described briefly in **Section 2.2.i.b**. In addition to flowering plant SI (reviewed by Charlesworth et al. 2005), SI is known in fungi (Casselton 1997; Kronstad and Staben 1997; Yun et al. 1999) and in some sessile animals, including bryozoans and sponges (Grosberg 1988) and ascidians (Grosberg 1991; Pemberton et al. 2004; Harada et al. 2008).

In most self-incompatible plants, the plants and the flowers of all types are morphologically indistinguishable (*homomorphic* systems), and incompatibility types at the loci controlling incompatibility (*S* loci) can be determined only by laborious testing of family members with pollen from others. There are two kinds of homomorphic SI systems, differing in the genetic control of the pollen types. In *gametophytic* systems, the pollen types are controlled by the haploid genotypes of the pollen grains themselves, i.e., by the gametophyte genotypes (Kao and Tsukamoto 2004). A plant heterozygous for two alleles at the S locus produces pollen of two incompatibility types, i.e., $S_1 S_2$ plants produce S_1 and S_2 pollen. The S_1 type of pollen is rejected by plants of genotypes such as $S_1 S_2$ or $S_1 S_3$, but is accepted by $S_2 S_3$ (but this genotype rejects the S_2 half of the pollen from an $S_1 S_2$ plant). In *sporophytic* systems, the incompatibility type of the pollen, as well as the stigma, is controlled by the diploid S locus genotype of the plant that produces it (Nasrallah 2002); if a given S allele is dominant over another allele, the pollen produced by a heterozygote for these alleles will all have the same incompatibility type. These SI systems maintain large numbers of alleles at the S loci, because of the frequency-dependent selection in favor of new S alleles (see **Sections 2.2.i.b** and **8.3.i**).

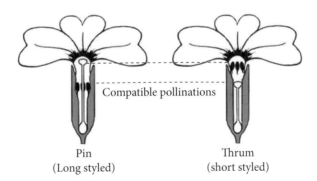

Compatible pollinations

Pin
(Long styled)

Thrum
(short styled)

FIGURE 9.1 Drawings of longitudinal sections of long- and short-styled flowers of the distylous plant, the primrose *Primula vulgaris*, corresponding to the photographs in **Plate 4,** back endpaper. The drawing shows the positions of the flower parts (note the anther and pistil heights) and indicates the pollen transfers that are compatible. [Adapted from Figure 1 of Darwin (1877).]

Apples, cherries, some potato and tomato species (family Solanaceae), poppies (*Papaver rhoeas*), and clovers (e.g., *Trifolium repens*) have gametophytic SI (Igic and Kohn 2001), and members of the Brassica (cabbage) and daisy families have sporophytic systems (Nasrallah 2002; Hiscock and McInnis 2003). Homomorphic SI probably exists in fewer than 80 of the more than 300 flowering plant families, but its distribution is still imperfectly known (Igic et al. 2008).

Another type of plant incompatibility system is *heteromorphism*: flowers of the different incompatibility types are morphologically different (reviewed by Barrett 1992). These are called *heterostylous* species. Some of these have two flower morphs (*distyly*), as in primroses (*Primula vulgaris*), which are controlled by a pair of alleles that segregate at a single locus (see **Plate 4**, back endpaper and **Figure 9.1**). Others have three morphs (*tristyly*), as in purple loosetrife, *Lythrum salicaria* (Darwin 1877). Each "morph" can fertilize the other two, because a sporophytic incompatibility system is associated with the flower morphs. This is the closest that any organism comes to having three morphologically distinct sexes, although these plants are all hermaphrodites.

9.1.iii.d. *Detecting outcrossing and estimating outcrossing rates in natural populations*

If a population is cosexual, and if individuals are not self-incompatible, establishing the mating system involves estimating the rates of outcrossing versus self-fertilization. Simply establishing that some outcrossing occurs can be valuable information. For instance, in many fungi it is very difficult to find the mating stages, which are often microscopic, so the occurrence of sexual reproduction can be hard to detect. However, population genetic analyses can be used to detect recombination in DNA sequences, implying outcrossing (Cooper et al. 2007; Douhan et al. 2007).

Some organisms that reproduce largely by inbreeding were mentioned in **Section 1.3.ii.c**. The most extreme inbreeding occurs when a haploid individual produces male and female gametes capable of mating with each other. This can occur in mosses and ferns, where in some species the haploid gametophyte stage produces both sperm and eggs cells, which are compatible and can undergo syngamy and produce zygotes. Intra-gametophyte selfing generates completely homozygous progeny. Inbreeding can also occur in organisms other than hermaphrodites. For instance, some species of *Daphnia* or aphids are cyclically parthenogenetic (i.e., during part of the year reproduction is asexual, with females producing asexual daughters, but a sexual phase occurs, often in the early winter, after females switch to producing some males as well as females). Male and female individuals produced by the same asexual female lineage may then mate, which is genetically equivalent to self-fertilization (Haag and Ebert 2007). Similarly, in malaria parasites (*Plasmodium*) the haploid stage male and female gametes produced by the same diploid genotype in an infection involving a single parasite genotype can mate (Read et al. 1992).

Outcrossing rates can be estimated from the segregation of genetic markers in the progenies of individuals sampled from natural populations (Ritland 1990b; Jarne and David 2008). Alternatively, one can assume that the population is at equilibrium under partial self-fertilization in the mixed-mating model of **Section 1.3.ii.c** (which assumes that individuals arise either from self-fertilization of the female gametes, or else by fertilization with male gametes randomly drawn from the parental population of hermaphrodites). The mating system of such a population can be described in terms of just one parameter, the self-fertilization frequency S; according to **Equation 1.4**, this results in an equilibrium inbreeding coefficient, $f^* = S/(2 - S)$, and thus determines the equilibrium genotype frequencies of neutral variants according to **Equation 1.3**. The genotype frequencies of genetic markers in populations, together with this equation, can be used to estimate the value of S (see **Problem 1.5**).

The distribution of self-fertilization rates among species of flowering plants estimated in these ways includes nearly all possible levels (**Figure 9.2**), but few, if any, species reproduce entirely by self-fertilization. Even plants whose flowers self-fertilize in the bud stage (cleistogamous species; see Schoen and Lloyd 1984) also have open flowers that are visited by insects. Intermediate selfing rates are common in both plants (Baker 1959; Lloyd 1979) and hermaphrodite animals (Jarne and Charlesworth 1993; Jarne and Staedler 1995), but extreme values are overrepresented, i.e., the distribution of selfing rates is somewhat bimodal (**Figure 9.2**).

In a plant, self-fertilization may involve either within-flower selfing (*autogamy*), or the transfer of pollen by pollinators from one flower to another on

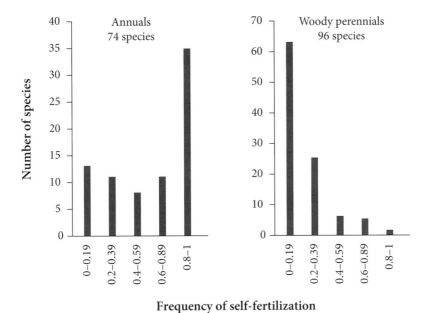

FIGURE 9.2 Distributions of self-fertilization rate estimates in plant species, illustrating the different distributions in annual versus long-lived plants. [Redrawn from Figure 2 of Barrett et al. (1996).]

the same plant (*geitonogamy*, which must involve pollinators). Autogamous self-fertilization can be tested for by removing all flowers except one. If selfed progeny are still produced, within-flower selfing must occur (Schoen and Lloyd 1992).

9.2. THE EVOLUTION OF INBREEDING AND OUTBREEDING

9.2.i. Introduction

Evolutionary shifts from partial inbreeding to full outcrossing are common in plants, fungi, and some animal taxa. Flowering plants have evolved separate sexes from an initially cosexual condition, independently in many different taxonomic groups, sometimes also evolving sex chromosome systems. These are usually male XY, female XX, like the human or *Drosophila* system, although there are a few species with male ZW, female ZZ systems, as is frequently found in non-mammalian vertebrates (Bull 1983, Chapter 1). Evolutionary changes in breeding systems can also reverse, although very highly selfing species probably rarely become outcrossing again (Barrett 2002).

Most plant genera with self-incompatible members also include some self-compatible species, and self-compatible populations of the same species may differ in their outcrossing rates. Many inbreeding plant species and populations evolved recently from outcrossers (e.g., Brauner and Gottlieb 1987; Holtsford and Ellstrand 1989; Macnair et al. 1989; Barrett and Husband 1990); a familiar example is *Arabidopsis thaliana*, whose closest relatives are self-incompatible (Charlesworth and Vekemans 2005). Self-incompatibility is rare in the floras of oceanic islands such as Hawaii, whose floras are mainly recent colonists (Baker 1955) (this is often called "Baker's rule"); similarly, self-compatibility is associated with the successful colonization of North America by European plants (Kleunen and Johnson 2007).

The direction of such changes can sometimes be determined from phylogenies estimated from DNA sequence data. In the nematode genus *Caenorhabditis*, *C. elegans* is an inbreeder, but other species, such as *C. remanei*, are dioecious, and phylogenetic analyses show that dioecy is the ancestral condition (Kiontke et al. 2004). In several plant genera, inbreeding has been shown to be derived from outcrossing, with the selfers being of relatively recent origin (e.g., Wyatt 1990; Schoen et al. 1997; Bena et al. 1998). In the genus *Linanthus*, for example, the incompatible ancestral state has been lost several times (Goodwillie and Stiller 2001). Cases of taxa with separate sexes reverting to hermaphroditism are also known among animals. Selfing animals are often colonizers, e.g., slugs (McCracken and Selander 1980) and a species of fish that lives in ephemeral habitats (Mackiewicz et al. 2006a,b). Many parasites, whose life cycles include the colonization of hosts, can self-fertilize (e.g., Milinski 2006).

We will next discuss some of the major concepts that help us understand these changes and counter-changes, particularly two important questions:

1. Why do cosexual species go to the trouble of mating with other individuals (outcrossing), instead of self-fertilizing?
2. Why have cosexual species often evolved two sexes, instead of remaining cosexual (this is the topic of **Section 9.3** below)?

Table 9.2 lists some of the main factors that exert selection on breeding systems, and their relevance to different known breeding systems.

9.2.ii. The evolution of selfing rates

9.2.ii.a. *Introduction*

A self-compatible plant could evolve a higher level of self-fertilization merely by having smaller flowers (Dole 1992), because (as mentioned in **Section 9.1.iii.b**) anther–stigma spatial separation often affects outcrossing rates (Barrett and Shore 1987; Brunet and Eckert 1998). Mutations causing loss of self-incompatibility are also common in plants (e.g., self-compatible apple culti-

TABLE 9.2 Four major factors that exert selection on breeding systems

| Breeding system | Factor[1] | | | |
	Transmission of genes	Relative survival of inbred versus outbred progeny	Allocation of reproductive resources	Mate availability
Apomixis	+	Not involved	Not involved	+
Self-fertilization	+	+ (inbreeding vs. outbreeding depression)	Not involved	+
Separate sexes (dioecy)	+	+ (avoidance of inbreeding depression)	+	+

[1]The factors involved in each case (indicated by +) are discussed and explained in the text.

vars). Why might changes in breeding systems to increased selfing be advantageous in nature?

In order to answer this question, it is important to understand that there are several possible advantages to self-fertilization. We have already mentioned the fact that reproduction may fail in colonizing situations, because self-incompatible plants then have no compatible neighbors, and pollination may be inefficient when plant densities are low. Such situations cause selection for self-fertilization, because it provides *reproductive assurance* (listed in the right-hand column of **Table 9.2**). Apomixis and autogamy indeed tend to occur in plant populations at the edges of species' ranges (Bierzychudek 1990), as does breakdown of dioecy by invasion of cosexual forms (e.g., Huff and Wu 1992; Pannell 2002; Pannell et al. 2008).

Selfing can also allow locally adapted genotypes to persist, because it reduces gene flow between different populations (see **Section 4.1**). This can favor selfing in some special situations, such as populations adapted to lead or copper mines (Antonovics 1968; Macnair et al. 1989). Similarly, when there is local adaptation, outcrossing between distant populations may reduce progeny fitness (*outbreeding depression*; see Galloway 1989).

9.2.ii.b. *The transmission advantage of selfing (the cost of outcrossing)*
A further important advantage of inbreeding is increased transmission of genes to the next generation. As with asexuality (**Section 9.1.ii.b**), selfing has a strong advantage, due to the relatedness between parents and their offspring. A cosexual individual producing zygotes by selfing provides both alleles at each locus to the resulting progeny, and can additionally potentially sire offspring by fertilizing other individuals. A rare selfing genotype thus has a 50% transmission advantage. **Box 9.2** shows how this arises, using the method we introduced for analyzing the maintenance of polymorphisms in **Section 2.1.ii.d**, and which we will employ extensively throughout this chapter.

The simple model in **Box 9.2** can be generalized to partial selfing (Lloyd 1979; Charlesworth 1980a) and other forms of inbreeding. For instance, *biparental inbreeding* (crosses between relatives, rather than inbreeding due to selfing) reduces the transmission advantage of selfing relative to crossing, and can lead to the maintenance of intermediate levels of selfing (Uyenoyama 1986).

9.2.ii.c. *Why do many organisms outcross? Inbreeding depression and the evolution of selfing rates*
Given the advantages of selfing just outlined, a strong counter-selection force is needed in order to account for the prevalence of outcrossing species and for the evolutionary origin of outcrossing systems, such as self-incompatibility, whose phylogenetic distribution and genetic basis indicate that it has originated several different times, presumably from partially selfing ancestors (**Section 9.1.iii.c**). The effective lack of genetic recombination in highly inbreeding species, due to their high degree of homozygosity (which reduces the effective-

Box 9.2 SELECTION ON A VARIANT THAT CAUSES SELF-FERTILIZATION

This model is based on Fisher (1941) and Nagylaki (1976b). It examines the invasibility of an outcrossing population by a dominant allele, A_2, causing complete self-fertilization, taking into account only success in reproducing (i.e., the transmission of gametes to the progeny generation), without any direct effects of the allele on survival or fertility, just as for the case of an asexual mutant described in **Box 9.1** of **Section 9.1.ii.b**. Because two of the genotypes in the system can self, we cannot assume Hardy–Weinberg genotype frequencies, and the system must be described in terms of genotype rather than allele frequencies. The parameters of the model are as follows:

Genotype	A_1A_1	A_1A_2	A_2A_2
Selfing rate	0	1	1
Frequency	x	y	z

The frequency of the A_1 allele in male gametes is $p = x + y/2$, and the frequency of A_2 is $q = z + y/2$. We can now work out the genotype frequencies in the progeny of such a population and find out whether the frequency of the allele causing selfing will increase. Because A_1A_2 individuals and A_2A_2 parents are assumed to reproduce by complete self-fertilization, the progeny of A_2A_2 individuals are all A_2A_2, and A_1A_2 parents produce the three genotypes in the Mendelian ratio 1 A_1A_1:2 A_1A_2:1 A_2A_2. The progeny of A_1A_1, however, are produced by outcrossing, i.e., there is random fertilization of A_1A_1 by gametes produced by the population as a whole, with frequencies p and q of A_1 and A_2, respectively. These progeny all receive the A_1 allele from their maternal parent, and will thus be A_1A_1 with frequency p and A_1A_2 with frequency q.

These genetic results yield the following genotype frequencies in the next generation:

$$A_1A_1 \qquad x' = xp + (y/4)$$
$$A_1A_2 \qquad y' = xq + (y/2)$$
$$A_2A_2 \qquad z' = (y/4) + z$$

The new frequency of the A_2 allele, q', is equal to $z' + (y'/2)$, so that:

$$q' = \frac{y}{4} + z + \frac{xq}{2} + \frac{y}{4} = \frac{y}{2} + z + \frac{xq}{2} = q\left(1 + \frac{x}{2}\right) \qquad (B9.2.1)$$

It follows that $q'/q = 1 + x/2$. When the A_2 allele is very rare, so that $x \approx 1$, q'/q is approximately 3/2. This means that a rare allele causing complete selfing increases in frequency by 50% in one generation.

ness of recombination, as explained in **Section 8.2.ii.b**), means that they suffer long-term fitness disadvantages similar to those of asexual populations that will be discussed in **Sections 10.2.iv–10.2.vi**. Selection acting through the differential success of different lineages, like that suggested to explain the phylogenetic distribution of asexuality (**Section 9.1.ii.c**), also applies to selfers, and probably explains why selfing species do not often persist for long evolutionary times.

However, the evolution of complex adaptations to avoid selfing, such as the SI mechanisms described in **Section 9.1.iii.c**, cannot be explained by appealing to the differential extinction of selfing lineages, because they require mutations that reduce the selfing rate to spread within partially selfing populations. A resolution of this puzzle is that inbreeding depression is a powerful force, which can cause natural selection to favor mechanisms that ensure outcrossing, as mentioned already (**Section 9.1.iii.b**). The 50% transmission advantage of selfing explained above suggests that, to a rough approximation, an increased rate of outcrossing will be favored only if the mean fitness of the progeny of selfing is less than half that of outcrossed progeny, i.e., the inbreeding depression, δ, must exceed 0.5 (Lloyd 1979; Charlesworth 1980a; Lande and Schemske 1985), where δ is defined as the difference between fitnesses of outcrossed and selfed progeny, divided by the outcrossed fitness value. Conversely, with $\delta < 0.5$, increased selfing should evolve, at least on the simple assumptions of **Box 9.2**, despite causing the population to have more inbred individuals with their reduced fitness.

This result illustrates the important principle that selection acting on individuals within a population does not necessarily lead to species having the characteristics that are optimal for the population, but only to characteristics that compete successfully with their alternatives within the population (Fisher 1941). This difference between the expected outcomes of individual selection and group selection was shown by Lloyd (1979) to imply that individual rather than group selection usually determines the outcome of the evolution of mating systems, contrary to the beliefs then current.

The model of **Box 9.2** is the simplest possible one; it assumes that individuals with different selfing rates do not differ in any other way. For instance, identical pollen output is assumed for a selfing mutant in a plant (or male gamete output for an animal), regardless of the selfing rate, and this may not be biologically plausible. As mentioned in **Section 9.2.i**, selfing may provide reproductive assurance by allowing fertilization to occur when pollination by other individuals is unsuccessful. Outcrossing individuals may then produce fewer offspring than selfers; this has the same effect as reducing δ. Ecological conditions that lead to reproductive assurance by selfing can thus overcome the disadvantage of inbreeding and favor the evolution of selfing (see **Section 9.2.iii** below).

Conversely, selfing in plants often evolves through a reduction in flower size; for instance, inbreeding species in the genus *Mimulus* have much smaller flowers than their more outcrossing relatives (Dole 1992). If this evolved during the initial change to selfing (rather than after selfing evolved), it might also reduce the amount of pollen available for outcrossing, as well as attractiveness to pollinators. If selfing is geitonogamous, for example, the loss of pollen carried to other individuals can be very high. Reduced pollen export in the selfing form (termed *pollen discounting*) reduces its advantage relative to individuals that can fertilize other plants. Pollen discounting thus reduces the threshold δ value that prevents selfing from evolving, making self-fertilization less likely to evolve (Lloyd 1979; Charlesworth 1980a). This quantitative difference does not, however, alter the basic principles underlying the evolution of selfing rates. Cleistogamous flowers have complete pollen discounting, so their within-flower selfing gives no transmission advantage. The only advantage of this form of selfing is reproductive assurance.

Our simple model assumes that each genotype has a fixed selfing rate. However, some species' flowers have mechanisms that allow flowers to self-pollinate if pollen is not received from other individuals. Such "delayed selfing" can evolve by providing reproductive assurance, and is always advantageous, even if there is strong inbreeding depression (Lloyd 1979).

Box 9.3 shows how we can obtain results for some of these more general models, using an approach known as the *phenotypic selection* method (Lloyd 1977). Instead of deriving recursion relations for all the genotype frequencies (which are often hard to handle when there is nonrandom mating), this method examines the different phenotypes' relative genetic contributions to the next generation. Although this method is subject to some limitations (discussed in **Box 9.3**), it greatly simplifies the calculations, and we will use it repeatedly from now onwards. It can be shown to yield the same conclusions as the full genotype recursion relations, provided that certain conditions are satisfied, such as complete dominance of one of the alleles at a biallelic locus controlling variation in the phenotype (Lloyd 1977; Charlesworth and Charlesworth 1978; Slatkin 1978).

Box 9.3 PHENOTYPIC SELECTION MODELS AND THE EVOLUTION OF SELF-FERTILIZATION

B9.3.i. A general model

This method was formalized by Lloyd (1977), Charlesworth and Charlesworth (1978), and Slatkin (1978). The idea is that we can define a fitness measure for individuals who share a common phenotype, i, by the fraction of the gametes present among adults of the next generation that are contributed by individuals of class i, divided by the total number of gametes contained in individuals of this class in the present generation. The mean fitness of the population on this definition is always 1. Contributions through female and male gametes must be taken into account separately, and we denote these by w_i^f and w_i^m, respectively. In most situations considered later in this chapter, either or both of these are affected by the frequencies of the different phenotypes in the population.

In the case of autosomal inheritance, there are equal contributions to each zygote from male and female gametes, so that:

$$w_i = \frac{1}{2}\left(w_i^f + w_i^m\right) \tag{B9.3.1}$$

Different relations apply to different modes of inheritance; for example, with purely maternal transmission, only w_i^f contributes to w_i.

For the autosomal case, assume for simplicity that there are three genotypes, A_1A_1, A_1A_2, and A_2A_2, but only two phenotypes, 1 and 2, i.e., the number of phenotypes in the system is less than the number of genotypes at the locus controlling them. (This is a general requirement for the utility of this method; see Lloyd 1977; Slatkin 1978). In the present case, this corresponds to assuming that either A_2 or A_1 is dominant, with A_1A_1 individuals being associated with phenotype 1 and A_2A_2 individuals with phenotype 2. An expression for the change in frequency of A_2 can then be derived (e.g., Charlesworth and Charlesworth 1978) and shows that the frequency of A_2, and hence of phenotype 2, will increase whenever $w_2 > w_1$ and will decrease when $w_2 < w_1$. There is an equilibrium with both phenotypes present when $w_1 = w_2$.

B9.3.ii. The evolution of increased self-fertilization

These general results can be applied as follows to the case of a dominant allele, A_2, which causes an increase in the selfing rate of a cosexual plant from S_1 to S_2. In order to incorporate possible effects of pollen discounting (see main text), we assume that the amount of pollen contributed by the more highly selfing phenotype 2 individuals to the pool available for fertilizing outcrossed ovules is multiplied by a factor of $1 - \lambda$, compared with that for phenotype 1 individuals. To model inbreeding depression, let the fitness of selfed individuals be $1 - \delta$, relative to that of outcrossed individuals. The viabilities and seed outputs of the two phenotypes are assumed to be equal. Let the frequencies of individuals with phenotypes 1 and 2 be X and $1 - X$, respectively.

The total contribution of female gametes to the next generation, assuming equal fertility of the two phenotypes, is thus proportional to:

$$W = X[1 - S_1 + S_1(1 - \delta)] + (1 - X)[1 - S_2 + S_2(1 - \delta)]$$
$$= X(1 - S_1\delta) + (1 - X)(1 - S_2\delta) \qquad (B9.3.2)$$

The total contribution of male gametes to the next generation is given by the same expression, because the two sexes contribute equal numbers of gametes.

The female fitness terms are thus given by:

$$w_1^f = \frac{(1 - S_1\delta)}{W}, \; w_2^f = \frac{(1 - S_2\delta)}{W} \qquad (B9.3.3)$$

There are two contributions to the male fitnesses, one through the fertilization of outcrossed ovules, and the other via self-pollination. For the whole population, the first of these is proportional to $V = X(1 - S_1) + (1 - X)(1 - S_2)$. The size of the pollen pool used in outcrossing is proportional to $U = X + (1 - X)(1 - \lambda) = 1 - (1 - X)\lambda$. The per capita contributions through outcrossing to male fitness are thus proportional to V/U and $(1 - \lambda)V/U$ for phenotypes 1 and 2, respectively. The contributions through self-pollination for the two phenotypes are the same as for female gametes, i.e., $S_1(1 - \delta)$ and $S_2(1 - \delta)$, respectively. Normalizing by the total gamete contribution to the next generation, W, we obtain:

$$w_1^m = \frac{1}{W}\left[\frac{V}{U} + S_1(1 - \delta)\right] \qquad (B9.3.4a)$$

$$w_2^m = \frac{1}{W}\left[\frac{V(1 - \lambda)}{U} + S_2(1 - \delta)\right] \qquad (B9.3.4b)$$

Equations B9.3.3 and B9.3.4 can be substituted into Equation B9.3.1 to obtain the net fitnesses of the two classes. This enables us to ask whether A_2 can increase when rare by comparing w_1 and w_2 when X is set to 1, which gives $V = 1 - S_1$ and $U = 1$. We obtain:

$$w_1 = \frac{1}{2W}\left[1 - S_1\delta + (1 - S_1) + S_1(1 - \delta)\right] = 1 \qquad (B9.3.5a)$$

and

$$w_2 = \frac{1}{2W}\left[1 - S_2\delta + (1 - S_1)(1 - \lambda) + S_2(1 - \delta)\right]$$

$$= \frac{1}{2(1 - S_1\delta)}\left[2 - S_1 - (1 - S_1)\lambda + S_2(1 - 2\delta)\right] \quad \text{(B9.3.5b)}$$

The difference, $w_2 - w_1$, is thus proportional to:

$$(S_2 - S_1)(1 - 2\delta) - (1 - S_1)\lambda \quad \text{(B9.3.6)}$$

If there is no pollen discounting ($\lambda = 0$), this shows that a dominant or recessive allele that increases the selfing rate will spread if $\delta < 1/2$. If there is complete discounting, so that $\lambda = S_2 - S_1$, the condition is more severe, i.e., we need $S_1 > 2\delta$ for an allele increasing selfing to spread. This is impossible to satisfy if the initial population is wholly outcrossing ($S_1 = 0$), but can be satisfied for a sufficiently high initial selfing rate.

9.2.ii.d. *More realistic models of inbreeding depression; purging of deleterious mutations*

The models above included the inbreeding depression level as a fixed parameter. This is unrealistic, because it ignores its genetic basis (discussed in **Section 4.4.iv**), which implies that δ changes as the breeding system evolves. If inbreeding depression is mainly caused by deleterious mutations with recessive or partially recessive effects, inbreeding exposes the mutations to selection and their frequency will be reduced (**Section 4.3.ii.b**), causing δ to decrease. This is called *purging* of the genetic load; under some circumstances, this may permit mutations with major effects on the selfing rate to spread, even if $\delta > 0.5$ on average, because they can become associated with a genetic background with fewer mutations and thus higher fitness than average (Charlesworth and Charlesworth 1998). Loss of genetic load due to loci with overdominance also occurs, because different alleles are less likely to be maintained in populations with high inbreeding levels (Charlesworth and Charlesworth 1990). Conversely, if outcrossing evolves, the population will build up a higher mutational load. Outcrossing will thus be increasingly favored as a population evolves higher outcrossing, and vice versa (Lande and Schemske 1985; Charlesworth and Charlesworth 1998).

Accordingly, it has been predicted that breeding systems should evolve to be either highly outcrossing or highly inbreeding (Lande and Schemske 1985), although purging proceeds slowly unless the deleterious mutations involved have large effects and are highly recessive (Charlesworth and Charlesworth 1998; Wang et al. 1999). As we have seen, however, intermediate outcrossing rates are common in plants (**Section 9.1.iii.d**). It turns out that the existence of intermediate selfing rates is not difficult to explain in principle, once we extend the limits of the rather oversimplified theories outlined above (Johnston et al. 2009). We already mentioned biparental inbreeding as a factor. Some other situations that lead to intermediate selfing rates are summarized in **Table 9.3**.

9.2.ii.e. *Origins and evolution of self-incompatibility systems*

The models for the evolution of outcrossing rates discussed so far have specified the selfing rate without including any biological details of the mechanisms controlling it, but these details may sometimes affect the evolution of outcrossing. Inbreeding avoidance by becoming unisexual is discussed later (**Section 9.3.ii**); here, we examine the conditions for the origin of self-incompatibility. Sporophytic incompatibility (*S*) alleles (**Section 9.1.iii.c**) cannot spread unless inbreeding depression exceeds 0.5. Imagine a population with no SI, in which individuals can potentially self-fertilize. If a new *S* allele arises, with sporophytic expression affecting both stigma and pollen, pollen of the carrier of this allele will be unable to fertilize the ovules of the carrier plant, i.e., it prevents self-fertilization, but pollination of other plants will be unaffected. This kind of

TABLE 9.3 Summary of some models that allow intermediate selfing rates

Factor or situation	Example	References
Increasing disadvantage of selfing with successive generations of inbreeding	Inbreeding depression, when caused by overdominance	Maynard Smith (1978, Chapter 8); Charlesworth and Charlesworth (1990)
Local adaptation		Lloyd (1980)
Trade-offs between male and female reproduction	Pollen competition for ovules (if greater pollen output produces higher outcross fertilization rates)	Charlesworth and Charlesworth (1978); Lloyd (1979); Gregorius et al. (1982); Holsinger (1988)
Biparental inbreeding		Uyenoyama (1986)

SI allele can therefore invade populations under similar conditions to those explained above.

A gametophytic S allele arising by a mutation in a previously self-compatible population is much less advantageous. This new pollen genotype would be unable to fertilize the plant that produces it, but the other half of the plant's pollen, carrying the other allele at the locus, will be able to do so. The S allele therefore prevents its own transmission to selfed progeny, yet it does not wholly prevent selfing (although it may reduce selfing to some extent, because it halves the quantity of compatible pollen). It is thus a "suicide allele" and suffers an even greater cost than do other ways of causing outcrossing. In the simplest models, inbreeding depression greater than 2/3 is needed for gametophytic SI to evolve, and to maintain SI if self-fertile mutants arise (Charlesworth and Charlesworth 1979b; Uyenoyama 2000; Porcher and Lande 2005).

9.2.iii. The breakdown of self-incompatibility systems

9.2.iii.a. *The breakdown of heterostyly*

For an understanding of the distribution and ecological patterns of plant breeding systems, it is just as important to understand how and when outbreeding systems break down as to know how they evolve. Although it is not yet fully understood how heterostyly evolves (Charlesworth and Charlesworth 1979a; Lloyd and Webb 1992a,b; Kohn et al. 1996; Barrett 2002), plants with this breeding system have been very important in studies of their breakdown and reversion to selfing (Charlesworth and Charlesworth 1979; Barrett 1992, 2002).

The genetic determination of distyly involves at least three different components, which in primroses are probably controlled by a closely linked cluster of separate loci that form a "supergene" (for further examples of supergenes, see **Section 10.2.iii.b**). Recombination between these can generate a self-compatible "homostyle" selfing form, with the female parts of the long-styled and the male parts of the short-styled morphs (**Figure 9.1** in **Section 9.1.iii.c**). These are observed in some natural populations of the primrose in England (Crosby 1949) and are highly selfing (Piper et al. 1984). Their spread into initially purely heterostyled populations is probably partly due to the transmission advantage discussed in **Section 9.2.ii.b** above (Crosby 1949). Homostyle primroses also enjoy reproductive assurance, because their anthers are close to the stigmas of their flowers and their pollen is compatible with their female reproductive tissues; in seasons with poor weather, the seed set of homostyles is higher than that of heterostyles (Piper et al. 1986). When pollinator abundance is low, the overall advantage from reproductive assurance is presumably sufficient to overcome the disadvantage of inbreeding depression. It is not understood why homostyles in these populations have not spread to fixation, but many other species of *Primula* have become secondarily fixed for homostyly (Mast et al.

2006), as has also happened in Jamaican populations of *Turnera*, whereas Brazilian populations of this group have remained heterostyled (Barrett and Shore 1987).

9.2.iii.b. *The breakdown of homomorphic SI*

The loss of homomorphic SI is also a common change (Barrett 2002), again presumably triggered by the advantage of reproductive assurance to selfers overcoming the disadvantage of inbreeding depression, and it is of interest to know how long ago such changes occurred. We mentioned in **Section 9.2.i** that selfing lineages of plants appear to have short evolutionary life-spans, i.e., they tend to be found on the external branches of phylogenetic trees. It is difficult, however, to estimate the time when a selfing lineage evolved. One situation where an estimate may be possible arises when self-incompatibility is lost. The *S* alleles will then become neutral, and genetic drift will, over time, lead to loss of all but one allele; if many alleles remain, the time involved can be delimited (Charlesworth and Vekemans 2005). If the *S* alleles are non-functional, their DNA sequences, and those of any other genes involved only in the incompatibility mechanism, will evolve like pseudogenes, because they are free to accumulate substitutions that would impair their functions in a self-incompatible species. This permits another estimate of the time since selective constraints were removed (Bechsgaard et al. 2006). For the plant *Arabidopsis thaliana*, this method estimates that selective constraints on the self-incompatibility locus ceased less than half a million years ago, about one-tenth of the estimated time since its split from the ancestor of the most closely related SI species, *A. lyrata* and *A. halleri* (Bechsgaard et al. 2006). The pattern of linkage disequilibrium in *A. thaliana* suggests a time of 1 million years (Tang et al. 2007). Further data are required for better estimates.

9.3. THE EVOLUTION OF SEPARATE SEXES

9.3.i. Introduction

Separate sexes have evolved many times, in a wide variety of animal and plant taxa (Bull 1983). The scattered taxonomic distribution of dioecy in flowering plants suggests that their ancestral state was cosexuality (Charlesworth 1985), consistent with the fact that, as mentioned in **Section 9.1.iii.a**, the flowers of dioecious plants often have opposite sex rudiments (Darwin 1877). Genetically determined dioecy can evolve from hermaphroditism, or from monoecy or from environmental sex determination, i.e., systems with separate male and female developmental systems already established.

Models for the evolution of dioecy involve finding the conditions for invasion of cosexual populations by unisexual types, and so we usually consider dioecy together with gynodioecy and androdioecy. One obvious potential advantage for unisexuals (females and males) is that their progeny must come from outcrossing and will not suffer from inbreeding depression. Unisexuals may also differ from cosexuals in their allocation of resources to reproduction. Female plants may, for example, deploy resources to seed production that cosexuals devote to anthers and pollen. Understanding why unisexuals may invade cosexual populations thus involves considering the allocation of resources between male and female functions, i.e., *trade-offs* between these two functions. We will focus attention here on flowering plants, because much of the comparative and genetic data come from plant studies, but we will also refer occasionally to animals.

9.3.ii. The evolutionary stability of cosexuality

Most species of flowering plants are cosexual, which is clearly their ancestral state. A major factor stabilizing cosexuality in flowering plants is likely to be the way in which resources are allocated between male and female functions (**Figure 9.3**); we discuss the evolution of allocation patterns in **Section 9.4.v**. With complete outcrossing, cosexuality is likely to be stable against invasion by unisexuals when the relationship between realized male and female fertility is *concave* (**Figure 9.3B**), i.e., there is a diminishing-returns relationship (trade-off) between the two aspects of reproductive function (Charnov et al. 1976; Maynard Smith 1978, pp. 130–133). With outcrossing, the net fitness of individuals with a given phenotype is proportional to the sum of the male and female fitness contributions (**Box 9.3** in **Section 9.2.ii.c**), so that a cosexual can have a higher fitness than either a male or female if this sum exceeds the fitnesses of pure males and females. In contrast, cosexuality is unstable if the curve is *convex* (i.e., there is an increasing-returns relationship).

One important factor that might generate a concave relationship is the fact that flowering plants often invest considerable resources in fruits, which develop after fertilization of the ovules (Maynard Smith 1978, pp. 132–133). A switch of a bisexual flower to all-female function may thus not proportionately increase successful offspring production, because this is limited by the resources available for fruit production. In addition, unless seeds are widely dispersed, competition is likely to occur among seeds produced by the same plant, so that successful offspring production has a diminishing-returns relation with seed output. There are therefore many factors that tend to stabilize cosexuality in flowering plants.

9.3.iii. Pathways to dioecy from cosexuality

Some plants have, however, evolved two sexes, despite the constraints just discussed. This evolutionary transition involves two important factors. The first is

A.

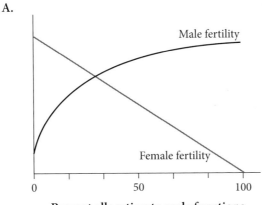

Percent allocation to male functions

B.

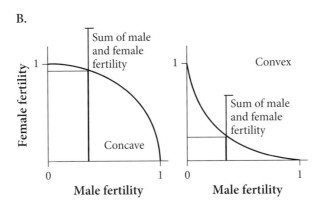

FIGURE 9.3 Gain curves for the allocation of reproductive resources between male and female functions, and the effect of this on the fitness of unisexual mutants. **A.** How male and female fertility may relate to the amount of resources allocated to male function, showing a trade-off between the two functions. Female fertility is assumed to increase linearly with resources allocated to female function, but male fertility shows a diminishing-returns (concave) relation with male resource allocation. **B.** The fitness gains in an outcrossing population that can be achieved by cosexual individuals with both sex functions, relative to the fitnesses of individuals that allocate all resources to either male or female functions. Both concave and convex (increasing-returns) relationships are shown. The vertical lines show the sum of the male and female contributions to fitness for a cosexual individual, relative to a fitness of one for pure males or females. [Modified from Charnov et al. (1976).]

inbreeding avoidance. Many cosexual species are not fully outcrossing, including most close relatives of dioecious plant species (Charlesworth 1985). Inbreeding depression can therefore help promote the spread of unisexual mutants, even when the relation between male and female function is concave, which would act to stabilize dioecy (see above). In addition, the genetics of the changes involved turns out to have profound effects on the evolution of unisexuality and dioecy, as we explain next.

Figure 9.4 shows the possible genetic changes that can lead to invasion by unisexuals, starting from an initially monomorphic, cosexual population. Two points are important. First, unisexual females can arise by mutations in maternally transmitted cytoplasmic genomes; male sterility in plants is often inherited as a mitochondrial mutation (Chase 2007). Second, the evolution of two genetically determined sexes from cosexuality, or from environmental sex determination (Bull 1983, Chapter 10), requires at least two genetic changes:

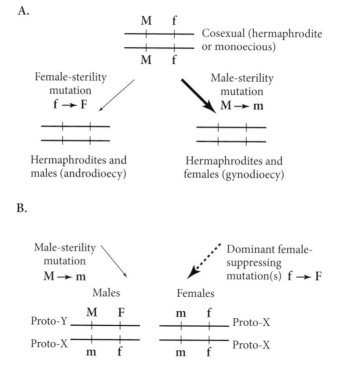

FIGURE 9.4 Genetic steps in two possible pathways in the evolution of separate sexes from a cosexual ancestral state (and reversion to cosexuality). This shows how, in the first stage (**A**), a mutation (M → m) can create unisexual females (right) or, more rarely, unisexual males (thinner arrow on the left), and lead to a sexually dimorphic population. Such a polymorphic mutation could be followed in a second stage (**B**) by a mutation in a linked gene leading to the appearance of the other sex (e.g., males appear in a gynodioecious population, as in the pathway diagrammed on the right-hand side). If the two genes are tightly linked, the spread of the second mutation would result in a population largely consisting of females and males. The chromosome carrying these two genes is termed a proto-sex chromosome.

one creating genetic females and the other forming genetic males. In the case of a cosexual ancestor, these could simply be male and female sterility mutations. It is impossible for a single mutation to cause the evolution of dioecy.

As shown in **Figure 9.4**, a first step might involve invasion of cosexual populations either by females (leading to gynodioecy) or by males (leading to androdioecy), followed by alteration of the remaining cosexuals to the complementary unisexual type. Alternatively, cosexuals might evolve a bias towards one sex or the other; these might be termed "partial sterility" morphs. We will pay most attention to models with male or female sterility mutations with major effects as the first steps, because these are the simplest to model. There is genetic evidence for the involvement of such mutations in the evolution of gynodioecy and dioecy (e.g., Kohn 1988; Weller and Sakai 1991; Dorken and Barrett 2004).

9.3.iv. When will unisexuals invade cosexual populations?

9.3.iv.a. *The establishment of cytoplasmic male sterility factors in populations* As already mentioned, male sterility (femaleness) in flowering plants can be caused by mutations in mitochondrial genes that are maternally transmitted; these are usually called *cytoplasmic male-sterility* (or CMS) factors (Chase 2007).

To ask whether a female mutant will invade a population of cosexuals (i.e., to determine when rare females have higher fitness than the cosexuals that are present initially), we can compare the fitnesses of females and cosexuals by counting the number of genes transmitted to the next generation (**Box 9.3** in **Section 9.2.ii.c**). If a female produces $1 + k$ times as many seeds as cosexuals on average, and if the mutation is transmitted exclusively via seeds, she will transmit $1 + k$ genes, compared with 1 for a cosexual individual (this ratio includes any effect of inbreeding depression on the progeny of cosexuals). Only if both sex phenotypes produce the same seed numbers will a cytoplasmic male-sterility mutation have no advantage, even if the mutation completely abolishes male functions (Lewis 1941; Lloyd 1975; Charlesworth and Ganders 1979). When male functions are abolished, some increase in seed output to a male sterile form is, however, likely to occur through re-allocation of reproductive resources to female functions. Males cannot, of course, spread in the same way, because an increase in male fertility, however large, gives no advantage to cytoplasmic factors that are entirely maternally transmitted.

Once females reach high frequencies in the population, the supply of pollen diminishes, which will ultimately reduce females' seed output. This prevents the population becoming all female (and going extinct), provided that the cosexuals are capable of selfing, and can maintain seed production when pollen from other individuals is scarce (Lloyd 1975). Cytoplasmic male sterility factors can thus be viewed as "selfish" genetic elements (similar to those discussed in

Section 10.3): in this example, they spread by virtue of an advantage to their female gametes, making use of the male gametes produced by others in the population. This, of course, reduces the population's seed output, and thus can be viewed as reducing the population's fitness.

Once gynodioecy is established by invasion by a cytoplasmic male sterility factor, the population may evolve further. If females become common, the male fertility of individuals producing male gametes becomes a major contributor to their fitness (similar to the principles governing the evolution of the sex ratio; see **Section 9.4** below). *Restorer* alleles at nuclear loci may appear, causing individuals with sterility cytoplasm to become functional cosexuals (**Table 9.4**). Such restorers gain a fitness advantage because of their male fertility and can invade a population with CMS (Charlesworth 1981; Couvet et al. 1986; Frank 1989; Jacobs and Wade 2003).

Restorer mutations have been found that act on male sterile mutants in many plants (Chase 2007). As expected from the theoretical considerations, they are common in gynodioecious populations of several wild plants with cytoplasmic control of male sterility, including ribwort plantain *Plantago lanceolata* (vanDamme 1983), the bladder campion, *Silene vulgaris* (Charlesworth and Laporte 1998; Andersson-Ceplitis and Bengtsson 2002; Olson and McCauley 2002), and other *Silene* species (Städler and Delph 2002). Another possibility is that males invade the population and convert it to dioecy; this can happen if the sterility cytoplasm goes to fixation and leaves the restorer allele segregating, so that male sterility is now determined by nuclear inheritance (Schultz 1994). We next consider the dynamics of nuclear male sterility.

9.3.iv.b. *The establishment of nuclear male sterility factors*
As we saw when thinking about the evolution of selfing rates, nuclear genes acquire fitness through both male and female functions, so that (unlike the case of cytoplasmic male sterility) a nuclear mutation causing the loss of male function has a strong disadvantage. This can be outweighed only by a correspond-

TABLE 9.4 A simple case of joint nuclear and cytoplasmic male sterility

| | Cytoplasmic genotype[1] | |
| | Sterility | Non-sterility |
Nuclear genotype	M_S	M_N
RR or *Rr*	H	H
rr	F	H

[1]H means hermaphrodite and F means female.

ingly strong increase in female performance. **Box 9.4** analyzes a simple model of an initial cosexual population reproducing by a mixture of selfing and random mating. As in **Box 9.3** of **Section 9.2.ii.c**, the model assumes that cosexuals have a selfing rate, S, and that the progeny produced by selfing suffer a reduction in fitness δ relative to the fitness of outcrossed progeny. In addition, females produce $1 + k$ times as many seeds as cosexuals. The analysis shows that females have a higher net fitness than cosexuals if:

$$k > 1 - 2S\delta \tag{9.1}$$

Box 9.4 A NUCLEAR GENE MODEL OF MALE STERILITY (GYNODIOECY)

We can use the method of **Box 9.3** of **Section 9.2.ii.c** to examine the case when a female (i.e., male sterile) phenotype caused by a nuclear gene mutation is introduced into a cosexual population, which is assumed to be partially selfing, with selfing rate S. As shown in **Box 9.3**, the total contributions of female and male gametes to the next generation for the cosexual phenotype are each $1 - S\delta$, where δ is the inbreeding depression associated with selfing.

To determine whether a female mutant will invade this population, we need only compare the fitnesses of females and cosexuals, using **Equation B9.3.1**. As explained in the text, there is likely to be a trade-off between resources invested in male and female reproduction, so that females may have a higher female fertility than cosexuals by a factor of $1 + k$, say. All the progeny of females are outcrossed, and hence suffer no inbreeding depression; using the equivalent of **Equations B9.3.5a** and **B9.3.5b**, their fitness in a population predominantly composed of cosexuals is therefore $(1 + k)/2(1 - S\delta)$, and so the condition for invasion is:

$$1 + k > 2(1 - S\delta) \tag{B9.4.1}$$

which yields **Equation 9.1** of the main text.

As females increase in frequency in the population, their fitness, relative to that of the cosexuals, will become lower, and the contribution to cosexual fitness through the fertilization of other individuals (i.e., through male functions) will exceed that through their own seeds, particularly once females become common. There will be an equilibrium when female and cosexual fitnesses are equal (**Problem 9.2.i**).

With no inbreeding depression ($\delta = 0$), this model therefore predicts that females can invade provided that $k > 1$, i.e., their fertility is at least double that of cosexuals (Lloyd 1975), consistent with the analysis in **Section 9.3.ii**. Even if inbreeding depression is severe, a large increase in female fertility is usually also required, as shown in **Figure 9.5**. For example, a population of cosexuals with selfing rate S and $\delta = 0.5$ would require an increase in fertility of $1 - S$. If there were no increase in fertility ($k = 0$), females could invade only if inbreeding depression were so severe that the product of the cosexuals' selfing rate and inbreeding depression exceed 0.5, which seems implausible (for $S = 0.7$, for instance, δ must exceed 0.714). It is therefore most biologically plausible that both advantages to females are present simultaneously, i.e., both the quantity and quality of females' seeds are higher (Charlesworth and Charlesworth 1978). Furthermore, a mutation that completely abolishes male function will often prevent self-fertilization, while one that slightly reduces male fertility may not, because a small amount of pollen is often sufficient for successful self-pollination (**Section 9.2.ii.e**). We are therefore more likely to see polymorphisms for fully male sterile individuals and cosexuals than for mutations with minor effects on male fertility.

9.3.iv.c. *Testing the theory*

Experiments on inbreeding depression (comparing cosexuals' progeny from selfing and outcrossing) in gynodioecious populations have found the pre-

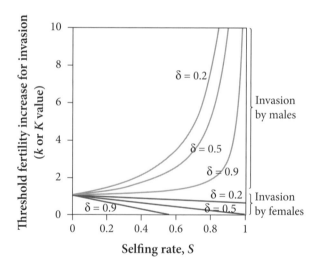

F I G U R E 9.5 Conditions for invasion of a cosexual population by females (leading to gynodioecy) or by males (leading to androdioecy) for various inbreeding depression values, δ. The black curves show how the minimum increase in female fertility, k, for a male-sterility mutation depends on the self-fertilization rate, S (x axis). The grey curves show the minimum increase in male fertility for a female-sterility mutation, K, in the same manner.

dicted intense inbreeding depression and the expected higher fertility of females (comparing progeny produced by outcrossing hermaphrodites and females).

If a male sterility gene can invade a population, it will reach an equilibrium (because a population cannot become all female), and the frequency of females can be calculated from the condition that the fitnesses of the females and cosexuals are equal (**Problem 9.2.i**). Such polymorphic populations are ideal for testing the theory. A few suitable gynodioecious populations have been studied and the facts fit the models well. **Figures 9.6** and **9.7** show some results for the gourd *Cucurbita foetidissima*, one of the few species in which male sterility is probably inherited as a single nuclear gene mutation, without cytoplasmic sterility factors (Kohn 1988, 1989). **Figure 9.7** uses the estimated values of the parameters k and δ (**Box 9.4** in **Section 9.3.iv.b**), and the selfing rate of *C. foetidissima* hermaphrodites (Kohn 1995), all estimated using plants growing in

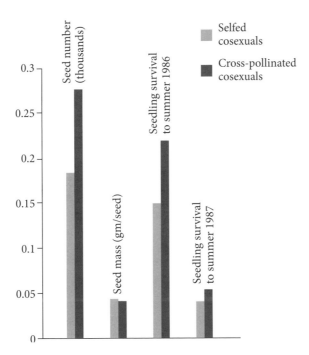

FIGURE 9.6 Inbreeding depression in the plant, *Cucurbita foetidissima*, a gynodioecious plant where females are caused by nuclear gene male sterility. Inbreeding depression (δ) values were estimated by comparing the relative numbers of seeds and survival of the progeny seedlings produced by cross-fertilizing or self-pollination of cosexual plants. Comparisons of the progeny of females with those of cross-fertilized cosexuals show that females have higher seed output, and allow the female fertility parameter k to be estimated as about 0.5. [Plotted from data in Kohn (1988).]

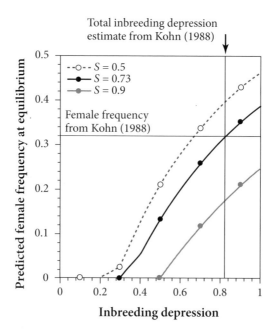

FIGURE 9.7 Testing the theory of gynodioecy, using data from a gynodioecious *Cucurbita foetidissima* population. The plot assumes the value estimated by Kohn (1988) for the female fertility parameter k (see **Figure 9.6**), and it plots the predicted frequency of females (y axis) for a range of inbreeding depression values (x axis); see **Problem 9.2.i**. Three different biologically plausible values of the self-fertilization rate, S, are assumed. The estimated inbreeding depression is indicated by a vertical line and the observed female frequency by a horizontal line. The line with the black symbols, which corresponds to the prediction using the estimated S for the population (0.73), agrees closely with the observed female frequency. [Data taken from Kohn (1988, 1989, 1995).]

their natural environment. The agreement is remarkably good, particularly as many ecological factors can affect the parameters of the model (e.g., Delph and Carroll 2001).

Figure 9.7 also shows that both increased inbreeding depression and increased selfing rates of the cosexuals lead to higher female frequencies. In several plant species, female frequencies indeed increase with the female fertility advantage over cosexuals (Webb 1979) and, in Hawaiian *Bidens* species, higher female frequencies were found with increasing selfing rates of cosexuals (Sun and Ganders 1986), which is also consistent with the model's predictions.

9.3.iv.d. *Androdioecy*

Now consider mutations producing males. It is easy to use the phenotypic selection method to derive the increase in male fertility, K, needed for males to have higher fitness than cosexuals when invading a cosexual population (**Problem 9.2.ii**):

$$K > \frac{2(1 - S\delta)}{(1 - S)} - 1 \tag{9.2}$$

Just as for females (**Equation 9.1** of **Section 9.3.iv.b**), males can invade only if their fertility is more than doubled ($K > 1$) compared with an outcrossing, initial cosexual form. Other things being equal, invasion of partially selfing populations by males requires even higher fertility differences because, with partial selfing, ovules are less available to pollen from other individuals, which makes it harder to gain fitness by increasing pollen output (see **Figure 9.5** of **Section 9.3.iv.c**). Female sterility is therefore less likely to evolve than male sterility, and males are also unlikely to reach high frequencies (Lloyd 1975; Charlesworth and Charlesworth 1978).

This theory thus predicts that androdioecy should be rare. A few androdioecious plant populations have been discovered, but they turn out to be exceptions that prove the rule. Most of them are not truly androdioecious, but are functionally dioecious populations in which the females are morphologically cosexual, probably because pollinators perceive the absence of pollen and discriminate against female plants (Charlesworth 1984; Anderson and Symon 1989); both pollinators and botanists have been deceived by this "floral mimicry." A few well-authenticated cases of functional androdioecy have, however, been found. In *Datisca glomerata*, there is enough information to show that the conditions are satisfied for polymorphism of males and cosexuals to be stably maintained, assuming a single recessive female sterility gene; **Figure 9.8** shows that male frequencies in two populations studied are close to those predicted, given the relative male fertility values, the inbreeding depression estimated in progeny produced by self-fertilizing or outcrossing cosexuals, and the outcrossing rates in the natural populations (Fritsch and Rieseberg 1992). This figure also illustrates the very high increase in male fertility relative to cosexuals that is necessary for males to be present.

D. glomerata is probably a case of breakdown of dioecy, because the most closely related species are dioecious. The same is true of another androdioecious plant, *Mercurialis annua* (Pannell 2002; Pannell et al. 2008). Most likely, in these cases, a dioecious population was invaded by a modified female that has reverted to partial male fertility, yielding cosexuals with low male fertility, thus satisfying the necessary condition for the population to remain polymorphic for cosexuals and males (see above). Cosexuals can replace females because of their ability to self-fertilize, i.e., through the advantage of reproductive assurance (**Section 9.2.ii.a**), for which there is evidence from the ecology of these species (Pannell 2002).

Loss of dioecy and reversion of females to partial male fertility may also explain some cases of androdioecy in animals, such as the mangrove killifish *Kryptolebias marmoratus* (Mackiewicz et al. 2006a,b) and the brine shrimp *Eulimnadia texana* (Sassaman and Weeks 1993). In these species, like the few

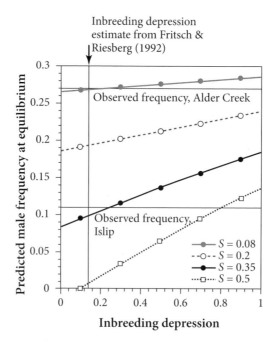

FIGURE 9.8 Testing the theory of androdioecy, using data from two populations of the plant *Datisca glomerata*. The plot assumes the value estimated by Fritsch and Rieseberg (1992) for the male fertility parameter K (see text), and it plots the predicted frequency of males (y axis) for a range of inbreeding depression values (x axis) , found by the method of **Problem 9.2.i**. The two populations have different selfing rates, and the figure plots the predictions using the estimated values of S for each population (black and grey lines and symbols) and a different (higher) value. The estimated inbreeding depression value is indicated by a vertical line. The observed female frequencies are indicated by horizontal lines. The predictions for both populations agree closely with the observed male frequencies. [Data taken from Fritsch and Rieseberg (1992).]

other androdioecious animals known, the hermaphrodites are largely self-fertilizing. This may seem to contradict the theory above (which suggests that self-fertilization makes the maintenance of males very unlikely). However, these hermaphrodite animals cannot mate with other hermaphrodites, so this is a case of complete male gamete discounting (**Section 9.2.ii.c**), which means that hermaphrodites are less likely to completely displace males (Otto et al. 1992; Weeks et al. 2000).

9.3.v. The transition from gynodioecy to dioecy

9.3.v.a. *The effect of the presence of females*

Given the stringent conditions for males to invade cosexual populations, it might seem that females can evolve, but not males. However, the establishment of unisexual females (which is the most likely intermediate stage in the evolution of dioecy, given the rarity of androdioecy; **Figure 9.5** of **Section 9.3.iv.c**) increases the fitness returns on the cosexual morphs' investment in male gamete output; the cosexuals in a gynodioecious population will thus be selected to reallocate reproductive resources and invest more in male functions (Charlesworth and Charlesworth 1978). Mutations that cause the cosexuals in gynodioecious populations to become fully or partially female sterile can therefore invade more easily than in the absence of females, i.e., they require a smaller increase in male fertility than for the invasion of pure cosexual populations (**Figure 9.9**). Invasion still requires a greater increase in male fertility than any associated reduction in female fertility. Reallocation of resources from female to male fertility is thus necessary for the evolution of dioecy via gynodioecy, i.e., dioecy does not evolve purely to avoid inbreeding. Furthermore, because invasion of gynodioecious populations by mutations that reduce female fertility is not driven primarily by inbreeding avoidance, mutations with only partial effects on fertility can invade, unlike the conditions for mutations reducing male fertility.

These conclusions about evolution from gynodioecy to dioecy are only partial, however. The trade-offs between male and female functions that are required for the invasion of a mutation creating males or partial males create *sexual antagonism* (Rice 1984), which we discussed briefly in **Section 2.2.iii**. In this case, there is selection against the spread of mutations that increase male (and decrease female) fertility when they are combined with the original mutation that causes male sterility (because this combination yields individuals with zero male fertility and low female fertility). **Table 9.5** summarizes the genetic essentials of **Figure 9.4** of **Section 9.3.iii**, and shows how such sexually antagonistic mutations will be favored in only one of the original two sexual phenotypes.

This has two interesting implications. First, evolution from gynodioecy to full dioecy is not inevitable, because selection for mutations making cosexuals more male-biased is opposed by negative effects on females, particularly if there is recombination between the two loci involved. Second, mutations that only partly suppress female functions are expected to be able to invade gynodioecious populations. Populations may thus often remain gynodioecious, or become sub-dioecious, with cosexuals that are somewhat male-biased in their gender, along with females. Many plants are indeed sub-dioecious (Burrows 1960; Humeau et al. 1999), e.g., the spindle tree (*Euonymus europaeus*), whose cosexuals range from fully male to hermaphrodite, and are highly variable in fruit and seed output, often producing fruits only in good years (Darwin 1877). If there were no cost to increased male function, there seems to be no selective reason why gynodioecy should evolve into full dioecy; instead, cosexuals could simply increase their male function while remaining cosexual. These examples thus provide evidence for the trade-off assumption of allocation models.

TABLE 9.5 The genetic situation in a gynodioecious population when a mutation arises that increases the maleness of the hermaphrodites

Second mutation (female-sterility locus)	First mutation $M \rightarrow m$ (male-sterility locus) Genotype at male-sterility locus		
	M/M or M/m		m/m
$f \rightarrow F$ f/f	Hermaphrodite	$\rightarrow$	Female
	$\downarrow$		$\downarrow$
F/f or F/F	Male	$\rightarrow$	Neuter

9.3.v.b. *The importance of linkage*

The first of the two conclusions just outlined suggests that a major factor determining whether a modifier producing a more male form will invade a gynodioecious population is its degree of linkage to the initial male sterility locus. Close linkage reduces the frequency of progeny carrying both male sterility and female sterility alleles and preserves genotypes with the advantageous combinations of alleles at the loci concerned (see **Table 9.5**). There is a threshold recombination rate between a suppressor of femaleness and the locus controlling the male sterility polymorphism, and this threshold depends on the k and K values in **Equations 9.1** and **9.2** of **Sections 9.3.iv.b** and **9.3.iv.d**, as shown in **Figure 9.9**. For the female suppressor to invade a gynodioecious population,

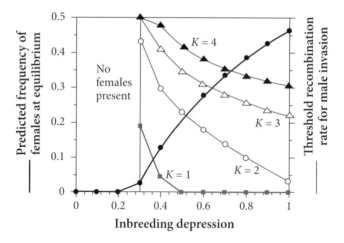

FIGURE 9.9 The conditions for invasion of a gynodioecious population by a female-sterility mutation that could create male individuals. The heavy line with filled circles shows how the predicted female frequency depends on the inbreeding depression values (*x* axis). The female fertility parameter, *k*, was set to 0.5, and the selfing rate, *S*, to 0.9. The thin lines show the maximum recombination rate between the male- and female-sterility loci that permits invasion of the female-sterility mutation, for different values of the increase in its male fertility, *K*.

the recombination rate must be below the threshold, otherwise the production of recombinants with lower fitness than the males and females eliminates it (Charlesworth and Charlesworth 1978).

Once this situation is established, with polymorphisms at linked male sterility and one or more female suppressor loci, there is also selection for reducing recombination between the two loci (following the rules for selection on modifiers of recombination rates that we discuss in **Section 10.2.iii**). This has important implications. It suggests that the evolution of dioecy will probably lead to a group of linked loci in one chromosomal region, which may start the evolution of a primitive sex chromosome system, with *proto-X* and *proto-Y* chromosomes (**Figure 9.10**).

There is also scope for further evolution of this proto-sex chromosome system. As pointed out by Bull (1983, Chapter 18) and Rice (1987a), recombination suppression might expand to include much of the proto-Y chromosome, because of the action of selection on other genes that are polymorphic for alleles with sexually antagonistic effects. This is because of the advantage of ensuring that alleles with beneficial effects on males, but harmful effects in females, remain associated with the female suppressor allele. This could lead eventually to restriction of crossing over across most of the proto-Y chromosome and to the evolution of 'classical' heteromorphic sex chromosomes (**Figure 9.10**).

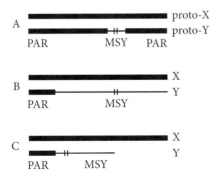

FIGURE 9.10 Different types of sex chromosomes (male heterogamety is assumed). Thick lines indicate regions where recombination occurs, and thin ones indicate non-recombining regions. **A.** "Proto-sex chromosomes" with the sex-determining genes indicated by vertical lines on the proto-Y chromosome, as in **Figure 9.4** of **Section 9.3.iii**. These are located within a small region in which recombination does not occur, so that there is a male-specific region (MSY) of the proto-Y chromosome, while the rest of the chromosome recombines and thus forms a large *pseudoautosomal* region, or PAR (as in papaya; see Yu et al. 2008). **B.** "Classical" sex chromosomes, in which recombination is suppressed throughout most of the chromosome pair in the heterogametic sex. There is thus a large MSY region. **C.** Heteromorphic sex chromosomes in which the MSY region has degenerated and large deletions of genetic material from the Y chromosome have occurred (as in humans, where the X and Y recombine only in small pseudoautosomal regions).

Genetic and physical mapping in organisms that may recently have evolved separate sexes show that the non-recombining chromosome region containing the sex-determining genes may indeed be physically small, as predicted for proto-sex chromosomes. In the plant papaya, the region is about 9 kb, only about 10% of chromosome 1 (Yu et al. 2008). In sex chromosomes at this stage of their evolution, divergence between the sequences of X- and Y-linked alleles should be small, reflecting the recent origin of the non-recombining region. This may be the case in papaya (Yu et al. 2008). If further regions of the sex chromosomes later stop recombining, older sex chromosomes will contain distinct regions (**Figure 9.11**). In the "old" sex chromosome region the genes' sequences

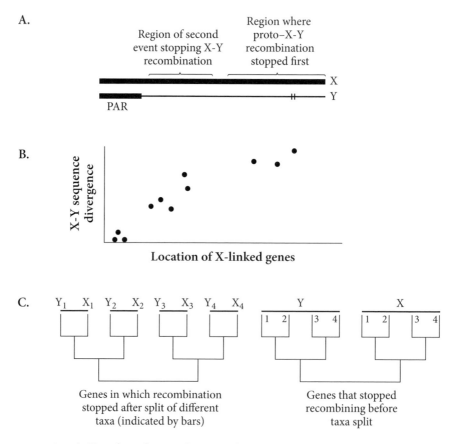

FIGURE 9.11 **A.** Hypothetical times when recombination was suppressed between a pair of sex chromosomes. **B.** The corresponding levels of divergence between sequences of X- and Y-linked homologues. **C.** The topologies of the phylogenetic trees that would be found for genes in the different regions, if the second step in suppressing recombination occurred independently in different species. [Adapted from Lawson-Handley et al. (2004).]

will be highly diverged between the X and Y, showing that they have not been recombining for a long time. If the same sex chromosome system is shared by a group of related species, the sequences for any gene in this region should indicate one cluster formed of X-linked sequences of all the species and another of their Y-linked sequences. In the "young" sex chromosome regions that stopped recombining recently, compared with the divergence times of the species, the X–Y divergence should be low, and the phylogenetic relationships for genes in this region show clusters of taxa instead of sex chromosomes (**Figure 9.11**). This is clearly seen in birds, for example (Lawson-Handley et al. 2004).

9.4. EVOLUTIONARILY STABLE STRATEGIES, SEX RATIOS, AND SEX ALLOCATION

9.4.i. Introduction

The 1:1 sex ratio of many animal species with separate sexes, and also of dioecious plants (Delph 1999), is a familiar fact of biology (though deviations from 1:1 are also observed—for recent reviews, see West et al. 2002) and West (2009). Darwin wondered why this happens, and suggested that, if one sex is rarer than the other, it will have a larger number of offspring per individual, compared to the commoner sex (Darwin 1871, p. 316). In population genetics terms, it can be shown that selection indeed acts on genes that control the sex ratio produced by a parent, in the following way. Fisher (1930b, pp. 141–143) pointed out that, because each individual has one mother and one father, in a diploid organism, the total genetic contribution at autosomal loci of all the parent generation males to the progeny generation must therefore equal the total contribution of all the females. An excess of males therefore implies that each male has fewer offspring, on average, than a female. A parental autosomal genotype producing excess female offspring in such a population would thus have more offspring per head, i.e., selection pushes the sex ratio towards 1:1. This argument was originally presented in an algebraic form (without a genetic formulation) by Carl Düsing in the 19th century (Edwards 2000), and demonstrates that frequency-dependent selection acts on the sex ratio.

Notice that, in most organisms, males can fertilize many females, so that a sex ratio with many females and few males would give the population the highest reproductive capacity, because offspring numbers are generally limited by the number of progeny or seeds that females can produce. Sex ratio theory is another case in which individual selection usually determines the evolutionary outcome, not group selection advantages to populations or species.

The population genetics theory supporting this idea has subsequently been considerably developed (e.g., Bulmer 1994, Chapter 10; West 2009). The argument given above simplifies the theoretical analysis by using the invasibility approach we have already used several times. This approach asks whether a population with one strategy (in this case, a given sex ratio) can be invaded by a different strategy (i.e., a genotype producing a slightly different sex ratio from the prevailing one), which is introduced at a low frequency into the population. Invasion can be studied either by calculating the changes in frequencies of alleles giving different sex ratios, or by using the fitnesses of different sex ratio phenotypes.

9.4.ii. Evolutionarily stable strategies

A population's phenotype that cannot be invaded by a different one is called an *Evolutionarily Stable Strategy* (*ESS*); in the case of the sex ratio, this was called the "unbeatable sex ratio" by Hamilton (1967). The ESS concept was formally introduced by Maynard Smith and Price (1973) in the context of animal conflict behavior. It is a powerful general method for predicting the outcome of phenotypic evolution when fitnesses are frequency dependent (with constant fitnesses, the ESS simply corresponds to the phenotype with maximum fitness). **Box 9.5** shows how to apply the phenotypic selection model to the determination of an ESS.

The ESS clearly represents a necessary, but not sufficient, condition for a population to be at an evolutionarily stable equilibrium. As discussed in **Box 9.5**, it is reasonable to think that the population will often eventually arrive at this "unbeatable" ESS state. There are, of course, situations in which genetic constraints mean that the sex ratio at equilibrium will not coincide with the ESS (detailed genetic models of sex ratio evolution are discussed by Karlin and Lessard 1986), but the ESS probably often reliably indicates the outcome of sex ratio evolution, and can usually be found by much simpler calculations than are needed for a full population genetic analysis (Maynard Smith 1982).

Box 9.5 EVOLUTIONARILY STABLE STRATEGIES

B9.5.i. General principles

This is a powerful method for determining the outcome of selection on alternative phenotypes or *strategies* that affect fitness. We will consider the case when the phenotype is a continuous variable, such as the sex ratio or the proportion of resources invested in female versus male reproduction in a cosexual. For a more general treatment, see Maynard Smith

(1982, Chapter 2 and appendices) and Otto and Day (2007, Chapter 12). In general, the fitness of an individual with phenotypic value y will depend not only on y, but also on the phenotypic value for the population as a whole, denoted by z. This fitness can be written as $w(y, z)$, where w is given by a formula such as **Equation B9.3.1** from **Box 9.3** in **Section 9.2.ii.c**. The phenotype y will have an advantage when rare if $w(y, z) > 1$.

The *Evolutionarily Stable Strategy* (ESS) is defined as the value of z that ensures that $w(y, z) \leq 1$ for all $y \neq z$. This value is denoted by z^*. If a value (or set of values) of z^* can be found, it is reasonable to regard it as corresponding to a likely outcome of evolution, because rare mutations that perturb the phenotype away from z^* are either neutral or deleterious.

B9.5.ii.* Determination of the ESS

We can determine the value of z^* as follows. To find the condition for $w(y, z^*) < 1$, we can treat $w(y, z^*)$ for a fixed value of z^* as a function of y, and ask if a small perturbation in y from z^*, δy, leads to a decrease in $w(y, z^*)$. This requires the standard mathematical conditions for a local maximum (**Appendix A1.i**):

$$\left(\frac{\partial w(y, z^*)}{\partial y}\right)_{z^*} = 0 \tag{B9.5.1a}$$

$$\left(\frac{\partial^2 w(y, z^*)}{\partial y^2}\right)_{z^*} < 0 \tag{B9.5.1b}$$

If all the derivatives of $w(y, z^*)$ are equal to 0 at z^*, as is sometimes the case in applications of this method (e.g., **Box 9.6** of **Section 9.4.iii.a** below), then $w(y, z^*) = 1$ for all y, so that z^* is neutral to perturbations. To determine whether z^* is a stable equilibrium point in this case, it is then necessary to ask whether a mutation, whose phenotype deviates from z^* by a small amount δz, will invade if it creates a phenotype that is closer to z^* than $z^* + \delta z$. If $\delta z < 0$, this requires:

$$\left(\frac{\partial w(y, z^* + \delta z)}{\partial y}\right)_{z^* + \delta z} > 0 \tag{B9.5.2}$$

with the opposite inequality when $\delta z > 0$.

9.4.iii. The ESS sex ratio in a panmictic population

9.4.iii.a. *Equal costs of male and female offspring*

Box 9.6 shows how to find the ESS sex ratio in a panmictic population, assuming autosomal gene control of the zygotic sex ratio produced by a female (similar results apply to paternal control of the sex ratio with autosomal inheritance, and to X-linked inheritance with maternal control). The model assumes that the total number of progeny of a mother is independent of the sex ratio; this is reasonable for species that provide no parental care, but it needs modification if there is parental care, with differences in the amount of resources provided to sons and daughters (see **Section 9.4.iii.b** below). The theory in **Box 9.6** is based on the principle that the fitness acquired through the reproduction of male progeny is equal to that acquired through the reproduction of female progeny (because each progeny individual has a male and female parent). The ESS sex ratio is then 1:1 males and females among the zygotes.

The primary sex ratio (the ratio in the zygotes) can be changed by subsequent survival differences of male and female zygotes. In mammals, differences in male and female prenatal mortality rates may often obscure the primary sex ratio (Toro et al. 2006). In dioecious plants, sex ratios among mature, flowering individuals are often male-biased because males may flower more regularly, or start flowering earlier than females, or may suffer lower mortality (Allen and Antos 1993).

When care is taken to determine the sex of all individuals, however, such as by following all plants until they eventually flower, or by genotyping plants with sex-linked markers specific for the male-determining chromosome, zygotic sex ratios close to 1:1 are often found when there is genetic sex determination. In flowering plants, there is generally some bias towards females, which may be due to greater success of pollen carrying X chromosomes than Y-bearing pollen (Lloyd 1974), as occurs in several plants (Correns 1928; Stehlik and Barrett 2005). In most animals and plants, however, segregation of the sex chromosomes appears to be generally very close to 1:1. In addition, there is little evidence for genetic variation in the primary sex ratio in species with chromosomal sex determination (Toro et al. 2006), other than through the "meiotic drive" mechanisms that distort sex ratios, which we discuss in **Section 10.3.ii**. This makes it hard to test the ESS theory rigorously in these organisms; situations where the sex ratio can be adjusted easily, such as with haplodiploid sex determination, are much more favorable (see **Section 9.4.iv** below). These cases have provided some of the best examples of theoretical predictions from evolutionary genetics surviving detailed empirical tests.

9.4.iii.b. *Unequal costs of sons and daughters*

Fisher (1930b, pp. 141–143) generalized the argument above to include situations where different amounts of resources are needed to produce a male

Box 9.6 PHENOTYPIC SELECTION ANALYSIS OF THE ESS SEX RATIO IN A DIPLOID POPULATION

The model imagines that females are of two kinds, both producing the same numbers of progeny, but with one type producing a proportion c_1 of male progeny, while the other produces a proportion c_2 of male progeny (these correspond to phenotypes 1 and 2 of **Box 9.3** of **Section 9.2.ii.c**). We want to ask what sex ratio is evolutionarily stable to invasion by other ratios. We therefore consider a population, all of whose females produce the sex ratio c_1, and we want to find the value of this sex ratio, such that any other value (c_2) has a lower fitness than that of the predominant kind of females; this c_1 value is the ESS.

When the phenotype with sex ratio c_2 is rare, the expression for its fitness can be found as follows, using a similar argument to that in **Box 9.3.ii**. The number of females in the population is approximately proportional to $(1 - c_1)$ and these contribute one-half of the total gametes to the next generation. Phenotype 2 females each produce a fraction of $(1 - c_2)$ female progeny, so that the per capita contribution of gametes through females by these is $w_2^f = (1 - c_2)/(1 - c_1)$. Similarly, their per capita contribution of gametes through males is $w_2^m = c_2/c_1$. Using **Equation B9.3.1**, the fitness of phenotype 2 is:

$$w_2 = \frac{1}{2}\left[\frac{(1 - c_2)}{(1 - c_1)} + \frac{c_2}{c_1}\right] \tag{B9.6.1}$$

It is easy to see that $w_2 = w_1 = 1$ for all c_2 when $c_1 = 1/2$, i.e., a sex ratio of one-half is neutral with respect to invasion by any other sex ratio. Write $\delta c = c_2 - c_1$; we then find that:

$$w_2 = 1 + \frac{(\delta c)}{2}\left[\frac{1}{c_1} - \frac{1}{(1 - c_1)}\right] \tag{B9.6.2}$$

If $c_1 < 1/2$, the term in large brackets is positive, so that a value of c_2 that exceeds c_1 gives $w_2 > 1$; if $c_1 > 1/2$, the condition is $c_2 < c_1$. A succession of fixations of mutations with small effects on the sex ratio can thus cause the population sex ratio to converge to one-half.

versus a female. If there is a linear trade-off between resources invested in males and females, the ESS is to devote equal total amounts of resources to male and female offspring (**Problem 9.3**). This does not produce a 1:1 sex ratio. For example, if male progeny cost 50% more than female offspring, the ESS is 1.5 females per male (a male frequency of 0.67). This principle has wider importance than for the sex ratio alone, and also applies to sex allocation in cosexuals (the proportion of resources invested in male vs. female functions)— see **Section 9.4.v** below.

9.4.iv. Extraordinary sex ratios

Hamilton (1967) drew attention to other situations in which sex ratios are very different from 1:1, and developed the first models to explain these, which we will now describe.

9.4.iv.a. *Local mate competition*

The models above assume that mating opportunities are not limited by availability of the opposite sex, but this is often not true. Matings of some animals occur in "patches," each of which contains the broods of a small number of mated females. For instance, the pollinator wasps mating within a single fig are the progeny of the few females that laid their eggs in the inflorescence that gives rise to the fig (Herre 1985); the male and female offspring emerge and mate, after which the males die, and the females leave the fig, starting the next generation by dispersing to newly available figs in the flowering stage. Within a fig, a small number of males can fertilize all the females, and therefore the males compete with one another for matings with the available females. A similar situation also occurs in *parasitoid* wasps that lay their eggs in other insects; again, the progeny of a female wasp that develop within a host insect often mate with one another before the females disperse to lay their eggs in new hosts.

In these examples of *local mate competition* (LMC), females transmit the highest number of genes to the offspring generation if they produce just enough male offspring to fertilize their female progeny. Excess males beyond this number are effectively wasted resources. **Box 9.7** shows that this situation leads to an ESS sex ratio with less than half males.

The models in **Boxes 9.6** and **9.7** assume, however, that the inheritance of the sex ratio depends on maternally acting autosomal genes in a diploid population, modifying the 1:1 segregation ratio of the progeny, with the ESS sex ratios being derived by counting one gamete transmitted per progeny individual of either sex. In animals whose sex determination is different from this, the equations must be modified by counting the appropriate numbers of gametes transmitted to progeny of the two sexes. For example, haplodiploid species such as Hymenopterans (to which wasps belong) have much more opportunity

Box 9.7 PHENOTYPIC FITNESS ANALYSIS OF THE ESS SEX RATIO WITH LOCAL MATE COMPETITION

B9.7.i. Fitnesses with local mate competition

As before, females can be of two kinds, both producing the same number of progeny: a predominant type producing a progeny sex ratio of c_1, or a rare mutant producing c_2. We now assume that each male mates within a patch containing only n females, for whom all males in the patch compete on an equal basis (i.e., mating is random within patches). As in the previous model (**Box 9.6** of **Section 9.4.iii.a**), the fitness of phenotype 1 approaches 1 as its frequency approaches 1.

The number of gametes transmitted to female progeny per female parent of the mutant phenotype is proportional to $(1 - c_2)$, so the female fitness contribution is $w_2^f = (1 - c_2)/(1 - c_1)$, as before. The male fitness of the mutant is, however, different in this model from that in **Box 9.6**. Because of their rarity in the population, the type 2 females will be single individuals in patches whose other members are $n - 1$ females of the c_1 type. In such patches, the number of female progeny produced by a patch is proportional to $(n - 1)(1 - c_1) + (1 - c_2)$, and the number of male progeny is proportional to $(n - 1)c_1 + c_2$. The males mate with the females from the same patch, and hence the number of offspring fathered by males in such a patch is the same as the number of female progeny produced by the patch. Among these, the type 2 males contribute a fraction $c_2/[(n - 1)c_1 + c_2]$. Normalizing the output of offspring from a patch by the number of female offspring produced by the population as a whole (which is proportional to $1 - c_1$), the male fitness contribution w_2^m is thus:

$$w_2^m = \frac{\left[(n - 1)(1 - c_1) + (1 - c_2)\right]c_2}{(1 - c_1)\left[(n - 1)c_1 + c_2\right]} \tag{B9.7.1}$$

and the net fitness of the mutant is given by $w_2 = (w_2^f + w_2^m)/2$.

B9.7.ii.* Determining the ESS

As described in **Box 9.5** of **Section 9.4.ii**, the ESS sex ratio can be found by differentiating w_2 with respect to c_2 and setting the derivative to 0 at $c_2 = c_1$. In this case, the ESS sex ratio is:

$$c^* = (n - 1)/(2n) \tag{B9.7.2}$$

(See **Problem 9.4***.) This is close to 0.5 when n is large, but can be considerably lower if n is small.

to control their sex ratios than species with X/Y chromosomal sex determination; females can lay unfertilized eggs, which develop as males, or fertilized eggs, which develop into females (Bull 1983, Chapter 11). Half of the gametes in each female progeny individual therefore come from their female parent, but all of the gametes in each male progeny are maternal in origin. To take haplodiploidy into account, one must weight male and female progeny differently; this reduces the ESS sex ratio below the ratio for diploidy when there is local mate competition (Bulmer 1994, Chapter 10).

Inbreeding also affects the numbers of gametes contributed to the progeny generation per female of the parental generation. Intuitively, one would predict that, when the mating males and females are related, the fitness advantage of producing more males to compete for matings will be weakest, because females in the parent generation can gain fitness by transmitting genes through their female progeny; their number of male progeny (which will transmit the same genes) should therefore be reduced to the number needed to fertilize the female gametes. Assuming nuclear genetic variation for the sex ratio produced, inbreeding therefore favors genotypes that produce an excess of females.

Situations with local mate competition also often involve inbreeding, particularly when there are few founding females in a given patch (Herre 1985). The theory has been modified to include both haplodiploidy and inbreeding (Herre 1985; Bulmer 1994, pp. 223–229; West 2009), and there is good quantitative agreement with data from both fig wasps (see **Figure 9.12** and West and Herre 1998) and the parasitoid wasp, *Nasonia vitripennis* (Werren 1980).

In *N. vitripennis*, patch sizes are equivalent to the mean numbers of females laying eggs per host. Not only are male proportions lowest in hosts in which few females lay their eggs, but female wasps also modify the proportions of unfertilized eggs to produce more males when eggs are already present from a different female. This is as the theory predicts, because the transmission benefit from producing males which will be able to mate with the unrelated female progeny from these eggs (exploiting eggs containing a gamete from an unrelated female) is higher than from fertilizing eggs from one's own progeny (Werren 1980). Temporal variation in the extent of local mate competition within the same host has a similar effect (Shuker et al. 2006).

9.4.iv.b. *Haplodiploid eusocial insects*

Deviations from an ESS 1:1 sex ratio or expenditure on males and females can also occur in eusocial insects with haplodiploid sex determination, notably the Hymenoptera (Trivers and Hare 1976; Bulmer 1994, pp. 216–221). This is because, if a single queen establishes a colony, sterile female workers produced by the queen may be able to control the sex ratio by allocating resources preferentially between males and females. If the queen mates only once, the coefficients of relatedness of the workers to their sisters and brothers are 3/4

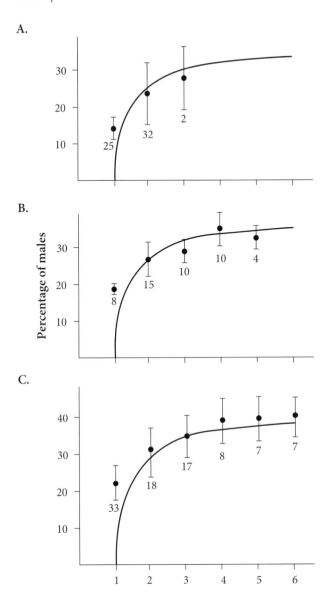

FIGURE 9.12 The good agreement between the predicted (lines) and observed (points, with error bars) sex ratios (percent males) with different numbers of foundresses in three different species (A, B, and C) of haplodiploid fig wasps with different levels of inbreeding. [Adapted from Figure 1 of Herre (1985).]

and 1/4, respectively, and so they transmit more of their genes indirectly through their sisters than their brothers (**Section 3.1.v.d**). Inclusion of this factor in calculations of the ESS expenditure of reproductive female and male offspring of the colony predicts that worker control of the sex ratio will produce a

3:1 ratio in favor of females. This is consistent with data showing substantial biases towards biomass of reproductive females versus males in groups such as ants (Trivers and Hare 1976); multiple matings by the queen reduce the extent of this bias (Bulmer 1994, pp. 216–221). This basic theory has subsequently been elaborated in relation to several specific features of eusocial Hymenoptera, and can explain a wide range of observations (e.g., Meunier et al. 2008; West 2009).

9.4.v. Sex allocation in cosexuals

9.4.v.a. *Theoretical models*

The factors determining how cosexuals should allocate resources to male and female functions are very similar to those determining sex ratio evolution in dioecious species, discussed in **Sections 9.4.i–9.4.iv** above (Charnov 1982; Lloyd 1984). In cosexual populations, genetic variation may often affect allocation to female versus male functions, in contrast to the situation we discussed for dioecious species with chromosomal sex determination (**Section 9.4.iii** above). Thus, sex allocation patterns in hermaphrodites are excellent material for testing the theory, just as for the sex ratio with haplodiploid sex determination.

Here, it is particularly easy to see that fitness through male functions (defined as the number of genes transmitted to the progeny generation by male gametes) is limited by the number of female gametes available in the population. This makes it evident, for example, that, if some female gametes are fertilized by male gametes from the same individual, and are thus unavailable for externally produced male gametes to fertilize, the "genetic value" of male gamete production is reduced. This is simply an extreme form of local mate competition, and it means that more reproductive resources should be allocated to female than to male functions, other things being equal (Charlesworth and Charlesworth 1981; Lloyd 1984, 1987). With a simple model of a linear relation between the amount of reproductive resources allocated to male function and the number of gametes produced, and between female gamete production and the amount of resources allocated to female functions, resources at the ESS are divided equally between the two sex functions when there is complete outcrossing, but are weighted increasingly towards female functions with higher selfing rates (**Problem 9.5.i***).

These assumptions are implausible for many organisms, however, as already mentioned in **Section 9.3.ii**. The number of progeny to which female gametes are transmitted may sometimes depend linearly on the resources allocated to female functions (eggs of an animal or ovules of a plant). Male success through outcrossing, however, is often limited by the availability of pollinators for transporting pollen; the saturation of pollen vectors with pollen can cause diminishing fitness returns for male gamete production, i.e., doubling the allocation to male function (and thus doubling male gamete output) may lead to much less than a doubling of the number of offspring sired. A simple way to model such situations is to assume that the number of male gametes transmitted follows a function such as a^y, where a is the allocation to male function and $y < 1$. Similarly, competition between the seedlings produced by the same maternal plant may cause the effective female fertility of the plant to be less than linearly related to its seed output. The ESS allocations to different functions then depend on the ratios of the powers for each different function (Lloyd 1984, 1987). For example, diminishing returns on investment in pollen lead to ESS allocations that are more biased towards femaleness than with linearity (Charlesworth and Charlesworth 1981; **Problem 9.5.ii***).

The same general principle can be applied to several different plant functions that can affect realized male or female fertility, including pollinator attractants such as petals and nectar, and investment in fruit to attract animal seed dispersal agents, as well as the pollen and ovules themselves (Lloyd 1984, 1987). Taking these factors into account, equal allocations to male and female functions are not the expectation, even for outcrossing populations, and extensive knowledge of the reproductive biology of the species is required to make detailed predictions.

9.4.v.b. *Empirical tests*

As with the theory for sex ratio, sex allocation theory is open to empirical testing, and several of the major predictions are upheld. Although we cannot easily measure allocation to different functions, or the shapes of the gain curves translating them into realized male or female fertility, this is often not necessary, because the model can be tested using comparisons between related species or populations with known differences in their ecology, especially their selfing rates (this implicitly assumes that the curves do not differ greatly between the taxa being compared).

An important pattern observed in natural plant populations, which the models can readily explain, is the generally low pollen production and smaller petals and nectar output in selfing plants compared with outcrossers (Ornduff 1969; Cruden 1977; Cruden and Lyon 1985). Ideally, related outcrossing species or populations should be compared, so that other factors are unlikely to differ greatly; examples where this has been done successfully include the plant genera *Leavenworthia* (Lloyd 1965), *Gilia achilleifolia* (Schoen 1982), and *Monochoria* (which does not have nectar, so that pollen output is a good basis for comparing male allocation; see Tang and Huang 2007). Similar comparisons can be made using sperm production in self-compatible hermaphroditic animals whose populations have different outcrossing rates (Johnston et al. 1998). Similarly, in wind-pollinated plants, where increased male gamete production may not be subject to diminishing fitness returns, the expected unusually high allocations to male functions are often observed, for example in some grasses (McKone et al. 1998).

9.5. AGE-STRUCTURED POPULATIONS AND THE EVOLUTION OF LIFE HISTORIES

9.5.i. Introduction

So far in this book, we have studied evolutionary processes in populations with discrete generations, in which the offspring are produced simultaneously by the adults, who then die. While this is characteristic of many species of annually breeding plants and insects, it applies to only a fraction of all species. In temperate regions, most species of birds, large mammals, and seed plants live for several years, reproducing in successive breeding seasons; reproduction, moreover, is often deferred for several years after birth. In the tropics, many species have no discrete breeding seasons, but reproduce more or less continuously throughout adult life. Familiar examples of this include *Drosophila melanogaster* and humans. Such populations are thus *age-structured*, with overlapping generations: individuals of different ages have different sizes, mortality rates, and rates of reproduction, and can reproduce with one another.

Are the population genetic results derived earlier still valid for age-structured populations, or do some of our major conclusions have to be revised for such populations? In the next sections, we extend the models of selection developed in **Chapters 2**, **3**, and **8** to age-structured populations, and show how most of the old results still apply, at least as approximations, while some new ones emerge. In addition, we show how understanding selection in age-structured populations sheds light on how selection shapes life histories, and, in particular, why aging evolves.

9.5.ii. Demographic models

9.5.ii.a. *Basic concepts and notation*
The theory of selection in age-structured populations rests on standard concepts from demography, e.g., Stearns (1992, Chapter 2), Charlesworth (1994b, Chapter 1), Caswell (2000, Chapter 1), and Roff (2002, Chapter 2), and so we start by introducing these. The science of demography relates the growth over time in the size of a population to the age-specific survival and reproductive rates of its constituent individuals—their *demographic parameters*. The popu-

lation is assumed to be sufficiently large that its behavior can be treated deterministically. For simplicity, we initially ignore sex differences in the case of a
species with separate sexes, and consider only the female component of the
population. The justification for this is that female fertility is rarely limited by
the number of males in the population, so that population growth can be realis-
tically modeled by considering females only. Similarly, for a cosexual population of animals or plants, we consider only the output of eggs or seeds, respectively.

We assume initially that the life history can be partitioned into discrete time
intervals, e.g., the years separating successive annual breeding seasons of temperate zone species. Ignoring differences between genotypes for the moment,
the state of the population at a given time t can be described by the number of
individuals in each *age class* (**Figure 9.13**), such that $n(x, t)$ is the number of
individuals of age class x at time t; in an organism with an annual breeding season, these would be counted at the time of breeding. It is convenient to assign
new zygotes to age 0. $n(0, t)$ is then the number of zygotes produced at time t,
$n(1, t)$ is the number of surviving individuals that were in age class 0 at time $t -$
1, and so on. Individuals of a given age x have an age-dependent probability,
$P(x, t)$, of survival to the next time interval, $t + 1$, when they will be aged $x + 1$
(**Figure 9.13**). The value of x has an upper limit, d, above which there are no
surviving individuals in the population, so we can represent the state of the
population by a vector $\boldsymbol{n}(t)$, with components $n(1, t), \ldots, n(d, t)$.

Female fecundity at age x, $m(x, t)$, is defined as the expected number of
daughters produced by a female aged x at time t. The age at which individuals
start breeding (the age of reproductive maturity) is denoted by b, which is ≥ 1
(**Figure 9.13**). The chance of survival of a newborn to the next time interval,
when they are aged 1, is $P(0, t)$. The effective fecundity of females aged x at time
t, in terms of daughters contributed to the population at time $t + 1$, is thus $f(x, t)$
$= m(x, t)P(0, t)$.

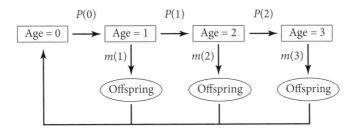

FIGURE 9.13 Number of individuals in each age class of a population with age structure
(overlapping generations). In this example, individuals survive to a maximum age of three and
start reproducing when they reach age one. The transitions between different age classes caused
by survival and fecundity are indicated by the P and m values, respectively.

We initially assume that the demographic parameters are constant over time, so that the dependence on t of P and m can be dropped. The probability of survival from age x to $x + 1$ can then be related to an underlying rate of mortality, $\mu(x)$, per unit time over this interval, so that we can write $P(x) = \exp -\mu(x)$, applying the argument used in **Section 6.1.ii.b**. It follows that $\ln P(x) = -\mu(x)$.

The net probability of surviving from conception to age x is given by the *survival function*: $l(x) = P(0) P(1) \dots P(x - 1)$. Because there is always some mortality at each age, $l(x)$ declines with age. Aging implies an increase in the rate of decline of $l(x)$ with age, but this is often detectable only when organisms are raised in sheltered conditions, because high rates of age-independent causes of death in the wild (caused by predation, disease, or starvation) conceal the underlying tendency for $P(x)$ for adult individuals to decline with age (Finch 1990, Chapter 1). An example is shown in **Figure 9.14**.

The life history can be summed up by the compound *reproductive function*: $k(x) = l(x)m(x)$, which is the expected number of daughters produced at age x by a newly formed female individual aged 0. The reproductive function, $k(x)$, is 0 for pre-reproductive ages ($x < b$), and rises after the start of reproduction. Because $l(x)$ declines with age, $k(x)$ may also start to decline, unless the decline in $l(x)$ is compensated for by an increase in $m(x)$. Such an increase happens in species that continue to grow after reaching reproductive maturity, and where fecundity is related to body size, as in trees and many cold-blooded vertebrates.

9.5.ii.b. *Predicting population growth*

With constant demographic parameters, the population rapidly approaches a constant exponential rate of growth in size, corresponding to a constant *age structure* [i.e., subject to some mild restrictions on the $k(x)$ function, the proportions of individuals in the different age classes remain unchanged over time]. The argument is outlined in **Box 9.8***. This growth rate, r, is the *intrinsic rate of increase* or *Malthusian parameter* (Fisher 1930b, p. 25), and is the unique real number that satisfies the *Euler–Lotka* equation, named after the eighteenth-century mathematician and the twentieth-century mathematical biologist who independently discovered it (Euler 1760; Sharpe and Lotka 1911). We have the following relation:

$$\sum_{x=b}^{d} \exp(-rx)k(x) = 1 \tag{9.3}$$

This representation is exact for species that reproduce at discrete time intervals. It is an approximation for species where reproduction occurs continuously in time, but it can be made arbitrarily close by increasing the number of age classes into which the life history is divided. Integration over a continuous range of ages then replaces summation (**Appendix A1.iii**). Given the values of

A.

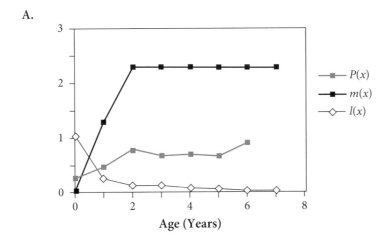

B.

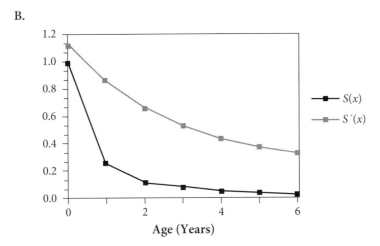

FIGURE 9.14 **A.** Demographic parameters for a wild population of gray squirrels (*Sciurus carolinensis*) in North Carolina. **B.** The corresponding sensitivities of fitness to changes in ln $P(x)$ and $m(x)$, $S(x)$ and $S'(x)$, respectively, from **Equations B9.13.3b** and **B9.13.4** in **Section 9.5.v.a**. [Data from Barkalow et al. (1970).]

$k(x)$ for each x, r can be determined numerically from **Equation 9.3** by Newton's method (**Appendix A1.ii.b**).

Demographic theory shows that the state of constant growth rate and age structure is approached quickly for biologically realistic life histories. Once this state is reached, some important quantities can be derived, which are described in **Box 9.9**; as we show in **Sections 9.5.iii** and **9.5.v** below, these are very useful for analyzing the effects of selection on life histories. But we first deal with some population genetics that includes age structure.

Box 9.8* PREDICTING CHANGES IN AN AGE-STRUCTURED POPULATION OVER TIME

B9.8.i. The population projection matrix

The state of the population at time t can be represented by a column vector $n(t)$, whose components are the numbers of individuals in age classes 1, 2, ..., d at time t, i.e., $n(1, t)$, $n(2, t)$, ..., $n(d, t)$. The numbers of individuals in age classes other than 1 at time $t + 1$ represent the survivors over one time interval, among individuals aged one unit less at time t, i.e.,

$$n(x + 1, t + 1) = n(x, t)P(x) \qquad (x = 1, 2, ..., d) \qquad \text{(B9.8.1)}$$

The individuals in age class 1 at time $t + 1$ are the survivors of the zygotes produced by the population at time t, whose number is denoted by $n(0, t)$.

This number is given by the sum over age classes b to d of $n(x, t)m(x)$; the number of survivors to stage 1 at time $t + 1$, $n(1, t + 1)$, is the product of this sum and $P(0)$. If we write $f(x) = m(x)P(0)$, we have:

$$n(1, t + 1) = \sum_{x=b}^{d} n(x,t)f(x) \qquad \text{(B9.8.2)}$$

This relation can be expressed in matrix notation (**Appendix A1.iv.a**) as:

$$n(t + 1) = An(t) \qquad \text{(B9.8.3)}$$

where A is the *population projection* or *Leslie* matrix (P. H. Leslie was one of the ecologists who introduced this representation into population biology; see Leslie 1945).

B9.8.ii. The asymptotic rate of change in population size

Based on these equations, A is a matrix whose elements are all 0, except (potentially) the members of the first row from column b onward, and for the sub-diagonal elements: $a_{1i} = f(i)$ $(i = b, ..., d)$ and $a_{i,i+1} = P(i)$. It is thus a *non-negative* matrix, because its elements are positive or 0 in value. It is known from the general theory of non-negative matrices that the largest eigenvalue, λ_1, of A (**Appendix A1.iv.c**) is unique and real, unless the fecundities $f(x)$ are periodic, i.e., all the $f(x)$ are 0 except for x values that are fixed multiples of an integer greater than 1 (Charlesworth 1994b, p. 23).

In all other cases, the population will reach a state in which $\mathbf{n}(t + 1) = \lambda_1 \mathbf{n}(t)$ (**Appendix A1.iv.d**), so that $n(x, t + 1) = \lambda_1 n(x, t + 1)$ for all x. If we take natural logarithms, the asymptotic rate of change in $\ln n(x, t)$ for all x is equal to $\ln \lambda_1$, which we can write as r, the *intrinsic rate of increase* of the population.

The *characteristic equation* that determines λ and r can be derived very easily by the following argument. Let $B(t)$ be the number of new zygotes formed at time t; the number of females aged x at time t is then $B(t - x)l(x)$, each of which produces $m(x)$ daughters. For $t \geq d$, so that no individuals alive at time 0 contribute to the population, we have:

$$B(t) = \sum_{x=b}^{d} B(t - x)k(x) \qquad (B9.8.4)$$

If the population has reached the asymptotic rate of growth, we have:

$$B(t - x)/B(t) = \lambda_1^{-x} = \exp(-rx) \qquad (B9.8.5)$$

Dividing both sides of **Equation B9.8.4** by $B(t)$ and using **Equation B9.8.5**, we obtain **Equation 9.3** of the main text. This corresponds to the characteristic equation of A, as can be confirmed by a direct evaluation.

Box 9.9 PROPERTIES OF A POPULATION WITH A CONSTANT AGE STRUCTURE

B9.9.i. The stable age distribution

The ratio of the number of individuals in two age classes (x and y) at time t, is $n(x, t)/n(y, t)$. This represents the *age distribution* of the population, i.e., the relative frequencies of individuals in different age classes. We have $n(x, t) = B(t - x)l(x)$, so that $n(x, t)/n(y, t) = B(t - x)l(x)/B(t - y)l(y)$. If the population has reached its asymptotic rate of growth, **Equation B9.8.5** implies that:

$$\frac{n(x, t)}{n(y, t)} = \frac{\exp(-rx)l(x)}{\exp(-ry)l(y)} \qquad (B9.9.1)$$

i.e., the age distribution is constant over time. This equilibrium age distribution is often referred to as the *stable age distribution*. Other things being equal, **Equation B9.9.1** implies that a high intrinsic rate of population growth causes the stable age distribution to be weighted towards a high frequency of young individuals.

B9.9.ii. Reproductive value

Fisher (1930b, pp. 27–28) introduced the concept of the *reproductive value* of age x, $v(x)$. This measures the expected net contribution of zygotes to the population from a female aged x, counting both the offspring she contributes at age x and the number she is expected to contribute from all future ages. Because the population increases by a factor of $\lambda_1 = \exp(r)$ each time interval, her fecundity at age $y = x + z$ is discounted by $\exp(-rz)$, multiplied by her probability of survival to age y, $l(y)/l(x)$. Summing over all ages from x on, this gives the following expression:

$$v(x) = \frac{\exp(rx)}{l(x)} \sum_{y=x}^{d} \exp(-ry)k(y) \qquad \text{(B9.9.2)}$$

As discussed in **Section 9.5.v**, reproductive value plays an important role in the theory of life history evolution. If this equation is applied to a new zygote, for which $x = 0$, its reproductive value is $v(0) = 1$.

B9.9.iii.* Mathematical interpretation of reproductive value and stable age distribution

It can be shown that the $v(x)$ values correspond to the components of the leading left eigenvector, $\mathbf{v}_1$, of A (**Appendices A1.iv.c** and **A1.iv.d**). **Equations B9.8.1–B9.8.3** give:

$$\lambda_1 v(x) = f(x)v(1) + P(x)v(x + 1) \qquad \text{(B9.9.3)}$$

(See **Problem 9.6.i***.) As is always the case with an eigenvector, only the relative values of the $v(x)$ are defined by this relation. Some simple algebra shows that this expression is equivalent to **Equation B9.9.2** (**Problem 9.6.ii***).

Similarly, the relative values of the components of $\mathbf{n}(t)$ for large t are given by μ_1, the leading right eigenvector of A. **Equation B9.9.1** thus implies that the xth component of μ_1 is proportional to $\exp(-rx)l(x)$, i.e., the components of the leading right eigenvector are proportional to the relative numbers of individuals in the different age classes.

B9.9.iv. Generation time

Formally, there is no such thing as a generation in an age-structured population. However, for purposes of comparison with discrete generation models, it is useful to define a generation time. The most appropriate definition is the mean age of mothers of a set of zygotes at a given time t. In general, this is given by:

$$\frac{\sum_x xn(x, t)m(x)}{B(t)} = \frac{\sum_x xB(t - x)k(x)}{B(t)}$$

Because $B(t - x)/B(t) = \exp(-rx)$ for a population at the stable age distribution, this allows us to write the generation time for such a population as:

$$T = \sum_x x\exp(-rx)k(x) \qquad (B9.9.4)$$

This formula brings out clearly that a high rate of population growth is associated with a short generation time, because a large value of r causes $\exp(-rx)$ to be weighted towards low values of x.

B9.9.v. Net reproduction rate

Another quantity that plays an important role in demography is the *net reproduction rate* or *lifetime reproductive success*, R. This is simply the sum over all ages (x) of fecundity at age x, discounted by the probability of surviving to age x:

$$R = \sum_x k(x) \qquad (B9.9.5)$$

If the population is stationary in size (i.e., $r = 0$), this is equivalent to $v(0)$. An increase in population size requires $R > 1$, and a decrease corresponds to $R < 1$.

9.5.ii.c. *Population genetics for age-structured populations*
Consider a single autosomal locus with two alleles A_1 and A_2. A given genotype $A_i A_j$ (i and j = 1 or 2) has its own set of demographic parameters for females, i.e., a set of values of $k_{ij}(x) = l_{ij}(x)m_{ij}(x)$. We can apply the Euler–Lotka equation to a hypothetical population exclusively of genotype $A_i A_j$, and define an intrinsic rate of increase r_{ij} for this genotype. This is often used as a measure of fitness, equivalent to the Malthusian parameter used for microbial populations in **Section 3.1.iii.d**.

If the population reproduces asexually, so that females produce only daughters, each genotype constitutes a separate sub-population that reproduces independently of the other genotypes. The genotype with the highest value of r_{ij} will thus eventually replace all of the others; similar principles apply to completely self-fertilizing populations of cosexuals (Pollak and Kempthorne 1970). This result has often been used by evolutionary ecologists to justify the use of r as a measure of fitness, initially by Cole (1954). The result applies to a population of any ploidy level.

For a sexually reproducing population, however, it is by no means obvious whether such a result is valid. Consider a population with separate sexes, where males and females contribute equally to the new zygotes. Both male and female demographic parameters must then appear in a selection model. The reproductive success of males (the number of offspring that they father) must in general depend on the age structure and the genotypic composition of the female population, because the number of offspring produced by a male depends on the ages and genotypes of the females with whom he mates. If some simplifying assumptions are made, however, a set of recursion relations for allele frequencies can be derived (**Box 9.10**). These assumptions include random mating with respect to age and genotype, progeny sex ratio independent of age and genotype, and the same ratios of male and female $k(x)$ values for a given genotype at each age x. For details, see Charlesworth (1994b, pp. 108–109).

9.5.iii. Selection equations

9.5.iii.a. *The general recursion relations with no sex differences*
The assumptions just mentioned mean that allele frequencies will be the same for the male and female gametes produced by the population at time t. As shown in **Box 9.10**, this yields a set of recursion equations that resemble those for selection at a single locus with discrete generations (**Section 2.1.ii**). This is the special case for a single age class with reproduction starting at age 1.

These results do not, however, immediately help us find a fitness measure analogous to that used in discrete generation models, because we have an intractable set of high-dimensional equations with no simple exact solution.

Box 9.10 THE SELECTION EQUATIONS
FOR A RANDOMLY MATING
AGE-STRUCTURED POPULATION

Let $p(t)$ and $q(t)$ be the frequencies of alleles A_1 and A_2 among the gametes produced at time t, where $q(t) = 1 - p(t)$. The simplifying assumptions described in the text mean that the frequencies of A_1A_1 and A_2A_2 individuals among the new zygotes produced at time t are $p(t)^2$ and $q(t)^2$, respectively, and the frequency of heterozygotes is $2p(t)q(t)$. This implies that the number of A_1A_1 individuals aged x at time t is equal to $B(t-x)p_1(t-x)^2l_{11}(x)$; the total number of offspring that they contribute to the population at time t is $m_{11}(x)/(1-c)$, where c is the proportion of males among the zygotes (the primary sex ratio of **Section 9.4.i**). Homozygotes A_1A_1 aged x thus contribute $2B(t-x)p(t-x)^2$ $l_{11}(x)m_{11}(x)/(1-c)$ to the number of A_1 alleles among the zygotes produced at time t. Similarly, heterozygotes A_1A_2 contribute $B(t-x)p(t-x)q(t-x)l_{12}(x)m_{12}(x)/(1-c)$.

The total number of A_1 alleles among the zygotes produced at time t, $B_1(t)$, is obtained by adding these terms and summing over x. A similar expression can be written for the total number of A_2 alleles, $B_2(t)$, substituting subscript 2 for 1. The total number of alleles entering the population at time t is $2B(t) = B_1(t) + B_2(t)$.

The frequency of A_1 among the zygotes at time t is thus $p(t) = B_1(t)/2B(t)$. The term $(1 - c)$ cancels from the top and bottom of this expression, and we can replace $l_{ij}(x)m_{ij}(x)$ by $k_{ij}(x)$. This yields the following expressions:

$$p(t) = \frac{\sum_x B(t-x)\left[p(t-x)k_{11}(x) + q(t-x)k_{12}(x)\right]}{B(t)} \tag{B9.10.1a}$$

$$q(t) = \frac{\sum_x B(t-x)\left[p(t-x)k_{12}(x) + q(t-x)k_{22}(x)\right]}{B(t)} \tag{B9.10.1b}$$

$$B(t) = \sum_x B(t-x)\left[p(t-x)^2 k_{11}(x) + 2p(t-x)q(t-x)k_{12}(x) + q(t-x)^2 k_{22}(x)\right] \tag{B9.10.1c}$$

9.5.iii.b. *Equilibrium conditions*

Several approaches have been used to solve this difficulty. A simple one is to consider the conditions for a polymorphic equilibrium, such that $p(t)$ and $q(t)$ have reached constant values, p^* and $q^* = 1 - p^*$, respectively. **Box 9.11** shows that the following expression determines p^* and q^* in the same way as the fitness w_{ij} with discrete generations:

$$w_{ij}(r) = \sum_x \exp(-rx)k_{ij}(x) \tag{9.4}$$

where r is the intrinsic rate of increase of the population as a whole.

To determine the genotypic composition of an equilibrium population with overlapping generations, we can use $w_{ij}(r)$ instead of the corresponding discrete generation fitness. This result applies quite generally, including equilibria under mutation–selection balance and multi-locus systems (Charlesworth

---

Box 9.11 CONDITIONS FOR A POLYMORPHIC EQUILIBRIUM

An equilibrium population can be treated as genetically homogeneous as far as its demography is concerned. It will thus achieve a constant growth rate, r, given by the equivalent of **Equation 9.3** of **Section 9.5.ii.b**. This means that $B(t - x)/B(t) = \exp(-rx)$, and so **Equation B9.10.1c** of **Section 9.5.iii.a** gives an Euler–Lotka equation:

$$\sum_x \exp(-rx)\left[p^{*2}k_{11}(x) + 2p^*q^*k_{12}(x) + q^{*2}k_{22}(x) \right] = 1 \tag{B9.11.1}$$

Equations B9.10.1a and **B9.10.1b** become the following:

$$\sum_x \exp(-rx)\left[p^*k_{11}(x) + q^*k_{12}(x) \right] = 1 \tag{B9.11.2a}$$

$$\sum_x \exp(-rx)\left[p^*k_{12}(x) + q^*k_{22}(x) \right] = 1 \tag{B9.11.2b}$$

Comparison with the equivalent equations for the discrete generation case (**Section 2.1.ii.c**) shows that the discrete generation fitness w_{ij} for genotype ij plays the same role in determining p^* as the quantity $w_{ij}(r)$ defined by **Equation 9.4** of the main text.

1994b, Chapter 3). It is worth noting that $w_{ij}(r)$ is the reproductive value at age 0 for genotype ij (**Box 9.9**).

A difficulty, however, is that the equation that determines r (i.e., **Equation B9.11.1**) itself involves the allele frequencies that are to be found. One solution to this problem is to regard r as a quantity that can be determined empirically. The other is to solve the set of simultaneous equations (**Equations B9.11.2a and B9.11.2b**), or their equivalents for more complex genetic models. There is a known solution for two alleles (Charlesworth 1994b, p. 122). For mutation–selection balance at a single site with two variants, r is well approximated by the value for a population homozygous for wild type (Charlesworth 1994b, p. 126). In general, however, the equations must be solved numerically.

A reasonable assumption for many natural populations is that the population growth rate is close to 0, due to density-dependent mortality or fertility (Begon et al. 2006, Chapter 5). **Equation 9.4** then reduces to the lifetime reproductive success or net reproduction rate R (**Box 9.9.v** of **Section 9.5.ii.b**) of genotype ij, which is often used by evolutionary ecologists as a fitness measure (e.g., Clutton-Brock 1988).

9.5.iii.c. *Stability and invasion conditions*

The second approach is to consider the conditions for the stability of equilibria, either global or local. The most complete treatment is that of Norton (1928) for the continuous-time version of **Equations B9.10.1a–B9.10.1c**. (Norton was the mathematician who also first investigated the speed of change of allele frequencies under selection for the discrete generation model; see **Section 3.1.iii.b**.) By a complicated argument, he showed that the ultimate outcome of selection is determined entirely by the relations between the r_{ij}. Allele A_2 tends to fixation if r_{22} is the largest of the set $\{r_{11}, r_{12}, r_{22}\}$. If there is heterozygote superiority in the r_{ij}, the allele frequencies tend to the neighborhood of the equilibrium discussed above; with heterozygote inferiority, the allele whose initial frequency lies between the equilibrium value and 1 tends to fixation. A local stability analysis of the polymorphic equilibrium can also be carried out for the discrete time interval model (Charlesworth 1994b, pp. 162–166) and gives corresponding results.

The local stability method can also be applied to the case of the invasion of a population that is fixed for allele A_1 by a rare, nonrecessive variant A_2 (Charlesworth 1973). The argument is displayed in **Box 9.12** and shows that the rate of increase in frequency of A_2 is determined by the difference between the intrinsic rate of increase r of its heterozygous carriers, and that of the predominant homozygous genotype, A_1A_1.

This type of analysis can be extended to the case of an initial population held at a stationary population size by density dependence (Charlesworth 1973); in this case, the demographic parameters for the invading mutant heterozygote

Box 9.12 CONDITIONS FOR INITIAL INCREASE OF A RARE ALLELE

We use the same approach as for the initial increase of a rare allele in the discrete generation case, i.e., we ignore second-order terms in its frequency (**Section 3.1.iii.c**). **Equation B9.10.1b** of **Section 9.5.iii.a** then yields the following relation:

$$q(t) \approx \sum_x q(t - x)\exp(-r_{11}x)k_{12}(x) \qquad (B9.12.1)$$

This leads to an Euler–Lotka equation (**Section 9.5.ii.b**), which gives the asymptotic rate of change of the natural logarithm of $q(t)$ as the real root of the equation:

$$\sum_x \exp\left[-\left(r_{11} + r\right)x\right]k_{12}\left(x\right) = 1 \qquad (B9.12.2)$$

This implies that the rate of change is $r_{12} - r_{11}$, which provides a measure of the selective advantage of A_2 when rare. A_2 will thus invade a population fixed for A_1 when $r_{12} > r_{11}$. By symmetry, A_1 will invade a population fixed for A_2 when $r_{12} > r_{22}$, consistent with the results on heterozygote superiority described above.

The case of invasion by a recessive mutant allele A_2 is harder to analyze, because second-order terms in mutant allele frequency can no longer be neglected (see **Section 3.1.iii.c**), but it can be shown that A_2 will invade an A_1 population if $r_{22} > r_{11}$. Its selective advantage, however, is not equal to $r_{22} - r_{11}$, except as an approximation when selection is weak (Charlesworth 1994b, p. 152).

are evaluated under the conditions for the stationary initial population. The predominant homozygote necessarily has an intrinsic rate of 0, so that the variant will invade only if the heterozygote has a positive growth rate under the prevailing intensity of density-dependent factors.

9.5.iii.d. *Weak selection approximations*

The final method is to assume weak selection and to seek approximations for rates of change of allele frequencies. This was first done by Haldane (1927a, 1962) and was later generalized by Charlesworth (1994b, Chapter 4). We can select an arbitrary genotype as a standard, with life-history parameters $k_S(x)$.

Weak selection implies that any other genotype ij has a $k_{ij}(x)$ function differing from $k_S(x)$ by an amount that is smaller in magnitude than a measure of the strength of selection, ε, where ε is sufficiently small that second-order terms can be neglected.

Under these conditions, the rate of change per time interval in the frequency of allele A_2 in **Equations B9.10.1b–B9.10.1c** rapidly approaches:

$$\Delta q \approx pq(r_{2.} - r_{1.}) \tag{9.5}$$

where $r_{1.} = pr_{11} + qr_{12}$, $r_{2.} = pr_{12} + qr_{22}$, and the error is on the order of ε^2.

This is identical in form to the corresponding discrete generation equation with weak selection (**Section 3.1.iii.b**), except that the rate of change is per time interval, not per generation. This approach can also be extended to the case of density-dependent populations, yielding results similar to those for the discrete generation case (Charlesworth 1994b, pp. 147–149).

9.5.iii.e. *Relationships between different fitness measures*

The results of the exact analysis of equilibrium populations give one possible interpretation of what is meant by fitness in an age-structured population. The analyses of conditions for local and global stability give a somewhat different one, but suggest that we can use differences in intrinsic rates of increase among genotypes to predict rates of change of allele frequencies, at least as a first-order approximation when selection is weak or when a rare allele is invading a population. However, the equilibrium fitness expression of **Equation 9.4** in **Section 9.5.iii.b** is not obviously related to the intrinsic rate of increase.

These discrepancies disappear, however, with sufficiently weak selection, as can be seen as follows. As noted above, the weak selection **Equation 9.5** gives a rate of change per time interval, not per generation. To make it comparable to the discrete generation case, we need to convert to a time scale of generations. This can be done using the formula for the generation time with age structure, T (**Box 9.9.iv** of **Section 9.5.ii.b**). When the genotypes have different life-history parameters, we can again select a standard genotype and use its generation time T_S as an approximate measure for the population as a whole. If **Equation 9.5** is multiplied by T_S, we obtain an expression for the rate of change of allele frequency that is on the same order of accuracy as the original equation, but expressed as a rate per generation. This shows that we can use $T_S r_{ij}$ as a measure of the fitness of genotype ij that is approximately equivalent to the familiar discrete generation measure w_{ij}.

This is confirmed by the fact that, with weak selection, genotypic differences in the equilibrium fitness function $w_{ij}(r)$ are approximately proportional to the corresponding differences in the intrinsic rates of increase, with a proportionality constant equal to the generation time for a standard genotype (**Problem 9.7***).

9.5.iii.f. *More general situations*

These results have all been derived using simplifying assumptions, notably random mating with respect to age and genotype, and proportionality of the male and female $k(x)$ functions for different genotypes. Modifications can be made to deal with more complicated situations, especially when selection is weak (Charlesworth 1994b, pp. 114–116, 120–121). For example, with weak selection, **Equation 9.4** can still be applied as a first-order approximation, even when there is nonrandom mating with respect to age and sex differences in the life history, by replacing $k_{ij}(x)$ by the arithmetic mean of the corresponding male and female functions (Charlesworth and Charlesworth 1973a). Selection in varying environments can also be modeled (Tuljapurkar 1990; Charlesworth 1994b, pp. 154–159).

Multi-locus systems are, however, even more intractable than the single-locus case. When selection is weak in relation to the recombination frequency, the quasi-linkage equilibrium approximation of **Section 8.4.ii.d** applies to two-locus systems with age structure (Charlesworth 1994b, pp. 146–147). Similar results to those for discrete generations should presumably apply to multi-locus systems with age structure (though this has not been formally proved). Assuming this to be true, equations can be derived for selection on a multivariate set of additively inherited quantitative traits; these are similar to the equations for discrete generations, but differences in w_{ij} are replaced by differences in $T_S r_{ij}$ (Lande 1982; Charlesworth 1993b).

9.5.iv. Implications for population genetic studies

One important conclusion from the study of selection in age-structured populations is that most of the basic results from models of discrete generation populations (or continuous-time models that ignore age structure—**Section 3.1.iii.d**) apply, especially if selection is weak. This also applies to other aspects of population genetics; results such as the equality of the neutral mutation rate and the rate of substitution of neutral mutations, and concepts such as effective population size, can also be extended to age-structured populations (Charlesworth 1994b, Chapter 2; Rousset 1999; Charlesworth 2001a; Nordborg and Krone 2002).

There are, however, some important qualifications. While differences in genotypic intrinsic rates of increase can be used to predict the initial increase in frequency of a nonrecessive rare allele, this does not hold true at or near polymorphic equilibria (**Section 9.5.iii.c**). Rather, the frequencies of genotypes in such equilibrium populations are determined by the fitness measure of **Equation 9.4**, as was first shown for the case of a biallelic locus by Norton (1928). For experimentalists dealing with examples of strong selection, it is important to be clear about this, because otherwise erroneous inferences of frequency-dependent selection might be made (Charlesworth, 1994b, p. 180).

Equation 9.4, and its extension to the case of sex differences in demographic parameters (Charlesworth, 1994b, p. 120), also provide a rigorous basis for estimating the relative fitnesses of different genotypes, which is relevant to studies of human populations (Charlesworth and Charlesworth 1973a). This formula has been applied in recent years to characterize the fitness effects of mutations in other organisms, such as *Caenorhabditis elegans* (e.g., Peters and Keightley 2000). It has the important implication that, in general, the relative fitnesses of genotypes depend on the demographic state of the population. Allele and genotype frequencies may thus be sensitive to environmental changes that alter the population's growth rate and age structure, even if these changes do not affect the relative values of the age-specific mortality and reproductive rates of different genotypes. This follows from the fact that genetic equilibrium is formally impossible until the population has attained a state of constant growth and age structure, unless the $k_{ij}(x)$ functions for different genotypes have the same relative values for each value of x (Charlesworth 1994b, pp. 117–119).

It follows that changes in the environment that lead to different values of the equilibrium population growth rate, r, will generally lead to different relative fitness values as determined by **Equation 9.4**, such that genotypes with relatively high $k_{ij}(x)$ functions early in life are favored if r is positive, and genotypes with relatively low $k_{ij}(x)$ functions early in life are favored if r is negative (Charlesworth 1994b, pp. 129–131). Fluctuations in population density, such as occur in microtine rodents, can thus drive changes in genotype frequencies (Charlesworth 1994b, pp. 126–129). Similar effects can occur if changes in mortality patterns change the relative proportions of young and old individuals in the population, even if the population growth rate remains constant.

The recent declines in population growth and mortality rates in many human populations mean that selection places a greater emphasis on late-life survival and reproduction than previously, so that diseases with a late age of onset, such as Huntington's disease, are now more disadvantageous in terms of their net fitness effects than they were a century ago, at least in the affluent section of the world. The selection coefficient against carriers of the dominant mutation responsible for this disease is estimated to have changed from 9% to 15% with the transition from a high-mortality to a low-mortality demographic regime, assuming that nothing else has changed with respect to their demographic parameters (Charlesworth 1994b, pp. 132–134). The same type of change will apply to the many late-onset diseases with complex genetic control, such as cancers and cardiovascular disease (Wright et al. 2003). If this demographic situation persists, a gradual decline in the frequency of such diseases is expected. This sensitivity of relative fitnesses to the growth rate and age structure of the population is the main novel feature introduced by age structure into selection theory, and forms the core of theories of life history evolution, as we discuss in the next section.

9.5.v. Principles of life history evolution

9.5.v.a. *The sensitivity of fitness to changes in demographic parameters*
A basic tool for life history theory is provided by the *fitness sensitivities*, which measure the effects on fitness of small changes in age-specific survival probabilities or fecundities. We saw a similar use of calculating the fitness effects of small phenotypic changes when we studied ESSs for the sex ratio in **Section 9.4.ii**. Fitness sensitivities were introduced by Hamilton (1966) in the context of the evolution of aging, and have been widely used in life history evolution, e.g., Caswell (2000). Here, we use the results discussed in **Section 9.5.iii.e**, and measure fitness, w, by the product of generation time, T, and intrinsic rate, r. Expressions for the fitness sensitivities are derived in **Box 9.13**, and an example from a wild population is shown in **Figure 9.14 of Section 9.5.ii.a**.

Equation B9.13.3b shows that the sensitivity of w to a change in log survival probability, $S(x)$, is equal to 1 for all pre-reproductive ages, and then decreases steadily throughout adult life, becoming 0 when reproduction ceases (**Figure 9.14**). The rate of decrease is highest when mortality or rate of population growth is high; it is least when these are low, or when fecundity increases with age, as is the case for organisms that continue to grow throughout adult life and in which fecundity increases with size, e.g., in many perennial plants and cold-blooded vertebrates. This suggests that selection puts the greatest weight on survival early in life in the first type of situation, and the least in the second. The implications of this for the evolution of aging will be discussed in **Section 9.5.v.d** below.

Box 9.13 FITNESS SENSITIVITIES

We first show how to calculate the effect on r of a small change in mortality at a given age x, denoted by $\delta P(x)$, leaving all other ages unchanged. This change will affect the survival and reproductive functions only for ages $y > x$. For these ages, we have $l(y) = P(0)P(1) \ldots P(x) \ldots P(y-1)$, so that the associated change in $l(y)$ is $\delta l(y) = P(0)P(1) \ldots \delta P(x) \ldots P(y) = l(y)\delta P(x)/P(x)$.

Let the change in r associated with the change in $P(x)$ be δr. The Euler–Lotka equation (**Equation 9.3**) of **Section 9.5.ii.b** implies that:

$$\sum_{y=b}^{d} \exp\left[-(r + \delta r)y\right]\left[l(y) + \delta l(y)\right]m(y) = 1 \qquad \text{(B9.13.1)}$$

We can write $\exp[-(r + \delta r)y] = \exp(-ry) \exp(-\delta ry)$; in addition, $\exp[-(\delta r)y]$ can be approximated by $1 - (\delta r)y$ (**Appendix A1.ii.c**). Ignoring the product of δr and $\delta l(y)$, we obtain:

$$\sum_{y=b}^{d} \exp(-ry)\left[l(y)m(y) + \delta l(y)m(y) - (\delta r)yl(y)m(y)\right] \approx 1 \quad \text{(B9.13.2)}$$

The first term on the left is equal to 1, by the definition of r from the Euler–Lotka equation, and thus cancels the 1 on the right-hand side. Rearranging, we get:

$$(\delta r)\sum_{y=b}^{d} y\exp(-ry)k(y) \approx \sum_{y=x+1}^{d} \exp(-ry)\delta l(y)m(y)$$

The multiplicand of δr is simply the generation time T, defined by **Equation B9.9.4** in **Section 9.5.ii.b**, and so the left-hand term represents the change in fitness expressed as a change per generation, which we can write as $\delta w \approx T\delta r$. Substituting the above expression for $\delta l(y)$, we obtain:

$$\delta w \approx \frac{\delta P(x)}{P(x)} \sum_{y=x+1}^{d} \exp(-ry)k(y) = \frac{\delta P(x)}{P(x)}S(x) \qquad \text{(B9.13.3a)}$$

$\delta P(x)/P(x)$ is approximately equal to $\delta \ln P(x)$ (**Appendix A1.ii.c**), so that the function $S(x)$ defined by this expression measures the sensitivity of fitness to a small change in log survival at age x. We can express this in terms of the mortality rate at age x, writing $\ln P(x) = -\mu(x)$ (**Section 9.5.ii.a**), and so we have:

$$\delta w \approx -\delta \mu(x)S(x) \qquad \text{(B9.13.3b)}$$

A similar procedure can be carried out for $\delta m(x)$, the effect of a small change in fecundity at age x, and we obtain:

$$\delta w \approx \delta m(x)\exp(-rx)l(x) = \delta m(x)S'(x) \qquad \text{(B9.13.4)}$$

where $S'(x)$ is the sensitivity of fitness to a change in fecundity.

$S(x)$ and $S'(x)$ are in fact the partial derivatives of w with respect to $\ln P(x) = -\mu(x)$ and $m(x)$, i.e., $-\partial w/\partial \mu(x)$ and $\partial w/\partial m(x)$, respectively (**Appendix A1.i**).

Provided that r is non-negative (i.e., the population is not declining in size), the sensitivity of w to a change in fecundity, $S'(x)$, always declines with age, because $\exp(-rx)l(x)$ decreases with age. This implies that, other things being equal, there is always a premium on early reproduction, and that this is greatest with high mortality and a high rate of population growth.

9.5.v.b. *Trade-off models*

At first sight, these results suggest that selection should favor a life history in which reproduction starts as early as possible in life, even at the expense of subsequent survival. This occurs in some types of *semelparous* species, which have a short juvenile period, followed by a burst of reproductive activity and death, e.g., annual plants or many species of insects. However, even many semelparous species defer reproduction for one or more years, the most extreme examples being species that wait many years before reproducing, like the century plant *Agave americana* (Schaffer and Schaffer 1979) and the periodic cicadas of the United States that spend 13 or 17 years as a nymph in the soil (Lloyd and Dybas 1966). In addition, many organisms are *iteroparous*, i.e., they reproduce repeatedly during a prolonged adult life.

An important goal of life history theory is to understand why these different life histories have evolved, in the light of the demographic theory of selection we outlined in **Sections 9.5.ii.c** and **9.5.iii**. We provide a few indications of how this can be done; for a detailed review of this topic, see Roff (2002). The key concept is that of a trade-off between different components of the life history, similar to those we discussed earlier in this chapter (**Sections 9.3.ii** and **9.4.v**), i.e., resources available at any one time in the life history are finite and investment in one aspect of the life history reduces resources available for others.

The simplest model of a life history trade-off is between reproduction and survival (or growth) at a given age; this is the *reproductive effort* model, first proposed by Williams (1966). An increase in reproduction at age x is assumed to result in a reduced probability of survival to $x + 1$ and/or a reduced rate of growth in body size. We will focus on survival here; let an increase $\delta m(x)$ in fecundity at age x be accompanied by a small reduction in survival probability, which we represent by $\delta P(x) \approx -D_{m(x)}\delta m(x)$, where $D_{m(x)}$ is the negative of the slope of the curve of $P(x)$ against $m(x)$, which is given by the derivative of $P(x)$ with respect to $m(x)$ (see **Appendix A1.i**).

From the general theory of an ESS (**Section 9.4.ii**), there will be an evolutionary equilibrium when a mutation that alters the life history by a small amount is associated with a zero change in w. From the argument in **Box 9.13** of **Section 9.5.v.a**, by ignoring second-order terms in $\delta m(x)$, we obtain:

$$\delta w \approx S'(x)\delta m(x) + S(x)\delta P(x)/P(x) = [S'(x) - D_{m(x)}S(x)/P(x)]\,\delta m(x) \qquad (9.6a)$$

Substituting the expressions for $S(x)$ and $S'(x)$ from **Box 9.13**, and using the equation for reproductive value in **Box 9.9** of **Section 9.5.ii.b**, we get:

$$\delta w \approx \exp(-rx)l(x)[1 - D_{m(x)}\, v(x + 1)\exp(-r)]\delta m(x) \qquad (9.6b)$$

The ESS for an intermediate value of reproductive effort [where neither $P(x)$ nor $m(x)$ is 0] is obtained by setting this expression to 0 (see **Box 9.5** of **Section 9.4.ii**), because this corresponds to the derivative of w being equal to 0 (**Appendix A1.i**). It follows that an ESS life history with a level of investment in reproductive effort that permits survival beyond age x must satisfy $v(x + 1) = \exp(r)/D_{m(x)}$. There is thus no possibility of maintaining nonzero $P(x)$ unless subsequent reproductive success and survival are sufficiently high that this condition can be met.

We also need to consider the additional condition that w must be at a maximum for an ESS life history whose reproductive effort is intermediate between 0 and its upper limit; this is necessary for an iteroparous life history to be an ESS. This can be shown to require the second derivative of $P(x)$ with respect to $m(x)$ to be negative (Schaffer 1974; Charlesworth 1994b, p. 215); i.e., $P(x)$ is a concave function of $m(x)$, declining faster as $m(x)$ increases (as in the left-hand panel of **Figure 9.3B** of **Section 9.3.ii**).

Numerical investigations of the case when the function relating $m(x)$ and $P(x)$ is the same for all ages show that the ESS values of $m(x)$ and $P(x)$ are often independent of age for all adult ages (Charlesworth 1990b), and can thus be written as m^* and P^*, respectively. If the upper limit to survival, d, is indefinitely large, a simple calculation shows that reproductive value for adult ages is also independent of age, and is given by $v^* = m^*/[1 - P^*\exp(-r)]$ (**Problem 9.8.i**). In addition, the Euler–Lotka equation in this case gives $1 = l(b)\, v^*\exp(-rb)$ (**Problem 9.8.ii**). As shown above, the ESS condition is $v^* = \exp(r)/D_{m^*}$. We thus have:

$$D_{m^*} = l(b)\exp{-r(b - 1)} \qquad (9.7)$$

Higher mortality of juveniles leads to lower values of $l(b)$, so that a lower value of D_{m^*} is required for this ESS when juvenile mortality is high. Given the requirement for $D_{m(x)}$ to increase with $m(x)$, this implies a lower value of m^* and a higher value of P^*. This is the justification for the idea that high juvenile mortality favors the evolution of low reproductive effort and high adult survival (Schaffer 1974). A similar argument shows that, if P for a given value of m is reduced to a new value, $P' = CP$ (where $C < 1$), as a result of an increase in mortality imposed by the external environment, an increased ESS reproductive effort will ensue (Charlesworth 1994b, p. 220), i.e., higher adult mortality favors increased reproductive effort.

This analysis leaves untouched the question of the ESS age of reproductive maturity, b. In the context of the reproductive effort model, this can be examined by noting that, for selection to resist the spread of a variant that increases reproductive effort above 0 at an age $x < b$, δw in **Equation 9.6b** must be negative when $\delta m(x) > 0$. Combining this with the formula for reproductive value (**Equation B9.9.2** in **Section 9.5.ii.b**), this leads to the relation:

$$D_{m(x)} > l(x) \exp -r(x - 1) \qquad (1 \leq x < b) \qquad (9.8)$$

where $D_{m(x)}$ is evaluated at $m(x) = 0$. To satisfy this relation jointly with **Equation 9.7**, P for all juvenile ages must be more sensitive to an increase in reproductive effort than P at later ages. This could arise, for example, because young individuals require resources for growth as well as survival, so that diversion of resources into reproduction has a larger effect on the survival of young individuals than it does on older individuals. Other factors influencing the optimal age of maturity are discussed by Charlesworth (1994b, pp. 205–212) and Roff (2002, pp. 181–188).

If $P(x)$ is a convex function, declining fastest for low $m(x)$ and more slowly when $m(x)$ is high, then no ESS is possible with values of $m(x)$ and $P(x)$ that lie between the extremes. If the same type of relation between $m(x)$ and $P(x)$ applies to all ages, then reproductive effort must be 0 for all ages except b, i.e., there is a semelparous life history. This situation probably arises when a considerable investment of resources is needed for any successful reproduction at all. This is likely to be the case for the century plant, for example, which produces a large and costly inflorescence stalk (Schaffer and Schaffer 1979; Howell and Roth 1981).

9.5.v.c. *Other life history models*

The discussion of the reproductive effort model is intended only to give an idea of how the properties of an optimal life history can be determined in the light of selection theory. This model makes many restrictive assumptions, notably a constant environment and a trade-off function that is restricted simply to survival and reproduction at the same age. Many more elaborate models of optimal life histories have also been studied, often based on "optimal control theory," under which a life history is optimized under a set of multiple constraints, e.g., Houston and McNamara (1996), Kozlowski and Teriokhin (1999), and Baudisch (2008, Chapters 4 and 5). The effects of temporally fluctuating environments are discussed by Tuljapurkar (1990), Charlesworth (1994b, pp. 203–205), Roff (2002, Chapter 5), and Metcalf and Koons (2007).

Nevertheless, some of the broad conclusions from the reproductive effort model, such as the conclusion that high juvenile mortality favors low reproductive effort, are likely to be robust to the details of trade-off models, and are supported by a large body of comparative and experimental data. For overviews of the empirical evidence, see Charlesworth (1994b, pp. 249–265) and Roff (2002, pp. 126–149, 188–220).

9.5.v.d. *The evolution of aging*

As stated above, aging is characterized demographically as increased mortality and declining reproductive success with age. While patterns of age-specific fecundity are very variable among species, a common pattern, at least for captive or sheltered populations, is for the logarithm of age-specific mortality, $\mu(x)$, (defined in **Section 9.5.ii.a**) to begin increasing approximately linearly with x, from a minimum level at around the age of reproductive maturity; this is called the *Gompertz–Makeham relation* (Finch 1990, Chapter 1; Gavrilov and Gavrilova 1991, Chapter 2); see **Figure 9.15**. At advanced ages (from the 90s on in humans), the rate of increase in mortality tends to level off (so-called mortality leveling; see Gavrilov and Gavrilova 1991, pp. 127–130; Vaupel 1997; Rose et al. 2007); this is visible in the *Drosophila* example in **Figure 9.15**.

Senescent increase in mortality or decrease in reproductive success is, at least in part, an evolved response to the greater selective impact of genes that affect survival or fecundity early in life, relative to ones with effects later in life (**Section 9.5.v.a** above). This idea was introduced by Fisher (1930b, p. 29) and Haldane (1941, pp. 192–194), and was further developed by Medawar (1946, 1952) and Williams (1957). Hamilton (1966) provided the first correct mathematical expressions for the effects on fitness of the age of action of a genetic change in mortality or fecundity, based on the intrinsic rate of increase as a

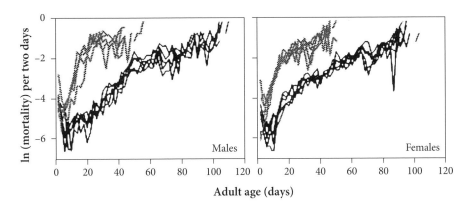

FIGURE 9.15 The plots of the natural logarithms of the age-specific mortality rates over two days, for adult males and females from two sets of laboratory populations of *Drosophila melanogaster*. The upper curves are for replicate populations maintained under standard conditions; the lower curves are for populations that had been selected for increased longevity. Early in reproductive life, there is a good fit to a linear increase in mortality rate with age, which is slower for the late-selected populations. Later on, the curves for both populations reach plateaus. [Adapted from Figure 5 of Rose et al. (2002).]

fitness measure (see **Sections 9.5.ii.b** and **9.5.iii.e** above). His fitness sensitivity measures, described in **Section 9.5.v.a** above, suggest that, other things being equal, a genetic variant that changes age-specific mortality by a given amount early in reproductive life will affect fitness more than one that acts later (see **Equation B9.13.3b** and **Figure 9.14**). If we start with a hypothetical life history in which aging is absent [i.e., $P(x)$ and $\mu(x)$ are independent of x], then the stronger selection in favor of genetic variants that increase survival early in adult life, and against variants that reduce it, will cause the evolution of higher survival rates early compared with late, i.e., the adult life history will start to show evidence of aging at the level of survival. Similar principles apply to changes in age-specific fecundity (see **Equation B9.13.4**). Subsequent theoretical work has elaborated on this idea (reviewed in Baudisch 2008, Chapter 1).

Two specific versions of this general theory are currently considered to be especially important by evolutionary geneticists interested in aging. The first is the "antagonistic pleiotropy" model of Medawar (1946, 1952) and Williams (1957). This resembles the reproductive effort model described in **Section 9.5.v.b**, and postulates that a genetic variant with positive effects on survival or fecundity early in life, but negative effects later on, is more likely to be established by selection than a variant with the opposite pattern of effects. Formal models of antagonistic pleiotropy can reproduce the general properties of observed age-specific mortality and fecundity patterns (Mueller and Rose 1996; Rose et al. 2007). One version of this idea is the "disposable soma" hypothesis (Kirkwood 1977, 1990; Abrams and Ludwig 1995), which assumes a trade-off between investment in the repair of cellular damage (which reduces survival), and investment in reproduction. Under suitable assumptions, the lower sensitivity of fitness to change in survival late in life leads to a decrease with age in allocation to repair versus reproduction. It is unclear, however, whether this model accurately predicts observed patterns of relations between age-specific survival and age, or between fecundity and age (Abrams and Ludwig 1995; Baudisch 2008, Chapter 1). Furthermore, as we have already seen with the reproductive effort model, optimization models do not necessarily predict an increase in mortality with age (Baudisch 2008, Chapters 4 and 5).

The second possibility is the "mutation accumulation" theory, first proposed by Medawar (1952). This refers to the fact that deleterious alleles with effects restricted to late stages of life equilibrate at higher frequencies under mutation–selection balance (**Section 4.2.ii.a**) than alleles that act earlier. In the simplest version of this model, mutations have effects on survival or fecundity confined to specific ages. We can then use **Equations B9.13.3** and **B9.13.4** to calculate their selection coefficients; these can be substituted into **Equation 4.3** to determine the equilibrium frequencies of the mutations. For example, if mutations in a given gene affecting only age x arise at rate $u(x)$ per generation, causing an increase in mortality rate of $s(x)$ over the wild-type value, their equilibrium frequency is given by:

$$q(x) \approx \frac{u(x)}{s(x)S(x)} \qquad (9.10a)$$

where $S(x)$ is the fitness sensitivity for mortality at age x for wild-type individuals.

Using the same argument as used for mutational load in **Section 4.3.ii.a**, this shows that there is an increase in mortality at age x over that for the wild-type individuals of:

$$\Delta\mu(x) \approx \frac{2u(x)}{S(x)} \qquad (9.10b)$$

If the effects of independent mutations on survival are multiplicative, as assumed in the discrete generation case in **Section 4.3.ii.c**, their effects on mortality rates will be additive. The net increase in mortality from a set of independent mutations is then obtained by summing across all mutable sites that affect age x, so that the net increase in mortality is given by the mean number of new mutations per generation that affect age x alone, divided by $S(x)$. If the mutation rate is the same for mutations affecting different ages, this implies that the increase in mortality is proportional to $1/S(x)$. The same argument for mutations affecting fecundity shows that the reduction in mean fecundity relative to that for mutation-free individuals is inversely proportional to $1/S'(x)$.

For a life history where reproduction starts at age b and continues at a constant level, with an age-independent adult survival probability of P_a (mortality rate μ_a), it is easy to show (**Problem 9.9**) that the fitness sensitivity S for a stationary population has the following form:

$$S(y) = P_a^y = \exp{-\mu_a(y)} \qquad (9.11)$$

where $y = x + 1 - b$.

In other words, the fitness sensitivity declines exponentially with age after maturity, at rate μ_a. Clearly, a life history that shows a senescent increase in mortality does not fit this model, but (as mentioned in **Section 9.5.ii.a**) age-independent sources of mortality due to predation, disease, and accidents in the wild often overwhelm senescent mortality (Finch 1990, Chapter 1). This relation may, therefore, often give a good indication of the general form of the decline in S with age under natural conditions. This suggests that, at least under this simple version of the mutation–accumulation model, the excess mortality rate should increase exponentially with age during the adult part of the life history, i.e., the log mortality rate should increase linearly, with a slope equal to the externally caused, age-independent mortality rate. In *Drosophila melanogaster*, for example, this slope is approximately equal to 0.09 per day (estimated from Table 1.1 of Finch 1990), which is consistent with what little is known about mortality rates in nature (Charlesworth 2001a).

This finding thus provides a natural explanation for the Gompertz–Makeham curve, and is consistent with the fact that, in captivity, the rates of increase in mortality with age for different species of birds and mammals are correlated positively with the species' age-independent components of mortality in the wild (e.g., Ricklefs 1998). The model is, however, oversimplified in several respects. First, it ignores the fact that female fecundity and male reproductive success often start to decline with age, after reaching a peak shortly after reproductive maturity. Second, mutations are unlikely to have effects confined to single ages. More elaborate models that include these complications have been developed (Mueller and Rose 1996; Charlesworth 2001b; Baudisch 2008, Chapter 3), but the broad conclusions remain valid. Finally, if deleterious mutations have effects spread across a wide range of ages, as well as peak effects on mortality at a specific age or group of ages, leveling of mortality rates late in life evolves (Mueller and Rose 1996; Charlesworth 2001b). This reflects the fact that fitness sensitivities are effectively 0 late in life, so that the timing of the peak age of effect late in life does not influence the overall selection intensity.

A great deal of experimental work has been carried out on *Drosophila*, with the aim of testing for the contributions of antagonistic pleiotropy versus mutation accumulation to the evolution of age-specific mortality patterns; for reviews, see Partridge and Gems (2002) and Rose et al. (2007). There is evidence for a trade-off between investment in reproduction early in life and survival late in life, with selection for survival and reproduction late in life not only producing a decrease in the rate of increase in mortality with age (**Figure 9.15**), but also causing an associated decline in female fecundity early in life, in accord with the antagonistic pleiotropy model. Evidence from the age-specific properties of quantitative genetic parameters also supports the mutation–accumulation theory (Reynolds et al. 2007; Escobar et al. 2008, but see Moorad and Promislow 2009 for a critique of this approach). The two models are, of course, not mutually exclusive.

PROBLEMS

9.1 Apply the method of **Box 9.1** of **Section 9.1.ii.b** to the spread of a dominant mutation, A_2, that causes seeds (but not pollen) to be produced asexually, with no other phenotypic effects. Assume that homozygotes for the wild-type allele, A_1, mate randomly, and that the mutation is initially carried in an A_1A_2 individual.

9.2. **i.** Apply the phenotypic fitness method of **Section 9.2.ii.c** (**Box 9.3**) to determine the equilibrium frequency of a dominant or recessive male-sterility mutation introduced into an initially cosexual population, using the model developed in **Section 9.3.iv.b** (**Box 9.4**).

ii. Use the same general framework to determine the conditions for invasion of a cosexual population by a dominant or recessive female-sterility mutation, discussed in **Section 9.3.iv.d**.

9.3 Modify the ESS sex ratio model described in **Section 9.4.iii.a** (**Box 9.6**) to allow unequal costs of male and female offspring. Assume a fixed total amount, R, of a resource, available to a mother for the purpose of raising offspring, and let each female and male offspring require one unit and α units of resource, respectively, to be raised successfully. Determine the ESS proportion of resources allocated to females and males, and the corresponding ESS sex ratio.

9.4* Use the model of local mate competition developed in **Section 9.4.iv.a** (**Box 9.7**) to show that the ESS sex ratio is given by **Equation B9.7.2**.

9.5* **i.** Use the framework of **Sections 9.2.ii.c** (**Box 9.3**) and **9.4.iii.a** (**Box 9.6**) to determine the ESS allocation of resources to male versus female functions in a partially selfing cosexual plant. Assume that the selfing rate is independent of allocation, that male pollen output is directly proportional to the fraction of resources allocated to male functions, a, and that female fertility is proportional to $1 - a$. Show that the ESS value of a, a^*, is one-half when the selfing rate $S = 1$, and that a^* decreases with increasing S.

ii. Extend this analysis to the case when male pollen output is related to a by the function a^y, and show that a diminishing-returns relation ($y < 1$) leads to a reduced ESS allocation to male function.

9.6* **i.** Show how **Equation B9.9.3** in **Section 9.5.ii.b** can be derived from the properties of the Leslie matrix of **Box 9.8***.

ii. Show that **Equation B9.9.3** for reproductive value is equivalent to **Equation B9.9.2**.

iii. Using the results in **Box 9.9**, show that the total reproductive value of a population, defined as the sum $v_T(t) = \Sigma_x v(x)n(x, t)$, increases by a factor λ_1 over each time interval (as originally derived using a continuous-time model by Fisher 1930b, p. 27).

9.7* Use a Taylor's series approximation to $w_{ij}(r)$, defined by **Equation 9.4** of **Section 9.5.iii.b**, to show that, with weak selection, differences among genotypes in the $w_{ij}(r)$ are approximately proportional to differences in r_{ij}, the genotypic intrinsic rates of increase.

9.8. **i.** Use the result in **Box 9.9** (**Section 9.5.ii.b**) to derive an expression for the reproductive value for adult age classes when fecundity (m) and survival across one time interval (P) are independent of age and the population growth rate is r. (Assume an indefinitely large number of age classes.)

ii. Show how the Euler–Lotka equation (**Equation 9.3**) of **Section 9.5.ii.b** can be expressed in terms of adult reproductive value in this case.

9.9 Derive **Equation 9.11** of **Section 9.5.v.d** for the sensitivity of fitness to a change in age-specific mortality, assuming constant adult survival and fecundity as in **Problem 9.8**, for the case when the population size is stationary ($r = 0$).

CHAPTER TEN

Some Topics in Genome Evolution

CHAPTER SUMMARY

This chapter deals with the evolutionary processes that shape three major features of genomes. The first is the composition of genomes, as described by the patterns in DNA-based genomes of the frequencies of GC versus AT base pairs, and by codon usage bias—the extent to which the frequencies of synonymous codons depart from those expected under the random use of nucleotide bases. The evolutionary forces that may affect these patterns are discussed; these include biases in mutation rates that favor GC to AT mutations, biased gene conversion in favor of GC versus AT, and selection for preferred codons. We describe population genetic methods for detecting these forces and estimating their intensities.

Another major feature of genomes is the frequency of occurrence of genetic recombination, with certain regions of eukaryote genomes, such as centromeres and Y chromosomes, that largely or completely lack recombination. We discuss how epistatic selection in a constant environment favors the evolution of reduced rates of genetic recombination and how this may explain some examples of suppressed recombination. This raises the question of why recombination occurs in most regions of eukaryote genomes, and we discuss how the evolution of increased recombination can be promoted by selection in changing environments, the interaction of mutation and selection, and the interaction between selection and finite population size. These factors mean that selection is relatively inefficient in asexual species or non-recombining regions of the genome such as Y chromosomes, leading to reduced levels of adaptation. This probably explains the rarity of fully asexual species, and the genetically degenerate nature of Y chromosomes.

The last feature we consider is the existence of components of the genome that propagate themselves within the genome, even at the expense of reduced fitness of the individuals that carry them—selfish DNA. We first discuss segregation distortion, where one allelic form of a gene or gene complex, when present in heterozygotes, causes defects in gametes carrying the alternative allele. Naturally occurring segregation distortion systems seem to involve at least two linked loci: a distorter and a responder, providing an example of epistatic selection. We show how the population dynamics of these systems favors the restriction of recombination between the loci involved, consistent with what is observed. Second, we discuss transposable elements, which have the property of being able to make new copies of themselves that can insert into new locations in the genome (transposition). Theoretical analyses and experimental observations on *Drosophila* suggest that these are maintained in the genome as a result of a balance between increased copy number through transposition and individual selection against their presence in the genome.

> The discovery of an agency which tends constantly to increase the intensity of linkage, naturally stimulates inquiry into the existence of other agencies having an opposite effect . . . .
>
> R. A. Fisher, *The Genetical Theory of Natural Selection*, Chapter 5

INTRODUCTION

Up to now, we have been discussing the evolution of phenotypes controlled by genetic variants, or the evolution of DNA and protein sequences. However, the genome itself is subject to variation and evolution. Data on genome organization have been accumulating since the era of classical genetics, starting with the discovery that genes are carried in the chromosomes of cells and can be mapped genetically and physically to specific locations in the chromosomes. The pioneers of population genetics initiated discussions of evolutionary processes affecting genomic features, such as genetic recombination (Fisher, 1930b, pp. 102–104), gene duplications (Haldane 1933), and the arrangement of genes along chromosomes (Wright 1941; Wright and Dobzhansky 1946). Molecular genetics has subsequently revealed many other aspects of genomes, such as the organization of eukaryote genes into introns and exons, and the large contributions from repetitive DNA sequences to the genomes of multicellular organisms. Whole genome sequences are now providing an unprecedented wealth of data on genomes and the evolution of their organization, especially from comparative genome sequencing projects, e.g., *Drosophila* 12 Genomes Consortium (2007).

This vast increase in information on genomes has revived interest in genome evolution, and two books have recently been devoted to it, but with only a brief coverage of the relevant population genetics theory (Burt and Trivers 2006; Lynch 2007). In this chapter, we focus on the population genetics aspects of three major topics: base composition and codon usage, genetic recombination, and selfish DNA. These topics illustrate some of the ways in which genetic approaches help us to understand the mechanisms of genome evolution, and this chapter frequently refers back to results explained in previous chapters. In particular, it is tempting to assume that the content and organization of genomes are shaped by natural selection acting on genome features that affect individuals' fitnesses, just as selection affects other phenotypes. Before concluding that any feature of a genome is adaptive, however, we need firm evidence, and non-selective alternatives must be considered, as was argued forcefully in the two books we have just mentioned. As we shall show, several deterministic forces that act within genomes (rather than on individuals) have roles in controlling genome evolution. In addition, base composition and codon usage are probably subject to very weak deterministic forces, so that to understand these features we need to study their interactions with genetic drift, using the approaches developed in **Sections 6.3** and **8.3**.

10.1. EVOLUTION OF NONCODING SEQUENCES AND CODON USAGE

10.1.i. Introduction

We first discuss how to characterize base composition, and how to study the frequencies of different synonymous codons used in coding sequences.

10.1.i.a. *Base composition*

For organisms with double-stranded DNA genomes, the nucleotide base composition can be measured by the fraction of base pairs in the genome that are GC (vs. AT). This can be studied for the genome as a whole, for coding sequences (possibly discriminating between first, second, and third coding positions), or for noncoding sequences (possibly discriminating between intergenic, intronic, and untranscribed regions of genes). We shall see that GC content can vary regionally in genomes and often differs between regions with low and high recombination rates.

Partial or complete genome sequences have shown that the GC contents of genomes vary widely among species. The overall GC content for different bacterial species, determined from complete genome sequences, varies from 22% to 72% (**Figure 10.1**). Because bacteria have little noncoding DNA, their GC content largely reflects that of coding sequences; GC3 (the GC content at the third coding position) ranges from 9% to 93% in these species (Sharp et al. 2005). Most mutations at first and second coding positions are nonsynonymous, whereas most mutations at third positions are synonymous. Because selective constraints are much weaker on synonymous than nonsynonymous mutations (**Section 6.2.iv**), variation between species is greatest at third positions. The possible causes of these species differences are discussed in **Section 10.1.ii.b** below.

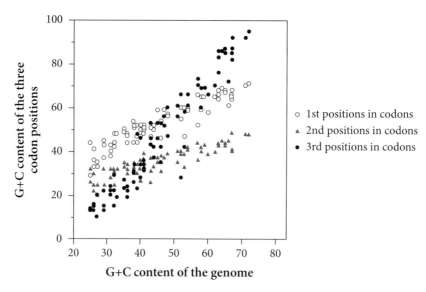

FIGURE 10.1 The GC content (in percent) of first, second, and third positions of codons in bacterial species, plotted against the GC content of the genome as a whole. [Adapted from Figure 2 of Sharp et al. (2005).]

In eukaryote nuclear genomes, a large proportion of the genome is noncoding, and the GC content of noncoding, highly repetitive sequences is generally much lower than that of coding sequences (Miklos and Gill 1982). The GC content of introns is also often lower than GC3 in the adjacent coding sequences (**Figure 10.2**). A third pattern, found in bird and mammalian genomes, is that GC content fluctuates over a scale of a few hundred kilobases (Graur and Li 2000, pp. 417–422; International Human Genome Sequencing Consortium 2001). Regions with high GC content are referred to as *isochores* (Bernardi 1993). Finally, the GC content of both coding and noncoding sequences tends to be reduced in genome regions with low levels of recombination, such Y or W chromosomes, and the regions around the centromeres and near the telomeres mentioned in **Section 8.3.i**. This effect has been best studied in *Drosophila* (**Figure 10.2**; Marais et al. 2001; Díaz-Castillo and Golic 2007), but is also seen in mammalian genomes (International Human Genome Sequencing Consortium 2001).

10.1.i.b. *Codon usage bias*

Another important feature of genome sequences that is related to base composition is the pattern of synonymous codons used in coding sequences. The genetic code is *degenerate*, i.e., several codons correspond to the same amino acid, as we discussed in **Section 6.1.ii.e**. Many mutations, especially at the third

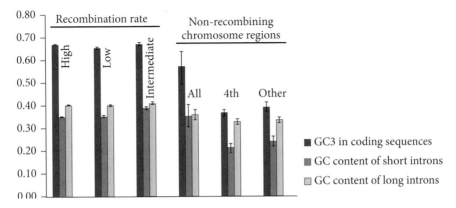

FIGURE 10.2 The GC content of third positions of codons (GC3) in coding sequences, and in short and long introns of the same genes of *Drosophila melanogaster*; the genes studied were sampled from genome regions with high and low recombination rates, and these are plotted separately. (4th refers to the small 4th chromosome) [Adapted from Figure 2 of Haddrill et al. (2007).]

positions in codons, are thus synonymous, as we have just mentioned. One might think that these would be neutral, with no effect on fitness, and that the frequencies of codons corresponding to each amino acid would be determined by random combinations of the four nucleotides, i.e., by the local base composition. DNA sequencing shows that this is not the case. Some species have considerable *codon usage bias*. In **Section 10.1.iii.a** below, we describe how to measure codon usage bias. Here, we simply note that bias varies between organisms (Ikemura 1982; Shields et al. 1988; Duret and Mouchiroud 1999; Comeron 2004), and that this feature of genomes demands an explanation.

10.1.ii. Evolutionary forces affecting base composition

10.1.ii.a. *Introduction*

We now discuss how to interpret some of the patterns just outlined, first considering base composition. Three possible processes exist that could explain the low frequency of GC base pairs observed in genome regions with low rates of recombination:

1. Higher GC → AT mutational bias in genome regions where crossing over is infrequent.
2. Biased gene conversion in favor of GC versus AT, with gene conversion occurring less often in regions where low crossing over is infrequent.
3. Low efficacy of selection (or biased gene conversion) in favor of GC, due to low N_e.

We discuss each of these processes, and the evidence for them, but it is often difficult to distinguish them, and they are not exclusive alternatives. All three probably contribute to some degree, and current data cannot reliably estimate their relative importance.

10.1.ii.b. *Mutation and genetic drift*

We expect selective constraints generally to be weak for noncoding sequences and synonymous variants in coding sequences (compared with nonsynonymous mutations), consistent with their faster sequence evolution (**Section 6.1.ii.e**). The GC content of noncoding sequence and synonymous sites should thus be strongly affected by mutation pressure and genetic drift. This can be modeled using the framework developed in **Section 6.3.ii.b**, where we introduced the idea of *mutational bias*. If we classify mutations at individual base pairs as either GC → AT (with a mean mutation rate v) or AT → GC (mean mutation rate u), we can define a mutational bias parameter κ as the ratio v/u. It is possible to extend this model to include mutations among all possible pairs of nucleotides (Crow and Kimura 1970, p. 263; see **Problem 1.7***), but we focus here on the more tractable case of mutations between GC and AT base pairs, which is of considerable biological importance.

The results in **Section 6.3.ii.b** show that, assuming neutrality, the base composition of a sequence of nucleotides will evolve under mutation pressure towards an equilibrium fraction of GC base pairs, $x^* = 1/(1 + \kappa)$. With $\kappa > 1$, this predicts a GC content much lower than 50% for neutrally evolving sequences. Such mutational bias probably explains the observed high AT content of sequences such as the untranscribed, highly repetitive satellite sequences of eukaryote nuclear genomes (**Section 10.1.i.a**). We explain in **Section 10.1.ii.e** below how to test whether equilibrium has been reached.

Mutation rates for predicting the neutral equilibrium base composition can be estimated from sequence comparisons between species, or among duplicated genes within the same genome, assuming neutrality. To ensure neutrality, such estimates often involve comparisons of sequences of ancient transposable element insertions that have become fixed in the population (see **Section 10.3.iv.f**), called "ancestral repeats" (Singh et al. 2005), or pseudogenes derived from insertions into the nuclear genome of chloroplast or mitochondrial DNA whose initial sequences are known (Huang et al. 2005; Keller et al. 2007). The rates of substitution from GC to AT, and vice versa, for such sequences can be used as estimates of the corresponding mutation rates. κ is often substantially greater than 1. Mutational bias can differ between different regions of a species' genome. For example, sequence comparisons of independent, ancient, nonfunctional insertions of the *DINE-1* transposable element fixed in *D. melanogaster* suggest a higher GC → AT mutational bias in genome regions with low rates of crossing over, with a κ value of about 2.1 (Singh et al. 2005).

Mutational bias may also vary among species. The extreme GC content differences of bacterial species, for example, may simply reflect different κ values among species (Muto and Ozawa 1987). However, mutation rates in DNA sequences have been estimated experimentally in very few species, so that there is little direct evidence on this point. A difficulty with this very simple interpretation is that bacterial species with low genomic GC content tend to be pathogens or symbionts (Rocha and Danchin 2002; Woolfit and Bromham 2003; Pallen and Wren 2007). This raises the possibility that host-dependent species might have unusually high mutational bias, due to misincorporation of AT instead of GC when metabolism is relatively inefficient (assuming a greater metabolic cost of using GC compared with AT nucleotides in DNA synthesis; see Rocha and Danchin 2002). However, other possible explanations cannot be excluded, including (i) selection for use of AT instead of GC in organisms with poor energy metabolism (Rocha and Danchin 2002), or (ii) lower N_e, reducing the efficacy of selection favoring GC over AT nucleotides (Woolfit and Bromham 2003), as we explain next.

10.1.ii.c. *Lower effectiveness of selection with reduced recombination*
We saw in **Section 8.3.i** that selection acting on a particular site in the genome affects linked sites more strongly in genomic regions with low levels of genetic recombination, reducing the effective population size, N_e, in such regions. The theory of the interaction between selection, mutation, and drift (**Section 6.3.ii.c**) thus predicts a lower effectiveness of selection in these regions, so that x^* will approach the neutral value if N_e is sufficiently reduced. We discuss this effect in more detail later (**Section 10.2.v**), but it can help to explain the much lower GC content of genes in low recombination regions than elsewhere in the genome (**Figure 10.2**), because selective forces will be relatively ineffective in these regions, and mutation and drift will dominate.

10.1.ii.d. *Biased gene conversion*
Another process complicating the understanding of GC content is biased gene conversion (BGC). Heterozygotes for GC and AT at a site may produce more than 50% of the GC variant in their gametes as a result of biased repair of DNA heteroduplexes produced in meiosis by exchanges of strands between maternal and paternal chromosomes (see **Section 1.3.i.b.**, Marais 2003 and Duret et al. 2006). **Box 10.1** shows that this process increases the frequency in a population of GC base pairs at a site. The equation for the rate of change in frequency is the same as for weak selection favoring a semidominant variant, with the rate of BGC, ω, playing the same role as the selection coefficient, s (this should not be confused with the use in **Section 6.2.v** of ω to denote the ratio of nonsynonymous to synonymous mutations). At statistical equilibrium under mutation, BGC, and drift, we can thus use **Equation 6.10** of **Section 6.3.ii.c** to predict the equilibrium GC content simply by substituting ω for s. If gene conversion rates

Box 10.1 THE EFFECTS OF BIASED GENE CONVERSION

Assume that the alternative GC and AT variants at a nucleotide site are selectively neutral, but the frequency of GC-carrying gametes among the gametes produced by GC/AT heterozygotes is k, where $k > 1/2$. This parameter can be related as follows to the frequency of biased gene conversion events at the site (Gutz and Leslie 1976). We assume that only one such event occurs in a given meiosis, between a maternal and paternal homologue at the four-strand stage of the first division. As a result, one of the two AT chromatids becomes GC instead of AT. Let the probability of this event be ω. There is thus a probability ω that a given meiosis results in a 3:1 ratio of GC:AT among its four products; with probability $1 - \omega$, the ratio is 1:1. Hence,

$$k = \frac{3}{4}\omega + \frac{1}{2}(1 - \omega) = \frac{1}{2}\left(1 + \frac{1}{2}\omega\right) \qquad \text{(B10.1.1)}$$

Let q be the population frequency of GC variants at this site. With random mating, the frequency of heterozygotes is $2pq$ (where $p = 1 - q$), and the change in variant frequency is given by the following expression (**Problem 10.1**):

$$\Delta q = \frac{\omega}{2}pq \qquad \text{(B10.1.2a)}$$

Comparison with the equation for selection on a semidominant, favorable mutation (**Section 3.1.iii.b**) shows an exact equivalence, with ω replacing the selection coefficient s.

If the population is inbred, with inbreeding coefficient f, the frequency of heterozygotes is reduced by a factor of $1 - f$ (**Section 1.3.ii.b**), and so this equation becomes:

$$\Delta q = \frac{\omega(1 - f)}{2}pq \qquad \text{(B10.1.2b)}$$

This means that biased gene conversion will be a much weaker force in highly inbred species than in outcrossers.

Given that ω is likely to be very small in most organisms (Marais 2003), it is necessary to include the effects of mutation and drift in predictions of the effects of BGC. This can be done by using the results of **Section 6.3.ii.c**, as expressed in **Equation 6.10**, which gives the expected equilibrium frequency of GC bases in a long sequence, substituting $N_e\omega$ for $N_e s$. In inbreeders with inbreeding coefficient f, the effect of gene conversion is further reduced by the fact that N_e is reduced by a factor of $1/(1 + f)$ relative to a random mating population with a similar number of breeding individuals (**Section 5.2.iii.a**).

are lower in genome regions with low rates of crossing over, the results in **Box 10.1**, together with **Equation 6.10**, predict a lower than average GC content in such regions, consistent with the observations just mentioned (Marais 2003).

However, the frequency of gene conversion may not be greatly reduced in genome regions with low levels of crossing over. Crossover and non-crossover events following a double strand break at a site of recombination seem to involve alternative pathways, which may behave quite differently (e.g., Baudat and De Massy 2007). Indeed, analyses of DNA sequence polymorphism data from genes in regions of the *D. melanogaster* genome that are known from genetic maps to have near-zero rates of crossing over nevertheless reveal signs of recombination. This probably involves gene conversion events, which may be occurring at near normal rates in these regions (Langley et al. 2000; Sheldahl et al. 2003; Gay et al. 2007).

Moreover, the equilibrium GC content x^* under BGC depends on $N_e\omega$ (**Box 10.1**), so that the rate of fixation of "disfavored" GC → AT mutations increases as N_e gets smaller if ω is kept constant. The proportion of GC → AT substitutions will thus be elevated in low recombination regions, relative to other genome regions, even if the rate of BGC, ω, is not lower. Higher rates of GC → AT substitutions in low recombination regions (Singh et al. 2005; Díaz-Castillo and Golic 2007), therefore, do not necessarily imply a mutational bias towards AT.

In highly inbreeding species, the effective population size is reduced (**Section 5.2.iii.a**), and, in addition, the reduced frequency of heterozygotes causes lower evolutionarily effective recombination rates over the entire genome (**Section 8.2.ii.b**), so that the effectiveness of BGC will be reduced. **Box 10.1** shows that inbreeders should have lower GC content in their non-coding sequences, compared with outcrossing relatives. At present, too few comparative genomic data are available to test this prediction rigorously, although data from comparisons of members of the wheat family with different breeding systems are consistent with this prediction (Haudry et al. 2008).

10.1.ii.e. *Departures from equilibrium base composition*

Base composition changes very slowly after a change in selection, BGC, or effective population size; the difference between x and its equilibrium value, x^*, under mutation and drift alone decreases at a rate $(u + v)(x - x^*)$ per generation in the absence of selection (Marais et al. 2004). By applying the method that we have employed previously to study linear processes (e.g., **Problem 8.3.ii**), it follows that it takes approximately $\ln(2)/(u + v) = 0.69/(u + v)$ generations to halve $x - x^*$, i.e., many millions of generations are needed to produce a noticeable change. This very slow rate of approach to equilibrium raises the question of how to test whether a species' GC content is in equilibrium under the various forces acting on it.

At equilibrium, the number of GC → AT and AT → GC fixations along a branch separating two species of interest should be the same (see **Section 6.3.ii.c**). One can thus compare two closely related species with a more distant outgroup species, and determine the changes fixed on the branches leading to one of the closely related species from its common ancestor with its partner (using the principle of parsimony, similar to the approach shown in **Figure 6.5** of **Section 6.3.i**; ideally, multiple sequences should be obtained from the species of interest to count fixed differences accurately and distinguish them from polymorphisms). If the numbers of GC → AT and AT → GC fixations along the branch leading to the species of interest differ significantly (e.g., on the basis of a χ^2 test for a 1:1 ratio of observed numbers of the two substitution types), one can conclude that equilibrium has not been reached.

Analyses of substitution patterns in the *D. melanogaster* group of species, using more sophisticated methods involving multiple species comparisons, indeed show an excess of GC → AT relative to AT → GC fixations of silent mutations along the branch leading to *D. melanogaster* from its ancestor with *D. simulans*, but not in the branch leading to *D. simulans*, suggesting that *D. melanogaster* is not at equilibrium, whereas *D. simulans* is close to equilibrium (Akashi et al. 2006). This may reflect a reduction in the efficiency of selection or of BGC, caused by a reduction in N_e in the *D. melanogaster* lineage.

In mammals, comparisons of genomes of different species suggest a long-term decline in the GC content of the GC-rich isochores mentioned in **Section 10.1.i.a** (Duret et al. 2002, 2006; Comeron 2006). Again, this may be caused by a reduction in N_e.

10.1.iii. Evolutionary forces affecting codon usage

10.1.iii.a. *Measuring codon usage bias*

To understand the evolutionary forces affecting codon usage, we need a quantitative measure of codon usage for individual genes, which we can compare across different genes in the same genome or between the same gene in different species. The most important of the various methods devised are described briefly in **Box 10.2**. These methods can identify one or more *preferred* codons for a given amino acid, i.e., codons that are more frequently used than other (non-preferred) ones, at least in genes with strong bias.

Box 10.2 MEASUREMENT OF CODON USAGE BIAS

Various methods for measuring codon bias have been proposed. They are all fairly *ad hoc* and present different problems of interpretation.

(1) The simplest is the index of *Relative Synonymous Codon Usage*, RSCU (Sharp et al. 1986). Let n be the level of degeneracy for a given amino acid, and X_i be the number of times the ith codon for the amino acid is used in the sequence under consideration. If all n codons were used equally frequently, the expected number of occurrences of a given codon, e, would be the expected proportion of this codon among all n possible ones, multiplied by the total number of occurrences of the amino acid in question, i.e.:

$$e = \frac{1}{n}\sum_i X_i \tag{B10.2.1}$$

The RSCU value for the ith codon is then given by X_i/e. This is equal to 1 for all codons, if they are used equally frequently; it is larger than 1 for codons that are more frequently used than average. This can be useful in picking out particular codons that are preferred or unpreferred, but does not in itself give a summary measure of codon usage.

(2) A useful, frequently used measure of overall bias is given by the *Effective Number of Codons*, ENC (Wright 1990). Let f_i be the frequency of the ith codon for an n-fold degenerate amino acid in a given sequence; ENU for this amino acid is defined as $1/(\Sigma_i f_i^2)$. This has a maximum value of n when all codons are equally frequent ($f_i = 1/n$); if only one codon out of n is used, it takes its minimum value of 1. The value of ENU is therefore *inversely* related to the strength of codon bias. The ENU value for a gene is obtained by summing the values for all 18 nondegenerate amino acids.

(3) An alternative overall bias measure is the *Scaled Chi-Squared Index* (Shields et al. 1988); this is simply obtained by calculating a chi-squared goodness-of-fit measure (**Appendix A2.v.f**) by comparing the observed numbers of each codon within a synonymous group with the numbers expected with equal codon usage.

The measure uses the standard chi-squared formula, i.e.:

$$\chi^2 = \sum_i \frac{\left(O_i - E_i\right)^2}{E_i} \tag{B10.2.2}$$

where O_i is the observed number of instances of the ith type of codon for the amino acid in question, and E_i is the expected number, calculated by multiplying the total number of codons for the amino acid in question by the frequency of the codon expected with equal frequencies of alternative codons.

By summing this over all amino acids other than Trp or Met (which are encoded by only one codon each), and dividing by the total number of codons in the sequence, we get a measure that lies between 0 and 1. This method, however, does not take into account the frequencies of different nucleotides in the genome. Improved versions of this method, which use such information, have been developed (Urrutia and Hurst 2001; Novembre 2002; Fuglsang 2006)

(4) Another method is to pick out codons for a given amino acid that are significantly more frequently used (based on a chi-squared test) in the genes with the highest bias compared with genes with the lowest bias, using one of the above measures, by analyzing sequences from a large number of genes (Ikemura 1982; Akashi 1994). The proportion of codons in a gene that are *preferred* under this criterion (the *major codons*) provides another index of overall codon bias, called *Major Codon Usage* or MCU (Akashi 1996).

A variant of this method, which has become popular with the advent of databases of levels of gene expression, is to identify codons that are more frequently used in genes with high levels of expression (Duret and Mouchiroud 1999; Sharp et al. 2005). These are often called *optimal codons*, and the frequency of optimal codons in a gene is known as F_{op}. This term is now often used interchangeably with MCU.

Patterns of codon usage tend to be fairly consistent across different genes in the genome of a species and in related species, i.e., the same codons are preferred in different genes, although the level of bias varies considerably among genes. Thus, analyses of the genomes of *Drosophila* species mostly show remarkable constancy in patterns of codon usage over 60 million years of evolution since the ancestor of the 12 species so far completely sequenced (*Drosophila* 12 Genomes Consortium 2007). This suggests that the forces that determine codon bias operate across the whole genome, rather than being specific for individual genes. In *Drosophila*, codons ending in G or C are preferred over ones ending in U (in mRNA; T in the corresponding DNA) or A. Such preference for GC-ending codons causes the higher value of GC3 compared with the GC content of introns, noted in **Section 10.1.i.a** above. Such GC3–GC$_{intron}$ differences could be due to different mutational biases between coding and noncoding sequences within the same gene, but this seems unlikely. It is more likely that selection acts on codon usage, leading to different CG content in neighboring coding and noncoding sequences. Additional evidence for selection is provided by the fact that, in a number of taxa, the synonymous site divergence values of genes between species decrease with the genes' degree of codon usage bias (e.g., Bierne and Eyre-Walker 2003). **Figure 10.3** shows an example of this.

10.1.iii.b. *Causes of selection for codon usage bias*

One factor that is likely to affect codon usage is selection for translational efficiency, i.e., the speed with which the messenger RNA of a gene is translated. Different codons might have different effects on this speed, due to differences in their chemical interactions with the anticodons of their transfer RNAs (tRNAs). Selection for translational efficiency predicts the following patterns:

1. More frequent use of codons corresponding to more abundant transfer tRNAs.

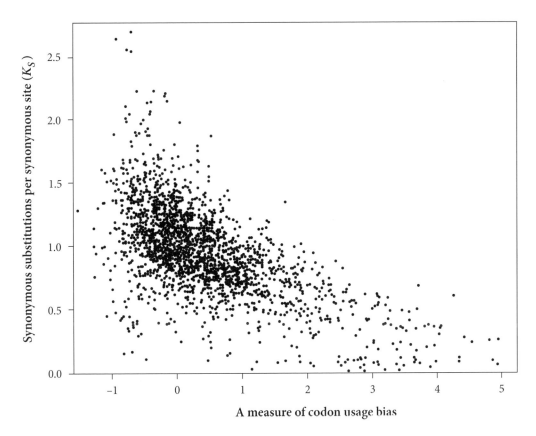

FIGURE 10.3 The estimated synonymous site divergence values of genes between the bacterial species *Escherichia coli* and *Salmonella enterica*, plotted against the genes' degree of codon usage bias. [Adapted from Figure 1 of Retchless and Lawrence (2007).]

2. Experimental manipulation of codon composition of a gene in favor of unpreferred codons should reduce gene expression. This has been demonstrated, for example, with the *Adh* gene of *D. melanogaster* (Carlini and Stephan 2003).

3. When efficient translation is selectively important, codon usage should be more biased. For instance, one might expect species with faster rates of cell division to fall into this category.

4. The most highly expressed genes, which presumably require more rapid mRNA translation, should have stronger codon bias. In addition, these genes draw more tRNA molecules from the available pool. Thus, if highly expressed genes make use of common tRNAs, they will deplete the pool less than if they used rare ones.

In unicellular organisms, data can relatively easily be obtained on expression levels and tRNA abundances. In yeast and *E. coli*, predictions 1 and 3 are confirmed (Ikemura 1982; Sharp et al. 1988, 2005). Furthermore, as predicted by 3 and 4, there is indeed a strong correlation between codon usage and level of gene expression, and between codon usage and rates of cell division in bacteria (Sharp et al. 1986, 1988, 2005). In multicellular organisms, it is harder to get good data of this kind, largely because of tissue-specific differences in gene expression and in tRNA levels. One measure of overall gene expression in whole multicellular organisms or specific tissues uses the frequency in cDNA libraries of clones corresponding to each gene (*Expressed Sequence Tags*: ESTs). Strong correlations are found between levels of codon bias and such estimates of gene expression (confirming prediction 4) in *C. elegans*, *D. melanogaster*, and *Arabidopsis thaliana* (Duret and Mouchiroud 1999), as shown in **Figure 10.4**. Expression levels estimated from microarray experiments reveal similar patterns, even in humans, where selection for codon usage bias previously seemed to be absent (Comeron 2004; Qin et al. 2004). These patterns strongly suggest that codon usage is under natural selection, at least in some organisms.

A further possible reason for such selection is *translational accuracy*, which favors codons that interact strongly with the anticodons, and are less likely to bind to the wrong tRNA and introduce the wrong amino acid during translation of the mRNA (Akashi 1994). Amino acid sites that are highly constrained by selection should then show higher codon usage bias than less conserved sites, and this is indeed found in sequence comparisons within *E. coli* (Stoletzki and Eyre-Walker 2007) and between species in several other taxonomic groups, including *Drosophila* and mammals (Drummond and Wilke 2008). Selection may also act against codons that cause inappropriate splicing of the mRNA precursor, or impair stability of the mRNA (Parmley and Hurst 2007). Maintenance of correct splicing may be the major factor in determining codon usage in mammals (Parmley and Hurst 2007).

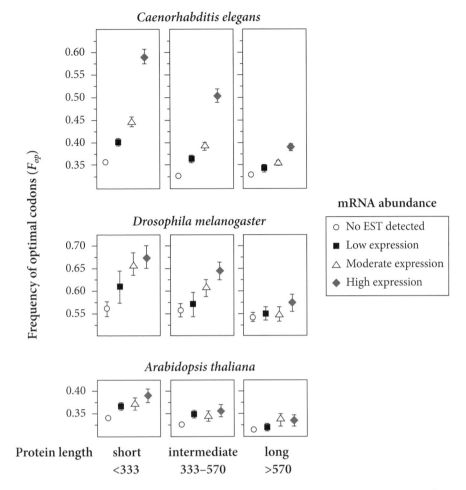

FIGURE 10.4 Correlations between levels of codon bias and gene expression in genomes of *Caenorhabditis elegans*, *D. melanogaster*, and *Arabidopsis thaliana*. For each species, three panels show genes with coding sequences of different lengths, and in each panel different symbols show genes with different levels of expression, as measured by the abundances of expressed sequence tags (see key). [Adapted from Figure 1 of Duret and Mouchiroud (1999).]

10.1.iv. Estimating the intensity of selection or biased gene conversion

The evidence we have just discussed suggests that base composition and codon usage are likely to be affected by selection and biased gene conversion, as well as mutation and drift. It sheds no light, however, on the important question of the strength of these forces, which forms the subject of this section.

10.1.iv.a. *The intensity of selection and BGC on noncoding sites*
We first consider the problem of estimating the intensity of selection for GC versus AT acting on noncoding sites (or BGC in favor of GC); because these forces produce similar effects, it is hard to distinguish them. The higher nucleotide diversity and between-species divergence for noncoding sites, compared with nonsynonymous sites, suggest that the product of N_e and s or ω cannot be much greater than 1 for most noncoding sites. Indeed, if $N_e\omega$ were much above 1, the predicted equilibrium frequency of GC base pairs would be close to 100% (unless mutational bias in favor of AT is implausibly high; this can be seen from **Figure 6.6** of **Section 6.3.ii.c**, substituting $N_e\omega$ for $N_e s$). We will illustrate the evidence on the product of N_e and s or ω, using data from *D. melanogaster*. This example demonstrates some of the difficulties in determining the nature and magnitude of the evolutionary forces acting on noncoding sequences.

Case Study 1
Is There Evidence for Biased Gene Conversion
in *Drosophila Melanogaster*?

As mentioned in **Section 10.1.ii.b** above, the κ value for mutational bias in favor of GC $\rightarrow$ AT mutations in regions of low recombination is about 2.1 (Singh et al. 2005). To find out what happens with neutrality, we use **Equation 6.10** of **Section 6.3.ii.c** with γ set to 0. This predicts an equilibrium GC content of 32%. Comparable nonfunctional sequences in regions with normal recombination have a predicted equilibrium GC content of about 35% (Singh et al. 2005). If this difference reflects selection or BGC, rather than a difference in mutational bias alone, we can use the predicted value of 35% to estimate $N_e\omega$ or $N_e s$ (equivalent to $\gamma/2$) in favor of GC nucleotides in these regions. By rearranging **Equation 6.10**, we obtain:

$$\gamma = \ln(\kappa x^*/(1 - x^*)) \tag{10.1}$$

Substituting $\kappa = 2.1$ and the neutral equilibrium value $x^* = 0.35$ into **Equation 10.1**, we obtain $\gamma = 0.12$, so that $N_e s$ or $N_e\omega = 0.06$. But some non-coding sequences in *Drosophila*, especially long introns, have a much higher GC content than 35%, as much as 50% or more (Haddrill et al. 2005b; Galtier et al. 2006). With $\kappa = 2.1$, **Equation 10.1** requires $N_e\omega$ to be about 0.37 for an equilibrium GC content of 50%.

Galtier et al. (2006) proposed a population genetics method for estimating the intensity of BGC (or, perhaps less likely, selection). This uses the frequency distribution of GC versus AT variants at a set of segregating sites, as described in **Section 5.3.iii.d**. Their approach resembles the method described for amino acid variants in **Section 6.4.iv.b**. In the case of base composition, the equilibrium distribution of the frequency of GC variants (assumed to be favored over AT) is given by the equivalent of **Equation 5.23** of **Section 5.3.iii.d**. Letting the mutational terms approach 0, in order to satisfy the infinite sites assumption, we have:

$$\phi(q) = C \exp(\gamma q) q^{-1}(1-q)^{-1} \qquad (10.2)$$

We can use **Equation 10.2** to generate an equation for the probability of seeing a GC variant at a given frequency in a sample from the population, using a similar method to that used in **Box 6.9*** of **Section 6.4.iv.b**; if different sites are assumed to be independent of each other, the log-likelihood for the frequencies for all sites can be obtained by summation across sites (**Appendix A2.vi.b**). Galtier et al. (2006) obtained $N_e\omega$ estimates of 0.15, 0.35, and 0.60 for low, medium, and high GC noncoding sequences, respectively, in a sample from an African population of *D. melanogaster*. The value of κ can be estimated for these sequences by substituting the corresponding γ estimates and observed GC contents into **Equation 10.1**, giving values of 2.7, 3.1, and 3.6, respectively. The κ values required to explain the observed GC contents are thus much higher than 2.1.

The most likely explanation for this discrepancy is that the base composition of *D. melanogaster* is not at equilibrium, either because of a long-term reduction in N_e or because of a change in κ. As mentioned in **Section 10.1.ii.e** above, the base composition of *D. melanogaster* is probably not at equilibrium, with GC base pairs decreasing in frequency in the genome (Akashi et al. 2006). This means that many AT variants have arisen relatively recently, and will thus be present at lower frequencies than expected at equilibrium under mutation and drift (Akashi 1997); in other words, the frequencies of GC variants will be higher than expected for the true value of γ. This difficulty also applies to attempts to estimate γ for GC versus AT noncoding variants in humans (Lercher et al. 2002), whose base composition is also not at equilibrium (**Section 10.1.ii.e**). BGC or selection in favor of GC base pairs may thus be acting at a much weaker intensity than is estimated by this method.

D. simulans seems to be much closer to equilibrium than *D. melanogaster* (Akashi et al. 2006), and provides evidence for selection or BGC in favor of GC versus AT in noncoding sequences, with estimated $N_e\omega$ or $N_e s$ values that are much more consistent with the observed base composition (Haddrill and Charlesworth 2008).

10.1.iv.b. *The intensity of selection on codon usage*
We saw that the frequency of preferred codons in a gene rarely exceeds 80%, even in genes with highly biased codon usage. Just as for GC content, this indicates that $N_e s$ values (where s is the selection coefficient favoring preferred

codons) cannot greatly exceed 1, unless there is a high mutational bias in favor of unpreferred codons. If we use a simple model in which the preferred codon for each amino acid, established by the criteria described in **Box 10.2** of **Section 10.1.iii.a**, is denoted by P and the unpreferred codon by NP, we can use the same model of drift, selection, and reversible mutation as for GC versus AT base pairs. Genes with low expression levels and no codon usage bias allow us to estimate any mutational bias in favor of NP mutations for each amino acid, by using the proportion of P codons. Using **Equation 10.1** of **Section 10.1.iv.a**, we can then estimate $N_e s$ for the preferred codon in bacterial highly expressed genes that do show a bias (Sharp et al. 2005); the mean $N_e s$ estimates range from 0 to 0.75 for 40 highly expressed genes.

Case Study 2
Estimating Selection on Codon Usage Bias from Polymorphism Data

Polymorphism data can also be used to estimate $N_e s$ values for mutations affecting codon usage, and various methods have been developed, based on the evolutionary model just described (Akashi 1995, 1999; Akashi and Schaeffer 1997; Maside et al. 2004; Comeron and Guthrie 2005; Cutter and Charlesworth 2006). It is usually assumed that codon usage is at statistical equilibrium under mutation, selection, and drift, and the frequencies of polymorphisms for P versus NP mutations in a sample of alleles are then fitted to the model's predictions.

If the system is in equilibrium, a simple method exists for estimating $N_e s$ from polymorphism data, using the relative numbers of P → NP and U → NP polymorphisms. Sequences from a related species can be used to distinguish polymorphisms from fixed differences on the lineage leading to the species of interest, and to determine the ancestral state (P or NP) of the segregating sites, allowing P → N and NP → P polymorphisms to be counted. The assumption of equilibrium for codon usage can be tested by the approach described in **Section 10.1.ii.e** above. For codon usage, equilibrium implies equal numbers of sites where P → NP fixations have occurred in a lineage and sites at which NP → P fixations have occurred (otherwise codon usage would change over time).

The rationale underlying the estimation of selection is as follows. Selection against deleterious mutations (P → NP) does not prevent them from becoming polymorphic, but they will generally not become fixed; conversely, advantageous mutations contribute less to polymorphism than to divergence (**Section 6.4.ii.b**). The lower effectiveness of selection on polymorphisms compared with fixation means that there should be more P → NP than NP → P polymorphisms, provided that the system is at equilibrium (Akashi 1995). The difference between the numbers of the two types of polymorphisms depends on the strength of selection. Data on these can then be used to obtain estimates of $N_e s$, the scaled strength of selection in favor of preferred codons, using the equations that describe the probability distribution of the frequencies of P → NP and NP → P mutations (**Equation B6.7.1** of **Section 6.3.ii.d**) to determine the expected numbers of P → NP and NP → P variants in a sample from the population, once again by a similar method to that used in **Box 6.9*** of **Section 6.4.iv.b**. These expectations can then be used to obtain an expression for the likelihood of the observed fraction of segregating sites that are P → U, among sites segregating for P → NP and NP → P in the sample (Maside et al. 2004).

Methods such as this have been applied to coding sequences from several species of *Drosophila* (Maside et al. 2004; Bartolomé et al. 2005; Comeron and Guthrie 2005) and humans (Comeron 2006; Kondrashov et al. 2006). For example, in *D. americana*, there is no evidence for departure from statistical equilibrium for mutations affecting codon usage on the basis of comparisons between the three related species *D. americana*, *D. virilis*, and *D. ezoana* (Maside et al. 2004). In a set of 17 genes, 124 P → NP and 20 NP → P polymorphisms were detected. These data yield a maximum likelihood estimate of $N_e s = 1.3$, with a 95% confidence interval of approximately 1 to 1.7. With an N_e of approximately one million (estimated from the *D. americana* nucleotide site diversity, using the method described in **Section 5.2.iii.e**), this corresponds to a selection coefficient of about 1.3×10^{-6} in favor of preferred codons.

Because preferred codons end in G or C, BGC in favor of GC, rather than selection, could potentially be responsible for the departure from neutral expectation, rather than selection. However, analysis of the numbers of GC → AT versus AT → GC polymorphisms found a significant excess only for synonymous sites, but not in noncoding sequences, suggesting at most only minor effects of BGC (which should act on both types of sequence). Thus, selection probably acts on codon usage in this species. A similar comparison of coding and noncoding sequences also suggests selection on codon usage in humans (Comeron 2006).

10.2. THE EVOLUTIONARY CAUSES AND CONSEQUENCES OF GENETIC RECOMBINATION

10.2.i. Introduction

We next turn to examining the evolutionary forces affecting the origin and maintenance of sexual reproduction and genetic recombination. The questions of what maintains genetic recombination, and what controls the differences in recombination rates between regions of the genome of a single species and between different species, have clear connections to the main topic of this chapter, genome evolution, particularly as recombination affects characteristics such as GC content (as described above). As described in **Section 9.1.ii.a**, most eukaryotes reproduce sexually. Sexual reproduction has two genetic consequences:

1. Mendelian segregation among the products of meiosis of a heterozygous diploid cell.
2. Recombination, so that each product of meiosis contains a patchwork of genetic information from each of the two gametes that contributed to the diploid parental genotype. This involves three types of processes: the independent segregation of genes on different chromosomes, reciprocal crossing over, and gene conversion.

While segregation and recombination both contribute to the evolutionary advantages of sexual reproduction, we will mainly discuss recombination here. Very few multicellular organisms lack effective recombination. Notable exceptions are several highly self-fertilizing plant species like some members of the genus *Oenothera*, where all individuals are heterozygous for translocation complexes that suppress crossing over throughout most of the genome (Cleland 1972; De Waal Malefijt and Charlesworth 1979) (**Figure 10.5**). The normal situation is, however, to have crossovers within chromosomes, as well as independent segregation of different chromosomes (although a few eukaryote species have only a single chromosome—see Crosland and Crozier 1986—and some chromosome regions do not recombine, as we have already mentioned in **Sections 8.3.i** and **10.1.i.a**).

Because the mechanisms of recombinational exchange between homologous DNA sequences are similar in prokaryotes and eukaryotes, including many of the basic proteins involved, it seems clear that they evolved even before the invention of meiosis and eukaryote chromosomes. A basic question, therefore, involves the possible selective advantages of recombination. Ideally, we would also like to understand the large variation in both chromosome numbers and in rates of crossing over among different species, as well as the strong regional patterning of crossing over rates within eukaryote genomes.

10.2.ii. The cost of recombination

To understand the evolutionary causes of sexual reproduction and recombination, we first consider some selective disadvantages of these processes. If sex and recombination are maintained by natural selection, they must have advantages that are large enough to outweigh these disadvantages.

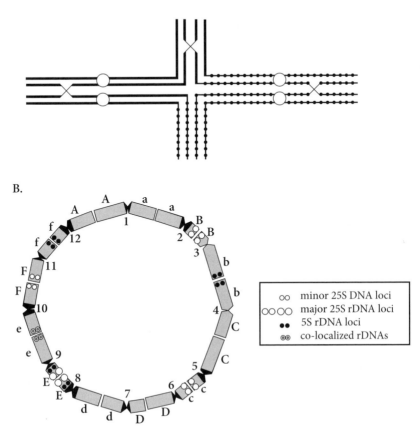

FIGURE 10.5 Reciprocal translocations between chromosomes. **A.** Diagram of a reciprocal translocation between two nonhomologous chromosomes, indicated by black and dotted symbols. The non-translocated chromosomes can be seen at the top left and bottom right parts of the diagram. The top right and bottom left parts show chromosomes (formed by the translocation event) that have the centromere (circles) and surrounding chromosome regions of one chromosome and the end of the long arm of the other chromosome. The diagram shows how pairing between homologous chromosome regions in meiosis produces a four-armed structure. Three crossover events between homologous chromosome regions are shown, localized to chromosome ends. At anaphase, when the centromeres separate, the crossover regions hold the chromosomes together until a late stage, often forming a ring structure. **B.** A plant species (*Rhoeo spathacea*) with multiple reciprocal translocations, for which the plants are permanently heterozygous. The homologous chromosome arms (either side of the centromeres, shown in black) are labeled with letters and some "landmarks" at some chromosome ends (dots and small circles), indicated in the key, showing that each of the 12 numbered chromosomes consists of parts of two different ancestral chromosomes (e.g., chromosome 1 has an **A** and an **a** arm; chromosome 12 has an **A** arm and an **f** arm, while chromosome 2 has **a** and **B** arms). The diagram has gaps where the two ends of each chromosome arm are each paired with the end of another chromosome with which it is homologous. The ring structure is very large, because it includes multiple chromosomes. [Adapted from Figure 10 of Golczyk et al. (2005).]

We have already discussed the vulnerability of sexual populations with male and female gametes to invasion by asexual mutants—the cost of sex (**Section 9.1.ii.b**). There is a smaller, and less obvious, cost of recombination, first pointed out by Fisher (1930b, p. 103). He noted that, when two loci are each polymorphic for two alleles with epistatic fitness effects leading to linkage disequilibrium (LD), as described in **Section 8.4.ii.b**, recombination reduces the frequencies of the selectively favorable combinations of alleles in an equilibrium population. He therefore concluded that, if there is genetic variability in the frequency of recombination between the loci, this

> . . . will always tend to diminish recombination, and therefore to increase the intensity of linkage in the chromosomes . . . .

Much theoretical work has been devoted to putting Fisher's verbal argument on a firm theoretical basis and generalizing it to multiple selected sites. To gain insight into this process, first consider the effect of recombination on a population's mean fitness. We saw in **Section 8.4.ii.a** that, with constant fitnesses, the mean fitness of a polymorphic population in which selection creates linkage disequilibrium (requiring nonadditive fitness effects across loci) is always highest when there is no recombination (Lewontin 1972b). This suggests that recombination is disfavored by selection under these conditions.

However, changes in frequencies of genetic variants that affect recombination frequency (called *recombination modifiers*) are not always correctly predicted by the effect of recombination on the population's mean fitness. Such modifiers are genetic variants that are postulated to be selectively neutral in themselves, but which affect the recombination frequencies between variants under selection. Evolutionary change in recombination rates as a result of the spread of recombination modifiers was first studied by Kimura (1956) and Nei (1967). Lewontin's result, just mentioned, implies that, if a modifier mutation that causes some recombination arises in a non-recombining population at equilibrium under epistatic selection, haplotypes with low fitness will be produced. Parents carrying the modifier will thus have lower mean offspring fitnesses than non-recombining genotypes, and the modifier will be eliminated from the population.

Studying the evolutionary fate of recombination modifiers is difficult, and not all possibilities have been explored. In general, however, the above conclusion holds—rare modifiers that slightly increase the frequency of recombination are eliminated, when introduced into a two-locus system at equilibrium under epistatic selection in a randomly mating population with LD, like the system described in **Section 8.4.ii.b**. Conversely, modifiers reducing recombination will always spread (Feldman 1972; Feldman and Liberman 1986; Altenberg and Feldman 1987).

Why do these effects occur? When LD is maintained by epistatic selection in the face of recombination, the haplotypes present in excess of random expectation have higher marginal fitnesses than the mean fitness of the population (and those in deficiency have lower marginal fitnesses than the mean). A modifier that reduces recombination becomes associated with the high fitness haplotypes, so it will increase in frequency by hitchhiking with them.

This establishes that, under the restrictive conditions assumed here, the state with no recombination is an Evolutionarily Stable Strategy (ESS), similar to those discussed for sex ratios in **Section 9.4.iii**. It is difficult to generalize this result to multi-locus systems, except for some special cases, e.g., modifiers that completely eliminate recombination when heterozygous (Charlesworth and Charlesworth 1973b) or modifiers with small effects on recombination in populations with weak epistatic selection (Zhivotovsky et al. 1994; Barton 1995a). The result is, however, valid even when the selected loci are subject to frequency-dependent selection or selection involving spatial heterogeneity within a panmictic population (for some exceptions, see **Section 10.2.iv.a** below).

10.2.iii. The evolution of close linkage

10.2.iii.a. *Inversion polymorphisms*

Some genetic systems do indeed show evidence that selection has reduced recombination. There is no doubt that many *Drosophila* inversions are maintained by natural selection (**Section 2.2.ii**). Inversions suppress crossing over only when heterozygous, so that epistatic selection of the type we have just described could maintain them as polymorphisms. The empirical evidence and theoretical models concerning the causes of inversion polymorphism are reviewed by Krimbas and Powell (1992b). The most convincing evidence for epistatic selection involving inversions comes from the inversions associated with segregation distorter systems (see **Section 10.3.iii.d** below). In general, however, it has proved remarkably difficult to rule out alternative possibilities (see Kirkpatrick and Barton 2006).

10.2.iii.b. *Supergenes*

There are several examples of close linkage among genes that are polymorphic for variants with major phenotypic effects, and the evolution of these systems of linked polymorphic genes (or *supergenes*) probably involved selection for close linkage. The classical examples of supergenes are the loci controlling polymorphisms for Batesian mimicry in butterflies (Sheppard 1959b). Tropical species of *Papilio*, such as *P. dardanus* and *P. memnon*, exhibit a variety of mimetic types, which appear to be controlled by a single locus with multiple alleles (**Sections 1.1.i**, **2.2.i.b**, and **8.1.iii.a**, **Plate 1** [front endpaper]). The morphs differ in several aspects of the butterflies' colors and patterns, yet genetic evidence suggests that the mimicry "locus" is in reality a complex of several tightly linked genes (Clarke and Sheppard 1960a, 1971). In *P. memnon*, five genes seem to control distinct aspects of the phenotype, and rare variant types found in natural populations appear to be recombinants among these genes (Clarke and Sheppard 1971). These phenotypes are poor mimics, because they combine features from the mimics of different model species; this is likely to be strongly disadvantageous.

Models for the evolution of mimicry supergenes have been developed (Charlesworth and Charlesworth 1975b; Turner 1977). The principle is similar to that described in **Section 9.3.v.b** for the early evolution of sex chromosomes. The further evolution of restricted recombination between the sex chromosomes, which we discussed briefly there, exemplifies the general principles discussed in **Section 10.2.ii** above (Charlesworth et al. 2005).

Plant self-incompatibility (SI) loci (**Sections 2.2.i.b** and **8.3.i**) provide another example of supergenes. In several taxa, the *S* loci consist of two separate, but physically close, genes; one controls the stigma incompatibility type, and the other the type of the pollen (Schopfer et al. 1999; Takayama et al. 2000; Charlesworth et al. 2005). To maintain incompatibility, the pollen and stigma genes must have the correct allele combinations, i.e., LD among them must be maintained, because genotypes with recombinant allelic combinations are self-compatible, leading to a loss in fitness due to inbreeding depression (**Section 9.2.ii.c**). In the *Arabidopsis lyrata* SI region, recombination rates appear to be low compared with the genome-wide average (Kamau et al. 2007).

10.2.iii.c. *Centromeric and telomeric suppression of crossing over*
Selection for reduced recombination does not necessarily require epistatic selection. We briefly describe one such situation here and another in **Section 10.2.iv.a** below. In **Section 8.3.i**, we mentioned the suppression of crossing over close to centromeres and telomeres. Centromeres in most species are compound structures, containing repetitive sequences around which the structure that binds the spindle fiber microtubules forms. Telomeres are also made up of repetitive units, of a different type from the centromere (Chan and Blackburn 2004). Unequal crossing over between the repeat units (**Figure 10.6**) would produce aberrant numbers of repeats, which could impair chromosome segregation in mitosis and meiosis, resulting in aneuploid cells and reduced fitness (Charlesworth et al. 1986). This could result in a selective advantage to reduced crossing over in or near centromeres and telomeres (Charlesworth et al. 1986). Additionally, exchanges near centromeres (even without unequal crossing over) may directly interfere with centromere disjunction in meiosis, again leading to aneuploidy. There is evidence for this from the smut fungus *Microbotryum violaceum* (Cattrall et al. 1978), yeast, *Drosophila*, and humans (Rockmill et al. 2006).

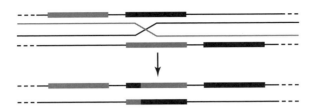

FIGURE 10.6 Diagram to show unequal crossing over between tandemly duplicated genes, resulting in one product with three copies of the gene and the other with only one copy. The two duplicated copies are shown as black and grey, and the diagram also shows that chimeric products, with part of the sequence from the left-hand gene and part from the right-hand copy, are produced by such events.

10.2.iii.d. *Congealing of the genome versus selection for recombination*
We have shown that selection for decreased recombination occurs in many natural situations, and epistatic selection should therefore cause the genome to "congeal", i.e., evolve towards zero recombination, with just a single, non-recombining chromosome (Turner 1967). There is evidence for widespread genetic variation within species in the frequency of recombination (Brooks 1988; Korol et al. 1994; Coop and Przeworski 2007), suggesting that, if selection generally favors reduced recombination, genomes could easily congeal. However, only a very few multicellular organisms effectively lack recombination, notably the plants with multiple translocations mentioned in **Section 10.2.i** above. This implies that selection must also often favor increased recombination.

One explanation for why genomes do not congeal is widely accepted among biologists working on the mechanism of meiosis, i.e., crossovers between homologous chromosomes help to ensure proper disjunction at the first division of meiosis by preventing premature separation of the centromeres of synapsed homologous chromosomes (Darlington 1931). Absence of crossovers thus causes the production of aneuploid gametes, resulting in inviable progeny (Page and Hawley 2003). There are, however, two reasons why this does not fully explain the maintenance of crossing over throughout large regions of most eukaryote genomes. First, crossing over is sometimes absent in one sex (invariably the heterogametic sex), such as the males of higher Diptera and the females of Lepidoptera (White 1973, pp. 476–490). Evolution can thus evidently produce meiotic mechanisms that achieve regular chromosome disjunction without crossovers. Furthermore, crossovers can be localized to small regions of the chromosomes, allowing disjunction, but without any crossing over in much of the genome (White 1973, pp. 169–173); this is illustrated by the terminal crossing over of the chromosomes in the *Oenothera* complexes (see **Figure 10.5** of **Section 10.2.i**), and the pseudoautosomal regions of mammalian X and Y chromosomes (the only regions where crossovers occur between the two chromosomes: Koller and Darlington 1934; Burgoyne 1982; Ross et al. 2005).

Second, the rate of crossing over per chromosome can vary considerably among related species, with some species having far more exchanges per bivalent than is needed to ensure disjunction. For example, *D. melanogaster* chromosomes disjoin accurately with a total map length only around one-quarter of that of *D. virilis* (Gubenko and Evgen'ev 1984; Lindsley and Zimm 1992); in the budding yeast, *Saccharomyces cerevisiae*, there are 5.2 crossovers per bivalent; this is equivalent to an average of 0.3–0.5 centiMorgans per kilobase, depending on the chromosome (www.yeastgenome.org), compared with about one centiMorgan per megabase in *D. melanogaster*.

Several ecological factors also correlate with rates of crossing over. For example, mammalian species with long development times tend to have higher rates of crossing over per chromosome than fast-developing species (Burt and Bell 1987; Sharp and Hayman 1988), and highly self-fertilizing species of plants tend to have higher rates of cytologically detected crossovers than related outcrossing species (Roze and Lenormand 2005). This evidence strongly suggests that the recombination rates vary among species due to selective pressures, and that we therefore need to search for processes in which recombination is associated with higher fitness.

10.2.iv. Possible advantages of sex and recombination: deterministic mechanisms

10.2.iv.a. *Introduction*

Many models have been studied that produce a selective advantage to recombination. In some cases, it has been shown that population mean fitness is higher than without recombination; this often, but not always, predicts when selection favors recombination modifiers that increase recombination rates, a process that is much more difficult to analyze theoretically (Otto and Lenormand 2002; Agrawal 2006b). A bewildering plethora of models exists, and we only briefly survey the main possibilities. Models can be classified into *deterministic* mechanisms that can act in infinitely large populations and *stochastic* processes that depend on finite population size (Kondrashov 1993). We first consider deterministic mechanisms, and will deal with stochastic mechanisms in **Section 10.2.v**.

A common property of nearly all of the models is that the populations must depart from equilibrium under selection and recombination alone, as might be expected from the results of **Section 10.2.ii** for equilibrium populations. We begin, however, with two situations that are exceptions to this rule. Then, after outlining other models, we review the evidence for the likely importance of the various possibilities in **Section 10.2.vi**.

The *sib-competition* model (Williams 1975) imagines an environment divided spatially into local patches, with variability being maintained by fitness differences among patches under density-dependent mortality, as in Levene's model (**Section 2.2.iii**). In addition, within each patch, the offspring of a single female are assumed to compete with each other; in the most extreme case, only individuals with the highest fitness within a patch survive. Under such conditions, sex and recombination are favored because they allow the production of progeny with mixed genotypes, so that a sibship includes at least some of the locally fittest genotypes. These conditions effectively ensure that the number of surviving progeny of a multiply heterozygous parent (its fitness) depends directly on its recombination frequency, in contrast to the equilibrium models described in **Section 10.2.ii**, where there was no such relationship between recombination and fitness. This leads to selection for modifiers that increase recombination above zero (Maynard Smith 1976b). It can also create an advantage to sexual versus asexual reproduction purely because sex allows Mendelian segregation (Barton and Post 1986).

In inbreeding populations, recombination can also be advantageous in a population at selection–recombination equilibrium. For partially self-fertilizing populations, some types of epistatic fitness effects create a selective advantage to modifiers that increase recombination (Charlesworth et al. 1979; Holden 1979; Roze and Lenormand 2005). For example, in a two-locus system with heterozygote advantage and the same selection coefficients at each locus, recombination modifiers spread when the fitnesses of double homozygotes are below those predicted from the additive effects of homozygosity at each locus alone. Higher recombination then increases the mean fitness of the selfed progeny of double heterozygotes.

In contrast, if the fitnesses of the double homozygotes are higher than predicted with additivity (for example, if the fitness effects of different loci are multiplicative), there is selection for reduced recombination in inbreeding populations (Charlesworth et al. 1977). This may contribute to the spread of the translocations in highly selfing species of plants, such as *Oenothera*, which we described in **Section 10.2.i** (De Waal Malefijt and Charlesworth 1979). These selection pressures on recombination in inbreeding populations (either for or against recombination) arise from identity disequilibrium, i.e., the correlated probability of homozygosity of linked variants at different sites. As discussed in **Section 8.3.iii.c**, closer linkage increases this probability, and thus can affect the outcome of selection without any LD.

We next describe the main processes that can generate an advantage to recombination when populations are not at equilibrium under selection.

10.2.iv.b. *Fluctuating epistasis*

Intuitively, one might expect temporal changes in fitnesses across generations to favor recombination. The most straightforward situation involves fluctuations in the direction of fitness epistasis between two loci (Sturtevant and Mather 1938; Maynard Smith 1971b). In some generations, haplotypes A_1B_1 and A_2B_2 have higher marginal fitnesses than A_1B_2 and A_2B_1, but this is reversed in other generations. Recombination allows transitions between these genotypes; computer models (Charlesworth 1976) and analytical results (Gandon and Otto 2007) show that recombination modifiers that push recombination frequencies above 0 are then favored. However, the selection pressure is usually weak, and the population generally reaches only a very low recombination frequency; modifiers that increase recombination frequency further will then not spread, i.e., there is a low ESS recombination frequency (Charlesworth 1976; Gandon and Otto 2007). Host–pathogen interactions of the kind discussed in **Section 2.2.i.b** can potentially generate the required type of cycles of haplotype frequencies (Jaenike 1978; Hamilton 1980); this is called the "Red Queen" model (Bell 1982, p. 157). However, a detailed theoretical investigation (Otto and Nuismer 2004) suggests that this occurs only under restricted conditions (but see Agrawal 2006a, who showed that maternal transmission of parasites makes it more likely that recombination will be favored).

10.2.iv.c. *Moving optima*

A changing environment can also cause the optimum for a quantitative trait to change over time. Consider, for example, the model introduced in **Section 4.5.i.b**, with a nor-optimal fitness function with respect to a trait, and additive genetic effects of variants at multiple sites on the trait value, z. The analysis in **Section 8.4.iv.a** showed that selection within a generation reduces the genetic variance of z, due to negative LD that builds up among the alleles that affect z in the same direction, whether or not the mean of z coincides with its optimum, θ. When $\bar{z} = \theta$, the mean fitness $\bar{w}$ of the population is higher, the lower the variance in z (see **Equation B4.8.4** of **Section 4.5.i.b**). Furthermore, the lower the harmonic mean recombination frequency, the smaller the genetic variance of z (**Equation B8.12.3** of **Section 8.4.iv.a**). Together, these two results suggest that pure stabilizing selection selects for reduced recombination, as was pointed out by Mather (1943), in line with the general analysis in **Section 10.2.i** above.

What happens when the environment changes, so that $\bar{z} \neq \theta$? Then $\bar{w}$ is influenced by both the extent of deviation of the mean from the optimum and the variance in z. The larger the additive genetic variance in z, the faster the mean will approach the optimum (**Section 3.3.iii.a**), and the faster $\bar{w}$ will increase over time. This suggests that increased recombination could be favored if the advantage of an increased response to selection outweighs the advantage of reduced variance.

These suggestions have been confirmed by modeling the spread of recombination modifiers (Maynard Smith 1988; Bergman and Feldman 1990; Charlesworth 1993a; Barton 1995a). Both smoothly changing and fluctuating optima can produce situations in which nonzero recombination is evolutionarily stable, although the ESS recombination is usually much less than 50%. This result is much more general than the nor-optimal model; it holds for any situation in which the relationship between phenotype and fitness means that higher recombination increases genetic variance (Felsenstein 1965). As mentioned in **Section 8.4.iv.a**, for a normally distributed quantitative trait, this requires a negative second derivative of $\ln w(z)$ with respect to z (Shnol and Kondrashov 1993).

10.2.iv.d. *Mutation and selection*

The effect just described depends on the population being perturbed away from its equilibrium under selection and recombination, due to a changed selective optimum. An advantage to recombination can occur in other situations where the population is not at equilibrium under selection alone, e.g., with mutation and selection. In a randomly mating population, autosomal deleterious mutations are mostly selected against when heterozygous (**Section 4.2.ii.a**). If different mutations affect fitness similarly, the number of heterozygous mutations carried by an individual can be treated as an additive quantitative trait with an approximately normal distribution.

The condition for recombination to increase the additive variance, just explained, is satisfied when mutations exhibit *synergistic epistasis*, i.e., log fitness declines faster, the more deleterious mutations are present (**Figure 10.7**).

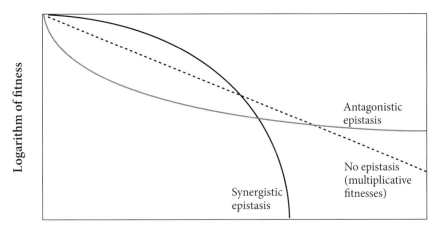

Number of deleterious mutations

FIGURE 10.7 Different forms of epistasis between deleterious mutations. For simplicity, all mutations are assumed to have the same effect on fitness, so that the fitness is a function of the number of mutations carried by an individual.

This happens, for example, when there is truncation selection of the type discussed in **Section 3.3.iii.a**, in which only the section of the population with the lowest number of mutations contributes to the next generation (Kondrashov 1984, 1988). Recombination then helps selection to remove deleterious mutations from the population, and the mean fitness of a population increases with the recombination frequency (Kimura and Maruyama 1966; Crow 1970; Kondrashov 1982, 1984, 1988; Charlesworth 1990a). There is also a large advantage to segregation without recombination over purely asexual reproduction; see Charlesworth 1990a. If mutations interact in the opposite way, there is *antagonistic epistasis* (i.e., log fitness declines more slowly than linearly as a function of the number of mutations under the opposite regime). In this case, recombination reduces mean fitness. Recombination modifiers that increase recombination away from zero have been shown to be favored if there is synergistic epistasis between mutations, and disfavored under antagonistic epistasis (Feldman et al. 1980; Kondrashov 1984; Charlesworth 1990a; Barton 1995a). If fitnesses are multiplicative, as we have often assumed for simplicity (**Section 4.3.ii.c**), differences in recombination are neutral.

In the models described here and in **Section 10.2.iv.c**, recombination in one generation increases genetic variance in fitness, and this often *reduces* the mean fitness of the progeny population, compared with no recombination (Charlesworth and Barton 1996). Modifiers increasing recombination are nevertheless favored under many conditions, because of their longer-term effects in promoting the response of the population to selection. Paradoxically, therefore, the observation of reduced fitness caused by the occurrence of recombination, which has been detected experimentally in both *Drosophila* (Charlesworth and Barton 1996) and *Daphnia* (Allen and Lynch 2008), is consistent with selection in favor of recombination, as well as with selection for reduced recombination of the type discussed in **Section 10.2.i**.

10.2.v. Stochastic processes

10.2.v.a. *General considerations*

The other main class of processes giving an advantage to increased recombination requires finite population size. Felsenstein (1974) described these cases under the generic term of the *Hill–Robertson effect*. Hill and Robertson (1966) showed that selection at one site in the genome impedes the action of selection at another site when recombination between them is rare or absent (**Figure 10.8**). This is because a finite population does not contain all possible combinations of variants that can be produced by mutation. If mutations (either favorable or deleterious) arise in different individuals, a favorable variant at one site will generally be present in a genotype with deleterious variants at other sites,

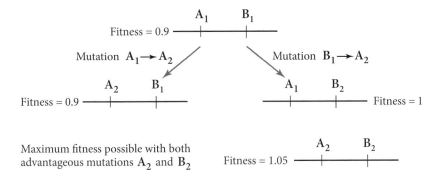

FIGURE 10.8 Diagram to illustrate the idea of Hill–Robertson interference between the action of selection for advantageous mutations at two sites in a genome when recombination between them is rare or absent. The diagram shows two loci or sites, A and B, at which advantageous mutations arise (causing increased fitness). The highest fitness can be achieved if both advantageous mutations are present in an individual.

i.e., there is negative linkage disequilibrium between selectively favorable variants, just as in the models described in **Sections 10.2.iv.c** and **10.2.iv.d** above. Because recombination breaks down this LD, it enhances the population's ability to respond to selection.

Another way of understanding this is to consider effective population size, N_e. We have already described several ways in which selection reduces neutral variability at linked sites; this effect can be viewed, at least in part, as a reduction in N_e. Selection implies heritable variance in fitness among individuals, and this reduces N_e (Robertson 1961). A site that is linked to a selected variant thus experiences a reduced N_e, as we discussed in **Sections 8.3.iii** and **8.3.v**, especially as the effect of variance in fitness at one site on the behavior of linked sites is maintained for many generations when there is close linkage (Santiago and Caballero 1995, 1998). Without recombination, selection is thus less effective than when recombination occurs. Almost any form of directional selection has such effects (no epistasis is necessary), in contrast to the deterministic mechanisms described in **Sections 10.2.iv** above. The following sections extend these ideas to the effects of selection at one site on other sites under selection, with some examples of specific biological situations where these effects occur.

10.2.v.b. *Selective sweeps*

Fisher (1930b, pp. 103–104) and Muller (1932) pointed out that two favorable mutations that arise at different sites cannot both become fixed in a non-recombining population unless both mutations occur in one lineage (**Figure 10.8**). This is often referred to as *clonal interference* (Gerrish and Lenski 1998), and it implies that recombination should facilitate adaptive evolution. The importance of this effect will depend on the rate of adaptive evolution driven by environmental change; if evolution is very slow, substitutions at different sites can occur in succession, and recombination is irrelevant. If, however, change is so rapid that two or more selectively favorable variants are spreading through the population at the same time, recombination improves the population's ability to fix the combination of best variants. In addition, population size affects the importance of recombination; in large populations, very rare events, like independent favorable mutations at two loci, are more likely to occur, so that recombination will be less advantageous than in small populations.

Clonal interference can generate an evolutionary advantage to recombination; both increased mean fitness of the population as a result of recombination and the spread of recombination modifiers have been demonstrated theoretically (Crow and Kimura 1965; Felsenstein 1974; Maynard Smith 1978, Chapter 2; Gerrish and Lenski 1998; Barton 1995b; Otto and Barton 2001; Martin et al. 2006; Desai and Fisher 2007).

10.2.v.c. *Background selection*

As described in **Section 8.3.iii**, neutral (or even weakly selected) diversity is reduced by the presence of strongly selected deleterious mutations at linked sites present at frequencies close to those expected under mutation–selection balance in an infinitely large population (background selection; see **Box 8.6***). This selective elimination of mutations reduces N_e for the linked neutral sites. From the theory in **Section 6.2.iii.a**, we also expect the chance of fixation of favorable variants to be reduced in proportion to the reduction in N_e, and the chance of fixation of deleterious mutations to be enhanced; this has been confirmed by detailed models in which the sites affected by background selection are more weakly selected than the mutations causing the background selection effects (Charlesworth 1994a; Peck 1994; Barton 1995b; Stephan et al. 1999). However, if a favorable mutation has a substantially larger selection coefficient than those of any linked deleterious mutations, it can spread to fixation, carrying these mutations with it (**Figure 10.9**) (Rice 1987b; Johnson and Barton 2002; Hadany and Feldman 2005).

The hitchhiking effects of relatively strongly selected favorable or deleterious mutations can thus undermine the level of adaptation at more weakly selected, linked sites, such as synonymous sites subject to selection for codon usage bias (Kim 2004); recombination alleviates these effects.

10.2.v.d. *Muller's ratchet*

The background selection model assumes mutations that are sufficiently deleterious that their expected frequencies remain close to their low equilibrium values in an infinite population. In non-recombining genomes or genomic

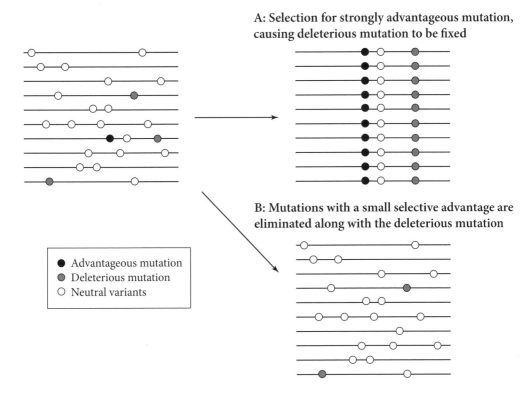

A: Selection for strongly advantageous mutation, causing deleterious mutation to be fixed

B: Mutations with a small selective advantage are eliminated along with the deleterious mutation

● Advantageous mutation
● Deleterious mutation
○ Neutral variants

FIGURE 10.9 Different possible outcomes of selection for a favorable mutation. The left-hand figure shows diagrammatically the state of a population when an advantageous mutation (indicated by the black dot) has occurred. There are disadvantageous mutations (grey dots) in the genome region near this in some individuals, and there are also neutral variants that differ between individuals (open circles). **A.** The result when the selection coefficient for the favorable mutation is substantially larger than those of the deleterious mutation in the same haplotype, with fixation of the favorable mutation leading to fixation of the haplotype's disadvantageous mutation(s) and neutral variants. **B.** The outcome when the haplotype with the favorable mutation carries a more strongly disadvantageous mutation that is eliminated from the population, also causing elimination of the favorable mutation, and the associated neutral variants.

regions, however, deleterious mutations will not necessarily behave like this, but instead may accumulate; this process was originally described by Muller (1964) and called *Muller's ratchet* by Felsenstein (1974), and has often been considered as potentially generating an advantage to recombination. For simplicity, we consider a population of haploid organisms (this also describes the evolution of a Y chromosome, except that the selection coefficients against deleterious mutations in a haploid are replaced by ones for heterozygotes).

A simple way to understand the process is to recall that the equilibrium distribution of the number of mutations per genome follows a Poisson distribution, if fitness effects are multiplicative (**Section 4.2.ii.c**). The equilibrium frequency of the zero mutation class, f_0, is given by the following expression:

$$f_0 = \exp(-\bar{n}) \tag{10.3}$$

where $\bar{n}$ is the mean number of deleterious mutations per individual at equilibrium under mutation and selection (**Equation 4.7c**). As we saw in **Section 6.4.iv.c**, $\bar{n}$ is likely to be very large for the genome as a whole, so that f_0 will be small, and at most only a small number of individuals are mutation-free.

The "least-loaded" class is therefore easily lost from the population by genetic drift; with no recombination, the assumption that back mutation is rare (i.e., mutation is one way, from good to bad alleles) means that the zero mutation class cannot be regenerated (**Figure 10.10**). This event is called a "click of the ratchet." In contrast, recombination among chromosomes carrying one or

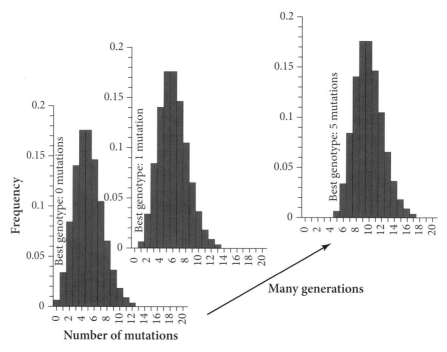

FIGURE 10.10 The progression of Muller's ratchet. The left-hand plot shows the distribution of numbers of deleterious mutations in a large equilibrium population, showing that only a few individuals lack such mutations even when the mean number of mutations is small. If there is no recombination, the loss of this class of individuals is irreversible, and the number of deleterious mutations changes to a new distribution, with higher numbers than previously (middle plot). Over evolutionary time, the individuals in the population will come to have high numbers of such mutations (right-hand plot).

more mutations can regenerate the zero mutation class. Once the zero class has been lost, the distribution of numbers of mutations reshapes itself and eventually approaches the deterministic equilibrium distribution again, now with the least-loaded class represented by individuals carrying one mutation. This process is repeated indefinitely in a ratchet-like manner, at an approximately constant click rate, which depends in a complex way on the effective population size, the value of f_0, and the strength of selection (Haigh 1978; Gessler 1995; Gordo and Charlesworth 2000; Stephan and Kim 2002; Söderberg and Berg 2007; Rouzine et al. 2008; Jain 2008). Modifiers that increase recombination away from 0 in populations subject to the ratchet experience a selective advantage (Keightley and Otto 2006; Gordo and Campos 2008).

The ratchet causes a steady increase in the mean number of mutations per individual, at a rate that can be several orders of magnitude faster than with free recombination. Each click of the ratchet is followed by the fixation of a single mutation somewhere in the genome region affected (Rice 1987b; Charlesworth and Charlesworth 1997). This happens because, after the loss of a least-loaded class, the new least-loaded class contains a mix of haplotypes, each carrying a mutation that was absent from the previous least-loaded class. Because this class consists of few individuals, one of the haplotypes will rapidly become fixed in this class by drift. The absence of back mutations means that, as with background selection, the least-loaded class eventually becomes the ancestor of all of the other classes in the population, so that the mutation that was fixed in the least-loaded class is eventually present throughout the entire population.

The results just described assume that the product of population size and f_0 for the region of the genome in question exceeds 1, so that some mutation-free individuals are present in the initial population before the ratchet starts to operate. As discussed in **Section 6.4.iv.c**, however, the mean number of deleterious mutations per haploid genome is likely to be several hundred in *Drosophila* or humans. With $\bar{n} = 50$, $f_0 = 1.9 \times 10^{-22}$. Unless the population size is impossibly large ($1/1.9 \times 10^{-22} = 5.2 \times 10^{21}$, in this example) no mutation-free individuals will be present in the population, and no quasi-stable distribution of the number of mutations can exist. The ratchet then moves very fast, accompanied by fixations of deleterious mutations (Gessler 1995; Rouzine et al. 2008).

This may be close to reality for an asexual population derived from a diploid sexual ancestor, where the whole genome contributes to the accumulation of deleterious mutations. In the limit (as the selection coefficient approaches 0), mutations accumulate at almost their rate of occurrence, so that the mean fitness will decline at a rate equal to the product of the deleterious mutation rate and the mean selection coefficient against a mutation. The mean fitness of an asexual population of multicellular eukaryotes, with a deleterious mutation rate as high as 1 or more (**Section 4.2.i**), will thus decline, and the population may be doomed to extinction. Similarly, an evolving Y chromosome may accumulate large numbers of mutations that reduce its gene functions, leading to its degeneration over time (**Section 10.2.vi.a** below).

10.2.v.e. *Reverse mutation and Hill–Robertson interference*

An unrealistic assumption of the ratchet model is one-way mutation from good to bad. This is reasonable for an initial process of departure from the deterministic equilibrium under mutation and selection (for example, if recombination stops in a previously recombining genome region), when the wild-type variant will be much more frequent than the mutant at each site. In such situations, reverse mutations can therefore be neglected. But once deleterious mutations become fixed at many nucleotide sites, reverse mutations will become increasingly frequent, slowing the ratchet down; ultimately, the rate of fixation of deleterious mutations will be balanced by fixations of favorable, reverse mutations. Unless the population has gone extinct, a new equilibrium, associated with a drastically reduced mean fitness, will be established. Thus, although Muller's ratchet is often described as an irreversible process of accumulation of deleterious mutations, this does not accurately describe what happens after a long period of time has elapsed.

The ratchet also assumes that the deleterious mutations present in the population at any one time are quite strongly selected and mostly held close to the deterministic equilibrium frequencies under mutation and selection. However, if many sites under relatively strong selection are packed into a non-recombining genome, Hill–Robertson interference among them will cause them all to behave as weakly selected variants. Thus, deleterious variants can reach high frequencies, or fixation. Again, recombination alleviates this and non-zero recombination is maintained by selection (Charlesworth et al. 2009).

For a non-recombining genome or genomic region with many sites under weak selection, however, this effect can be very large (for example, an incipient Y chromosome, or even the small dot chromosome of *Drosophila*, which has only 80 genes, and does not undergo crossing over). Simulations show that the decline in the overall level of adaptation is accompanied by an increased rate of substitution of deleterious variants, and a reduction in DNA sequence variability at both the selected sites and neutral sites located in the same genomic region (Comeron et al. 2008); the reduction in variability, reflecting a reduction in N_e, can be on the order of 100-fold in very large regions (Charlesworth et al. 2009).

10.2.vi. Empirical evidence concerning the consequences of recombination

10.2.vi.a. *Reduced levels of adaptation in regions with low recombination*

In **Section 10.1.i**, we described patterns in the genome that suggest that the efficacy of selection is reduced in genomic regions where recombination is rare or absent, e.g. the reduced frequencies of optimal codons in genes (as reflected in their GC3 content) on the dot (fourth) chromosome of *D. melanogaster* (**Figure 10.2**). This chromosome also has an accelerated rate of amino acid substitution, compared with the rest of the genome (Haddrill et al. 2007). Similar findings have been noted in recently evolved asexual lineages of *Daphnia* (Paland and Lynch 2006), in the ancient asexual Bdelloid rotifers (Barraclough et al. 2007), and in endosymbiotic bacteria, with their reduced opportunities for genetic exchange and lower effective population sizes compared with free-living species (Herbeck et al. 2003).

The short evolutionary persistence times of asexual and self-fertilizing species mentioned in **Section 9.1.ii.a** are consistent with much other evidence that genetic recombination and sexual reproduction confer advantages in terms of population mean fitness. In addition, the degeneration of Y and W chromosomes provides compelling examples of the evolutionary consequences of suppressed recombination, as we now describe.

Case Study 3
The Evolution of Y or W Chromosomes

The initial steps in the evolution of sex chromosome systems were discussed in **Section 9.3.v.b**. The Y or W chromosomes are confined to the heterogametic sex (males for Y and females for W), and do not recombine over part or all of their length with their partner X or Z chromosomes. Y or W chromosomes often have few remaining active genes, whereas gene densities on the X or Z are normal. The Y or W are therefore "degenerate" descendants of chromosomes originally carrying similar complements of genes to their partners (Charlesworth et al. 2005). For such degenerate chromosomes, the theory developed above implies that the evolutionary forces causing degeneration must now be comparatively weak, because these forces involve the action of selection at multiple sites in the genome. To detect the signature of this selection, we need to examine systems with recently evolved Y or W chromosomes, which still retain many active loci.

One such system is provided by the *neo-X* and *neo-Y* chromosomes of some *Drosophila* species, which formed when a fusion or translocation of an autosome onto an X or Y chromosome spreads to fixation within the species (**Figure 10.11**). Initially, the new components of the neo-X and neo-Y cannot have been any more different than a random pair of autosomes from the same population. The lack of crossing over in male *Drosophila* means that the neo-Y chromosome is immediately placed in a genetic environment that is identical to that of the true Y chromosome (Lucchesi 1978; Steinemann and Steinemann 1998).

D. miranda has a neo-Y chromosome that was formed by the fusion of the Y chromosome to an autosome (the homologue of the right arm of chro-

A.

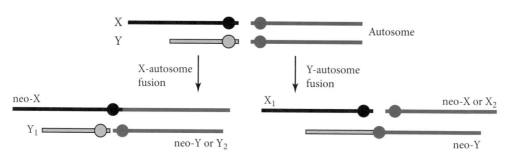

B.

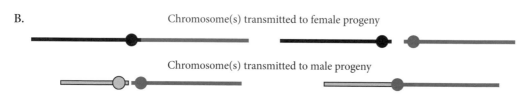

FIGURE 10.11 Chromosome fusions. **A.** A fusion or translocation of an autosome to a sex chromosome, creating neo-X and neo-Y chromosomes. **B.** In *Drosophila*, there is no recombination in males. Thus, both kinds of fusion create non-recombining neo-sex chromosomes (the two autosomal copies in males segregate with the sex chromosomes in the first division of meiosis, in such a way that one always accompanies the X into a sperm and the other accompanies the Y).

mosome 2 of *D. melanogaster*) (MacKnight 1939), less than two million years ago, i.e., since its split from its close relative *D. pseudoobscura*. The silent-site divergence between homologous genes on the neo-Y and neo-X chromosomes suggests that these diverged about 1.5 million years ago (Bartolomé and Charlesworth 2006). Silent-site variability within *D. miranda* on the neo-Y chromosome is about 1% of the value for the neo-X, indicating a greatly reduced effective population size (Bartolomé and Charlesworth 2006).

About one-half of the neo-Y genes so far sequenced are fixed for major mutations, such as frame shifts and deletions (Bachtrog et al. 2008). In addition, use of *D. pseudoobscura* as an outgroup (see **Section 10.1.ii.e**) shows that the rate of nonsynonymous substitutions is approximately 60% higher on the branch leading to the neo-Y from its common ancestor with the neo-X than on the neo-X branch (Bachtrog 2004; Bachtrog et al. 2008). The ratio of nonsynonymous to synonymous polymorphism levels in neo-Y chromosome genes is significantly higher than for genes on the neo-X (Bartolomé and Charlesworth 2006). The divergence and polymorphism data both suggest that selection against nonsynonymous mutations is less effective on the neo-Y chromosome, consistent with Hill–Robertson effects reducing N_e for the neo-Y chromosome, allowing deleterious mutations to accumulate.

However, other possibilities need to be excluded before this interpretation can be accepted. In particular, it is possible that the expression levels of neo-Y genes have been greatly reduced, so that they are effectively nonfunctional and the accumulation of coding sequence changes simply reflects neutral evolution. Measurements of the relative messenger RNA levels produced by neo-X and neo-Y gene copies show that about two-thirds of the genes have significantly lower mRNA production from neo-Y copies than from their neo-X counterparts; occasionally, Y-linked mRNA production is significantly higher (Bachtrog 2006). However, neo-Y genes with low expression levels do not have faster than average nonsynonymous site evolution, which suggests that reduced gene functionality probably does not explain the excess nonsynonymous fixations on the neo-Y. Indeed, the fact that neo-Y expression is sometimes elevated suggests that misregulation of neo-Y gene activity is common, with some substitutions causing higher than optimal expression (Bachtrog 2006).

10.2.vi.b. *Discriminating between the different possible advantages of recombination*

These examples of the consequences of turning off recombination or sexual reproduction indicate very clearly that these processes provide a long-term fitness benefit to a population. This raises the question of which of the various processes described above are most important for maintaining sex and nonzero levels of recombination; these processes are not mutually exclusive. At present, there is no conclusive answer to this question, and different authors have expressed very different views. The older literature used evidence from comparative studies, such as the association between asexuality and low population density, which has widely been interpreted as evidence for the "Red Queen" model of **Section 10.2.iv.b** (e.g., Bell 1982). However, this evidence is open to alternative interpretations, and so is inconclusive. It is also very difficult to obtain direct experimental evidence to discriminate among the specific models described in **Sections 10.2.iv** and **10.2.v**, although a number of experimental evolution studies suggest that sexual reproduction facilitates adaptive evolution (Colegrave 2002; Goddard 2002). Similarly, genetic analyses of the results of artificial selection experiments suggest that a response to selection is often associated with increased rates of crossing over, but the population genetic causes of these effects are difficult to determine (reviewed by Otto and Barton 2001).

Accordingly, we mainly have to rely on evidence concerning the occurrence and intensity of the processes involved in the different models. Processes involving the effects of deleterious mutations, such as Muller's ratchet (**Section 10.2.v.d**) and selection against synergistic mutations (**Section 10.2.iv.d**), are effective in protecting populations against invasion by asexual mutants or mutations that restrict recombination if the genome-wide deleterious mutation rate is high. Recent estimates of this parameter suggest that, even in a multicellular species with a relatively small genome such as *Drosophila*, there is at least one deleterious new mutation per individual per generation (Haag-Liautard et al. 2007), and many more in large-genome organisms such as humans (see **Section 4.2.i**). The evidence from mutation–accumulation experiments of the type described in **Section 4.4.iv.b** is conflicting on whether deleterious mutations tend to have synergistic effects on log fitness, as is required by the deterministic mutational theory (Charlesworth et al. 2004). Studies of mutations with large deleterious effects in microbial systems suggest a wide distribution of the direction and magnitude of interactions, with roughly 50% having synergistic effects and 50% having antagonistic effects (Kouyous et al. 2007). This pattern is similar to that predicted by the multivariate quantitative genetics model of Martin and Lenormand (2006, 2008) discussed in **Section 3.3.iv**. It is seemingly inconsistent with an important role for deterministic effects of deleterious mutations in favoring sex or recombination, but it is not known whether it applies to the much larger class of mutations with minor effects on fitness, which may have the major effect on the outcome of selection for increased recombination under this model (Kouyous et al. 2007). A recent mutation–accumulation study in yeast suggests that synergistic epistasis is in fact widespread in this species (Dickinson 2008).

The importance of processes that generate fluctuating epistasis in fitness, such as host–parasite "Red Queen" coevolution, is also difficult to assess, with empirical evidence for the occurrence of genetic variation in host resistance and parasite virulence, but little direct evidence for covarying changes in frequencies of resistance and virulence genotypes in host and parasite, respectively (Otto and Nuismer 2004). A recent study of *Daphnia* has provided evidence of this kind (Decaestecker et al. 2007), but it is still unclear whether these changes are of the kind that will generate an advantage to increased recombination.

The plausibility of Hill–Robertson effects associated with the spread of selectively favorable mutations (**Section 10.2.v.b**) is highly dependent on the frequency with which selective sweeps occur within the genome; if they are sufficiently infrequent that it is rare for more than one selectively favorable mutation to be spreading through the population at one time, then this process can be ruled out. Attempts have recently been made to estimate the frequency of such events in *Drosophila* and humans, suggesting that multiple simultaneous sweeps are occurring in the genome (Andolfatto 2007; Hawks et al. 2007) and may be reducing the effective population size at closely linked sites. The elimination of deleterious mutations may also have similar effects (Loewe and Charlesworth 2007). This suggests that these types of Hill–Robertson effects may well have an important role in generating an advantage to recombination, even if they are not sufficient to overcome the two-fold cost of sex.

10.3. SELFISH DNA

10.3.i. Introduction

We have mostly assumed that natural selection acts either among individuals within a population or (probably less often) among populations of the same species (**Section 7.3.iv.b**). However, the process of biased gene conversion (**Section 10.1.ii.d**) illustrates the possibility that genetic variants can also compete with each other within the same individual. Such selection involves "selfish" behavior, and it plays a considerable role in genome evolution (Burt and Trivers 2006; Lynch 2007). We will discuss two examples of selfish DNA for which population genetic models are important: genetic entities that subvert the rules of Mendelian segregation to their own advantage (*segregation distorters*), and transposable elements (TEs), which can replicate themselves within the host genome. TEs are widely distributed throughout the genomes of virtually all organisms, whereas segregation distorters appear to be limited to specific regions of the genome and have only been discovered in a few dozen species. We concentrate on illustrating how the unorthodox behavior of these systems can be understood by using population genetics models and only describe enough of the biological details to understand the models, because these details are well reviewed in the books just cited.

10.3.ii. Gametic selection and segregation distortion

10.3.ii.a. *Gametic selection*
It has long been realized that selection can act at the level of gametes as well as individuals (Heribert-Nilsson 1920; Onodera et al. 2008). In flowering plants, the pollen grains are miniature haploid individuals (*male gametophytes*) with two genetically identical nuclei. A substantial fraction of a plant's genes are expressed in these nuclei (Tanksley et al. 1981; Honys and Twell 2003; Onodera et al. 2008), and selection can occur on allelic variants among the pollen grains that have landed on a stigma, resulting in a departure from the expected Mendelian 1:1 ratio (Heribert-Nilsson 1920; Snow et al. 2000). As Haldane

pointed out (1932, p. 123), with intense competition between pollen grains for success in fertilizing the ovules:

> A gene which greatly accelerates pollen tube growth will spread through a species even if it causes moderately disadvantageous changes in the adult plant.

Genes are also expressed in the haploid *female gametophyte*, which in flowering plants is represented by a set of eight haploid nuclei produced by meiosis and a subsequent mitotic division. In a given ovule, only one of these nuclei forms the egg nucleus, which contributes the maternal genome to the developing embryo. There can thus be competition among nuclei to form the egg nucleus. In flowering plants, selection can thus act in both the male and female gametophyte stages (Onodera et al. 2008). This occurs in an extreme form in some species of *Oenothera*, whose members are all heterozygous for two haploid genomes (α and β), which are maintained distinct by a complex of translocations involving all the chromosomes, between which there is little or no crossing over (see **Section 10.2.i** above). The α genome is lethal in pollen, and the β genome is lethal to egg nuclei, so that the plants in the next generation are always α/β heterozygotes (Renner 1925; Cleland 1972). The evolution of this system from the probable ancestral state of zygotic lethality of the translocation complexes is discussed by Charlesworth (1979b).

10.3.iii. Meiotic drive and segregation distortion

10.3.iii.a. *Introduction*

In sperm cells of animals, the packaging of the chromosomes into a very small space requires replacement of histones by protamines, and this is accompanied by repression of most gene activity. Genetic manipulations of *Drosophila* and mouse show that sperm cells with much of their genome deleted are capable of normal fertilization (Lindsley and Grell 1969; Ford and Evans 1973), implying that active genes are not required for sperm function. However, recent data on gene expression in the post-meiotic, haploid precursors of sperm cells in *Drosophila* and mammals, and evidence that disruption of some of the genes concerned affects male fertility, imply that there may be a few hundred genes where selection can act in the haploid stage (Joseph and Kirkpatrick 2004), although sperm length in Drosophila is not influenced by haploid gene expression (Pitnick et al. 2009). Much more commonly, the diploid genotypes of males determine the functionality and competitive ability of their sperm (Parker 1970; Fiumera et al. 2007).

In female meiosis, only one of the four products will form an egg nucleus; the remainder are discarded. Competition for inclusion in the egg nucleus among the products of female meiosis is therefore possible in both plants and animals (Malik and Henikoff 2002). This has, for example, been observed in a cross between two species of the monkey flower *Mimulus* (Fishman and Willis 2005; Fishman and Saunders 2008). A similar competitive situation occurs with *centric fusions* between autosomes in the house mouse (these are similar to the sex chromosome–autosome fusion shown in **Figure 10.11** of **Section 10.2.vi.a**); heterozygous mothers transmit a higher frequency of the unfused chromosomes to the progeny (De Villena and Sapienza 2002). These situations represent a form of *meiotic drive*, defined as a distorted segregation ratio that is not caused by gametic selection (Sandler and Novitski 1957). Other examples of meiotic drive include the behavior of heterochromatic knob-bearing copies of chromosome 10 in maize and many examples of supernumerary (B) chromosomes (reviewed by Burt and Trivers 2006, Chapters 8 and 9).

10.3.iii.b. *Segregation distortion*

Despite the lack of evidence for true gametic selection among sperm in animals, another phenomenon can occur: *segregation distortion*. This involves the post-meiotic destruction of developing sperm cells, or malfunctioning of mature sperm, that contain the "wrong" variant D_1 at a locus in males heterozygous for D_1 and the "driving variant" D_2. The functional sperm of $D_1 D_2$ males are therefore mainly D_2. Many systems of this kind have been described (Burt and Trivers 2006, Chapters 2 and 3). Distorter systems involving X-linked loci that cause inactivation of Y-bearing sperm are particularly easy to identify, because this results in female-biased sex ratios. These are thus disproportionately represented among the known cases of segregation distortion.

If male fertility were proportional to the number of sperm produced, a D_2 variant would have no selective advantage in a large population, because the number of D_2 copies transmitted to their progeny by $D_1 D_2$ males would be unaffected by loss of the D_1 sperm. However, male fertility usually increases more slowly than linearly with sperm production; in *D. melanogaster*, for example, the number of progeny produced by mating a female to a single male is barely affected even by a halving of the number of sperm under normal conditions, and a decrease in the number of fertilized eggs due to a high proportion of dysfunctional sperm is detectable only by special breeding experiments (Hartl et al. 1967). The mean number of D_2-carrying eggs laid by females fertilized by $D_1 D_2$ males is therefore likely to be increased by segregation distortion, and so D_2 will invade a D_1 population, unless the loss in fertility due to the production of dysfunctional sperm is very severe (**Box 10.3**).

Segregation distortion systems are usually detected as polymorphisms in natural populations. What prevents them from spreading to fixation? In the two best-studied systems, the *Segregation Distorter* (SD) system of *D. melanogaster* and the *t-haplotype* system of the house mouse, the primary factor opposing fixation is sterility of males homozygous for the distorter (Wu

Box 10.3 SEGREGATION DISTORTION:
POPULATION DYNAMICS OF AN AUTOSOMAL
DISTORTER CAUSING HOMOZYGOUS
MALE STERILITY

In a given generation, let the frequency in males of the driving D_2 variant be q_m and the frequency in females be q_f. The overall frequency, q, is the mean of these. Let the frequency of D_2 among the sperm of D_1D_2 males be k ($k > 0.5$ in the presence of drive); this measures the level of distortion. Let the fertilities of D_1D_2 and D_2D_2 males relative to D_1D_1 males be $1 - s$ and $1 - t$, respectively. Because drive in males causes differences in frequencies between eggs and sperm, we need to consider the genotype frequencies in zygotes (these are independent of sex).

Assuming random mating, let the frequencies of D_1D_1, D_1D_2, and D_2D_2 be $x = p_f p_m$, $y = p_f q_m + p_m q_f$ and $z = q_f q_m$, respectively (these sum to 1). The new frequencies of D_2 in eggs and sperm are:

$$q_f' = \frac{1}{2}y + z = \frac{1}{2}\left(q_f + q_m\right) \tag{B10.3.1a}$$

$$q_m' = \frac{k(1 - s)y + (1 - t)z}{(1 - ys - zt)} \tag{B10.3.1b}$$

If q_f and q_m are close to 0, we can use the standard method for approximating the conditions for the initial spread of a variant, by ignoring the second-order terms in q_f and q_m (**Section 3.1.iii.c**). The new frequency of D_2, averaged over males and females, is then given by:

$$q' = \left[\frac{1}{2} + k(1 - s)\right]q \tag{B10.3.2}$$

There will be an increase over the old frequency if $2k(1 - s) > 1$. With complete drive ($k = 1$), s need be only slightly less than 1/2 to allow D_2 to spread.

The corresponding condition for D_1 to increase when rare is $2(1 - k)$ $(1 - s) < 1 - t$, i.e., $2k(1 - s) - 1 < t - 2s$. This shows that a sufficiently large loss in homozygous male fertility will prevent a distorter from spreading to fixation if it invades the population, implying the existence of a protected polymorphism (see **Section 2.3.ii**).

The composition of the equilibrium population can be found as follows. **Equation B10.3.1a** implies that, at equilibrium, $q_f^* = q^*$. Because q is the mean of the male and female frequencies, this implies that we also have $q_m^* = q^*$. Somewhat counterintuitively, there is thus no difference in the equilibrium frequency of D_2 between eggs and sperm in the absence of selection on females. This simplifies the analysis of the equilibrium frequency, which can be found by equating the variant frequencies on both sides of **Equation B10.3.1b** (**Problem 10.2**):

$$q^* = \frac{2k(1-s)-1}{(t-2s)} \qquad\qquad \text{(B10.3.3)}$$

and Hammer 1991; Burt and Trivers 2006, Chapter 2). Probably the non-distorter alleles of these genes lead to better sperm development (see **Section 10.3.iii.c** below). The nature of the counter-selection pressure is not always known in other cases, although homozygotes for the X-linked sex-ratio distorter of *D. recens* are known to be female-sterile (Dyer et al. 2007). **Box 10.3** derives the conditions under which sterility of D_2D_2 males can lead to a stable polymorphism; not surprisingly, a small distortion coefficient can produce a high frequency of the distorter in a population.

Counter-selection may often be weak, and unable to prevent a driving variant spreading and becoming fixed in the population. However, the deleterious fitness effects of homozygous distorters may give advantages to suppressors of distorters. A suppressed distorter system would become apparent only in crosses to individuals from populations that lack suppressors. A system of this kind has been found in *D. simulans*. Here, the X-linked distorter populations also have suppressors of distortion, located on the Y chromosome and autosomes (Atlan et al. 1997; Jutier et al. 2004). Other examples are reviewed by Presgraves (2008).

Non-Mendelian ratios for genetic markers are often reported in crosses between populations or species (Hall and Willis 2005), but there is usually not enough information to tell whether these are caused by true segregation distorter systems, or by hybrid incompatibilities causing fitness differences among genotypes (**Section 8.4.vi**).

10.3.iii.c. *Examples of segregation distorter systems*
The best understood example of segregation distortion is *SD* in *D. melanogaster*, which was discovered by Yuichiro Hiraizumi when he crossed wild-caught males to females carrying the recessive second chromosome mark-

ers *cn* and *bw* (Sandler and Hiraizumi 1959). On backcrossing F_1 + +/*cn bw* males to *cn bw* females, nearly all the progeny in some crosses were wild-type, not the expected 50% wild-type and 50% *cn bw*. These males carried the *Segregation Distorter* (*SD*) second chromosome. *SD*/+ males from natural populations will usually produce sperm with over 95% *SD*, because sperm carrying + chromosomes fail to develop properly (Wu and Hammer 1991; Burt and Trivers 2006, Chapter 2). The system involves two loci, *Sd* and *Rsp*, located on either side of the centromere of chromosome 2 (**Figure 10.12**). Molecular genetic analysis has shown that the driving version of *Sd* is a duplicated copy of a gene, *RanGAP*, located in the proximal euchromatin of the left arm of chromosome 2, which is involved in transporting proteins across the nuclear membrane (Kusano et al. 2003). The duplication encodes a truncated version of the protein, causing its mislocalization in cells. The *Rsp* (*responder*) locus is a tandem array of 120 bp AT-rich repeats, present in the centromeric heterochromatin of the right arm of chromosome 2 (Wu and Hammer 1991).

Driving chromosomes carry the mutant form of *Sd* and an *Rsp* locus with a small or zero number of repeats, denoted by *Rsp^i*; i.e., the *SD* haplotype is *Sd Rsp^i*. The frequency of this haplotype is usually around 1–2% in natural populations. Nondriving chromosomes fall into two classes, with roughly equal frequencies of the drive-sensitive *Sd^+ Rsp^s* haplotype and the insensitive haplotype with the constitution *Sd^+ Rsp^i*. The fourth possible haplotype is *Sd Rsp^s*. The presence of *Sd* causes *Sd Rsp^s* to have a high probability of spermatid development failure, and so this haplotype is strongly disadvantageous; it is a "suicide

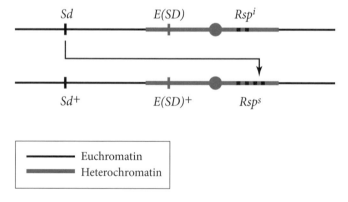

FIGURE 10.12 Component genetic factors and organization of the SD region of *D. melanogaster*. The factors shown are the segregation distortion (*Sd*) and Responder (*Rsp*) genes and the enhancer of segregation distortion, symbolized by *E*(*SD*). The arrow indicates that the product of the *Sd* allele interacts with the sensitive *Rsp* sequence (*Rsp^s*) of *Sd^+* chromosomes, as described in the text. [Adapted from Wu and Hammer (1991).]

combination" and is not found in population samples (it can be constructed artificially by obtaining recombinants from $Sd\ Rsp^i/Sd^+\ Rsp^s$ female parents). Modeling shows that an equilibrium with a low frequency of $Sd\ Rsp^i$, and approximately equal frequencies of $Sd^+\ Rsp^s$ and $Sd^+\ Rsp^i$, can be produced if $Sd^+\ Rsp^i/Sd^+\ Rsp^s$ suffers a fitness loss compared with $Sd^+\ Rsp^s/Sd^+\ Rsp^s$. There is a highly oscillatory approach to equilibrium (Charlesworth and Hartl 1978), because the eigenvalues of the stability matrix for the internal equilibrium with all three haplotypes are complex (see **Box 8.9*** of **Section 8.4.ii.c**). A population cage experiment has detected an apparent disadvantage to $Sd\ Rsp^i$ when competed against $Sd^+\ Rsp^s$ (Wu and Hammer 1991).

Other loci in the *D. melanogaster* genome carry alleles that modify the degree of distortion in $SD/+$ males, i.e., the fraction of SD sperm that they produce. One such locus, $E(SD)$, is located between Sd and Rsp, to the left of the centromere of chromosome 2; SD chromosomes carry an allele at this locus that enhances the drive. The other major chromosomes are also polymorphic for modifier loci that affect the drive strength (Wu and Hammer 1991). Finally, most SD chromosomes are associated with chromosome 2 inversions that suppress crossing over in the SD region when heterozygous over wild-type.

The classic X-linked distorter system, discovered by Sturtevant and Dobzhansky (1936) in *D. pseudoobscura*, involves an X chromosome denoted by X_r. The Y chromosome bearing sperm of X_r/Y males mostly fail to develop, resulting in an excess of female progeny (Wu and Hammer 1991). The X_r frequency varies geographically, with some populations completely lacking it and others having up to ~20% of X_r chromosomes. The *D. pseudoobscura* X_r is associated with three inversions; in the closely related species, *D. persimilis*, X_r has the *D. pseudoobscura* non-inverted arrangement, while the non-X_r chromosomes carry an inversion. Little is so far known about the genetic and molecular basis of X_r, although crossing experiments indicate that the *D. persimilis* X_r contains several genes needed for distortion (Wu and Hammer 1991). *D. recens*, mentioned in **Section 10.3.iii.b** above, also has an X-linked sex ratio distorter associated with a complex of inversions that suppress crossing over across almost the entire X chromosome in heterozygotes (Dyer et al. 2007).

The mouse *t-haplotype* system is similar to the SD system. By crossing wild mice to lab strains carrying a dominant gene, T, that reduces tail length (Dunn and Suckling 1956; Silver 1993; Burt and Trivers 2006, Chapter 2), it was found that wild mice are often heterozygous for alleles (t) that cause loss of tails in the presence of T; > 95% of the offspring of $+/t$ males have the t chromosome, whereas $+/t$ females show normal segregation. Most t chromosomes are homozygous lethal; the lethals fall into a number of different complementation groups, i.e., some crosses between mice both carrying t alleles yield viable t homozygotes, which must be heterozygous for different lethals, l_1/l_2.

Males homozygous for nonlethal *t* chromosomes, or heterozygous for two *t* haplotypes with different lethals, are usually sterile. Therefore (as in the case of *SD*) the segregation distortion mechanism seems to be associated with abnormal sperm function. +/*t* males produce both types of sperm, but *t* sperm are nonfunctional. The *t* system maps to chromosome 17, near the *MHC* complex; driving chromosomes are differentiated from wild-type chromosomes by a set of four inversions, so that +/*t* mice have a widespread reduction of crossing over on chromosome 17. Genetic and molecular analyses of this system indicate that the lethality and tailless phenotypes are separable from segregation distortion and from each other (Silver 1993). As in *SD*, + chromosomes carry a sensitive allele at a responder locus, *Tcr*, and distorting chromosomes carry an insensitive allele; the *Tcr* gene encodes a protein kinase (Herrmann et al. 1999). At least three loci contribute to segregation distortion (*Tcd-1, 2, 3*, two of which have now been characterized; see Bauer et al. 2007). In the *t* system, unlike *SD*, there is no evidence for wild-type chromosomes insensitive to distortion.

10.3.iii.d. *The evolution of segregation distorter systems*

A common feature of the best-studied segregation distortion systems just reviewed is their complex, multi-locus genetic basis, with close linkage between several components, i.e., they are supergenes (**Section 10.2.iii.b**). The conditions for the initial evolution of a system such as *SD* can be modeled by considering a starting population initially with the *Sd*$^+$ *Rsp*^{s} genotype. Both single mutations (*Sd* or *Rsp*^{i}) will be selected against, as explained in **Section 10.3.iii.c** above, so the evolution of such a system is puzzling. Suppose, however, that an *Sd Rsp*^{i} double mutant arises. **Box 10.4** shows that, for this to spread, the two loci must be closely linked, because otherwise the effectively lethal *Sd*$^+$ *Rsp*^{s} haplotype is produced. There is thus a linkage constraint on the evolution of systems like these, although it differs somewhat from that described for sex chromosomes in **Section 9.3.v.b** (Charlesworth and Hartl 1978). The special mutations necessary must be rare events, which may explain why segregation distorter polymorphisms rarely evolve.

Once polymorphisms have been established at both the *Sd* and *Rsp* loci, the extreme fitness epistasis between them implies that selection will favor reduced recombination (according to the principles outlined in **Section 10.2.ii** above) to avoid producing the selectively inferior haplotypes. This explain the observation, noted in **Section 10.3.iii.c**, that segregation distorter systems tend to be associated with inversions.

It is likely that a newly evolved distorter system will be inefficient (i.e., the *k* coefficients in **Box 10.4** will be much lower than 1). Theoretical analyses show that a mutation that enhances distortion can be favored by selection if it is

Box 10.4 CONDITIONS FOR THE INITIAL SPREAD
OF A TWO-LOCUS SEGREGATION DISTORTER

We consider the case when the initial frequency of the *Sd Rsp^i* haplotype
is very low. If the population is randomly mating, it will then largely be
carried in heterozygotes with wild-type, i.e., in the *Sd Rsp^i/Sd^+ Rsp^s* genotype. If the drive loci are on the same chromosome, recombination occurs
only through crossing over. In *Drosophila*, there is no crossing over in
males, whereas segregation distortion occurs exclusively in males.

Crossing over with frequency *c* in *Sd Rsp^i/Sd^+ Rsp^s* females causes them
to produce a fraction $(1 - c)/2$ of *Sd Rsp^i* eggs; these combine with *Sd^+ Rsp^s*
gametes to produce a fraction $(1 - c)/2$ of new *Sd Rsp^i/Sd^+ Rsp^s* individuals. In males, if the fertility disadvantage is *s* and the drive coefficient is *k*,
the *Sd Rsp^i/Sd^+ Rsp^s* genotype produces a fraction $k(1 - s)$ of new *Sd Rsp^i/
Sd^+Rsp^s* individuals. The net change in frequency of *Sd Rsp^i/Sd^+ Rsp^s* per
individual of this genotype is thus:

$$\frac{(1 - c)}{2} + k(1 - s) - 1 = k(1 - s) - \frac{c}{2} - \frac{1}{2} \qquad \text{(B10.4.1)}$$

and so *Sd Rsp^i* will increase in frequency if:

$$k(1 - s) > \frac{(1 + c)}{2} \qquad \text{(B10.4.2)}$$

It is clear from this result that there is no constraint imposed by the
level of crossing over when distortion is complete and there is no fertility loss to males carrying the distorter haplotype. However, as discussed
in the text, it is likely that distortion will initially be incomplete ($k < 1$),
in which case there is a critical value of *c* above which the distorter complex will fail to spread. If $k = 0.6$ and $s = 0$, for example, *c* must be less
than 0.2.

This constraint is more severe if recombination occurs in both sexes, as
in the case of the *t* haplotype system, or for *Drosophila* with the two loci
on separate chromosomes. In this case, recombination in males means
that only a fraction $(1 - c)k$ of the sperm produced by a *Sd Rsp^i/ Sd^+ Rsp^s*
male are *Sd Rsp^i*. Thus, **Equation B10.4.1** is replaced by:

$$(1 - c)\left[\frac{1}{2} + k(1 - s)\right] - 1 = (1 - c)k(1 - s) - \frac{(1 + c)}{2} \qquad \text{(B10.4.3)}$$

and *Sd Rsp^i* will increase if:

$$k(1 - s) > \frac{(1 + c)}{2(1 - c)} \qquad \text{(B10.4.4)}$$

Even with complete distortion and no fertility loss, it is impossible for a freely recombining pair of loci to invade, because the right-hand side is equal to 3/2 when $c = 1/2$.

closely linked to the primary distorter locus, because it receives a transmission advantage through being associated with the distorter locus, similar to the hitchhiking benefits to a recombination modifier (see **Section 10.2.ii**).

On the other hand, if a distorter system is kept polymorphic because of fitness disadvantages of the kind described in **Section 10.3.iii.b** above, loosely linked mutations suppressing distortion may be favored; these mutations will not be associated with the distorter locus, so that they gain no transmission advantage, and suppressing distortion gives their carriers higher than average mean fitness (Wu and Hammer 1991; Burt and Trivers 2006, Chapter 2). This is particularly important for sex-ratio distorters, because a highly distorted sex ratio leads to individuals of the minority sex having much higher average numbers of offspring per capita than the opposite sex (**Section 9.4.i**). Restoring a 1:1 sex ratio thus gives a large fitness gain. The widespread existence of both suppressors and enhancers of segregation distortion (**Section 10.3.iii.c**) is thus explicable. Selection to suppress drive may be a further reason for the rarity of observable drive systems (Eshel 1985; Crow 1991), in addition to the rarity of the mutations necessary for such systems to evolve. These considerations suggest the distorter systems arise rarely and may not persist for long.

A final question is why recessive lethal genes are associated with the *t*-haplotype system. One plausible explanation is suggested by the fact that, in species like mice, producing lethal embryos may not greatly reduce female fertility, because more embryos are implanted than can survive to birth. Consider a polymorphic distorter system that is stabilized by male sterility of homozygous *t* haplotypes. In such a situation, an early-acting lethal mutation linked to *t* in a +/*t* female can be advantageous, because *t/t* individuals eliminated early in development will cause almost no reduction in female fertility, and their loss enhances the transmission of *t* via the +/*t* siblings (Charlesworth 1994c). This is an extreme example of kin selection (**Section 3.1.v.d**). Other possibilities are discussed by Burt and Trivers (2006, Chapter 2).

10.3.iv. Transposable elements

10.3.iv.a. *Introduction*

Both prokaryote and eukaryote genomes contain repetitive DNA sequences that have the ability to make new copies of themselves (**Figure 10.13**). These can insert elsewhere in the genome, in contrast to the stable behavior of most genome features. These transposable elements or TEs were first discovered in maize using classical cytogenetic methods (McClintock 1950, 1984). The implications of these findings were first understood when the molecular details of some of the maize TEs were worked out, and similar entities were found in the genomes of bacteria and *Drosophila* (Shapiro 1983). TEs from a large variety of species have now been studied (Craig et al. 2002; Burt and Trivers 2006, Chapter 7; Lynch 2007, Chapter 7).

New insertions that occur in the germ line or gamete-forming cell lineages, as well as the insertions previously present, are transmitted to the progeny along with the host genome (*vertical transmission*), so that the number of TEs

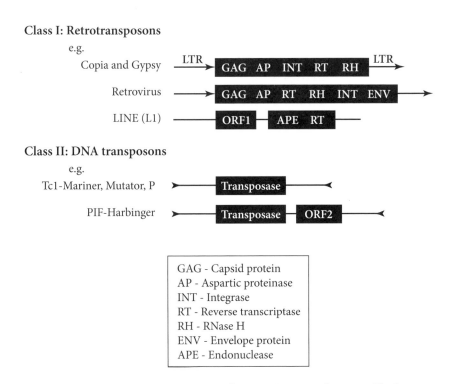

FIGURE 10.13 Diagrammatic representation of some major types of transposable elements (TEs). [Adapted from Figure 1 of Wicker et al. (2007).]

TABLE 10.1 Types of TEs and some important properties[1]

Replication method	Names	Categories	Examples in the human and *Drosophila* genomes
Reverse transcription of their mRNA into DNA, followed by re-insertion into the genome (copy and paste)	Retrotransposons (Class I elements)	Elements with Long Terminal Repeats (LTRs)	Endogenous retroviruses; MALR1
		Elements lacking LTRs	Long, autonomously replicating elements (LINEs); Short, non-autonomous elements (SINEs), which depend on the LINEs to provide enzymes to catalyze their replication
Direct replication, without RNA intermediates (e.g., cut-and-paste mechanisms)	DNA elements (transposons) (Class II elements)		Mariner, hAT group *Drosophila* P element

[1]For more details, see International Human Genome Sequencing Consortium (2001) and Wicker et al. (2007).

in genomes can increase across the generations. TEs are divided into different major classes (**Table 10.1** explains these and defines some of the terminology used in describing TEs). The TE classes are further subdivided into *families* according to their structures and DNA sequences. In addition, the mammalian retroviruses are a category of LTR retrotransposons that have acquired the ability to infect new host individuals (*horizontal transmission*).

Family members can be recognized because they share a common structure and gene content (**Figure 10.13**) and often have similar DNA sequences. Families may also include defective elements, lacking parts of the full sequence. Defective transposons usually cannot transpose unless intact elements are present elsewhere in the genome to provide transposase enzyme functions, and many defective retrotransposons are probably completely disabled (Burt and Trivers 2006, Chapter 7; Lynch 2007, Chapter 7). Defective elements may nevertheless reach high abundances in genomes. Most of the TE-derived component of the human genome is made up of defective elements (International Human Genome Sequencing Consortium 2001), and in *D. melanogaster* about

TABLE 10.2 TEs in the human and *D. melanogaster* sequenced euchromatic genomes[1]

Type	Number of copies in haploid genome	Number of families	% of the genome
Humans			
LTR elements	443×10^3	100	8
SINES	1558×10^3	3	13
LINES	868×10^3	3	20
DNA elements	294×10^3	60	3
D. melanogaster			
LTR elements	682	49	2.6
LINES	486	27	0.9
DNA elements	404	20	0.3

[1]For more details, see International Human Genome Sequencing Consortium (2001) and Kaminker et al. (2002).

65% of the approximately 1500 TEs in the sequenced genome are defective (Kaminker et al. 2002).

Complete genome sequences provide unprecedented details about the abundance and organization of TEs, albeit mostly for a single individual per species. **Table 10.2** shows the abundances of different classes of TEs in the human and *D. melanogaster* genomes. TEs or sequences derived from TEs represent substantial fractions of the hosts' DNA—in total, over 45% of the human genome (International Human Genome Sequencing Consortium 2001) and even more for plants like maize with very large genomes (SanMiguel et al. 1996). Differences in TE abundances play a major part in differences in genome sizes among eukaryotes, and TEs or TE-derived sequences make up only 5–10% of the genomes of organisms with small genomes compared with humans or maize, such as *Arabidopsis thaliana* and *D. melanogaster* (**Table 10.2**).

10.3.iv.b. *The evolution of TEs in genomes*
Understanding the forces responsible for the maintenance of TEs in genomes is critical for understanding genome organization and evolution. There are two major views about why TEs exist and the forces regulating their abundance. One view is that TEs persist in host genomes because they can benefit them by causing favorable mutations, thus aiding adaptive evolution (e.g., Nevers and Saedler 1977). This suffers from the difficulties already explained in accepting that evolution can work effectively through differences in the fitnesses of populations or species, discussed in **Section 7.3.iv.b.**

The other view of TEs is that they are genomic parasites, which maintain themselves by replicating within their host genomes (the *selfish DNA* hypothesis of Doolittle and Sapienza 1980 and Orgel and Crick 1980). This hypothesis is open to empirical tests and is therefore widely used as a working hypothesis by evolutionary geneticists interested in TEs. The selfish DNA hypothesis does not deny that TE insertions may occasionally induce favorable mutations and benefit the host, like any other kind of mutation. For example, in *Drosophila*, some insecticide-resistance mutations have been caused by the recent spread of TEs associated with increased expression of genes involved in detoxification pathways (Schlenke and Begun 2004; Chung et al. 2007). Moreover, a nearly neutral TE that becomes fixed in the population at a particular site in the genome, by genetic drift or by hitchhiking with a linked, advantageous mutation, may later evolve a role in regulating gene activity. There are many examples in which TE-derived sequences have been "domesticated," and fixed in the host genome, and now have a role in host functions (Brookfield 2003). But we shall see in **Section 10.3.iv.c** below that new TE insertions mostly reduce host fitness. This is a serious difficulty for the view that TEs generally benefit their hosts.

10.3.iv.c. *TE behavior in genomes and populations*

We shall focus on data from *D. melanogaster*, the species whose TEs have been best studied by population geneticists. As can be seen from **Table 10.2**, the sequenced genome contains over 90 different families of TEs. For a given TE family, individuals sampled from a population can be characterized in terms of *n*, the number of elements of each family. One way of doing this is to score the presence or absence of TE at a band in the polytene chromosomes of larval salivary glands by *in situ* hybridization of a chromosome preparation using a chemically labeled probe for the TE.

This has usually been done using individual chromosomes isolated from a population and made homozygous by means of the balancer chromosome procedure described in **Section 1.1.iii.b**. The number of TE insertions in a given chromosome is then estimated by counting every location where hybridization is observed (Charlesworth and Langley 1989). An alternative approach is based on PCR with primers for sequences flanking a site of an element insertion in the sequenced genome, or for a flanking site and a site internal to the TE. On amplifying DNA from samples from a population, a TE insertion at the site is indicated by a product of the expected length (Petrov et al. 2003; Bartolomé and Maside 2004; González et al. 2008). This technique gives information on the presence of TEs at precise locations, unlike the *in situ* method, and is less laborious. However, it is biased in favor of elements present at high frequencies (because rare ones are less likely to be found in the strain used for sequencing the genome).

We can summarize the state of the population for a given TE family by the mean and variance of copy number. Different TE families have different haploid copy numbers in the sequenced genome, ranging from two or three to over 100 (**Table 10.2** in **Section 10.13.iv.a**), but copy numbers usually vary between individuals in a natural population (Charlesworth and Langley 1989). A more complete description of the population includes the frequencies of TE insertions at each site in the genome (Charlesworth and Langley 1989; Charlesworth et al. 1994; Le Rouzic and Deceliere 2005), and we can describe the state at the ith site by q_i, the frequency at which the site is occupied by a TE. The mean diploid copy number for autosomes and for the X chromosome in females is given by:

$$\bar{n} = 2\sum_i q_i \qquad (10.4)$$

A given $\bar{n}$ value can be caused either by a few sites with high insertion frequencies or fixation, or by many sites with rare insertions (i.e., low q_i values).

Using both approaches for estimating TE frequencies described above, TE insertions in *D. melanogaster* are generally found at very low frequencies in populations (Charlesworth and Langley 1989; Charlesworth et al. 1994; Le Rouzic and Deceliere 2005; González et al. 2008), except in genome regions with suppressed crossing over (see **Section 10.3.iv.d** below), with some exceptions that may either represent selection for favorable mutations induced by TE insertions or the effect of nearby selective sweep (González et al. 2008).

What controls TE numbers? Clearly, only events that occur in the germ line matter. In a germ-line cell, n can increase when new copies arise by *transposition,* and decreases through *excision* of elements. The frequencies of transposition, and excision events have been estimated in *D. melanogaster* by experiments using mutation accumulation lines (like those described in **Section 1.3.iii.b**), in which many independent lines are started from an isogenic stock, so that all individuals initially have the same TE insertions in the same locations. After many generations, the locations of members of a set of element families can be determined by *in situ* hybridization, thus revealing how many new insertions and excisions have occurred.

One can then estimate the per generation rate of transposition, u (the probability that an element present at a given location in an individual produces a newly transposed element in its germ line), and of excision, v (the probability that an element initially present at a given location in an individual excises from this location in its germ line), just like the mutation rate estimation procedure. In *D. melanogaster*, the estimated mean u value over all element families is on the order of 10^{-4}, and v is less than one-tenth of this. Overall, approximately 0.2 new transposition events occur per individual per generation (Maside et al. 2000).

Unusually high transposition rates (u) sometimes occur due to failure to repress transposition. In the progeny of crosses between *D. melanogaster* females that lack *P* elements and males carrying *P* elements, elements move at very high rates (Rio 2002), causing a set of effects called *hybrid dysgenesis*. The *hobo* and *I* element families behave similarly (Blackman 1987; Bucheton et al. 2002). In addition, transposition rates of some TE families, such as the *copia* and *gypsy* elements, vary considerably among inbred lines, showing that host genomes vary genetically for factors that regulate element movement (Nuzhdin et al. 1998; Mevel-Ninio et al. 2007).

There is good evidence that eukaryote genomes have evolved mechanisms to suppress TE movement, including the methylation of TE sequences (Zilberman and Henikoff 2004), silencing by small interfering RNAs in plants and *Drosophila* (Lippman and Martienssen 2004; Aravin et al. 2007), and specialized systems (such as Repeat Induced Point-mutation, RIP, in the fungus *Neurospora* and other fungi; see Galagan and Selker 2004) that detect and destroy elements. In *Drosophila*, recent work shows that small transposon-derived RNA molecules, processed by the host's PiWi-piRNA pathway, regulate TE activity (Aravin et al. 2007) and may play a key role in hybrid dysgenesis (Brennecke et al. 2009). Genetic variation in the proteins involved in this pathway, as well as in the copy number of the degenerate elements that provide the source of the TE-derived small RNAs (Aravin et al. 2007; Mevel-Ninio et al. 2007), may contribute to host-controlled genetic variation in TE activity. The existence of genetic variation in host control mechanisms suggests the possibility of "arms races" between hosts and their intragenomic parasites similar to those for other parasites (**Section 6.2.iv**). Elements are selected to resist host-repression mechanisms, and hosts are selected to overcome this resistance (Charlesworth and Langley 1986). A possible example of such coevolution has recently been described (Sawyer and Malik 2006).

10.3.iv.d. *Population dynamics of TEs*

The behavior of TEs in a population of host organisms depends strongly on whether the population is sexual or asexual. With asexuality, a transposition event affects just the clone descended from the individual in which it occurred, and clones not carrying TEs cannot acquire them; thus, there cannot be a deterministic increase in the frequency in the population with which TEs are carried by clones (Hickey 1982). This is consistent with the dearth of elements in the non-recombining, maternally transmitted mitochondrial genome of animals (Hickey 1982).

In a sexually reproducing population, however, transpositions and excisions in the germ lines of either mothers or fathers are transmissible to their offspring, so that an individual without an insertion can produce progeny that carry it. Hence, the number of TE copies per individual in the population can

Box 10.5 DYNAMICS OF A TE FAMILY IN THE ABSENCE OF SELECTION

The state of the population can be summarized by the mean and variance of copy number, $\bar{n}$ and V_n, respectively. If elements are scattered over the genome, neighboring insertions will mostly be loosely linked, and we can assume that TEs at polymorphic sites are in linkage equilibrium. In the absence of transposition and excision, $\bar{n}$ and V_n will then remain unchanged over the generations.

The effects of element mobility on $\bar{n}$ can be described as follows. In some cases, transposition rates are negatively regulated by copy number, so that in general we should write the transposition rate for an individual with n copies as a decreasing function of copy number, u_n, giving a net increase in copy number of nu_n due to transposition. Even when there is no regulation, if there is a finite number of sites into which elements can insert, the effective rate of transposition will decline as sites fill up with TEs, so that u_n is again a decreasing function of n. The addition to the mean copy number in the next generation from new transpositions is $E\{nu_n\}$, the expectation of nu_n over all individuals in the population. Provided the dependence of u on n is not too strong, this can be approximated by $nu_{\bar{n}}$ (**Problem 10.3***).

Similarly, the reduction in mean copy number due to excision is $\bar{n}v$. The contribution of element movement to the change in copy number per generation is thus:

$$\Delta \bar{n}_M = E\{nu_n\} - v\bar{n} \approx \bar{n}\left(u_{\bar{n}} - v\right) \qquad \text{(B10.5.1)}$$

This equation shows that an element family can invade an infinite population only if the transposition rate for low copy numbers exceeds the rate of excision; its initial rate of change away from a copy number of 1 is approximately $\bar{n}(u_1 - v)$.

Copy number will cease to change when the right-hand side of **Equation B10.5.1** is equal to 0. This will happen either when all occupiable sites are filled up with elements, or self-regulation reduces the overall rate of production of new copies by transposition to the rate of their removal by excision. Given the fact that, for most element families, there are many thousands of occupiable sites in the genome, this suggests that element numbers for a newly arisen TE family will increase for very long periods of time in the absence of self-regulation of transposition or of other forces that oppose their spread.

change. If all elements of a given family have the same transposition and excision rates, u and v, respectively, the mean number of copies per individual in a large panmictic population changes by an amount $\bar{n}(u - v)$ each generation. **Box 10.5** gives a more general description of the dynamics of TEs, allowing transposition and excision rates to differ in individuals with different copy numbers, but assuming that no opposing forces operate to prevent high element numbers. The theory shows that elements will then spread and be maintained in the population if the rate of transposition for individuals with low copy numbers exceeds the excision rate.

In *D. melanogaster*, u is much greater than v (**Section 10.3.iv.c**). The theory in **Box 10.5** thus predicts that copy numbers in this species should increase, and all sites in the genome that are capable of being occupied by TEs should eventually be filled up, if no opposing forces operate. Although the estimated rate of increase of TEs by transposition in *D. melanogaster* seems low (about one new element per individual every five generations), copy number should increase rapidly over evolutionary time scales; for example, a 10-fold increase would take only 23,000 *Drosophila* generations, i.e., less than 2000 years. In fact, however, copy numbers in organisms like *Drosophila* are quite low; the most abundant family still active in *D. melanogaster* is the retrotransposon *roo*, with about 60 full-length copies per haploid genome in the euchromatin and many sites occupied at low frequencies (Kaminker et al. 2002). This supports the hypothesis that copy numbers are regulated.

The predicted potential for rapid increase can readily explain large differences in TE abundances among related species, if the pressures restricting their spread vary; *D. simulans*, for example, seems to have about three times fewer TEs than its close relative, *D. melanogaster* (*Drosophila* 12 Genomes Consortium 2007). Sequence comparisons of members of several TE families in *D. melanogaster* suggest that they have increased in abundance quite recently, possibly after horizontal transfer from a related species (Bergman and Bensasson 2007; Bartolomé et al. 2009). The operation of forces keeping TE numbers in check is further supported by the very low frequencies of insertions in samples from natural populations of *D. melanogaster* (see above).

Overall, therefore, the *Drosophila* evidence strongly suggests that the increase in element frequencies expected from the excess of transposition over excision is resisted by some force or forces. Clearly, elements do not often cause beneficial mutations. **Box 10.6*** describes models of TE behavior when the fitness of host individuals decreases with increasing number of copies of a TE family. This form of selection can stabilize copy numbers, provided that the fitness effect increases with the number of elements in the genome (because the rate of addition of new elements increases in proportion to the number of elements that are already present, a greater-than-linear dependence of the fitness effect on copy number is required to counteract the additions). Stabilizing ele-

Box 10.6* DYNAMICS OF TES WITH SELECTION

We will consider a simple model in which fitness depends only on copy number. We can then write the fitness of an individual with n copies as w_n, which we assume to be a decreasing function of n. This is a caricature of reality, because there is considerable heterogeneity among genomic sites with respect to the fitness effects of element insertions. However, selection will rapidly eliminate elements that insert into sites where they severely reduce fitness, such as coding sequences. Only sites where insertions have very small deleterious effects, such as sites in intergenic or intron sequences, will contribute much to the TE content of a genome.

With this model of selection, we can use the theory of selection on additive quantitative traits (**Section 3.3.iii.a**) to obtain the following expression for the effect of selection on the mean copy number:

$$\Delta \bar{n}_s = V_n \frac{\partial \ln \bar{w}}{\partial \bar{n}} \approx V_n \left(\frac{\partial \ln w_n}{\partial n} \right)_{\bar{n}} \qquad \text{(B10.6.1)}$$

To use this equation, we need an explicit expression for V_n. If element frequencies at all polymorphic sites are low, as in the case of *D. melanogaster* populations (see main text), we can apply the argument used for the number of deleterious mutations in a genome (**Section 4.2.ii.c**) to establish that the distribution of the number of elements among individuals follows a Poisson distribution. (This is not quite accurate, because it ignores linkage disequilibrium induced by selection—see **Section 8.4.iv.a**. A correction for this is given by Charlesworth and Barton 1996.)

With a Poisson distribution, $V_n = \bar{n}$. Adding **Equation B10.5.1** to the approximate version of **Equation B10.6.1**, and assuming for convenience that there is no dependence of transposition rate on copy number, we obtain the following expression for the net change in mean copy number:

$$\Delta \bar{n} \approx \bar{n} \left[\left(\frac{\partial \ln w_n}{\partial n} \right)_{\bar{n}} + u - v \right] \qquad \text{(B10.6.2)}$$

At equilibrium, we thus have:

$$-\left(\frac{\partial \ln w_n}{\partial n} \right)_{\bar{n}} \approx u - v \qquad \text{(B10.6.3)}$$

The left-hand side of this expression measures the net proportional reduction in fitness caused by adding an additional element to the genome in individuals with the mean copy number, and so we obtain the intuitively satisfying result that the spread of elements is checked when this is equal to the net rate at which an element generates progeny copies.

Given a functional relation between fitness and copy number, this relation yields an explicit formula for $\bar{n}$. For example, let $\ln(w_n)$ be a quadratic function of n, such that $\ln(w_n) = -(\alpha n + \beta n^2)$. (This is the simplest form that allows an explicit solution to **Equation B10.6.3**.) We then have:

$$\bar{n} \approx \frac{(u - v - \alpha)}{2\beta} \tag{B10.6.4}$$

Importantly, examination of **Equation B10.6.2** shows that stability requires the derivative of $-\ln(w_n)$ to be smaller than $u - v$ when n is below $\bar{n}$, and larger when n is above $\bar{n}$. The derivative of $\ln(w_n)$ must thus be a decreasing function of n in the neighborhood of equilibrium, i.e., a necessary condition for stability is:

$$\left(\frac{\partial^2 \ln w_n}{\partial n^2} \right)_{\bar{n}} < 0 \tag{B10.6.5}$$

This is satisfied by the quadratic function used above.

ment abundance by selection thus requires synergistic epistasis in the fitness effects of TE number, similar to that proposed for the effects of deleterious mutations in **Section 10.2.iv.d** above.

10.3.iv.e. *Discriminating among possible regulators of element abundance* **Table 10.3** lists five kinds of processes that may be involved in regulating element abundance. These mechanisms are not mutually exclusive, and it is hard to discriminate among them. However, the following points seem clear. First, self-regulated transposition alone (model 1) cannot fully explain the data. While some non-LTR retrotransposons and DNA-based transposons have mechanisms for regulating their transposition rate (**Section 10.3.iv.c** above), there is no evidence for such a process for retrotransposons with LTRs, the majority TE component of many genomes, including *Drosophila*. These have been well studied in the population surveys mentioned in **Section 10.3.iv.c**, and the accumulation experiments described in that section indicate that u for genomes with copy numbers typical of natural populations is always much

TABLE 10.3 Processes that may be involved in regulating TE abundance

	Process	Action	Evidence and references
1a	Self-regulation of transposition rates	Decrease in u with the number of elements in the genome. At high copy number, the mean value of u may become equal to v, and copy number will stabilize (Box 10.5).	Self-regulation of the rate of transposition is known to occur with certain eukaryotic transposons, such as the P elements of *Drosophila*, which produce repressor proteins that inhibit transposition, and many types of bacterial transposons (Craig et al. 2002).
1b	Host regulation of transposition rates	Hosts respond to increase in element abundance by intensifying the activity of repressor mechanisms.	No direct evidence
2	Selection against mutations	Insertions into coding sequence of genes, or into regulatory sequences	Most element insertions detected in genome sequences (e.g., International Human Genome Sequencing Consortium 2001; Kaminker et al. 2002) or in population surveys (Charlesworth et al. 1994) are in intergenic or intron regions. In contrast, many laboratory mutations are due to insertions of this kind, indicating that TEs can jump into these regions of the genome.
3	Ectopic exchange	Selection against deleterious chromosome rearrangements induced by recombination between homologous elements located in different places	Observed in the laboratory in yeast, *Drosophila*, and in some human genetic diseases (Montgomery et al. 1991; Charlesworth et al. 1994; Stankiewicz and Lupski 2002).
4	Direct negative fitness effects of transposition on host fitness	Induction of chromosome breaks during transposition	Observed for P elements (Rio 2002); no firm evidence for such effects in non-dysgenic crosses or for retrotransposons
5	Indirect effects of copy number on fitness	High copy number reduces the rate of cell division by requiring more time and energy for DNA replication, which prolongs development.	Comparative data showing correlations between life histories with slow development times and large amounts of repetitive DNA in the genome (Charlesworth et al. 1994; Burt and Trivers 2006, Chapter 7)

larger than v. Under the self-regulation model, however, they should be approximately the same, because the accumulation lines have copy numbers similar to those for randomly chosen natural genomes.

Selection against mutations caused when TEs insert (Model 2) is also inadequate as the sole means of regulating element abundance, although, as mentioned already, many insertions are probably removed from populations by this mechanism. The disruption of gene function by insertion of a TE into a coding sequence is severe, and is likely to be associated with a significant fitness effect even in heterozygotes (see **Section 4.2.ii.a**). This is confirmed by direct measurements of the effects of insertions in *D. melanogaster*, which suggest a mean heterozygous fitness effect of a few percent (e.g., Mackay et al. 1992). Insertions into noncoding sequences are likely to have much smaller or even zero mutational effects on fitness, consistent with the fact that most TEs are found in noncoding sequences (Bartolomé et al. 2002; Kaminker et al. 2002; Fontanillas et al. 2007).

At equilibrium between selection and transposition, the probability per generation that a new insertion is removed from the population by selection must obviously equal $u - v$ (about 10^{-4} in *Drosophila*); elements will quickly be lost from sites where selection is stronger than this, such as coding sequences, but may be able to spread at sites where elements are selected less strongly. This mechanism thus has great difficulty in explaining the almost universally low population frequencies of elements at euchromatic sites outside coding sequences in organisms with low copy numbers, such as *Drosophila*.

The ectopic exchange (model 3) is free from these difficulties (**Figure 10.14**). On a simple model of exchanges arising from random encounters between members of the same element family, the probability that elements of the same family located in different genomic locations engage in an ectopic exchange event is approximately proportional to the square of their abundance in the genome, independent of whether they are in genic or intergenic sequences. Thus, as the abundance of an element rises in the genome, the frequency of ectopic exchanges rises disproportionately (Langley et al. 1988). This is exactly what is required to stabilize element numbers, as shown by the model in **Box 10.6***. It is also consistent with the fact that element densities (per megabase of genome) tend to be highest in euchromatic genome regions that have low crossover rates (Charlesworth et al. 1994; Bartolomé et al. 2002; Fontanillas et al. 2007), and elements in these regions are also found at higher frequencies in populations (Bachtrog 2003; Bartolomé and Maside 2004); the same applies to heterochromatin, which lacks crossing over (Maside et al. 2005; Bergman et al. 2006). However, these effects can also be explained by other models, particularly by a reduced efficacy of selection in low recombination regions (**Section 10.2.vi.a**).

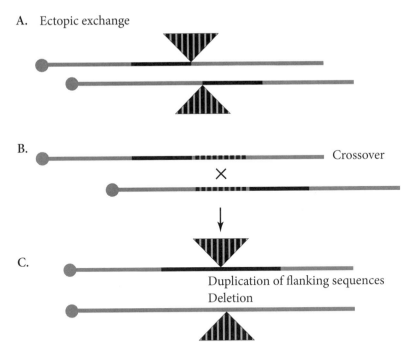

A. Ectopic exchange

B.   Crossover

C.   Duplication of flanking sequences
                                Deletion

FIGURE 10.14 Ectopic exchange between similar sequences, such as TEs, in different genome locations (indicated in part **A** by triangles at the insertion locations of TEs and in part **B** by striped lines) generates duplications and deletions of nearby regions (indicated by black lines). **C.** The products of an ectopic exchange event.

There is no evidence for direct fitness effects of transposition (model 4) for most TEs, but model 5 (indirect fitness effects) is supported by comparative data that show that TE abundances tend to be highest in species that are large and relatively slowly developing (Burt and Trivers 2006, Chapter 7). Many of these species are flowering plants, in which TEs make up a large fraction of the repetitive portion of the genome, suggesting that slow development time is permissive for the accumulation of TEs. There is also evidence from the distribution of TEs among long and short noncoding sequences in *D. melanogaster* that regions of the genome that are apparently under selection for compactness contain less TE-derived sequence than other regions (Fontanillas et al. 2007). Both models 4 and 5 are capable of explaining the concentration of TEs in intergenic, rather than coding, sequences, but do not obviously lead to synergistic fitness effects, in contrast to model 3, so it is hard to see how these processes could stabilize element frequencies at low levels, as seen in *Drosophila*.

10.3.iv.f. *Genomes and genomic regions with high TE contents*

If ectopic exchange is an important regulator of TE abundance, there is a difficulty in understanding how elements can ever become very abundant in host genomes or reach high frequencies. The presence of hundreds of thousands of copies of a family within a genome would be expected to generate a huge rate of ectopic exchange, creating an intolerable load of deleterious chromosomal rearrangements. Clearly, this does not happen in species with high TE copy numbers. In the human genome, for instance, the vast majority of TEs at individual genomic sites are fixed (International Human Genome Sequencing Consortium 2001; Deininger et al. 2003).

A possible resolution of this difficulty is to note that ectopic exchange is unlikely to involve sites with homozygous elements, because the two homologous copies at a site can pair with each other; there is experimental evidence for this (Montgomery et al. 1991). The results described in **Section 6.2.iii.a** imply that, if the effective population size is low relative to the intensity of selection against an insertion, an element initially inserted into a particular site in a single individual can spread to a high frequency or fixation. Once all individuals are homozygous for the TE at this site, it will be almost immune to ectopic exchange, and this form of selection will cease to act on it. In the absence of other factors regulating the abundance of TEs, copy numbers could thus grow indefinitely, unless checked by forces such as those postulated in model 5.

Another factor that may counter the tendency for indefinite increase in TE numbers is inactivation by deleterious mutations of elements that have remained in the same site for a long time. Up to now our models have assumed that all members of a TE family in a host individual transpose at similar rates. However, once a TE becomes inserted into the genome, it is vulnerable to inactivating mutations. Selection against mutations that reduce a TE's ability to transpose acts only during transposition. If excision rates are low, elements fixed by drift will therefore gradually become defective and, possibly even eliminated from the genome, as a result of the accumulation of deletions (Petrov et al. 1996). Members of the *melanogaster* sub-group of *Drosophila* species indeed possess some old-established, defective, and fixed elements, such as *DINE-1*, distributed throughout the genome (Kaminker et al. 2002; Yang et al. 2006).

These considerations suggest that species and genome regions with low effective population sizes should tend to have high TE abundances, with many sites at which elements reach high frequencies or fixation; many of these elements will have mutations that impede transposition. This is what is seen for most families of TEs in the human genome. Similarly, low recombination regions of the *D. melanogaster* genome, such as the boundary between the euchromatin and the centromeric heterochromatin, not only have higher TE densities than regions with normal recombination frequencies, but TE insertions are often at high frequencies or near fixation, and many of these elements are defective, unlike those found at low frequencies (Bartolomé and Maside 2004; Maside et al. 2005). The build-up of old insertions at the euchromatin–heterochromatin boundary is so extreme that often elements are inserted into other, older elements (Di Nocera and Dawid 1983; Maside et al. 2005; Bergman et al. 2006).

Some species, however, have high TE abundances throughout their genomes, even though they do not have very low effective population sizes. Maize (and, to a lesser extent, its closest relatives) is a striking example (San-Miguel et al. 1996). There seems to have been a very large expansion of TEs in the maize lineage since its divergence from its common ancestor with its relative *Sorghum* (Meyers et al. 2001). Its causes are unknown. There is still much to learn about the factors controlling TE abundances in genomes.

PROBLEMS

10.1   Derive **Equation B10.1.2a** (**Box 10.1** of **Section 10.1.ii.d**) for the effect of biased gene conversion on the frequency of a GC variant in a randomly mating population, by using **Equation B10.1.1**.

10.2   Use the result in **Section 10.3.iii.b** (**Box 10.3**) to derive **Equation B10.3.3** for the equilibrium frequency of a segregation distorter that causes male sterility when homozygous. Note that it helps to write $2k(1 - s)$ as $2k(1 - s) - 1 + 1$.

10.3*   Use a Taylor's series expansion around $\bar{n}$ to show that $\mathrm{E}\{nu_n\}$ in **Section 10.3.iv.d** (**Box 10.5**) can be approximated by $nu_{\bar{n}}$.

Appendix

A1. BASIC MATHEMATICS

A1.i. Differentiation

There are many situations in which we need to know how fast $f(x)$, a function of a variable x, changes as x changes. For instance, this can be used to determine the maximum point of a curve, when the curve stops increasing and starts decreasing again, i.e., it passes through a point of zero change. The rate of change is described by the *differential coefficient* or *derivative*, $df(x)/dx$. In geometrical terms, this corresponds to the tangent of the curve produced by graphing $f(x)$ against x, i.e., the slope of the straight line that just touches $f(x)$ at a point x (**Figure A1.1**). It is defined more formally as the limit of the difference between the value of the function at $x + \delta x$, (where δx is a small increment) and its value at x, divided by δx, as δx approaches 0. In other words, $df(x)/dx$ is the limit of $[f(x + \delta x) - f(x)]/\delta x$ as δx is made arbitrarily small. Expressions for the derivatives of a large number of functions can be found in standard texts on calculus and in mathematical software packages such as *Mathematica*. In particular, for an arbitrary power of x, x^i, we have $dx^i/dx = ix^{i-1}$. In addition, the derivative of the product of two functions, $f(x) = g(x)h(x)$, is given by $g(dh/dx)$

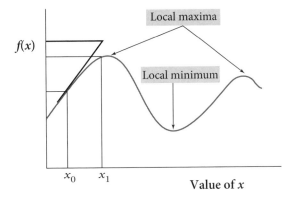

FIGURE   A.1.1. The graph of a function $f(x)$ of a variable x, showing the tangent to the curve representing $f(x)$ at the point x_0; this is the straight line that touches the curve at $f(x_0)$. When $x_1 - x_0$ is small, the difference between the values of $f(x)$ for $x = x_1$ and $x = x_0$ is approximated by the corresponding difference between the values given by the tangent line at x_1 and x_0. The graph also allows one to see that the local maxima and minima correspond to points where the tangents to the curve are horizontal.

+ $h(dg/dx)$. Another useful result is the *chain rule*: the derivative with respect to x of a function $f(y)$ of a variable y that is a function of x is given by $df(x)/dx =$ $(df/dy)(dy/dx)$.

The concept of a derivative can be extended to functions of more than one variable, the simplest case being a function of two variables, x and y. The *partial derivative* of $f(x, y)$ with respect to x is the derivative obtained by holding y constant and differentiating with respect to x. It is denoted by $\partial f(x, y)/\partial x$.

Higher-order derivatives can be obtained by successive rounds of differentiation, e.g., by differentiating $df(x)/dx$, we obtain the *second derivative* $d^2f(x)/dx^2$. The nth derivative is obtained by n successive differentiations and is denoted by $d^nf(x)/dx^n$. A similar notation can be applied to partial derivatives: e.g., the partial derivative of $\partial f(x, y)/\partial x$ with respect to y is denoted by $\partial^2 f(x, y)/\partial x \partial y$. The condition for a point x_0 with $df(x)/dx = 0$ to be a local maximum of $f(x)$ is that a change to a slightly different x value causes $f(x)$ to decrease, so that $df(x)/dx$ must be negative for $x > x_0$ and positive for $x < x_0$, implying that $d^2f(x)/dx^2$ is negative at x_0. For $f(x)$ to be at a minimum at x_0, $d^2f(x)/dx^2$ must be positive.

A1.ii. Representing functions by series

A1.ii.a. *Taylor's series*

Many useful mathematical functions can be expressed as infinite series of terms involving successive powers of the variable, x, on which they depend. We used this in **Box 2.4*** of **Section 2.1.ii.c** to show that the equilibrium state with heterozygote advantage is a stable equilibrium, by considering the effects of slight perturbations of the allele frequencies from the equilibrium and by asking whether the system changes back towards it. It was also used in **Box 2.6*** of **Section 2.2.i.b**. This approach is based on a remarkable theorem, which states that a suitably "well-behaved" function (e.g., a continuous function) at a point $x + z$ can be written as a *Taylor's series*:

$$f(x + z) = f(z) + x\frac{df(z)}{dz} + \frac{x^2}{2}\frac{d^2 f(z)}{dz^2} + \frac{x^3}{3!}\frac{d^3 f(z)}{dz^3} + \ldots \quad (A1.1)$$

This makes sense in intuitive terms, because we expect to be able to find the value of a function close to a given point by going a small distance away, and adding the quantity indicated by the slope of the function at the given point (its first derivative). This will be incorrect for larger distances, unless the function is a straight line, so we expect to need further "corrections" to take into account its curvature, i.e., derivatives higher than the first (the derivatives are all evaluated at z). In the ith term in the series, the quantity by which the derivative is multiplied has the form $x^i/i!$, in which $i!$ (*factorial i*) is $i(i - 1)(i - 2) \ldots 1$. If the ith derivative is 0, the series terminates at this point (e.g., if the curve is a straight line, only the first derivative is needed). If z is set equal to 0, the series represents the value of the function at x, $f(x)$.

A1.ii.b. *Approximate numerical solutions of equations*

One use of this method is that, for small absolute values of x, $|x|$, the series can be well approximated by its first two or three terms, because the higher powers of x are small relative to the lower powers. This can often allow one to obtain a numerical solution to an equation for which there is no simple algebraic solution. Suppose we have a function $f(x)$ and we wish to know the value of x, x^*, that satisfies the equation $f(x^*) = 0$. We can use a trial value of x, x_0, which yields a value of the function $f(x_0) \neq 0$, where x_0 is equivalent to z in **Equation A.1.1.1** and x^* is equivalent to $x + z$. Using Taylor's series, we have:

$$f(x^*) = 0 \approx f(x_0) + \left(x^* - x_0\right)\left(\frac{df}{dx}\right)_{x_0} \tag{A1.2}$$

where the subscript for the derivative indicates that it is evaluated at $x = x_0$. We can thus obtain an approximation to x^*, x_1, by rearranging this expression:

$$x_1 = x_0 - \frac{f(x_0)}{(df/dx)_{x_0}} \tag{A1.3}$$

If we replace x_0 by x_1 on the right-hand side of this expression, we should obtain a better approximation, x_2. This process can be repeated until the difference between successive values becomes negligibly small, yielding a final value x_n that approximates x^*. This is known as *Newton's method*, or *Newton–Raphson iteration*.

A1.ii.c. *Some useful series*

The *exponential function*, exp (x), is one of the most commonly used infinite series, and has the following form:

$$\exp(x) = 1 + x + \frac{x^2}{2} + \frac{x^3}{3!} + \cdots \tag{A1.4}$$

where the $(i + 1)$th term in the series is $x^i/i!$. (It is sometimes convenient to omit the brackets around the argument x, so that the function is written as exp x.)

This series always *converges* (i.e., it approaches a limiting value as the number of terms increases indefinitely), but it cannot be represented by an algebraic formula. Its utility arises from the fact that its differential coefficient with respect to x is equal to exp (x); this can be verified by differentiating it one term at a time, using the formula for the differential coefficient of a power. Similarly, the differential coefficient of exp(ax) with respect to x, where a is an arbitrary constant, is equal to $a \exp(ax)$. Also, $\exp(x + y) = \exp(x) \exp(y)$. This justifies the commonly used notation $\exp(x) = e^x$, where the constant e is the value of $\exp(x)$ for $x = 1$ (approximately equal to 2.718). For small values of x, $\exp(x)$ is approximately $1 + x$; hence, a product of the form $(1 + x_1)(1 + x_2) \ldots (1 + x_n)$ is well approximated by $\exp(x_1 + x_2 + \cdots x_n)$ when each of the x_i is small.

Another function for which a power series representation is useful is the *natural logarithm* of x, $\ln(x)$. This is defined by the relation $\exp \ln(x) = x$. When the absolute value of x, $|x|$, is less than 1, we can use the fact that $d \ln(x)/dx = 1/x$, and substitute this and the higher derivatives of $\ln(x)$ into the Taylor's series for $\ln(1 + x)$. This gives the following:

$$\ln(1 + x) = x - \frac{x^2}{2} + \frac{x^3}{3} - \frac{x^4}{4} + \cdots \qquad (A1.5)$$

The ith term is $(-1)^{i-1}x^i/i$. For $|x| \ll 1$, we obtain the useful approximation $\ln(1 + x) \approx x$ by neglecting the second and higher powers of x. This relation implies that a small change in x, δx, divided by x is approximately equal to the corresponding change in $\ln(x)$, $\delta \ln(x)$.

Another important series is the *geometric series*. For small values of x, when $|x| < 1$, we can represent $1/(1 - x)$ by:

$$\frac{1}{(1 - x)} = 1 + x + x^2 + x^3 + \cdots \qquad (A1.6)$$

Here the ith term is x^i. If $|x| \ll 1$, then $1/(1 - x) \approx 1 + x$.

A1.iii. Integration

The *indefinite integral* of a function $f(x)$ is defined as the function, $g(x)$, whose derivative is equal to $f(x)$. The indefinite integrals of some familiar functions are as follows:

$f(x)$	$g(x)$
$x^i \ (i \neq -1)$	$x^{i+1}/(i + 1)$
$1/x$	$\ln(x)$
$\exp(ax)$	$[\exp(ax)]/a$

Because the derivative of a constant is 0, a given $f(x)$ corresponds to an infinite number of $g(x)$ functions; by convention, the simplest form of expression is chosen for $g(x)$, and the integral is often written as $g(x) + C$, where C is called a *constant of integration*. For example, the indefinite integral of $1/x$ is written as $\ln(x) + C$.

In general, we write the following:

$$g(x) = \int f(x)dx \qquad (A1.7)$$

The reason for this notation is that the integral can be related to a summation. Consider values of x in an interval $x_0 \leq x \leq x_1$. The *definite integral* measures the area under the curve represented by $f(x)$ between x_0 and x_1. We can divide this interval into n sub-intervals of width $\Delta x = (x_1 - x_0)/n$ (**Figure A1.2**). Use the notation $f(x_i)$ to denote the value of the function in the middle of the ith sub-interval. The area for the ith sub-interval is approximately the area of the rectangle with height $f(x_i)$ and width Δx. Summing over these areas [i.e., summing up $f(x_i)\Delta x$ over the interval] gives the approximate area under the curve. By taking the limit as n is increased indefinitely, we obtain the area, denoted by:

$$\int_{x_0}^{x_1} f(x)dx \tag{A1.8}$$

The connection with the indefinite integral is that this expression can be shown to be equivalent to $g(x_1) - g(x_0)$; an alternative way of looking at this result is that the derivative of **Equation A1.8** with respect to x_1 (keeping x_0 constant) is equal to $f(x_1)$.

A1.iv. Vectors and matrices

A1.iv.a. *Some definitions*

A *vector x* is a set of n numbers, x_i ($i = 1, 2, \ldots, n$), each of which represents the value of a single component of an n-dimensional system—for instance, the probability that an allele sampled from a subdivided population is present in

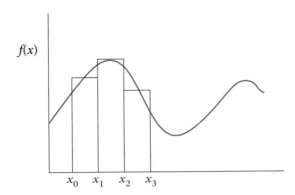

FIGURE A.1.2. This shows how the area under the curve of a function $f(x)$ between the points x_0 and x_3 is approximated by the sum of the areas of the rectangles whose heights are given by the values of $f(x)$ at $x_0 + (x_1 - x_0)/2$, $x_1 + (x_2 - x_1)/2$, and $x_2 + (x_3 - x_2)/2$. By increasing the number of sub-intervals between x_0 and x_3 (i.e. decreasing the widths of the rectangles), the approximation becomes more accurate.

deme i, as in **Box 7.3*** of **Section 7.2.ii.d**. A very useful shorthand is to write $x = (x_1, x_2, \ldots, x_n)$. In many applications, the state of the system after the lapse of one unit of time, x', will be a linear function of its current state. We can then write a_{ij} for the probability that state j gives rise to state i. (An alternative description of the system is in terms of the probability that state i is derived from state j; this is what is used in **Box 7.3***.) The overall new probability of being in state i depends on the transitions from all possible current states, so that $x_i' = \Sigma_j a_{ij} x_j$. The set of values a_{ij} can be represented by a *matrix* of numbers, $A = \{a_{ij}\}$, with the index i referring to the ith row of the matrix and j to its jth column. For example, we might have a two-dimensional matrix of transition probabilities:

$$A = \begin{pmatrix} a_{11} & a_{12} \\ a_{21} & a_{22} \end{pmatrix}$$

In this case, the A matrix is *square*, i.e., it has n rows and n columns.

The recursion relation for the x' values (i.e., the new x values after one application of the transition probabilities) can be conveniently represented by the following shorthand notation:

$$x' = Ax \qquad\qquad (A1.9)$$

The vector x can also be regarded as a matrix with n rows and one column (it is a *column vector*).

In this example, the matrix A is said to be *post-multiplied* by the vector x. To get the correct value for each x_i', the rule for multiplication involves going across the ith row of the matrix and down the column vector, adding up the products of the corresponding elements $a_{ij} x_j$. This rule can be generalized to define the product C when a matrix A is multiplied by another matrix B, i.e., $C = AB$, where $c_{ij} = \Sigma_k a_{ik} b_{kj}$.

We could equally well describe the system by a *row vector*, y. In this case, we would represent the probability that j gives rise to i by b_{ji}, so that $y_i' = \Sigma_j y_j b_{ji}$. If we use the same rule for multiplication as above (move across the rows and down the columns), then we must *pre-multiply* the matrix $B = \{b_{ji}\}$ by y:

$$y' = yB \qquad\qquad (A1.10)$$

In this example, the matrices A and B are said to be *stochastic*, meaning that they refer to probabilities of transitions between states. Because state j has probability one of giving rise to some other state (including itself), we must have $\Sigma_i a_{ij} = 1$; similarly $\Sigma_i b_{ji} = 1$.

A matrix that has no effect when it multiplies another vector or matrix is the *unit matrix*, I. According to the multiplication rule, the unit matrix has diagonal elements of 1 and all other elements of 0, so that $IA = A$ and $BI = B$.

A1.iv.b. *The inverse of a matrix and the solution of linear equations*
The concept of the unit matrix leads to the definition of the *inverse* of a matrix, which is useful in the following context. Suppose that we have a set of linear equations, of the form $z_i = \sum_j a_{ij} x_j$, where the z_i might represent a set of measurements and the x_j are the values of unknown variables that determine the z_i. We then wish to determine the x_j values. The entire set of equations can be represented by the following matrix relation:

$$z = Ax \qquad (A1.11)$$

To determine the values of the vector x, we define the inverse of A as A^{-1}, such that:

$$A^{-1}A = AA^{-1} = I \qquad (A1.12)$$

We then have:

$$A^{-1}z = A^{-1}Ax = Ix = x \qquad (A1.13)$$

If we can determine A^{-1}, we can find x.

There is an explicit formula for A^{-1}. It requires the use of the matrix's *determinant*. The determinant of a matrix with n columns is the sum of all possible products of the elements of the matrix of the form $\pm a_{1\alpha} a_{2\beta} a_{3\gamma} \cdots a_{n\omega}$, where each of the column subscripts $\alpha, \beta, \ldots, \omega$ is a permutation of the numbers $1, 2, \ldots, n$, and the resulting number is given a plus sign when the given permutation $\alpha, \beta, \ldots, \omega$ has an even (or zero) number of switches of terms from the order $1, 2, \ldots, n$, and a minus sign otherwise. The determinant of A is denoted by det A or $|A|$. For example, in the case of a two-dimensional matrix, we have:

$$\det A = a_{11} a_{22} - a_{12} a_{21}$$

The *minor* of an element of a matrix is the determinant of the matrix that results when the row and column in which the element resides are deleted from the matrix. For a two-dimensional matrix, for example, the minor of a_{11} is a_{22}, and the minor of a_{12} is a_{21}. The *cofactor* of element a_{ij} is its minor multiplied by $(-1)^{(i+j)}$, so that the sign is unchanged when $i + j$ is even and is changed when $i + j$ is odd. We denote the cofactor by A_{ij}. The determinant of A can represented in the following way:

$$\det A = \sum_j a_{ij} A_{ij} \qquad (A1.14)$$

That is, we move along a row and multiply the successive elements by their cofactors. (The same principle applies to movement down a column.)

The *adjugate* of the matrix A, or adj A, is the *transpose* of the matrix formed from the cofactors of A, i.e., the matrix formed by exchanging the rows and columns of a matrix. In the case of a two-dimensional matrix, we have:

$$\text{adj}\,A = \begin{pmatrix} a_{22} & -a_{12} \\ -a_{21} & a_{11} \end{pmatrix}$$

The inverse of A is given by the following relation:

$$A^{-1} = \frac{\text{adj}\,A}{\det A} \qquad\qquad (A1.15)$$

In the case of a two-dimensional system, it can be verified that the solution to a set of linear equations, given by substituting **Equation A1.15** into **Equation A1.13**, is the same as that obtained directly.

If det $A = 0$, then A^{-1} is undefined. In this case, no solution can be found unless the values of the components of z in **Equation A1.9** are all 0. Conversely, if these are all 0 (this is called a *homogeneous* set of equations), a finite solution for x exists only when det $A = 0$; in this case, only the relative values of the components of x can be determined, as can be verified for the two-dimensional case.

A1.iv.c. *Eigenvalues and eigenvectors*
We now return to the use of matrix notation to describe a linear recursion relation for a set of n variables (**Equation A1.9**). As we will show more formally below, in many applications, the system will eventually reach a constant rate of change, such that the ratio of each component of x to its previous value becomes constant. This constant is called the *leading eigenvalue* of the matrix A, denoted by λ_1. It is defined by the following relation:

$$x' = Ax = \lambda_1 x = \lambda_1 Ix \qquad\qquad (A1.16a)$$

This can be rearranged to give:

$$(A - \lambda_1 I)\,x = 0 \qquad\qquad (A1.16b)$$

where 0 is the vector whose elements are all 0 (here we are using the rule for matrix addition, i.e., the elements of the matrix formed as the sum of two matrices are simply the sums of the elements in the corresponding rows and columns in the two matrices).

From what was said in **Appendix A1.iv.b** above for the homogeneous case, this relation implies that λ_1 satisfies the following relation:

$$\det(A - \lambda I) = 0 \qquad\qquad (A1.17)$$

This expression provides a means of calculating the constant, λ_1, because it is a polynomial of degree n (i.e., it involves terms in powers of λ up to λ^n). It is known as the *characteristic equation* of the matrix. Because a polynomial of degree n has n possible solutions, there are n different eigenvalues corresponding to **Equation A1.16**, which we can denote by $\lambda_1, \lambda_2, \ldots, \lambda_n$. In the two-dimensional case, for example, the characteristic equation is the quadratic equation:

$$\lambda^2 - \lambda(a_{11} + a_{22}) + \det A = 0 \qquad (A1.18)$$

whose solution (using the equation for the solution of a quadratic equation) is:

$$\lambda = \frac{(a_{11} + a_{22}) \pm \sqrt{(a_{11} + a_{22})^2 - 4\det A}}{2} \qquad (A1.19)$$

From the general theory of polynomials, the eigenvalues must be either *real* or *complex* numbers. If the term inside the square root in **Equation A1.19** is positive, then the two solutions (λ_1 and λ_2) are both real numbers. If it is negative, the solutions must be complex numbers, because the square root of a negative quantity is undefined for real numbers.

A complex number can be regarded as a pair of real numbers representing the position of a point in two-dimensional space; their use allows us to represent the solutions of polynomials of any degree. Complex numbers obey their own set of rules of addition, multiplication, and division. They can be written in the form $z = x + iy$, where i is the square root of -1. In this formulation, x is the *real part* of the number and y is the *imaginary part*. The position of a point in two-dimensional space can always be represented by *polar coordinates*, r and θ, where r is the length of the line connecting the origin to the point and θ is its the angle with the x axis. We can therefore write $z = r(\cos\theta + i\sin\theta)$. A different way of writing the same relationship is $z = r\exp(i\theta)$, which can be expressed as a power series in exactly the same way as for a real number (**Equation A1.4**). The roots of a polynomial are either real numbers or pairs of *complex conjugates*, with the same real parts and opposite signs for their imaginary parts.

To each eigenvalue λ_i, there corresponds an *eigenvector* $\boldsymbol{\mu}_i$, such that:

$$A\boldsymbol{\mu}_i = \lambda_i\boldsymbol{\mu}_i \qquad (A1.20a)$$

This is the *right* (column) *eigenvector*; we can also define a *left* (row) *eigenvector* by the following relation:

$$\boldsymbol{v}_i A = \lambda_i \boldsymbol{v}_i \qquad (A1.20b)$$

A1.iv.d. *The spectral expansion of a matrix*

Under suitable conditions (e.g., when all of the eigenvalues of the matrix have distinct values), it is possible to represent a matrix by its *spectral expansion*:

$$A = \lambda_1 P_1 + \lambda_2 P_2 + \cdots + \lambda_n P_n \qquad \text{(A1.21a)}$$

where the matrices P_i are defined by the following relation:

$$P_i = \mu_i v_i \qquad \text{(A1.21b)}$$

and the elements of each eigenvector are normalized so that the product $v_i \mu_i$ is 1. This has the useful property that, if we multiply the matrix A by itself t times, so that we form the matrix A^t, we have:

$$A^t = \lambda_1^t P_1 + \lambda_2^t P_2 + \cdots + \lambda_n^t P_n \qquad \text{(A1.22)}$$

It follows that, for sufficiently large t, the expression is dominated by the leading eigenvalue or eigenvalues and the corresponding eigenvectors; contributions from all the others can then be neglected. The local stability analysis in **Section 8.4.ii.c (Box 8.9*)** is an application of this principle.

We also have:

$$x_t = A^t x_0 \qquad \text{(A1.23)}$$

where x_t is the value of the vector at time t.

If the leading eigenvalue is real, **Equation A1.23** implies that the rate of change of each component of the system asymptotically approaches a constant value, such that each component of the vector at time t is equal to λ_1 multiplied by its value at time $t-1$. If there is a pair of complex conjugate leading eigenvalues, λ_1 and λ_2, the contributions from their imaginary parts cancel each other out, and the asymptotic ratio of each of the vector components at times t and $t-1$ is equal to $r \cos t\theta/[\cos (t-1)\theta]$, where r is the absolute value of the leading eigenvalue and $r \cos \theta$ is its real part. This implies that the ratio fluctuates over time with a period of $\theta/2\pi$. If $r > 1$, the magnitude of the components of the vector increases over time, so that the system undergoes oscillations of increasing size; if $r < 1$, the oscillations damp down and equilibrium is approached.

Consider a recursion relation of the form of **Equation A1.9**. If the leading eigenvalue is real, **Equation A1.20a** implies that the system approaches a state in which the relative values of the vector components reach a constant value, given by the ratios of the corresponding components of the right leading eigenvector. In the case of a stochastic matrix, these represent the ratios of the probabilities of the different states of the system; in this case, the leading eigenvalue can be shown to be equal to 1, so these probabilities become constant and an equilibrium is reached. If the system is represented by a row vector instead of a column vector, this result applies to the row eigenvector (as used in **Box 7.3*** of **Section 7.2.ii.d**).

A2. BASIC PROBABILITY AND STATISTICS

A2.i. Introduction

Biological data of the type encountered in population and quantitative genetics, such as the frequency of a variant at a nucleotide site or a set of measurements of a quantitative trait, are usually based on only a small sample from a population. The properties of the sample are thus drawn from a probability distribution that reflects the way in which the sample is taken, and hence we only have a more or less uncertain estimate of the true population values of any parameters of interest. In addition, the population values of these parameters may also be drawn from a probability distribution, as in the case of the frequency of a nucleotide variant subject to genetic drift. The basic concepts of probability theory and statistics are therefore a core foundation of our subject.

A2.ii. The rules of probability

The probability of occurrence of a discrete event, such as the presence of an allele A_1 at a biallelic locus in a haploid genome sampled from a population, can be thought of as the frequency with which this event occurs in a sample as the sample size is increased indefinitely. For a continuous variable, x, there is evidently no meaning to the probability of a given value of x, because there is an infinitely small chance that we draw a specific value of x. Nevertheless, the probability that x lies between two defined values, x_0 and x_1, is meaningful. We can then use **Equation A1.8** to define the *probability density*, $f(x)$. This is the function whose definite integral between x_0 and x_1 defines the relevant probability (**Figure A1.2**). The *cumulative probability function*, $F(z)$, gives the probability that x lies in the integral between 0 and z, and is the special case of **Equation A1.8** in **Appendix A1.iii** when $x_0 = 0$ and $x_1 = z$.

It is often of interest to ask about the probabilities of multiple events. If two events are *mutually exclusive*, as is the case when sampling two alleles, A_1 and A_2, at a biallelic locus, the net probability of either of the two events occurring is the sum of the probabilities of each event (in this case, this probability is 1). If we have two *independent* events, the probability that both events occur is the product of the individual probabilities. More generally, this probability is given by:

$$P(1 \text{ and } 1) = P(1|2)P(2) = P(2|1)P(1) \qquad (A2.1)$$

where $P(i|j)$ is the probability of event i occurring given that j has occurred (this is called the probability of event i *conditional* on event j; the vertical bar denotes the value of i conditional on the value of j). If the two events are independent, $P(i|j)$ is simply $P(i)$, and the formula above yields the rule for the multiplication of independent probabilities.

A2.iii. The moments of a distribution

The probability distribution or probability density is a complete description of the properties of a random variable. However, it is often convenient to use quantities such as the *mean* and *variance* that summarize the main properties of the distribution.

For a discrete variable, x, the mean of a distribution, $\bar{x}$, is defined as:

$$\bar{x} = \sum_x xP(x) \tag{A2.2a}$$

and the variance, V_x, is defined by:

$$V_x = \sum_x (x - \bar{x})^2 P(x) \tag{A2.2b}$$

The variance measures the scatter of a variable around its mean. To express this scatter on the same scale of measurement as the variable itself, the *standard deviation* (defined as the square root of the variance) is used; it is often denoted by σ_x (the variance is then denoted by σ_x^2).

It is often convenient to write these expressions as the *expectations* of the relevant terms. The expectation, $E\{f(x)\}$, of a function $f(x)$ of a discrete random variable x is obtained by multiplying each of the possible values of $f(x)$ by the corresponding probability of x, $P(x)$, and summing the product, i.e., $E\{f(x)\} = \Sigma_x f(x)P(x)$. Thus,

$$\bar{x} = E\{x\} \quad \text{and} \quad V_x = E\{(x - \bar{x})^2\} \tag{A2.2c}$$

For a continuous variable, the probabilities are replaced by probability densities and summation is replaced by integration. In what follows, we mostly present formulae only for discrete variables, but the equivalent integrals for continuous distributions can be easily by written down.

These are the definitions of the mean and variance for the distribution as a whole; the estimates of these parameters for a sample of n values of x drawn independently from the distribution are usually obtained as follows:

$$\hat{\bar{x}} = \frac{1}{n}\sum_{i=1}^{n} x_i \tag{A2.3}$$

$$\hat{V}_x = \frac{1}{(n-1)} \sum_{i=1}^{n} (x_i - \bar{x})^2 \qquad (A2.4)$$

where x_i is the value of the ith observation in the sample and the carets indicate estimates of the parameters in question.

In the formula for the sample estimate of the variance, there is a divisor of $n - 1$ instead of n. Use of this divisor means that the population variance V_x is equal to the expectation of the sample variance over a large number of repeated samples of n independent observations. The expectation of the sample mean is the same as the population mean (this is true even if the x_i are not independent of each other).

More generally, we can define the nth moment μ_n of degree n of x as the expectation of x^n. This can be defined as the moment either about 0 or about the mean, μ_1. For second moments, these are $E\{x^2\}$ and $E\{(x - \bar{x})^2\}$, respectively. The variance is thus the second moment about the mean. Simple algebra shows that the variance of a distribution is related to the second moment about 0 by:

$$V_x = \mu_2 - \mu_1^2 \qquad (A2.5)$$

It is often convenient to use the *moment generating function (m.g.f.)* of a distribution of a variable x when attempting to obtain explicit expressions for its moments. This is defined in terms of an arbitrary variable, θ, such that the m.g.f. is the following function:

$$M(\theta) = E\{\exp(\theta x)\} \qquad (A2.6)$$

The utility of this is that the expansion of the exponential function, as in **Equation A1.4** in **Appendix A.1.ii.c**, implies that μ_n is the coefficient of $\theta^n/n!$ in the Taylor's series representation of $M(\theta)$, since this is multiplied by $E\{x^n\}$ (**Appendix A1.ii.a**). If we can use the formula for the probability distribution to obtain an expression for $M(\theta)$, we can immediately write down all of the moments. This is facilitated by the fact that the nth-order derivative of $M(\theta)$ with respect to θ, evaluated at $\theta = 0$, is equal to μ_n. Some examples of m.g.f.'s are given below.

A2.iv. Pairs of variables

These concepts can be extended to more than one variable; here, we only consider two variables, x and y. If these are non-independent, the probability of y is conditional on the value of x and vice versa (see **Appendix A2.i** above). The extent of interdependence of the variables can be measured by their *covariance*, which is defined as:

$$C_{xy} = E\{(x - \bar{x})(y - \bar{y}\} = E\{xy\} - \bar{x}\,\bar{y} \qquad (A2.7)$$

In addition, we have:

$$E\{x\,y\} = E\{(x|y)y\} \qquad (A2.8)$$

where (as explained above) the vertical bar denotes the value of x conditional on the value of y.

If x and y are independent of each other, $E\{x|y\} = E\{x\}$ and $E\{y|x\} = E\{y\}$, so that $E\{xy\} = \bar{x}\,\bar{y}$. Substituting this into **Equation A2.7** shows that $C_{xy} = 0$ in this case. A positive value of C implies that large values of x tend to be associated with large values of y; a negative value means that large values of x are associated with small values of y and vice versa. In order to remove the effect of the scale of the measurements, we can use the *correlation coefficient, r*, which is calculated by dividing the covariance by the product of the standard deviations of x and y (see **Appendix A2.iii** above). This lies between -1 and $+1$.

Another way of quantifying the relation between two variables is to assume that they are linearly related, with random variability in either or both that obscures their underlying relationship. This is most commonly applied to continuous variables, where the *dependent* variable y is assumed to be a linear function of the *independent* variable x:

$$y = \alpha + \beta x + \varepsilon \qquad (A2.9)$$

Here ε is a random variable with an expectation of 0, which represents the distribution of y around the value predicted by the equation. In most applications, the distribution of ε is assumed to be the same for each value of x. This implies that, if x is held constant, there is a single variance of y, called the *residual variance, V_ε*.

Geometrically, the constant α is the intercept of the line with the y axis when x is 0, and β is the slope of the line. It follows from **Equation A2.9** that:

$$\bar{y} = \alpha + \beta\bar{x} \quad \text{or} \quad \alpha = \bar{y} - \beta\bar{x} \qquad (A2.10a)$$

$$C_{xy} = \beta V_x \quad \text{or} \quad \beta = C_{xy}/V_x \qquad (A2.10b)$$

In order to estimate the values of α and β from a set of data, the method of *least squares* is often used. This means that the sum of the squared deviations of the observed values y from their values predicted from the corresponding x values is minimized. This yields the *regression line*. It turns out that the estimates a and b of α and β are then given by the expressions corresponding to **Equations A2.10a** and **A2.10b**, using the sample estimates of the means, the variance of x, and the covariance. The estimate of β is known as the *regression coefficient, b*.

By taking the expectation of y^2 in **Equation A2.9**, the variance of y is seen to be related to the variance of x by:

$$V_y = \beta^2 V_x + V_\varepsilon \qquad (A2.11)$$

The proportion of the total variance in y explained by its relation with x is thus $\beta^2 V_x/V_y$. Using the definition of the correlation coefficient, this can be seen to be equal to r^2.

There is no guarantee that two variables being studied are related linearly, or that the variance of y conditioned on x is independent of x. More generally, we can always write the variance of y as:

$$V_y = V_{\bar{y}|x} + E\left\{V_{y|x}\right\} \qquad (A2.12)$$

It is also of interest to consider the variance of a variate y that is made of the sum of n other variates, x_i ($i = 1, 2, \ldots, n$). The expectation of y is simply the sum of the expectations of the x_i, i.e., $\Sigma_i \bar{x}_i$. The variance of y is given by:

$$V_y = \sum_i V_{x_i} + 2 \sum_{i<j} C_{x_i x_j} \qquad (A2.13)$$

When the x_i are independent of each other, the covariance terms vanish and the variance of the sum of n independent random variables is the sum of the variances of each variable. This is the basis for partitioning variance into components associated with different causes, such as genotypic and environmental effects—the *analysis of variance* (**Section 3.3.ii**).

When the x_i represent the values for a sample formed by independent draws from a population, all the x_i have the same variance, V_x, and the variance of their sum is simply nV_x. The variance of the corresponding sample mean is obtained by dividing this by n^2, and is thus equal to V_x/n.

A2.v. Some important probability distributions

The next sections (**Appendix 2.v.a–2.v.f**) describe some of the most commonly used distributions in elementary probability and statistics.

A2.v.a. *The binomial distribution*

We are often interested in the situation when two alternatives are possible, with the probability of one alternative being p in the population from which a sample is drawn (the alternative having probability $q = 1 - p$). The probability of observing i events of the first alternative in a sample of size n is denoted by $P(i)$; the distribution that gives $P(i)$ is called the *binomial distribution* because $P(i)$ is the ith term in the *binomial series* representation of $(p + q)^n$. This term is given by:

$$\binom{n}{i} p^i q^{n-i} = \frac{n!}{i!(n-i)!} p^i q^{n-i} \tag{A2.14}$$

and the sum of all of the terms equals $(p + q)^n$. The expression in brackets on the left-hand side of the above equation is the number of ways of choosing i objects out of a set of n (the *number of combinations*); the "!" symbols denote factorials (defined in **Appendix A1.ii.a**).

The mean and variance of i can be shown to be np and npq, respectively. The frequency of the event in the sample is i/n, which thus has mean and variance p and pq/n, respectively.

A2.v.b. *The multinomial distribution*

The binomial distribution deals with just two alternatives. The multinomial distribution generalizes this to the case when there are k alternatives with probabilities $p_1, p_2, \ldots, p_k$. The probability of observing a sample with i_1 events of type 1, i_2 of type 2, etc., is:

$$\frac{n!}{i_1! i_2! \ldots i_k!} p_1^{i_1} p_2^{i_2} \cdots p_k^{i_k} \tag{A2.15}$$

The expectation of i_j is np_j, its variance is $np_j(1 - p_j)$, and the covariance of i_j and i_k $(j \neq k)$ is $-np_j p_k$.

A2.v.c. *The geometric and exponential distributions*

The *geometric distribution* arises when we consider an event that has a constant probability p of occurring in a given interval of time (e.g., the death of an individual over the course of a year). If there is independence between events in different time intervals, the probability of an event failing to occur over $i - 1$ time intervals, and then happening in the ith time interval, is $p(1 - p)^{(i-1)}$. The waiting time to the occurrence of the event (i) has mean $1/p$.

If time is continuous rather than discrete, the equivalent process is one in which there is a constant rate, λ, of events occurring, as in the case of radioactive decay (λ is called the *rate constant*). The probability that an event occurs over a sufficiently small time interval, i.e., between t and $t + \delta t$ (conditional on the event not having happened previously), is $\lambda \delta t$, plus terms on the order of $(\delta t)^2$. The probability that no event has occurred by time t, $P(t)$, thus changes by approximately $-P(t)\lambda \delta t$, so that we have:

$$\frac{dP(t)}{dt} = -\lambda P(t) \tag{A2.16}$$

The initial value of P is 1. From the properties of the exponential function (**Equation A1.4** in **Appendix A1.ii.c**), this equation has the solution $P(t) = \exp$

$(-\lambda t)$. The probability of a event happening between t and $t + \delta t$ is thus $P(t)\lambda\delta t$ = $\lambda\exp(-\lambda t)\,\delta t$, where $\lambda\exp(-\lambda t)$ is the *probability density* of the waiting time to an event. This is the *exponential distribution*. It provides an example of the usefulness of the m.g.f. (**Appendix A2.iii**), because the m.g.f. in this case is defined by the following expression:

$$M(\theta) = \lambda\int_0^\infty \exp(\theta - \lambda)\mathrm{d}t = \frac{1}{\left(1 - \dfrac{\theta}{\lambda}\right)} \tag{A2.17}$$

By evaluating the derivatives of this expression at $\theta = 0$, or by using **Equation A1.6**, we find that the mean is $1/\lambda$, and the second moment about 0 is $2/\lambda^2$. The variance is thus $1/\lambda^2$. Higher moments can be obtained in a similar way.

The exponential distribution has the useful property that, if there are n different mutually exclusive events that may occur within a short time interval, and the rate constant for the ith type of event is λ_i, the probability that the ith event occurs, conditional on one of the events having taken place, is $\lambda_i/(\lambda_1 + \lambda_2 + \cdots + \lambda_n)$.

A2.v.d. *The Poisson distribution*

The binomial distribution is often used in situations when the sample size n is very large relative to the mean number of events of interest in the sample (i.e., the probability p of an event is $\ll 1$). The probability of observing i events is then well approximated by:

$$P(i) = \frac{\mu^i}{i!}\exp(-\mu) \tag{A2.18}$$

where μ is the expected number of events, pn. This is the *Poisson* distribution of the number of occurrences of rare events. Its variance can be obtained by taking the limit of the variance of the binomial distribution, $np(1 - p)$, as p approaches 0; this is $np = \mu$. The variance is thus the same as the mean.

In the case of events occurring at a constant rate per unit time, λ, where the waiting time is described by the exponential distribution, we can apply the Poisson distribution to the probabilities of the numbers of events that occur over a finite time interval t, with the expected number of events in this case given by λt. Another useful property of the Poisson distribution is that a sum of n independent Poisson variates is itself a Poisson variate, with mean and variance equal to the sum of the means of each variate.

A2.v.e. *The normal distribution*

The binomial distribution for large n and intermediate p can be approximated by the *normal distribution*, sometimes called the *Gaussian* distribution. If the observed frequency of the event of interest in a sample is x (which is now a continuous variable), the probability density of x is then given by:

$$\phi(x) = \frac{1}{\sqrt{2\pi}\sigma}\exp\left[-\frac{(x-\mu)^2}{2\sigma^2}\right] \tag{A2.19}$$

where μ and σ are the mean and standard deviation, respectively, of the distribution of the continuous variate, x. From the results in **Appendix A2.v.a**, $\mu = np$ and $\sigma^2 = np(1-p)$.

The normal distribution also arises when x is the sum of n independent variables, as an approximation to the probability density of x as n becomes sufficiently large. This is known as the *central limit theorem* of statistics. Because it is reasonable to assume that variables subject to many different influences can be represented in this way (e.g., a phenotype subject to variation caused by several different loci and environmental effects), there is an expectation that many metrical traits will be distributed normally, at least if a scale can be found on which the underlying causes of variability are additive (see **Section 3.3.ii**).

It is often convenient to work with a *standardized normal variate*, $z = (x - \mu)/\sigma$, with mean 0 and standard deviation 1. Integration shows that the m.g.f. of z is:

$$M(\theta) = \exp\left(\frac{\theta^2}{2}\right) \tag{A2.20}$$

This implies that the odd moments about the mean of the normal distribution are all 0, because all of the powers of θ in the expansion of $\exp(\theta^2/2)$ are even. The coefficent of θ^{2i} in the expansion is $1/(2^i i!)$, so that the $(2i)$th moment of x about the mean is equal to $\sigma^{(2i)}(2i)!/2^i i!$.

The normal distribution can be generalized to the case of an arbitrary number of variables, described by a vector $x = (x_1, x_2, ..., x_m)$. The *multivariate normal* distribution is described by the corresponding vector of means, $\mu = (\mu_1, \mu_2, ..., \mu_m)$, and the matrix of variances and covariances, $V = \{C_{ij}\}$, where C_{ij} is the covariance between x_i and x_j (for $i = j$, C_{ij} is the variance of x_i). If the variables are standardized in the same way as for the univariate normal distribution, we have:

$$\phi(z) = (2\pi)^{-(m+1)/2} \mid \det A \mid^{1/2} \exp(-Q/2) \tag{A2.21}$$

where A is the inverse of the matrix of correlation coefficients r_{ij} between each pair of variates (with diagonal elements of 1), and $Q = z^T A z$, with z being the column vector of standardized variates, and z^T being the corresponding row vector.

In particular, this distribution approximates the multinomial distribution of **Appendix A2.v.b**, just as the univariate normal distribution approximates the binomial distribution. It also plays an important role in the theory of selection on multivariate quantitative traits (**Box 3.8*** of **Section 3.3.iv**).

A2.v.f. *Other distributions*

Some more specialized distributions, such as the beta and gamma distributions, arise in specific contexts, and their properties are described in the relevant sections of the main text (i.e., **Box 5.9*** of **Section 5.3.iii.c** and **Box 6.9*** of **Section 6.4.iv.b**).

Another useful distribution is the *chi-squared* distribution, defined by:

$$\phi(\chi^2) = \left(\frac{\chi^2}{2}\right)^{(n/2)-1} \frac{\exp(-\chi^2/2)}{2\Gamma(n/2)} \qquad (A2.22)$$

where n is the *number of degrees of freedom* and Γ is the gamma integral defined in **Box 5.9***.

This is used in the test for the goodness-of-fit for data represented by discrete classes (such as the genotypes at a locus), where the sum of squares of the deviations of the observed numbers in each class from their expected values, divided by the expected values, is the measure of the fit of the data. In this case, n is one less than the number of classes. If m parameters are estimated from the data in order to determine the goodness-of-fit, the degrees of freedom are reduced by m.

The use of the chi-squared distribution in this case is justified by the fact that, when the numbers in the different classes follow the multinomial distribution (**Equation A2.15** of **Appendix A2.iv.b**), the resulting goodness-of-fit statistic approaches the chi-squared distribution, provided that the expected numbers for each class are sufficiently large (usually taken to mean five or above).

A2.vi. Statistical estimation

A2.vi.a. *Introduction*

We briefly consider the problem of how to relate the properties of a sample to the properties of the population from which it is drawn. The common-sense approach is to find a statistic whose expectation over a large number of repeated samplings is equal to the population value of the corresponding parameter, referred to as an *unbiased estimate* of the parameter. For example, **Equations A2.3** and **A2.4** in **Appendix A2.iii** above describe the unbiased estimates of the mean and variance, respectively. Similarly, given the assumptions mentioned after **Equation A2.9**, the least squares estimates, a and b, of the parameters, α and β, of a linear regression equation (**Appendix A2.iv** above) are also unbiased.

A2.vi.b. *Maximum likelihood*

Estimates of parameters are necessarily subject to sampling errors, caused by the fact that samples represent only a small part of the population. It is desirable for the sampling variances of estimates to be as low as possible. Unbiased estimates, however, do not necessarily achieve this. For this and other reasons, it is often desirable to use the method of *maximum likelihood*. This approach uses the idea that a good estimator of a parameter, θ, should yield a high probability of obtaining the sample that is observed. This suggests that we should determine the value of θ that yields the maximum value of the probability of the sample, conditional on θ. If the state of the sample is described by a set of measurements, **S**, therefore, we need to maximize the conditional probability or probability density $P(S|\theta)$. This is the definition of the *likelihood* of θ, denoted by $L(\theta, S)$. This principle can be generalized to a situation in which we wish to estimate several different parameters, in which case θ is replaced by a vector, $\boldsymbol{\theta}$.

The principle of *maximum likelihood* estimation can be illustrated by the case when we wish to estimate the mean and variance (or standard deviation) of a normally distributed trait x from a set of n independent values, $x_1, x_2, \ldots, x_n$. These constitute the data set **S**. The two parameters are μ and σ^2 of **Equation A2.19**, so that $\boldsymbol{\theta} = (\mu, \sigma^2)$. The likelihood of **S** is the product of the probability densities for each x_i, conditional on the assumed value of $\boldsymbol{\theta}$:

$$L(\theta, S) = \phi(x_1)\phi(x_2) \cdots \phi(x_n) \tag{A2.23}$$

In practice, it is more convenient to work with the natural logarithm of L, the *log-likelihood*; a maximum of $\ln(L)$ corresponds to a maximum in the logarithm, and we can combine independent sets of observations by summing their log-likelihoods. In the present example, this gives:

$$\ln(L) = -n \ln\left(\sqrt{2\pi}\sigma\right) - \sum_{i=1}^{n} \frac{\left(x_i - \mu\right)^2}{2\sigma^2} \tag{A2.24}$$

From basic calculus (**Appendix A1.i**), a maximum or minimum of $\ln L$ requires its partial derivatives with respect to μ and σ^2 to be equal to 0, i.e., we have:

$$\frac{\partial \ln(L)}{\partial \mu} = \sum_{i=1}^{n} \frac{\left(x_i - \mu\right)}{\sigma^2} = 0 \tag{A2.25a}$$

$$\frac{\partial \ln(L)}{\partial \sigma^2} = -\frac{n}{2\sigma^2} + \sum_{i=1}^{n} \frac{\left(x_i - \mu\right)^2}{2\sigma^4} = 0 \tag{A2.25b}$$

These yield the following maximum likelihood estimates of μ and σ^2:

$$\hat{\mu} = \sum_{i=1}^{n} x_i / n \qquad \text{(A2.26a)}$$

$$\hat{\sigma}^2 = \sum_{i=1}^{n} \left(x_i - \hat{\mu} \right)^2 / n \qquad \text{(A2.26b)}$$

In this example, the estimate of μ is the same as the unbiased estimate given by **Equation A2.3** of **Appendix A2.iii**, while the estimate of σ^2 differs from **Equation A2.4** by having a denominator of n instead of $n - 1$. For large sample sizes, the difference is unimportant, but this example shows that maximum likelihood estimates are not necessarily unbiased.

A very useful property of likelihoods is that it is possible to evaluate the goodness-of-fit of a given set of parameter values ($\boldsymbol{\theta}_0$) to the data by using the ratio of the maximum likelihood for the data set to the likelihood for this alternative set of parameter values. This is called a *likelihood ratio* test. The log-likelihood at the maximum and the log-likelihood for $\boldsymbol{\theta}_0$ are calculated, and the difference between them, Δ, is computed. If the sample size is sufficiently large, the distribution of 2Δ approaches a chi-squared distribution, with a number of degrees of freedom equal to the number of parameters whose values are estimated. The principle of the *likelihood ratio* test can be extended to more complex situations, such as comparisons between multiple possible sets of parameters, $\boldsymbol{\theta}_0$ versus $\boldsymbol{\theta}_1$, say.

It is often not possible to find simple analytic expressions to calculate maximum likelihood estimates, especially when several parameters need to be simultaneously estimated. Various numerical algorithms are available for finding maxima (see http://www.nr.com/). These methods can be difficult to apply when the data set is highly multidimensional, as is the case for DNA sequence data sets, where the sample is represented by the states of hundreds or thousands of nucleotide sites. In these cases, stochastic methods are used. Small perturbations are made to the parameter set $\boldsymbol{\theta}$, the new values of the likelihood are computed, and perturbations are accepted if they increase it by a pre-set amount. This is called a *Monte Carlo Markov chain* (*MCMC*). In principle, this procedure should eventually lead to the overall maximum likelihood of the whole parameter space.

With very high dimensional systems, however, almost all parameter sets are associated with very low likelihoods and it may be impossibly time-consuming to find the maximum. Various computational and statistical tricks have been devised to overcome this problem. These are beyond the scope of this account, but are described in several chapters of the *Handbook of Statistical Genetics*

(Balding et al. 2003). One helpful approach that is often used is to reduce the dimensionality of the data by using a small number of *summary statistics* to describe the sample, rather than the raw data itself. In the case of polymorphism data, for example, the diversity estimates π and θ_w (**Section 1.2.iv.b**) can be used, instead of the full list of haplotypes that make up the sample. This greatly simplifies the computations, although information is lost compared with the full likelihood approach.

A2.vi.c. *Bayesian methods*

An increasingly common extension to likelihood methods is based on the idea that we may have some prior information about $\boldsymbol{\theta}$, independent of a particular data set. For example, in mapping human genes, from the known map lengths and numbers of genes for the 23 different chromosomes, we can compute the *prior probability* that a random pair of genes have a given map distance. This is relevant to interpreting pedigree data that yield an estimate of the likelihood of the data, given an assumed map distance. It would seem sensible to combine the information from the likelihood calculations with the prior probability when estimating the map distances.

Bayesian statistical methods are intended to use information on such prior probabilities of $\boldsymbol{\theta}$, which we can denote by $P(\boldsymbol{\theta})$ for discrete variables (or a corresponding probability density, for continuous variables). The method uses the following principle. Using **Equation A2.1** of **Appendix A2.ii** above, the joint probability of the data set and a given value of $\boldsymbol{\theta}$, $\boldsymbol{\theta}_0$, is $P(S \text{ and } \boldsymbol{\theta}_0) = P(S|\boldsymbol{\theta}_0)P(\boldsymbol{\theta}_0)$ $= L(\boldsymbol{\theta}_0, S)P(\boldsymbol{\theta}_0)$. But we also have $P(S \text{ and } \boldsymbol{\theta}_0) = P(\boldsymbol{\theta}_0|S)P(S)$, where $P(\boldsymbol{\theta}_0|S)$ is the *posterior probability* of $\boldsymbol{\theta}_0$ conditioned on the data, and $P(S)$ is the overall probability of the data. The posterior probability can be viewed as a statement of our confidence in $\boldsymbol{\theta}_0$, given the result of the observations.

Equating the two expressions for $P(S \text{ and } \boldsymbol{\theta}_0)$, we obtain:

$$P(\boldsymbol{\theta}_0|S) = L(\boldsymbol{\theta}_0, S)P(\boldsymbol{\theta}_0)/P(S) \tag{A2.27}$$

The remaining problem is to find $P(S)$. We again apply the rule for conditional probabilities, summing the product $P(S|\boldsymbol{\theta})P(\boldsymbol{\theta})$ over all values of $\boldsymbol{\theta}$ to obtain $P(S)$.

We can then compare the posterior probabilities for different $\boldsymbol{\theta}_0$ and use the $\boldsymbol{\theta}$ value with maximum posterior probability as our estimator. The problem with this method is that the choice of prior probabilities is usually arbitrary. Uniform and independent distributions over pre-assigned intervals are often used for each component of $\boldsymbol{\theta}$. There is then little practical difference from a maximum likelihood procedure in which the parameters are restricted to defined intervals, as is often done in practice, and the Bayesian estimate corresponds to such a *restricted maximum likelihood* estimate.

Answers to Problems

Chapter 1

1.1.i. There are six monomorphic loci (loci with the commonest allele with a frequency > 0.99). Hence, the proportion of polymorphic loci is $2/8 = 0.25$.

ii. The gene diversity can be taken to be 0 for the monomorphic loci; for the polymorphic loci, we have the following values:

6-Pgd: The gene diversity is $2 \times 0.79 \times 0.21 = 1 - 0.79^2 - 0.21^2 = 0.332$.

Pt-8: The gene diversity is $2 \times (0.020 \times 0.876 + 0.020 \times 0.104 + 0.876 \times 0.104)$
$= 1 - 0.020^2 - 0.876^2 - 0.104^2 = 0.221$.

The mean gene diversity over all loci is thus $(0.332 + 0.221)/8 = 0.069$.

1.2 Let the frequencies of A_1 and A_2 be p and $q = 1 - p$, respectively. These are assumed to be the same for eggs and sperm. (If the frequencies differed in the initial generation, in the next generation males and females are both formed from the same pool of eggs and sperm, so they will have the same frequencies of A_1 and A_2. From then on, we can assume the same allele frequencies for eggs and sperm.)

The probability that a zygote is formed by the fusion of an A_1 egg and an A_1 sperm is p^2, because random mating implies that the genotype of the mother is independent of the genotype of the father, and each has a probability p of producing an A_1 gamete. The frequency of A_1A_1 individuals in the progeny generation is thus p^2. A similar argument shows that the frequency of A_2A_2 individuals is q^2. The frequency of A_1A_2 individuals is given by $1 - p^2 - q^2 = 1 - p^2 - (1 - p)^2 = 1 - p^2 - 1 + 2p - p^2 = 2(p - p^2) = 2pq$. Alternatively, the probability that an A_1 egg is fertilized by an A_2 sperm is pq, which is also the probability that an A_2 egg is fertilized by an A_1 sperm, so that the net frequency of A_1A_2 individuals is $2pq$.

1.3 Among the female zygotes in the next generation, the frequency of A_1 is $(p_f + p_m)/2$, because half of their alleles are maternally derived and half are paternally derived. Among the male zygotes, the frequency of A_1 is p_f, because their alleles are maternally derived. The difference between female and male frequencies is thus $(p_f - p_m)/2$. In the following generation, the frequency of A_1 among female zygotes is $[(p_f + p_m)/4] + (p_f/2) = (3p_f + p_m)/4$, and the frequency among males is $(p_f + p_m)/2$. The difference between female and male frequencies is thus $[3p_f + p_m - 2(p_f + p_m)]/4 = (p_f - p_m)/4$. This implies that the difference is of the opposite sign and half of the magnitude of the value in the previous generation.

Because the values of p_f and p_m are arbitrary, this conclusion is quite general: each generation, the difference between male and female frequencies changes sign and halves in magnitude.

1.4 The appropriate test is to compare observed and expected numbers for each genotype. To obtain the expected numbers, we first estimate the frequency of the G variant as $(2 \times 46 + 127)/(2 \times 239) = 0.458$. The frequency of T is $1 - 0.458 = 0.542$. On the hypothesis of Hardy–Weinberg equilibrium, the expected frequencies of the three genotypes are $(0.458)^2 = 0.210$, $2 \times 0.458 \times 0.542 = 0.496$, and $(0.542)^2 = 0.294$. The corresponding expected numbers are obtained by multiplying these by 239, i.e., 50.2, 118.5, and 70.3, respectively.

There is thus a slight deficit of heterozygotes compared with H–W expectation. The significance of this can be tested by calculating the sum of $(O - E)^2/E$ over the three genotypes, where O and E are the observed and expected numbers for a given genotype. This yields the sum $0.351 + 0.610 + 0.263 = 1.224$. This is distributed approximately as χ^2 (**Appendix 2.v.f**). There is only one degree of freedom, because one parameter (the G variant frequency) has been estimated from the data. The table of the χ^2 distribution with one degree of freedom shows that a value as small as 1.22 has a probability of more than 0.20 of arising by chance. There is thus no reason to believe that there is any deviation from Hardy–Weinberg equilibrium in this case.

1.5.i. The allele frequencies are $(90 + 2)/[2(45 + 52 + 2)] = 92/198 = 0.465$ for F and $1 - 0.465 = 0.535$ for S. The gene diversity is $2 \times 0.465 \times 0.535 = 0.498$. The frequency of heterozygotes is $2/99 = 0.020$.

ii. Using **Equation 1.3**, the predicted frequency of heterozygotes is $0.498(1 - f)$. We thus have $1 - f = 0.020/0.498 = 0.040$. Hence, $f = 0.960$.

iii. From **Equation 1.4**, $0.960 = S/(2 - S)$, where S is the frequency of self-fertilization. Hence, $1.920 - 0.960S = S$, so that $1.920 = 1.960S$, giving $S = 0.980$.

1.6 From **Equation 1.5**, there is no change in variant frequencies when $v(1 - p^*) = up^*$, corresponding to equal net rates of change in each direction. This gives $v = up^* + vp^*$, so that $p^* = v/(u + v)$.

For arbitrary p, we can write $p = p^* + \delta p$, where δp measures the departure of p from its equilibrium value. From **Equation 1.5**, $\Delta p = v (1 - p^* - \delta p) - u (p^* + \delta p) = - (u + v)\delta p$. Hence, if $\delta p < 0$, we must have $\Delta p > 0$, and the frequency of the GC variant will increase. The opposite is true if $\delta p > 0$. The variant frequency thus approaches the equilibrium value from either side.

1.7* We can represent the frequencies of the four nucleotides on a single strand of DNA by a four-dimensional vector column x, where x_1, x_2, x_3, and x_4 represent the frequencies of G, C, A and T nucleotide sites along the strand, respectively. The mutation rates can be represented by a 4×4 matrix A, where the diagonal elements represent the probabilities of no mutation for each of the four nucleotides, and the off-diagonal element a_{ij} $(i \neq j)$ is the probability that nucleotide j mutates to i in a given generation (**Appendix A1.iv.a**). The value of the vector in the next generation, x', is given by the matrix equation $x' = Ax$. The equilibrium value of the vector is thus given by $x^* = Ax^*$.

A is a stochastic matrix, because the elements in each of its columns must sum to 1. This means that it has a leading eigenvalue of 1 (**Appendix A1.iv.c**), and x^* is equal to the right eigenvector of A that corresponds to this eigenvalue, v_1, scaled such that its elements sum to 1, so that $x_1^* = 1 - x_2^* - x_3^* - x_4^*$. This allows us to write the individual terms in the matrix equation as an inhomogeneous set of linear equations. Choose the three that correspond to x_2^*, x_3^*, and x_4^* on the left-hand side. For the row involving x_2^*, we have:

$$x_2^* = a_{21}(1 - x_2^* - x_3^* - x_4^*) + a_{22}x_2^* + a_{23}x_3^* + a_{24}x_4^*$$

so that

$$-a_{21} = (a_{22} - a_{21} - 1)x_2^* + (a_{23} - a_{21})x_3^* + (a_{24} - a_{21})x_4^*$$

Similarly,

$$-a_{31} = (a_{32} - a_{31})x_2^* + (a_{33} - a_{31} - 1)x_3^* + (a_{34} - a_{31})x_4^*$$
$$-a_{41} = (a_{42} - a_{41})x_2^* + (a_{43} - a_{41})x_3^* + (a_{44} - a_{41} - 1)x_4^*$$

This can be written as a standard matrix equation of the form $z = Bx^*$, where the elements of z are $-a_{i1}$ $(i = 2$ to $4)$, $b_{ij} = a_{ij} - a_{i1}$ $(i \neq j)$, and $b_{ii} = a_{ii} - a_{i1} - 1$. The components of x^* can be found by solving this equation, e.g., using the formula for the inverse of a matrix (**Equations A1.13** and **A1.15** of **Appendix A1.iv.b**).

Chapter 2

2.1.i. From **Equation B2.3.1**, there is an equilibrium when $p^*s = q^*t$, i.e., the reductions below 1 in the marginal fitnesses of each allele due to homozygosity are equal. Writing $q^* = 1 - p^*$, this is equivalent to $p^*s = t - p^*t$, so that $p^*(s + t) = t$, i.e., $p^* = t/(s + t)$. Subtracting this from 1, we have $q^* = 1 - p^* = 1 - t/(s + t) = s/(s + t)$.

ii. In general, from **Equation B2.3.1** we have $w_2 - w_1 = ps - qt = s - q(s + t)$. When $q = q^*$, this expression is equal to 0. Because it decreases with q, it must be > 0 for $q < q^*$ and < 0 when $q > q^*$.

2.2.i. Because the balancer suppresses crossing over when heterozygous, we can treat Cy and $+$ as alleles with frequencies p and q, respectively. Because Cy is homozygous lethal, the selection coefficient s against Cy/Cy homozygotes is equal to 1. The selection coefficient against +/+ homozygotes is t, which is to be estimated.

We have $p^* = t/(s + t) = t/(1 + t)$ from **Equation 2.3** of **Section 2.1.ii.c**. But we only have an estimate of the frequency of Cy among surviving adults, whereas p^* applies to the new zygotes. Let the viability of +/+ relative to Cy/+ be V, where $V = 0.8$. The frequency of Cy/+ among the adults is thus $P = 2p^*q^*/(2p^*q^* + Vq^{*2}) = 2p^*/(2p^* + Vq^*)$. Substituting $p^* = t/(1 + t)$ and $q^* = 1/(1 + t)$, we have $P = 2t/(2t + V)$, i.e., $2tP + VP = 2t$, so that $t = VP/2(1 - P)$. With $V = 0.8$ and $P = 0.6$, $t = 0.48/0.8 = 0.6$.

ii. The net selection coefficient against +/+ homozygotes is 0.6, whereas the reduction in viability of homozygotes relative to that of heterozygotes is 0.2. This implies that some other aspect of fitness must be affected; this might be female fertility, male reproductive success, or adult longevity.

2.3 With heterozygote disadvantage, we can write the fitnesses of A_1A_1 and A_2A_2 relative to that of A_1A_2 as $1 + s$ and $1 + t$, respectively. **Equation B.2.2.3** in **Section 2.1.ii.a** shows that Δq is proportional to $w_{2.} - w_{1.}$. Using the same argument as in **Problem 2.1.ii**, $w_{2.} - w_{1.} = qt - ps = q(s + t) - s$. This is an increasing function of q, and so Δq is positive when $q > q^*$ and negative when $q < q^*$. The frequency of A_2, therefore, always moves away from q^*.

2.4* In **Equation B.2.4.1** we can write $f(q) = q + \Delta q$. This gives $df(q)/dq = 1 + d(\Delta q)/dq$. From **Equation B2.2.3**, $\Delta q = pq(w_{2.} - w_{1.})/\overline{w}$. Using the rule for differentiating a product (**Appendix A1.i**), $d(\Delta q)/dq = (pq/\overline{w})[d(w_{2.} - w_{1.})/dq] + (w_{2.} - w_{1.})[d(pq/\overline{w})/dq]$. At equilibrium, $(w_{2.} - w_{1.}) = 0$, so that the equilibrium value of $d(\Delta q)/dq$ is simply:

$$\frac{p^*q^*}{\overline{w}^*}\left(\frac{d(w_{2.} - w_{1.})}{dq}\right)_{q^*}$$

We have $p^* = t/(s + t)$, $q^* = s/(s + t)$, and $\overline{w}^* = 1 - p^{*2}s - q^{*2}t = 1 - st/(s + t)$, so that $p^*q^*/\overline{w}^* = [st/(s + t)^2]/[1 - st/(s + t)]$.

From **Problem 2.1.ii**, $w_{2.} - w_{1.} = ps - qt$. Because $dp/dq = -1$, $d(w_{2.} - w_{1.})/dq = -(s + t)$. Hence, $(p^*q^*/\overline{w}^*)d(w_{2.} - w_{1.})/dq^* = -[st/(s + t)]/[1 - st/(s + t)]$. We thus have $\lambda(q^*) = 1 - [st/(s + t)]/[1 - st/(s + t)] = [1 - 2st/(s + t)]/[1 - st/(s + t)]$.

2.5.i.* For equilibrium, we must have $s(q) = 0$ at $q = q^*$ (**Box 2.6***), i.e., $a = bq^*$. This gives the equilibrium frequencies as $q^* = a/b$ and $p^* = (b - a)/b$. (This requires $a < b$ for the equilibrium to exist.) We have $ds/dq = b$, so that the formula for $\lambda(q^*)$ (**Equation B2.6.2**) gives:

$$\lambda(q^*) = 1 - b(a/b)^2(b - a)/b = 1 - (a/b)^2(b - a)$$

ii. For stability rather than limit cycles or chaos, $(a/b)^2(b - a) < 1$, or $(b - a) < (b/a)^2$. Writing $a = kb$, the existence of limit cycles or chaos requires $b > 1/k^2(1 - k)$.

2.6 The assumption of hard selection means that a similar argument to that in **Box 2.8** can be applied, but with the total contribution to the gene pool from niche k now being proportional to $c_k \bar{w}^{(k)}$ instead of c_k, where $\bar{w}^{(k)}$ is the mean fitness in niche k. Let p_k' be the frequency of A_1 in niche k after selection. The overall frequency of A_1 in the next generation is then given by:

$$\bar{w}p' = \sum_k p_k' c_k \bar{w}_k$$

where $\bar{w} = \Sigma_k c_k \bar{w}_k$.

Under the assumptions of the model, the allele frequencies at the start of each generation are the same for each niche, so that (using the argument in **Box 2.2** of **Section 2.1.ii.a**):

$$p_k' = \frac{p\left[pw_{11}^{(k)} + qw_{12}^{(k)}\right]}{\bar{w}_k}$$

Based on these two equations, the factor of $\bar{w}^{(k)}$ cancels from the numerator and the denominator. We can then rewrite the first equation as $p' = pw_1./\bar{w}$, where $w_1.$ is given by:

$$\sum_k c_k \left[pw_{11}^{(k)} + qw_{12}^{(k)}\right]$$

This is identical in form to the standard selection equation (**Equations 2.2a** of **Section 2.1.ii.a**), except that fitness w_{ij} is replaced by its mean over all niches, with the value for niche k being weighted by c_k. It follows from the arguments in **Boxes 2.3** and **2.4*** that heterozygote advantage in this mean is required for a stable polymorphic equilibrium.

2.7.i.* Write fitness as a function of x, $w(x)$. First consider the homozygote A_1A_1, for which the activity in a given environment is x_1 and the fitness is $w(x_1)$. Let the mean and variance of x_1 be $\bar{x}_1$ and V_{x_1}. Using a similar argument to that in **Box 2.9.i***, the mean of the fitness of A_1A_1 over the set of environments is:

$$w_{11} \approx w(\bar{x}_1) + \frac{V_{x_1}}{2}\left(\frac{d^2 w(x)}{dx^2}\right)_{\bar{x}_1}$$

Similarly, for A_1A_1 we have activity x_2 and fitness $w(x_2)$ in a given environment, with mean fitness:

$$w_{22} \approx w(\bar{x}_2) + \frac{V_{x_2}}{2}\left(\frac{d^2 w(x)}{dx^2}\right)_{\bar{x}_2}$$

The fitnesses of the heterozygote, $A_1 A_2$, is given by $w(z)$, where $z = (x_1 + x_2)/2$. Using the same Taylor's expansion as before, we have:

$$w_{12} \approx w(\bar{z}) + \frac{V_z}{2}\left(\frac{d^2 w(z)}{dz^2}\right)_{\bar{z}}$$

By the chain rule (**Appendix A1.i**), however, $\partial w/\partial x_i = (dw/dz)/2$ ($i = 1$ or 2) and $\partial^2 w/\partial x_i \partial x_j = (d^2 w/dz^2)/4$, so that this can be written as:

$$w_{12} \approx w\left(\frac{\bar{x}_1 + \bar{x}_2}{2}\right) + \frac{V_{x_1}}{8}\left(\frac{\partial^2 w(z)}{\partial x_1^2}\right)_{\bar{x}_1} + \frac{V_{x_2}}{8}\left(\frac{\partial^2 w(z)}{\partial x_2^2}\right)_{\bar{x}_2} + \frac{C_{x_1 x_2}}{4}\left(\frac{\partial^2 w(z)}{\partial x_1 \partial x_2}\right)_{\bar{x}_1, \bar{x}_2}$$

because $V_{(x_1 + x_2)} = V_{x_1} + V_{x_2} + 2C_{x_1 x_2}$, where $C_{x_1 x_2}$ is the covariance between x_1 and x_2 (**Appendix A2.iv**).

ii. If we assume that $\bar{x}_1 = \bar{x}_2 = \bar{x}$ and $Vx_1 = Vx_2 = V_x$, this simplifies to:

$$w_{12} \approx w(\bar{x}) + \frac{V_x}{4}\left(\frac{\partial^2 w(z)}{\partial x_1^2}\right)_{\bar{x}} + \frac{C_{x_1 x_2}}{4}\left(\frac{\partial^2 w(z)}{\partial x_1 \partial x_2}\right)_{\bar{x}, \bar{x}}$$

Concavity of w implies that all of the second derivatives are negative. Comparing this with the expressions for w_{11} and w_{22}, there is net heterozygote advantage provided that the covariance $C_{x_1 x_2} < V_x$, i.e., the values of x_1 and x_2 are not completely correlated. From the results in **Box 2.9***, such net heterozygote advantage is sufficient, but not necessary, for protected polymorphism in temporally and spatially variable environments.

Chapter 3

3.1.i. By substituting the fitness model $w_{11} = w_{12} = 1$ and $w_{22} = 0$ into **Equation B2.2.2b** with $q = q_0$, we find that the marginal fitness of A_2 is $w_2 = 1 - q_0$ and $\bar{w} = 1 - q_0^2$. This gives $q_1 = q_0(1 - q_0)/(1 - q_0^2)$. But $(1 - q_0^2) = (1 - q_0)(1 + q_0)$, so that this expression simplifies to $q_1 = q_0/(1 + q_0)$. Similarly, we have $q_2 = q_1/(1 + q_1) = q_0/\{[1 + q_0][1 + q_0/(1 + q_0)]\}$. The denominator of the right-hand term can be written as $(1 + q_0)(1 + 2q_0)/(1 + q_0) = 1/(1 + 2q_0)$, so that $q_2 = q_0/(1 + 2q_0)$.

This suggests the general relation $q_t = q_0/(1 + tq_0)$. The validity of this can be established using *mathematical induction*. Assume that it is true for t; we then have $q_{t+1} = q_t/(1 + q_t)$. The denominator of this expression is, by assumption, equal to $1 + q_0/(1 + tq_0) = [1 + (t + 1)q_0]/(1 + tq_0)$, and the numerator is $q_0/(1 + tq_0)$. Canceling $(1 + tq_0)$ from the numerator and the denominator gives $q_{t+1} = q_0/[1 + (t + 1)q_0]$. Because the relation is known to be true for $t = 1$, this establishes that it is true for all generations.

ii. In this case, the marginal fitnesses are $w_{1.} = p + q(1 + s) = (1 + qs)$ and $w_{2.} = p(1 + s) + q(1 + s)^2 = (1 + s)(1 + qs)$. From **Equations 2.2a** and **2.2b** of **Section 2.1.ii.a**, $p' = pw_{1.}/\overline{w} = p(1 + qs)/\overline{w}$, and $q' = qw_{2.}/\overline{w} = q(1 + s)(1 + q s)/\overline{w}$. Writing $u = q/p$, we have $u' = u(1 + s)$, which is the same as **Equation B3.1.1**. This means that **Equation B3.1.2** for the haploid case is also the solution for this case.

3.2.i.* Using **Equation B3.2.2**, we can set $h = 0$ for a favorable recessive mutation, giving:

$$\left(t_1 - t_0\right) \approx \frac{1}{s} \int_{q_0}^{q_1} \frac{dq}{pq^2}$$

The integrand can be written as:

$$\frac{1}{q}\left(\frac{1}{p} + \frac{1}{q}\right) = \frac{1}{pq} + \frac{1}{q^2} = \frac{1}{p} + \frac{1}{q} + \frac{1}{q^2}$$

The integrals of $dq/p = -dp/p$ and dq/q are $-\ln(p)$ and $\ln(q)$, respectively, and the integral of dq/q^2 is $-1/q$. It follows that:

$$\left(t_1 - t_0\right) \approx \frac{1}{s}\left[\ln\left(\frac{p_0 q_1}{p_1 q_0}\right) + \left(\frac{1}{q_0} - \frac{1}{q_1}\right)\right]$$

The case of a favorable dominant can be studied by setting $h = 1$ in **Equation B3.2.2**. This is equivalent to interchanging p and q in the denominator of the above integral, and hence in the expression for the integrand we have just derived. The only difference in the final result is that $1/q_0$ and $1/q_1$ in the second term in the final expression are replaced by $1/p_1$ and $1/p_0$, respectively.

ii. In the case of a favorable recessive, $1/q_0$ will be very large and $1/q_1$ will be close to 1 when q_0 is low. At a low value of the initial frequency of A_2, the time to fixation is greatly increased over that for intermediate dominance (**Equation 3.3**), i.e., the initial spread of the mutation is very slow. For a favorable dominant, the low value of p_1 (the final frequency of the recessive, disfavored allele) inflates the value of the expression for the time, so that the approach to fixation of the favored allele is very slow.

3.3 In both cases, we can write $\bar{w} = 1 + f(q)$, where $f(q)$ approaches 0 as q becomes small, i.e., $f(q)$ is of order q. In addition, we can use the result that, when x is small, $1/(1 + x) \approx 1 - x$ (**Appendix 1.ii.c**), to write $1/\bar{w} = 1 + g(q)$, where $g(q)$ is also of order q. From **Equation B2.2.3**, with recessivity we thus have $\Delta q = spq^2/\bar{w} = spq^2[1 + g(q)]$. Because $g(q)$ is of order q, the product $spq^2g(q)$ can be neglected in comparison with spq^2 as q becomes small; also, $p = 1 - q$, so that spq^2 can be approximated by sq^2, as in **Equation 3.4a**, with the error becoming negligible relative to this term as q approaches 0. For the non-recessive case, a similar argument shows that $\Delta q \approx hspq/\bar{w} \approx hsq$, where again the relative error becomes negligible as q approaches 0.

3.4* The frequency of A_2, q, is given by $q = N_2/(N_1 + N_2)$. Writing $N = N_1 + N_2$, the rule for differentiating a product (**Appendix A1.i**) gives:

$$\frac{dq}{dt} = \frac{N\dfrac{dN_2}{dt} - N_2\left(\dfrac{dN_1}{dt} + \dfrac{dN_2}{dt}\right)}{N^2}$$

Substituting the relation $dN_i/dt = r_iN_i$, and noting that $p = N_1/N$, this becomes:

$$\frac{dq}{dt} = \frac{N_2}{N}r_2 - \left(\frac{N_2}{N}\right)\left(\frac{N_1}{N}r_1 + \frac{N_2}{N}r_2\right) = qr_2 - pqr_1 - q^2r_2 = q(r_2 - \bar{r})$$

where $\bar{r} = pr_1 + qr_2$. But $r_2 - \bar{r} = pr_2 - pr_1$, so that $dq/dt = pq(r_2 - r_1)$.

3.5 In the case of an autosomal mutation, **Equation 3.7** implies that $\Delta q > 0$ if $s_m + s_f > 0$. Thus, if $s_m < 0$ and $s_f > 0$, we require the absolute value of s_m to be $< s_f$. If $s_m > 0$ and $s_f < 0$, the condition is that the absolute value of s_f is $< s_m$. For the X-linked case, the equivalent conditions are that the absolute value of s_m is $< 2hs_f$, and that the absolute value of $2hs_f$ is $< s_m$, respectively.

With recessivity ($h = 0$), the conditions for the autosomal case remain unchanged, but for X-linkage the mutation cannot spread if $s_m < 0$, but will spread if $s_m > 0$, regardless of the value of s_f.

3.6 From **Equations B3.3.2a** and **B3.3.2b**, the change in frequency of A_2 is $\Delta q = (n_2'/n') - q = (n_2' - qn')/n'$, where $n' = n_1' + n_2'$. Using **Equations B3.3.2** and **3.11**, we note that $q + pqa_{12}(rb_{12} - c_{12}) + q^2a_{22}(rb_{22} - c_{22}) = q(p\tilde{w}_{12} + q\tilde{w}_{22})$, and $p + p^2a_{11}(rb_{11} - c_{11}) + pqa_{12}(rb_{12} - c_{12}) = p(p\tilde{w}_{11} + q\tilde{w}_{12})$. This gives:

$$n_2' - q(n_1' + n_2') \approx 2nw\left[q(p\tilde{w}_{12} + q\tilde{w}_{22}) - pq(p\tilde{w}_{11} + q\tilde{w}_{12}) - q^2(p\tilde{w}_{12} + q\tilde{w}_{22})\right]$$

because the terms in $(1 - r)(p^2a_{11}b_{11} + 2pqa_{12}b_{12} + q^2a_{22}b_{22})$ cancel each other.

Dividing by *2nw* and simplifying, we obtain:

$$\Delta q \approx pq\left(p\tilde{w}_{12} + q\tilde{w}_{22} - p\tilde{w}_{11} - q\tilde{w}_{12}\right)$$

which is identical in form to the standard selection equation with weak selection, when the denominator $\bar{w}$ can be ignored (**Section 3.1iii.b**).

3.7 In this case, we assume that A_2 is a rare variant that causes its heterozygous carriers to act as altruists, whereas the prevailing $A_1 A_1$ homozygotes do not behave altruistically, i.e., $a_{11} = 0$. Their inclusive fitness is thus 1. For $A_1 A_2$ individuals to have higher inclusive fitness, we require $1 + a_{12}(rb_{12} - c_{12})$ > 1, i.e., $a_{12}(rb_{12} - c_{12}) > 0$, which is equivalent to $rb_{12} - c_{12} > 0$, i.e., Hamilton's rule.

3.8* Equation B3.4.1 can be written as $E - f(E) = 0$, where $f(E) = \Sigma_i P_i E^i$. $E = 1$ is always a possible solution to this equation, because $\Sigma_i P_i = 1$. However, differentiating $E - f(E)$ with respect to E gives $1 - (df(x)/dx)_E = 0$. Thus, $(df(x)/dx)_E = 1$ is a condition for a solution with $E \leq 1$. But $(df(x)/dx)_1 = \Sigma_i i P_i$, the mean offspring number, m. If $m = 1$, $E = 1$ is the only solution to the equation for E. Because $df(x)/dx$ increases with x, a value of $m = (df(x)/dx)_1 > 1$ means that a value of $E < 1$ is required for the condition $(df(x)/dx)_E = 1$ to hold.

3.9 A probability of survival greater than 0.95 is equivalent to a probability of loss of less than 0.05. If n independent mutations enter the population, the probability that they are all lost is $(1 - 2hs)^n$. We therefore need to solve the equation $(1 - 2hs)^n = 0.05$, i.e., $n = \ln(0.05)/\ln(1 - 2hs)$. For $hs = 0.01$, this gives $n = 148$; for $hs = 0.05$, we get $n = 28$.

3.10* First, convert the trait to a standardized normal deviate (**Appendix A2.v.e**) by transforming to the scale of deviations from the mean, divided by the standard deviation, $\tilde{z} = (z - \bar{z})/\sigma$. The cumulative distribution function corresponding to a probability of choosing the proportion P of the population with the largest values of z is $\phi(a) = 1 - P$, where a is the cut-off point on the standardized scale, such that only individuals with values $> a$ are allowed to contribute to the next generation. On the standardized scale, the mean of the individuals chosen for breeding is given by:

$$\tilde{z}^\star = \int_a^\infty \tilde{z}\phi(\tilde{z})\,d\tilde{z}/P$$

where $\phi(\tilde{z})$ is the standardized normal probability density, $\exp(-\tilde{z}^2/2)$. This expression can be evaluated by noting that $d[\exp(-x^2/2)]/dx = -x\exp(-x^2/2)$, so that:

$$\int_a^\infty \tilde{z}\phi(\tilde{z})d\tilde{z} = -\int_a^\infty d\phi(\tilde{z}) = \phi(a)$$

Because $\tilde{z}^*$ is equal to the selection intensity, i, and $\phi(a)$ is the height of the normal distribution at the truncation point, t, this establishes **Equation 3.13**.

3.11 Write $w'(z) = w(z)/\bar{w}$. The mean value of $w'(z)$ is 1. By definition, $S = (\Sigma_z zw'(z) - \bar{z})/n$, and $\bar{z} = \Sigma_z z/n$, where Σ_z denotes summation over all values of z in the population and n is the number of individuals. The covariance between z and $w'(z)$ is equal to $\Sigma_z(z - \bar{z})(w'(z) - 1)/n$. We have:

$$\Sigma_z\left(z - \bar{z}\right)\left(w'(z) - 1\right) = \Sigma_z\left(z - \bar{z}\right)w'(z) - \Sigma_z\left(z - \bar{z}\right) = \Sigma_z\left(z - \bar{z}\right)w'(z)$$

Because $\bar{z}\Sigma_z w'(z) = n\bar{z}$, this expression reduces to nS; dividing by n yields the desired relation between S and the covariance between z and $w(z)/\bar{w}$.

Chapter 4

4.1.i. We can use **Equation B4.1.3**, with $q_1 = 0.1$, to obtain the relation $0.09s - 0.008m \approx 0$, so that $s \approx 0.008/0.09 = 0.089$. When $m = 0.05$, $s \approx 0.040/0.09 = 0.444$.

ii. The speed of advance of the mutation in this case is given by $v = \sigma\sqrt{s}$, because the rate of change of allele frequency for a semidominant autosomal allele is half that for the haploid case. The rate of spread of the mutation implies that $v = 7/55 = 0.127$ km per year, and $v/\sigma = 0.127/0.5 = 0.254$. This give $s = (v/\sigma)^2 = 0.0645$.

4.2 From **Equations 4.1** and **4.2**, the net change in frequency of the deleterious allele A_2 is given approximately by $\Delta q \approx u - hsq$, provided that $q \ll 1$. The equilibrium is $q^* = u/hs$, so that Δq can be written as $u - hsq + hsq^* - hsq^* = u - hsq + hsq^* - u = -hs(q - q^*)$. But Δq is the same as the change in $q - q^*$. Hence, $\Delta(q - q^*) = -hs(q - q^*)$. Thus, if $q > q^*$, $\Delta q < 0$, and if $q > q^*$, $\Delta q > 0$, implying that q^* is stable.

Writing the value of q in the next generation as $q' = q + \Delta q$, we have $q' - q^* = (q - q^*)(1 - hs)$. The value of $q - q^*$ after t generations is therefore $(1 - hs)^t$ multiplied by its initial value. The time for $q - q^*$ to reach one-half of its initial value is thus $\ln(0.5)/\ln(1 - hs) = -0.69 \ln(1 - hs)$. If hs is small, this is approximately $0.69/hs$ (**Appendix A1.ii.c**).

4.3.i. The approximate expression for change in frequency of the deleterious allele A_2 due to mutation is u, as before. The argument used in **Problem 3.3**

gives $\Delta q \approx -sq^2$ (we simply change the sign of the selection coefficient). Thus, at equilibrium $u \approx sq^2$, and so $q^2 \approx u/s$, giving $q^* \approx \sqrt{u/s}$.

ii. With $h = 0.01$, the equilibrium frequency of a homozygous lethal mutation is $u/0.01 = 100u = 10^{-4}$; with $h = 0$, $q^* \approx \sqrt{u} = 10^{-3}$.

4.4 The fertility of dwarfs was $27/108 = 0.25$ children per individual, whereas the fertility of normal individuals was $582/457 = 1.27$. The fitness reduction to carriers of mutations relative to normal individuals, hs, is thus $1 - 0.25/1.27 = 0.803$. The frequency of dwarfs was $10/94,075 = 1.06 \times 10^{-4}$. With random mating, this is approximately equal to $2q$, where q is the frequency of the mutant allele, so that $q = 0.53 \times 10^{-4}$. At equilibrium, $q = u/hs$ (**Equation 4.3**), so that $u = qhs = 0.53 \times 10^{-4} \times 0.803 = 4.26 \times 10^{-5}$.

4.5.i. In the case of a recessive deleterious mutation, the mean fitness relative to wild-type is $1 - sq^{*2}$, so that the load $L = sq^{*2}$. The answer to **Problem 4.3.i** gave $q^{*2} \approx u/s$, so that $L \approx u$.

ii. With equal selection on the two sexes, **Equation 4.5** gives $q^* \approx 3u/(2h + 1)s$. The load in females comes mainly from heterozygotes and is thus approximately equal to $2q^*hs \approx 6uh/(2h + 1)$. The load in males is $q^*s \approx 3u/(2h + 1)$. Averaging over the two sexes, the net load is thus $3u(2h + 1)/2(2h + 1) = 3u/2$.

   If selection acts only on males, **Equation 4.5** gives $q^* \approx 3u/s_m$ and so the male load $= q^*s_m \approx 3u$. If selection acts only on females, $q^* \approx 3u/(2hs_f)$ and the female load $= 2q^*hs_f \approx 3u$.

4.6.i. From **Equation 2.3** of **Section 2.1.ii.c**, the equilibrium frequencies of A_1 and A_2 are $p^* = t/(s + t)$ and $q^* = s/(s + t)$, respectively. The load is $L = p^{*2}s + q^{*2}t = (st^2 + ts^2)/(s + t)^2 = st(s + t)/(s + t)^2 = st/(s + t)$.

ii. In this case, the load for each locus is $s^2/2s = s/2$. Using **Equation B4.4.2** of **Box 4.4** in **Section 4.2.ii.c**, we have $1 - \bar{w} = L \approx 1 - \exp(-ms/2)$, where m is the number of sites under selection.

iii. The maximum number of offspring that a female can produce in one generation is 100, because survival probability from egg to adult cannot exceed 1. For the population to remain in existence, each female must produce on average two offspring that survive to become adults. If $\bar{w}$ is the mean fitness relative to the fitness of the optimal genotype, we thus need $50\,\bar{w} \geq 1$ for the population to sustain itself, i.e., $\bar{w} \geq 0.02$, or $\ln(\bar{w}) \geq \ln(0.02) = -3.91$. From the above formula, this is equivalent to $m \leq 7.82/s$. Thus, with $s = 0.1$, a maximum of 78 loci with heterozygote advantage is compatible with population survival; with $s = 0.001$, 7820 loci could segregate.

4.7* The cost C is the sum of $sp = s(1 - q)$ over all generations when the favorable mutation is spreading through the population. If allele frequency change is slow, this can be approximated by an integral:

$$s \int_{t_0}^{t_1} (1 - q)\,dt$$

where t_0 and t_1 are the initial and final generations of the process of fixation of A_2. From **Equation B3.2.1**, we have $dq \approx sq(1 - q)dt/2$. We can substitute this into the above integral, changing the integration limits to q_0 and q_1, the initial and final frequencies of A_2, respectively, where q_1 can be set to 1. This gives:

$$C \approx 2s \int_{q_0}^{q_1} \frac{(1 - q)\,dq}{sq(1 - q)} = 2 \int_{q_0}^{1} \frac{dq}{q} = -2\ln(q_0)$$

4.8 Let the time taken for an individual substitution of a favorable mutation be t_s. In a given generation, we thus have K sites at which a substitution has just completed, K in which a substitution was initiated $t_s - 1$ generations ago, and so on. Thus, at any one time, there will be Kt_s sites segregating for mutations that are on their way to fixation. The mean load per generation for any one site is C/t_s. If fitnesses act multiplicatively, the mean fitness of the population, relative to the fitness of a population that is fixed for all currently segregating mutations, is $\left(1 - C/t_s\right)^{Kt_s} \approx \exp(-Kt_s C/t_s) = \exp(-KC)$, by the argument used to obtain **Equation B4.4.2** in **Section 4.3.ii.c**. The lag load is thus $L \approx 1 - \exp(-KC)$.

For $L = 0.1$ and $C = 30$, we have $0.9 = \exp(-30K)$, i.e., $K = -\ln(0.9)/30 = 3.51 \times 10^{-3}$. This is equivalent to one new substitution starting every 285 generations.

4.9 We can apply the standard notation for the fitnesses of heterozygotes and homozygotes to any trait, where s now specifies the reduction in trait value for A_2A_2 homozygotes relative to A_1A_1 rather than a selection coefficient. The trait mean for a randomly mating population can be written as $1 - 2pqhs - q^2s = 1 - qs(2ph + q) = 1 - qs[2h + q(1 - 2h)]$. The inbred population is made up of a fraction $1 - f$ of genotypes with this mean value (formed by combining non-i.b.d. gametes), and a fraction f with mean $1 - qs$ (formed by combining i.b.d. gametes). The difference in mean between the outbred and inbred populations is thus $-fqs[2h + q(1 - 2h)] + fqs = fqs[1 - 2h - q(1 - 2h)] = fqs(1 - 2h)(1 - q)$. Thus, if $h = 0.5$, there is no effect of inbreeding. If $h < 0.5$, then inbreeding reduces the mean, but increases it if $h > 0.5$.

4.10 The mean fitness of the outbred population is $1 - st/(s + t)$, from the answer to **Problem 4.6.i.** As in **Problem 4.9**, the inbred population is made up of two components. A fraction $1 - f$ has the same mean fitness as the outbred population; a fraction f is i.b.d., with fraction $p^* = t/(s + t)$ being homozygous for allele A_1 and $q^* = s/(s + t)$ being homozygous for allele A_2. The mean fitness of this component is thus $1 - p^*s - q^*t = 1 - 2st/(s + t)$. The difference in mean fitness between the outbred and inbred populations is $[-st/(s + t)] + [(1 - f)st/(s + t)] + 2fst/(s + t) = fst/(s + t)$.

The value of B with respect to fitness for a single site is thus $st/(s + t)$; the value for a set of sites is the sum of this over all sites. The value for a fitness component z is obtained by multiplying the value for each site by the coefficient c_{zi} that measures the relation between the effect of a site on fitness and its effect on z. This yields **Equation 4.17**.

Chapter 5

5.1.i. The results in **Section 5.1.i.b** imply that the ultimate state of the set of populations is to approach $F = 1$, where each population is fixed for one of the two variants. The mean frequency of A_1 over the set of populations is unchanged by drift, and thus remains equal to p_0. This corresponds to the probability that a population is fixed for A_1, i.e., it has a frequency 1 of A_1. The probability that a population is fixed for A_2 is $1 - p_0 = q_0$.

ii. Because a proportion p_0 of populations have a frequency of A_1 of 1, and the others have a frequency of 0, the variance (**Appendix A2.iii**) in the frequency of A_1 is equal to $p_0 - p_0^2 = p_0 q_0$.

5.2.i. Set $H_{t-1} = H_t = H^*$. We then have $1 - H^* \approx (1/2N) + 1 - 2u - (1/2N) - H^* + (2u + 1/2N)H^*$; canceling terms from both sides, we get $(2u + 1/2N)H^* \approx 2u$, so that $(1 + 4Nu)H^* \approx 4Nu$, i.e., $H^* \approx 4Nu/(1 + 4Nu) = \theta/(1 + \theta)$ (**Equation 5.4**).

When $\theta \ll 1$, then $\theta/(1 + \theta) \approx \theta(1 - \theta) \approx \theta$, where terms in θ^2 and higher powers of θ have been neglected (**Appendix A1.ii.c**). Replacing H^* with π^*, this gives **Equation 5.5**.

ii. With $Nu = 0.05$, **Equation 5.4** gives $H^* = 0.20/1.20 = 0.167$, and **Equation 5.5** gives $\pi^* = 0.20$. With $\theta = 0.005$, $H^* = 0.02/1.02 = 0.0196$, and **Equation 5.5** gives $\pi^* = 0.020$.

5.3 Let the nucleotide diversity at site i be π_i, which we assume to be drawn from a distribution with the same expected value π^* and variance V_π for each site. The mean over m sites is given by $\Sigma_i \pi_i/m$. The expected value of this is $\Sigma_i E\{\pi_i\}/m = \pi^*$.

The variance is $V\{\Sigma_i \pi_i/m)\} = V\{\Sigma_i \pi_i\}/m^2$. But $V\{\Sigma_i \pi_i\} = \Sigma_i V_\pi + 2\Sigma_{i \neq j} C_{ij}$, where C_{ij} is the covariance between the diversities at sites i and j (**Appendix A2.iv**). If the correlation coefficient for diversities at site i and j is r_{ij}, we have $C_{ij} = r_{ij}V_\pi$ and so $V\{\Sigma_i \pi_i\}$ can thus be written as $(m + 2\Sigma_{i \neq j} r_{ij})V_\pi$. The variance of mean diversity is $(1 + 2\Sigma_{i \neq j} r_{ij}/m)(V_\pi/m)$, which depends on the magnitude of the r_{ij}.

If the correlations are all positive, because nearby sites have similarly genealogies, the variance is greater than when sites are independent. In general, therefore, we expect a greater variance when sites are not independent of each other.

5.4* The mean of the exponential distribution is the integral of $\lambda x \exp(-\lambda x)$ between 0 and ∞. We can integrate this by the method of integration by parts; we can write $d(fg) = fdg + gdf$, so that $fdg = d(fg) - gdf$, where we put $\lambda x = f$ and $g = -\exp(-\lambda x)/\lambda$, because the integral of $\exp(-\lambda x) = -\exp(-\lambda x)/\lambda$ (see **Appendix A1.iii**). The indefinite integral of $\lambda x \exp(-\lambda x)$ is thus equal to:

$$-x\exp(-\lambda x) + \int \exp(-\lambda x)\mathrm{d}x = -x\exp(-\lambda x) - \frac{\exp(-\lambda x)}{\lambda}$$

The definite integral between 0 and ∞ is thus $1/\lambda$, because the values of all of the terms are 0, except for $\exp(-\lambda x)/\lambda$ at $x = 0$. The mean of the exponential distribution is therefore $1/\lambda$.

The variance can be determined by evaluating the integral of $\lambda x^2 \exp(-\lambda x)$. Integrating this by parts gives the indefinite integral:

$$-x^2\exp(-\lambda x) + 2\int x\exp(-\lambda x)\mathrm{d}x$$

The first term again contributes nothing to the definite integral between 0 and ∞; evaluating the second term in the same way as above, we get a contribution of $2/\lambda^2$. The variance is obtained by subtracting the square of the mean from this, yielding $1/\lambda^2$.

In the case of the coalescent process, $\lambda = 1/2N$, so that the mean is $2N$ and the variance is $(2N)^2$.

5.5 The mean time taken for i alleles to coalesce to $i - 1$ alleles is $2N/[i(i - 1)/2]$ $= 4N/i(i - 1)$ (**Section 5.1.iii.c**). The mean time to the most recent common ancestor is the sum of this from $i = k$ to $i = 2$. We can use the relation $1/(i - 1) - 1/i = 1/i(i - 1)$ to express the sum of the terms in i as:

$$\sum_{i=2}^{k} \left(\frac{1}{i - 1} - \frac{1}{i} \right) = -\frac{1}{k} + 1 = \frac{(k - 1)}{k}$$

The mean time to the most recent common ancestor is thus $4N(k - 1)/k$.

5.6. For a given tree length L, let the number of segregating mutations be S_L. This is distributed as a Poisson variate with mean and variance muL. The expectation of S_L^2 conditioned on L is thus $muL + (muL)^2$ (using **Equation A2.5** of **Appendix A2.iii**). The unconditional expectation of S_L^2 is thus $muE\{L\} + (mu)^2(V_L + E\{L\}^2)$. Because the unconditional expectation of S is $muE\{L\}$, the unconditional variance of S is $muE\{L\} + (mu)^2[V_L + E\{L\}^2] - (mu)^2E\{L\}^2 = muE\{L\} + (mu)^2V_L$.

5.7 We have $k = 5$, so that $a_k = 1 + 0.5 + 0.333 + 0.25 = 2.083$ and $\Sigma_i(1/i^2) = 1 + 0.25 + 0.111 + 0.0625 = 1.424$.

With no recombination, from **Equation B5.6.2**, the variance of θ_w for $m = 100$ is $0.01/(100 \times 2.083) + (0.01/2.083)^2 \times 1.424 = 4.80 \times 10^{-5} + 3.28 \times 10^{-5} = 8.08 \times 10^{-5}$, so that the standard deviation is 8.99×10^{-3}. For $m = 1000$, this becomes $0.01/(1000 \times 2.083) + (0.010/2.083)^2 \times 1.424 = 4.80 \times 10^{-6} + 3.28 \times 10^{-5} = 3.76 \times 10^{-5}$, and the standard deviation is 6.13×10^{-3}.

With independence between sites, from the discussion following **Equation B5.6.2**, the variance of θ_w for $m = 100$ is $4.80 \times 10^{-5} + [(0.01/2.083)^2 \times 1.424/100] = 4.80 \times 10^{-5} + 3.28 \times 10^{-7} = 4.83 \times 10^{-5}$, so that the standard deviation is 6.95×10^{-3}. For $m = 1000$, this becomes $4.80 \times 10^{-6} + [(0.01/2.083)^2 \times 1.424/1000] = 4.80 \times 10^{-6} + 3.28 \times 10^{-8} = 4.83 \times 10^{-6}$, and the standard deviation is 2.20×10^{-3}.

5.8 The population size in the first case is 100; from **Equation 5.11**, $1/N_e \approx 1/200 + 1/200 = 1/100$, giving $N_e \approx 100$. Population size N and N_e are thus approximately the same in this case. In the second case, $1/N_e \approx 1/380 + 1/20 = 0.0026 + 0.050 = 0.0526$, so that $N_e \approx 19$. N_e is thus much smaller than N (which is equal to 100 again), and is close to four times the number of males. In the third case, $c = 0.5$ and $\Delta V_m = 8$, and $1/N_e \approx [2 + (0.25 \times 8)]/200 = 0.02$, so that $N_e \approx 50$. The excess variance has a similar effect to a sex inequality in the number of breeding individuals, reflecting the fact that genes are passed to the next generation through a smaller number of parents than with purely random variation.

5.9 Replacing N with N_e in **Equation 5.5**, we have $\pi = 4N_e u$, so that $N_e = \pi/4u$. Equating observed and expected values, $N_e = 0.023/(2 \times 10^{-8}) = 1.15 \times 10^6$ for the African population. For the U.S. population, $N_e = 0.011/(2 \times 10^{-8}) = 550,000$.

Either the African population has recently expanded in size or the U.S. population has shrunk, compared with the ancestral population. Because the relatives of *D. melanogaster* live in Africa, the latter possibility seems more likely, with variation being lost when the U.S. population was founded. If this is the case, then the U.S. population may be far from equilibrium and its current N_e is thus much larger than this estimate.

5.10* Substituting **Equation B5.9.1** into **Equation 5A.7**, we have:

$$\phi(q) = \frac{C2N_e}{pq} \exp \int 2N_e \left(\frac{d\ln \bar{w}}{dq} + \frac{u}{q} - \frac{v}{p} \right) dq$$

$$= \frac{C2N_e}{pq} \exp 2N_e (\ln \bar{w} + u\ln q + v\ln p)$$

$$= C2N_e \bar{w}^{2N_e} p^{4N_e v - 1} q^{4N_e u - 1}$$

5.11* The kth moment about 0 of the beta distribution is given by:

$$\frac{1}{B(\alpha,\beta)} \int_0^1 x^{\alpha+k-1}(1-x)^{\beta-1}\,dx = \frac{B(\alpha+k,\beta)}{B(\alpha,\beta)} = \frac{\Gamma(\alpha+k)\Gamma(\alpha+\beta)}{\Gamma(\alpha)\Gamma(\alpha+k+\beta)}$$

For $k = 1$ (the mean), by using the factorial-like property of the Γ function, we find that this becomes:

$$\frac{\alpha\Gamma(\alpha)\Gamma(\alpha+\beta)}{\Gamma(\alpha)(\alpha+\beta)\Gamma(\alpha+\beta)} = \frac{\alpha}{(\alpha+\beta)}$$

For $k = 2$, we obtain:

$$\frac{(\alpha+1)\alpha\Gamma(\alpha)\Gamma(\alpha+\beta)}{\Gamma(\alpha)(\alpha+\beta+1)(\alpha+\beta)\Gamma(\alpha+\beta)} = \frac{(\alpha+1)\alpha}{(\alpha+\beta+1)(\alpha+\beta)}$$

The variance is obtained by subtracting the square of the mean from this, yielding:

$$\frac{\alpha}{(\alpha+\beta)}\left[\frac{\alpha+1}{(\alpha+\beta+1)} - \frac{\alpha}{(\alpha+\beta)} \right] = \frac{\alpha\beta}{(\alpha+\beta)^2(\alpha+\beta+1)}$$

Using **Equation 5.19**, we can set $4N_e v = \alpha$ and $4N_e u = \beta$. The mean of $p = 1 - q$ is thus $\alpha/(\alpha+\beta) = v/(u+v) = p^*$, and the mean of q is $\beta/(\alpha+\beta) = u/(u+v) = q^*$. The variance normalized by p^*q^* is thus $1/(\alpha+\beta+1) = 1/[1 + 4N_e(u+v)]$.

Chapter 6

6.1.i. For the example in question, $D = 59/5022 = 0.0117$, and so **Equation B6.1.3** gives the estimate of K as $-\ln(0.9883) = 0.0118$. We have $\lambda = K/2T$, so that the estimate of λ is $0.0118/(2 \times 5 \times 10^6) = 1.18 \times 10^{-9}$.

ii. The variance of the number of changes per site is equal to the variance of the number of differences in the whole sequence, divided by m^2 (**Appendix A2.iv**). Assuming a Poisson distribution, the estimated standard deviation is thus $\left(\sqrt{Km}\right)/m = \sqrt{K/m}$. The standard deviation for the estimate of D in **Problem 6.1.i** is thus given by the square root of $0.0118/5022 = 0.00153$. The corresponding 95% confidence interval for the estimate of K is thus $\pm(1.96 \times 0.00153) = \pm0.00300$. The 95% confidence interval for the estimate of λ is thus $\pm(0.00300/(2 \times 5 \times 10^6) = \pm3 \times 10^{-10}$.

6.2.i. Males and females each contribute 1/2 of the genes transmitted to the next generation. The initial frequency in the gene pool of a mutation initially present in a breeding female is thus $1/4N_p$ and that for a male is $1/4N_m$. These represent the respective probabilities of fixation, conditioned on the sex of the individual in which the mutation is carried. A new autosomal mutation has an equal chance of appearing in a female or male zygote, so the probabilities that it is found in a breeding female or male are N_f/N and N_m/N, respectively. The overall probability of fixation is thus $(N_f/N) \times 1/4N_f + (N_m/N) \times 1/4N_m = 1/2N$.

ii. For an X-linked mutation, 2/3 of the genes transmitted to the next generation come from females and 1/3 come from males. The initial frequency of a mutation present in a breeding female is thus $2/6N_f = 1/3N_p$ and that for a male is $1/3N_m$, because females are diploid and males are haploid in this case. The probability that a new mutation is present in a breeding female is $2N_f/(2N_f + N_m)$, and the probability that it is present in a breeding male is $N_m/(2N_f + N_m)$. The overall probability of fixation is thus $1/(2N_f + N_m)$, i.e., the reciprocal of the number of allele copies present in the population.

iii. If j copies of a mutation at a given site are present in the population, there are j mutually exclusive possible fixation events, because only one of the j mutations can become fixed. Using the rule for the summation of probabilities of mutually exclusive events, the net probability of the fixation of a mutation is j multiplied by the fixation probability of a single mutation.

6.3.i.* Using the Taylor's series expansion of the exponential function (**Appendix A1.ii**), the numerator of **Equation B6.4.2a** can be approximated by $1 - (1 - N_e s/N) = N_e s/N$, when $N_e s/N \ll 1$; this yields **Equation B6.4.2b**. This result will be valid if $s \ll N/N_e$.

ii. Again using the Taylor's series expansion of the exponential function, the denominator of **Equation B6.4.2b** can be written as $1 - (1 - \gamma + \gamma^2/2 - \cdots) \approx \gamma$ if $\gamma \ll 1$. The terms in γ thus cancel from the numerator and the denominator, leaving $Q \approx 1/2N$ (the neutral value), to an accuracy of order γ^2.

6.4* Consider a large number of independent nucleotide sites that are each polymorphic for the class of mutation in question. The segregating mutations observed at a given time arose at various, unknown times in the past. The longer a mutation spends at a given frequency, the greater the chance that it is observed at that frequency at a random moment in its history. By definition, a mutation spends a mean amount of time $t(q)\delta q$ in the frequency range q to $q + \delta q$. The chance of observing a mutation with a frequency in the range q to $q + \delta q$ is thus proportional to $t(q)\delta q$. The probability density of frequency q for a polymorphic site can thus be approximated by $t(q)/\int t(q)dq$, where the integral is taken over the interval $1/2N$ to $1 - 1/2N$.

6.5.i. For mutations of the type A_1 to A_2 (mutation rate u), we obtain, by summing **Equation B6.6.2a** over all values of i between 1 and $k - 1$, the probability that a given site is segregating for mutations of the type A_1 to A_2 as $4N_e u a_k$, where $a_k = 1 + 1/2 + \cdots + 1/3 + 1/(k - 1)$. The expected number of segregating sites in the sample is thus $4N_e u m a_k$.

ii. For mutations in the reverse direction (A_2 mutating to A_1), a similar argument shows that the expected number of segregating sites is $4N_e v m a_k$. As shown in **Section 6.3.ii.b**, the equilibrium frequencies of sites fixed for A_1 and A_2 are approximately $v/(u + v)$ and $u/(u + v)$, respectively. The overall expected number of segregating sites is obtained by weighting the expressions for each type of mutation by the frequencies of the corresponding fixed sites, giving $8N_e u v m a_k/(u + v) = 4N_e \tilde{u} m a_k$, where $\tilde{u} = 2uv(u + v)$.

6.6 From **Equations 6.6a** and **B6.4.2b** of **Section 6.2.iii.a**, we have $\lambda_1 \approx \gamma u/[1 - \exp(-\gamma)]$, where $\gamma = 2N_e s$. Similarly, $\lambda_2 \approx -\gamma v/[1 - \exp(\gamma)]$. At equilibrium, a fraction of sites x^* are fixed for A_2. If there is equality of substitutions in each direction, as is required for equilibrium, we must thus have $\gamma u(1 - x^*)/[1 - \exp(-\gamma)] \approx -\gamma v x^*/[1 - \exp(\gamma)]$. The right-hand term can be rewritten as $\gamma v x^* \exp(-\gamma)/[1 - \exp(-\gamma)]$, and so we have $x^*/(1 - x^*) \approx u \exp(\gamma)/v = \exp(\gamma)/\kappa$, where $\kappa = v/u$. Rearranging, we obtain $x^* \approx \exp(\gamma)/[\kappa + \exp(\gamma)] = 1/[1 + \kappa \exp(-\gamma)]$.

6.7.i.* The first term in the denominator of **Equation 6.11** can be approximated as follows:

$$1 + \kappa \exp(-\gamma) \approx 1 + \kappa\left(1 - \gamma + \frac{\gamma^2}{2}\right) = (1 + \kappa)\left[1 - \frac{\kappa\left(\gamma - \frac{1}{2}\gamma^2\right)}{(1 + \kappa)}\right]$$

The second term in the denominator can be approximated by (see **Problem 6.3***):

$$\gamma + \frac{1}{2}\gamma^2 + \frac{1}{6}\gamma^3$$

These approximations imply that:

$$\lambda \approx \frac{2\kappa u}{(1 + \kappa)\left[1 - \dfrac{\kappa\left(\gamma - \dfrac{1}{2}\gamma^2\right)}{(1 - \kappa)}\right]\left(1 + \dfrac{\gamma}{2} + \dfrac{\gamma^2}{6}\right)}$$

Expanding $1/(1 + x)$ as $1 - x + x^2 - \cdots$ (**Appendix A1.ii.c**), and neglecting powers of γ higher than 2, we get:

$$\lambda \approx 2au\left[1 + a\left(\gamma - \frac{1}{2}\gamma^2\right) + a^2\gamma^2\right]\left[1 - \frac{1}{2}\gamma + \frac{1}{12}\gamma^2\right]$$

where $a = \kappa/(1 + \kappa)$.

This reduces to:

$$\lambda \approx 2au\left[1 + \left(a - \frac{1}{2}\right)\gamma - \left(a - a^2 - \frac{1}{12}\gamma^2\right)\right]$$

ii. Examining the behavior of this expression when γ is close to 0, and ignoring the squared terms in γ, we have:

$$\lambda \propto 1 + \left(a - \frac{1}{2}\right)\gamma$$

so that λ increases with γ when $a > 1/2$, i.e., $\kappa > 1$.

If the squared terms in γ are included, we can find the approximate condition for a maximum in λ by differentiating the quadratic expression for λ with respect to γ and setting it equal to 0:

$$\frac{\partial \lambda}{\partial \gamma} \propto \left(a - \frac{1}{2}\right) - 2\left(a - a^2 - \frac{1}{12}\right)\gamma = 0$$

This gives the value of γ that maximizes λ, γ^*, as approximately $(a - 1/2)/2(a - a^2 - 1/12)$. With $\kappa = 2$, we have $a = 0.667$, so that we have $\gamma^* = 0.601$. With $\kappa = 4$, we have $a = 0.8$ and $\gamma^* = 1.96$.

6.8.i.* From the results in **Box 6.4*** of **Section 6.2.iii.a**, we have $\psi(x) = \exp(-\gamma x)$ in **Equation 6A.13**. The indefinite integral of this is $-\exp(-\gamma x)/\gamma$, and so we have:

$$\frac{\int_q^1 \psi(x)\,dx}{\int_0^1 \psi(x)\,dx} = \frac{\left[\exp(-\gamma q) - \exp(-\gamma)\right]}{1 - \exp(-\gamma))}$$

Noting that $V_{\delta q} = pq/2N_e$ and that $\psi(q) = \exp(-\gamma q)$, **Equation 6A.13** becomes:

$$\frac{4N_e u}{p\,q} \frac{\exp(\gamma q)\left[\exp(-\gamma q) - \exp(-\gamma)\right]}{\left[1 - \exp(-\gamma)\right]}$$

which yields **Equation B6.7.1** by multiplying the term in square brackets in the numerator by $\exp(\gamma q)$.

ii. We can use the same approach as in **Problem 6.3.ii***. As γ approaches 0, the denominator of **Equation B6.7.1** approaches $pq\gamma$ and the numerator approaches $4N_e up\gamma$, neglecting terms in γ^2. The overall expression thus approaches $4N_e u/q$, which is equivalent to the neutral **Equation B6.6.1** of **Section 6.3.ii.b**.

Similarly, for **Equation B6.7.3**, the denominator approaches $s\gamma$ as γ tends to 0, and the numerator approaches $4\mu\gamma^2/2$. The net result is thus $2\mu\gamma/s = 4N_e u$, the neutral value given by **Equation B6.6.3**.

6.9 Equation 4.18 (when applied to fitness) implies that, with semidominance, the additive genetic variance in fitness contributed by a single nucleotide site i is approximately $u_i s_i$, where u_i is the mutation rate to a deleterious amino acid mutation and s_i is the homozygous selection coefficient at site i. Summing over all sites, and assuming independence between u_i and s_i, this gives a total additive variance of $U\bar{s}$, where $\bar{s}$ is the mean of the s values that are sufficiently large for the deterministic approximation to apply and U is the sum of the mutation rates to deleterious amino acid variants. The results of **Section 4.5.i.a** imply that this is close to the total genetic variance in fitness. The results of **Section 6.4.iv.b** suggested a mean s of 0.066 for all mutations, which will be somewhat lower than the mean for those that satisfy the conditions for the deterministic equation to apply. U was estimated as 0.335 in **Section 6.4.iv.b**, so that $U\bar{s}$ = 0.022.

The corresponding standard deviation is 0.148; with independence between sites, the distribution of fitnesses should be approximately normal, so that the upper and lower 2.5% percentiles of the distribution are ±0.29 around the mean fitness, which was estimated as 0.55 in **Section 6.4.iv.c**. Expressing fitnesses relative to the mean, the upper and lower 2.5% percentiles are ±0.53.

Chapter 7

7.1.i. G_{ST} is defined as $(H_T - H_S)/H_T$, where H_T is the gene diversity for a pair of alleles sampled randomly from the set of populations. For a biallelic locus, $H_T = 2\bar{p}\bar{q}$, where $\bar{p}$ and $\bar{q}$ are the mean frequencies of the two alleles over the set of populations. H_S is the mean diversity for each population, i.e., the mean of $2pq$ over all populations. The argument used in **Section 5.1.i.c** shows that this mean is equal to $2\bar{p}\bar{q} - 2V_q$, where V_q is the variance in allele frequencies among populations. From **Equation 5.3**, $V_q = F_{ST}\bar{p}\bar{q}$; it follows that $H_T - H_S = 2F_{ST}\bar{p}\bar{q}$, so that $G_{ST} = (H_T - H_S)/H_S = F_{ST}$.

ii. The mean frequency of the SNP is $(0.1 + 0.15 + 0.2 + 0.3)/4 = 0.1875$, and the variance is $[(0.1)^2 + (0.15)^2 + (0.2)^2 + (0.3)^2]/4 - (0.1875)^2 = 0.00547$. Hence, $F_{ST} = 0.00547/(0.1875 \times 0.8125) = 0.0359$. The diversities within each population, $2pq$, are 0.18, 0.255, 0.32, and 0.42, respectively, giving the mean within-population diversity as $H_S = (0.18 + 0.255 + 0.32 + 0.42)/4 = 0.2938$. The total diversity is $H_T = 2 \times 0.1875 \times 0.8125 = 0.3047$, so that $G_{ST} = (0.3047 - 0.2938)/0.3047 = 0.0358$.

7.2 Neglecting the second-order terms as indicated in the question, **Equation B7.2.1b** for T_{ii} can be approximated by:

$$T_{ii} \approx 1 + m_{ii}^2 T_{ii} - P_{ci} T_{ii} + 2\sum_{k \neq i} m_{ik} T_{ik}$$

Because $m_{ii}^2 \approx 1 - 2\sum_{k \neq i} m_{ik}$ and $P_{ci} \approx 1/2N_{ei}$, this can be further approximated by:

$$1 \approx \left(2\sum_{k \neq i} m_{ik} + \frac{1}{2N_{ei}}\right) T_{ii} - 2\sum_{k \neq i} m_{ik} T_{ik}$$

Similarly, for $i \neq j$:

$$T_{ij} \approx 1 + m_{ii} m_{jj} T_{ij} + \sum_{k \neq i} m_{ik} T_{jk} + \sum_{l \neq j} m_{jl} T_{il}$$

which reduces to:

$$1 \approx \left(\sum_{k \neq i} m_{ik} + \sum_{l \neq j} m_{jl}\right) T_{ij} - \sum_{k \neq i} m_{ik} T_{ik} + \sum_{l \neq j} m_{jl} T_{il}$$

7.3 In this case, $T_{ii} = T_S$ and $T_{ij} = T_B$ $(i \neq j)$, because all demes are equivalent. Furthermore, $m_{ij} = m/(d-1)$ when $i \neq j$, where m is the net migration rate. The simplified equation for T_{ii} gives:

$$1 \approx \left(2m + \frac{1}{2N_e} \right) T_S - 2mT_B$$

and the simplified equation for T_{ij} gives:

$$1 \approx 2mT_B - \frac{2m}{(d-1)} T_S - \frac{2m(d-2)}{(d-1)} T_B$$

so that:

$$1 \approx \frac{2m}{(d-1)} T_B - \frac{2m}{(d-1)} T_S \; ; \; T_B \approx T_S + \frac{(d-1)}{(2m)}$$

Substituting this expression for T_B into the first equation, we get:

$$1 \approx 2mT_S + \frac{1}{2N_e} T_S - 2mT_S - (d-1); \; T_S \approx 2dN_e$$

The mean coalescent time for a random pair of alleles is given by $T_T = (T_S/d) + (1 - 1/d)T_B$. Using the above results, we thus have $T_T = T_S + (d-1)^2/(2dm)$.

7.4.i.* The variable m_{ij} is defined as the probability that an allele sampled from deme i was derived from deme j in the previous generation. If we write $x_i(t)$ for the probability that an allele was present in deme i at a given time t, the probability that it was in deme j at time $t - 1$ is $x_j(t-1) = \Sigma_i \, x_i(t)m_{ij}$. This is equivalent to the matrix equation $x(t-1) = x(t)M$. The equilibrium value of x thus satisfies $x^* = x^*M$. Because the leading eigenvalue of M is 1, x^* must correspond to the leading left eigenvector v of M, normalized such that its components sum to 1.

ii. If we multiply both sides of **Equation B7.2.1b** by the product of the ith and jth elements of v, and sum over all i and j, we have:

$$\sum_{ij} v_i v_j T_{ij} \approx \sum_{ij} v_i v_j + \sum_{ijkl} v_i v_j m_{ik} m_{jl} T_{kl} - \sum_i v_i^2 P_{ci} T_{ii}$$

The eigenvector v can be normalized such that its elements sum to 1, without loss of generality, so that $\Sigma_{ij} v_i v_j = (\Sigma_i v_i)(\Sigma_j v_j) = 1$. In addition, by virtue of the relation $v = vM$, we have $\Sigma_i v_i m_{ik} = v_k$ and $\Sigma_j v_j m_{jl} = v_l$. The first two terms on the right-hand side thus reduce to 1 and $\Sigma_{kl} v_k v_l T_{kl}$, respectively. Because the subscripts are arbitrary, the second of these terms is the same as the left-hand side, and we therefore obtain **Equation B7.3.2**.

7.5.i.[*] Let the allele frequency in deme i in generation t be $p_i(t)$. From the definition of M, we have $p_i(t) = \Sigma_k p_k(t-1)m_{ik}$. Taking the weighted mean allele frequency across demes, we have:

$$\bar{p}(t) = \sum_i v_i p_i(t) = \sum_{ik} v_i m_{ik} p_k(t-1) = \sum_k v_k p_k(t-1) = \bar{p}(t-1)$$

ii. Let the number of individuals in deme i before migration be N_i. Denote the fraction of individuals in deme i that migrate from deme j to deme i by $\tilde{m}_{ij}$ (the value for $i = j$ indicates the fraction that stay in deme j). The number of individuals in deme i after migration is thus $N_i' = \Sigma_j \tilde{m}_{ij} N_j$. But the migration probabilities m_{ij} that we have used before are related to the $\tilde{m}_{ij}$ in the following way, using the fact that the number of individuals entering population i from population j is equal to the product $N_j \tilde{m}_{ij}$: $m_{ij} = \tilde{m}_{ij} N_j / (\Sigma_k \tilde{m}_{ik} N_k)$. If deme sizes are unchanged by migration, the denominator must be equal to N_i, so that $m_{ij} = \tilde{m}_{ij} N_j / N_i$. This implies that $\Sigma_i N_i m_{ij} = \Sigma_j N_j \tilde{m}_{ij} = N_j$. The vector of N_i values is thus the leading left eigenvector of M, normalized such that its components sum to $N_T = \Sigma_i N_i$.

iii. If the values of the leading left eigenvector of M are normalized to sum to 1, the result from **Problem 7.5.ii** for conservative migration implies that the ith component of the eigenvector is $v_i = N_i/N_T = c_i$. Use of the result from **Problem 7.5.i** shows that the use of weights of c_i for each deme yields constancy of the corresponding mean allele frequency.

7.6[*] In the one-dimensional case, $m(y)$ is a normal variate, with mean 0 and variance σ^2. It follows that:

$$\int_{-\infty}^{\infty} m(y)^2 \, dy = \frac{1}{2\pi\sigma^2} \int_{-\infty}^{\infty} \exp\left(-\frac{y^2}{\sigma^2}\right) dy$$

Let $x = y/\sigma$, so that $y^2/\sigma^2 = x^2/2$ and $dy = \sigma dx/\sqrt{2}$. Changing variables in the integral from y to x, and using the fact that the integral of a standardized normal variate, $\left(1/\sqrt{2\pi}\right)\exp\left(-x^2/2\right)$, is 1, we obtain:

$$\int_{-\infty}^{\infty} m(y)^2 \, dy = \frac{1}{2\pi\sigma\sqrt{2}} \int_{-\infty}^{\infty} \exp\left(-\frac{x^2}{2}\right) dx = \frac{1}{2\sigma\sqrt{\pi}}$$

Substituting this expression into **Equation B7.5.1** yields **Equation B7.5.2**.
  Similarly, we have:

$$\int\limits_{-\infty}^{\infty}\int\limits_{-\infty}^{\infty} m(y)^2 m(z)^2 \, dy \, dz = \frac{1}{4\pi^2\sigma^4}\left[\int\limits_{-\infty}^{\infty}\exp\left(-\frac{y^2}{\sigma^2}\right)dy\right]^2 = \frac{1}{8\pi^2\sigma^2}\left[\int\limits_{-\infty}^{\infty}\exp\left(-\frac{x^2}{2}\right)dx\right]^2$$

$$= \frac{1}{4\pi\sigma^2}$$

Substituting this expression into **Equation B7.5.3** yields **Equation B7.5.4**.

7.7* Write $a = N_e/n_e$ and $b = 2N_e[1 - \phi(1 - 1/2n_e)]$. The derivative of F_{ST} in **Equation 7.11** with respect to e is proportional to:

$$(1 + 4N_e m + eb)a - (1 + ea)b = a - b + 4N_e ma$$
$$= -2N_e(1 - \phi) + N_e(1 - \phi)/n_e + 4N_e^2 m/n_e$$

If $\phi = 1$ (the propagule-pool model), then the derivative is always positive, so that F_{ST} increases with e.

In the general case, we can divide this expression by $2N_e$, giving the condition that the derivative is positive when $(1 - \phi)(1 - 2n_e) + 4N_e m > 0$, which yields **Equation 7.12**.

7.8 With no extinction, the probability of coalescence within deme i is $P_{ci} = 1/2N_{ei}$, so that $F_i = P_{ci}/(P_{ci} + 2m_i)$ and $1 - F_i = 2m_i/(P_{ci} + 2m_i)$. From **Equation B7.6.4**, we have $\tilde{T}_T = 1/P_{Cc} \approx 1/[2\Sigma_i v_i^2 m_i P_{ci}/(P_{ci} + 2m_i)]$. Using weights of $v_i^2 P_{ci}/\Sigma_j v_j^2 P_{cj}$ for the contribution to $\tilde{T}_S$ from each deme, from **Equation 7.16** we have $\tilde{T}_S \approx \tilde{T}_T \Sigma_i v_i^2 P_{ci}(1 - F_i)/\Sigma_j v_j^2 P_{cj} = \tilde{T}_T 2[\Sigma_i v_i^2 P_{ci} m_i/(P_{ci} + 2m_i)]/\Sigma_j v_j^2 P_{cj} = 1/\Sigma_j v_j^2 P_{cj}$, which is identical to **Equation B7.3.2**.

7.9* Exchangeability implies that deme effective sizes, migration rates, and extinction rates are independent of deme identity, so that subscripts indicating deme identity can be omitted. For the migrant-pool model, we can use **Equation B7.6.2** to write the probability of coalescence within a deme as $P_c = (n_e + N_e e)/2n_e N_e$. From **Equation B7.6.3**, we have $F = (n_e + N_e e)/2N_e n_e [P_c + 2m + e(1 - 1/2n_e)] = (n_e + N_e e)/(n_e + N_e e + 4N_e n_e m + 2N_e n_e e - N_e e) = (n_e + N_e e)/(n_e + 4N_e n_e m + 2N_e n_e e)$. **Equation 7.11** for the case of $\phi = 0$ is identical to this expression.

For the propagule-pool model, the same procedure can be used, except that we now do not include colonization events in the denominator of the equivalent of **Equation B7.6.3**, because these are treated as equivalent to the alleles in question staying in the same deme. The denominator of the expression for F thus becomes $2N_e n_e(P_c + 2m) = (n_e + 4N_e n_e m + N_e e)$, so that $F = (n_e + N_e e)/(n_e + 4N_e n_e m + N_e e)$. This is the same as **Equation 7.11** with $\phi = 1$.

7.10* Exchangeability implies that the values of the components of the leading left eigenvector of the B matrix in **Box 7.6*** must all be equal to $1/d$, where d is the number of demes. The probability that an allele is derived from a different deme in the previous generation is $m + e$ for both types of extinction/colonization model. **Equation B7.6.4** thus implies that $T_T \approx d/2(m + e)F$, and **Equation 7.16** gives $T_S \approx d(1 - F)/2(m + e)F$.

From the expression for F with the migrant-pool model, we thus have $T_T \approx d(n_e + 4N_e n_e m + 2N_e n_e e)/2(m + e)(n_e + N_e e)$. In addition, $1 - F = [4N_e n_e m + N_e e(2n_e - 1)]/(n_e + 4N_e n_e m + 2N_e n_e e)$, so that $T_S \approx d[4N_e n_e m + N_e e(2n_e - 1)]/2(m + e)(n_e + N_e e)$. These expressions for T_T and T_S are the same as **Equations 7.14a** and **7.14b**.

From the expression for F with the propagule-pool model, we have $T_T \approx d(n_e + 4N_e n_e m + N_e e)/2(m + e)(n_e + N_e e)$. In addition, $1 - F = 4N_e n_e m/(n_e + 4N_e n_e m + N_e e)$, so that $T_S \approx 2dN_e n_e m/(m + e)(n_e + N_e e)$. These are the same as **Equations 7.13a** and **7.13b**.

Chapter 8

8.1 If we substitute the expressions for the haplotype frequencies given by **Equations B8.1.1a–B8.1.1d** into **Equation B8.1.2**, we obtain $D = (p_A p_B + D)(q_A q_B + D) - (p_A q_B - D)(q_A p_B - D) = p_A p_B q_A q_B - p_A q_B q_A p_B + D(p_A p_B + q_A q_B + p_A q_B + q_A p_B) + D^2 - D^2 = D(p_A p_B + q_A q_B + p_A q_B + q_A p_B) = D[p_A(p_B + q_B) + q_A(q_B + p_B)] = D$.

8.2.i. From **Equation B8.1.2**, in the first case we have $D = (0.6)(0.1) - (0.1)(0.2) = 0.06 - 0.02 = 0.04$. We have $p_A = 0.8$, $q_A = 0.2$; $p_B = 0.7$, $q_B = 0.3$. Hence, $p_A q_B = 0.24$ and $q_A p_B = 0.14$; using the results following **Equation B8.2.1**, $D' = D/|D|_{MAX} = 0.04/0.14 = 0.286$. In addition, the product of the four variant frequencies is $0.8 \times 0.2 \times 0.7 \times 0.3 = 0.0336$, with square root 0.183. **Equation B8.2.3** thus gives $r = 0.219$.

In the second case, $D = (0.8)(0.05) - (0.1)(0.05) = 0.035$. We have $p_A = 0.9$, $q_A = 0.1$; $p_B = 0.85$, $q_B = 0.05$, so that $p_A q_B = 0.045$ and $q_A p_B = 0.085$, and $D' = 0.778$. The product of the four allele frequencies is 0.00382, with square root 0.0618, so that $r = 0.566$.

ii. The value of D is $(0.55)(0.17) - (0.16)(0.12) = 0.0743$. We have $p_A = 0.71$, $q_A = 0.29$; $p_B = 0.67$, $q_B = 0.33$. The expected numbers in each cell of the 2×2 table of **Box 8.3**, under the null hypothesis of no LD, are given by the products of the corresponding allele frequencies and the sample size (60), and are all greater than 5. We can thus use the χ^2 test for agreement between expected and observed numbers, given by **Equation B3.1.1**. This is equal to $(60)(0.0743)^2/(0.71 \times 0.29 \times 0.67 \times 0.33) = 7.28$. The table of χ^2 for one degree of freedom shows that the probability of exceeding this value is less than 1%, so that there is evidence for significant LD.

8.3.i. Substitute the formula for D in **Equation B8.1.2** in **Section 8.1.ii.a** into **Equations B8.4.2a–B8.4.2d**. The value of D in the new generation is thus given by $x_1'x_4' - x_2'x_3' = (x_1 - cD)(x_4 - cD) - (x_2 + cD)(x_3 + cD) = x_1x_4 - (x_1 + x_4)cD - x_2x_3 - (x_2 + x_3)cD = D - (x_1 + x_2 + x_3 + x_4)cD = (1 - c)D$.

ii. Equation B8.4.4 implies that the time for D to reach half of its initial value is given by $0.5 = (1 - c)^t$, so that $0.5 = t\ln(1 - c)$. For small c, $\ln(1 - c) \approx -c$ (**Appendix A1.ii.c**). The half-life is thus given by $\ln(0.5) = -ct$, so that $t = 0.69/c$.

8.4* We can use **Equation 7.16** to obtain a mean time T_S to coalescence within a deme, using weights of v_i for each deme, where v_i is the ith element of the left leading eigenvector of the migration matrix (scaled so that the v_i sum to 1). In the case of conservative migration with deme size proportional to effective deme size, the results in **Box 7.3*** imply that T_S defined in this way is equal to $2N_{eT}$. From **Equation 7.16**, the right-hand side of **Equation 8.3** is equivalent to T_S/T_T. This implies that $T_T\tilde{c} \approx T_Sc = 2N_{eT}c$. By definition, $T_T = 2N_{eM}$, so this implies that $4N_{eM}\tilde{c} \approx 4N_{eT}c$.

8.5* We can treat the wild-type and mutant alleles as defining two sub-populations, with effective sizes $N_e(1 - q)$ and N_eq, respectively. Using **Equation B7.4.1**, the rate of coalescence for a pair of alleles from the population as a whole is approximated by the sum $v_1{}^2P_{c1} + v_2{}^2P_{c2}$, where P_{c1} is given by $1/2N_e(1 - q)$ and $P_{c2} = 1/2N_eq$. From **Box 8.6***, we have $v_1 = b_{21}/(b_{12} + b_{21})$ and $v_2 = b_{12}/(b_{12} + b_{21})$. In addition, $b_{12} = qc(1 - t)$ and $b_{21} = (1 - q)[c(1 - t) + t]$, so that $b_{12} + b_{21} = c(1 - t) + t - qt$. The ratio of the rate of coalescence to $1/2N_e$ is thus equal to:

$$\frac{1}{\left[c(1 - t) + t - qt\right]^2}\left\{(1 - q)\left[c(1 - t) + t\right]^2 + qc^2(1 - t)^2\right\}$$

$$= \frac{\left\{\left[c(1 - t) + t\right]^2 - 2qct(1 - t) - qt^2\right\}}{\left[c(1 - t) + t - qt\right]^2}$$

This can be further simplified by approximating the denominator by ignoring terms in q^2, and writing it as $[c(1 - t) + t]^2\{1 - 2qt/[c(1 - t) + t]\}$. Similarly, the numerator can be written as $[c(1 - t) + t]^2\{1 - [2qct(1 - t) - qt^2]/[c(1 - t) + t]^2\}$. But $1/\{1 - 2qt/[c(1 - t) + t]\}$ is approximately $1 + 2qt/[c(1 - t) + t]$, so that the full expression can be approximated by:

$$1 + \frac{2qt}{\left[c(1 - t) + t\right]} - \frac{2qct(1 - t) + qt^2}{\left[c(1 - t) + t\right]^2}$$

$$= 1 + \frac{2qct(1-t) + 2qt^2 - 2qct(1-t) - qt^2}{\left[c(1-t) + t\right]^2} = 1 + \frac{qt^2}{\left[c(1-t) + t\right]^2}$$

Inverting this and ignoring terms in q^2 and higher, we obtain the expression for coalescence time in **Equation B8.6.2**.

8.6.i.* We have $q_{A1} = x_3/p_B = (q_Ap_B - D)/p_B$ and $q_{A2} = x_4/q_B = (q_Aq_B + D)/q_B$. Hence, $p_Bq_{A1} = q_Ap_B - D$ and $q_Bq_{A2} = q_Aq_B + D$, so that $p_Bq_B(q_{A1} - q_{A2}) = q_Ap_Bq_B - q_BD - p_Bq_Aq_B - p_BD = -D$. Hence, $p_Bq_B(q_{A2} - q_{A1}) = D$.

ii. Let $y = 2c/s$ and $x = q_B$. Then we need to evaluate the integral of $(1 - x)^y x^{-y}$ between 0 and 1, because q_{B0} is close to 0. We can expand $(1 - x)^y$ as $1 - yx$ plus higher order terms in x. But if $y \ll 1$, this series is dominated by its first term, so that we need integrate only the term x^{-y}, whose indefinite integral is $x^{1-y}/(1 - y)$, which is approximately equal to x. The definite integral between 0 and 1 is thus 1, which establishes the desired result.

8.7.i. The initial gamete frequencies are $A_1B_1 = 0.2$ and $A_2B_1 = 0.8$; immediately after the rise in frequency of B_2, a fraction 0.25 of all gametes are A_1B_2. There is no change in frequency of A_1 versus A_2 among the B_1 gametes, whose frequency is 0.75. Hence, the new A_1B_1 frequency is $0.75 \times 0.2 = 0.15$, and the new A_2B_1 frequency is $0.75 \times 0.8 = 0.60$. The new allele frequencies are thus: $p_A = 0.25 + 0.15 = 0.40$, and $q_B = 0.25$. The initial linkage disequilibrium is $D_0 = -0.25 \times 0.60 = 0.15$.

ii. The state under investigation is such that A_1B_2 and A_2B_2 have equal frequencies, i.e., $x_2 = x_4$. The allele frequencies are unchanged by recombination. Using the relations given in **Box 8.1** of **Section 8.1.ii.a**, we have $q_B = x_2 + x_4 = 2x_2 = 0.25$, so that $x_2 = x_4 = 0.125$. Similarly, $p_A = x_1 + x_2 = x_1 + 0.125 = 0.40$, so that $x_1 = 0.275$. In addition, $x_3 = 1 - 2 \times 0.125 - 0.275 = 0.475$. The final linkage disequilibrium is thus $D_t = 0.125 \times 0.275 - 0.125 \times 0.475 = -0.025$.

iii. The final linkage disequilibrium is $D_t = -0.025$. Hence, $0.025 = 0.15 \times (1 - 0.001)^t$, from **Equation B.8.4.4** of **Section 8.2.ii.a**. This implies that $\ln(0.025/0.15) = t \ln(1 - 0.001)$, i.e. $1.79 = t \times 0.001$, so that $t = 1790$.

8.8 Using **Equations B8.8.1**, we have $\bar{w}^2 x_1' x_4' = (x_1w_1 - cDw_{14})(x_4w_4 - cDw_{14}) = x_1w_1 x_4w_4 - (x_1w_1 + x_4w_4)cDw_{14} + (cDw_{14})^2$ and $\bar{w}^2 x_2' x_3' = (x_2w_2 + cDw_{14})(x_3w_3 + cDw_{14}) = x_2w_2 x_3w_3 + (x_2w_2 + x_3w_3)cDw_{14} + (cDw_{14})^2$. Hence, $\bar{w}^2 D' = \bar{w}^2(x_1'x_4' - x_2'x_3') = (x_1w_1 x_4w_4 - x_2w_2 x_3w_3) - (x_1w_1 + x_4w_4 + x_2w_2 + x_3w_3)cDw_{14} + (cDw_{14})^2 - (cDw_{14})^2$, so that $\bar{w}^2 D' = (x_1w_1 x_4w_4 - x_2w_2 x_3w_3) - \bar{w}cDw_{14}$. This is equivalent to **Equation B8.8.2c**.

8.9.i. Use the fitness matrix in **Table 8.1**. If fitnesses are additive, we have $w_1 = 1 - (s_A + s_B)x_1 - s_Ax_2 - s_Bx_3$; $w_2 = 1 - s_Ax_1 - (s_A + t_B)x_2 - t_Bx_4$; $w_3 = 1 - s_Bx_1 - (s_B$

$+ t_A)x_3 - t_A x_4$; $w_4 = 1 - t_B x_2 - t_A x_3 - (t_A + t_B)x_4$. These give $w_{1.} + w_{4.} = 2 - (s_A + s_B)x_1 - (s_A + t_B)x_2 - (s_B + t_A)x_3 - (t_A + t_B)x_4$; similarly, $w_{2.} + w_{3.} = 2 - (s_A + s_B)x_1 - (s_A + t_B)x_2 - (s_B + t_A)x_3 - (t_A + t_B)x_4$. Hence, $\bar{e} = w_{1.} + w_{4.} - w_{2.} - w_{3.} = 0$.

ii. If the equilibrium value of D is 0, the equilibrium allele frequencies at each locus are the same as in the single-locus case. This can be verified by assuming it is true and showing that the results are consistent with the conditions for equilibrium at each locus. The single-locus equilibrium formulae give $p_A = t_A/(s_A + t_A)$, $q_A = s_A/(s_A + t_A)$; $p_B = t_B/(s_B + t_B)$, $q_B = s_B/(s_B + t_B)$ (**Section 2.1.ii.c**). At equilibrium with $D = 0$, **Equations B8.8.1** show that all four marginal fitnesses must equal the population mean fitness, $\bar{w}$. With $D = 0$, the reasoning used in **Section 4.3.iii** shows that the equilibrium $\bar{w} = [1 - s_A t_A/(s_A + t_A)][1 - s_B t_B/(s_B + t_B)]$. It is thus sufficient to show that $w_{1.}$, the marginal fitness of $A_1 B_1$, is equal to $\bar{w}$ under these conditions. With $D = 0$, the haplotype frequencies are given by the products of the allele frequencies, so that $w_{1.} = 1 - (s_A + s_B - s_A s_B)p_A p_B - s_A p_A q_B - s_B q_A p_B = 1 - s_A p_A - s_B p_B + s_A s_B p_A p_B = 1 - s_A t_A/(s_A + t_A) - s_B t_B/(s_B + t_B) + s_A s_B t_A t_B/(s_A + t_A)(s_B + t_B)$. This is equivalent to the expression for $\bar{w}$.

8.10 Under the multiplicative fitness model with heterozygote advantage, a non-recombining population segregating for $A_1 B_1$ and $A_2 B_2$ is equivalent to a single locus system with homozygous fitnesses for $A_1 B_1/A_1 B_1$ and $A_2 B_2/A_2 B_2$ of $(1 - s_A)(1 - s_B) = 1 - (s_A + s_B - s_A s_B)$ and $(1 - t_A)(1 - t_B) = 1 - (t_A + t_B - t_A t_B)$, respectively. Symmetrical fitness effects at each locus mean that $s_A = t_A$ and $s_B = t_B$, so that the homozygous fitnesses are both equal to $1 - (s_A + s_B - s_A s_B)$. The equilibrium frequencies of $A_1 B_1$ and $A_2 B_2$ are each 1/2. The mean fitness of this population, $\bar{w}$, is thus equal to $1 - (s_A + s_B - s_A s_B)/2$. An $A_1 B_2$ haplotype introduced at low frequency into this population has marginal fitness $w_{2.} = 1 - (s_A + s_B)/2$, because it is heterozygous with $A_1 B_1$ and $A_2 B_2$ with equal frequencies. The difference $\bar{w} - w_{2.}$ is thus equal to $s_A s_B/2$. From the definition of the selection coefficients, this is necessarily positive; **Equations B8.8.1** imply that this haplotype will therefore decrease in frequency in the absence of recombination. An identical result holds for the other possible new haplotype, $A_2 B_1$.

8.11* **Equations B8.8.1** imply that we can write the change in frequency of haplotype i in the form $\Delta x_i = [x_i(w_{i.} - \bar{w}) \pm cDw_{14}]/\bar{w}$, where $w_{i.} = \Sigma_j x_j w_{ij}$ ($i = 1, \dots, 4$) and the signs of the terms in cD are negative for $i = 1$ or 4 and positive for $i = 2$ or 3. The first term on the right-hand side of **Equation B8.11.1** is thus equal to $\Sigma_i w_{i.}[x_i(w_{i.} - \bar{w}) \pm cDw_{14}]/\bar{w} = [\Sigma_i x_i(w_{i.} - \bar{w})^2 \pm \Sigma_i w_{i.}cDw_{14}]/\bar{w}$, because $\Sigma_i x_i w_{i.} = \bar{w}$. By the definition of $\bar{e}$, this reduces to $[\Sigma_i x_i(w_{i.} - \bar{w})^2 + \bar{e}cD]/\bar{w}$, neglecting terms of order s^3, noting that $w_{14}/\bar{w} = 1 + O(s)$, and D is $O(s)$. From the definition of $w_{i.}$, we have $\Delta w_{i.} = \Sigma_j w_{ij}[x_j(w_{j.} - \bar{w}) \pm cDw_{14}]/\bar{w}$, where j is summed from 1 to 4. Summing $x_i \Delta w_{i.}$ over i from 1 to 4 and collecting terms, we find that $\Sigma_i x_i \Delta w_{i\cdot} \approx [\Sigma_i x_i (w_{i\cdot} - \bar{w})^2 + \bar{e}\, cD]/\bar{w}$, to the same accuracy as above. Substituting these results into **Equation B8.11.1** yields **Equation B8.11.2**.

Chapter 9

9.1 All carriers of the mutation in each generation must be A_1A_2, because A_1A_2 individuals cannot mate with each other. Let the frequency of carriers of the mutation in a given generation be $2q$; a fraction $1 - 2q$ of the population is A_1A_1. The fraction of the pollen that carries A_2 will be $q/(1 - 2q + 2q) = q$. This fertilizes the ovules of sexual plants in the next generation, which are all A_1A_1, giving us a net contribution of $(1 - 2q)q$ to the pool of A_1A_2 individuals from seed produced by A_1A_1 parents. There is a similar contribution of $2q$ from seed produced by A_1A_2 parents, so that the new frequency of A_1A_2 is $2q' = q[2 + (1 - 2q)]$. This gives $q' = q(3/2 - q)$. For small q, this reduces to $q' = 3q/2$.

9.2.i. The parameters needed are described in **Box 9.4**. Using the approach of **Box 9.3**, we can treat the cosexuals as phenotype 1, with frequency X and selfing rate S. The females are phenotype 2, with no selfing or male fertility, whose frequency is $1 - X$. Their female fertility is $1 + k$ relative to that of the cosexuals. The condition for equilibrium is that the total fitnesses of each phenotype, as given by **Equation B9.3.1**, are the same. The quantity W in **Equations B9.3.3** and **B9.3.4** is irrelevant, because it is the same for both phenotypes. The female fitness components are proportional to $1 - S\delta$ for phenotype 1, and to $1 + k$ for phenotype 2. There is no male fitness component for phenotype 2; for phenotype 1, the size of the pool of outcrossed seeds is proportional to $V = X(1 - S) + (1 - X)(1 + k)$, and the size of the pollen pool is proportional to $U = X$. From **Equation B9.3.4a**, the male fitness component for phenotype 1 is thus proportional to $S(1 - \delta) + [X(1 - S) + (1 - X)(1 + k)]/X$, so that the net fitness of phenotype 1 is proportional to $1 - S\delta + S(1 - \delta) + [X(1 - S) + (1 - X)(1 + k)]/X$. The condition for equilibrium is thus $1 + k = 1 - S\delta + S(1 - \delta) + [X^*(1 - S) + (1 - X^*)(1 + k)]/X^*$. Let $Y^* = (1 - X^*)/X^*$; then $1 + k = 1 - S\delta + S(1 - \delta) + (1 - S) + (1 + k)Y^*$, giving $Y^* = [(k - 1) + 2S\delta]/(1 + k)$.

ii. In this case, phenotype 2 lacks female fertility and thus makes no contribution to the pool of outcrossed seeds; its male fertility is increased by a factor of $1 + K$ compared with the cosexuals. Because the males are assumed to be rare, the sizes of the pool of outcrossed seeds and the pollen pool are approximately proportional to $V = (1 - S)$ and $U = 1$, respectively. The male fitness component for phenotype 1 is thus proportional to $S(1 - \delta) + (1 - S)$, so that the approximate net fitness of phenotype 1 is proportional to $1 - S\delta + S(1 - \delta) + (1 - S) = 2(1 - S\delta)$. The approximate net fitness of phenotype 2 is proportional to $(1 + K)(1 - S)$. The condition for invasion is thus $1 + K > [2(1 - S\delta)]/(1 - S)$, which is equivalent to **Equation 9.2**.

9.3 Denote the proportion of maternal resource devoted to males as a_1 for the initial population and a_2 for the rare mutants. The numbers of males produced per female are then Ra_1/α and Ra_2/α, respectively. The corresponding numbers of females are $R(1 - a_1)$ and $R(1 - a_2)$, respectively. The argument in **Box 9.6** can be applied using these expressions for the numbers of males and females produced by each type; this implies that **Equations B9.6.1** can be used, simply by substituting a_i for c_i ($i = 1$ or 2), because R and α cancel from the numerator and the denominator. This implies that the ESS is $a^* = 1/2$, i.e., an equal fraction of resources is expended on male and female offspring that are raised successfully by the mother. Because a male costs α units of resource to raise compared with 1 unit for a female, we have $c^*\alpha = (1 - c^*)$, where c^* is the ESS proportion of males raised successfully. This implies that the ESS ratio of the numbers of female and male offspring raised is α to 1.

9.4* The rules for evaluating an ESS (**Box 9.5**) imply that we need to evaluate the derivative of the total fitness of the mutant, w_2, with respect to its associated proportion of males, c_2. Using the rule for differentiating a product (**Appendix A1.i**), the expressions for $w_2{}^f$ and $w_2{}^m$ in **Box 9.7** imply that the derivative of their sum with respect to c_2 is proportional to:

$$-1 + \frac{\left[(n-1)c_1 + c_2\right]\left[(n-1)(1 - c_1) + (1 - c_2) - c_2\right] - \left[(n-1)(1 - c_1) + (1 - c_2)\right]c_2}{\left[(n-1)c_1 + c_2\right]^2}$$

If we set this to 0, and equate c_2 to c_1, the overall sex ratio, we obtain an expression that determines the ESS sex ratio, c^*. With $c_2 = c_1 = c^*$, the above expression is proportional to:

$$-(nc^*)^2 + (nc^*)[n(1 - c^*) - c^*] - n(1 - c^*)c^* = -2(nc^*)^2 + n^2c^* - nc^*$$

Equating this to 0 and simplifying, we obtain $2nc^* = n - 1$, which yields **Equation B9.7.2**.

9.5.i.* Using the formulation in **Box 9.3**, and assuming that the invading phenotype (i.e., 2) is at low frequency, we can set $X = 1$ in the equivalents of **Equations B9.3.2–B9.3.4**. The selfing rate S is assumed to be independent of allocation. The female component of fitness for phenotype 2 is thus proportional to $(1 - a_2)S(1 - \delta)$. We have $U = a_1$ and $V = (1 - a_1)(1 - S)$. From **Equations B9.3.4**, the net fitness of phenotype 2 is proportional to $(1 - a_2)(1 - S\delta) + (1 - a_2)S(1 - \delta) + (1 - a_1)(1 - S)a_2/a_1$. Differentiating this with respect to a_2, we have $dw_2/da_2 \propto -[1 + S(1 - 2\delta)] + (1 - a_1)(1 - S)/a_1$. The necessary condition for an ESS is for this to be 0 at $a_1 = a^*$, which yields $a^*/(1 - a^*) = (1 - S)/[1 + S(1 - 2\delta)]$. It can be seen from the expression for dw_2/da_2 that this derivative is positive for $a_1 < a^*$

and negative for $a_1 > a^*$, so that a^* represents a stable point according to the criterion of **Box 9.5.ii***.

With $S = 0$, $a^*/(1 - a^*) = 1$, so that there is equal allocation to male and female function. The derivative of $a^*/(1 - a^*)$ with respect to S is proportional to $-[1 + S(1 - 2\delta)] - (1 - S)(1 - 2\delta) = -2(1 - \delta) < 0$. Unless $\delta = 1$, so that selfing is lethal, a^* therefore decreases with S.

ii. We now have $U = a_1{}^y$ and the male component of fitness for phenotype 2 is proportional to $(1 - a_2)S(1 - \delta) + (1 - a_1)(1 - S)a_2{}^y/a_1{}^y$. Adding the female component of fitness from the answer to **Problem 9.5.i** and differentiating, we now have $dw_2/da_2 = -[1 + S(1 - 2\delta)] + (1 - a_1)(1 - S)ya_2{}^{y-1}/a_1{}^y$. We require this to be 0 at $a_1 = a_2 = a^*$, which gives $[1 + S(1 - 2\delta)] = y(1 - a^*)(1 - S)/a^*$. We thus have $a^*/(1 - a^*) = y(1 - S)/[1 + S(1 - 2\delta)]$. If $y < 1$, as is the case for a diminishing-returns function, then the ratio is smaller than with a linear function by a factor of y.

9.6.i.* The xth column of the Leslie matrix described in **Box 9.8*** has its first component equal to $f(x)$; the only other potentially nonzero component of this column is the $(x + 1)$th element, the survival probability $P(x)$. By the definition of a left eigenvector (**Equation A1.20b**), we have $v(1)f(x) + v(x + 1)P(x) = \lambda_1 v(x)$; this is the same as **Equation B9.9.3**.

ii. Equation B9.9.2 can be written as $v(x) = m(x) + \exp(rx)l(x + 1)[\Sigma_{y=x+1} k(y) \exp(-ry)]/l(x)l(x + 1)$; this is the same as $v(x) = m(x) + \lambda_1{}^{-1}P(x)v(x + 1)$. From **Equation B9.9.2**, the $v(x)$ values are defined on a scale such that $v(1) = \lambda_1/P(0)$, so that $v(x) = P(0)m(x)v(1)\lambda_1{}^{-1} + \lambda_1{}^{-1}P(x)v(x + 1)$. Noting that $f(x) = P(0)m(x)$, this is equivalent to **Equation B9.9.3**.

iii. The total reproductive value can be written as the matrix product $v_T(t) = \mathbf{v}_1 n(t)$. Because $n(t + 1) = An(t)$ (**Equation B9.8.3**), we have $v_T(t + 1) = \mathbf{v}_1 An(t) = \lambda_1 \mathbf{v}_1 n(t) = \lambda_1 v_T(t)$.

9.7* Using the first two terms of the Taylor's series expansion of $w_{ij}(r)$ at r_{ij}, the intrinsic rate for genotype ij, we have $\mathrm{w}_{ij}(r) \approx w_{ij}(r_{ij}) + (r - r_{ij})(dw_{ij}(r)/dr)_{r_{ij}}$. But $w_{ij}(r_{ij}) = 1$, and $(dw_{ij}(r)/dr)_{r_{ij}} = -\Sigma_x x \exp(-r_{ij}x)k(x) = -T_{ij}$, where T_{ij} is the generation time in a population consisting solely of genotype ij (**Box 9.9.iv** in **Section 9.5.i.b**). Hence, $w_{ij}(r) - 1 \approx (r_{ij} - r) T_{ij}$. If selection is weak, we can replace T_{ij} for each genotype with the generation time of a standard genotype, T_S, neglecting second-order terms in the strength of selection, as was done in the initial approximation. It follows that the fitness difference between genotypes ij and kl, $w_{ij} - w_{kl}$, is equal to $(w_{ij} - w_{kl})T_S$ plus second-order terms in the strength of selection.

9.8.i. From **Equation B9.9.2**, we have $v(x) = \Sigma_y ml(y) [\exp -r(y - x)]/l(x)$. For adult ages, $l(y)/l(x) = P^{(y-x)}$. Writing $z = y - x$, we thus have $v(x) = m\Sigma_z P^z$

exp($-rz$), where z runs from 0 to infinity. Using the expression for the sum of a geometric series (**Appendix A1.ii.c**), noting that exp($-rz$) = [exp($-r$)]^z and assuming that P exp($-r$) < 1, we have Σ_z P^z exp($-rz$) = 1/[1 − P exp($-r$)]. This implies that $v(x)$ for adult ages is independent of age and can be written as $v = m/[1 − P\exp(-r)]$.

ii. Let the probability of survival to the age of first reproduction be $l(b)$. Then the Euler–Lotka equation with constant adult fecundity and survival becomes 1 = $l(b)$ exp($-rb$)Σ_z $P^z m$ exp($-rz$) = $l(b)$ exp($-rb$)m/[1 − P exp($-r$)]. Hence, 1 = $vl(b)$exp($-rb$).

9.9 In this case, $S(x)$ for an adult age x is given by $m\Sigma_y$ $l(y)$, where y runs from x + 1 to infinity. This can be written as $ml(b)P^{x+1-b}\Sigma_z P^z$, where $z = y - x$. From the results in **Problem 9.8**, this is equal to $ml(b)$ P^{x+1-b}/(1 − P). From **Problem 9.8.ii**, we have 1 = $ml(b)$/(1 − P), so that $S(x) = P^{x+1-b}$.

Chapter 10

10.1 Let the frequency of the GC variant be q. Biased gene conversion has an effect only in heterozygotes, whose frequency is $2pq$. From **Equation B10.1.1**, the new frequency of GC among the gametes produced by heterozygotes is $k = (1 + \omega/2)/2$. The difference in frequency from the value of one-half among the heterozygotes at the start of a generation is thus $\omega/4$, so that the net change in frequency in the population is $2pq\omega/4 = pq\omega/2$.

10.2 From the assumption of distortion only in males, the equilibrium frequency of the distorter is the same in eggs and sperm. Denote this frequency by q^*. Then **Equation B10.3.1b** can be written as $q^* = [2p^*q^*k(1 - s) + q^{*2}(1 - t)]/(1 - 2p^*q^*s - q^{*2}t)$. Canceling q^* from both sides gives $1 - 2p^*q^*s - q^{*2}t = 2p^*k(1 - s) + q^*(1 - t)$. The right-hand side can be written as $p^*[2k(1 - s) - 1] + p^* + q^*(1 - t) = 1 - q^*t + (1 - q^*)[2k(1 - s) - 1]$, and the left-hand side can be written as $1 - 2q^*s - q^{*2}(t - 2s)$. Subtracting the left-hand side from the right-hand side and collecting terms, we obtain the quadratic equation $(t - 2s)q^{*2} - \{t - 2s + [2k(1 - s) - 1]\}q^* + 2k(1 - s) - 1 = 0$. Write $C = 2k(1 - s) - 1$. The solution to this quadratic equation is then:

$$q^* = \frac{(t - 2s) + C \pm \sqrt{[t - 2s + C]^2 - 4(t - 2s)C}}{2(t - 2s)}$$

Collecting terms in the square root term, this reduces to:

$$q^* = \frac{(t - 2s) + C \pm [(t - 2s) - C]}{2(t - 2s)}$$

The only meaningful solution (with $q^* < 1$) is $q^* = [2k(1-s) - 1]/(t - 2s)$.

10.3★ The Taylor's series expansion of $E\{nu_n\}$ around $\bar{n}$ is:

$$\bar{n}u_{\bar{n}} + E\left\{(n - \bar{n})\left(\frac{\mathrm{d}u_n}{\mathrm{d}n}\right)_{\bar{n}} + \frac{(n - \bar{n})^2}{2}\left(\frac{\mathrm{d}^2u_n}{\mathrm{d}n^2}\right)_{\bar{n}} + \cdots\right\}$$

Because $E\{n\} = \bar{n}$, this is approximately equal to $\bar{n}u_{\bar{n}}$, provided that the higher-order derivatives are sufficiently small, as would be the case for a nearly linear relation between u and n.

References

Abbott, R.J. and M.F. Gomes. 1988. Population genetic structure and outcrossing rate of *Arabidopsis thaliana* (L.) Heynh. *Heredity* **62**: 411–418.

Abdallah, F., F. Salamini, and D. Leister. 2000. A prediction of the size and evolutionary origin of the proteome of chloroplasts of Arabidopsis. *Trends Plant Sci.* **5**: 141–142.

Abrams, P.A. and D. Ludwig. 1995. Optimality theory, Gompertz's law and the disposable soma theory of senescence. *Evolution* **49**: 1055–1066.

Agrawal, A.F. 2006a. Similarity selection and evolution of sex: revisiting the Red Queen. *PLoS Biol.* **4**: 1364–1371.

Agrawal, A.F. 2006b. Evolution of sex: why do organisms shuffle their genotypes? *Curr. Biol.* **16**: R696–R704.

Akashi, H. 1994. Synonymous codon usage in *Drosophila melanogaster*: natural selection and translational accuracy. *Genetics* **136**: 927–935.

Akashi, H. 1995. Inferring weak selection from patterns of polymorphism and divergence at "silent" sites in *Drosophila* DNA. *Genetics* **139**: 1067–1076.

Akashi, H. 1996. Molecular evolution between *Drosophila melanogaster* and *D. simulans*: reduced codon bias, faster rates of amino acid substitution, and larger proteins in *D. melanogaster*. *Genetics* **144**: 1297–1307.

Akashi, H. 1997. Distinguishing the effects of mutational biases and natural selection on DNA sequence variation. *Genetics* **147**: 1989–1991.

Akashi, H. 1999. Inferring the fitness effects of DNA polymorphisms and divergence data: statistical power to detect directional selection under stationarity and free recombination. *Genetics* **151**: 221–238.

Akashi, H. and S. Schaeffer. 1997. Natural selection and the frequency distributions of "silent" DNA polymorphism in *Drosophila*. *Genetics* **146**: 289–307.

Akashi, H., W.-Y. Ko, S. Piao, A. John, P. Goel, et al. 2006. Molecular evolution in the *Drosophila melanogaster* species subgroup: frequent parameter fluctuations on the timescale of molecular divergence. *Genetics* **172**: 1711–1726.

Akey, J.M. 2009. Constructing genomic maps of positive selection in humans: where do we go from here? *Genome Res.* **19**: 711–722.

Akey, J.M., G. Zhang, K. Zhang, L. Jin, and M.D. Shriver. 2002. Interrogating a high-density SNP map for signatures of natural selection. *Genome Res.* **12**: 1805–1814.

Akin, E. 1982. Cycling in simple genetic systems. *J. Math. Biol.* **13**: 305–324.

Allard, R.W., G.R. Babbel, M.T. Clegg, and A.L. Kahler. 1972. Evidence for coadaptation in *Avena barbata*. *Proc. Natl. Acad. Sci. U.S.A.* **69**: 3043–3048.

Allen, D.E. and M. Lynch. 2008. Both costs and benefits of sex correlate with relative frequency of asexual reproduction in cyclically parthenogenic *Daphnia pulicaria* populations. *Genetics* **179**: 1497–1502.

Allen, G.A. and J.A. Antos. 1993. Sex ratio variation in the dioecious shrub *Oemleria cerasiformis*. *Am. Nat.* **141**: 537–553.

Allison, A.C. 1964. Polymorphism and natural selection in human populations. *Cold Spring Harbor Symp. Quant. Biol.* **29**: 137–149.

Allison, A.C. 2004. Two lessons from the interface of genetics and medicine. *Genetics* **166**: 1591–1599.

Altenberg, L. and M.W. Feldman. 1987. Generalized reduction principle and the evolution of modifier genes. I. The reduction principle. *Genetics* **117**: 559–572.

Alter, S.E., E. Rynes, and S.R. Palumbi. 2007. DNA evidence for historic population size and past ecosystem impacts of gray whales. *Proc. Natl. Acad. Sci. U.S.A.* **104**: 15162–15167.

Anderson, G.J. and D.E. Symon. 1989. Functional dioecy and andromonoecy in Solanum. *Evolution* **43**: 204–219.

Anderson, L. K., G. G. Doyle, B. Brigham, J. Carter, K. D. Hooker, et al. 2003. High-resolution crossover maps for each bivalent of *Zea mays* using recombination nodules. *Genetics* **165**: 849–865.

Anderson, R.M. and R.M. May. 1991. *Infectious Diseases of Humans: Dynamics and Control*. Oxford University Press, Oxford, U.K.

Anderson, W.W. 1973. Genetic divergence in body size among experimental populations of *Drosophila pseudoobscura* kept at different temperatures. *Evolution* **27**: 278–284.

Andersson, M. 1994. *Sexual Selection*. Princeton University Press, Princeton, NJ.

Andersson-Ceplitis, H. and B.O. Bengtsson. 2002. Transmission rates and phenotypic effects of mitochondrial plasmids and cytotypes in *Silene vulgaris*. *Evolution* **56**: 1586–1591.

Andolfatto, P. 2001. Contrasting patterns of X-linked and autosomal nucleotide variation in African and non-African populations of *Drosophila melanogaster* and *D. simulans*. *Mol. Biol. Evol.* **18**: 279–290.

Andolfatto, P. 2005. Adaptive evolution of non-coding DNA in *Drosophila*. *Nature* **437**: 1149–1152.

Andolfatto, P. 2007. Hitchhiking effects of recurrent beneficial amino acid substitutions in the *Drosophila melanogaster* genome. *Genome Res.* **17**: 1755–1762.

Andolfatto, P. 2008. Controlling type-I error of the McDonald–Kreitman test in genomewide scans for selection on noncoding DNA. *Genetics* **180**: 1767–1771.

Andolfatto, P. and M. Przeworski. 2000. A genome wide departure from the standard neutral model in natural populations of *Drosophila*. *Genetics* **156**: 257–268.

Antonovics, J. 1968. Evolution in closely adjacent plant populations. V. Evolution of self-fertility. *Heredity* **23**: 219–238.

Aquadro, C.F., S.F. Deese, M.M. Bland, C.H. Langley, and C.C. Laurie-Ahlberg. 1986. Molecular population genetics of the alcohol dehydrogenase gene region of *Drosophila melanogaster*. *Genetics* **114**: 1165–1190.

Aravin, A.A., G.J. Hannon, and J. Brennecke. 2007. The Piwi-piRNA pathway provides an adaptive defense in the transposon arms race. *Science* **318**: 761–764.

Arndt, P.H. and T. Hwa. 2005. Identification and measurement of neighbor-dependent nucleotide substitution processes. *Bioinformatics* **21**: 2322–2328.

Arnold, J. and W.W. Anderson. 1983. Density-regulated selection in a heterogeneous environment. *Am. Nat.* **121**: 656–668.

Ashburner, M., K.G. Golic, and R.S. Hawley. 2005. *Drosophila. A Laboratory Handbook*. Cold Spring Harbor Press, Cold Spring Harbor, NY.

Asthana, S., S. Schmidt, and S. Sunyaev. 2005. A limited role for balancing selection. *Trends Genet.* **21**: 30–32.

Asthana, S., W. S. Noble, G. Kryukov, C. E. Grant, S. Sunyaev, and J. A. Stamatoyannopoulos. 2007. Widely distributed noncoding purifying selection in the human genome. *Proc. Natl. Acad. Sci. U.S.A.* **104**: 12410–12415.

Atlan, A., H. Mercot, C. Landre, and C. Montchamp-Moreau. 1997. The sex-ratio trait in *Drosophila simulans*: geographical distribution of distortion and resistance. *Evolution* **51**: 1886–1895.

Atwood, K.C., L.K. Schneider, and F.J. Ryan. 1951. Periodic selection in *Escherichia coli. Proc. Natl. Acad. Sci. U.S.A.* **37**: 146–155.

Austerlitz, F., S. Mariette, N. Machon, P.H. Gouyon, and B. Goddelle. 2000. Effects of colonization processes on genetic diversity: differences between annual plants and tree species. *Genetics* **154**: 1309–1321.

Avent, N.D. and M.E. Reid. 2000. The Rh blood group system: a review. *Blood* **95**: 375–387.

Avila, V., D. Chavarrias, E. Sanchez, A. Manrique, C. Lopez-Fanjul, and A. Garcia-Dorado. 2006. Increase of the spontaneous mutation rate in a long-term experiment on *Drosophila melanogaster*. *Genetics* **173**: 267–277.

Avise, J.C. 2000. *Phylogeography: The History and Formation of Species*. Harvard University Press, Cambridge, MA.

Axelsson, E., N.G.C. Smith, H. Sundstrom, S. Berlin, and H. Ellegren. 2004. Male-biased mutation rate and divergence in autosomal, Z-linked and W-linked introns of chicken and turkey. *Mol. Biol. Evol.* **21**: 1538–1547.

Bachtrog, D. 2003. Accumulation of Spock and Worf, two novel non-LTR retrotransposons, on the neo-Y chromosome of *Drosophila miranda*. *Mol. Biol. Evol.* **20**: 173–181.

Bachtrog, D. 2004. Evidence that positive selection drives Y-chromosome degeneration in *Drosophila miranda*. *Nat. Genet.* **36**: 518–522.

Bachtrog, D. 2006. Expression profile of a degenerating neo-Y chromosome in *Drosophila*. *Curr. Biol.* **16**: 1694–1699.

Bachtrog, D., E. Hom, K.M. Wong, X. Maside, and P. De Jong. 2008. Genomic degradation of a young Y chromosome in *Drosophila miranda*. *Genome Biol.* **9**: R30.

Baer, C.F. 1999. Among-locus variation in F_{st}: fish, allozymes and the Lewontin–Krakauer test revisited. *Genetics* **152**: 653–659.

Bahlo, M. and R.C. Griffiths. 2000. Inferences from gene trees in a subdivided population. *Theor. Popul. Biol.* **57**: 79–95.

Baines, J.F. and B. Harr. 2007. Reduced X-linked diversity in derived populations of house mice. *Genetics* **175**: 1911–1921.

Baines, J.F., S.A. Sawyer, D.L. Hartl, and J. Parsch. 2008. Effects of sex-linkage and sex-biased gene expression in the rate of adaptive protein evolution in *Drosophila*. *Mol. Biol. Evol.* **25**: 1639–1650.

Baker, H.G. 1955. Self-compatibility and establishment after "long-distance" dispersal. *Evolution* **9**: 347–348.

Baker, H.G. 1959. Reproductive methods as a factor in speciation in flowering plants. *Cold Spring Harbor Symp. Quant. Biol.* **24**: 177–191.

Balding, D.J., M. Bishop, and C. Cannings. 2003. *Handbook of Statistical Genetics.* John Wiley, Chichester, U.K.

Barkalow, F.S., R.B. Hamilton, and R.F. Soots. 1970. The vital statistics of an unexploited gray squirrel population. *J. Wildl. Man.* **34**: 489–500.

Barraclough, T.G., D. Fontaneto, C. Ricci, and E.A. Herniou. 2007. Evidence for inefficient selection against deleterious mutations in cytochrome oxidase I of asexual bdelloid rotifers. *Mol. Biol. Evol.* **24**: 1952–1962.

Barrett, S.C.H. 1992. Heterostylous genetic polymorphisms: model systems for evolutionary analysis. In *Evolution and Function of Heterostyly* (ed. S.C.H. Barrett), pp. 1–29. Springer-Verlag, Berlin.

Barrett, S.C.H. 2002. The evolution of plant sexual diversity. *Nat. Rev. Genet.* **3**: 274–284.

Barrett, S.C.H. and B.C. Husband. 1990. Variation in outcrossing rates in *Eichhornia paniculata*: the role of demographic and reproductive factors. *Plant Species Biology* **5**: 41–55.

Barrett, S.C.H. and J. Shore. 1987. Variation and evolution of breeding systems in the *Turnera ulmifolia* L. complex (Turneraceae). *Evolution* **41**: 340–354.

Barrett, S.C.H., L.D. Harder, and A.C. Worley. 1996. Comparative biology of plant reproductive traits. *Phil. Trans. R. Soc. B* **351**: 1272–1280.

Barrière, A. and M.-A. Félix. 2005. High local genetic diversity and low outcrossing rate in *Caenorhabditis elegans* natural populations. *Curr. Biol.* **15**: 1176–1184.

Barrière, A., S. Yang, E. Pekarek, C. Thomas, E. Haag, and I. Ruvinsky. 2009. Detecting heterozygosity in shotgun genome assemblies: Lessons from obligately outcrossing nematodes. *Genome Research* **19**: 470–480.

Bartolomé, C. and B. Charlesworth. 2006. Evolution of amino-acid sequences and codon usage on the *Drosophila miranda* neo-sex chromosomes. *Genetics* **174**: 2033–2044.

Bartolomé, C. and X. Maside. 2004. The lack of recombination drives the fixation of transposable elements on the fourth chromosome of *Drosophila melanogaster*. *Genet. Res.* **83**: 91–100.

Bartolomé, C., X. Bello, and X. Maside. 2009. Widespread evidence for horizontal transfer of transposable elements across *Drosophila* genomes. *Genome Biol.* **10**: R22.

Bartolomé, C., X. Maside, and B. Charlesworth. 2002. On the abundance and distribution of transposable elements in the genome of *Drosophila melanogaster*. *Mol. Biol. Evol.* **19**: 926–937.

Bartolomé, C., X. Maside, S. Yi, A.L. Grant, and B. Charlesworth. 2005. Patterns of selection on synonymous and non-synonymous variants in *Drosophila miranda*. *Genetics* **169**: 1495–1507.

Barton, N.H. 1993. The probability of fixation of a favoured allele in a subdivided population. *Genet. Res.* **62**: 149–157.

Barton, N.H. 1995a. A general model for the evolution of recombination. *Genet. Res.* **65**: 123–144.

Barton, N.H. 1995b. Linkage and the limits to natural selection. *Genetics* **140**: 821–841.

Barton, N.H. 1996. Natural selection and random genetic drift as causes of evolution on islands. *Phil. Trans. R. Soc. B* **351**: 785–795.

Barton, N.H. 2000. Genetic hitchhiking. *Phil. Trans. R. Soc. B* **355**: 1553–1562.

Barton, N.H. and B.O. Bengtsson. 1986. The barrier to genetic exchange between hybridising populations. *Heredity* **56**: 357–376.

Barton, N.H. and B. Charlesworth. 1984. Genetic revolutions, founder effects and speciation. *Ann. Rev. Syst. Ecol.* **15**: 133–164.

Barton, N.H. and A.M. Etheridge. 2004. The effect of selection on genealogies. *Genetics* **166**: 1115–1131.

Barton, N.H. and G.M. Hewitt. 1989. Adaptation, speciation and hybrid zones. *Nature* **341**: 497–503.

Barton, N.H. and R.J. Post. 1986. Sibling competition and the advantage of mixed families. *J. Theor. Biol.* **120**: 381–387.

Barton, N.H. and M. Turelli. 1987. Adaptive landscapes, genetic distance and the evaluation of quantitative characters. *Genet. Res.* **49**: 157–173.

Barton, N.H. and M. Turelli. 1991. Natural and sexual selection on many loci. *Genetics* **127**: 229–255.

Barton, N.H. and I. Wilson. 1995. Genealogies and geography. *Phil. Trans. R. Soc. B* **349**: 49–59.

Barton, N.H. and I. Wilson. 1996. Genealogies and geography. In *New Uses for Old Phylogenies* (Eds. P.H. Harvey, A.J. Leigh Brown, and J. Maynard Smith), pp. 23–56. Oxford University Press, Oxford, U.K.

Barton, N.H., F. Depaulis, and A.M. Etheridge. 2002. Neutral evolution in spatially continuous populations. *Theor. Popul. Biol.* **61**: 31–48.

Bateson, W. 1909. Heredity and variation in modern lights. In *Darwin and Modern Science* (Ed. A.C. Seward), pp. 85–101. Cambridge University Press, Cambridge, U.K.

Baudat, F. and B. De Massy. 2007. Regulating double-stranded DNA break repair towards crossover or non-crossover during mammalian meiois. *Chrom. Res.* **15**: 565–577.

Baudisch, A. 2008. *Inevitable Aging?* Springer, Berlin.

Bauduer, F., J. Feingold, and D. Lacombe. 2005. The Basques: review of population genetics and Mendelian disorders. *Hum. Biol.* **77**: 619–637.

Bauer, H., N. Veron, J. Willert, and B.G. Herrmann. 2007. A *t*-complex-encoded guanine nucleotide exchange factor *Fgd2* reveals that two opposing signaling pathways promote transmission ratio distortion in the mouse. *Genes Dev.* **21**: 143–147.

Baysal, B.E., E.C. Lawrence, and R.E. Ferrell. 2007. Sequence variation in human succinate dehydrogenase genes: evidence for long-term balancing selection on *SDHA*. *BMC Biology* **5**: 12.

Bazykin, A.D. 1969. Hypothetical mechanism of speciation. *Evolution* **23**: 685–687.

Beaumont, M.A. 1999. Detecting population expansion and decline using microsatellites. *Genetics* **153**: 2013–2029.

Beaumont, M.A. 2003. Estimation of population growth or decline in genetically monitored populations. *Genetics* **164**: 1139–1160.

Beaumont, M.A. and D.J. Balding. 2004. Identifying adaptive genetic divergence among populations from genome scans. *Mol. Ecol.* **13**: 969–980.

Beaumont, M.A. and R.A. Nichols. 1996. Evaluating loci for use in the genetic analysis of population structure. *Proc. R. Soc. B* **263**: 1619–1626.

Beaumont, M.A., W. Zhang, and D.J. Balding. 2003. Approximate Bayesian computation in population genetics. *Genetics* **2002:** 2025–2035.

Bechsgaard, J.S., V. Castric, D. Charlesworth, X. Vekemans, and M.H. Schierup. 2006. The transition to self-compatibility in *Arabidopsis thaliana* and evolution within S-haplotypes over 10 Myr. *Mol. Biol. Evol.* **23:** 1741–1750.

Bedford, T., I. Wapinski, and D.L. Hartl. 2008. Overdispersion of the molecular clock varies between yeast, *Drosophila* and mammals. *Genetics* **179:** 977–984.

Beerli, P. 2004. Effect of unsampled populations on the estimation of population sizes and migration rates between sampled populations. *Mol. Ecol.* **13:** 827–836.

Beerli, P. and J. Felsenstein. 2001. Maximum likelihood estimation of a migration matrix and effective population sizes in *n* subpopulations by using a coalescent approach. *Proc. Natl. Acad. Sci. U.S.A.* **98:** 4563–4568.

Begon, M., J.L. Harper, and C.R. Townsend. 2006. *Ecology: From Individuals to Ecosystems*, 4th ed. Blackwell, Oxford, U.K.

Begun, D.J. and C.F. Aquadro. 1992. Levels of naturally occurring DNA polymorphism correlate with recombination rate in *Drosophila melanogaster*. *Nature* **356:** 519–520.

Bell, G. 1982. *The Masterpiece of Nature*. Croom-Helm, London.

Bell, G. 2008. *Selection*, 2nd ed. Oxford University Press, Oxford, U.K.

Bena, G., J.M. Prosperi, B. Lejeune, and I. Olivieri. 1998. Evolution of annual species of the genus Medicago: a molecular phylogenetic approach. *Mol. Phylogenet. Evol.* **9:** 552–559.

Bennett, J.H. 1954. On the theory of random mating. *Ann. Eugen.* **18:** 311–317.

Bergman, A. and M.W. Feldman. 1990. More on selection for and against recombination. *Theor. Popul. Biol.* **38:** 68–92.

Bergman, C.M. and D. Bensasson. 2007. Recent LTR retrotransposon insertion contrasts with waves of non-LTR insertion since speciation in *Drosophila melanogaster*. *Proc. Natl. Acad. Sci. U.S.A.* **104:** 11340–11345.

Bergman, C.M. and M. Kreitman. 2001. Analysis of conserved noncoding DNA in *Drosophila* reveals similar constraints in intergenic and intronic sequences. *Genome Res.* **11:** 1135–1345.

Bergman, C.M., H. Quesneville, D. Anxolabehere, and M. Ashburner. 2006. Recurrent insertion and duplication generate networks of transposable element sequences in the *Drosophila melanogaster* genome. *Genome Biol.* **7:** R112.

Berlin, S., M. Quintela, and J. Hoglund. 2008. A multilocus assay reveals high nucleotide diversity and limited differentiation among Scandinavian willow grouse (*Lagopus lagopus*). *BMC Genetics* **9:** 89.

Bernardi, G. 1993. The vertebrate genome: isochores and evolution. *Mol. Biol. Evol.* **10:** 186–220.

Berry, A.J., J.W. Ajioka, and M. Kreitman. 1991. Lack of polymorphism on the *Drosophila* fourth chromosome resulting from selection. *Genetics* **129:** 1111–1117.

Betancourt, A.J. and J.P. Bollback. 2006. Fitness effects of beneficial mutations: the mutational landscape model in experimental evolution. *Curr. Op. Genet. Dev.* **16:** 618–623.

Betancourt, A.J., J.J. Welch, and B. Charlesworth. 2009. Reduced effectiveness of selection caused by lack of recombination. *Curr. Biol.* **19**: 655–660.

Bielawski, J.P. and Z.H. Yang. 2001. Positive and negative selection in the DAZ gene family. *Mol. Biol. Evol.* **18**: 523–529.

Bierne, N. and A. Eyre-Walker. 2003. The problem of counting sites in the estimation of the synonymous and nonsynonyous substitution rates: implications for the correlation between the synonymous substitution rate and codon usage. *Genetics* **165**: 1587–1597.

Bierne, N. and A. Eyre-Walker. 2004. The genomic rate of adaptive amino-acid substitutions in *Drosophila. Mol. Biol. Evol.* **21**: 1350–1360.

Bierne, N., S. Launey, Y. Naciri-Graven, and F. Bonhomme. 2000. Early effect of inbreeding as revealed by microsatellite analyses on *Ostrea edulis* larvae. *Genetics* **148**: 1893–1906.

Bierzychudek, P. 1987. Resolving the paradox of sexual reproduction: a review of experimental tests. In *The Evolution of Sex and its Consequences* (Ed. S.C. Stearns), pp. 163–174. Birkhaüser, Basel.

Bierzychudek, P. 1990. The demographic consequences of sexuality and apomixis in Antennaria. In *Biological Approaches and Evolutionary Trends in Plants* (Ed. S. Kawano), pp. 293–307. Academic Press, New York.

Bishop, J.G., A.M. Dean, and T. Mitchell-Olds. 2000. Rapid evolution in plant chitinases: molecular targets of selection in plant–pathogen coevolution. *Proc. Natl. Acad. Sci. U.S.A.* **97**: 5322–5327.

Bittles, A.H. and J.V. Neel. 1994. The costs of human inbreeding and their implications for variations at the DNA level. *Nat. Genet.* **8**: 117–121.

Blackman, R.K., R. Grimaila, M.M.D. Koehler, and W.M. Gelbart. 1987. Mobilization of hobo elements residing within the decapentaplegic gene complex: suggestion of a new hybrid dysgenesis system in *Drosophila melanogaster. Cell* **49**: 497–505.

Blattner, F.R., G. Plunkett, C.A. Bloch, N.T. Perna, V. Burland, et al. 1997. The complete genome sequence of *Escherichia coli* K-12. *Science* **277**: 1453–1462.

Blows, M.W. and A.A. Hoffmann. 2005. A reassessment of genetic limits to evolutionary change. *Ecology* **86**: 1371–1384.

Bodmer, W.F. and J. Felsenstein. 1967. Linkage and selection: theoretical analysis of the deterministic two locus random mating model. *Genetics* **57**: 237–265.

Boer, R.J.D., J.A.M. Borghans, M. van Boven, C. Kesmir, and F.J. Weissing. 2004. Heterozygote advantage fails to explain the high degree of polymorphism of the MHC. *Immunogenetics* **55**: 725–731.

Boorman, S.A. and P.R. Levitt. 1973. Group selection at the boundary of a stable population. *Theor. Popul. Biol.* **4**: 85–128.

Bourguet, D. and M. Raymond. 1998. The molecular basis of dominance relationships: the case of some recent adaptive genes. *J. Evol. Biol.* **11**: 103–122.

Boyko, A., S.H. Williamson, A.R. Indap, J.D. Degenhardt, R.D. Hernandez, et al. 2008. Assessing the evolutionary impact of amino-acid mutations in the human genome. *PLOS Genet.* **5**: e1000083.

Bradshaw, H.D. and D.W. Schemske. 2003. Allele substitution at a flower colour locus produces a pollinator shift in monkeyflowers. *Nature* **426**: 176–178.

Bradshaw, H.D., K.G. Otto, B.E. Frewen, J.K. McKay, and D.W. Schemske. 1998. Quantitative trait loci affecting differences in floral morphology between two species of monkeyflower. *Genetics* **149:** 367–382.

Brauner, S. and L.D. Gottlieb. 1987. A self-compatible plant of *Stephanomeria exigua* subsp. *coronaria* (Asteraceae) and its relevance to the origin of its self-pollinating derivative *S. malheurensis*. *Syst. Bot.* **12:** 299–304.

Braverman, J.M., R.R. Hudson, N.L. Kaplan, C.H. Langley, and W. Stephan. 1995. The hitchiking effect on the site frequency spectrum of DNA polymorphism. *Genetics* **140:** 783–796.

Brennecke, J., J.G. Malone, A.A. Aravin, R. Sachidanandam, A. Stark, and G.J. Hannon. 2009. An epigenetic role for maternally inherited piRNAs in transposon silencing. *Science* **322:** 1387–1392.

Brideau, N.J., H.A. Flores, J. Wang, S. Maheshwari, X. Wang, and D.A. Barbash. 2006. Two Dobzhansky–Muller genes interact to cause hybrid lethality in *Drosophila*. *Science* **314:** 1292–1295.

Brookfield, J.F.Y. 2003. Mobile DNAs: the poacher turned gamekeeper. *Curr. Biol.* **13:** R846–R847.

Brooks, L.D. 1988. The evolution of recombination rates. In *The Evolution of Sex* (Eds. R.E. Michod and B.R. Levin), pp. 87–105. Sinauer, Sunderland, MA.

Brotherstone, S. and Goddard. 2005. Artificial selection and maintenance of genetic variance in the global dairy cow population. *Phil. Trans. R. Soc. B* **360:** 1479–1148.

Brown, A.H.D. and R.W. Allard. 1970. Estimation of the mating system in open-pollinated maize populations using isozyme polymorphisms. *Genetics* **66:** 133–145.

Brown, J.K.M. 2003. A cost of disease resistance: paradigm or peculiarity? *Trends Genet.* **19:** 667–671.

Brunet, J. and C.G. Eckert. 1998. Effects of floral morphology and display on outcrossing in Blue Columbine, *Aquilegia caerulea* (Ranunculaceae). *Funct. Ecol.* **12:** 596–606.

Bubb, K.L., D. Bovee, D. Buckley, E. Haugen, M. Kibukawa, et al. 2006. Scan of human genome reveals no new loci under ancient balancing selection. *Genetics* **173:** 2165–2177.

Bucheton, A., I. Busseau, and D. Teninges. 2002. *I* elements in *Drosophila melanogaster*. In *Mobile DNA II* (Eds. N.L. Craig, R. Craigie, M. Gellert, and A.M. Lambowitz), pp. 796–812. American Society of Microbiology, Washington, DC.

Bull, J.J. 1983. *Evolution of Sex Determining Mechanisms*. Benjamin Cummings, Menlo Park, CA.

Bulmer, M.G. 1971. The effect of selection on genetic variability. *Am. Nat.* **105:** 201–211.

Bulmer, M.G. 1974. Linkage disequilibrium and genetic variability. *Genet. Res.* **23:** 281–289.

Bulmer, M.G. 1980. *The Mathematical Theory of Quantitative Genetics*. Oxford University Presss, Oxford, U.K. (Reprinted and revised, 1985.)

Bulmer, M.G. 1991. The selection-mutation-drift theory of synonomous codon usage. *Genetics* **129:** 897–907.

Bulmer, M.G. 1994. *Theoretical Evolutionary Ecology*. Sinauer, Sunderland, MA.

Bürger, R. 2000. *The Mathematical Theory of Selection, Recombination and Mutation.* John Wiley, Chichester, U.K.

Burgoyne, P.S. 1982. Genetic homology and crossing over in the X and Y chromosomes of mammals. *Hum. Genet.* **61:** 85–90.

Buri, P. 1956. Gene frequency in small populations of mutant *Drosophila. Evolution* **10:** 367–402.

Burrows, C.J. 1960. Studies in Pimelea. I. The breeding system. *Trans. R. Soc. New Zealand* **88:** 29–45.

Burt, A. 1995. The evolution of fitness. *Evolution* **49:** 1–8.

Burt, A., Bell, G. 1987. Mammalian chiasma frequencies as a test of two theories of recombination. *Nature* **326:** 803–805.

Burt, A. and R.L. Trivers. 2006. *Genes in Conflict.* Harvard University Press, Cambridge, MA.

Burton, O.J. and J.M.J. Travis. 2008. Landscape structure and boundary effects determine the fate of mutations occurring during range expansions. *Heredity* **101:** 329–340.

Bustamante, C.D., R. Nielsen, S.A. Sawyer, K.M. Olsen, M.D. Purugganan, and D.L. Hartl. 2002. The cost of inbreeding in *Arabidopsis. Nature* **416:** 531–534.

Caballero, A. 1994. Developments in the prediction of effective population size. *Heredity* **73:** 657–679.

Cai, J.J., J.M. Macpherson, G. Sella, and D.A. Petrov. 2009. Pervasive hitchiking at coding and regulatory sites in humans. *PLoS Genetics* **5:** e1000336.

Caicedo, A. L., S. Williamson, R. Hernandez, A. Boyko, A. Fledel-Alon,, et al. 2007. Genome-wide patterns of nucleotide polymorphism in domesticated rice. *PLoS Genetics* **3:** 1745–1756.

Cargill, M., D. Altschuler, J. Ireland, P. Sklar, K. Ardlie, et al. 1999. Characterization of single-nucleotide polymorphisms in coding regions of human genes. *Nat. Genet.* **22:** 231–238.

Carlini, D.B. and W. Stephan. 2003. In vivo introduction of unpreferred synonymous codons into the *Drosophila Adh* gene results in reduced levels of ADH protein. *Genetics* **163:** 239–243.

Carroll, S. 2008. Evo-devo and the expanding evolutionary synthesis: a genetic theory of morphological evolution. *Cell* **134:** 25–36.

Carson, H.L. and A.R. Templeton. 1984. Genetic revolutions in relation to speciation phenomena. *Ann. Rev. Syst. Ecol.* **15:** 97–131.

Casselton, L.A. 1997. Molecular recognition in fungal mating. *Endeavour* **21:** 159–163.

Caswell, H. 2000. *Matrix Population Models,* 2nd ed. Sinauer, Sunderland, MA.

Cattrall, M.E., M.L. Baird, and E.D. Garber. 1978. Genetics of *Ustilago violacea.* III. Crossing over and nondisjunction. *Bot. Gaz.* **13:** 266–270.

Cavalli-Sforza, L.L. and M.W. Feldman. 1978. Darwinian selection and "altruism." *Theor. Popul. Biol.* **14:** 268–280.

Chan, S.R.W.L. and E.H. Blackburn. 2004. Telomeres and telomerase. *Phil. Trans. R. Soc. B* **359:** 109–121.

Charlesworth, B. 1973. Selection in populations with overlapping generations. V. Natural selection and life histories. *Am. Nat.* **107:** 303–311.

Charlesworth, B. 1976. Recombination modification in a fluctuating environment. *Genetics* **83**: 181–195.

Charlesworth, B. 1979a. A note on the evolution of altruism in structured demes. *Am. Nat.* **113**: 601–605.

Charlesworth, B. 1979b. Selection for gamete lethals and S-alleles in complex heterozygotes. *Heredity* **43**: 159–164.

Charlesworth, B. 1979c. Evidence against Fisher's theory of dominance. *Nature* **278**: 848–849.

Charlesworth, B. 1980a. The cost of sex in relation to mating system. *J. Theor. Biol.* **84**: 655–671.

Charlesworth, B. 1980b. Models of kin selection. In *Evolution of Social Behavior: Hypothesis and Empirical Tests* (Ed. H. Markl), pp. 11–26. Dahlem Workshops, Berlin.

Charlesworth, B. 1984. The cost of phenotypic evolution. *Paleobiology* **10**: 319–327.

Charlesworth, B. 1990a. Mutation–selection balance and the evolutionary advantage of sex and recombination. *Genet. Res.* **55**: 199–221.

Charlesworth, B. 1990b. Optimization models, quantitative genetics, and mutation. *Evolution* **44**: 520–538.

Charlesworth, B. 1992. Evolutionary rates in partially self-fertilizing species. *Am. Nat.* **140**: 126–148.

Charlesworth, B. 1993a. Directional selection and the evolution of sex and recombination. *Genet. Res.* **61**: 205–224.

Charlesworth, B. 1993b. Natural selection on multivariate traits in age-structured populations. *Proc. R. Soc. B* **251**: 47–52.

Charlesworth, B. 1994a. The effect of background selection against deleterious alleles on weakly selected, linked variants. *Genet. Res.* **63**: 213–228.

Charlesworth, B. 1994b. *Evolution in Age-structured Populations*, 2nd ed. Cambridge University Press, Cambridge, U.K.

Charlesworth, B. 1994c. The evolution of lethals in the *t*-haplotype system of the mouse. *Proc. R. Soc. B* **258**: 101–107.

Charlesworth, B. 1996. Background selection and patterns of genetic diversity in *Drosophila melanogaster*. *Genet. Res.* **68**: 131–150.

Charlesworth, B. 2001a. The effect of life-history and mode of inheritance on neutral genetic variability. *Genet. Res.* **77**: 153–166.

Charlesworth, B. 2001b. Patterns of age-specific means and genetic variances of mortality rates predicted by the mutation-accumulation theory of ageing. *J. Theor. Biol.* **210**: 47–66.

Charlesworth, B. and N.H. Barton. 1996. Recombination load associated with selection for increased recombination. *Genet. Res.* **67**: 27–41.

Charlesworth, B. and D. Charlesworth. 1973a. The measurement of fitness and mutation rate in human populations. *Ann. Hum. Genet.* **37**: 175–187.

Charlesworth, B. and D. Charlesworth. 1973b. Selection of new inversions in multilocus genetic systems. *Genet. Res.* **21**: 167–183.

Charlesworth, B. and D. Charlesworth. 1978. A model for the evolution of dioecy and gynodioecy. *Am. Nat.* **112**: 975–997.

Charlesworth, B. and D. Charlesworth. 1979. The maintenance and breakdown of distyly. *Am. Nat.* **114:** 499–513.

Charlesworth, B. and D. Charlesworth. 1997. Rapid fixation of deleterious alleles by Muller's ratchet. *Genet. Res.* **70:** 63–73.

Charlesworth, B. and D. Charlesworth. 1998. Some evolutionary consequences of deleterious mutations. *Genetica* **102/103:** 3–19.

Charlesworth, B. and D.L. Hartl. 1978. Population dynamics of the segregation distorter polymorphism of *Drosophila melanogaster. Genetics* **89:** 171–192.

Charlesworth, B. and K.A. Hughes. 2000. The maintenance of genetic variation in life-history traits. In *Evolutionary Genetics from Molecules to Morphology* (Eds. R.S. Singh and C.B. Krimbas), pp. 369–392. Cambridge University Press, Cambridge, U.K.

Charlesworth, B. and C.H. Langley. 1986. Evolution of self-regulated transposition of transposable elements. *Genetics* **112:** 359–383.

Charlesworth, B. and C.H. Langley. 1989. The population genetics of *Drosophila* transposable elements. *Ann. Rev. Genet.* **23:** 251–287.

Charlesworth, B., D. Charlesworth, and N.H. Barton. 2003. The effects of genetic and geographic structure on neutral variation. *Ann. Rev. Ecol. Evol. Syst.* **34:** 99–125.

Charlesworth, B., J.A. Coyne, and N.H. Barton. 1987. The relative rates of evolution of sex chromosomes and autosomes. *Am. Nat.* **130:** 113–146.

Charlesworth, B., R. Lande, and M. Slatkin. 1982. A neo-Darwinian commentary on macro-evolution. *Evolution* **36:** 474–498.

Charlesworth, B., C.H. Langley, and W. Stephan. 1986. The evolution of restricted recombination and the accumulation of repeated DNA sequences. *Genetics* **112:** 947–962.

Charlesworth, B., M.T. Morgan, and D. Charlesworth. 1993. The effect of deleterious mutations on neutral molecular variation. *Genetics* **134:** 1289–1303.

Charlesworth, B., M. Nordborg, and D. Charlesworth. 1997. The effects of local selection, balanced polymorphism and background selection on equilibrium patterns of genetic diversity in subdivided populations. *Genet. Res.* **70:** 155–174.

Charlesworth, B., P. Sniegowski, and W. Stephan. 1994. The evolutionary dynamics of repetitive DNA in eukaryotes. *Nature* **371:** 215–220.

Charlesworth, B., A.J. Betancourt, V.B. Kaiser, and I. Gordo. 2009. Genetic recombination and molecular evolution. *Cold Spring Harb. Symp. Quant. Biol.* **74:** In press.

Charlesworth, B., H. Borthwick, C. Bartolomé, and P. Pignatelli. 2004. Estimates of the genomic mutation rate for detrimental alleles in *Drosophila melanogaster. Genetics* **168:** 1105–1109.

Charlesworth, D. 1981. A further study of the problem of the maintenance of females in gynodioecious species. *Heredity* **46:** 27–39.

Charlesworth, D. 1984. Androdioecy and the evolution of dioecy. *Biol. J. Linn. Soc.* **23:** 333–348.

Charlesworth, D. 1985. Distribution of dioecy and self-incompatibility in angiosperms. In *Evolution. Essays in Honour of John Maynard Smith* (Eds. P.H. Harvey, M. Slatkin, and P.J. Greenwood), pp. 237–269. Cambridge University Press, Cambridge, U.K.

Charlesworth, D. 2003. Effects of inbreeding on the genetic diversity of plant populations. *Phil. Trans. R. Soc. B* **358:** 1051–1070.

Charlesworth, D. 2006. Balancing selection and its effects on sequences in nearby genome regions. *PLoS Genet.* **2:** e2.

Charlesworth, D. and B. Charlesworth. 1975a. Theoretical genetics of Batesian mimicry. I. Single-locus models. *J. Theor. Biol.* **55:** 283–303.

Charlesworth, D. and B. Charlesworth. 1975b. Theoretical genetics of Batesian mimicry. II. Evolution of supergenes. *J. Theor. Biol.* **55:** 305–324.

Charlesworth, D. and B. Charlesworth. 1978. The population genetics of partial male-sterility and the evolution of monoecy and dioecy. *Heredity* **41:** 137–153.

Charlesworth, D. and B. Charlesworth. 1979a. A model for the evolution of distyly. *Am. Nat.* **114:** 467–498.

Charlesworth, D. and B. Charlesworth. 1979b. The evolution and breakdown of S-allele systems. *Heredity* **43:** 41–55.

Charlesworth, D. and B. Charlesworth. 1981. Allocation of resources to male and female functions in hermaphrodites. *Biol. J. Linn. Soc.* **15:** 57–74.

Charlesworth, D. and B. Charlesworth. 1987. Inbreeding depression and its evolutionary consequences. *Ann. Rev. Ecol. Syst.* **18:** 237–268.

Charlesworth, D. and B. Charlesworth. 1990. Inbreeding depression with heterozygote advantage and its effect on selection for modifiers changing the outcrossing rate. *Evolution* **44:** 870–888.

Charlesworth, D. and B. Charlesworth. 1995. Quantitative genetics in plants: the effect of the breeding system on genetic variability. *Evolution* **49:** 911–920.

Charlesworth, D. and F.R. Ganders. 1979. The population genetics of gynodioecy with cytoplasmic male-sterility. *Heredity* **43:** 213–218.

Charlesworth, D. and V. Laporte. 1998. The male-sterility polymorphism of *Silene vulgaris*. Analysis of genetic data from two populations, and comparison with *Thymus vulgaris*. *Genetics* **150:** 1267–1282.

Charlesworth, D. and X. Vekemans. 2005. How and when did *Arabidopsis thaliana* become highly self-fertilising? *BioEssays* **27:** 472–476.

Charlesworth, D., B. Charlesworth, and G. Marais. 2005. Steps in the evolution of heteromorphic sex chromosomes. *Heredity* **95:** 118–128.

Charlesworth, D., B. Charlesworth, and M.T. Morgan. 1995. The pattern of neutral molecular variation under the background selection model. *Genetics* **141:** 1619–1632.

Charlesworth, D., B. Charlesworth, and C. Strobeck. 1977. Effects of selfing on selection for recombination. *Genetics* **86:** 213–226.

Charlesworth, D., B. Charlesworth, and C. Strobeck. 1979. Selection for recombination in partially self-fertilizing species. *Genetics* **93:** 237–244.

Charlesworth, D., E. Kamau, J. Hagenblad, and C. Tang. 2006. Trans-specificity at loci near the self-incompatibility loci in Arabidopsis. *Genetics* **172:** 2699–2704.

Charlesworth, D., X. Vekemans, V. Castric, and S. Glémin. 2005. Plant self-incompatibility systems: a molecular evolutionary perspective. *New Phytol.* **168:** 61–69.

Charlesworth, J. and A. Eyre-Walker. 2006. The rate of adaptive evolution in enteric bacteria. *Mol. Biol. Evol.* **23:** 1348–1356.

Charlesworth, J. and A. Eyre-Walker. 2007. The other side of the nearly neutral theory, evidence of slightly advantageous back-mutations. *Proc. Natl. Acad. Sci. U.S.A.* **104:** 16992–16997.

Charlesworth, J. and A. Eyre-Walker. 2008. The McDonald–Kreitman test and slightly deleterious mutations. *Mol. Biol. Evol.* **25:** 1007–1015.

Charnov, E.L. 1982. *The Theory of Sex Allocation.* Princeton University Press, Princeton, NJ.

Charnov, E.L., J. Maynard Smith, and J.J. Bull. 1976. Why be an hermaphrodite? *Nature* **263:** 125–126.

Chase, C. 2007. Cytoplasmic male sterility: a window to the world of plant mitochondrial–nuclear interactions. *Trends Genet.* **23:** 81–90.

Cherry, J.L. 2003a. Selection in a subdivided population with dominance or local frequency dependence. *Genetics* **163:** 1511–1518.

Cherry, J.L. 2003b. Selection in a subdivivided population with local extinction and recolonization. *Genetics* **164:** 789–779.

Cherry, J.L. and J. Wakeley. 2003. A diffusion approximation for selection and drift in a subdivided population. *Genetics* **163:** 421–428.

Chesser, R.K., O.E. Rhodes, D.W. Sugg, and A. Schnabel. 1993. Effective sizes for subdivided populations. *Genetics* **135:** 1221–1232.

Chimpanzee Sequencing and Analysis Consortium. 2005. Initial sequence of the chimpanzee genome and comparison with the human genome. *Nature* **437:** 69–87.

Chung, H., M.R. Bogwitz, C. McCart, A. Andrianopoulos, R.H. Ffrench-Constant, et al. 2007. Cis-regulatory elements in the *Accord* retrotransposon result in tissue-specific expression of the *Drosophila melanogaster* insecticide resistance gene *Cyp6g1*. *Genetics* **175:** 1071–1077.

Christiansen, F.B. 2000. *Population Genetics of Multiple Loci.* John Wiley, Chichester, U.K.

Clarke, B.C. 1969. The evidence for apostatic selection. *Heredity* **24:** 347–352.

Clarke, B.C. 1979. The evolution of genetic diversity. *Proc. R. Soc. B* **205:** 453–474.

Clarke, C.A. and P.M. Sheppard. 1960a. The evolution of mimicry in the butterfly *Papilio dardanus. Heredity* **14:** 163–173.

Clarke, C.A. and P.M. Sheppard. 1960b. The genetics of *Papilio dardanus.* III. Race antinorii from Abyssinia and race meriones from Madagascar. *Genetics* **45:** 683–698.

Clarke, C.A. and P.M. Sheppard. 1971. Further studies on the genetics of the mimetic butterfly *Papilio memnon. Phil. Trans. R. Soc. B* **263:** 35–70.

Cleland, R.E. 1972. *Oenothera. Cytogenetics and Evolution.* Academic Press, New York.

Clutton-Brock, T.H. 1988. *Reproductive Success.* University of Chicago Press, Chicago, IL.

Cockerham, C.C. and B.S. Weir. 1968. Sib-mating with two loci. *Genetics* **60:** 629–640.

Cole, L.C. 1954. The population consequences of life history phenomena. *Quart. Rev. Biol.* **29:** 103–137.

Colegrave, N. 2002. Sex releases the speed limit on evolution. *Nature* **420:** 664–666.

Comeron, J.M. 2004. Selective and mutational patterns associated with gene expression in humans: influences on synonymous composition and intron presence. *Genetics* **167:** 1293–1304.

Comeron, J.M. 2006. Weak selection and recent mutational changes influence poly-morphic synonymous mutations in humans. *Proc. Natl. Acad. Sci. U.S.A.* **103:** 6940–6945.

Comeron, J.M. and T.B. Guthrie. 2005. Intragenic Hill–Robertson interference influ-ences selection on synonymous mutations in *Drosophila. Mol. Biol. Evol.* **22:** 2519–2530.

Comeron, J.M., A. Williford, and R.M. Kliman. 2008. The Hill–Robertson effect: evolu-tionary consequences of weak selection in finite populations. *Heredity* **100:** 19–31.

Cook, L.M. 2003. The rise and fall of the *Carbonaria* form of the peppered moth. *Quart. Rev. Biol.* **78:** 399–417.

Coop, G. and R.C. Griffiths. 2004. Ancestral inference on gene trees under selection. *Theor. Popul. Biol.* **66:** 219–232.

Coop, G. and M. Przeworski. 2007. An evolutionary view of human recombination. *Nat. Rev. Genet.* **8:** 23–24.

Coop, G., J.K. Pickrell, J. Novembre, S. Kudaravalli, J. Li, et al. 2009. The role of geog-raphy in human evolution. *PLoS Genetics* **5:** e1000500.

Cooper, M.A., R.D. Adam, M. Worobey, and C.R. Sterling. 2007. Population genetics provides evidence for recombination in Giardia. *Curr. Biol.* **17:** 1984–1988.

Copenhaver, G.P., W.E. Browne, and D. Preuss. 1998. Assaying genome-wide recom-bination and centromere functions with Arabidopsis tetrads. *Proc. Natl. Acad. Sci. U.S.A.* **95:** 247–252.

Corander, J., P. Waldmann, P. Marttinen, and M.J. Sillanpaa. 2004. BAPS 2: enhanced possibilities for the analysis of genetic population structure. *Bioinformatics* **20:** 2363–2369.

Correns, C. 1928. Bestimmung, Vererbung und Verteilung des Geschlechtes bei den höheren Pflanzen. *Handb. Vererbungswissenschaft* **2:** 1–138.

Cotterman, C.W. 1940. A calculus for statistico-genetics. (Ph.D Dissertation.) Ohio State Univ., Columbus, OH.

Couvet, D., F. Bonnemaison, and P.-H. Gouyon. 1986. The maintenance of females among hermaphrodites: the importance of nuclear-cytoplasmic interactions. *Heredity* **57:** 325–330.

Coyne, J.A. and H.A. Orr. 1997. "Patterns of speciation in *Drosophila*" revisited. *Evolu-tion* **51:** 295–303.

Coyne, J.A. and H.A. Orr. 2004. *Speciation*. Sinauer, Sunderland, MA.

Coyne, J.A., I.A. Boussy, S.H. Bryant, J.S. Jones, and J.A. Moore. 1982. Long-distance migration of *Drosophila. Am. Nat.* **119:** 589–595.

Coyne, J.A., N.H. Barton, and M. Turelli. 1997. Perspective: a critique of Wright's shift-ing balance theory of evolution. *Evolution* **51:** 643–671.

Coyne, J.A., N.H. Barton, and M. Turelli. 2000. Is Wright's shifting balance process important in evolution? *Evolution* **54:** 306–317.

Craig, N.L., R. Craigie, M. Gellert, and A.M. Lambowitz. 2002. *Mobile DNA II*. Ameri-can Society of Microbiology, Washington, DC.

Crandall, K.A., C.R. Kelsey, H. Imamichi, H.C. Lane, and N.P. Salzman. 1999. Parallel evolution of drug resistance in HIV: failure of nonsynonymous/synonymous sub-stitution rate ratio to detect selection. *Mol. Biol. Evol.* **16:** 372–382.

Crawford, T.J. 1984. What is a population? In *Evolutionary Ecology* (Ed. B. Shorrocks), pp. 135–173. Blackwell, Oxford, U.K.

Crispo, E., P. Bentzen, D. Reznick, M. Kinnison, and A. Hendry. 2006. The relative influence of natural selection and geography on gene flow in guppies. *Mol. Ecol.* **15**: 49–62.

Crnokrak, P. and D.A. Roff. 1995. Dominance variance associations with selection and fitness. *Heredity* **75**: 530–545.

Cronin, J.E., N.T. Boaz, C.B. Stringer, and Y. Rak. 1981. Tempo and mode in hominid evolution. *Nature* **292**: 113–122.

Crosby, J.L. 1949. Selection of an unfavourable gene-complex. *Evolution* **3**: 212–230.

Crosland, M.W.J. and R.H. Crozier. 1986. *Myrmecia pilosula*, an ant with only one pair of chromosomes. *Science* **231**: 1278.

Crow, J.F. 1954. Breeding structure of populations. II. Effective population number. In *Statistics and Mathematics in Biology* (Eds. O. Kempthorne, T.A. Bancroft, J.W. Gowen, and J.L. Lush), pp. 543–556. Iowa State University Press, Ames, IA.

Crow, J.F. 1958. Some possibilities for measuring selection intensities in man. *Hum. Biol.* **30**: 1–13.

Crow, J.F. 1970. Genetic loads and the cost of natural selection. In *Mathematical Topics in Population Genetics* (Ed. K. Kojima), pp. 128–177. Springer-Verlag, Berlin.

Crow, J.F. 1991. Why Mendelian segregation is so exact. *Bioessays* **13**: 305–312.

Crow, J.F. 1993. Mutation, mean fitness, and genetic load. *Oxf. Surv. Evol. Biol.* **9**: 3–42.

Crow, J.F. and K. Aoki. 1984. Group selection for a polygenic behavioral trait—estimating the degree of population subdivision. *Proc. Natl. Acad. Sci. U.S.A.* **81**: 6073–6077.

Crow, J. F. and C. Denniston. 1988. Inbreeding and variance effective numbers. *Evolution* **43**: 482–495.

Crow, J.F. and M. Kimura. 1965. Evolution in sexual and asexual populations. *Am. Nat.* **99**: 439–450.

Crow, J.F. and M. Kimura. 1970. *An Introduction to Population Genetics Theory*. Harper and Row, New York.

Crow, J.F. and N.E. Morton. 1955. Measurement of gene-frequency drift in small populations. *Evolution* **9**: 202–214.

Crow, J.F. and R.G. Temin. 1964. Evidence for the partial dominance of recessive lethal genes in natural populatons of *Drosophila*. *Am. Nat.* **98**: 21–33.

Crozier, R.H. and P. Pamilo. 1996. *Evolution of Social Insect Colonies: Sex Allocation and Kin Selection*. Oxford University Press, Oxford, U.K.

Cruden, R.W. 1977. Pollen–ovule ratios: a conservative index of breeding systems in flowering plants. *Evolution* **31**: 32–46.

Cruden, R.W. and D.L. Lyon. 1985. Patterns of biomass allocation to male and female functions in plants with different mating systems. *Oecologia (Berlin)* **66**: 299–306.

Currat, M., G. Trabuchet, D. Rees, P. Perrin, R. M. Harding, et al. 2002. Molecular analysis of the beta-globin gene cluster in the Niokholo Mandenka population reveals a recent origin of the beta(S) Senegal mutation. *Am. J. Hum. Genet.* **70**: 207–223.

Cutler, D.J. 2000. Understanding the overdispersed molecular clock. *Genetics* **154:** 1403–1417.

Cutter, A.D. 2005. Nucleotide polymorphism and linkage disequilibrium in wild populations of the partial selfer *Caenorhabditis elegans*. *Genetics* **172:** 171–184.

Cutter, A.D. and B. Charlesworth. 2006. Selection intensity on preferred codons correlates with overall codon usage bias in *Caenorhabditis remanei*. *Curr. Biol.* **16:** 2053–2057.

Cutter, A.D., S.E. Baird, and D. Charlesworth. 2006. High nucleotide polymorphism and rapid decay of linkage disequilibrium in wild populations of *Caenorhabditis remanei*. *Genetics* **174:** 901–913.

Dahlgaard, J. and A.A. Hoffmann. 2000. Stress resistance and environmental dependency of inbreeding depression in *Drosophila melanogaster*. *Conserv. Biol.* **14:** 1187–1192.

Darlington, C.D. 1931. Meiosis. *Biol. Rev.* **6:** 221–264.

Darwin, C.R. 1859. *The Origin of Species*. John Murray, London.

Darwin, C.R. 1862. *The Various Contrivances by which Orchids are Fertilised by Insects*. John Murray, London.

Darwin, C.R. 1868. *The Variation of Animals and Plants under Domestication*. 2 Vols. John Murray, London.

Darwin, C.R. 1871. *The Descent of Man and Selection in Relation to Sex*. John Murray, London.

Darwin, C.R. 1876. *The Effects of Cross and Self Fertilisation in the Vegetable Kingdom*. John Murray, London.

Darwin, C.R. 1877. *The Different Forms of Flowers on Plants of the Same Species*. John Murray, London.

Darwin, C.R. and A.R. Wallace. 1858. On the tendency of species to form varieties; and on the perpetuation of varieties and species by natural selection. *J. Proc. Linn. Soc. (Zoology)* **3:** 45–62.

David, P. 1998. Heterozygosity-fitness correlations: new perspectives on old problems. *Heredity* **80:** 531–537.

Dawson, K.J. and K. Belkhir. 2001. Bayesian approach to the identification of panmictic populations and the assignment of individuals. *Genet. Res.* **78:** 69–77.

De, A. and R. Durrett. 2007. Stepping-stone spatial structure causes slow decay of linkage disequilibrium and shifts the site frequency spectrum. *Genetics* **176:** 969–981.

De Villena, F. and C. Sapienza. 2002. Female meiosis drives karyotypic evolution in mammals. *Genetics* **160:** 1651–1659.

De Waal Malefijt, M. and B. Charlesworth. 1979. A model for the evolution of translocation heterozygosity. *Heredity* **43:** 315–331.

Decaestecker, E., S. Gaba, J.A.M. Raeymaekers, R. Stoks, L. Van Kerckhoven, et al. 2007. Host–parasite "Red Queen" dynamics archived in pond sediment. *Nature* **450:** 870–873.

Deininger, P.L., J.V. Moran, M.A. Batzer, and H.H. Kazazian. 2003. Mobile elements and mammalian genome evolution. *Curr. Opin. Genet. Dev.* **13:** 651–658.

Delph, L. 1999. Sexual dimorphism in life history. In *Gender and Sexual Dimorphism in Flowering Plants* (Eds. M.A. Geber, T.E. Dawson, and L.F. Delph), pp. 149–174. Springer-Verlag, Berlin.

Delph, L. and S. Carroll. 2001. Factors affecting relative seed fitness and female frequency in a gynodioecious species, *Silene acaulis*. *Evol. Ecol. Res.* **3:** 487–505.

DeLuna, A., N.S.K. Vetsigian, M. Hegreness, M. Colón-González, S. Chao, and R. Kishony. 2008. Exposing the fitness contribution of duplicated genes. *Nat. Genet.* **40:** 676–681.

Dempster, E.R. 1955. Maintenance of genetic heterogeneity. *Cold Spring Harbor Symp. Quant. Biol.* **20:** 25–32.

Denver, D.R., K. Morris, M. Lynch, I.L. Vassilieva, and W.K. Thomas. 2000. High direct estimate of the mutation rate in the mitochondrial genome of *Caenorhabditis elegans*. *Science* **289:** 2342–2344.

Denver, D.R., P.C. Dolan, L.J. Wilhelm, W. Sung, J.I. Lucas-Lledó, et al. 2009. A genome-wide view of *Caenorhabditis elegans* base-substitution mutation processes. *Proc. Natl. Acad. Sci. U.S.A.* **106:** 16310–16314.

Depaulis, F. and M. Veuille. 1998. Neutrality tests based on the distribution of haplotypes under an infinite-site model. *Mol. Biol. Evol.* **15:** 1788–1790.

Depaulis, F., S. Mousset, and M. Veuille. 2001. Haplotype tests using coalescent simulations conditional on the number of segregating sites. *Mol. Biol. Evol.* **18:** 1136–1138.

DePristo, M.A., D.L. Hartl, and D.M. Weinreich. 2007. Mutational reversions during adaptive protein evolution. *Mol. Biol. Evol.* **24:** 1608–1610.

Desai, M. and D.S. Fisher. 2007. Beneficial mutation–selection balance and the effect of linkage on positive selection. *Genetics* **176:** 1759–1798.

Dettman, J., C.C. Sirjusingh, L. Kohn, and J. Anderson. 2007. Incipient speciation by divergent adaptation and antagonistic epistasis in yeast. *Nature* **447:** 585–589.

Deutschbauer, A.M., D.F. Jaramillo, M. Proctor, J. Kumm, M.E. Hillenmeyer, et al. 2005. Mechanisms of haploinsufficiency revealed by genome-wide profiling in yeast. *Genetics* **169:** 1915–1925.

Díaz-Castillo, C. and K.G. Golic. 2007. Evolution of gene sequence in response to chromosomal location. *Genetics* **177:** 359–374.

Di Cesnola, A.P. 1907. A first study of natural selection in *Helix arbustorum* (Helicogena). *Biometrika* **5:** 387–399.

Di Nocera, P.P. and I.B. Dawid. 1983. Interdigitated arrangement of two oligo(a)-terminated DNA sequences in *Drosophila*. *Nucleic Acids Res.* **11:** 5475–5482.

Di Rienzo, A., P. Donnelly, C. Toomajian, B. Sisk, A. Hill, et al. 1998. Heterogeneity of microsatellite mutations within and between loci, and implications for human demographic histories. *Genetics* **148:** 1269–1284.

Dickinson, W.J. 2008. Synergistic fitness interactions and a high frequency of beneficial changes among mutations accumulated under relaxed selection in *Saccharomyces cerevisiae*. *Genetics* **178:** 1571–1578.

Ding, X.D., Q. Zhang, C. Flury, and H. Simianer. 2006. Haplotype reconstruction and estimation of haplotype frequencies from nuclear families with only one parent available. *Hum. Hered.* **62:** 12–19.

Dobzhansky, T. 1943. Genetics of natural populations. IX. Temporal changes in the composition of populations of *Drosophila pseudoobscura*. *Genetics* **28:** 1621–1186.

Dobzhansky, T. 1955. A review of some fundamental concepts and problems of population genetics. *Cold Spring Harbor Symp. Quant. Biol.* **20:** 1–15.

Dobzhansky, T. and C.C. Tan. 1936. Studies on hybrid sterility. III. A comparison of the gene order in two species, *Drosophila pseudoobscura* and *Drosophila miranda. Z. Ind. Abstamm. Vererb.* **72:** 88–114.

Dobzhansky, T. and S. Wright. 1941. Genetics of natural populations. V. Relations between mutation rate and accumulation of lethals in populations of *Drosophila pseudoobscura. Genetics* **26:** 23–51.

Dobzhansky, T. and S. Wright. 1947. Genetics of natural populations. XV. Rate of diffusion of a mutant gene through a population of *Drosophila pseudoobscura. Genetics* **32:** 303–339.

Dole, J.A. 1992. Reproductive assurance mechanisms in three taxa of the *Mimulus guttatus* complex (Scrophulariaceae). *Am. J. Bot.* **79:** 650–659.

Donnelly, P. 2008. Progress and challenges in genome-wide association studies in humans. *Nature* **456:** 728–731.

Donnelly, P. and S. Tavaré. 1995. Coalescents and genealogical structure under neutrality. *Ann. Rev. Genet.* **29:** 410–421.

Doolittle, W.F. and C. Sapienza. 1980. Selfish genes, the phenotype paradigm, and genome evolution. *Nature* **284:** 601–603.

Dorken, M.E. and S.C.H. Barrett. 2004. Sex determination and the evolution of dioecy from monoecy in *Sagittaria latifolia* (Alismataceae). *Proc. R. Soc. B* **271:** 213–219.

Douhan, G.W., D.P. Martin, and D.M. Rizzo. 2007. Using the putative asexual fungus *Cenococcum geophilum* as a model to test how species concepts influence recombination analyses using sequence data from multiple loci. *Curr. Genet.* **52:** 191–201.

Drake, J.W., B. Charlesworth, D. Charlesworth, and J.F. Crow. 1998. Rates of spontaneous mutation. *Genetics* **148:** 1667–1686.

Drosophila 12 Genomes Consortium. 2007. Evolution of genes and genomes on the *Drosophila* phylogeny. *Nature* **450:** 203–218.

Drummond, A., O.G. Pybus, and A. Rambaut. 2003. Inference of viral evolutionary rates from molecular sequences. *Adv. Parasit.* **54:** 331–358.

Drummond, A.J., S.Y.W. Ho, M.J. Phillips, and A. Rambaut. 2006. Relaxed phylogenetics and dating with confidence. *PLoS Biol.* **4:** 699–710.

Drummond, D.A. and C.O. Wilke. 2008. Mistranslation-induced protein misfolding as a dominant constrain on coding-sequence evolution. *Cell* **13:** 341–352.

Dunn, L.C. and J. Suckling. 1956. Studies of the genetic variability in populations of wild house mice. I. Analysis of several alleles at locus *t. Genetics* **41:** 344–352.

Duret, L. and D. Mouchiroud. 1999. Expression pattern and, surprisingly, gene length shape codon usage in *Caenorhabditis*, *Drosophila*, and *Arabidopsis. Proc. Natl. Acad. Sci. U.S.A.* **96:** 4482–4487.

Duret, L., M. Semon, G. Piganeau, and D. Mouchiroud. 2002. Vanishing GC-rich isochores in mammalian genomes. *Genetics* **162:** 1837–1847.

Duret, L., A. Eyre-Walker, and N. Galtier. 2006. A new perspective on isochore evolution. *Gene* **385:** 71–74.

Durrett, R. and J. Schweinsberg. 2004. Approximating selective sweeps. *Theor. Popul. Biol.* **66:** 129–138.

Dye, C. and B.G. Williams. 1997. Multigenic drug resistance among inbred malaria parasites. *Proc. R. Soc. B* **264**: 61–67.

Dyer, K.A., B. Charlesworth, and J. Jaenike. 2007. Chromosome-wide linkage disequilibrium as a consequence of meiotic drive. *Proc. Natl. Acad. Sci. U.S.A.* **104**: 1587–1592.

Dykhuizen, D.E. 1990. Experimental studies of natural selection in bacteria. *Ann. Rev. Ecol. Evol. Syst.* **21**: 373–398.

Dykhuizen, D.E. and A.M. Dean. 1990. Enzyme-activity and fitness-evolution in solution. *Trends Ecol. Evol.* **5**: 257–262.

Eanes, W.F. 1999. Analysis of selection on enzyme polymorphisms. *Ann. Rev. Ecol. Syst.* **30**: 301–326.

Eanes, W.F., J. Hey, and D. Houle. 1985. Homozygous and hemizygous viability variation on the X chromosome of *Drosophila melanogaster*. *Genetics* **111**: 831–844.

East, E. M. 1910. A Mendelian interpretation of variation that is apparently continuous. *Am. Nat.* **44**: 65–82.

East, E. M. 1916. Studies on size inheritance in *Nicotiana*. *Genetics* **1**: 164–176.

Easton, D.F., K.A. Pooley, A.M. Dunning, P.D. Pharoah, D. Thompson, et al. 2007. Genome-wide association study identifies novel breast cancer susceptibility loci. *Nature* **447**: 1087–1093.

Ebert, D., C. Haag, M. Kirkpatrick, M. Riek, J.W. Hottinger, and V.I. Pajunen. 2002. A selective advantage to immigrant genes in a *Daphnia* metapopulation. *Science* **295**: 485–488.

Edwards, A.W.F. 2000. Carl Düsing (1884) on the regulation of the sex-ratio. *Theor. Popul. Biol.* **58**: 255–257.

Eldon, B. and J. Wakeley. 2006. Coalescent processes when the distribution of offspring number among individuals is highly skewed. *Genetics* **172**: 2621–2633.

Eldon, B. and J. Wakeley. 2009. Coalescence times and F_{ST} under a skewed offspring distribution among individuals in a population. *Genetics* **181**: 615–629.

Emerson, J.J., M. Cardoso-Moreira, J.O. Borevitz, and M. Long. 2008. Natural selection shapes genome wide patterns of copy number polymorphism in *D. melanogaster*. *Science* **320**: 1629–1631.

ENCODE Project Consortium. 2007. Identification and analysis of functional elements in 1% of the human genome by the ENCODE pilot project. *Nature* **447**: 800–816.

Endler, J.A. 1977. *Geographic Variation, Speciation and Clines*. Princeton University Press, Princeton, NJ.

Endler, J.A. 1986. *Natural Selection in the Wild*. Princeton University Press, Princeton, NJ.

Eshel, I. 1985. Evolutionary genetic stability of Mendelian segregation and the role of free recombination in the chromosomal system. *Am. Nat.* **125**: 412–420.

Escobar, J.S., P. Jarne, A. Charmantier, and P. David. 2008. Outbreeding alleviates senescence in hermaphroditic snails as expected from the mutation-accumulation theory. *Curr. Biol.* **18**: 906–910.

Escobar, J.S., A. Nicot, and P. David. 2008. The different sources of variation in inbreeding depression, heterosis and outbreeding depression in a metapopulation of *Physa acuta*. *Genetics* **180**: 1593–1608.

Estes, S. and S.J. Arnold. 2007. Resolving the paradox of stasis: models with stabilizing selection explain evolutionary divergence on all timescales. *Am. Nat.* **169**: 227–244.

Estoup, A., M. Beaumont, F. Sennedot, C. Moritz, and J.M. Cornuet. 2004. Genetic analysis of complex demographic scenarios: spatially expanding populations of the cane toad, *Bufo marinus*. *Evolution* **58**: 2021–2036.

Ethier, S. and T. Nagylaki. 1980. Diffusion approximations of Markov chains with two time scales and applications to population genetics. *Adv. Appl. Prob.* **12**: 14–49.

Euler, L. 1760. Recherches generales sur la mortalité et la multiplication. *Mém. Acad. R. Sci. Bell. Lett.* **16**: 144–164.

Evans, S.N., Y. Shvets, and M. Slatkin. 2006. Non-equilibrium theory of the allele frequency spectrum. *Theor. Popul. Biol.* **71**: 109–119.

Ewens, W.J. 1964. The pseudo-transient distribution and its uses in genetics. *J. App. Prob.* **1**: 141–156.

Ewens, W.J. 2004. *Mathematical Population Genetics. 1. Theoretical Introduction.* Springer, New York.

Excoffier, L. 2003. Analysis of population subdivision. In *Handbook of Statistical Genetics* (Eds. D.J. Balding, M. Bishop, and C. Cannings), pp. 713–750. John Wiley, Chichester, U.K.

Excoffier, L. and M. Slatkin. 1995. Maximum-likelihood estimation of molecular haplotype frequencies in a diploid population. *Mol. Biol. Evol.* **12**: 921–927.

Excoffier, L., P.E. Smouse, and J.M. Quattro. 1992. Analysis of molecular variance inferred from metric distances among DNA haplotypes: application to human mitochondrial DNA restriction data. *Genetics* **131**: 479–491.

Eyre-Walker, A. 1992. The effect of constraint on the rate of evolution in neutral models with biased mutation. *Genetics* **131**: 233–234.

Eyre-Walker, A. 2002. Changing effective population size and the McDonald–Kreitman test. *Genetics* **162**: 2017–2024.

Eyre-Walker, A. 2006. The genomic rate of adaptive evolution. *Trends Ecol. Evol.* **21**: 569–575.

Eyre-Walker, A. and P.D. Keightley. 1999. High genomic deleterious mutation rates in hominids. *Nature* **397**: 344–347.

Eyre-Walker, A. and P.D. Keightley. 2009. Estimating the rate of adaptive mutations in the presence of slightly deleterious mutations and population size change. *Mol. Biol. Evol.* **26**: 2097-2108.

Eyre-Walker, A., M. Woolfit, and T. Phelps. 2006. The distribution of fitness effects of new deleterious amino-acid mutations in humans. *Genetics* **173**: 891–900.

Falconer, D.S. and T.F.C. Mackay. 1996. *An Introduction to Quantitative Genetics*, 4th ed. Longman, London.

Falush, D., M. Stephens, and J.K. Pritchard. 2003. Inference of population structure using multilocus genotype data: linked loci and correlated allele frequencies. *Genetics* **164**: 1567–1587.

Fang, D.Q., M.L. Roose, R.R. Krueger, and C.T. Federici. 1997. Fingerprinting trifoliate orange germ plasm accessions with isozymes, RFLPs, and inter-simple sequence repeat markers. *Theor. Appl. Genet.* **95**: 211–219.

Fay, J.C. and C.I. Wu. 2000. Hitchhiking under positive Darwinian selection. *Genetics* **155:** 1405–1413.

Fay, J.C., G.J. Wyckoff, and C.I. Wu. 2001. Positive and negative selection on the human genome. *Genetics* **158:** 1227–1234.

Fay, J., G.J. Wyckhoff, and C.I. Wu. 2002. Testing the neutral theory of molecular evolution with genomic data from *Drosophila*. *Nature* **415:** 1024–1026.

Fearnhead, P. 2006. Perfect simulation from nonneutral population genetic models: variable population size and population subdivision. *Genetics* **174:** 1397–1406.

Fearnhead, P. and P. Donnelly. 2001. Estimating recombination rates from population genetic data. *Genetics* **159:** 1299–1318.

Feldman, M.W. 1972. Selection for linkage modification. 1. Random mating populations. *Theor. Popul. Biol.* **3:** 324–346.

Feldman, M.W. and U. Liberman. 1986. An evolutionary reduction principle for genetic modifiers. *Proc. Natl. Acad. Sci. U.S.A.* **83:** 4824–4827.

Feldman, M.W., F.B. Christiansen, and L.D. Brooks. 1980. Evolution of recombination in a constant environment. *Proc. Natl. Acad. Sci. U.S.A.* **77:** 4824–4827.

Felsenstein, J. 1965. The effect of linkage on directional selection. *Genetics* **52:** 349–363.

Felsenstein, J. 1974. The evolutionary advantage of recombination. *Genetics* **78:** 737–756.

Felsenstein, J. 1975. A pain in the torus: some difficulties with the model of isolation by distance. *Am. Nat.* **109:** 359–368.

Felsenstein, J. 2004. *Inferring Phylogenies*. Sinauer, Sunderland, MA.

Felsenstein, J. 2006. Accuracy of coalescent likelihood estimates: do we need more sites, more sequences, or more loci? *Mol. Biol. Evol.* **23:** 691–700.

Ffrench-Constant, R.H., B. Pittendrigh, A. Vaughan, and N. Anthony. 1998. Why are there so few resistance-associated mutations in insecticide target genes? *Phil. Trans. R. Soc. B* **353:** 1685–1693.

Ffrench-Constant, R.H., P.J. Daborn, and G. Le Goff. 2004. The genetics and genomics of insecticide resistance. *Trends Genet.* **20:** 163–170.

Filatov, D.A. and D. Charlesworth. 1999. DNA polymorphism, haplotype structure and balancing selection in the Leavenworthia PgiC locus. *Genetics* **153:** 1423–1434.

Finch, C.E. 1990. *Longevity, Senescence, and the Genome*. University of Chicago Press, Chicago, IL.

Fisher, R.A. 1922. On the dominance ratio. *Proc. R. Soc. Edinburgh* **52:** 312–341.

Fisher, R.A. 1928. The possible modification of the wild-type in response to recurrent mutations. *Am. Nat.* **62:** 115–126.

Fisher, R.A. 1930a. The distribution of gene ratios for rare mutations. *Proc. R. Soc. Edinburgh* **50:** 205–220.

Fisher, R.A. 1930b. *The Genetical Theory of Natural Selection*. Oxford University Press, Oxford, U.K.

Fisher, R.A. 1931. The evolution of dominance. *Biol. Rev.* **6:** 345–368.

Fisher, R.A. 1937. The wave of advance of advantageous genes. *Ann. Eugen.* **7:** 355–369.

Fisher, R.A. 1938. Dominance in poultry. Feathered feet, rose comb, internal pigment and pile. *Proc. R. Soc. B* **125:** 25–48.

Fisher, R.A. 1941. Average excess and average effect of a gene substitution. *Ann. Eugen.* **11**: 31–38.

Fisher, R.A. 1950. Gene frequencies in a cline determined by selection and migration. *Biometrics* **6**: 353–361.

Fisher, R.A. 1954. A fuller theory of junctions in inbreeding. *Heredity* **8**: 187–199.

Fishman, L. 2001. Inbreeding depression in two populations of *Arenaria uniflora* (Caryophyllaceae) with contrasting mating systems. *Heredity* **86**: 184–189.

Fishman, L. and A. Saunders. 2008. Centromere-associated female meiotic drive entails male fitness costs in monkeyflowers. *Science* **322**: 1559–1562.

Fishman, L. and J.H. Willis. 2005. A novel meiotic drive locus almost completely distorts segregation in *Mimulus* (monkeyflower) hybrids. *Genetics* **169**: 347–353.

Fishman, L., and J. H. Willis. 2006. A cytonuclear incompatibility causes anther sterility in *Mimulus* hybrids. Evolution 60:1372–1381.

Fitzpatrick, M.J., E. Feder, L. Rowe, and M.B. Sokolowski. 2007. Maintaining a behaviour polymorphism by frequency-dependent selection on a single gene. *Nature* **447**: 210–213.

Fiumera, A.C., B.L. Dumont, and A.G. Clark. 2007. Associations between sperm competition and natural variation in male reproductive genes on the third chromosome of *Drosophila melanogaster*. *Genetics* **176**: 1245–1260.

Flint, J. and T.F.C. Mackay. 2009. Genetic architecture of quantitative traits in mice, flies, and humans. *Genome Res.* **19**: 723–733.

Flint, J. and R. Mott. 2008. Applying mouse complex-trait resources to behavioural genetics. *Nature* **456**: 724–727.

Fontanillas, P., D.L. Hartl, and M. Reuter. 2007. Genome organization and gene expression shape the transposable element distribution in the *Drosophila melanogaster* euchromatin. *PLoS Genet.* **3**: e210.

Ford, C.E. and E.P. Evans. 1973. Robertsonian fusions in mice: segregational irregularities in male heterozygotes and zygotic imbalance. *Chromosomes Today* **4**: 387–397.

Frank, S. A. 1989. The evolutionary dynamics of cytoplasmic male sterility. *Am. Nat.* **133**: 345–576.

Frank, S.A. 1996. Statistical properties of polymorphism in host–parasite genetics. *Evol. Ecol.* **10**: 307–317.

Frank, S.A. 1998. *Foundations of Social Evolution*. Princeton University Press, Princeton, NJ.

Frankham, R. 1995. Effective population-size adult-population size ratios in wildlife—a review. *Genet. Res.* **66**: 95–107.

Frankham, R., J.D. Ballou, and D.A. Briscoe. 2002. *Introduction to Conservation Genetics*. Cambridge University Press, Cambridge, U.K.

Franklin, I.R. and R.C. Lewontin. 1970. Is the gene the unit of selection? *Genetics* **65**: 701–734.

Fritsch, P. and L.H. Rieseberg. 1992. High outcrossing rates maintain male and hermaphrodite individuals in populations of the flowering plant *Datisca glomerata*. *Nature* **359**: 633–636.

Fu, Y.-X. 1995. Statistical properties of segregating sites. *Theor. Popul. Biol.* **48**: 172–197.

Fu, Y.-X. 1997. Statistical tests of neutrality for samples from a population. *Genetics* **143:** 557–570.

Fu, Y.-X. and W.-H. Li. 1993. Statistical tests of neutrality of mutations. *Genetics* **133:** 693–709.

Fuglsang, A. 2006. Accounting for background nucleotide composition when measuring codon usage bias: brilliant idea, difficult in practice. *Mol. Biol. Evol.* **23:** 1345–1347.

Galagan, J.E. and E.U. Selker. 2004. RIP: the evolutionary cost of genome defense. *Trends Genet.* **20:** 417–423.

Galen, C. 1997. Rates of floral evolution: adaptation to bumblebee pollination in an alpine wildflower, *Polemonium viscosum*. *Evolution* **50:** 120–125.

Galloway, L.F. 1989. Population differentiation and hybrid success in *Campanula americana*: geography and genome size. *J. Evol. Biol.* **18:** 81–89.

Galtier, N., E. Bazin, and N. Bierne. 2006. GC-biased segregation of non-coding polymorphisms in *Drosophila*. *Genetics* **172:** 221–228.

Ganders, F.R. 1978. The genetics and evolution of gynodioecy in *Nemophila menziesii* (Hydrophyllaceae). *Can. J. Bot.* **56:** 1400–1408.

Gandon, S. and S.P. Otto. 2007. The evolution of sex and recombination in response to abiotic or coevolutionary fluctuations in epistasis. *Genetics* **175:** 1835–1853.

Garcia-Dorado, A. and A. Caballero. 2000. On the average coefficient of dominance of deleterious spontaneous mutations. *Genetics* **155:** 1991–2001.

García-Dorado, A. and A. Gallego. 2003. Comparing analysis methods for mutation-accumulation data: a simulation study. *Genetics* **164:** 807–819.

Gardner, A., S.A. West, and N.H. Barton. 2007. The relation between multilocus population genetics theory and social evolution theory. *Am. Nat.* **169:** 207–226.

Garrigan, D. and P.W. Hedrick. 2003. Perspective: detecting adaptive molecular polymorphism: lessons from the MHC. *Evolution* **57:** 1707–1722.

Gavrilets, S. 2004. *Fitness Landscapes and the Origin of Species*. Princeton University Press, Princeton, NJ.

Gavrilov, L.A. and N.S. Gavrilova. 1991. *The Biology of Life-Span: A Quantitative Approach*. Harwood, Chur, Switzerland.

Gay, J., S. Myers, and G. McVean. 2007. Estimating meiotic gene conversion rates from population genetic data. *Genetics* **177:** 881–894.

Gelehrter, T.D., F.S. Collins, and D. Ginsburg. 1998. *Principles of Medical Genetics*. Williams and Wilkins, Baltimore, MD.

Gerrish, P.J. and R.E. Lenski. 1998. The fate of competing beneficial mutations in an asexual population. *Genetica* **102/3:** 127–144.

Gerton, J.L., J. DeRisi, R. Shroff, M. Lichten, P.O. Brown, and T.D. Petes. 2000. Global mapping of meiotic recombination hotspots and coldspots in the yeast *Saccharomyces cerevisiae*. *Proc. Natl. Acad. Sci. U.S.A.* **97:** 11383–11390.

Gessler, D.G. 1995. The constraints of finite size in asexual populations and the rate of the ratchet. *Genet. Res.* **66:** 241–253.

Giaever, G., A.M. Chu, C. Connelly, L. Riles, S. Véronneau, et al. 2002. Functional profiling of the *Saccharomyces cerevisiae* genome. *Nature* **418:** 387–391.

Gibson, G. and B.S. Weir. 2005. The quantitative genetics of transcription. *Trends Genet.* **21:** 616–623.

Gigord, L., M. Macnair, and A. Smithson. 2001. Negative frequency-dependent selection maintains a dramatic flower color polymorphism in the rewardless orchid *Dactylorhiza sambucina* (L.) Soo. *Proc. Natl. Acad. Sci. U.S.A.* **98:** 6253–6255.

Giles, B.E. 1997. A case study of genetic structure in a metapopulation. In *Metapopulation Biology: Ecology, Genetics, and Evolution* (Eds. I. Hanski and M.E. Gilpin), pp. 429–454. Academic Press, San Diego, CA.

Gillespie, J.H. 1977. A general model to account for enzyme variation in natural environments. III. Multiple alleles. *Evolution* **31:** 85–90.

Gillespie, J.H. 1983. A simple stochastic gene substitution process. *Theor. Popul. Biol.* **23:** 202–215.

Gillespie, J.H. 1984. Molecular evolution over the mutational landscape. *Evolution* **38:** 1116–1119.

Gillespie, J.H. 1986. Natural selection and the molecular clock. *Mol. Biol. Evol.* **3:** 138–155.

Gillespie, J.H. 1991. *The Causes of Molecular Evolution*. Oxford University Press, Oxford, U.K.

Gillespie, J.H. 1994. Substitution processes in molecular evolution. III. Deleterious alleles. *Genetics* **138:** 943–952.

Gillespie, J.H. 2000. Genetic drift in an infinite population: the pseudohitchiking model. *Genetics* **155:** 909–919.

Gillespie, J.H. and C.H. Langley. 1974. A general model to account for enzyme variation in natural populations. *Genetics* **76:** 837–884.

Gillespie, J.H. and C.H. Langley. 1979. Are evolutionary rates really variable? *J. Mol. Evol.* **13:** 27–34.

Gilligan, D.M., D.A. Briscoe, and R. Frankham. 2005. Comparative losses of quantitative and molecular genetic variation in finite populations of *Drosophila melanogaster*. *Genet. Res.* **85:** 47–77.

Gilpin, M.E. 1975. *Group Selection in Predator–Prey Communities*. Princeton University Press, Princeton, NJ.

Glémin, S. 2005. Lethals in subdivided populations. *Genet. Res.* **86:** 41–51.

Glémin, S., T. Gaude, M.L. Guillemin, M. Lourmas, I. Olivieri, and A. Mignot. 2005. Balancing selection in the wild: testing population genetics theory of self-incompatibility in the rare species *Brassica insularis*. *Genetics* **171:** 279–289.

Glinka, S., L. Ometto, S. Mousset, W. Stephan, and D. De Lorenzo. 2003. Demography and selection have shaped genetic variation in *Drosophila melanogaster*: a multilocus approach. *Genetics* **165:** 1269–1278.

Goddard, M.R., C.H. Godfray, and A. Burt. 2002. Sex increases the efficacy of natural selection in experimental yeast populations. *Nature* **434:** 636–640.

Golczyk, H., R. Hasterok, and A. Joachimiak. 2005. FISH-aimed karyotyping and characterization of Renner complexes in permanent heterozygote *Rhoeo spathacea*. *Genome* **48:** 145–153.

Goldschmidt, R.B. 1940. *The Material Basis of Evolution*. Yale University Press, New Haven, CT.

Goldschmidt, R.B. 1945. Mimetic polymorphism: a controversial chapter of Darwinisim. *Quart. Rev. Biol.* **20:** 147–164, 205–230.

Goldstein, D.B. and C. Schloetterer. 1999. *Microsatellites. Evolution and Applications.* Oxford University Press, Oxford, U.K.

Goldstein, D.B., A.R. Linares, L.L. Cavalli-Sforza, and M.W. Feldman. 1995. An evaluation of genetic distances for use with microsatellite loci. *Genetics* **139:** 463–471.

Gonder, M., T. Disotell, and J. Oates. 2006. New genetic evidence on the evolution of chimpanzee populations and implications for taxonomy. *Int. J. Primatol.* **27:** 1103–1127.

González, J., K. Lenkov, M. Lipatov, J.M. Macpherson, and D.A. Petrov. 2008. High rate of transposable-element-induced adaptation in *Drosophila melanogaster. PLoS Biol.* **6:** e251.

Goodwillie, C. and J.W. Stiller. 2001. Evidence for polyphyly in a species of Linanthus (Polemoniaceae): convergent evolution in self-fertilizing taxa. *Syst. Bot.* **26:** 273–282.

Gordo, I. and R.A. Campos. 2008. Sex and deleterious mutations. *Genetics* **179:** 621–626.

Gordo, I. and B. Charlesworth. 2000. On the speed of Muller's ratchet. *Genetics* **156:** 2137–2140.

Gordo, I., A. Navarro, and B. Charlesworth. 2002. Muller's ratchet and the pattern of variation at a neutral locus. *Genetics* **161:** 835–848.

Gorelick, R. and M.D. Laublicher. 2004. Decomposing multilocus linkage disequilibrium. *Genetics* **166:** 1581–1583.

Goudet, J. and G. Martin. 2007. Under neutrality, $Q_{ST} < F_{ST}$ when there is dominance in an island model. *Genetics* **176:** 1371–1374.

Goymer, P., S.G. Kahn, J.G. Malone, S.M. Gehrig, A.J. Spiers, and P.B. Rainey. 2006. Adaptive divergence in experimental populations of *Pseudomonas fluorescens.* II. Role of the GGDEF regulator WspR in evolution and development of the wrinkly spreader phenotype. *Genetics* **173:** 515–526.

Grafen, A. 2006. Optimization of inclusive fitness. *J. Theor. Biol.* **238:** 541–563.

Grahame, J.W., C.S. Wilding, and R.K. Butlin. 2006. Adaptation to a steep environmental gradient and an associated barrier to gene exchange in *Littorina saxatilis. Evolution* **60:** 268–278.

Grant, P.R. and B.R. Grant. 2002. Unpredictable evolution in a 30-year study of Darwin's finches. *Science* **296:** 707–711.

Graur, D. and W.-H. Li. 2000. *Fundamentals of Molecular Evolution*, 2nd ed. Sinauer, Sunderland, MA.

Green, R.E., J. Krause, S.E. Ptak, A.W. Briggs, M.T. Ronan, J.F. Simons, et al. 2006. Analysis of one million base pairs of Neanderthal DNA. *Nature* **444:** 330–336.

Gregorius, H.-R., M.D. Ross, and E. Gillet. 1982. Selection in plant populations of effectively infinite size. 3. Maintenance of females among hermaphrodites. *Heredity* **48:** 329–343.

Griffing, B. 1960. Theoretical consequences of truncation selection based on the individual phenotype. *Aust. J. Biol. Sci.* **13:** 307–343.

Griffiths, R.C. 1980. Lines of descent in the diffusion approximation of neutral Wright–Fisher models. *Theor. Popul. Biol.* **17:** 37–50.

Grosberg, R.K. 1988. The evolution of allorecognition specificity in clonal invertebrates. *Quart. Rev. Biol.* **63:** 377–412.

Grosberg, R.K. 1991. Sperm-mediated gene flow and the genetic-structure of a population of the colonial ascidian *Botryllus schlosseri*. *Evolution* **45**: 130–142.

Gruber, J.D., A. Genissel, S.J. MacDonald, and A.D. Long. 2007. How repeatable are
associations between polymorphisms in *achaete–scute* and bristle number variation in *Drosophila*? *Genetics* **175**: 1987–1997.

Gubenko, I.S. and M.B. Evgen'ev. 1984. Cytological and linkage maps of *Drosophila
virilis* chromosomes. *Genetica* **65**: 127–139.

Guillot, G., A. Estoup, F. Mortier, and J.F. Cosson. 2005. Bayesian clustering using
hidden Markov random fields in spatial population genetics. *Genetics* **174**: 805–
816.

Gutz, H. and J.F. Leslie. 1976. Gene conversion: a hitherto overlooked parameter in
population genetics. *Genetics* **83**: 861–866.

Haag, C. and D.D. Ebert. 2007. Genotypic selection in Daphnia populations consisting
of inbred sibships. *J. Evol. Biol.* **20**: 881–891.

Haag-Liautard, C., M. Dorris, X. Maside, S. Macaskill, D.L. Halligan, et al. 2007. Direct
estimation of per nucleotide and genomic deleterious mutation rates in
Drosophila. *Nature* **445**: 82–85.

Haag-Liautard, C., N. Coffey, D. Houle, M. Lynch, B. Charlesworth, and P.D. Keightley. 2008. Direct estimation of the mitochondrial DNA mutation rate in
Drosophila melanogaster. *PLoS Biol.* **6**: e204.

Hadany, L. and M.W. Feldman. 2005. Evolutionary traction: the cost of adaptation
and the evolution of sex. *J. Evol. Biol.* **18**: 309–314.

Haddrill, P.R. and B. Charlesworth. 2008. Non-neutral processes drive the nucleotide
composition of non-coding sequences in *Drosophila*. *Biol. Lett.* **4**: 438–441.

Haddrill, P.R., K.R. Thornton, B. Charlesworth, and P. Andolfatto. 2005a. Multilocus
patterns of nucleotide variability and the demographic and selection history of
Drosophila melanogaster populations. *Genome Res.* **15**: 790–799.

Haddrill, P.R., B. Charlesworth, D.L. Halligan, and P. Andolfatto. 2005b. Patterns of
intron sequence evolution in *Drosophila* are dependent upon length and GC content. *Genome Biol.* **6**: R67.61–68.

Haddrill, P.R., D.L. Halligan, D. Tomaras, and B. Charlesworth. 2007. Reduced efficacy
of selection in regions of the *Drosophila* genome that lack crossing over. *Genome
Biol.* **8**: R18.11–17.

Haigh, J. 1978. The accumulation of deleterious genes in a population. *Theor. Popul.
Biol.* **14**: 251–267.

Haldane, J.B.S. 1922. Sex-ratio and unisexual sterility in hybrid animals. *J. Genet.* **12**:
101–109.

Haldane, J.B.S. 1924. A mathematical theory of natural and artificial selection. Part I.
Trans. Camb. Philos. Soc. **23**: 19–41.

Haldane, J.B.S. 1927a. A mathematical theory of natural and artificial selection. Part IV.
Proc. Camb. Philos. Soc. **23**: 607–615.

Haldane, J.B.S. 1927b. A mathematical theory of natural and artificial selection. Part V.
Selection and mutation. *Proc. Camb. Philos. Soc.* **23**: 838–844.

Haldane, J.B.S. 1930a. A mathematical theory of natural and artificial selection. VI. Isolation. *Proc. Camb. Philos. Soc.* **26**: 220–230.

Haldane, J.B.S. 1930b. A mathematical theory of natural and artificial selection. VII. Selection intensity as a function of mortality rate. *Proc. Camb. Philos. Soc.* **27:** 131–136.

Haldane, J.B.S. 1932. *The Causes of Evolution.* Longmans Green, London.

Haldane, J.B.S. 1933. The part played by recurrent mutation in evolution. *Am. Nat.* **67:** 5–19.

Haldane, J.B.S. 1935. The rate of spontaneous mutation of a human gene. *J. Genet.* **31:** 317–326.

Haldane, J.B.S. 1937. The effect of variation on fitness. *Am. Nat.* **71:** 337–349.

Haldane, J.B.S. 1941. *New Paths in Genetics.* Allen and Unwin, London.

Haldane, J.B.S. 1942. Selection against heterozygosis in man. *Ann. Eugen.* **11:** 333–340.

Haldane, J.B.S. 1948. The theory of a cline. *J. Genet.* **48:** 277–284.

Haldane, J.B.S. 1949a. The association of characters as a result of inbreeding. *Ann. Eugen.* **15:** 15–23.

Haldane, J.B.S. 1949b. Disease and evolution. *La Ricerca Scient.* **19:** 1–11.

Haldane, J.B.S. 1949c. The rate of mutation of human genes. *Hereditas, Suppl. Vol.* **1949:** 267–273.

Haldane, J.B.S. 1954. The measurement of natural selection. *Caryologia* **6:** 480–487 (Suppl.).

Haldane, J.B.S. 1955. Population genetics. *Penguin New Biol.* **18:** 34–51.

Haldane, J.B.S. 1957. The cost of selection. *J. Genet.* **55:** 511–524.

Haldane, J.B.S. 1962. Natural selection in a population with annual breeding but overlapping generations. *J. Genet.* **58:** 122–124.

Haldane, J.B.S. 1964. A defense of beanbag genetics. *Perspect. Biol. Med.* **7:** 343–349.

Haldane, J.B.S. and S.D. Jayakar. 1963. Polymorphism due to selection of varying direction. *J. Genet.* **58:** 237–242.

Hall, M. and J.H. Willis. 2005. Transmission ratio distortion in intraspecific hybrids of *Mimulus guttatus*: implications for genomic divergence. *Genetics* **170:** 375–386.

Hamilton, W.D. 1963. The evolution of altruistic behavior. *Am. Nat.* **97:** 354–356.

Hamilton, W.D. 1964a. The genetical evolution of social behaviour. I. *J. Theor. Biol.* **7:** 1–16.

Hamilton, W.D. 1964b. The genetical evolution of social behaviour. II. *J. Theor. Biol.* **7:** 17–52.

Hamilton, W.D. 1966. The moulding of senescence by natural selection. *J. Theor. Biol.* **12**.

Hamilton, W.D. 1967. Extraordinary sex ratios. *Science* **156:** 477–488.

Hamilton, W.D. 1980. Sex versus non-sex versus parasite. *Oikos* **35:** 282–290.

Hamrick, J.L. and M.J.W. Godt. 1996. Effects of life history traits on genetic diversity in plant species. *Phil. Trans. R. Soc. B* **351:** 1291–1298.

Handley, L.J.L., A. Manica, J. Goudet, and F. Balloux. 2007. Going the distance: human population genetics in a clinal world. *Trends Genet.* **23:** 432–439.

Hanski, I. and M.E. Gilpin. 1997. *Metapopulation Biology: Ecology, Genetics, and Evolution.* Academic Press, San Diego, CA.

Hao, W. and G.B. Golding. 2006. The fate of laterally transferred genes: Life in the fast lane to adaptation or death. *Genome Res.* **16:** 636–643.

Harada, Y., Y. Takagaki, M. Sunagawa, T. Saito, L. Yamada, et al. 2008. Mechanism of self-sterility in a hermaphroditic chordate. *Science* **320:** 548–550.

Hardy, G.H. 1908. Mendelian proportions in a mixed population. *Science* **28:** 49–50.

Harris, H. 1966. Enzyme polymorphisms in man. *Proc. R. Soc. B* **164:** 298–310.

Hartl, D.L. and D.E. Dykhuizen. 1981. Potential for selection among nearly neutral allozymes of 6-phosphogluconate dehydrogenase in *Escherichia coli*. *Proc. Natl. Acad. Sci. U.S.A.* **78:** 62444–66348.

Hartl, D.L., Y. Hiraizumi, and J.F. Crow. 1967. Evidence for sperm dysfunction as the mechanism of segregation distortion in *Drosophila melanogaster*. *Proc. Natl. Acad. Sci. U.S.A.* **58:** 2240–2245.

Hasson, E., I.N. Wang, L.W. Zeng, M. Kreitman, and W.F. Eanes. 1998. Nucleotide variation in the triosephosphate isomerase (*Tpi*) locus of *Drosophila melanogaster* and *Drosophila simulans*. *Mol. Biol. Evol.* **15:** 756–769.

Hastings, A. 1981. Stable cycling in discrete-time genetic models. *Proc. Natl. Acad. Sci. U.S.A.* **78:** 7224–7225.

Haudry, A., A. Cenci, C. Ravel, T. Bataillon, D. Brunel, et al. 2007. Grinding up wheat: a massive loss of nucleotide diversity since domestication. *Mol. Biol. Evol.* **24:** 1506–1517.

Haudry, A., A. Cenci, C. Guilhaumon, E. Paux, S. Poirier, et al. 2008. Mating system and recombination affect molecular evolution in four Triticeae species. *Genet. Res.* **90:** 97–109.

Hawks, J., E.T. Wang, G. Cochran, M.H.C. Harpending, and R.K. Moyzis. 2007. Recent acceleration of human adaptive evolution. *Proc. Natl. Acad. Sci. U.S.A.* **104:** 20753–20758.

Hazel, L.N. 1943. The genetic basis for constructing selection indexes. *Genetics* **28:** 476–490.

Hedrick, P.W. 1987. Gametic disequilibrium measures: proceed with caution. *Genetics* **117:** 331–341.

Hedrick, P.W. 2007. Sex: differences in mutation, recombination, selection, gene flow, and genetic drift. *Evolution* **61:** 2750–2771.

Hein, J., M.H. Schierup, and C. Wiuf. 2005. *Gene Genealogies, Variation and Evolution*. Oxford University Press, Oxford, U.K.

Hellmann, I., I. Ebersberger, S.E. Ptak, S. Paabo, and M. Przeworski. 2003. A neutral explanation for the correlation of diversity with recombination rates in humans. *Am. J. Hum. Genet.* **72:** 1527–1535.

Henderson, I.R. and I.R. Jacobsen. 2007. Epigenetic inheritance in plants. *Nature* **447:** 418–424.

Herbeck, J.T., D.J. Funk, P.H. Degnan, and J.J. Wernegreen. 2003. A conservative test of genetic drift in the endosymbiotic bacterium Buchnera: slightly deleterious mutations in the chaperonin groEL. *Genetics* **165:** 1651–1660.

Heribert-Nilsson, N. 1920. Zuwachsgeschwindigkeit der Pollenschläuche und gestörte Mendelzahlen bei *Oenothera lamarckiana*. *Hereditas* **1:** 11–67.

Hermisson, J. and P.S. Pennings. 2005. Soft sweeps: molecular population genetics of adaptation from standing genetic variation. *Genetics* **169:** 2335–2352.

Herre, E. A. 1985. Sex-ratio adjustment in fig wasps. *Science* **228:** 896–898.

Herrmann, B.G., B. Koschorz, K. Wertz, K.J. McLaughlin, and A. Kispert. 1999. A protein kinase encoded by the *t* complex responder gene causes non-Mendelian inheritance. *Nature* **402**: 141–146.

Hershberg, R. and D.A. Petrov. 2008. Selection on codon bias. *Ann. Rev. Genet.* **42**: 287–299.

Hewitt, G.M. 2001. Speciation, hybrid zones and phylogeography—or seeing genes in space and time. *Mol. Ecol.* **10**: 537–549.

Hey, J. 1991. A multi-dimensional coalescent process applied to multiallelic selection models and migration models. *Theor. Popul. Biol.* **39**: 30–48.

Hey, J. and R. Nielsen. 2004. Multilocus methods for estimating population sizes, migration rates and divergence time, with applications to the divergence of *Drosophila pseudoobscura* and *D. persimilis. Genetics* **167**: 747–760.

Hickey, D.A. 1982. Selfish DNA: a sexually transmitted parasite. *Genetics* **101**: 519–531.

Hill, W.G. 1982. Predictions of response to selection from new mutations. *Genet. Res.* **40**: 255–278.

Hill, W.G. and A. Robertson. 1966. The effect of linkage on limits to artificial selection. *Genet. Res.* **8**: 269–294.

Hill, W.G. and A. Robertson. 1968. Linkage disequilibrium in finite populations. *Theor. Appl. Genet.* **38**: 226–231.

Hill, W.G., M.E. Goddard, and P.M. Visscher. 2008. Data and theory point to mainly additive genetic variance for complex traits. *PLoS Genetics* **4**: e100008.

Hilliker, A.J. and A. Chovnick. 1981. Further observations on intragenic recombination in *Drosophila melanogaster. Genet. Res.* **38**: 281–296.

Hilliker, A.J., G. Harauz, A.G. Reaume, M. Gray, S.H. Clark, and A. Chovnick. 1994. Meiotic gene conversion tract length distribution within the *rosy* locus of *Drosophila melanogaster. Genetics* **137**: 1019–1026.

Hiscock, S.J. and S.M. McInnis. 2003. Pollen recognition and rejection during the sporophytic self-incompatibility response: Brassica and beyond. *Trends Plant Sci.* **8**: 606–613.

Hoekstra, H.E. and J.A. Coyne. 2007. The locus of evolution: evo devo and the genetics of adaptation. *Evolution* **61**: 995–1016.

Hoekstra, H.E., K.E. Drumm, and M.W. Nachman. 2004. Ecological genetics of adaptive color polymorphism in pocket mice: geographic variation in selected and neutral genes. *Evolution* **58**: 1329–1341.

Hoekstra, H.E., R.J. Hirschmann, R.A. Bundey, P.A. Insel, and J.P. Crossland. 2006. A single amino acid mutation contributes to adaptive beachmouse color pattern. *Science* **313**: 101–104.

Holden, L.R. 1979. New properties of the two-locus partial selfing model with selection. *Genetics* **93**: 237–244.

Holsinger, K.E. 1988. Inbreeding depression doesn't matter: the genetic basis of mating-system evolution. *Evolution* **442**: 1235–1244.

Holsinger, K. 1999. Analysis of genetic diversity in geographically structured populations: a Bayesian perspective. *Hereditas* **130**: 245–255.

Holt, R.A., G.M. Subramanian, A. Halpern, G.G. Sutton, R. Charlab, et al. 2002. The genome sequence of the malaria mosquito *Anopheles gambiae. Science* **298**: 129–149.

Holtsford, T.P. and N.C. Ellstrand. 1989. Variation in outcrossing rate and population genetic structure of *Clarkia tembloriensis* (Onagraceae). *Theor. Appl. Genet.* **78:** 480–488.

Honys, D. and D. Twell. 2003. Comparative analysis of the *Arabidopsis* pollen transcriptome. *Plant Phys.* **132:** 640–652.

Houle, D. 1992. Comparing evolvability and variability of quantitative traits. *Genetics* **130:** 195–204.

Houston, A.I. and J.M. McNamara. 1996. State-dependent life histories. *Nature* **380:** 215–221.

Howell, N. 1976. Towards a uniformitarian theory of human paleodemography. In *The Demographic Evolution of Human Populations* (Eds. R.H. Ward and K.M. Weiss), pp. 25–40. Academic Press, London.

Howell, N. and B.S. Roth. 1981. Sexual reproduction in Agaves: the benefits of bats, the cost of semelparous reproduction. *Ecology* **62:** 1–7.

Howell, N., C.B. Smejkal, D.A. Mackey, P.F. Chinnery, D.M. Turnbull, et al. 2003. The pedigree rate of sequence divergence in the human mitochondrial genome: there is a difference between phylogenetic and pedigree rates. *Am. J. Hum. Genet.* **72:** 659–670.

Huang, C.Y., N. Grunheit, N. Ahmadinejad, J.N. Timmis, and W. Martin. 2005. Mutational decay and age of chloroplast and mitochondrial genomes transferred recently to angiosperm nuclear chromosomes. *Plant Physiol.* **138:** 1723–1733.

Hubby, J.L. and R.C. Lewontin. 1966. A molecular approach to the study of genic heterozygosity in natural populations. I. The number of alleles at different loci in *Drosophila pseudoobscura*. *Genetics* **54:** 577–594.

Hudson, R.R. 1983a. Properties of a neutral allele model with intragenic recombination. *Theor. Popul. Biol.* **23:** 183–201.

Hudson, R.R. 1983b. Testing the constant-rate neutral model with protein sequence data. *Evolution* **37:** 203–217.

Hudson, R.R. 1990. Gene genealogies and the coalescent process. *Oxf. Surv. Evol. Biol.* **7:** 1–45.

Hudson, R.R. 1998. Island models and the coalescent process. *Mol. Ecol.* **7:** 413–418.

Hudson, R.R. 2000. A new statistic for detecting differentiation. *Genetics* **155:** 2011–2014.

Hudson, R.R. 2001. Two-locus sampling distributions and their application. *Genetics* **159:** 1805–1817.

Hudson, R.R. and N.L. Kaplan. 1985. Statistical properties of the number of recombination events in the history of a sample of DNA sequences. *Genetics* **111:** 147–164.

Hudson, R.R. and N.L. Kaplan. 1988. The coalescent process in models with selection and recombination. *Genetics* **120:** 831–840.

Hudson, R.R. and N.L. Kaplan. 1995. Deleterious background selection with recombination. *Genetics* **141:** 1605–1617.

Hudson, R.R., D.D. Boos, and N.L. Kaplan. 1992. A statistical test for detecting geographic subdivision. *Mol. Biol. Evol.* **9:** 138–151.

Hudson, R.R., M. Kreitman, and M. Aguadé. 1987. A test of molecular evolution based on nucleotide data. *Genetics* **116:** 153–159.

Hudson, R.R., K. Bailey, D. Skarecky, J. Kwiatowski, and F.J. Ayala. 1994. Evidence for positive selection in the superoxide dismutase (*Sod*) region of *Drosophila melanogaster*. *Genetics* **136**: 1329–1340.

Huff, D.R. and L. Wu. 1992. Distribution and inheritance of inconstant sex forms in natural populations of dioecious buffalograss (*Buchloe dactyloides*). *Am. J. Bot.* **79**: 207–215.

Hughes, A.L., T. Ota, and M. Nei. 1990. Positive Darwinian selection promotes charge profile diversity in the antigen-binding cleft of class I major-histocompatibility-complex molecules. *Mol. Bio. Evol.* **7**: 515–524.

Humeau, L., T. Pailler, and J.D. Thompson. 1999. Cryptic dioecy and leaky dioecy in endemic species of Dombeya (Sterculiaceae) on La Réunion. *Am. J. Bot.* **86**: 1437–1447.

Hurst, L.D. 2002. The Ka/Ks ratio: diagnosing the form of sequence evolution. *Trends Genet.* **18**: 486–487.

Hutter, S., H.P. Li, S. Beisswanger, D. De Lorenzo, and W. Stephan. 2007. Distinctly different sex ratios in African and European populations of *Drosophila melanogaster* inferred from chromosome-wide nucleotide polymorphism data. *Genetics* **177**: 469–480.

Hyten, D.L., I.Y. Choi, Q. Song, R.C. Shoemaker, R.L. Nelson, et al. 2007. Highly variable patterns of linkage disequilibrium in multiple soybean populations. *Genetics* **175**: 1937–1944.

Iafrate, A.J., L. Feuk, M.N. Rivera, M.L. Listwenik, P.K. Donahoe, et al. 2004. Detection of large-scale variation in the human genome. *Nat. Genet.* **36**: 949–951.

Igic, B. and J.R. Kohn. 2001. Evolutionary relationships among self-incompatibility RNases. *Proc. Natl. Acad. Sci. U.S.A.* **98**: 13167–13171.

Igic, B., R. Lande, and J.R. Kohn. 2008. Loss of self-incompatibility and its evolutionary consequences. *Int. J. Plant Sci.* **169**: 93–104.

Ikemura, T. 1982. Correlation between the abundance of yeast transfer RNAs and the occurrence of the respective codons in protein genes; differences in synonymous codon choice patterns of yeast and *E. coli* with reference to the abundance of isoaccepting transfer RNAs. *J. Mol. Evol.* **158**: 389–409.

International Chicken Polymorphism Map Consortium. 2004. A genetic variation map for chicken with 2.8 million single nucleotide polymorphisms. *Nature* **432**: 717–722.

International Human Genome Sequencing Consortium. 2001. Initial sequencing and analysis of the human genome. *Nature* **409**: 861–921.

Ingvarsson, P.K. 2002. A metapopulation perspective on genetic diversity and differentiation in partially self-fertilizing plants. *Evolution* **56**: 2368–2373.

Ingvarsson, P.K. 2005. Nucleotide polymorphism and linkage disequilibrium within and among natural populations of European aspen (*Populus tremula* L., Salicaceae). *Genetics* **169**: 945–953.

Ingvarsson, P.K., K. Olsson, and L. Ericson. 1997. Extinction–recolonization dynamics in the mycophagous beetle *Phalacrus substriatus*. *Evolution* **51**: 187–195.

Innan, H. and M. Nordborg. 2003. The extent of linkage disequilibrium and haplotype sharing around a polymorphic site. *Genetics* **165**: 437–444.

Irwin, D.E. 2002. Phylogeographic breaks without geographic barriers to gene flow. *Evolution* **56**: 2383–2394.

Jacobs, M.S. and M.J. Wade. 2003. A synthetic review of the theory of gynodioecy. *Am. Nat.* **161**: 837–851.

Jaenike, J. 1978. An hypothesis to account for the maintenance of sex within populations. *Evol. Theor.* **3**: 191–194.

Jain, K. 2008. Loss of least-loaded class in asexual populations due to drift and epistasis. *Genetics* **179**: 2125–2134.

Jain, S.K. and A.D. Bradshaw. 1966. Evolutionary divergence among adjacent plant populations. I. The evidence and its theoretical analysis. *Heredity* **21**: 407–441.

James, T.Y., D. Porter, J.L. Hamrick, and R. Vilgalys. 1999. Evidence for limited intercontinental gene flow in the cosmopolitan mushroom *Schizophyllum commune*. *Evolution* **53**: 1665–1677.

Jarne, P. and D. Charlesworth. 1993. The evolution of the selfing rate in functionally hermaphrodite plants and animals. *Ann. Rev. Ecol. Syst.* **23**: 441–466.

Jarne, P. and P. David. 2008. Quantifying inbreeding in natural populations of hermaphroditic organisms. *Heredity* **100**: 431–439.

Jarne, P. and T. Staedler. 1995. Population genetic structure and mating system in freshwater pulmonates. *Experientia* **51**: 482–497.

Jenkin, F. 1867. Origin of Species. *N. Brit. Rev.* **46**: 277–318.

Jensen, J.D., Y. Kim, V. Bauer DuMont, C.F. Aquadro, and C.D. Bustamante. 2005. Distinguishing between selective sweeps and demography using DNA polymorphism data. *Genetics* **170**: 1401–1410.

Jensen, M.A., B. Charlesworth, and M. Kreitman. 2002. Patterns of genetic variation at a chromosome 4 locus of *Drosophila melanogaster* and *D. simulans*. *Genetics* **160**: 493–507.

Jimenez-Ambriz, G., C. Petit, I. Bourrie, S. Dubois, I. Olivieri, and O. Ronce. 2007. Life history variation in the heavy metal tolerant plant *Thlaspi caerulescens* growing in a network of contaminated and noncontaminated sites in southern France: role of gene flow, selection and phenotypic plasticity. *New Phytol.* **173**: 199–215.

Johannsen, W. 1909. *Elemente der exakten Erblichkeitslehre*. Fischer, Jena.

Johnson, T. and N.H. Barton. 2002. The effect of deleterious alleles on adaptation in asexual populations. *Genetics* **162**: 395–411.

Johnson, T. and N.H. Barton. 2005. Theoretical models of selection and mutation on quantitative traits. *Phil. Trans. R. Soc. B* **360**: 1411–1425.

Johnston, M.O., B. Das, and W.R. Hoeh. 1998. Negative correlation between male allocation and rate of self-fertilization in a hermaphroditic animal. *Proc. Natl. Acad. Sci. U.S.A.* **95**: 617–620.

Johnston, M.O., E. Porcher, P. Cheptou, C. Eckert, E. Elle, et al. 2009. Correlations among fertility components can maintain mixed mating in plants. *Am. Nat.* **173**: 1–11.

Jolley, K., D. Wilson, P. Kriz, G. McVean, and M. Maiden. 2005. The influence of mutation, recombination, population history, and selection on patterns of genetic diversity in *Neisseria meningitidis*. *Mol. Biol. Evol.* **22**: 562–569.

Jorde, P.E. and N. Ryman. 2007. Unbiased estimator for genetic drift and effective population size. *Genetics* **177**: 927–935.

Joseph, S.B. and M. Kirkpatrick. 2004. Haploid selection in animals. *Trends Ecol. Evol.* **19**: 592–597.

Jost, M.C., D. Hillis, Y. Lu, J. Kyle, H. Fozzard, and H. Zakon. 2008. Toxin-resistant sodium channels: parallel adaptive evolution across a complete gene family. *Mol. Biol. Evol.* **25**: 1016–1024.

Joyce, P., D. R. Rokyta, C. Beisel, and H. A. Orr. 2008. A general extreme value theory model for the adaptation of DNA sequences under strong selection and weak mutation. *Genetics* **180**: 1627–1643.

Jukes, T.H. and C.R. Cantor. 1969. Evolution of protein molecules. In *Mammalian Protein Metabolism III* (Ed. H.N. Munro), pp. 21–132. Academic Press, New York.

Jutier, D., N. Derome, and C. Montchamp-Moreau. 2004. The sex-ratio trait and its evolution in *Drosophila simulans*: a comparative approach. *Genetica* **120**: 87–99.

Kacser, H. and J.A. Burns. 1981. The molecular basis of dominance. *Genetics* **97**: 639–666.

Kaiser, V.B. and H. Ellegren. 2006. Nonrandom distribution of genes with sex-biased expression in the chicken genome. *Evolution* **60**: 1945–1951.

Kamau, E., B. Charlesworth, and D. Charlesworth. 2007. Linkage disequilibrium and recombination rate estimates in the self-incompatibility region of *Arabidopsis lyrata. Genetics* **176**: 2357–2369.

Kaminker, J.S., C.M. Bergman, B. Kronmiller, J. Carlson, R. Svirskas, et al. 2002. The transposable elements of the *Drosophila melanogaster* genome: a genomics perspective. *Genome Biol.* **3**: Research 0084

Kao, T.H. and T. Tsukamoto. 2004. The molecular and genetic bases of S-RNase-based self-incompatibility. *Plant Cell* **16**: S72–S83.

Kaplan, N.L., R.R. Hudson, and C.H. Langley. 1989. The "hitch-hiking" effect revisited. *Genetics* **123**: 887–899.

Karlin, S. 1968. Equilibrium behavior of population genetic models with non-random mating. *J. App. Prob.* **5**: 231–313.

Karlin, S. 1975. General two-locus selection models: some objectives, results and interpretations. *Theor. Popul. Biol.* **7**: 364–398.

Karlin, S. and S. Lessard. 1986. *Theoretical Studies of Sex Ratio Evolution.* Princeton University Press, Princeton, NJ.

Karlin, S. and U. Liberman. 1979. Representation of non-epistatic selection models and analysis of multilocus Hardy–Weinberg equilibrium configurations. *J. Math. Biol.* **7**: 353–374.

Karlin, S. and McGregor, J. 1974. Towards a theory of the evolution of modifier genes. *Theor. Popul. Biol.* **5**: 59–103.

Kauppi, L., A.J. Jeffreys, and S. Keeney. 2004. Where the crossovers are: recombination distributions in mammals. *Nat. Rev. Genet.* **5**: 413–424.

Kazazian, H., A. Chakravarti, S.H. Orkin, and S.E. Antonakis. 1983. DNA polymorphisms in the human β globin gene cluster. In *Evolution of Genes and Proteins* (Eds. M. Nei and R.K. Koehn), pp. 137–146. Sinauer, Sunderland, MA.

Keightley, P.D. 1996. A metabolic basis for dominance and recessivity. *Genetics* **143:** 621–625.

Keightley, P.D. 2004. Comparing analysis methods for mutation–accumulation data. *Genetics* **167:** 551–553.

Keightley, P.D. and A. Eyre-Walker. 1999. Terumi Mukai and the riddle of deleterious mutation rates. *Genetics* **153:** 515–523.

Keightley, P.D. and A. Eyre-Walker. 2007. Joint inference of the distribution of fitness effects of deleterious mutations and population demography based on nucleotide polymorphism frequencies. *Genetics* **177:** 2251–2261.

Keightley, P.D. and S.P. Otto. 2006. Interference among deleterious mutations favours sex and recombination in finite populations. *Nature* **443:** 89–92.

Keightley, P.D., G.V. Kryukov, S. Sunyaev, D.L. Halligan, and D.J. Gaffney. 2005. Evolutionary constraints in conserved nongenic sequences of mammals. *Genome Res.* **15:** 1371–1378.

Keightley, P.D., M. Trivedi, M. Thomson, F. Oliver, S. Kumar, and M.L. Blaxter. 2009. Analysis of the genome sequences of three *Drosophila melanogaster* spontaneous mutation accumulation lines. *Genome Res.* **19:** 1195–1201.

Keller, I., D. Bensasson, and R.A. Nichols. 2007. Transition–transversion bias is not universal: a counter example from grasshopper pseudogenes. *PLoS Genet.* **3:** 185–191.

Keller, L.F. 1998. Inbreeding and its fitness effects in an insular population of song sparrows (*Melospiza melodia*). *Evolution* **52:** 240–250.

Keller, L.F. and D.M. Waller. 2002. Inbreeding effects in wild populations. *Trends Ecol. Evol.* **17:** 230–241.

Kelly, C.K., M.G. Fowler, F. Breden, M. Fenner, and G.M. Poppy. 2005. An analytical model assessing the potential threat to natural habitats from insect resistance transgenes. *Proc. R. Soc. B* **272:** 1759–1767.

Kelly, J.K. 1999a. Response to selection in partially self-fertilizing populations. I. Selection on a single trait. *Evolution* **53:** 336–349.

Kelly, J.K. 1999b. Response to selection in partially self-fertilizing populations. II. Selection on multiple traits. *Evolution* **53:** 350–357.

Kelly, J.K. and S. Williamson. 2000. Predicting response to selection on a quantitative trait: a comparison between models for mixed-mating populations. *J. Theor. Biol.* **207:** 37–56.

Kettlewell, H.B.D. 1956. Further selection experiments on industrial melanism in the Lepidoptera. *Heredity* **10:** 287–301.

Keyfitz, N. 1968. *Introduction to the Mathematics of Population*. Addison-Wesley, Reading, MA.

Kidd, J.M., G.M. Cooper, W.F. Donahue, H.S. Hayden, N. Sampas, et al. 2008. Mapping and sequencing of structural variation from eight human genomes. *Nature* **453:** 56–64.

Kidwell, M.G. 1992. Horizontal transfer of *P* elements and other short inverted repeat transposons. *Genetica* **86:** 275–286.

Kim, S., V. Plagnol, T. Hu, C. Toomajian, R. Clark, et al. 2007. Recombination and linkage disequilibrium in *Arabidopsis thaliana*. *Nat. Genet.* **39:** 1151–1155.

Kim, S.H. and S.V. Yi. 2008. Mammalian nonsynonymous sites are not overdispersed: comparative genomic analysis of index of dispersion of mammalian proteins. *Mol. Biol. Evol.* **25:** 634–642.

Kim, S.H., N. Elango, C. Warden, E. Vigoda, and S.V. Yi. 2006. Heterogeneous genomic molecular clocks in primates. *PLoS Genet.* **2:** 1527–1534.

Kim, Y. 2004. Effect of strong directional selection on weakly selected mutations at linked sites: implication for synonymous codon usage. *Mol. Biol. Evol.* **21:** 286–294.

Kimura, M. 1953. "Stepping stone" model of population. *Ann. Rep. Nat. Inst. Genet.* **3:** 63–65.

Kimura, M. 1955. Stochastic processes and distribution of gene frequencies under natural selection. *Cold Spring Harbor Symp. Quant. Biol.* **20:** 33–53.

Kimura, M. 1956. A model of a system which leads to closer linkage by natural selection. *Evolution* **9:** 419–435.

Kimura, M. 1962. On the probability of fixation of a mutant gene in a population. *Genetics* **47:** 713–719.

Kimura, M. 1964. Diffusion models in population genetics. *J. App. Prob.* **1:** 177–223.

Kimura, M. 1965a. Attainment of quasi linkage equilibrium when gene frequencies are changing by natural selection. *Genetics* **52:** 875–890.

Kimura, M. 1965b. A stochastic model concerning the maintenance of genetic variability in quantitative characters. *Proc. Natl. Acad. Sci. U.S.A.* **54:** 731–736.

Kimura, M. 1968. Evolutionary rate at the molecular level. *Nature* **217:** 624–626.

Kimura, M. 1971. Theoretical foundations of population genetics at the molecular level. *Theor. Popul. Biol.* **2:** 174–208.

Kimura, M. 1980. A simple method for estimating evolutionary rate in a finite population due to mutational production of neutral and nearly neutral base substitution through comparative studies of nucleotide sequences. *J. Mol. Evol.* **16:** 111–120.

Kimura, M. 1983. *The Neutral Theory of Molecular Evolution*, Cambridge University Press, Cambridge, U.K.

Kimura, M. and J.F. Crow. 1963. The measurement of effective population size. *Evolution* **17:** 279–288.

Kimura, M. and J.F. Crow. 1964. The number of alleles that can be maintained in a finite population. *Genetics* **49:** 725–738.

Kimura, M. and T. Maruyama. 1966. The mutational load with epistatic gene interactions in fitness. *Genetics* **54:** 1303–1312.

Kimura, M. and T. Ohta. 1971. *Theoretical Aspects of Population Genetics*. Princeton University Press, Princeton, NJ.

King, J.L. and T.H. Jukes. 1969. Non-Darwinian evolution. *Science* **164:** 788–798.

Kingman, J.F.C. 1961. A matrix inequality. *Quart. J. Math.* **12:** 78–80.

Kingman, J.F.C. 1982. On the genealogy of large populations. *J. Appl. Prob.* **A19:** 27–43.

Kingsolver, J.G., H.E. Hoekstra, J.M. Hoekstra, D. Berrigan, S.N. Vignieri, et al. 2001. The strength of phenotypic selection in natural populations. *Am. Nat.* **157:** 245–261.

Kiontke, K., N.P. Gavin, Y. Raynes, C. Roehrig, F. Piano, and D.H.A. Fitch. 2004. *Caenorhabditis* phylogeny predicts convergence of hermaphroditism and extensive intron loss. *Proc. Natl. Acad. Sci. U.S.A.* **101:** 9003–9008.

Kirkpatrick, M. and N. Barton. 2006. Chromosome inversions, local adaptation and speciation. *Genetics* **173:** 419–434.

Kirkpatrick, M., T. Johnson, and N.H. Barton. 2002. General models of multilocus evolution. *Genetics* **161:** 1727–1750.

Kirkwood, T.B.L. 1977. Evolution of ageing. *Nature* **270:** 301–304.

Kirkwood, T.B.L. 1990. The disposable soma theory of aging. In *Genetic Effects on Aging II* (Ed. D.E. Harrison), pp. 9–19. Telford Press, Caldwell, NJ.

Kleunen, M.V. and S. Johnson. 2007. Effects of self-compatibility on the distribution range of invasive European plants in North America. *Conserv. Biol.* **21:** 1537–1544.

Kliman, R.M., P. Andolfatto, J.A. Coyne, F. Depaulis, M. Kreitman, et al. 2000. The population genetics of the origin and divergence of the *Drosophila simulans* complex of species. *Genetics* **156:** 1913–1931.

Klopfstein, S., M. Currat, and L. Excoffier. 2006. The fate of mutations surfing on the wave of a range expansion. *Mol. Biol. Evol.* **23:** 482–490.

Knight, R.D., S.J. Freeland, and L.F. Landweber. 2001. Rewiring the keyboard; evolvability of the genetic code. *Nat. Rev. Genet.* **2:** 49–58.

Knowles, L.L. and W.P. Maddison. 2002. Statistical phylogeography. *Mol. Ecol.* **11:** 2623–2635.

Koch, R., H.G.A.M. Van Luenen, M. Van der Horst, K.L. Thijssen, and R.H.A. Plasterk. 2000. Single nucleotide polymorphisms in wild isolates of *Caenorhabditis elegans*. *Genome Res.* **10:** 1690–1696.

Koelewijn, H.P., Koski, V., and Savolainen, O. 1999. Magnitude and timing of inbreeding depression in Scots pine (*Pinus sylvestris* L.). *Evolution* **53:** 758–768.

Kohn, J. 1988. Why be female? *Nature* **355:** 431–433.

Kohn, J. 1989. Sex ratio, seed production, biomass allocation, and the cost of male function in *Cucurbita foetidissima* HBK (Cucurbitaceae). *Evolution* **43:** 1424–1434.

Kohn, J. 1995. Outcrossing rates and inferred levels of inbreeding depression in gynodioecious *Cucurbita foetidissima* (Cucurbitaceae). *Heredity* **75:** 77–83.

Kohn, J., S. Graham, B. Morton, J. Doyle, and S.C.H. Barrett. 1996. Reconstruction of the evolution of reproductive characters in Pontederiaceae using phylogenetic evidence from chloroplast DNA restriction-site variation. *Evolution* **50:** 1454–1459.

Kojima, K.-I. and T.M. Kelleher. 1961. Changes of mean fitness in random-mating populations when epistasis and linkage are present. *Genetics* **46:** 527–540.

Kolkman, J.M., S.T. Berry, A.J. Leon, M.B. Slabaugh, S. Tang, et al. 2007. Single nucleotide polymorphisms and linkage disequilibrium in sunflower. *Genetics* **177:** 457–468.

Koller, P.C. and C.D. Darlington. 1934. The genetical and mechanical properties of the sex chromosomes. I. *Rattus norvegicus* male. *J. Genet.* **28:** 159–173.

Kondrashov, A.S. 1982. Selection against harmful mutations in large sexual and asexual populations. *Genet. Res* **44:** 325–332.

Kondrashov, A.S. 1984. Deleterious mutations as an evolutionary factor. 1. The advantage of recombination. *Genet. Res.* **44:** 199–217.

Kondrashov, A.S. 1988. Deleterious mutations and the evolution of sexual reproduction. *Nature* **336:** 435–440.

Kondrashov, A.S. 1993. A classification of hypotheses on the advantage of amphimixis. *J. Hered.* **84:** 372–387.

Kondrashov, A.S. 2003. Direct estimates of human per nucleotide mutation rates at 20 loci causing Mendelian diseases. *Hum. Mut.* **21:** 12–27.

Kondrashov, A.S. and J.F. Crow. 1993. A molecular approach to estimating the human deleterious mutation rate. *Hum. Mut.* **2:** 229–234.

Kondrashov, A.S. and M. Turelli. 1992. Deleterious mutations, apparent stabilizing selection and the maintenance of quantitative genetic variation. *Genetics* **132:** 603–618.

Kondrashov, F.A., A.Y. Ogurtsov, and A.S. Kondrashov. 2006. Selection in favor of nucleotides G and C diversifies evolution rates and levels of polymorphism at mammalian synonymous sites. *J. Theor. Biol.* **240:** 616–626.

Korol, A.B., I.A. Preygel, and S.I. Preygel. 1994. *Recombination, Variability and Evolution*. Chapman & Hall, London.

Kosakovsky Pond, S.L., S.D.W. Frost, and S.V. Muse. 2005. HyPhy: hypothesis testing using phylogenies. *Bioinformatics* **21:** 676–679.

Kouyos, R.D., O.K. Silander, and S. Bonhoeffer. 2007. Epistasis between deleterious mutations and the evolution of recombination. *Trends Ecol. Evol.* **22:** 308–315.

Kozlowski, J. and A.T. Teriokhin. 1999. Allocation of energy between growth and reproduction: the Pontryagin Maximum Principle solution for the case of age- and season-dependent mortality. *Evol. Ecol. Res.* **1:** 423–441.

Krause, J., C. Lalueza-Fox, L. Orlando, W. Enard, R. E. Green, et al. 2007. The derived FOXP2 variant of modern humans was shared with neandertals. *Curr. Biol.* **17:** 1908–1912.

Krebs, H.J. and N.B. Davies. 1997. *Behavioural Ecology: An Evolutionary Approach*. Blackwell, Oxford, U.K.

Kreitman, M. 1983. Nucleotide polymorphism at the alcohol dehydrogenase locus of *Drosophila melanogaster*. *Nature* **304:** 412–417.

Krimbas, C.B. and J.R. Powell. 1992a. *Drosophila Inversion Polymorphism*. CRC Press, Boca Raton, FL.

Krimbas, C.B. and J.R. Powell. 1992b. Introduction. In *Drosophila Inversion Polymorphism* (Eds. C.B. Krimbas and J.R. Powell), pp. 1–52. CRC Press, Boca Raton, FL.

Kronstad, J.W. and C. Staben. 1997. Mating type in filamentous fungi. *Ann. Rev. Genet.* **31:** 245–276.

Kruuk, L.E.B., B.C. Sheldon, and J. Merila. 2002. Severe inbreeding depression in collared flycatchers (*Ficedula albicollis*). *Proc. R. Soc. B* **269:** 1581–1589.

Kryukov, G.V., L.A. Pennachio, and S. Sunyaev. 2007. Most rare missense alleles are deleterious in humans: implications for complex disease and association studies. *Am. J. Hum. Genet.* **80:** 727–739.

Kuhner, M.K., J. Felsenstein, and J. Yamato. 2000. Maximum likelihood estimation of recombination rates from population data. *Genetics* **154:** 931–942.

Kuhner, M.K., J. Yamato, and J. Felsenstein. 1995. Estimating effective population size and mutation rate from sequence data using Metropolis–Hasting sampling. *Genetics* **140:** 1421–1430.

Kusano, A., C. Staber, and B. Ganetzky. 2003. Closing the (Ran) GAP on segregation distortion in *Drosophila melanogaster*. *Bioessays* **25**: 108–115.

Kwiatkowski, D.P. 2005. How malaria has affected the human genome and what human genetics can teach us about malaria. *Am. J. Hum. Genet.* **77**: 171–192.

Lamb, R.Y. and R.B. Willey. 1979. Are parthenogenetic and related bisexual insects equal in fertility? *Evolution* **33**: 774–775.

Lambie, E.J. and G.S. Roeder. 1986. Repression of meiotic crossing over by a centromere. *Genetics* **114**: 769–789.

Land, M.F. and D.-E. Nilsson. 2002. *Animal Eyes*. Oxford University Press, Oxford, U.K.

Lande, R. 1976. The maintenance of genetic variability by mutation in a polygenic character. *Genet. Res.* **26**: 221–235.

Lande, R.S. 1979a. Effective deme sizes during long-term evolution estimated from rates of chromosomal rearrangements. *Evolution* **33**: 234–251.

Lande, R. 1979b. Quantitative genetic analysis of multivariate evolution, applied to brain:body size allometry. *Evolution* **33**: 402–416.

Lande, R. 1982. A quantitative genetic theory of life history evolution. *Ecology* **63**: 607–615.

Lande, R.S. 1984. The expected fixation rate of chromosomal inversions. *Evolution* **38**: 743–752.

Lande, R. 2000. Quantitative genetics and phenotypic evolution. In *Evolutionary Genetics: From Molecules to Morphology* (Eds. R.S. Singh and C.B. Krimbas), pp. 335–350. Cambridge University Press, Cambridge, U.K.

Lande, R. and S.J. Arnold. 1983. The measurement of selection on correlated characters. *Evolution* **37**: 1210–1226.

Lande, R. and D.W. Schemske. 1985. The evolution of self-fertilization and inbreeding depression in plants. I. Genetic models. *Evolution* **39**: 24–40.

Langley, C.H., D.B. Smith, and F.M. Johnson. 1978. Analysis of linkage disequilibria between allozyme loci in natural populations of *Drosophila melanogaster*. *Genet. Res.* **32**: 215–229.

Langley, C.H., E.A. Montgomery, R.R. Hudson, N.L. Kaplan, and B. Charlesworth. 1988. On the role of unequal exchange in the containment of transposable element copy number. *Genet. Res.* **52**: 223–235.

Langley, C.H., B.P. Lazzaro, W. Phillips, E. Heikkinen, and J.M. Braverman. 2000. Linkage disequilibria and the site frequency spectra in the su(s) and su(wa) regions of the *Drosophila melamogaster* X chromosome. *Genetics* **156**: 1837–1852.

Laporte, V. and B. Charlesworth. 2002. Effective population size and population subdivision in demographically structured populations. *Genetics* **162**: 501–519.

Lapoumiéroulie, C., O. Dunda, R. Ducrocq, G. Trabuchet, M. Mony-Lobé, et al. 1992. A novel sickle cell mutation of yet another origin in Africa: the Cameroon type. *Hum. Genet.* **89**: 333–337.

Larracuente, A.M., T.B. Sackton, A.J. Greenberg, A. Wong, N.D. Singh, et al. 2008. Evolution of protein-coding genes in *Drosophila*. *Trends Genet.* **24**: 114–123.

Latter, B.D.H. and J.A. Sved. 1994. A reevaluation of data from competitive tests shows high levels of heterosis in *Drosophila melanogaster*. *Genetics* **137**: 509–511.

Lawson-Handley, L.J., H. Ceplitis, and H. Ellegren. 2004. Evolutionary strata on the chicken Z chromosome: implications for sex chromosome evolution. *Genetics* **167**: 367–376.

Le Rouzic, A. and G. Deceliere. 2005. Models of the population genetics of transposable elements. *Genet. Res.* **85**: 171–181.

Lees, D.R. 1981. Industrial melanism: genetic adaptation of animals to air pollution. In *Genetic Consequences of Man-Made Change* (Eds. L.M. Cook and J.A. Bishop), pp. 129–176. Academic Press, Orlando, FL.

Leigh, E.G. 1987. Ronald Fisher and the development of evolutionary theory. II. Influences of new variation on the evolutionary process. *Oxf. Surv. Evol. Biol.* **4**: 213–263.

Lenski, R.E. and M. Travisano. 1994. Dynamics of adaptation and diversification: a 10,000-generation experiment with bacterial populations. *Proc. Natl. Acad. Sci. U.S.A.* **91**: 6608–6818.

Lercher, M.J., N.G.C. Smith, A. Eyre-Walker, and L.D. Hurst. 2002. The evolution of isochores: evidence from SNP frequency distributions. *Genetics* **162**: 1805–1810.

Leslie, P.H. 1945. On the use of matrices in certain population mathematics. *Biometrika* **33**: 183–212.

Levene, H. 1953. Genetic equilibrium when more than one ecological niche is available. *Am. Nat.* **87**: 331–333.

Levin, B.R. and W.L. Kilmer. 1974. Interdemic selection and the evolution of altruism. *Evolution* **28**: 527–545.

Levin, S.A., H.C. Muller-Landau, R. Nathan, and J. Chave. 2003. The ecology and evolution of seed dispersal. *Ann. Rev. Ecol. Evol. Syst.* **34**: 575–604.

Lewis, D. 1941. Male sterility in natural populations of hermaphrodite plants. *New Phytol.* **40**: 56–63.

Lewontin, R.C. 1964. The interaction of selection and linkage. I. General considerations, heterotic models. *Genetics* **49**: 49–67.

Lewontin, R.C. 1971. The effect of genetic linkage on the mean fitness of a population. *Proc. Natl. Acad. Sci. U.S.A.* **68**: 984–986.

Lewontin, R.C. 1972a. The apportionment of human diversity. *Evol. Biol.* **6**: 381–398.

Lewontin, R.C. 1972b. The effect of genetic linkage on the mean fitness of a population. *Proc. Natl. Acad. Sci. U.S.A.* **68**: 984–986.

Lewontin, R.C. 1974. *The Genetic Basis of Evolutionary Change*. Columbia University Press, New York.

Lewontin, R.C. 1985. Population genetics. *Ann. Rev. Genet.* **19**: 81–102.

Lewontin, R.C. 1988. On measures of gametic disequilibrium. *Genetics* **120**: 849–852.

Lewontin, R.C. and J.L. Hubby. 1966. A molecular approach to the study of genic heterozygosity in natural populations. II. Amount of variation and degree of heterozygosity in *Drosophila pseudoobscura*. *Genetics* **54**: 595–609.

Lewontin, R.C. and J. Krakauer. 1973. Distribution of gene frequency as a test of neutrality. *Genetics* **74**: 175–195.

Lewontin, R.C., L.R. Ginzburg, and S.D. Tuljapurkar. 1978. Heterosis as an explanation for large amounts of genic polymorphism. *Genetics* **88**: 149–169.

Li, W.-H. 1987. Models of nearly neutral mutations with particular implications for non-random usage of synonymous codons. *J. Mol. Evol.* **24**: 337–345.

Lin, X.Y., S. Kaul, S. Rounsley, T.P. Shea, M. Benito, et al. 1999. Sequence and analysis of chromosome 2 of the plant *Arabidopsis thaliana*. *Nature* **402**: 761–768.

Lindsley, D.L. and E.H. Grell. 1969. Spermiogenesis without chromosomes in *Drosophila melanogaster*. *Genetics* **61 (suppl. 1)**: 69–78.

Lindsley, D.L. and L. Sandler. 1977. The genetic analysis of meiosis in female *Drosophila*. *Phil. Trans. R. Soc. B* **277**: 295–312.

Lindsley, D.L. and G.G. Zimm. 1992. *The Genome of Drosophila melanogaster*. Academic Press, San Diego, CA.

Linhart, Y.B. and M.C. Grant. 1996. Evolutionary significance of local genetic differentiation in plants. *Ann. Rev. Ecol. Syst.* **27**: 237–277.

Lippman, Z. and R. Martienssen. 2004. The role of RNA interference in heterochromatic silencing. *Nature* **431**: 364–370.

Lloyd, D.G. 1965. Evolution of self-compatibility and racial differentiation in Leavenworthia (Cruciferae). *Contrib. Gray Herbarium Harv. Univ.* **195**: 3–134.

Lloyd, D.G. 1974. Female-predominant sex ratios in angiosperms. *Heredity* **32**: 35–44.

Lloyd, D.G. 1975. The maintenance of gynodioecy and androdioecy in angiosperms. *Genetica* **45**: 325–339.

Lloyd, D.G. 1977. Genetic and phenotypic models of natural selection. *J. Theor. Biol.* **69**: 543–560.

Lloyd, D.G. 1979. Some reproductive factors affecting the selection of self-fertilization in plants. *Am. Nat.* **113**: 67–79.

Lloyd, D.G. 1980. Benefits and handicaps of sexual reproduction. *Evol. Biol.* **13**: 69–111.

Lloyd, D.G. 1984. Gender allocations in outcrossing cosexual plants. In *Perspectives on Plant Population Ecology* (Eds. R. Dirzo and J. Sarukhan), pp. 277–300. Sinauer, Sunderland, MA.

Lloyd, D.G. 1987. Allocations to pollen, seeds and pollination mechanisms in self-fertilizing plants. *Funct. Ecol.* **1**: 83–89.

Lloyd, D.G. and C.J. Webb. 1992a. The evolution of heterostyly. In *Evolution and Function of Heterostyly* (Ed. S.C.H. Barrett), pp. 151–178. Springer-Verlag, Heidelberg, Germany.

Lloyd, D.G. and C.J. Webb. 1992b. The selection of heterostyly. In *Evolution and Function of Heterostyly* (Ed. S.C.H. Barrett), pp. 179–207. Springer-Verlag, Heidelberg, Germany.

Lloyd, M. and H.S. Dybas. 1966. The periodical cicada problem. I. Population ecology. *Evolution* **20**: 133–149.

Loewe, L. and B. Charlesworth. 2007. Background selection in single genes may explain patterns of codon bias. *Genetics* **175**: 1381–1393.

Loewe, L., B. Charlesworth, C. Bartolomé, and V. Nöel. 2006. Estimating selection on nonsynonymous mutations. *Genetics* **172**: 1079–1092.

Lucchesi, J.C. 1978. Gene dosage compensation and the evolution of sex chromosomes. *Science* **202**: 711–716.

Lush, J.L. 1937. *Animal Breeding Plans.* Iowa State College Press, Ames, IA.

Lynch, M. 1988. The rate of polygenic mutation. *Genet. Res.* **51**: 137–148.

Lynch, M. 2007. *The Origins of Genome Architecture.* Sinauer, Sunderland, MA.

Lynch, M. and J.B. Walsh. 1998. *Genetics and Analysis of Quantitative Traits.* Sinauer, Sunderland, MA.

Lynch, M., B. Koskella, and S. Schaack. 2006. Mutation pressure and the evolution of organelle genomic architecture. *Science* **311**: 1727–1730.

Lynch, M., M. Pfrender, K. Spitze, N. Lehman, J. Hicks, et al. 1999. The quantitative and molecular genetic architecture of a subdivided species. *Evolution* **53**: 100–110.

Lynch, M., W. Sung, N. Coffey, C.R. Landry, E.B. Dopman, et al. 2008. A genome-wide view of the spectrum of spontaneous mutations in yeast. *Proc. Natl. Acad. Sci. U.S.A.* **105**: 9272–9277.

Mackay, T.F.C. 2004. The genetic architecture of quantitative traits: lessons from *Drosophila. Curr. Op. Genet. Dev.* **14**: 253–257.

Mackay, T.F.C. and R.F. Lyman. 2005. *Drosophila* bristles and the nature of quantitative genetic variation. *Phil. Trans. R. Soc. B* **360**: 1513–1527.

Mackay, T.F.C., R.F. Lyman, and M.S. Jackson. 1992. Effects of *P* element mutations on quantitative traits in *Drosophila melanogaster. Genetics* **130**: 315–332.

Mackiewicz, M., A. Tatarenkov, D.S. Taylor, B.J. Turner, and J.C. Avise. 2006a. Extensive outcrossing and androdioecy in a vertebrate species that otherwise reproduces as a self-fertilizing hermaphrodite. *Proc. Natl. Acad. Sci. U.S.A.* **103**: 9924–9928.

Mackiewicz, M., A. Tatarenkov, B.J. Turner, and J.C. Avise. 2006b. A mixed-mating strategy in a hermaphrodite vertebrate. *Proc. R. Soc. B* **273**: 2449–2452.

MacKnight, R.H. 1939. The sex-determining mechanism of *Drosophila miranda. Genetics* **24**: 180–201.

MacLeod, A.K., C.S. Haley, and J.A. Woolliams. 2005. Marker densities and the mapping of ancestral junctions. *Genet. Res.* **85**: 69–79.

Macnair, M.R., V.E. Macnair, and B.E. Martin. 1989. Adaptive speciation in Mimulus: an ecological comparison of *M. cupriphilus* with its presumed progenitor *M. guttatus. New Phytol.* **112**: 269–279.

Makalowski, W. and M. S. Boguski. 1998. Evolutionary parameters of the transcribed mammalian genome: an analysis of 2,280 orthologous rodent and human sequences. *Proc. Natl. Acad. Sci. U.S.A.* **95**: 9407–9412.

Malécot, G. 1948. *Les Mathématiques de l'Hérédité.* Masson, Paris.

Malécot, G. 1969. *The Mathematics of Heredity.* W.H. Freeman, San Francisco, CA.

Malécot, G. 1975. Heterozygosity and relationship in regularly subdivided populations. *Theor. Popul. Biol.* **8**: 212–241.

Malik, H.S. and S. Henikoff. 2002. Conflict begets complexity: the evolution of centromeres. *Curr. Op. Genet. Dev.* **12**: 711–718.

Mandel, S.P.H. 1959. The stability of a multiple allelic system. *Heredity* **13**: 289–302.

Manicacci, D., A. Atlan, and D. Couvet. 1997. Spatial structure of nuclear factors involved in sex determination in the gynodioecious *Thymus vulgaris* L. *J. Evol. Biol.* **10**: 889–907.

Mank, J.E., E. Axelsson, and H. Ellegren. 2007. Fast-X on the Z: rapid evolution of sex-linked genes in birds. *Genome Res.* **17:** 618–624.

Marais, G. 2003. Biased gene conversion: implications for genome and sex evolution. *Trends Genet.* **19:** 330–338.

Marais, G., B. Charlesworth, and S.I. Wright. 2004. Recombination and base composition: the case of the highly self-fertilizing plant *Arabidopsis thaliana. Genome Biol.* **5:** R45.

Marais, G., D. Mouchiroud, and L. Duret. 2001. Does recombination improve selection on codon usage? Lessons from nematode and fly genomes. *Proc. Natl. Acad. Sci. U.S.A.* **98:** 5688–5692.

Mark Welch, D.B., J.L. Mark Welch, and M. Meselson. 2008. Evidence for degenerate tetraploidy in bdelloid rotifers. *Proc. Natl. Acad. Sci. U.S.A.* **105:** 5145–5149.

Marshall, A.R., K.L. Knudsen, and F.W. Allendorf. 2004. Linkage disequilibrium between the pseudoautosomal PEPB-1 locus and the sex-determining region of chinook salmon. *Heredity* **93:** 85–97.

Martin, G. and T. Lenormand. 2006. A multivariate extension of Fisher's geometrical model and the distribution of fitness effects across species. *Evolution* **60:** 893–907.

Martin, G. and T. Lenormand. 2008. The distribution of beneficial and fixed mutation fitness effects close to an optimum. *Genetics* **179:** 907–916.

Martin, G., S.F. Elena, and T. Lenormand. 2007. Distributions of epistasis in microbes fit predictions from a fitness landscape model. *Nat. Genet.* **39:** 555–559.

Martin, G., S.P. Otto, and T. Lenormand. 2006. Selection for recombination in structured populations. *Genetics* **172:** 593–609.

Martinsen, G.D., T.G. Whitham, R.J. Turek, and P. Keim. 2001. Hybrid populations selectively filter gene introgression between species. *Evolution* **55:** 1325–1335.

Maruyama, T. 1970a. Analysis of population structure. I. One-dimensional stepping stone models of finite length and other geographically structured populations. *Ann. Hum. Genet.* **34:** 201–219.

Maruyama, T. 1970b. Effective number of alleles in a subdivided population. *Theor. Popul. Biol.* **1:** 273–306.

Maruyama, T. 1970c. On the fixation probabilities of mutant genes in a subdivided population. *Genet. Res.* **15:** 221–226.

Maruyama, T. 1971a. Analysis of population structure. II. Two-dimensional stepping stone models of finite length and other geographically structured populations. *Ann. Hum. Genet.* **35:** 179–196.

Maruyama, T. 1971b. An invariant property of a subdivided population. *Genet. Res.* **18:** 81–84.

Maruyama, T. 1974. A simple proof that certain quantities are independent of the geographic structure of population. *Theor. Popul. Biol.* **5:** 148–154.

Maruyama, T. 1977. *Lectures in Biomathematics. 17. Stochastic Problems in Population Genetics.* Springer-Verlag, Berlin.

Maruyama, T. and M. Kimura. 1980. Genetic variability and effective population size when local extinction and recolonization of populations are frequent. *Proc. Natl. Acad. Sci. U.S.A.* **77:** 6710–6714.

Maside, X., S. Assimacopoulos, and B. Charlesworth. 2000. Rates of movement of transposable elements on the second chromosome of *Drosophila melanogaster*. *Genet. Res.* **75**: 275–284.

Maside, X., S. Assimacopoulos, and B. Charlesworth. 2005. Fixations of transposable elements in the *D. melanogaster* genome. *Genet. Res.* **85**: 195–203.

Maside, X., A. Weishan Lee, and B. Charlesworth. 2004. Selection on codon usage in *Drosophila americana*. *Curr. Biol.* **14**: 150–154.

Mast, A., S. Kelso, and E. Conti. 2006. Are any primroses (Primula) primitively monomorphic? *New Phytol.* **171**: 605–616.

Mather, K. 1943. Polygenic inheritance and natural selection. *Biol. Rev* **18**: 32–64.

Matsen, F.A. and J. Wakeley. 2006. Convergence to the island-model coalescent process in populations with restricted migration. *Genetics* **172**: 701–708.

May, R.M. and R.M. Anderson. 1983. Epidemiology and genetics in the coevolution of parasites and hosts. *Proc. R. Soc. B* **219**: 281–313.

Maynard Smith, J. 1964. Group selection and kin selection. *Nature* **200**: 1145–1147.

Maynard Smith, J. 1966. Sympatric speciation. *Am. Nat.* **104**: 487–490.

Maynard Smith, J. 1968. "Haldane's dilemma" and the rate of evolution. *Nature* **219**: 1114–1116.

Maynard Smith, J. 1970. Natural selection and the concept of a protein space. *Nature* **225**: 563–564.

Maynard Smith, J. 1971a. The origin and maintenance of sex. In *Group Selection* (Ed. G.C. Williams), pp. 163–175. Aldine-Atherton, Chicago, IL.

Maynard Smith, J. 1971b. What use is sex? *J. Theor. Biol.* **30**: 319–355.

Maynard Smith, J. 1976a. Group selection. *Quart. Rev. Biol.* **51**: 277–283.

Maynard Smith, J. 1976b. A short-term advantage for sex and recombination through sib-competition. *J. Theor. Biol.* **63**: 245–258.

Maynard Smith, J. 1976c. What determines the rate of evolution? *Am. Nat.* **110**: 331–338.

Maynard Smith, J. 1978. *The Evolution of Sex*. Cambridge University Press, Cambridge, U.K.

Maynard Smith, J. 1982. *Evolution and the Theory of Games*. Cambridge University Press, Cambridge, U.K.

Maynard Smith, J. 1983. Models of evolution. *Proc. R. Soc. B* **219**: 315–325.

Maynard Smith, J. 1988. Selection for recombination in a polygenic model—the mechanism. *Genet. Res* **51**: 69–63.

Maynard Smith, J. and J. Haigh. 1974. The hitch-hiking effect of a favourable gene. *Genet. Res.* **23**: 23–35.

Maynard Smith, J. and G.R. Price. 1973. The logic of animal conflict. *Nature* **246**: 15–18.

Maynard Smith, J., R. Burian, S. Kaufman, P. Alberch, J. Campbell, et al. 1985. Developmental constraints and evolution. *Quart. Rev. Biol.* **60**: 265–287.

Maynard Smith, J., N.H. Smith, M. O'Rourke, and B.G. Spratt. 1993. How clonal are bacteria? *Proc. Natl. Acad. Sci. U.S.A.* **90**: 4384–4388.

Mayo, O. and R. Bürger. 1997. The evolution of dominance: a theory whose time has passed? *Biol. Rev.* **72**: 97–110.

Mayr, E. 1954. Change of genetic environment and speciation. In *Evolution as a Process* (Eds. J.S. Huxley, A.C. Hardy, and E.B. Ford), pp. 157–180. Allen and Unwin, London.

Mayr, E. 1963. *Animal Species and Evolution*. Harvard University Press, Cambridge, MA.

McCauley, D.E. and J.R. Ellis. 2008. Recombination and linkage disequilibrium among mitochondrial genes in structured populations of the gynodioecious plant *Silene vulgaris*. *Evolution* **62:** 823–832.

McClintock, B. 1950. The origin and behavior of mutable loci in maize. *Proc. Natl. Acad. Sci. U.S.A.* **36:** 344–355.

McClintock, B. 1984. The significance of responses of the genome to challenge. *Science* **226:** 792–800.

McCracken, G.F. and R.K. Selander. 1980. Self-fertilization and monogenic strains in natural populations of terrestrial slugs. *Proc. Natl. Acad. Sci. U.S.A.* **77:** 684–688.

McCune, A., R.C. Fuller, A.A. Aquilina, R.M. Dawley, J.M. Fadool, et al. 2002. A low genomic number of recessive lethals in natural populations of bluefin killifish and zebrafish. *Science* **296:** 2398–2401.

McDonald, J.H. and M. Kreitman. 1991. Accelerated protein evolution at the *Adh* locus in *Drosophila*. *Nature* **351:** 652–654.

McGuigan, K. and M.W. Blows. 2007. The phenotypic and genetic covariance structure of Drosophilid wings. *Evolution* **61:** 902–911.

McKay, J.K. and R.G. Latta. 2002. Adaptive population divergence: markers, QTL and traits. *Trends Ecol. Evol.* **53:** 285–291.

McKone, M., C. Lund, and J. O'Brien. 1998b. Reproductive biology of two dominant prairie grasses (*Andropogon gerardii* and *Sorghastrum nutans*, Poaceae): male-biased sex allocation in wind-pollinated plants? *Am. J. Bot.* **85:** 776–783.

McVean, G.A.T. 2002. A genealogical interpretation of linkage disequilibrium. *Genetics* **162:** 987–991.

McVean, G.A.T. and B. Charlesworth. 1999. A population genetic model for the evolution of synonymous codon usage: patterns and predictions. *Genet. Res.* **74:** 145–158.

McVean, G.A.T. and J. Vieira. 1999. The evolution of codon preferences in *Drosophila*: a maximum-likelihood approach to parameter estimation and hypothesis testing. *J. Mol. Evol.* **49:** 63–75.

McVean, G., P. Awadalla, and P. Fearnhead. 2002. A coalescent-based method for detecting and estimating recombination from gene sequences. *Genetics* **160:** 1231–1241.

McVean, G.A.T., C.C.A. Spencer, and R. Chaix. 2006. Perspectives on human genetic variation from the HapMap project. *PLoS Genet.* **1:** 413–414.

Meagher, S., D.J. Penn, and W.K. Potts. 2000. Male–male competition magnifies inbreeding depression in wild house mice. *Proc. Natl. Acad. Sci. U.S.A.* **97:** 3324–3329.

Medawar, P.B. 1946. Old age and natural death. *Mod. Quart.* **1:** 30–56.

Medawar, P.B. 1952. *An Unsolved Problem of Biology*. H.K. Lewis, London.

Mendel, G. 1866. Versuche über Pflanzenhybriden. *Ver. Naturforsch. Ver. Brünn* **4**: 3–47.

Metcalf, C.J.E. and D.N. Koons. 2007. Environmental uncertainty, autocorrelation and the evolution of survival. *Proc. R. Soc. B* **274**: 2153–2160.

Meunier, J., S.A. West, and M. Chapuisat. 2008. Split sex ratios in the social Hymenoptera: a meta-analysis. *Behav. Ecol.* **19**: 382–390.

Mevel-Ninio, M., A. Pelisson, J. Kinder, A.R. Campos, and A. Bucheton. 2007. The flamenco locus controls the *gypsy* and *ZAM* retroviruses and is required for *Drosophila* oogenesis. *Genetics* **175**: 1615–1624.

Meyer, K. and M. Kirkpatrick. 2008. Perils of parsimony: properties of reduced-rank estimates of genetic covariance matrices. *Genetics* **180**: 1153–1166.

Meyers, B.C., S.V. Tingey, and M. Morgante. 2001. Abundance, distribution, and transcriptional activity of repetitive elements in the maize genome. *Genome Res.* **11**: 1660–1676.

Miklos, G.L.G. and A.C. Gill. 1982. Nucleotide sequences of highly repeated DNAs; compilation and comments. *Genet. Res.* **39**: 1–30.

Milinski, M. 2006. Fitness consequences of selfing and outcrossing in the cestode *Schistocephalus solidus*. *Integ. Comp. Biol.* **46**: 373–380.

Misra, S., M.A. Crosby, C.J. Mungall, B.B. Matthews, K.S. Campbell, et al. 2002. Annotation of the *Drosophila melanogaster* euchromatic genome: a systematic review. *Genome Biol.* **3**: Research 0083.

Miyata, T., H. Hayashida, K. Kuma, K. Mitsuyasu, and T. Yasunaga. 1987. Male-driven molecular evolution: a model and nucleotide sequence analysis. *Cold Spring Harbor Symp. Quant. Biol.* **52**: 863–867.

Mizawa, K. and F. Tajima. 1997. Estimation of the amount of genetic variation when the mutation rate varies among sites. *Genetics* **147**: 1959–1964.

Moeller, D.A., M.I. Tenaillon, and P. Tiffin. 2001. Population structure and its effects on patterns of nucleotide polymorphism in Teosinte (*Zea mays* ssp. *parviglumis*). *Genetics* **176**: 1799–1809.

Montgomery, E.A., S.-M. Huang, C.H. Langley, and B.H. Judd. 1991. Chromosome rearrangement by ectopic recombination in *Drosophila melanogaster*: genome structure and evolution. *Genetics* **129**: 1085–1098.

Moorad, J.A. and D.E.L. Promislow. 2009. What can genetic variation tell us about the evolution of senescence? *Proc. R. Soc. B* **276**: 2271–2278.

Moran, P.A.P. 1958. Random processes in genetics. *Proc. Camb. Philos. Soc.* **54**: 60–71.

Moran, P.A.P. 1962. *The Statistical Processes of Evolutionary Theory*. Oxford University Press, Oxford, U.K.

Moran, P.A.P. 1964. On the nonexistence of adaptive topographies. *Ann. Hum. Genet.* **27**: 383–392.

Morgan, M.T. 2002. Genome-wide deleterious mutation favors dispersal and species integrity. *Heredity* **89**: 253–257.

Morgante, M., S. Brunner, G. Pea, K. Fengler, A. Zuccolo, and A. Rafalski. 2005. Gene duplication and exon shuffling by helitron-like transposons generate intraspecies diversity in maize. *Nat. Genet.* **37**: 997–1002.

Moriyama, E.N. and J.R. Powell. 1996. Intraspecific nuclear DNA variation in *Drosophila*. *Mol. Biol. Evol.* **13**: 261–277.

Morton, N.E. 1955. Sequential tests for the detection of linkage. *Am. J. Hum. Genet.* **7**: 277–318.

Morton, N.E. 1991. Parameters of the human genome. *Proc. Natl. Acad. Sci. U.S.A.* **88**: 7474–7476.

Morton, N.E., J.F. Crow, and H.J. Muller. 1956. An estimate of the mutational damage in man from data on consanguineous marriages. *Proc. Natl. Acad. Sci. U.S.A.* **42**: 855–863.

Mott, R. 2006. Finding the molecular basis of complex genetic variation in humans and mice. *Phil. Trans. R. Soc. B* **361**: 393–401.

Mu, J., P. Awadalla, J. Duan, K. McGee, J. Keebler, et al. 2007. Genome-wide variation and identification of vaccine targets in the *Plasmodium falciparum* genome. *Nat. Genet.* **39**: 126–130.

Mu, J., J. Duan, K.D. Makova, D.A. Joy, C.Q. Huynh, et al. 2002. Chromosome-wide SNPs reveal an ancient origin for *Plasmodium falciparum*. *Nature* **418**: 323–326.

Mueller, L.D. and M.R. Rose. 1996. Evolutionary theory predicts late-life mortality plateaus. *Proc. Natl. Acad. Sci. U.S.A.* **93**: 15249–15253.

Mukai, T. 1964. The genetic structure of natural populations of *Drosophila melanogaster*. I. Spontaneous mutation rate of polygenes controlling viability. *Genetics* **50**: 1–19.

Mukai, T. 1969. The genetic structure of natural populations of *Drosophila melanogaster*. VIII. Natural selection on the degree of dominance of viability polygenes. *Genetics* **63**: 476–478.

Mukai, T. 1988. Genotype-environment interaction in relation to the maintenance of genetic variability in populations of *Drosophila melanogaster*. In *Proceedings of the 2nd International Conference on Quantitative Genetics* (Eds. B.S. Weir, E.J. Eisen, M.M. Goodman, and G. Namkoong), pp. 21–31. Sinauer, Sunderland, MA.

Mukai, T., S.I. Chigusa, L.E. Mettler, and J.F. Crow. 1972. Mutation rate and dominance of genes affecting viability in *Drosophila melanogaster*. *Genetics* **72**: 335–355.

Mullen, L.M. and H.E. Hoekstra. 2008. Natural selection along an environmental gradient: A classic cline in mouse pigmentation. *Evolution* **62**: 1555–1569.

Muller, H.J. 1928. The measurement of mutation rate in *Drosophila*, its high variability and its dependence on temperature. *Genetics* **13**: 279–357.

Muller, H.J. 1932. Some genetic aspects of sex. *Am. Nat.* **66**: 118–138.

Muller, H.J. 1940. Bearing of the *Drosophila* work on systematics. In *The New Systematics* (Ed. J.S. Huxley), pp. 185–268. Oxford University Press, Oxford, U.K.

Muller, H.J. 1949. Redintegration of the symposium on genetics, paleontology, and evolution. In *Genetics, Paleontology and Evolution* (Eds. G.L. Jepsen, G.G. Simpson, and E. Mayr), pp. 421–445. Princeton University Press, Princeton, NJ.

Muller, H.J. 1950. Our load of mutations. *Am. J. Hum. Genet.* **2**: 111–176.

Muller, H.J. 1964. The relation of recombination to mutational advance. *Mut. Res.* **1**: 2–9.

Muller, H.J. and E. Altenburg. 1920. The genetic basis of truncate wing, an inconstant and modifiable character in *Drosophila*. *Genetics* **5**: 1–59.

Muse, S.V. and B.S. Gaut. 1994. A likelihood approach for comparing synonymous and non-synonymous substitution rates. *Mol. Biol. Evol.* **11**: 715–724.

Muto, A. and S. Ozawa. 1987. The guanine and cytosine content of genomic DNA and bacterial evolution. *Proc. Natl. Acad. Sci. U.S.A.* **84**: 166–169.

Myers, S., L. Bottolo, C. Freeman, G. McVean, and P. Donnelly. 2005. A fine-scale map of recombination rates and hotspots across the human genome. *Science* **310**: 312–324.

Nachman, M. 2005. The genetic basis of adaptation: lessons from concealing coloration in pocket mice. *Genetica* **123**: 125–136.

Nachman, M.W. and S.L. Crowell. 2000. Estimate of the mutation rate per nucleotide in humans. *Genetics* **156**: 297–304.

Nagylaki, T. 1975. Polymorphisms in cyclically varying environments. *Heredity* **35**: 67–74.

Nagylaki, T. 1976a. The evolution of one- and two-locus systems. *Genetics* **83**: 583–600.

Nagylaki, T. 1976b. A model for the evolution of self-fertilization and vegetative reproduction. *J. Theor. Biol.* **58**: 55–58.

Nagylaki, T. 1978. A diffusion model for geographically structured populations. *J. Math. Biol.* **6**: 375–382.

Nagylaki, T. 1979. Selection in dioecious populations. *Ann. Hum. Genet.* **14**: 143–150.

Nagylaki, T. 1980. The strong-migration limit in geographically structured populations. *J. Math. Biol.* **9**: 101–114.

Nagylaki, T. 1982. Geographical invariance in population genetics. *J. Theor. Biol.* **99**: 159–172.

Nagylaki, T. 1987. Evolution under fertility and under viability selection. *Genetics* **115**: 367–375.

Nagylaki, T. 1989. Gustave Malécot and the transition from classical to modern population genetics. *Genetics* **122**: 253–268.

Nagylaki, T. 1990. Models and approximations for random genetic drift. *Theor. Popul. Biol.* **37**: 192–212.

Nagylaki, T. 1992. *Introduction to Theoretical Population Genetics.* Springer-Verlag, Berlin.

Nagylaki, T. 1993. The evolution of multilocus systems under weak selection. *Genetics* **134**: 627–647.

Nagylaki, T. 1998a. Fixation indices in subdivided populations. *Genetics* **148**: 1325–1332.

Nagylaki, T. 1998b. The expected number of heterozygous sites in a subdivided population. *Genetics* **149**: 1599–1604.

Nagylaki, T., J. Hofbauer, and P. Brunovsky. 1999. Convergence of multilocus systems under weak epistasis or weak selection. *J. Math. Biol.* **38**: 103–133.

Nair, S., J.T. Williams, A. Brockman, L. Paiphun, M. Mayxay, et al. 2003. A selective sweep driven by pyrimethamine treatment in Southeast Asian malaria parasites. *Mol. Biol. Evol.* **20**: 1526–1536.

Nasrallah, J.B. 2002. Recognition and rejection of self in plant reproduction. *Science* **296**: 305–308.

Nassar, J.M., J.L. Hamrick, and T.H. Fleming. 2003. Population genetic structure of Venezuelan chiropterophilous columnar cacti (Cactaceae). *Am. J. Bot.* **90**: 1628–1637.

Navarro, A. and N.H. Barton. 2002. The effects of multilocus balancing selection on neutral variability. *Genetics* **161**: 849–863.

Nei, M. 1967. Modification of linkage intensity by natural selection. *Genetics* **57**: 625–641.

Nei, M. 1973. Analysis of gene diversity in subdivided populations. *Proc. Natl. Acad. Sci. U.S.A.* **70**: 3321–3323.

Nei, M. and T. Gojobori. 1986. Simple methods for estimating the numbers of synonymous and nonsynonymous nucleotide substitutions. *Mol. Biol. Evol.* **3**: 418–426.

Nei, M. and W.-H. Li. 1973. Linkage disequilibrium in subdivided populations. *Genetics* **75**: 213–219.

Neuhauser, C. and S.M. Krone. 1997. The genealogy of samples in models with selection. *Genetics* **145**: 519–534.

Nevers, P. and H. Saedler. 1977. Transposable genetic elements as agents of instability and chromosomal rearrangements. *Nature* **268**: 109–115.

Nicholson, A.J. 1927. A new theory of mimicry in insects. *Austral. Zool.* **5**: 10–104.

Nielsen, R. and J. Wakeley. 2001. Distinguishing migration from isolation: a Markov chain Monte Carlo approach. *Genetics* **158**: 885–896.

Nielsen, R., S. Williamson, Y. Kim, M.J. Hubisz, A.G. Clark, and C.D. Bustamante. 2005. Genomic scans for selective sweeps using SNP data. *Genome Res.* **15**: 1566–1575.

Nielsen, R., I. Hellmann, M. Hubisz, C. Bustamante, and A.G. Clark. 2007. Recent and ongoing selection in the human genome. *Nat. Rev. Genet.* **8**: 857–868.

Nilsson, D.-E. and S. Pelger. 1994. A pessimistic estimate of the time required for an eye to evolve. *Proc. R. Soc. B* **256**: 53–58.

Nilsson, E. and J. Agren. 2006. Population size, female fecundity, and sex ratio variation in gynodioecious *Plantago maritima*. *J. Evol. Biol.* **9**: 825–833.

Nilsson-Ehle, H. 1909. Kreuzungsuntersuchungen an Hafer und Weizen. *Lunds Univ. Aarskrift NS* **5**: 1–122.

Nordborg, M. 1997. Structured coalescent processes on different time scales. *Genetics* **146**: 1501–1514.

Nordborg, M. 2000. Linkage disequilibrium, gene trees and selfing: an ancestral recombination graph with partial self-fertilization. *Genetics* **154**: 923–929.

Nordborg, M. and P. Donnelly. 1997. The coalescent process with selfing. *Genetics* **146**: 1185–1195.

Nordborg, M. and H. Innan. 2003. The genealogy of sequences containing multiple sites subject to strong selection in a subdivided population. *Genetics* **163**: 1201–1213.

Nordborg, M. and S.M. Krone. 2002. Separation of time-scales and convergence to the coalescent in structured populations. In *Modern Developments in Population Genetics. The Legacy of Gustave Malécot* (Eds. M. Slatkin and M. Veuille), pp. 194–232. Oxford University Press, Oxford, U.K.

Nordborg, M., B. Charlesworth, and D. Charlesworth. 1996. The effect of recombination on background selection. *Genet. Res.* **67:** 159–174.

Nordborg, M., T.T. Hu, Y. Ishino, J. Jhaveri, C. Toomajian, et al. 2005. The pattern of polymorphism in *Arabidopsis thaliana. PLoS Biol.* **3:** 1289–1299.

Normark, B.B., O.P. Judson, and N.A. Moran. 2003. Genomic signatures of ancient asexual lineages. *Biol. J. Linn. Soc.* **79:** 69–84.

Norton, H.T.J. 1928. Natural selection and Mendelian variation. *Proc. London Math. Soc.* **28:** 1–45.

Notley-McRobb, L., S. Seeto, and T. Ferenci. 2002. Enrichment and elimination of *mutY* mutators in *Escherichia coli* populations. *Genetics* **162:** 1055–1062.

Novella, I., S. Zarate, D. Metzgar, and B. Ebendick-Corpus. 2004. Positive selection of synonymous mutations in vesicular stomatitis virus. *J. Mol. Biol.* **342:** 1415–1421.

Novembre, J.A. 2002. Accounting for background nucleotide composition when measuring codon usage bias. *Mol. Biol. Evol.* **19:** 1390–1394.

Novembre, J., J. Galvani, and M. Slatkin. 2005. The geographic spread of the CCR5 D32 HIV-resistance allele. *PLoS Biol.* **2:** 1954–1962.

Nunney, L. 1989. The maintenance of sex by group selection. *Evolution* **43:** 245–257.

Nunney, L. 1991. The influence of age structure and fecundity on effective population size. *Proc. R. Soc. B* **246:** 71–76.

Nunney, L. 1993. The influence of mating system and overlapping generations on effective population size. *Evolution* **47:** 1329–2341.

Nuzhdin, S.V., E.G. Pasyukova, E.A. Morozova, and A.J. Flavell. 1998. Quantitative genetic analysis of copia retrotransposon activity in inbred *Drosophila melanogaster* lines. *Genetics* **150:** 755–766.

O'Neill, S.L., A.A. Hoffmann, and J.H. Werren. 1997. *Influential Passengers.* Oxford University Press, Oxford, U.K.

Oakeshott, J.G., J.B. Gibson, P.R. Anderson, W.R. Knibb, and G.K. Chambers. 1982. Alcohol dehydrogenase and glycerol-3-phosphate dehydrogenase clines in *Drosophila melanogaster* on different continents. *Evolution* **36:** 86–96.

Ohta, T. 1971. Associative overdominance caused by linked detrimental mutations. *Genet. Res.* **18:** 277–286.

Ohta, T. 1982. Linkage disequilibrium with the island model. *Genetics* **101:** 139–155.

Ohta, T. 1992. The nearly neutral theory of molecular evolution. *Ann. Rev. Ecol. Syst.* **23:** 263–286.

Ohta, T. and M. Kimura. 1970. Development of associative overdominance through linkage disequilibrium in finite populations. *Genet. Res.* **18:** 277–286.

Ohta, T. and M. Kimura. 1971. Linkage disequilibrium between two segregating nucleotide sites under steady flux of mutations in a finite population. *Genetics* **68:** 571–580.

Olendorf, R., F.H. Rodd, D. Punzalan, A.E. Houde, C. Hurt, et al.. 2006. Frequency-dependent survival in natural guppy populations. *Nature* **441:** 633–636.

Olson, M.S. and D.E. McCauley. 2002. Mitochondrial DNA diversity, population structure, and gender association in the gynodioecious plant *Silene vulgaris. Evolution* **56:** 253–262.

Ometto, L., S. Glinka, D. De Lorenzo, and W. Stephan. 2005. Inferring the effects of demography and selection on *Drosophila melanogaster* populations from a chromosome-wide scan of DNA variation. *Mol. Biol. Evol.* **22**: 2119–2130.

Onodera, Y., K. Nakagawa, J.R. Haag, D. Pikaard, T. Tetsuo Mikami, et al. 2008. Sex-biased lethality or transmission of defective transcription machinery in Arabidopsis. *Genetics* **180**: 207–218.

Oosterhout, C.v., D. Joyce, S. Cummings, J. Blais, N. Barson, et al. 2006. Balancing selection, random genetic drift, and genetic variation at the major histocompatibility complex in two wild populations of guppies (*Poecilia reticulata*). *Evolution* **60**: 2562–2574.

Orgel, L.E. and F.H.C. Crick. 1980. Selfish DNA: the ultimate parasite. *Nature* **284**: 604–607.

Ornduff, R. 1969. Reproductive biology in relation to systematics. *Taxon* **18**: 121–133.

Orr, H.A. 1991. A test of Fisher's theory of dominance. *Proc. Natl. Acad. Sci. U.S.A.* **88**: 11413–11415.

Orr, H.A. 1995. The population genetics of speciation: the evolution of hybrid incompatibilities. *Genetics* **139**: 1805–1813.

Orr, H.A. 1998. The population genetics of adaptation: the distribution of factors fixed during adaptive evolution. *Evolution* **52**: 935–949.

Orr, H.A. 2002. The population genetics of adaptation: the adaptation of DNA sequences. *Evolution* **56**: 1317–1330.

Orr, H.A. 2003. The distribution of fitness effects among beneficial mutations. *Genetics* **163**: 1519–1526.

Orr, H.A. 2005a. The genetic theory of adaptation: a brief history. *Nat. Rev. Genet.* **6**: 119–127.

Orr, H.A. 2005b. The probability of parallel evolution. *Evolution* **59**: 216–220.

Orr, H.A. and A.J. Betancourt. 2001. Haldane's sieve and adaptation from the standing genetic variation. *Genetics* **157**: 875–884.

Orr, H.A. and M. Turelli. 2001. The evolution of postzygotic isolation: accumulating Dobzhansky–Muller incompatibilities. *Evolution* **55**: 1085–1094.

Orr, H.A., J.P. Masly, and D.C. Presgraves. 2004. Speciation genes. *Curr. Op. Genet. Dev.* **14**: 675–679.

Otto, S.P. and N.H. Barton. 2001. Selection for recombination in small populations. *Evolution* **55**: 1921–19231.

Otto, S.P. and T. Day. 2007. *A Biologist's Guide to Mathematical Modeling*. Princeton University Press, Princeton, NJ.

Otto, S.P. and T. Lenormand. 2002. Resolving the paradox of sex and recombination. *Nat. Rev. Genet.* **3**: 256–261.

Otto, S.P. and S.L. Nuismer. 2004. Species interactions and the evolution of sex. *Science* **304**: 1018–1020.

Otto, S.P., C. Sassaman, and M.W. Feldman. 1992. Evolution of sex determination in the conchostracan shrimp *Eulimnidiu texana*. *Am. Nat.* **141**: 329–337.

Ouborg, N.J. and R. Van Treuren. 1994. The significance of genetic erosion in the process of extinction. IV. Inbreeding load and heterosis in relation to population size in the mint *Salvia pratensis*. *Evolution* **48**: 996–1008.

Ouborg, N.J., Y. Piquot, and J.M.V. Groenendael. 1999. Population genetics, molecular markers and the study of dispersal in plants. *J. Ecol.* **87**: 551–568.

Padhukasahasram, B., J.D. Wall, P. Marjoram, and M. Nordborg. 2006. Estimating recombination rates from single-nucleotide polymorphisms using summary statistics. *Genetics* **174**: 1517–1528.

Page, S.L. and R.S. Hawley. 2003. Chromosome choreography: the meiotic ballet. *Science* **301**: 785–789.

Paland, S. and M. Lynch. 2006. Transitions to asexuality result in excess amino-acid substitutions. *Science* **311**: 990–992.

Pallen, M.J. and B.W. Wren. 2007. Bacterial pathogenomics. *Nature* **449**: 835–842.

Panhuis, T.M., N.L. Clark, and W.J. Swanson. 2006. Rapid evolution of reproductive proteins in abalone and *Drosophila*. *Phil. Trans. R. Soc. B* **361**: 261–268.

Pannell, J.R. 2002. The evolution and maintenance of androdioecy. *Ann. Rev. Ecol. Evol. Syst.* **33**: 397–425.

Pannell, J.R. 2003. Coalescence in a metapopulation with recurrent local extinction and recolonization. *Evolution* **57**: 949–961.

Pannell, J.R. and B. Charlesworth. 1999. Neutral genetic diversity in a metapopulation with recurrent local extinction and recolonization. *Evolution* **53**: 664–676.

Pannell, J.R. and B. Charlesworth. 2000. Effects of metapopulation processes on measures of genetic diversity. *Phil. Trans. R. Soc. B* **355**: 1851–1864.

Pannell, J.R., M.E. Dorken, B. Pujol, and R. Berjano. 2008. Gender variation and transitions between sexual systems in *Mercurialis annua* (Euphorbiaceae). *Int. J. Plant Sci.* **169**: 129–139.

Parker, G.A. 1970. Sperm competition and its evolutionary consequences in insects. *Biol. Rev.* **45**: 525–567.

Parmley, J.L. and L.D. Hurst. 2007. How do synonymous mutations affect fitness? *Bioessays* **29**: 515–519.

Partridge, L. and D. Gems. 2002. Mechanisms of ageing: public or private? *Nat. Rev. Genet.* **3**: 165–175.

Payne, F. 1920. Selection for high and low bristle number in the mutant strain "reduced." *Genetics* **5**: 501–542.

Pearson, K. 1903. Mathematical contributions to the theory of evolution. XI. On the influence of natural selection on the variability and correlation of organs. *Phil. Trans. R. Soc. A* **200**: 1–66.

Peck, J. 1994. A ruby in the rubbish: beneficial mutations, deleterious mutations, and the evolution of sex. *Genetics* **137**: 597–606.

Pemberton, A., A. Sommerfeldt, C. Wood, H. Flint, L. Noble, et al. 2004. Plant-like mating in an animal: sexual compatibility and allocation trade-offs in a simultaneous hermaphrodite with remote transfer of sperm. *J. Evol. Biol.* **17**: 506–518.

Perfeito, L., L. Fernandes, C. Mota, and I. Gordo. 2007. Adaptive mutations in bacteria: high rate and small effects. *Science* **317**: 813–815.

Peters, A.D. and P.D. Keightley. 2000. A test for epistasis among induced mutations in *Caenorhabditis elegans*. *Genetics* **156**: 1635–1647.

Petrov, D.A., E.R. Lozovskaya, and D.L. Hartl. 1996. High intrinsic rate of DNA loss in *Drosophila*. *Nature* **384**: 346–349.

Petrov, D.A., Y.T. Aminetzach, J.C. Davis, D. Bensasson, and A.E. Hirsh. 2003. Size matters: non-LTR retrotransposable elements and ectopic recombination in *Drosophila. Mol. Biol. Evol.* **20**: 880–892.

Pennings, P.S. and J. Hermisson. 2006. Soft sweeps III: the signature of positive selection from recurrent mutation. *PLoS Genet.* **2**: 1998–2012.

Pialek, J., H.C. Hauffe, and J.B. Searle. 2005. Chromosomal variation in the house mouse. *Biol. J. Linn. Soc.* **84**: 535–563.

Piganeau, G. and A. Eyre-Walker. 2003. Estimating the distribution of fitness effects from DNA sequence data: implications for the molecular clock. *Proc. Natl. Acad. Sci. U.S.A.* **100**: 10335–10340.

Piganeau, G., M. Gardner, and A. Eyre-Walker. 2004. A broad survey of recombination in animal mitochondria. *Mol. Biol. Evol.* **24**: 374–381.

Piper, J., B. Charlesworth, and D. Charlesworth. 1984. A high rate of self-fertilization and increased seed fertility of homostyle primroses. *Nature* **310**: 50–51.

Piper, J., B. Charlesworth, and D. Charlesworth. 1986. Breeding system evolution in *Primula vulgaris* and the role of reproductive assurance. *Heredity* **56**: 207–217.

Pitnick, S., R. Dobler, and D. Hosken. 2009. Sperm length is not influenced by haploid gene expression in the flies *Drosophila melanogaster* and *Scathophaga stercoraria. Proc. R. Soc. B* **276**: 4029-4034.

Planes, S. and C. Fauvelot. 2002. Isolation by distance and vicariance drive genetic structure of a coral reef fish in the Pacific Ocean. *Evolution* **56**: 378–399.

Pollak, E. 1987. On the theory of partially inbreeding populations. I. Partial selfing. *Genetics* **117**: 353–360.

Pollak, E. and O. Kempthorne. 1970. Malthusian parameters in genetic populations. Part I. Haploid and selfing models. *Theor. Popul. Biol.* **1**: 315–345.

Polley, S.D. and D.J. Conway. 2001. Strong diversifying selection on domains of the *Plasmodium falciparum* apical membrane antigen 1 gene. *Genetics* **158**: 1505–1512.

Porcher, E. and R. Lande. 2005. Loss of gametophytic self-incompatibility with evolution of inbreeding depression. *Evolution* **59**: 46–60.

Presgraves, D.C. 2005. Recombination enhances protein adaptation in *Drosophila melanogaster. Curr. Biol.* **15**: 1651–1656.

Presgraves, D.C. 2008. Sex chromosomes and speciation in *Drosophila. Trends Genet.* **24**: 336–343.

Presgraves, D.C., L. Balagopalan, S.M. Abmayr, and H.A. Orr. 2003. Adaptive evolution drives divergence of a hybrid inviability gene between two species of *Drosophila. Nature* **423**: 715–719.

Price, G.R. 1970. Selection and covariance. *Nature* **227**: 520–521.

Pritchard, J.K. 2001. Are rare variants responsible for susceptibility to complex diseases? *Am. J. Hum. Genet.* **69**: 124–137.

Pritchard, J.K., M. Stephens, and P. Donnelly. 2000. Inference of population structure using multilocus genotype data. *Genetics* **155**: 945–959.

Pröschel, M., Z. Zhang, and J. Parsch. 2006. Widespread adaptive evolution of *Drosophila* genes with sex-biased expression. *Genetics* **174**: 893–900.

Prout, T. 1968. Sufficient conditions for multiple niche polymorphism. *Am. Nat.* **102**: 494–496.

Prout, T. 1971a. The relation between fitness components and population prediction in *Drosophila*. I. The estimation of fitness components. *Genetics* **68**: 127–149.

Prout, T. 1971b. The relation between fitness components and population prediction in *Drosophila*. II. Population prediction. *Genetics* **68**: 151–167.

Prout, T. 2000. How well does opposing selection maintain variation? In *Evolutionary Genetics from Molecules to Morphology* (Eds. R.S. Singh and C.B. Krimbas), pp. 157–181. Cambridge University Press, Cambridge, U.K.

Prout, T. and J.S.F. Barker. 1993. *F* statistics in *Drosophila buzzatii*: selection, population size and inbreeding. *Genetics* **134**: 369–375.

Provine, W.B. 1971. *The Origins of Theoretical Population Genetics*. University of Chicago Press, Chicago, IL (reprinted with afterword, 2000).

Prusinkiewicz, P., Y. Erasmus, B. Lane, L.D. Harder, and E. Coen. 2007. Evolution and development of inflorescence architecture. *Science* **316**: 1452–1456.

Przeworski, M. 2002. The signature of positive selection at randomly chosen loci. *Genetics* **160**: 1179–1189.

Ptak, S.E. and M. Przeworski. 2002. Evidence for population growth in humans is confounded by fine-scale population structure. *Trends Genet.* **18**: 559–563.

Punnett, R.C. 1915. *Mimicry in Butterflies*. Cambridge University Press, Cambridge, U.K.

Qin, H., W.B. Wu, J.M. Comeron, M. Kreitman, and W.-H. Li. 2004. Intragenic spatial patterns of codon usage in prokaryotic and eukaryotic genomes. *Genetics* **168**: 2245–2260.

Race, H.L., R.G. Herrmann, and W. Martin. 1999. Why have organelles retained genomes? *Trends Genet.* **15**: 364–370.

Race, R.R. and R. Sanger. 1975. *Blood Groups in Man*, 6th ed. Blackwell, Oxford, U.K.

Ramsey, M., G. Vaughton, and R. Peakall. 2006. Inbreeding avoidance and the evolution of gender dimorphism in *Wurmbea biglandulosa* (Colchicaceae). *Evolution* **60**: 529–537.

Rand, D.M. and L.M. Kann. 1996. Excess amino acid polymorphisms in mitochondrial DNA: contrasts among genes from *Drosophila*, mice and humans. *Mol. Biol. Evol.* **13**: 735–748.

Rannala, B. and J.A. Hartigan. 1995. Identity by descent in island–mainland populations. *Genetics* **139**: 429–437.

Ray, N., M. Currat, P. Berthier, and L. Excoffier. 2005. Recovering the geographic origin of early modern humans by realistic and spatially explicit simulations. *Genome Res.* **15**: 1161–1167.

Raymond, C.K., M.P.A. Kas, R. Qiu, Y. Zhou, S.A. Subramanian, et al. 2005. Ancient haplotypes of the HLA class II region. *Genome Res.* **15**: 1250–1257.

Read, A., A. Narara, S. Nee, A. Keymer, and K. Day. 1992. Gametocyte sex ratios as indirect measures of outcrossing rates in malaria. *Parasitology* **104**: 387–395.

Reich, D.E., M. Cargill, S. Bollk, J. Ireland, P.C. Sabeti, D.J. Richter, et al. 2001. Linkage disequilibrium in the human genome. *Nature* **411**: 199–1024.

Reik, W. 2007. Stability and flexibility of epigenetic regulation in mammalian development. *Nature* **447**: 425–431.

Rendel, J.M. 1943. Variations in the weights of hatched and unhatched ducks' eggs. *Biometrika* **33**: 48–58.

Renner, O. 1925. Untersuchungen über die faktorielle Konstitution einiger komplex-heterozygotischer Oenotheren. *Bibl. Genet.* **9**: 1–168.

Retchless, A.C. and J.G. Lawrence. 2007. Temporal fragmentation of speciation in bacteria. *Science* **317**: 1093–1096.

Reynolds, R.M., S. Temiyasathit, M.M. Reedy, E.A. Ruedi, J.M. Drnevich, et al. 2007. Age specificity of inbreeding load in *Drosophila melanogaster* and implications for the evolution of late-life mortality plateaus. *Genetics* **177**: 587–595.

Rice, W.R. 1984. Sex chromosomes and the evolution of sexual dimorphism. *Evolution* **38**: 735–742.

Rice, W.R. 1987a. The accumulation of sexually antagonistic genes as selective agent promoting the evolution of reduced recombination between primitive sex chromosomes. *Evolution* **41**: 911–914.

Rice, W.R. 1987b. Genetic hitch-hiking and the evolution of reduced genetic activity of the Y sex chromosome. *Genetics* **116**: 161–167.

Richards, C.M. 2000. Inbreeding depression and genetic rescue in a plant metapopulation. *Am. Nat.* **155**: 383–394.

Richards, E.J. 2006. Inherited epigenetic variation: revisiting soft inheritance. *Nat. Rev. Genet.* **7**: 395–401.

Richman, A.D., M.K. Uyenoyama, and J.R. Kohn. 1996. Allelic diversity and gene genealogy at the self-incompatibility locus in the Solanaceae. *Science* **273**: 1212–1216.

Ricklefs, R.E. 1998. Evolutionary theories of aging: confirmation of a fundamental prediction, with implications for the genetic basis and evolution of life span. *Am. Nat.* **152**: 24–44.

Rieseberg, L. H., S. J. E. Baird, and A. M. Desrochers. 1998. Patterns of mating in wild sunflower hybrid zones. *Evolution* **52**: 713–726.

Rio, D.C. 2002. *P* transposable elements in *Drosophila melanogaster*. In *Mobile DNA II* (Eds. N.L. Craig, R. Craigie, M. Gellert, and A.M. Lambowitz), pp. 484–58. American Society of Microbiology, Washington, DC.

Ritland, K. 1990a. Inferences about inbreeding depression based on changes in the inbreeding coefficient. *Evolution* **44**: 1230–1241.

Ritland, K. 1990b. A series of FORTRAN computer programs for estimating plant mating systems. *J. Hered.* **81**: 235–237.

Robbins, R.B. 1918. Some applications of mathematics to breeding problems. *Genetics* **3**: 375–389.

Robertson, A. 1960. A theory of limits in artificial selection. *Proc. R. Soc. B* **153**: 234–249.

Robertson, A. 1961. Inbreeding in artificial selection programmes. *Genet. Res.* **2**: 189–194.

Robertson, A. 1967. The nature of quantitative genetic variation. In *Heritage from Mendel* (Ed. R.A. Brink), pp. 265–280. University of Wisconsin Press, Madison, WI.

Robertson, H.M. and D.J. Lampe. 1995. Distribution of transposable elements in arthropods. *Ann. Rev. Ent.* **40**: 333–357.

Rocha, E.P.C. and A. Danchin. 2002. Base competition might result from competition for metabolic resources. *Trends Genet.* **18**: 291–294.

Rockmill, B., K. Voelkel-Meiman, and G.S. Roeder. 2006. Centromere-proximal crossovers are associated with precocious separation of sister chromatids during meiosis in *Saccharomyces cerevisiae*. *Genetics* **174**: 1745–1754.

Roff, D.A. 1996. The evolution of threshold traits in animals. *Quart. Rev. Biol.* **71**: 3–35.

Roff, D.A. 2002. *Life History Evolution*. Sinauer, Sunderland, MA.

Rokyta, D.R., P. Joyce, S.B. Caudle, and H.A. Wichman. 2005. An empirical test of the mutational landscape model of adaptation using a single-stranded DNA virus. *Nat. Genet.* **37**: 441–444.

Roman, J. and S.R. Palumbi. 2003. Whales before whaling in the North Atlantic. *Science* **301**: 508–510.

Rose, L.E., P.D. Bittner-Eddy, C.H. Langley, E.B. Holub, R.W. Michelmore, and J.L. Beynon. 2004. The maintenance of extreme amino acid diversity at the disease resistance gene, RPP13, in *Arabidopsis thaliana*. *Genetics* **166**: 1517–1527.

Rose, M.R. 1982. Antagonistic pleiotropy, dominance, and genetic variation. *Heredity* **48**: 63–78.

Rose, M.R., M.D. Drapeau, P.G. Yazdi, K.H. Shah, D.B. Moise, et al. 2002. Evolution of late-life mortality in *Drosophila melanogaster*. *Evolution* **56**: 1982–1991.

Rose, M.R., C.L. Rauser, G. Benford, M. Matos, and L.D. Mueller. 2007. Hamilton's forces of natural selection after forty years. *Evolution* **61**: 1265–1276.

Roselius, K., W. Stephan, and T. Städler. 2005. The relationship of nucleotide polymorphism, recombination rate and selection in wild tomato species. *Genetics* **171**: 753–763.

Roskam, J.C. and P.M. Brakefield. 1996. A comparison of temperature-induced polyphenism in African Bicyclus butterflies from a seasonal savannah-rainforest ecotone. *Evolution* **50**: 2360–2372.

Ross, M.T., D.V. Grafham, A.J. Coffey, S. Scherer, K. McLay, et al. 2005. The DNA sequence of the human X chromosome. *Nature* **434**: 325–337.

Ross-Ibarra, J., S. Wright, J. Foxe, A. Kawabe, L. DeRose-Wilson, et al. 2008. Patterns of polymorphism and demographic history in natural populations of *Arabidopsis lyrata*. *PLoS One* **6**: 3.

Rousset, F. 1999. Genetic differentiation in populations with different classes of individuals. *Theor. Popul. Biol.* **55**: 297–308.

Rousset, F. 2003. Inferences from spatial population genetics. In *Handbook of Statistical Genetics* (Eds. D.J. Balding, M. Bishop, and C. Cannings), pp. 681–712. John Wiley, Chichester, U.K.

Rousset, F. 2004. *Genetic Structure and Selection in Subdivided Populations*. Princeton University Press, Princeton, NJ.

Roux, C.Z. 1974. Hardy–Weinberg equilibria in random mating-populations. *Theor. Popul. Biol.* **5**: 393–416.

Roux, F., S. Giancola, S. Durand, and X. Reboud. 2006. Building of an experimental cline with *Arabidopsis thaliana* to estimate herbicide fitness cost. *Genetics* **173**: 1023–1031.

Rouzine, I.M., E. Brunet, and C.O. Wilke. 2008. The traveling-wave approach to asexual evolution: Muller's ratchet and speed of adaptation. *Theor. Popul. Biol.* **73**: 24–46.

Rozas, J., J.C. Sanchez-DelBarrio, X. Messeguer, and R. Rozas. 2003. DnaSP, DNA polymorphism analyses by the coalescent and other methods. *Bioinformatics* **19**: 2496–2497.

Roze, D. and T. Lenormand. 2005. Self-fertilization and the evolution of recombination. *Genetics* **170:** 840–557.

Roze, D. and F. Rousset. 2003. Selection and drift in subdivided populations: a straightforward method for deriving diffusion approximations and applications involving dominance, selfing and local extinctions. *Genetics* **165:** 2153–2166.

Roze, D. and F. Rousset. 2004. Joint effects of self-fertilization and population structure on mutation load, inbreeding depression and heterosis. *Genetics* **167:** 1001–1015.

Sabeti, P.C., D.E. Reich, J.M. Higgins, H.Z.P. Levine, D.J. Richter, et al. 2002. Detecting recent positive selection in the human genome from haplotype structure. *Nature* **419:** 832–837.

Salvini-Plawen, L.V. and E. Mayr. 1977. On the evolution of photoreceptors and eyes. *Evol. Biol.* **10:** 207–263.

Sandler, L. and Y. Hiraizumi. 1959. Meiotic drive in natural populations of *Drosophila melanogaster*. II. Genetic variation at the segregation-distorter locus. *Proc. Natl. Acad. Sci. U.S.A.* **45:** 1412–1422.

Sandler, L. and E. Novitski. 1957. Meiotic drive as an evolutionary force. *Am. Nat.* **91:** 105–110.

SanMiguel, P., A. Tikhonov, N. Jin, N. Motchulskaia, A. Zakharov, et al. 1996. Nested retrotransposons in the intergenic regions of the maize genome. *Science* **273:** 765–769.

Santiago, E. and A. Caballero. 1995. Effective size of populations under selection. *Genetics* **139:** 1013–1030.

Santiago, E. and A. Caballero. 1998. Effective size and polymorphism of linked neutral loci in populations under selection. *Genetics* **149:** 2105–2117.

Santure, A.W. and J. Wang. 2009. The joint effects of selection and dominance on the $Q_{ST} - F_{ST}$ contrast. *Genetics* **181:** 259–276.

Sassaman, C. and S.C. Weeks. 1993. The genetic mechanism of sex determination in the conchostracan shrimp *Eulimnidia texana*. *Am. Nat.* **141:** 314–328.

Sato, T., T. Nishio, R. Kimura, M. Kusaba, G. Suzuki, et al. 2002. Coevolution of the S-locus genes SRK, SLG and SP11/SCR in *Brassica oleracea* and *B. rapa*. *Genetics* **162:** 931–940.

Saunders, M.A., M.F. Hammer, and M.W. Nachman. 2002. Nucleotide variability at *G6pd* and the signature of malarial selection in humans. *Genetics* **162:** 1849–1861.

Sawyer, S.A. and D.L. Hartl. 1992. Population genetics of polymorphism and divergence. *Genetics* **132:** 1161–1176.

Sawyer, S.A., R.J. Kulathinal, C.D. Bustamante, and D.L. Hartl. 2003. Bayesian analysis suggests that most amino acid replacements in *Drosophila* are driven by positive selection. *J. Mol. Evol.* **57:** S154–S164.

Sawyer, S.A., J. Parsch, Z. Zhang, and D.L. Hartl. 2007. Prevalence of positive selection among nearly neutral amino acid replacements in *Drosophila*. *Proc. Natl. Acad. Sci. U.S.A.* **104:** 6504–6510.

Sawyer, S.L. and H.S. Malik. 2006. Positive selection of yeast nonhomologous end-joining genes and a retrotransposon conflict hypothesis. *Proc. Natl. Acad. Sci. U.S.A.* **103:** 17614–17619.

Schacherer, J., J. A. Shapiro, D. M. Ruderfer, and L. Kruglyak. 2009. Comprehensive polymorphism survey elucidates population structure of *Saccharomyces cerevisiae*. *Nature* **458**: 342–345.

Schaeffer, S.W. and E.L. Miller. 1992. Estimates of gene flow in *Drosophila pseudoobscura* determined from nucleotide sequence analysis of the alcohol dehydrogenase region. *Genetics* **132**: 471–480.

Schaeffer, S.W. and E.L. Miller. 1993. Estimates of linkage disequilibrium and the recombination parameter determined from segregating nucleotide sites in the alcohol dehydrogenase region of *Drosophila pseudoobscura*. *Genetics* **135**: 541–552.

Schaffer, W.M. 1974. Selection for optimal life histories: the effect of age structure. *Ecology* **55**: 291–303.

Schaffer, W.M. and M.V. Schaffer. 1979. The adaptive significance of variations in reproductive habit in the Agavaceae. II. Pollinator foraging behavior and selection for increased reproductive expenditure. *Ecology* **60**: 1051–1059.

Schaffner, S.F., C. Foo, S. Gabriel, D. Reich, M.J. Daly, and D. Altschuler. 2005. Calibrating a coalescent simulation of human genome sequence variation. *Genome Res.* **15**: 1576–1583.

Scheet, P. and M. Stephens. 2006. A fast and flexible statistical model for large-scale genotype data: applications to inferring missing genotypes and haplotype phase. *Am. J. Hum. Genet.* **78**: 629–644.

Schemske, D.W. and H.D. Bradshaw. 1999. Pollinator preferences and the evolution of floral traits in monkeyflowers (*Mimulus*). *Proc. Natl. Acad. Sci. U.S.A.* **96**: 11910–11914.

Schierup, M.H., X. Vekemans, and V. Christiansen. 1998. Mate availability and fecundity selection in multi-allelic self-incompatibility systems in plants. *Evolution* **52**: 19–29.

Schierup, M.H., X. Vekemans, and D. Charlesworth. 2000. The effect of subdivision on variation at multi-allelic loci under balancing selection. *Genet. Res.* **76**: 51–62.

Schlenke, T.A. and D.J. Begun. 2004. Strong selective sweep associated with a transposon insertion in *Drosophila simulans*. *Proc. Natl. Acad. Sci. U.S.A.* **101**: 1626–1631.

Schlötterer, C. 2004. The evolution of molecular markers—just a matter of fashion? *Mol. Ecol.* **5**: 63–69.

Schmid, K. J., S. Ramos-Onsins, H. Ringys-Beckstein, D. Weisshaar, and T. Mitchell-Olds. 2005. A multilocus sequence survey in *Arabidopsis thaliana* reveals a genome-wide departure from a neutral model of DNA sequence polymorphism. *Genetics* **169**: 1601–1615.

Schoen, D.J. 1982. The breeding system of *Gilia achilleifolia*: variation in floral characteristics and outcrossing rate. *Evolution* **36**: 352–360.

Schoen, D.J. and D.G. Lloyd. 1984. The selection of cleistogamy and heteromorphic diaspores. *Bot. J. Linn. Soc.* **23**: 303–322.

Schoen, D.J. and D.G. Lloyd. 1992. Self- and cross-fertilization in plants. III. Methods for studying modes and functional aspects of self-fertilization. *Int. J. Plant Sci.* **153**: 381–393.

Schoen, D.J., A.-M. L'Heureux, J. Marsolais, and M.O. Johnston. 1997. Evolutionary history of the mating system in Amsinckia (Boraginaceae). *Evolution* **51**: 1090–1099.

Schopfer, C.R., M.E. Nasrallah, and J.B. Nasrallah. 1999. The male determinant of self-incompatibility in *Brassica. Science* **286**: 1697–1700.

Schultz, S. 1994. Nucleo-cytoplasmic male sterility and alternative routes to dioecy. *Evolution* **48**: 1933–1945.

Sea Urchin Genome Sequencing Consortium. 2006. The genome of the sea urchin *Strongylocentrotus purpuratus. Science* **314**: 941–952.

Sebat, J. 2005. Large-scale copy number polymorphism in the human genome. *Science* **305: 525–528.**

Seger, J. 1988. Dynamics of some simple host–parasite models with more than two genotypes in each species. *Phil. Trans. R. Soc. B* **319**: 541–555.

Sezgin, E., D.D. Duvernell, L.M. Matzkin, Y.H. Duan, C.T. Zhu, et al. 2004. Single-locus latitudinal clines and their relationship to temperate adaptation in metabolic genes and derived alleles in *Drosophila melanogaster. Genetics* **168**: 923–831.

Shapiro, B., A.J. Drummond, A. Rambaut, M.C. Wilson, P.R. Matheus, et al. 2004. Rise and fall of the Beringian steppe bison. *Science* **306**: 1561–1565.

Shapiro, J.A. 1983. *Mobile Genetic Elements.* Academic Press, New York.

Shapiro, J.A., W. Huang, C. Zhang, M. Hubisz, J. Lu, et al. 2007. Adaptive genic evolution in the *Drosophila* genome. *Proc. Natl. Acad. Sci. U.S.A.* **104**: 2271–2276.

Sharp, P.J., D.L. Hayman, 1988. An examination of the role of chiasma frequency in the genetic system of marsupials. *Heredity* **60**: 77–85.

Sharp, P.M., T.M.F. Tuohy, and K.R. Mosurski. 1986. Codon usage in yeast: cluster analysis clearly differentiates highly and lowly expressed genes. *Nuc. Acids. Res.* **14**: 5125–5143.

Sharp, P.M., E. Cowe, D.G. Higgins, D. Shields, K.E. Wolfe, and F. Wright. 1988. Codon usage patterns in *Escherichia coli, Bacillus subtilis, Saccharomyces cerevisiae, Schizosaccharomyce spombe, Drosophila melanogaster* and *Homo sapiens*: a review of the considerable within-species diversity. *Nuc. Acids. Res.* **16**: 8207–8211.

Sharp, P.M., E. Bailes, R.J. Grocock, J.F. Peden, and R.E. Sockett. 2005. Variation in the strength of selected codon usage bias among bacteria. *Nucleic Acids Res.* **33**: 1141–1153.

Sharpe, F.R. and A.J. Lotka. 1911. A problem in age distribution. *Philos. Mag.* **21**: 435–438.

Shaw, K.L. 2002. Conflict between nuclear and mitochondrial DNA phylogenies of a recent species radiation: what mtDNA reveals and conceals about modes of speciation in Hawaiian crickets. *Proc. Natl. Acad. Sci. U.S.A.* **99**: 16122–16127.

Sheldahl, L.E., D.M. Weinreich, and D.M. Rand. 2003. Recombination, dominance and selection on amino-acid polymorphisms in the *Drosophila* genome: contrasting patterns on the X and fourth chromosomes. *Genetics* **165**: 1195–1208.

Sheppard, P.M. 1959a. Blood groups and natural selection. *Brit. Med. Bull.* **15**: 134–139.

Sheppard, P.M. 1959b. The evolution of mimicry: a problem in ecology and genetics. *Cold Spring Harbor Symp. Quant. Biol.* **24**: 131–140.

Sheppard, P.M. 1975. *Natural Selection and Heredity*, 4th ed. Hutchinson, London.

Sherman, J.D. and S.M. Stack. 1995. Two-dimensional spreads of synaptonemal complexes from Solanaceous plants. VI. High-resolution recombination map for tomato (*Lycopersicon esculentum*). *Genetics* **141**: 683–708.

Sherman-Broyles, S., N. Boggs, A. Farkas, P. Liu, J. Vrebalov, et al. 2007. *S* Locus genes and the evolution of self-fertility in *Arabidopsis thaliana*. *Plant Cell* **19**: 94–106.

Shields, D.C., P.M. Sharp, D.G. Higgins, and F. Wright. 1988. "Silent" sites in *Drosophila* are not neutral: evidence of selection among synonymous codons. *Mol. Biol. Evol.* **5**: 704–716.

Shiina, T., M. Ota, S. Shimizu, Y. Katsuyama, N. Hashimoto, et al. 2006. Rapid evolution of major histocompatibility complex class I genes in primates generates new disease alleles in humans via hitchhiking diversity. *Genetics* **173**: 1555–1570.

Shnol, E.E. and A.S. Kondrashov. 1993. Effect of selection on the phenotypic variance. *Genetics* **134**: 995–996.

Shuker, D.M., I. Pen, and S.A. West. 2006. Sex ratios under asymmetrical local mate competition in the parasitoid wasp *Nasonia vitripennis*. *Behav. Ecol.* **17**: 345–352.

Siepel, A. and D. Haussler. 2004. Phylogenetic estimation of context-dependent subsitution rates by maximum likelihood. *Mol. Biol. Evol.* **21**: 468–488.

Silver, L.M. 1993. The peculiar journey of a selfish chromosome: mouse *t* haplotypes and meiotic drive. *Trends Genet.* **9**: 250–254.

Simmons, M.J. and J.F. Crow. 1977. Mutations affecting fitness in *Drosophila* populations. *Ann. Rev. Genet.* **11**: 49–78.

Simonsen, K.L., G.A. Churchill, and C.F. Aquadro. 1995. Properties of statistical tests of neutrality for DNA polymorphism data. *Genetics* **141**: 413–429.

Simpson, G.G. 1953. *The Major Features of Evolution*. Columbia University Press, New York.

Singh, N.D., P.F. Arndt, and D.A. Petrov. 2005. Genomic heterogeneity of background substitutional patterns in *Drosophila melanogaster*. *Genetics* **169**: 709–722.

Skellam, J.G. 1951. Gene dispersion in heterogeneous populations. *Heredity* **5**: 433–435.

Slatkin, M. 1973a. On treating the chromosome as the unit of selection. *Genetics* **72**: 157–168.

Slatkin, M. 1973b. Gene flow and selection in a cline. *Genetics* **75**: 733–756.

Slatkin, M. 1977. Gene flow and genetic drift in a species subject to frequent local extinction. *Theor. Popul. Biol.* **12**: 253–262.

Slatkin, M. 1978. Equilibration of fitnesses by natural selection. *Am. Nat.* **112**: 845–859.

Slatkin, M. 1981. Fixation probabilities and fixation times in subdivided populations. *Evolution* **35**: 477–488.

Slatkin, M. 1991. Inbreeding coefficients and coalescence times. *Genet. Res.* **58**: 167–175.

Slatkin, M. 1993. Isolation by distance in equilibrium and non-equilibrium populations. *Evolution* **47**: 264–279.

Slatkin, M. 1994. Linkage disequilibrium in stable and growing populations. *Genetics* **137**: 331–336.

Slatkin, M. 1995a. Hitchhiking and associative overdominance at a microsatellite locus. *Mol. Biol. Evol.* **12**: 473–480.

Slatkin, M. 1995b. A measure of population subdivision based on microsatellite allele frequencies. *Genetics* **139**: 457–462.

Slatkin, M. 2002. The age of alleles. In *Modern Developments in Theoretical Population Genetics* (Eds. M. Slatkin and M. Veuille), pp. 233–260. Oxford University Press, New York.

Slatkin, M. 2005. Seeing ghosts: the effect of unsampled populations on migration rates estimated for sampled populations. *Mol. Ecol.* **14**: 67–73.

Slatkin, M. 2008. Linkage disequilibrium—understanding the evolutionary past and mapping the medical future. *Nat. Rev. Genet.* **9**: 477–485.

Slatkin, M. and R.R. Hudson. 1991. Pairwise comparisons of mitochondrial DNA sequences in stable and exponentially growing populations. *Genetics* **143**: 579–587.

Slatkin, M. and T. Wiehe. 1998. Genetic hitch-hiking in a subdivided population. *Genet. Res.* **71**: 155–160.

Small, K.S., M. Brudno, M. Hill, and A. Sidow. 2007. Extreme genomic variation in a natural population. *Proc. Natl. Acad. Sci. U.S.A.* **104**: 5698–5703.

Smirnova, I., M.T. Hamblin, C. McBride, B. Beutler, and A. Di Rienzo. 2001. Excess of rare amino acid polymorphisms in the Toll-like receptor 4 in humans. *Genetics* **158**: 1657–1664.

Smirnova, I., N. Mann, A. Dols, H.H. Derkx, M.L. Hibberd, et al. 2003. Assay of locus-specific genetic load implicates rare Toll-like receptor 4 mutations in meningococcal susceptibility. *Proc. Natl. Acad. Sci. U.S.A.* **100**: 6075–6080.

Smith, C.D., S. Shu, C.J. Mungall, and G.H. Karpen. 2007. The release 5.1 annotation of *Drosophila melanogaster* heterochromatin. *Science* **316**: 1587–1591.

Smith, D., and R. Lee. 2008. Nucleotide diversity in the mitochondrial and nuclear compartments of *Chlamydomonas reinhardtii*: investigating the origins of genome architecture. *BMC Evol. Biol.* **8**: 156.

Smith, F. 1937. A discriminant function for plant selection. *Ann. Eugen.* **7**: 240–250.

Smith, N.G.C. and A. Eyre-Walker. 2002. Adaptive protein evolution in *Drosophila*. *Nature* **415**: 1022–1024.

Smith, T.B. 1993. Disruptive selection and the genetic basis of bill size polymorphism in the African finch *Pyrenestes*. *Nature* **363**: 618–620.

Smouse, P. 1974. Likelihood analysis of recombinational disequilibrium in multiple-locus gametic frequencies. *Genetics* **76**: 557–565.

Snow, A.A., T.P. Spira, and H. Liu. 2000. Effects of sequential pollination on the success of "fast" and "slow" pollen donors in *Hibiscus moscheutos* (Malvaceae). *Am. J. Bot.* **87**: 1656–1659.

Söderberg, R.J. and O.G. Berg. 2007. Mutational interference and the progression of Muller's ratchet when mutations have a broad range of deleterious effects. *Genetics* **177**: 971–986.

Sokal, R.R. and F.J. Rohlf. 1995. *Biometry*. W.H. Freeman, San Francisco.

Song, S., D. Dey, and K. Holsinger. 2006. Differentiation among populations with migration, mutation, and drift: implications for genetic inference. *Evolution* **60**: 1–12.

Spencer, C.C.A., P. Deloukas, S. Hunt, J. Mullikin, S. Myers, et al. 2006. The influence of recombination on human genetic diversity. *PLoS Genet.* **2**: 1375–1385.

Spencer, N., D. Hopkinson, and H. Harris. 1964. Quantitative differences and gene dosage in the human red cell acid phosphatase polymorphism. In *The Principles of Human Biochemical Genetics* (Ed. H. Harris), pp. 107–140. North Holland Press, Amsterdam.

Städler, T. and L.F. Delph. 2002. Ancient mitochondrial haplotypes and evidence for intragenic recombination in a gynodioecious plant. *Proc. Natl. Acad. Sci. U.S.A.* **99:** 11730–11735.

Stam, P. 1980. The distribution of the fraction of the genome identical by descent in finite random mating populations. *Genet. Res.* **35:** 131–155.

Stankiewicz, P. and J.R. Lupski. 2002. Genome architecture, rearrangements and genetic disorders. *Trends Genet.* **18:** 74–82.

Stearns, S.C. 1992. *The Evolution of Life Histories.* Oxford University Press, Oxford, U.K.

Stefansson, H., A. Helgason, G. Thorleifsson, V. Steinthorsdottir, G. Masson, et al. 2005. A common inversion under selection in Europeans. *Nat. Genet.* **37:** 129–137.

Stehlik, I. and S.C.H. Barrett. 2005. Mechanisms governing sex-ratio variation in dioecious *Rumex nivalis. Evolution* **59:** 814–825.

Steinemann, M. and S. Steinemann. 1998. Enigma of Y chromosome degeneration: neo-Y and neo-X chromosomes of *Drosophila miranda* a model for sex chromosome evolution. *Genetica* **102/103:** 409–420.

Steinmetz, L.M., C. Scharfe, A.M. Deutschbauer, D. Mokranjac, Z.S. Herman, et al. 2002. Systematic screen for human disease genes in yeast. *Nat. Genet.* **31:** 400–404.

Stephan, W. 1995. An improved method for estimating the rate of fixation of favorable mutations based on DNA polymorphism data. *Mol. Biol. Evol.* **12:** 959–962.

Stephan, W. and Y. Kim. 2002. Recent applications of diffusion theory to population genetics. In *Modern Developments in Theoretical Population Genetics* (Eds. M. Slatkin and M. Veuille), pp. 72–93. Oxford University Press, Oxford, U.K.

Stephan, W., T.H.E. Wiehe, and M.W. Lenz. 1992. The effect of strongly selected substitutions on neutral polymorphism: analytical results based on diffusion theory. *Theor. Popul. Biol.* **41:** 237–254.

Stephan, W., B. Charlesworth, and G.A.T. McVean. 1999. The effect of background selection at a single locus on weakly selected, partially linked variants. *Genet. Res.* **73:** 133–146.

Stewart, C.N., M.D. Halfhill, and S.I. Warwick. 2003. Transgene introgression from genetically modified crops to their wild relatives. *Nat. Rev. Genet.* **4:** 806–817.

Stoletzki, N. and A. Eyre-Walker. 2007. Synonymous codon usage in *Escherichia coli*: selection for translational accuracy. *Mol. Bio. Evol.* **24:** 374–381.

Streisfeld, M.A. and J.R. Kohn. 2005. Contrasting patterns of floral and molecular variation across a cline in *Mimulus aurantiacus. Evolution* **59:** 2548–2559.

Strobeck, C. 1987. Average number of nucleotide differences in a sample from a single subpopulation: a test for population subdivision. *Genetics* **117:** 149–153.

Sturgill, D., Y. Zhang, M. Parisi, and B. Oliver. 2007. Demasculinization of X chromosomes in the *Drosophila* genus. *Nature* **450:** 238–242.

Sturtevant, A.H. and T. Dobzhansky. 1936. Geographical distribution and cytology of "sex-ratio" in *Drosophila pseudoobscura* and related species. *Genetics* **21:** 473–490.

Sturtevant, A.H. and T. Dobzhansky. 1938. Inversions in the chromosomes of *Drosophila pseudoobscura*. *Genetics* **23**: 28–64.

Sturtevant, A.H. and K. Mather. 1938. The interrelations of inversions, heterosis and recombination. *Am. Nat* **72**: 447–452.

Subramanian, S. and S. Kumar. 2006. Higher intensity of purifying selection on > 90% of the human genes revealed by the intrinsic replacement mutation rates. *Mol. Biol. Evol.* **23**: 2283–2287.

Sueoka, N. 1962. On the genetic basis of variation and heterogeneity of DNA base composition. *Proc. Natl. Acad. Sci. U.S.A.* **48**: 582–592.

Sueoka, N. 1988. Directional mutation pressure and neutral molecular evolution. *Proc. Natl. Acad. Sci. U.S.A.* **85**: 2653–2657.

Sumner, F.B. 1929. The analysis of a concrete case of intergradation between two subspecies. *Proc. Natl. Acad. Sci. U.S.A.* **15**: 110–120.

Sumner, F.B. 1930. Genetic and distributional studies of three subspecies of Peromyscus. *J. Genet.* **23**: 275–376.

Sun, M. and F.R. Ganders. 1986. Female frequencies in gynodioecious populations correlated with selfing rates in hermaphrodites. *Am. J. Bot.* **73**: 1645–1648.

Sundström, H., M.T. Webster, and H. Ellegren. 2004. Reduced variation on the chicken Z chromosome. *Genetics* **167**: 377–385.

Sunyaev, S., W.C. Lathe, V.E. Ramensky, and P. Bork. 2000. SNP frequencies in human genes: an excess of rate alleles and differing modes of selection. *Trends Genet.* **16**: 335–337.

Suzuki, Y. and T. Gojobori. 1999. A method for detecting positive selection at single amino acid sites. *Mol. Bio. Evol.* **16**: 1315–1328.

Sved, J.A. 1968. The stability of linked systems of loci with a small population size. *Genetics* **59**: 543–563.

Sved, J.A. 1971. An estimate of heterosis in *Drosophila melanogaster. Genet. Res.* **18**: 97–105.

Sved, J.A. 1972. Heterosis at the level of the chromosome and at the level of the gene. *Theor. Popul. Biol.* **3**: 491–506.

Sved, J.A. and O. Mayo. 1970. The evolution of dominance. In *Mathematical Topics in Population Genetics* (Ed. K. Kojima), pp. 289–316. Springer-Verlag, Berlin.

Szymura, J.M. and N.H. Barton. 1991. The genetic structure of the hybrid zone between *Bombina bombina* and *B. variegata*—comparisons between transects and between loci. *Evolution* **45**: 237–261.

Tajima, F. 1983. Evolutionary relationship of DNA sequences in a finite population. *Genetics* **105**: 437–460.

Tajima, F. 1989a. DNA polymorphism in a subdivided population: the expected number of segregating sites in the two-subpopulation model. *Genetics* **123**: 229–240.

Tajima, F. 1989b. Statistical method for testing the neutral mutation hypothesis. *Genetics* **123**: 585–595.

Tajima, F. 1989c. The effect of change in population size on DNA polymorphism. *Genetics* **123**: 597–601.

Takahashi, A., Y. Liu, and N. Saitou. 2004. Genetic variation versus recombination rate in a structured population of mice. *Mol. Biol. Evol.* **21**: 404–409.

Takahata, N. 1991. Genealogy of neutral genes and spreading of selected mutations in a geographically structured population. *Genetics* **129**: 585–595.

Takahata, N. and Y. Satta. 1998. Footprints of intragenic recombination at HLA loci. *Immunogenetics* **47**: 430–441.

Takahata, N., S.-H. Lee, and Y. Satta. 2001. Testing multiregionality of modern human origins. *Mol. Biol. Evol.* **18**: 172–183.

Takano-Shimizu, T. 1999. Local recombination rate and mutation effects on molecular evolution. *Genetics* **153**: 1285–1296.

Takayama, S., H. Shiba, M. Iwano, H. Shimosato, F.-S. Che, et al. 2000. The pollen determinant of self-incompatibility in *Brassica campestris*. *Proc. Natl. Acad. Sci. U.S.A.* **97**: 1920–1925.

Takebayashi, N., P.B. Brewer, E. Newbigin, and M.K. Uyenoyama. 2003. Patterns of variation within self-incompatibility loci. *Mol. Biol. Evol.* **20**: 1778–1794.

Tang, C., C. Toomajian, S. Sherman-Broyles, V. Plagnol, Y. Guo, et al. 2007. The evolution of selfing in *Arabidopsis thaliana*. *Science* **317**: 1070–1072.

Tang, L. and S. Huang. 2007. Evidence for reductions in floral attractants with increased selfing rates in two heterandrous species. *New Phytol.* **175**: 588–595.

Tanksley, S.D., D. Zamir, and C.M. Rick. 1981. Evidence for extensive overlap of sporophytic and gametophytic gene expression in *Lycopersicon esculentum*. *Science* **213**: 453–455.

Tavaré, S. 1984. Lines-of-descent and genealogical processes, and their applications in population genetic models. *Theor. Popul. Biol.* **26**: 119–164.

Teschke, M., O. Mukabayire, T. Wiehe, and D. Tautz. 2008. Identification of selective sweeps in closely related populations of the house mouse based on microsatellite scans. *Genetics* **180**: 1537–1545.

Thornton, K.R. and P. Andolfatto. 2006. Approximate Bayesian inference reveals evidence for a recent, severe bottleneck in a Netherlands population of *Drosophila melanogaster*. *Genetics* **172**: 1607–1619.

Thornton, K.R., J.D. Jensen, C. Becquet, and P. Andolfatto. 2007. Progress and prospects in mapping recent selection in the genome. *Heredity* **98**: 340–348.

Tishkoff, S.A., F.A. Reed, A. Ranciaro, B.F. Voight, C.C. Babbitt, et al. 2007. Convergent adaptation of human lactase persistence in Africa and Europe. *Nat. Genet.* **39**: 31–40.

Toomajian, C., R.S. Ajioka, L.B. Jorde, J.P. Kushner, and M. Kreitman. 2003. A method for detecting recent selection in the human genome from allele age. *Genetics* **165**: 287–297.

Toro, M.A., A. Fernández, L.A. García-Cortés, J. Rodrigáñez, and L. Silió. 2006. Sex ratio variation in Iberian pigs. *Genetics* **173**: 911–917.

Trivers, R.L. and H. Hare. 1976. Haplodiploidy and the evolution of social insects. *Science* **191**: 249–263.

Tsitrone, A., F. Rousset, and P. David. 2001. Heterosis, marker mutational processes and population inbreeding history. *Genetics* **159**: 1845–1859.

Tuljapurkar, S. 1990. *Population Dynamics in Variable Environments*. Springer-Verlag, Berlin.

Turelli, M. 1984. Heritable variation via mutation–selection balance: Lerch's zeta meets the abdominal bristle. *Theor. Popul. Biol.* **25**: 138–193.

Turelli, M. and N.H. Barton. 1994. Genetical and statistical analyses of strong selection on polygenic traits—what me normal? *Genetics* **138**: 913–941.

Turelli, M. and H.A. Orr. 1995. The dominance theory of Haldane's rule. *Genetics* **140**: 389–402.

Turelli, M. and H.A. Orr. 2000. Dominance, epistasis and the genetics of postzygotic isolation. *Genetics* **154**: 1663–1679.

Turelli, M., D.W. Schemske, and P. Bierzychudek. 2001. Stable two-allele polymorphisms maintained by fluctuating fitnesses and seed banks: protecting the blues in *Linanthus parryae*. *Evolution* **55**: 1283–1298.

Turner, J.R.G. 1967. Why does the genome not congeal? *Evolution* **21**: 645–656.

Turner, J.R.G. 1977. Butterfly mimicry: the genetical evolution of an adaptation. *Evol. Biol.* **10**: 163–206.

Turner, J.R.G. 1981. Adaptation and evolution in Heliconius: a defense of neo-Darwinism. *Ann. Rev. Ecol. Syst.* **12**: 99–121.

Umina, P., A. Weeks, M. Kearney, S. McKechnie, and A. Hoffmann. 2005. A rapid shift in a classic clinal pattern in *Drosophila* reflecting climate change. *Science* **308**: 691–693.

Urrutia, A.O. and L.D. Hurst. 2001. Codon usage bias covaries with expression breadth and the rate of synonymous evolution in humans, but this is not evidence for selection. *Genetics* **159**: 1191–1199.

Uyenoyama, M.K. 1986. Inbreeding and the cost of meiosis: the evolution of selfing in populations practicing biparental inbreeding. *Evolution* **40**: 388–404.

Uyenoyama, M.K. 2000. Evolutionary dynamics of self-incompatibility. *Genetics* **156**: 351–359.

Valdar, W., L.C. Solberg, D. Guaguier, S. Burnett, P. Klenerman, et al. 2006. Genome-wide genetic association of complex traits in heterogeneous stock mice. *Nat. Genet.* **38**: 879–887.

Valdes, A.M., M. Slatkin, and N.B. Freimer. 1993. Allele frequencies at microsatellite loci: the stepwise mutation model revisited. *Genetics* **133**: 737–749.

Van Damme, J.M.M. 1983. Gynodioecy in *Plantago lanceolata* L. II. Inheritance of three male sterility types. *Heredity* **50**: 253–273.

Van Dijk, P.J. 2003. Ecological and evolutionary opportunities of apomixis: insights from *Taraxacum* and *Chondrilla*. *Phil. Trans. R. Soc. B* **358**: 1113–1129.

Van Treuren, R., R. Bijlsma, N.J. Ouborg, and W. VanDelden. 1993. The significance of genetic erosion in the process of extinction. IV. Inbreeding depression and heterosis effects caused by selfing and outcrossing in *Scabiosa columbaria*. *Evolution* **47**: 1669–1680.

Van Valen, L. 1975. Group selection, sex and fossils. *Evolution* **29**: 87–98.

Vaupel, J.W. 1997. Trajectories of mortality at advanced ages. In *Between Zeus and the Salmon: The Biodemography of Longevity* (Eds. K.W. Wachter and C.E. Finch), pp. 17–37. National Academy Press, Washington, DC.

Vekemans, X. and M. Slatkin. 1994. Gene and allelic genealogies at a gametophytic self-incompatibility locus. *Genetics* **137**: 1157–1165.

Vetukhiv, M. 1953. Viability of hybrids between local populations of *Drosophila pseudoobscura*. *Proc. Natl. Acad. Sci. U.S.A.* **39**: 30–34.

Vetukhiv, M. 1956. Fecundity of hybrids between geographic populations of *Drosophila pseudoobscura*. *Evolution* **10**: 139–146.

Via, S. 2001. Sympatric speciation in animals: the ugly duckling grows up. *Trends Ecol. Evol.* **16**: 381–390.

Via, S. and J. West. 2008. The genetic mosaic suggests a new role for hitchhiking in ecological speciation. *Mol. Ecol.* **17**: 4334–4345.

Vickery, R.K. 1978. Case studies in the evolution of species complexes in Mimulus. *Evol. Biol.* **11**: 405–508.

Vicoso, B. and B. Charlesworth. 2006. Evolution on the X chromosome: unusual patterns and processes. *Nat. Rev. Genet.* **7**: 645–653.

Vicoso, B. and B. Charlesworth. 2009. Recombination rates may affect the ratio of X to autosomal noncoding polymorphism in African populations of *Drosophila melanogaster*. *Genetics* **181**: 1699–1701.

Visel, A., S. Prabhakar, J. A. Akiyama, M. Shoukry, K. D. Lewis, et al. 2008. Ultraconservation identifies a small subset of extremely constrained developmental enhancers. *Nat. Genet.* **40**: 158–160.

Vitalis, R. 2002. Sex-specific genetic differentiation and coalescence times: estimating sex-biased dispersal rates. *Mol. Ecol.* **11**: 125–138.

Vitalis, R., S. Glémin, and I. Olivieri. 2004. When genes go to sleep: the population genetic consequences of seed dormancy and monocarpic perenniality. *Am. Nat.* **163**: 295–311.

Vogel, F. and A.G. Motulsky. 1997. *Human Genetics: Problems and Approaches*. Springer-Verlag, Berlin.

Vogler, D.W. and S. Kalisz. 2001. Sex among the flowers: the distribution of plant mating systems. *Evolution* **55**: 202–204.

Voight, B.F., A.M. Adams, L.A. Frisse, Y.D. Qian, R.R. Hudson, and A. Di Rienzo. 2005. Interrogating multiple aspects of variation in a full resequencing data set to infer human population size changes. *Proc. Natl. Acad. Sci. U.S.A.* **102**: 18508–18513.

Voight, B.F., S. Kudaravalli, X. Wen, and J.K. Pritchard. 2006. A map of recent positive selection in the human genome. *PLoS Biol.* **4**: 449–458.

Wade, M.J. 1978. A critical review of the models of group selection. *Quart. Rev. Biol.* **53**: 101–114.

Wade, M.J. and C.J. Goodnight. 1998. The theories of Fisher and Wright in the context of metapopulations: when nature does many small experiments. *Evolution* **52**: 1537–1553.

Wade, M.J. and C.J. Goodnight. 2000. The ongoing synthesis. *Evolution* **54**: 306–317.

Wahlund, S. 1928. Zusammensetzung von Populationen und Korrelationserscheinungen vom Standpunkt der Vererbungslehre aus betrachtet. *Hereditas* **11**: 65–106.

Wakeley, J. 1998. Segregating sites in Wright's island model. *Theor. Popul. Biol.* **53**: 166–174.

Wakeley, J. 1999. Nonequilibrium migration in human history. *Genetics* **153**: 1863–1871.

Wakeley, J. 2001. The coalescent in an island model of population subdivision with variation among demes. *Theor. Popul. Biol.* **59**: 133–144.

Wakeley, J. 2003. Polymorphism and divergence for island-model species. *Genetics* **163:** 411–420.

Wakeley, J. 2008. *Coalescent Theory. An Introduction.* Roberts & Co., Greenwood Village, CO.

Wakeley, J. and N. Aliacar. 2001. Gene genealogies in a metapopulation. *Genetics* **159:** 893–905.

Wakeley, J. and J. Hey. 1997. Estimating ancestral population parameters. *Genetics* **145:** 847–855.

Wakeley, J. and S. Lessard. 2003. Theory of the effects of population structure and sampling on patterns of linkage disequilibrium applied to genomic data from humans. *Genetics* **164:** 1043–1053.

Wakeley, J. and O. Sargsyan. 2009. Extensions of the coalescent effective population size. *Genetics* **181:** 341–345.

Wall, J.D. 1999. Recombination and the power of statistical tests of neutrality. *Genet. Res.* **74:** 65–80.

Wall, J.D. 2003. Estimating ancestral population sizes and divergence times. *Genetics* **163:** 395–404.

Wall, J.D. and M. Przeworski. 2000. When did the human population size start increasing? *Genetics* **155:** 1865–1874.

Walsh, J.B. 1982. Rate of accumulation of reproductive isolation in chromosome arrangements. *Am. Nat.* **120:** 510–532.

Wang, J. 2005. Estimation of effective population sizes from data on genetic markers. *Phil. Trans. R. Soc. B* **360:** 1395–1409.

Wang, J. and A. Caballero. 1999. Developments in predicting the effective size of subdivided populations. *Heredity* **82:** 212–226.

Wang, J. and W.G. Hill. 1999. Effect of selection against deleterious mutations on the decline in heterozygosity at neutral loci in closely inbreeding populations. *Genetics* **153:** 1475–1489.

Wang, J., W.G. Hill, D. Charlesworth, and B. Charlesworth. 1999. Dynamics of inbreeding depression due to deleterious mutations in small populations: mutation parameters and inbreeding rate. *Genet. Res.* **74:** 165–178.

Wang, J., P.D. Keightley, and T. Johnson. 2006. MCALIGN2: faster, accurate global pairwise alignment of non-coding DNA sequences based on explicit models of indel evolution. *BMC Bioinformatics* **7:** 292.

Wang, R.L., J. Wakeley, and J. Hey. 1997. Gene flow and natural selection in the origin of *Drosophila pseudoobscura* and close relatives. *Genetics* **147:** 1091–1106.

Waples, R.S. and M. Yokota. 2007. Temporal estimates of effective population size in species with overlapping generations. *Genetics* **175:** 219–233.

Waser, P.M., S.N. Austad, and B. Keane. 1986. When should animals avoid inbreeding? *Am. Nat.* **128:** 529–537.

Watterson, G.A. 1975. On the number of segregating sites in genetical models without recombination. *Theor. Popul. Biol.* **7:** 256–276.

Watterson, G.A. 1978. The homozygosity test of neutrality. *Genetics* **88:** 405–417.

Webb, C.J. 1979. Breeding systems and the evolution of dioecy in New Zealand apioid Umbelliferae. *Evolution* **33:** 662–672.

Weber, K.E. 1992. How small are the smallest selectable domains of form? *Genetics* **130:** 345–353.

Weber, K.E. 1996. Large genetic change at small fitness cost in large populations of *Drosophila melanogaster*: rethinking fitness surfaces. *Genetics* **144:** 205–213.

Weber, K.E. and L.T. Diggins. 1990. Increased selection response in large populations. II. Selection for ethanol vapor resistance in *Drosophila melanogaster* at two population sizes. *Genetics* **125:** 585–597.

Weeks, S.C., B.R. Crosser, R. Bennett, M. Gray, and N. Zucker. 2000. Maintenance of androdioecy in the freshwater shrimp, *Eulimnadia texana*: estimates of inbreeding depression in two populations. *Evolution* **54:** 878–887.

Weinberg, W. 1908. Über den Nachweis der Vererbung beim Menschen. *Jahresheftes Ver. Vät. Naturk. Würtemberg* **64:** 368–382.

Weinreich, D.M. and D.M. Rand. 2000. Contrasting patterns of nonneutral evolution in proteins encoded in nuclear and mitochondrial genomes. *Genetics* **156:** 385–399.

Weinreich, D.M., N.F. Delane, M.A. DePristo, and D.L. Hartl. 2006. Darwinian evolution can follow only very few mutational paths to fitter proteins. *Science* **312:** 111–114.

Weir, B.S. 1996. *Genetic Data Analysis II*. Sinauer, Sunderland, MA.

Weir, B.S. and W.G. Hill. 2002. Estimating F-statistics. *Ann. Rev. Genet.* **36:** 721–750.

Welch, J.J. 2006. Estimating the genomewide rate of adaptive protein evolution in Drosophila. *Genetics* **173:** 821–827.

Weldon, W.F.R. 1901. A first study of natural selection in *Clausilia laminata* (Montagu). *Biometrika* **1:** 109–124.

Wellcome Trust Case Control Consortium. 2007. Genome-wide association study of 14,000 cases of seven common diseases and 3,000 shared controls. *Nature* **447:** 661–678.

Weller, S.G. and A.K. Sakai. 1991. The genetic basis of male sterility in Schiedea (Caryophyllaceae), an endemic Hawaiian genus. *Heredity* **67:** 265–273.

Wenseleers, T. and F.L.W. Ratnieks. 2006. Enforced altruism in insect societies. *Nature* **444:** 50–50.

Werren, J.H. 1980. Sex ratio adaptation to local mate competition in a parasitic wasp. *Science* **208:** 1157–1159.

West, S.A. 2009. *Sex Allocation*. Princeton University Press, Princeton NJ.

West, S.A. and E.A. Herre. 1998. Partial local mate competition and the sex ratio: a study on non-pollinating fig wasps. *J. Evol. Biol.* **11:** 531–538.

West, S.A., A.S. Griffin, and A. Gardner. 2006. Social semantics: altruism, cooperation, mutualism, strong reciprocity and group selection. *J. Evol. Biol.* **20:** 415–432.

West, S.A., S.E. Reece, and B.C. Sheldon. 2002. Sex ratios. *Heredity* **88:** 117–124.

Wheat, C.W., W.B. Watt, S.D. Pollock, and P.M. Schulte. 2006. From DNA to fitness differences: sequences and structures of adaptive variants of *Colias* phosphoglucose isomerase (PGI). *Mol. Biol. Evol.* **23:** 499–512.

White, M.J.D. 1973. *Animal Cytology and Evolution*. Cambridge University Press, Cambridge, U.K.

White, M.D. 1978. *Modes of Speciation*. W.H. Freeman, San Francisco, CA.

Whitlock, M.C. 2000. Fixation of new alleles and the extinction of small populations: drift load, beneficial alleles, and sexual selection. *Evolution* **54:** 1855–1861.

Whitlock, M.C. 2002. Selection, load and inbreeding depression in a large metapopulation. *Genetics* **160:** 1191–1202.

Whitlock, M.C. 2003. Fixation probability and time in subdivided populations. *Genetics* **164:** 767–779.

Whitlock, M.C. 2008. Evolutionary inference from Q(ST). *Mol. Ecol.* **17:** 1885–1896.

Whitlock, M.C. and N.H. Barton. 1997. The effective size of a subdivided population. *Genetics* **146:** 427–441.

Whitlock, M.C. and D.E. McCauley. 1990. Some population genetic consequences of colony formation and extinction: genetic correlations within founding groups. *Evolution* **44:** 1717–1724.

Whitlock, M.C. and D.E. McCauley. 1999. Indirect measures of gene flow and migration: $F_{ST} \neq 1/(4Nm + 1)$. *Heredity* **82:** 117–125.

Whittam, T.S., H. Ochman, and R.K. Selander. 1983. Multilocus genetic structure in natural populations of *Escherichia coli*. *Proc. Natl. Acad. Sci. U.S.A.* **80:** 1751–1755.

Wichman, H.A., M.R. Badgett, L.A. Scott, C.M. Boulianne, and J.J. Bull. 1999. Different trajectories of parallel evolution during viral adaptation. *Science* **285:** 422–424.

Wichman, H.A., J. Wichman, and J.J. Bull. 2005. Adaptive molecular evolution for 13,000 phage generations: a possible arms race. *Genetics* **170:** 19–31.

Wicker, T., F. Sabot, J. L. B. A. Hua-Van, P. Capy, B. Chalhoub, et al. 2007. A unified classification system for eukaryotic transposable elements. *Nat. Rev. Genet.* **8:** 973–983.

Wickler, W. 1968. *Mimicry in Plants and Animals*. Weidenfeld & Nicolson, London.

Wild, G., A. Gardner, and S.A. West. 2009. Adaptation and the evolution of parasite virulence in a connected world. *Nature* **459:** 983–986.

Wilkins, J. 2004. A separation-of-timescales approach to the coalescent in a continuous population. *Genetics* **168:** 2227–2244.

Wilkins, J.F. and J. Wakeley. 2002. The coalescent in a continuous, finite, linear population. *Genetics* **161:** 873–888.

Wilkinson-Herbots, H.M. 1998. Genealogy and subpopulation differentiation under various models of population structure. *J. Math. Biol.* **37:** 535–585.

Willi, Y. and M. Fischer. 2005. Genetic rescue in interconnected populations of small and large size of the self-incompatible *Ranunculus reptans*. *Heredity* **95:** 437–443.

Williams, G.C. 1957. Pleiotropy, natural selection and the evolution of senescence. *Evolution* **11:** 398–411.

Williams, G.C. 1966. Natural selection, the costs of reproduction and a refinement of Lack's principle. *Am. Nat.* **108:** 705–709.

Williams, G.C. 1975. *Sex and Evolution*. Princeton University Press, Princeton, NJ.

Williamson, S.H., M.J. Hubisz, A.G. Clark, B.A. Payseur, C.D. Bustamante, and R. Nielsen. 2007. Localizing recent adaptive evolution in the human genome. *PLoS Genet.* **3:** 901–915.

Wilson, D.S. 1975. A theory of group selection. *Proc. Natl. Acad. Sci. U.S.A.* **72:** 143–146.

Wilson, D.S. 1980. *The Natural Selection of Populations and Communities*. Benjamin Cummings, Menlo Park, CA.

Wilson, D.S. 2007. Social semantics: toward a genuine pluralism in the study of social behaviour. *J. Evol. Biol.* **21:** 368–373.

Wilson, D.S. and M. Turelli. 1986. Stable underdominance and the evolutionary invasion of empty niches. *Am. Nat.* **127:** 835–850.

Wilson, D.S. and E.O. Wilson. 2007. Rethinking the theoretical foundation of sociobiology. *Quart. Rev. Biol.* **82:** 327–348.

Wilton, A.N. and J.A. Sved. 1979. X-chromosomal heterosis in *Drosophila melanogaster*. *Genet. Res.* **34:** 303–315.

Wiuf, C., K. Zhao, H. Innan, and M. Nordborg. 2004. The probability and chromosomal extent of trans-specific polymorphism. *Genetics* **168:** 2363–2372.

Wolf, J.B., E.B. Brodie, and M.J. Wade. 2000. *Epistasis and the Evolutionary Process.* Oxford University Press, New York.

Wolf, Y., C. Viboud, E. Holmes, E.V. Koonin, and D. Lipman. 2006. Long intervals of stasis punctuated by bursts of positive selection in the seasonal evolution of influenza A virus. *Biol. Direct* **1:** 34.

Wood, R.J. 1981. Insecticide resistance: genes and mechanisms. In *Genetic Consequences of Man-Made Change* (Eds. L.M. Cook and J.A. Bishop), pp. 53–96. Academic Press, Orlando, FL.

Woodruff, R.C. and J.N. Thompson. 1992. Have premeiotic clusters of mutations been overlooked in evolutionary theory? *J. Evol. Biol.* **5:** 457–464.

Woodruff, R.C. and J.N. Thomson. 2005. The fundamental theorem of neutral evolution: rates of substitution and mutation should factor in premeiotic clusters. *Genetica* **125:** 333–339.

Woolfit, M. and L. Bromham. 2003. Increased rates of sequence evolution in endosymbiotic bacteria and fungi with reduced effective population sizes. *Mol. Biol. Evol.* **20:** 1545–1555.

Woolfit, M. and L. Bromham. 2005. Population size and molecular evolution on islands. *Proc. R. Soc. B* **272:** 2277–2282.

Woolliams, J.A., N.R. Wray, and R. Thompson. 1993. Prediction of long-term contributions and inbreeding in populations undergoing mass selection. *Genet. Res.* **62:** 231–242.

Wray, G.A. 2007. The evolutionary significance of *cis*-regulatory mutations. *Nature Rev. Genet.* **8:** 206–216.

Wright, A., B. Charlesworth, I. Rudan, A. Carothers, and H. Campbell. 2003 A polygenic basis for late-onset disease. *Trends Genet.* **19:** 97–106.

Wright, F. 1990. The "effective number of genes" in a codon. *Gene* **87:** 23–29.

Wright, S. 1920. The relative importance of heredity and environment in determining the piebald pattern of guinea-pigs. *Proc. Natl. Acad. Sci. U.S.A.* **6:** 320–333.

Wright, S. 1921. Systems of mating. II. The effects of inbreeding on the genetic composition of a population. *Genetics* **6:** 124–143.

Wright, S. 1929. Fisher's theory of dominance. *Am. Nat.* **63:** 274–279.

Wright, S. 1931. Evolution in Mendelian populations. *Genetics* **16:** 97–159.

Wright, S. 1932. The role of mutation, inbreeding, crossbreeding and selection in evolution. *Proc. VI Intern. Cong. Genet.* **1:** 356–366.

Wright, S. 1934. Physiological and evolutionary theories of dominance. *Am. Nat.* **68:** 25–53.

Wright, S. 1937. The distribution of gene frequencies in populations. *Proc. Natl. Acad. Sci. U.S.A.* **23:** 307–320.

Wright, S. 1938. The distribution of gene frequencies under irreversible mutation. *Proc. Natl. Acad. Sci. U.S.A.* **24:** 253–259.

Wright, S. 1939. The distribution of self-sterility alleles in populations. *Genetics* **24:** 538–552.

Wright, S. 1940a. Breeding structure of species in relation to speciation. *Am. Nat.* **74:** 232–248.

Wright, S. 1940b. The statistical consequences of Mendelian heredity in relation to speciation. In *The New Systematics* (Ed. J.S. Huxley), pp. 161–183. Systematics Association, London.

Wright, S. 1941. On the probability of fixation of reciprocal translocations. *Am. Nat.* **74:** 513–522.

Wright, S. 1943. Isolation by distance. *Genetics* **28:** 114–138.

Wright, S. 1945. The differential equation of the distribution of gene frequencies. *Proc. Natl. Acad. Sci. U.S.A.* **31:** 381–389.

Wright, S. 1946. Isolation by distance under diverse systems of mating. *Genetics* **31:** 39–59.

Wright, S. 1951. The genetical structure of populations. *Ann. Eugen.* **15:** 323–354.

Wright, S. 1968. *Evolution and the Genetics of Populations. Vol. 1. Genetic and Biometric Foundations.* University of Chicago Press, Chicago, IL.

Wright, S. 1969. *Evolution and the Genetics of Populations. Vol. 2. The Theory of Gene Frequencies.* University of Chicago Press, Chicago, IL.

Wright, S. 1977. *Evolution and the Genetics of Populations. Vol. 3. Experimental Results and Evolutionary Deductions.* University of Chicago Press, Chicago, IL.

Wright, S. 1978. *Evolution and the Genetics of Populations. Vol. 4. Variability Within and Between Populations.* University of Chicago Press, Chicago, IL.

Wright, S. and T. Dobzhansky. 1946. Genetics of natural populations. XII. Experimental reproduction of some of the changes caused by natural selection in certain populations of *Drosophila pseudoobscura*. *Genetics* **31:** 125–156.

Wright, S., T. Dobzhansky, and W. Hovanitz. 1942. Genetics of natural populations. VII. The allelism of lethals in the third chromosome of *Drosophila pseudoobscura*. *Genetics* **27:** 363–394.

Wright, S.I. and B. Charlesworth. 2004. The HKA test revisited: a maximum-likelihood-ratio test of the standard neutral model. *Genetics* **168:** 1071–1076.

Wright, S.I., J.P. Foxe, L.D. Wilson, A. Kawabe, M. Looseley, et al. 2006. Testing for effects of recombination rate on nucleotide diversity in natural populations of *Arabidopsis lyrata*. *Genetics* **174:** 1421–1430.

Wu, C.-I. and M.F. Hammer. 1991. Molecular evolution of ultraselfish genes of meiotic drive systems. In *Evolution at the Molecular Level* (Eds. R.K. Selander, A.G. Clark, and T.S. Whittam), pp. 177–203. Sinauer, Sunderland, MA.

Wyatt, R. 1990. The evolution of self-pollination in granite outcrop species of Arenaria

(Caryophyllaceae). V. Artificial crosses within and between populations. *Syst. Bot.* **15**: 363–369.

Wynne-Edwards, V.C. 1962. *Animal Dispersion in Relation to Social Behaviour.* Oliver and Boyd, Edinburgh, U.K.

Yablokov, A.V. 1974. *Variability of Mammals.* Amerind Publishing Co., New Delhi.

Yamamoto, M.-T. and G.L.G. Miklos. 1978. Genetic studies on heterochromatin and their implications for the functions of satellite DNA. *Chromosoma* **66**: 71–98.

Yampolsky, L.Y., F.A. Kondrashov, and A.S. Kondrashov. 2005. Distribution of the strength of selection against amino-acid replacements in human proteins. *Hum. Mol. Genet.* **2005**: 3191–3201.

Yang, H.P., T.L. Hung, T.L. You, and T.H. Yang. 2006. Genomewide comparative analysis of the highly abundant transposable element *DINE-1* suggests a recent transpositional burst in *Drosophila yakuba. Genetics* **173**: 189–196.

Yang, Z. 1997. PAML: a program package for phylogenetic analysis by maximum likelihood. *Comput. Appl. Biosci.* **13**: 555–556.

Yang, Z. 2006. *Computational Molecular Evolution.* Oxford University Press, Oxford, U.K.

Yang, Z., and J.P. Bielawski. 2000. Statistical methods for detecting molecular adaptation. *Trends Ecol. Evol.* **15**: 496–503.

Yang, Z. R. Nielsen, N. Goldman, and A.-M.K. Pedersen. 2000. Codon-substitution models for heterogeneous selection pressure at amino acid sites. *Genetics* **155**: 431–449.

Yokoyama, S. and M. Nei. 1979. Population dynamics of sex-determining alleles in honeybees and self-incompatibility alleles in plants. *Genetics* **91**: 609–626.

Yu, N., F.-C. Chen, S. Ota, L.B. Jorde, P. Pamilo, et al. 2002. Larger genetic differences within Africans than between Africans and Eurasians. *Genetics* **161**: 269–274.

Yu, N., M.I. Jensen-Seaman, L. Chemnick, O. Ryder, and W.H. Li. 2004. Nucleotide diversity in gorillas. *Genetics* **166**: 1375–1383.

Yu, Q., S. Hou, A. Feltus, M.R. Jones, J.E. Murray, et al. 2008. Low X/Y divergence in four pairs of papaya sex-linked genes. *Plant J. Cell Mol. Biol.* **53**: 124–132.

Yu, Y.H., Y.W. Lin, J.F. Yu, W. Schempp, and P.H. Yen. 2008. Evolution of the DAZ gene and the AZFc region on primate Y chromosomes. *BMC Evol. Biol.* **8**: 96.

Yun, S.H., M.L. Berbee, O.C. Yoder, and B.G. Turgeon. 1999. Evolution of the fungal self-fertile reproductive life style from self-sterile ancestors. *Proc. Natl Acad. Sci. U.S.A.* **96**: 5592–5597.

Yunis, J.J., J.R. Sawyer, and K. Dunham. 1980. The striking resemblance of high-resolution G-banded chromosomes of man and chimpanzee. *Science* **208**: 1145–1148.

Zapata, C. 2000. The D′ measure of overall gametic disequilibrium between pairs of multiallelic loci. *Evolution* **54**: 1809–1812.

Zeng, K., X.-Y. Fu, S. Shi, and C.-I. Wu. 2006. Statistical tests for detecting positive selection by utilizing high-frequency variants. *Genetics* **174**: 1431–1439.

Zeyl, C., T. Vanderford, and M. Carter. 2003. An evolutionary advantage of haploidy in large yeast populations. *Science* **299**: 555–558.

Zhang, J.Z., R.Y. Nielsen, and Z. Yang. 2005. Evaluation of an improved branch-site likelihood method for detecting positive selection at the molecular level. *Mol. Biol. Evol.* **22**: 2472–2479.

Zhang, L. and W.-H. Li. 2005. Human SNPs reveal no evidence of frequent positive selection. *Mol. Biol. Evol.* **22**: 2504–2507.

Zilberman, D. and S. Henikoff. 2004. Silencing of transposons in plant genomes: kick them when they're down. *Genome Biol.* **5**: 249.

Zhivotovsky, L.A., M.W. Feldman, and F.B. Christiansen. 1994. Evolution of recombination among multiple selected loci—a generalized reduction principle. *Proc. Natl. Acad. Sci. U.S.A.* **91**: 1079–1093.

Zuckerkandl, E. and L. Pauling. 1965. Evolutionary divergence and convergence in proteins. In *Evolving Genes and Proteins* (Eds. V. Bryson and H.J. Vogel), pp. 97–166. Academic Press, New York.

Author Index

Subject Index

Mimulus lewisii   *Mimulus cardinalis*

PLATE 3 Flowers of the closely related monkey flower species, *Mimulus lewisii* and *M. cardinalis* (see **Section 3.4.iv.b**). Adapted from Schemske and Bradshaw (1999).